Benchmark Papers in Electrical Engineering and Computer Science

Series Editor: John B. Thomas
Princeton University

Published Volumes

SYSTEM SENSITIVITY ANALYSIS
J. B. Cruz, Jr.
RANDOM PROCESSES, Part I: Multiplicity Theory and Canonical Decompositions
Anthony Ephremides and John B. Thomas
ALGEBRAIC CODING THEORY: History and Development
Ian F. Blake
PATTERN RECOGNITION: Introduction and Foundations
Jack Sklansky
COMPUTER-AIDED CIRCUIT DESIGN: Simulation and Optimization
S. W. Director
ENVIRONMENTAL MODELING: Analysis and Management
Douglas Daetz and Richard H. Pantell
AUTOMATIC CONTROL: Classical Linear Theory
George J. Thaler
CIRCUIT THEORY: Foundations and Classical Contributions
M. E. Van Valkenburg
DATA COMMUNICATION: Fundamentals of Baseband Transmission
L. E. Franks
NONLINEAR SYSTEMS: Processing of Random Signals—Classical Analysis
A. H. Haddad
RANDOM PROCESSES, Part II: Poisson and Jump Point Processes
Anthony Ephremides
DIGITAL FILTERS AND THE FAST FOURIER TRANSFORM
Bede Liu

Additional Volumes in Preparation

Benchmark Papers
in Electrical Engineering
and Computer Science/12

A BENCHMARK® Books Series

DIGITAL FILTERS AND THE FAST FOURIER TRANSFORM

Edited by
BEDE LIU
Princeton University

Dowden, Hutchinson & Ross, Inc.
Stroudsburg, Pennsylvania

Distributed by
HALSTED PRESS *A Division of John Wiley & Sons, Inc.*

Benchmark Papers in Electrical Engineering and Computer Science, Volume 12
Library of Congress Catalog Card Number: 75–9533
ISBN: 0–470–54150–4

77 76 75 1 2 3 4 5
Manufactured in the United States of America.

LIBRARY OF CONGRESS CATALOGING IN PUBLICATION DATA

Main entry under title:

Digital filters and the fast Fourier transform.

(Benchmark papers in electrical engineering and computer science ; v. 12)
Includes indexes.
1. Digital filters (Mathematics) 2. Fourier transformations. I. Liu, Bede.
TK7872.F5D53 621.3815'32 75-9533
ISBN 0-470-54150-4

Exclusive Distributor: **Halsted Press**
A Division of John Wiley & Sons, Inc.

Permissions

The following papers have been reprinted with the permission of the authors and copyright holders.

AMERICAN FEDERATION OF INFORMATION PROCESSING SOCIETIES
AFIPS Proceedings of the 1966 Fall Joint Computer Conference
Fast Fourier Transforms—for Fun and Profit
AFIPS Proceedings of the 1966 Spring Joint Computer Conference
High-Speed Convolution and Correlation

AMERICAN MATHEMATICAL SOCIETY—*Mathematics of Computation*
An Algorithm for the Machine Calculation of Complex Fourier Series

AMERICAN TELEPHONE AND TELEGRAPH COMPANY—*The Bell System Technical Journal*
Design of Wideband Sampled-Data Filters
Floating-Point-Roundoff Accumulation in Digital-Filter Realizations
Overflow Oscillations in Digital Filters

ASSOCIATION FOR COMPUTING MACHINERY, INC.—*Journal of the Association for Computing Machinery*
Accumulation of Round-Off Error in Fast Fourier Transforms

INSTITUTE OF ELECTRICAL AND ELECTRONICS ENGINEERS, INC.
Proceedings of the IEEE
Effects of Finite Register Length in Digital Filtering and the Fast Fourier Transform
Error Analysis of Digital Filters Realized with Floating-Point Arithmetic
Proceedings of the 1974 IEEE EASCON
Some New Realizations of Dedicated Hardware Digital Signal Processors
IEEE Transactions on Acoustics, Speech, and Signal Processing
Equiripple and Minimax (Chebyshev) Approximations for Recursive Digital Filters
A Parallel Arithmetic Hardware Structure for Recursive Digital Filtering
IEEE Transactions on Audio and Electroacoustics
An Absolute Bound on Limit Cycles due to Roundoff Errors in Digital Filters
Application of the Fast Fourier Transform to Computation of Fourier Integrals, Fourier Series, and Convolution Integrals
An Approach to the Implementation of Digital Filters
A Bound on Limit Cycles in Fixed-Point Implementations of Digital Filters
Computer-Aided Design of Recursive Digital Filters
A Computer Program for Designing Optimum FIR Linear Phase Digital Filters
On the Design of Digital Filters with Coefficients of Limited Word Length
A Fixed-Point Fast Fourier Transform Error Analysis
Instability Thresholds in Digital Filters due to Coefficient Rounding
Limit-Cycle Oscillations in Floating-Point Digital Filters
A Method for Computing the Fast Fourier Transform with Auxiliary Memory and Limited High-Speed Storage
A New Approach to the Realization of Nonrecursive Digital Filters
Parallelism in Fast Fourier Transform Hardware
Roundoff-Noise Analysis for Fixed-Point Digital Filters Realized in Cascade or Parallel Form
A Simple Design of Maximally Flat Delay Digital Filters
What is the Fast Fourier Transform?

IEEE Transactions on Circuit Theory
Equal-Ripple Delay Recursive Digital Filters
A Unified Approach to the Design of Optimum FIR Linear-Phase Digital Filters
IEEE Transactions on Computers
A Fast Fourier Transform for High-Speed Signal Processing
A Pipeline Fast Fourier Transform

INSTITUTION OF ELECTRICAL ENGINEERS (LONDON)—*Proceedings of the IEE*
Effect of a Finite-Word-Length Computer in a Sampled-Data Feedback System

JOHN WILEY & SONS, INC.—*System Analysis by Digital Computer*
Digital Filters

Series Editor's Preface

The Benchmark Series in Electrical Engineering and Computer Science is aimed at sifting, organizing, and making readily accessible to the reader the vast literature that has accumulated. Although the series is not intended as a complete substitute for a study of this literature, it will serve at least three major critical purposes. In the first place, it provides a practical point of entry into a given area of research. Each volume offers an expert's selection of the critical papers on a given topic as well as his views on its structure, development, and present status. In the second place, the series provides a convenient and time-saving means for study in areas related to but not contiguous with one's principal interests. Last, but by no means least, the series allows the collection, in a particularly compact and convenient form, of the major works on which present research activities and interests are based.

Each volume in the series has been collected, organized, and edited by an authority in the area to which it pertains. To present a unified view of the area, the volume editor has prepared an introduction to the subject, has included his comments on each article, and has provided a subject index to facilitate access to the papers.

We believe that this series will provide a manageable working library of the most important technical articles in electrical engineering and computer science. We hope that it will be equally valuable to students, teachers, and researchers.

This volume, *Digital Filters and the Fast Fourier Transform,* has been edited by Professor Bede Liu of Princeton University. It contains thirty-four papers treating the design, implementation, error analysis, and application of digital filters and fast Fourier transforms. Professor Liu is extremely well-qualified as the editor of this volume, having been heavily involved in research in this area and having made many significant contributions to the development of the field.

John B. Thomas

Contents

Permissions v
Series Editor's Preface vii
Contents by Author xiii

Introduction 1

I. DIGITAL FILTERS: INTRODUCTION

Editor's Comments on Paper 1 4

1 **KAISER, J. F.:** Digital Filters 5
Systems Analysis by Digital Computer, F. F. Kuo and J. F. Kaiser, eds., John Wiley & Sons, Inc., 1966, pp. 218–285

II. DESIGN AND IMPLEMENTATION

Editor's Comments on Papers 2 Through 12 76

2 **GOLDEN, R. M., and J. F. KAISER:** Design of Wideband Sampled-Data Filters 80
Bell Syst. Tech. J., **43,** Pt. 2, 1533–1546 (July 1964)

3 **STEIGLITZ, K:** Computer-Aided Design of Recursive Digital Filters 94
IEEE Trans. Audio Electroacoustics, **AU–18**(2), 123–129 (1970)

4 **DECZKY, A. G.:** Equiripple and Minimax (Chebyshev) Approximations for Recursive Digital Filters 101
IEEE Trans. Acoustics, Speech, Signal Processing, **ASSP–22**(2), 98–111 (1974)

5 **McCLELLAN, J. H., and T. W. PARKS:** A Unified Approach to the Design of Optimum FIR Linear-Phase Digital Filters 115
IEEE Trans. Circuit Theory, **CT–20**(6), 697–701 (1973)

6 **McCLELLAN, J. H., T. W. PARKS, and L. R. RABINER:** A Computer Program for Designing Optimum FIR Linear Phase Digital Filters 120
IEEE Trans. Audio Electroacoustics, **AU–21**(6), 506–526 (1973)

7 **FETTWEIS, A.:** A Simple Design of Maximally Flat Delay Digital Filters 141
IEEE Trans. Audio Electroacoustics, **AU–20**(2), 112–114 (1972)

8 **THIRAN, J. P.:** Equal-Ripple Delay Recursive Digital Filters 144
IEEE Trans. Circuit Theory, **CT–18**(6), 664–669 (1971)

9 **JACKSON, L. B., J. F. KAISER, and H. S. McDONALD:** An Approach to the Implementation of Digital Filters 150
IEEE Trans. Audio Electroacoustics, **AU–16**(3), 413–421 (1968)

10 **GABEL, R. A.:** A Parallel Arithmetic Hardware Structure for Recursive Digital Filtering 159
IEEE Trans. Acoustics, Speech, Signal Processing, **ASSP–22**(4), 255–258 (1974)

11 **PELED, A., and B. LIU:** A New Approach to the Realization of Nonrecursive Digital Filters 163
IEEE Trans. Audio Electroacoustics, **AU–21**(6), 477–484 (1973)

12 **PELED, A., and B. LIU:** Some New Realizations of Dedicated Hardware Digital Signal Processors 171
Proc. 1974 IEEE EASCON, Institute of Electrical and Electronics Engineers, Inc., pp. 464–468

III. EFFECTS OF FINITE WORD LENGTH

Editor's Comments on Papers 13 Through 23 178

13 **OPPENHEIM, A. V., and C. J. WEINSTEIN:** Effects of Finite Register Length in Digital Filtering and the Fast Fourier Transform 182
Proc. IEEE, **60**(8), 957–976 (1972)

14 **KNOWLES, J. B., and R. EDWARDS:** Effect of a Finite-Word-Length Computer in a Sampled-Data Feedback System 202
Proc. IEE, **112**(6), 1197–1207 (1965)

15 **JACKSON, L. B.:** Roundoff-Noise Analysis for Fixed-Point Digital Filters Realized in Cascade or Parallel Form 213
IEEE Trans. Audio Electroacoustics, **AU–18**(2), 107–122 (1970)

16 **SANDBERG, I. W.:** Floating-Point-Roundoff Accumulation in Digital-Filter Realizations 229
Bell Syst. Tech. J., **46**(8), 1775–1791 (1967)

17 **LIU, B., and T. KANEKO:** Error Analysis of Digital Filters Realized with Floating-Point Arithmetic 246
Proc. IEEE, **57**(10), 1735–1747 (1969)

18 **AVENHAUS, E.:** On the Design of Digital Filters with Coefficients of Limited Word Length 259
IEEE Trans. Audio Electroacoustics, **AU–20**(3), 206–212 (1972)

19 **OTNES, R. K., and L. P. McNAMEE:** Instability Thresholds in Digital Filters due to Coefficient Rounding 266
IEEE Trans. Audio Electroacoustics, **AU–18**(4), 456–463 (1970)

20 **LONG, J. L., and T. N. TRICK:** An Absolute Bound on Limit Cycles due to Roundoff Errors in Digital Filters **274**
IEEE Trans. Audio Electroacoustics, **AU–21**(1), 27–30 (1973)

21 **SANDBERG, I. W., and J. F. KAISER:** A Bound on Limit Cycles in Fixed-Point Implementations of Digital Filters **278**
IEEE Trans. Audio Electroacoustics, **AU–20**(2), 110–112 (1972)

22 **EBERT, P. M., J. E. MAZO, and M. G. TAYLOR:** Overflow Oscillations in Digital Filters **281**
Bell Syst. Tech. J., **48**(9), 2999–3020 (1969)

23 **KANEKO, T:** Limit-Cycle Oscillations in Floating-Point Digital Filters **303**
IEEE Trans. Audio Electroacoustics, **AU–21**(2), 100–106 (1973)

IV. THE FAST FOURIER TRANSFORM: INTRODUCTION

Editor's Comments on Paper 24 **312**

24 **IEEE G–AE Subcommittee on Measurement Concepts:** What is the Fast Fourier Transform? **313**
IEEE Trans. Audio Electroacoustics, **AU–15**(2), 45–55 (1967)

V. ALGORITHMS, HARDWARE IMPLEMENTATION, AND SOME APPLICATIONS OF THE FAST FOURIER TRANSFORM

Editor's Comments on Papers 25 Through 32 **326**

25 **COOLEY, J. W., and J. W. TUKEY:** An Algorithm for the Machine Calculation of Complex Fourier Series **328**
Math. Comp., **19,** 297–301 (1965)

26 **GENTLEMAN, W. M., and G. SANDE:** Fast Fourier Transforms—for Fun and Profit **333**
AFIPS PROC. 1966 Fall Joint Computer Conf., Vol. 29, Spartan Books, 1966, pp. 563–578

27 **SINGLETON, R. C.:** A Method for Computing the Fast Fourier Transform with Auxiliary Memory and Limited High-Speed Storage **349**
IEEE Trans. Audio Electroacoustics, **AU–15**(2), 91–98 (1967)

28 **GOLD, B., and T. BIALLY:** Parallelism in Fast Fourier Transform Hardware **357**
IEEE Trans. Audio Electroacoustics, **AU–21**(1), 5–16 (1973)

29 **GROGINSKY, H. L., and G. A. WORKS:** A Pipeline Fast Fourier Transform **369**
IEEE Trans. Computers, **C–19**(11), 1015–1019 (1970)

30 CORINTHIOS, M. J.: A Fast Fourier Transform for High-Speed Signal Processing **374**
IEEE Trans. Computers, **C–20**(8), 843–846 (1971)

31 STOCKHAM, T. G.: High-Speed Convolution and Correlation **378**
AFIPS Proc. 1966 Spring Joint Computer Conf., Vol. 28, Spartan Books, 1966, pp. 229–233

32 COOLEY, J. W., P. A. W. LEWIS, and P. D. WELCH: Application of the Fast Fourier Transform to Computation of Fourier Integrals, Fourier Series, and Convolution Integrals **383**
IEEE Trans. Audio Electroacoustics, **AU–15**(2), 79–84 (1967)

VI. ROUNDOFF ERROR IN FAST FOURIER TRANSFORM

Editor's Comments on Papers 33 and 34 **390**

33 WELCH, P. D.: A Fixed-Point Fast Fourier Transform Error Analysis **391**
IEEE Trans. Audio Electroacoustics, **AU–17**(2), 151–157 (1969)

34 KANEKO, T., and B. LIU: Accumulation of Round-Off Error in Fast Fourier Transforms **398**
J. Assoc. Computing Machinery, **17**(4), 637–654 (1970)

Author Citation Index **417**

Subject Index **421**

Contents by Author

Introduction

A digital filter is a discrete time system which operates on an input sequence to produce an output sequence according to some computational algorithm. Digital filters can be constructed using electronic components, or the algorithms can be programmed on a general-purpose computer. Excellent methods have been developed to design these filters with desired characteristics. By means of analog-to-digital and digital-to-analog conversion devices, a digital filter can also be used to process analog signals. With the rapid technological advances in digital devices, circuits, and systems, signal processing by digital techniques has become increasingly attractive. Consequently, digital filtering has found applications in a growing number of fields of science and engineering.

Digital filters offer three distinct advantages:

1. Flexibility: the characteristics of a digital filter can be changed easily by simply reading in a new set of parameters which determines the characteristics of the filter. Thus, time-varying filters can be easily implemented. More importantly, a single filter structure can be used to realize a multiplicity of filtering functions on a time-shared basis.

2. Reliability: the use of digital components offers a more reliable system, since this avoids a number of problems, such as element tolerance, and large physical size of components for low-frequency operation, that arise in many analog systems.

3. Modularity: digital filters can be implemented with a highly modularized structure, making them suitable for large-scale integration (LSI). For example, it is now technically feasible to put one or two second-order digital filter sections with programmable characteristics on a single LSI chip.

A closely related topic to digital filters is that of the fast Fourier transform, or FFT, which refers to a class of highly efficient algorithms for computing Fourier coefficients of a finite sequence. Because of its substantial time saving over the conventional approaches of Fourier analysis, FFT has become an extremely useful tool in many applications that involve the processing of large quantities of data, including digital filtering. In addition, real-time processing becomes practical through using these algorithms, and special-purpose processors have been built for just such purposes.

In the short span of less than ten years, the fields of digital filtering and FFT have reached a certain degree of maturity. There is a need for a single reference volume that contains the basic works done in these fields. This book is intended to serve this function. It can also be used as a textbook on digital signal processing at the graduate level with minimal help from the course instructor.

The papers collected in this volume are divided into six parts. The first three parts deal largely with digital filters and the last three with FFT; some overlap exists between these two groups of papers.

It is obviously not possible to include in a single volume of this size all the excellent works done in such a broad field. No paper dealing with the many interesting applications of digital signal processing or with multidimensional filtering is included. In the selection of papers for each topic in this volume, my emphasis has been placed on coherence and continuity. Therefore, some important works are, regretfully, not included.

I
Digital Filters: Introduction

Editor's Comments on Paper 1

1 **Kaiser:** *Digital Filters*

Digital filtering has a long and interesting history, tracing back several centuries. However, in the opinion of many, the publication of the book *System Analysis by Digital Computer* in 1966 had a profound impact on the field of digital signal processing. In Chapter 7 of the book, J. F. Kaiser offered the first unifying comprehensive treatment on the subject from a modern viewpoint. Although almost ten years have elapsed since its publication, this article still remains an excellent introduction to digital filters. The list of references should also be of considerable value to active researchers in the field.

The subject of converting an analog signal into digital form and vice versa is not treated in many textbooks. For the benefit of some readers, I have included at the end of the article a supplement summarizing some of the basic relationships concerning digital processing of analog signals.

1

Reprinted by permission from *System Analysis by Digital Computer,* F. F. Kuo and J. F. Kaiser, eds., John Wiley & Sons, Inc., 1966, 218–285

Digital Filters

J. F. KAISER
Bell Telephone Laboratories
Murray Hill, New Jersey

The simulation of linear dynamic systems and continuous filter networks and the filtering or processing of data signals by means of digital computers require both the design and utilization of digital filters. The term *digital filter* refers to the computational process or algorithm by which a sampled signal or sequence of numbers (acting as an input) is transformed into a second sequence of numbers termed the output signal. The computational process may be that of low-pass filtering (smoothing), bandpass filtering, interpolation, the generation of derivatives etc. The process is assumed to be linear, that is that the principle of superposition applies to the input-output relationship. A detailed discussion of some of the digital filter design methods and considerations in the use of these methods is the primary intent of this chapter. The emphasis is centered more on the frequency domain methods and on the processing of long streams of data rather than on time domain methods and the processing of short sections of data. Conspicuously absent from our discussion will be a treatment of the digital filter designs resulting from optimal filter theory and the theory of signal estimation in the presence of noise. The work on these problems has been extensive and widely reported in the literature.

After the early development of digital filter theory is briefly summarized, the theory relating to the sampling process as applied both to signals and to signal processors is described. The class of linear digital filters is subdivided into two types: the nonrecursive filters and the recursive filters. Several methods for the design of filters of each type from frequency domain specifications are developed in detail. The advantages and shortcomings of the different design methods are then pointed out, with the aid of design examples.

The chapter concludes with a discussion of some of the problems associated with the realization of the digital filters, i. e., coefficient accuracy, quantization and rounding effects. A comparison is made between the different canonical realization forms and their related computational procedures to establish insight into the choice of a suitable form.

7.1 THE EARLY DEVELOPMENT OF DIGITAL FILTERING

The processing of discrete signals by linear filters or weighting sequences had its origin in the early 1600's. It began with the work of the mathematicians who were busy constructing mathematical tables and the astronomers who were concerned with the determination of the orbits of heavenly bodies from their observations. The work of Napier, Gregory, Newton, the Bernoullis, Euler, Taylor, Lagrange, Laplace and Gauss is clearly evident in the classical numerical analysis techniques used even today for numerical integration, interpolation, differentiation, and etc. From the frequency domain viewpoint the characteristic features of this work as yielding usable filtering methods clearly reflect the nature of these early applications; i. e., the methods work extremely well for non-noisy signals of low relative bandwidth. For example it is easy to show that for the classical symmetrical interpolating filters the first n successive derivatives of their magnitude-frequency characteristics exhibit zero values (a behavior at zero frequency much like that of the Butterworth low-pass filters). The treatise of Jordan [1] on the calculus of finite differences illustrates the generality and refinement of this fundamental work.

The utility of these classical methods however diminished when the data to be processed contained undesirable noise components. The development of fire control systems and radar equipment in the 1940's and 50's, for example, furnished a strong impetus for the development of more refined signal processing techniques. Out of this work sprang the extensive literature on sampled-data control systems beginning with the work of Hurewicz [2] and others, and culminating in the flood of texts on sampled-data systems [3, 4, 5, 6, 7, 8, 9, 10]. This work, while developing in great detail the theory of sampled-data systems via the z-transform, is confined primarily to control system analysis and optimization studies. The applicability of this literature to many of the problems in digital filtering was found wanting.

Paralleling the extensive development of the theory of sampled-data control systems was strong interest in processing numerically the data signals from radars and other transducers. The purposes here were, on one hand, to extract *cleaner* output signals representing desired quantities such as velocity and acceleration. On the other hand the purposes were also to obtain a characterization of the noise, signal, or system being measured. The advent of the digital computer accelerated the development of signal processing techniques by making it possible to perform quite rapidly more extensive computation procedures on more complex problems. The early and sustained work of Blackman clearly shows this development of technique; a compendium of this extensive and important work is only now readily accessible [11].

Other selected references through which digital filter development can be followed include Lanczos [12], Blackman and Tukey [13], Martin [14], and Ormsby [15]. Monroe [16] summarizes and outlines many different techniques for the synthesis of sampled-data filters; the bibliography of this work is one of the most extensive known on work done since 1950.

With the advent of larger and faster digital computers the role of the digital computer broadened. In addition to being used for data processing applications, the computer became a very versatile and powerful vehicle for the *simulation* of complex systems [17, 18]. This use greatly increased the need for finding effective ways not only to process sampled-data signals but to obtain meaningful sampled-data representations of the systems themselves. This representation problem is one of approximation and is far from trivial.

The first efforts at a solution of this representation problem were to approximate the basic operations of continuous integration and differentiation by discrete processes; Tustin [19], Salzer [20], Boxer [21, 22], and Baxter [23, 24] represent developments along this line. Another approach was to utilize the well developed z-transform theory of sampled-data control systems [3-10]. That the z-transform or *partial-fraction substitution* method was not always satisfactory for the digitalization process was recognized early by Blackman [11, p. 72] and in another context by Gibson [25] but this fact is repeatedly overlooked by the writers of current texts on sampled-data systems.

Steiglitz [26, 27] investigated in detail the theory of the digitalization process. Kaiser [28, 29] was concerned with extending the methods of digitalization both to make possible the simulation of complex continuous linear filters and to make possible the utilization of the extensive continuous filter art of the electrical engineer. That this filter art can be of great utility in many diverse data processing applications is continually being realized by statisticians [30, 31], engineers, and workers in other scientific fields.

In this chapter we shall place emphasis on the mechanics of the digitalization process primarily from a frequency domain viewpoint; the time domain methods are treated extensively by Blackman [11], Monroe [16], and others. In filter simulation work it is imperative that the user fully understand the nature of the approximations being made in the design method. The user should also be informed concerning the numerical problems involved in both the digitalization process and the use of the obtained filter. Therefore we begin by a review of the sampling process as applied to signals and signal processors.

7.2 THE SAMPLING PROCESS†

Linvill [33] showed that the operation of sampling can be viewed as a form of impulse modulation. Accordingly if a signal f(t) is sampled every T seconds then the sampled wave, f*(t), can be represented as a train of impulses given by

$$f^*(t) = T\sum_{n=0}^{\infty} f(nT)\,\delta(t-nT) \tag{7.1}$$

where f(t) has been assumed to be zero for $t < 0$ and $\delta(t-nT)$ is a unit impulse at time $t = nT$. The sampled signal f*(t) is purely mathematical as it is a string of impulse functions. The signal, f*(t), however is usually thought of in terms of the sequence of numbers, $\{f(nT)\}$, representing the values of the signal at the sampling instants nT. If the Laplace transform is taken of both sides of (7.1) there results

$$\mathcal{L}[f^*(t)] = F^*(s) = \sum_{n=-\infty}^{\infty} F(s+jn\omega_s) + \frac{T}{2} f(0^+) \tag{7.2}$$

or its equivalent form

$$F^*(z) = T\sum_{n=0}^{\infty} f(nT)\,z^{-n} \tag{7.3}$$

where $z = \epsilon^{sT}$ is the unit advance operator (the standard z-transform variable) and where $\omega_s = 2\pi/T$ is the radian sampling frequency. Inspection of (7.2) reveals that the spectrum of the sampled signal is periodic in frequency with period ω_s. Further, from (7.2) it can be seen that the original signal f(t) can be recovered exactly from f*(t) by passing f*(t) through

---

†Several excellent descriptions of the sampling process appear in the literature; for example see Ross [34] and Wilts [6, pp. 196-200]. Our development follows that of Wilts.

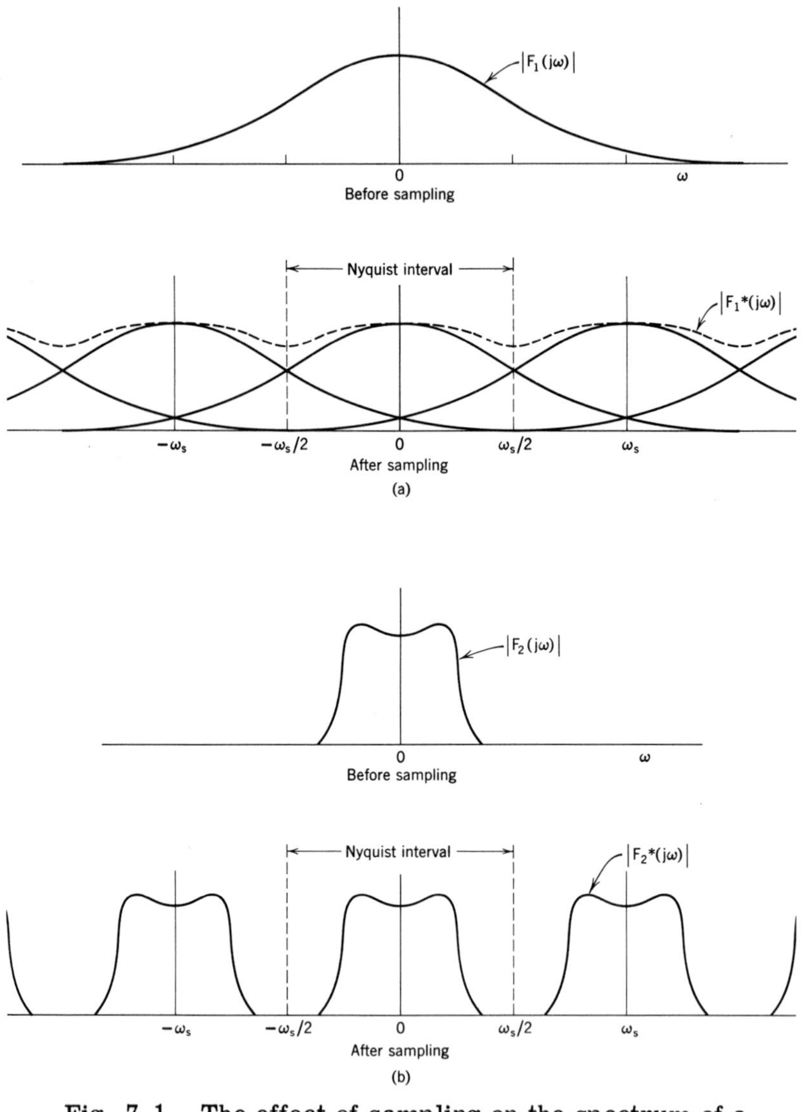

Fig. 7.1. The effect of sampling on the spectrum of a signal.

an ideal low-pass filter of bandwidth $\pm\omega_s/2$ only if F(s) is identically zero outside of the central strip, $s = \pm j\omega_s/2$. Figure 7.1 illustrates this fact for the bandlimited signal $F_2(j\omega)$ and shows the aliasing problems wrought by a signal $F_1(j\omega)$ not bandlimited to $\pm\omega_s/2$.

The signal, $f_0(t)$, recovered from passing $f^*(t)$ through an ideal low-pass filter of bandwidth $\pm\omega_s/2$ is given by

$$f_o(t) = \sum_{n=0}^{\infty} f(nT) \frac{\sin[\omega_s(t-nT)/2]}{\omega_s(t-nT)/2} \qquad (7.4)$$

Again $f_0(t) = f(t)$ only if f(t) is bandlimited to $\pm\omega_s/2$. Thus to prepare satisfactorily a continuous signal for a subsequent sampling operation the signal must first be bandlimited by filtering it with a good low-pass filter. A good low-pass filter is one having both a flat magnitude characteristic and a linear phase characteristic over the major portion of the Nyquist interval ($\pm\omega_s/2$) and having large attenuation at frequencies greater than $\omega_s/2$.

For simulation work this bandlimiting of the signal is most essential in order to minimize errors due to frequency aliasing. The general scheme of system simulation using digital filter representations for the continuous system dynamics is shown in Fig. 7.2. The simulation is said to be *wideband* if both the passband of the bandlimiting filter and the usable frequency range of the digital filter are an appreciable fraction of the Nyquist interval (i.e., 70% to 80%). This sampling rate

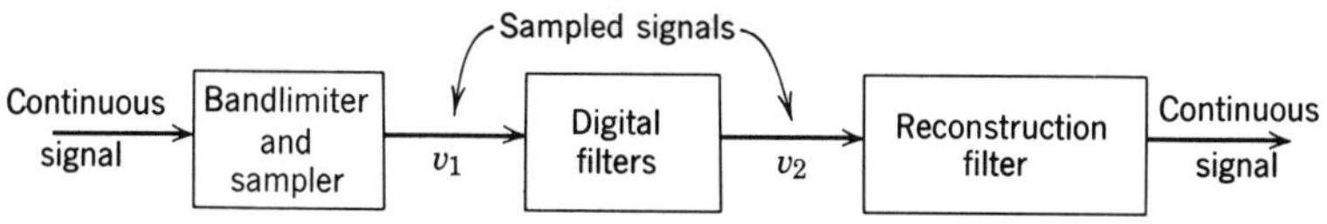

Fig. 7.2. The simulation of a continuous system with digital filters.

is in somewhat sharp contrast to that rate used in sampled-data control system design where the sampling rate is usually chosen to be five to twelve or more times the effective control system bandwidth so as not to introduce excessive amounts of delay in the control loop. The use of a minimal sampling rate will in general result in a minimum total amount of computation to be performed in any simulation or data processing operation unless, of course, the use of this sampling rate requires inordinately complex digital filter designs.

The filter used for reconstruction of a continuous signal from the sequence of output samples need only approximate the ideal low-pass filter over the usable frequency range of the simulation. Thus for low bandwidth simulations the zero-order hold filter, $H_0(s)$, may be satisfactory where

$$H_o(s) = (1-e^{-sT})/s \tag{7.5}$$

The frequency characteristics of H_0 are given by

$$H_o(j\omega) = \frac{\sin(\omega T/2)}{\omega T/2} \angle - (\omega T/2) \tag{7.6}$$

indicating good linear-phase characteristics but relatively poor magnitude characteristics. The effective bandwidth of this reconstruction filter can be measurably improved if it is followed by a linear-phase continuous low-pass filter whose amplitude characteristic tends to equalize the $[\sin(\omega T/2)]/(\omega T/2)$ droop in the range $0 < \omega T/2 < \pi$.

Having presented the basic ideas of the sampling process we now consider the design of the digital filters.

7.3 THE CLASSIFICATION OF DIGITAL FILTERS

It is assumed that the transfer characteristic, H(s), of the linear system for which a digital approximation is desired can be represented in the form

$$H(s) = \frac{\sum_{m=0}^{M} c_m s^m}{\sum_{n=0}^{N} d_n s^n} \tag{7.7}$$

where $M < N$ for $H(s)$ to be quasi-bandlimited.† Utilizing the well known relationships between linear differential and linear difference equations we see that the corresponding digital transfer characteristic will be of the general form

$$H^*(z) = \frac{\sum_{j=0}^{N-1} a_j z^{-j}}{1 + \sum_{j=1}^{N} b_j z^{-j}} \tag{7.8}$$

where z^{-1} is the unit delay operator.

Using (7.8) to relate the output signal or sequence, $v_2(nT)$, to the sampled input signal, $v_1(nT)$, and solving this equation for the current value of the output yields

$$v_2(kT) = \sum_{j=0}^{N-1} a_j v_1(kT-jT) - \sum_{j=1}^{N} b_j v_2(kT-jT) \tag{7.9}$$

If at least one b_j and one a_j value are nonzero the filter, as given by (7.8), is said to be of the recursive type, i.e., computation of the present value of the output v_2 depends not only on the present and (N-1) past values of the input but also on the previous N values of the output v_2. A second-order recursive filter is shown in block diagram form in Fig. 7.3.

If all the b_j are zero then the filter is said to be of the nonrecursive or transversal type, i.e., the output is a simple

†The development can easily be made to include the case where $M = N$.

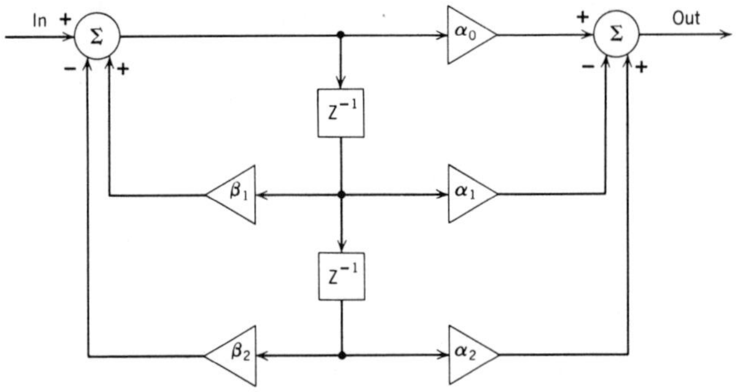

Fig. 7.3 A second-order digital filter of the recursive type.

linear weighting of the present and previous (N-1) samples of the input. Figure 7.4 is an example of an nth order nonrecursive filter.

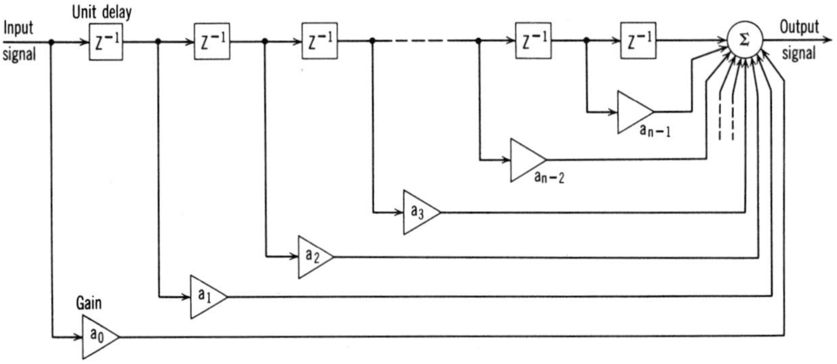

Fig. 7.4 An nth-order nonrecursive digital filter with $a_n = 1$.

The design methods for each of these two classes of filters differ markedly from one another and thus are considered separately. Each class of filters has its distinct properties. The nonrecursive filter has a *finite memory* and can have excellent phase characteristics. The nonrecursive filter however tends to require a large number of terms to obtain a relative sharp cutoff. On the other hand the recursive filter has *an infinite memory* and tends to have fewer terms.

With this economization of terms there usually exists relatively poorer phase characteristics. Sharp cutoff filters are much easier to design using a recursive structure.

The digital counterpart of the linear lumped parameter continuous system is the recursive filter of (7.8). The digital filter used for the approximation of frequency characteristics by trigonometric polynomials is the nonrecursive filter. We now turn our attention first to the nonrecursive filter design methods.

7.4 NONRECURSIVE DIGITAL FILTER DESIGN METHODS

The design methods for nonrecursive filters generally fall into three classes: first, those relating to the classical interpolation and differentiating formulae for equally spaced data, second, those deriving from a Fourier series approach, and third, those utilizing least-squares approximation in some form.

7.4.1 The Classical Numerical Analysis Approach

The classic interpolation and differenting formulae [1] for equally spaced data can be digitized directly by making the correspondence between the unit advance operator **E**, as used in numerical analysis, and the z-transform operator z. In general application of these formulae result in low-pass and differentiating filters with excellent amplitude characteristics only at the very low frequencies (small relative bandwidths). As can be easily shown these methods correspond to using the first few terms of the power series expansion around zero frequency for the desired magnitude characteristics. The use of this method is restricted primarily to filters whose function is integration, interpolation, differentiation, or some combination thereof.

7.4.2 The Fourier Series Approach

The Fourier series and related methods, however, offer considerably greater flexibility. The Fourier method consists first in expanding the magnitude-frequency characteristics of the desired continuous filter, H(s), in a Fourier series over the band $|\omega| < \omega_s/2$. The choice of a sine or cosine series is made on the basis of the behavior of H(s) at very small frequencies. If $H(s) \approx Ks^m$ for small s then a cosine series is chosen when m is even and a sine series when m is odd. Thus

$$H(\omega) = \sum_{n=0}^{\infty} a_n \cos n\omega T, \quad m \text{ even} \tag{7.10}$$

or

$$H(\omega) = \sum_{n=1}^{\infty} b_n \sin n\omega T, \quad m \text{ odd} \tag{7.11}$$

but by definition

$$z^{-1} \equiv \varepsilon^{-j\omega T} \tag{7.12}$$

hence by using the exponential forms for the trigonometric functions there results

$$H^*(z) = a_o + \tfrac{1}{2} \sum_{n=1}^{\infty} a_n \left(z^n + z^{-n}\right), \quad m \text{ even} \tag{7.13}$$

and

$$H^*(z) = \tfrac{1}{2} \sum_{n=1}^{\infty} b_n \left(z^n - z^{-n}\right), \quad m \text{ even} \tag{7.14}$$

which are the desired nonrecursive digital filters.

If the convergence of the Fourier series is not sufficiently rapid enough to make errors resulting from truncating the infinite series of (7.13) or (7.14) small, then in order to yield a satisfactory approximation with a finite number of terms the coefficients a_n or b_n must be modified. Martin [14] and

Ormsby [15] got around the difficulty of slow convergence by going back to the original $H(\omega)$ and modifying it so as to eliminate discontinuities in $H(\omega)$ and its first derivative, $H'(\omega)$. This modification improves the convergence somewhat by reducing the source of the Gibbs oscillations, [12, p. 217-219] but at the same time it makes the determination of the series coefficients more difficult. Their method is applicable to filters with magnitude characteristics that are piecewise specified with initially undefined transition region behavior as sketched in Fig. 7.5

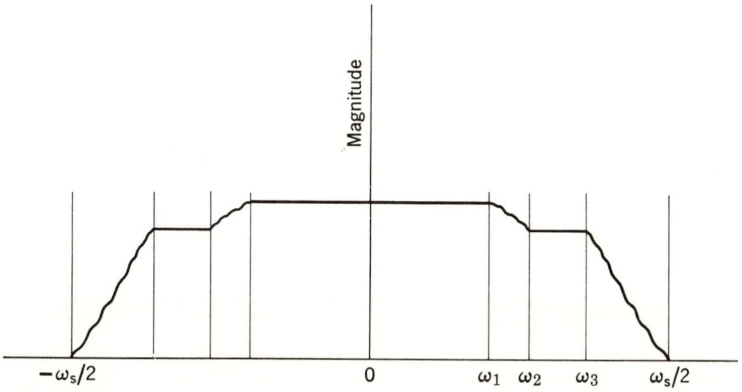

Fig. 7.5. A piecewise-specified magnitude characteristic with unspecified transition regions.

7.4.3 The Modified Fourier Series Method

An alternate approach to that of Ormsby and Martin is to modify the coefficients a_n or b_n in such a way that the error incurred by truncating the series (7.13) or (7.14) is greatly reduced. This can be conveniently accomplished by multiplying the time response $h(nT)$ by a time limited even function $w(t)$, i.e., $w(t) = w(-t)$ and $w(t) = 0$ for $|t| > \tau$. But multiplication in the time domain corresponds to convolution in the frequency domain. Thus, if the weighting function, $w(t)$, is chosen such that the frequency content of its transform is concentrated primarily in the central lobe, i.e., $|\omega| < \omega_o$ then the effect of this convolution on the frequency characteristics of H*(z) will be to smooth out the sharp transitions or slope discontinuities in H*(z). This procedure† is sketched in Fig. 7.6 where the desired filter is a wideband differentiator.

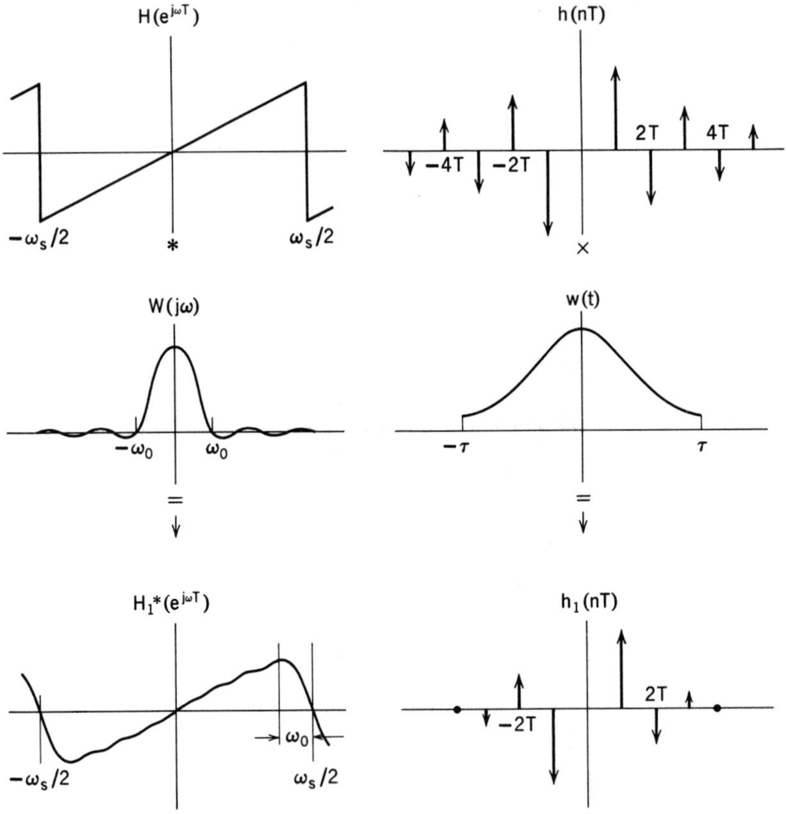

Fig. 7.6. The weighting function method applied to improve convergence.

The design equations thus become

†This technique has been used by Ross [34], Kaiser [28], Heyliger [35], Graham [36], Anders [37], and others.

$$H_1^*(z) = a_o w(o) + \frac{1}{2} \sum_{n=1}^{N} [a_n w(nT)][z^n + z^{-n}], \quad m \text{ even} \tag{7.15}$$

or

$$H_1^*(z) = \frac{1}{2} \sum_{n=1}^{N} [b_n w(nT)][z^n - z^{-n}], \quad m \text{ odd} \tag{7.16}$$

where N is the greatest integer in (τ/T).

An especially simple weighting function which may be used is Hamming's [13, p. 95-99] window function defined by

$$w_h(t) = \begin{cases} 0.54 + 0.46 \cos(\pi t/\tau), & |t| < \tau \\ 0, & |t| > \tau \end{cases} \tag{7.17}$$

For this function 99.96% of its energy lies in the band $|\omega| \leq 2\pi/\tau$ with the peak amplitude of the side lobes of $W_h(j\omega)$ being less than 1% of the peak.

Other specific window functions [38, 39] may be used to advantage. One especially flexible *family* of weighting functions with nearly optimum† characteristics is given by the Fourier cosine transform pair [40]

$$w(t) = \begin{cases} \dfrac{I_o\left[\omega_a \sqrt{\tau^2 - t^2}\right]}{I_o(\omega_a \tau)} & |t| < \tau \\ 0 & |t| > \tau \end{cases} \tag{7.18}$$

† This family of window functions was "discovered" by Kaiser in 1962 following a discussion with B. F. Logan of the Bell Telephone Laboratories.

$$W(j\omega) = \frac{2}{I_0(\omega_a\tau)} \frac{\sin\left[\tau\sqrt{\omega^2 - \omega_a^2}\right]}{\sqrt{\omega^2 - \omega_a^2}} \tag{7.19}$$

where I_0 is the modified Bessel function of the first kind and order zero. By varying the product $\omega_a\tau$ the energy in the central lobe and the amplitudes of the side lobes can be changed. The usual range on values of $\omega_a\tau$ is $4 < \omega_a\tau < 9$ corresponding to a range of side lobe peak heights of 3.1% down to 0.047%. Figure 7.7 shows w(t) and W(j ω) for $\omega_a\tau = 6.0$ and 8.5.

The flexibility of this set of window functions is now illustrated. Setting $\omega_a\tau = \pi\sqrt{3} = 5.4414$ places the first zero of W(jω) at 2π on a normalized scale, the same location as that of the cosine transform of the Hamming window [13, p. 95-99]. The closeness of the resulting windows is immediately apparent. The Hamming window, while having a slightly lower peak amplitude on the first two side lobes, continues to oscillate approximately sinusoidally with a slowly diminishing amplitude. On the other hand the I_0-sinh window with $\omega_a\tau = \sqrt{3}\pi$ has a slightly greater magnitude for the first two side lobes but the amplitude of the sinusoidal oscillations of the window tails diminishes much more rapidly. The Hamming window has all but 0.037% of its energy in the main lobe while the I_0-sinh window with $\omega_a\tau = \sqrt{3}\pi$ has all but 0.012% of its energy in the central lobe.

As a second example, the Blackman window, his "not very serious proposal" [13, p. 98], has its first zero at 3π and corresponds to the I_0-sinh window with $\omega_a\tau = 2\sqrt{2}\pi = 8.885$. The nearly equivalent value of $\omega_a\tau = 8.50$ can be used for comparison. This function has its first zero at $2\pi(1.44226)$ rather than at 3π. Again the closeness of these two window functions can be observed by comparing our Fig. 7.7c with

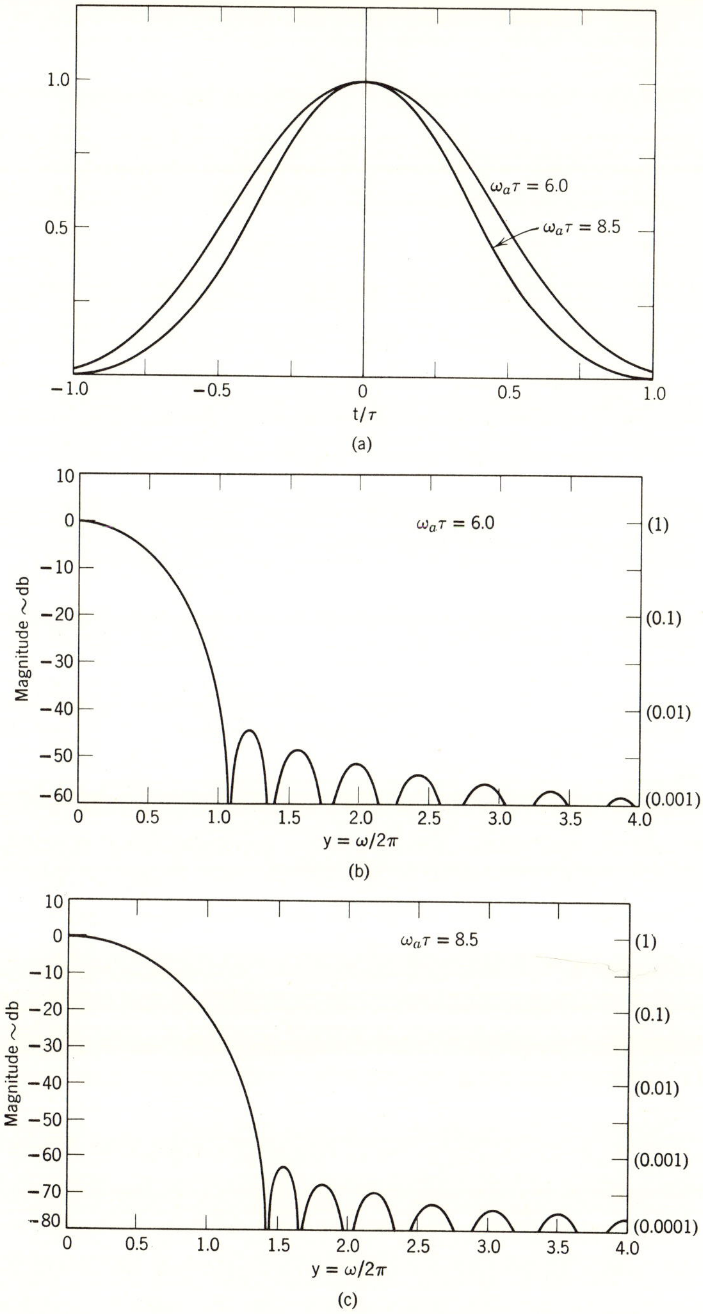

Fig. 7.7. The I_0-sinh window function (a) and its Fourier transform, (b) and (c), for $\omega_a\tau$ = 6.0, 8.5.

Fig. 15e, of Blackman and Tukey [13, p. 97]. It is seen that the I_0-sinh window, $\omega_a\tau = 8.5$, has measurably smaller side lobes than the Blackman window in spite of the smaller value of its first zero. The same general comments with regard to side lobe amplitude apply here as were presented with reference to the Hamming window.

Another valuable property of this family of modified Bessel type window functions is that they also approximate closely [41] the prolate spheroidal wave functions of order zero and whose bandlimiting properties [42] are well established. The remarkable closeness of the two sets of functions is illustrated by a plot of their difference, Fig. 7.8, for various values of $\alpha = \omega_a\tau$ and the corresponding prolate parameter c_0. Recall that $w(o) = 1$ and $S_{00}(c_0, 0) = 1$, and hence the difference is also the relative error. The approximate correspondence in parameter values of α and c_0 for the two sets of functions is shown in Fig. 7.9 as a plot of the ratio c_0/α against c_0.

Returning to the use of these weighting functions for reducing the amplitude of the Gibbs oscillations, the following table [41], Table 7.1, indicates the $\omega_a\tau$ parameter value necessary to reduce the amplitude of the first Gibbs overshoot for step discontinuities to the prescribed value.

TABLE 7.1

$\omega_a\tau$	overshoot amplitude %
0.	8.949
5.0000	0.197
5.4414	0.125
6.0000	0.070
6.5000	0.0415
7.0000	0.0250
7.5000	0.0145
8.0000	0.0085
8.5000	0.0050

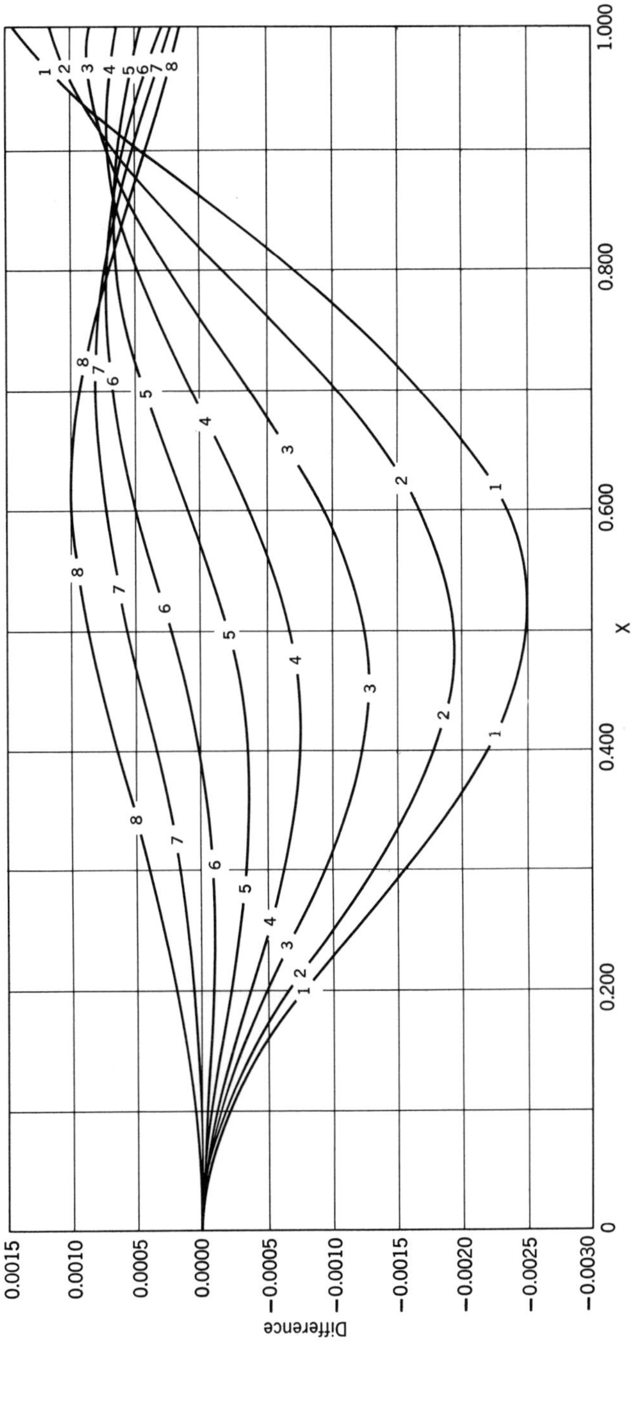

1 - c = 5.311, α = 5.000　　3 - c = 6.292, α = 6.0　　5 - c = 7.274, α = 7.0　　7 - c = 8.257, α = 8.0
2 - c = 5.744, α = 5.441　　4 - c = 6.783, α = 6.5　　6 - c = 7.765, α = 7.5　　8 - c = 8.749, α = 8.5

Fig. 7.8. Approximation of the prolate $S_{00}(c, x)$ by $I_0(\alpha, x)$. Difference = $S_{00}(c, x) - I_0(\alpha, x)$.

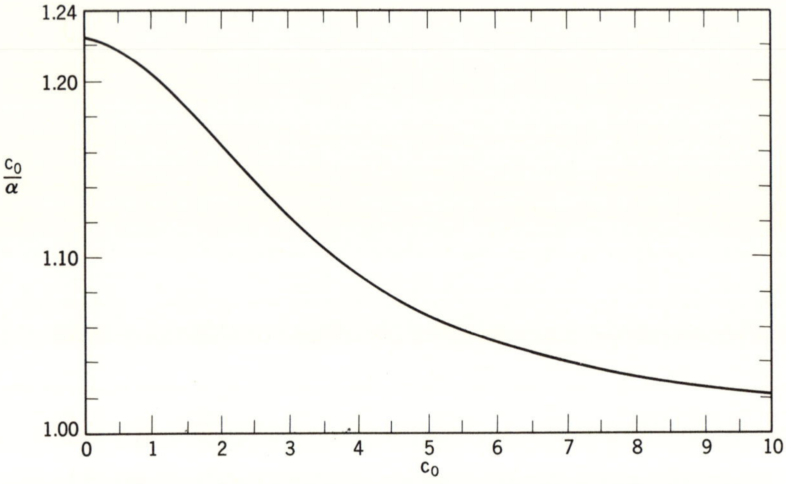

Fig. 7.9. The correspondence of parameter values for the prolate functions, $S_{00}(c, x)$, and the $I_0(\alpha, x)$ functions as shown by their ratio c/α.

By viewing this scheme for increasing the convergence of Fourier series at points removed from discontinuities, the relation between the number of terms required in the digital filter $H^*(z)$ and the usable frequency range of the filter becomes clear. Referring to Fig. 7.6 again, the halving of ω_b and the concomitant doubling of τ yield a filter design having twice as many terms as before.

An estimate for the number of terms (unit delays) required in the nonrecursive digital filter depends on both the allowable tolerance on the overshoot to step discontinuities and on the relative width (the fraction of the Nyquist frequency) for the transition regions. For example this relative width is given by $(\omega_2 - \omega_1)/(\omega_s/2)$ for the filter characteristic shown in Fig. 7.5, p. 230. A convenient expression for the number of terms, n_s, for 0.3% peak overshoot is

$$n_s \approx \frac{8}{\omega_r} \tag{7.20}$$

where ω_r is the per-unit width of the transition region relative to the Nyquist frequency (half the sampling frequency). If the overshoot specification is tightened to 0.01%, the factor *8* increases to *11.5*.

Thus we have a design procedure which really makes the Fourier expansion *useful* for precision designs. This nonrecursive filter design procedure is best applied to filters whose magnitude characteristics are fairly well behaved especially at very small frequencies and at frequencies approaching the Nyquist limit. For *wideband* filter designs this procedure is superior to the standard numerical analysis approach and has the additional advantage of being admirably suited to either machine or hand calculation.

7.4.4 Least-Square, Chebyshev, and Other Methods

The direct Fourier series method has the property that it gives the best least-squares fit to the desired transfer characteristic for equal weighting of errors over the entire frequency range of $-\frac{\omega_s}{2} < \omega \leq \frac{\omega_s}{2}$. But we have seen the objectionable Gibbs oscillations that result from discontinuities in the desired $H(j\omega)$ or some of its derivatives. An alternate approach then is to perform a least-squares design but to weight errors only in the frequency range where specific behavior is desired. For example, in Fig. 7.5, p. 230 the frequency ranges of interest are from 0 to ω_1 and ω_2 to ω_3. A later example, Fig. 7.10e, will show the utility of this approach. Fleischer [43] has utilized this generalized least-squares approach in the design of nonrecursive filters.

Another design method consists of the determination of a Chebyshev or min-max type approximation to the desired transfer characteristic. Martin [44] has used this technique to some advantage. The scheme, however, requires considerable calculation to arrive at the desired result, the computation being carried out usually by machine. The number of cases for which a Chebyshev type approximation can be determined analytically in closed form is vanishingly small.

Another interesting source of design techniques for nonrecursive or transversal filters is the literature on the design of phased arrays of antennae to produce desired beam patterns. See for example [45, 46, 47].

Blackman [11, p. 158-159] gives the outline of a transformation method for designing nonrecursive filters. The method assumes the transfer function is given in squared-magnitude form as a trigonometric polynomial.

Blackman and Tukey [13] and Blackman [11] have also shown the computational advantages to be gained by building up a nonrecursive filter from a cascade of simple sums and decimators. This method has considerable merit for yielding filter designs which are well suited for either hand computation or simple physical implementation in digital hardware.

7.4.5 Some Examples

In order to obtain a comparison among some of the nonrecursive design methods outlined in this chapter the design of a differentiating filter is carried out using six different methods. The comparison is made on the basis of the normalized magnitude-frequency characteristic plotted on a frequency scale normalized such that 1.0 unit of normalized frequency corresponds to the Nyquist frequency, $\omega_s/2$. The first five filter designs all have 19 taps each; the sixth filter has 21 taps. All filters are antisymmetrical with a zero center tap, i.e.

$$H^*(z) = \frac{\pi}{T} \sum_{n=1}^{\overset{(10)}{9}} a_n(z^n - z^{-n}), \quad a_o = 0 \tag{7.21}$$

Tables 7.2 and 7.3 give for each filter the coefficients, a_n, and the magnitude response based on $T = \pi$ seconds.

The filter obtained by the direct Fourier series method is shown in Fig. 7.10a; the Gibbs' oscillation is evident. Using the window method with an $\omega_a\tau$ value of 6 to weight the filter

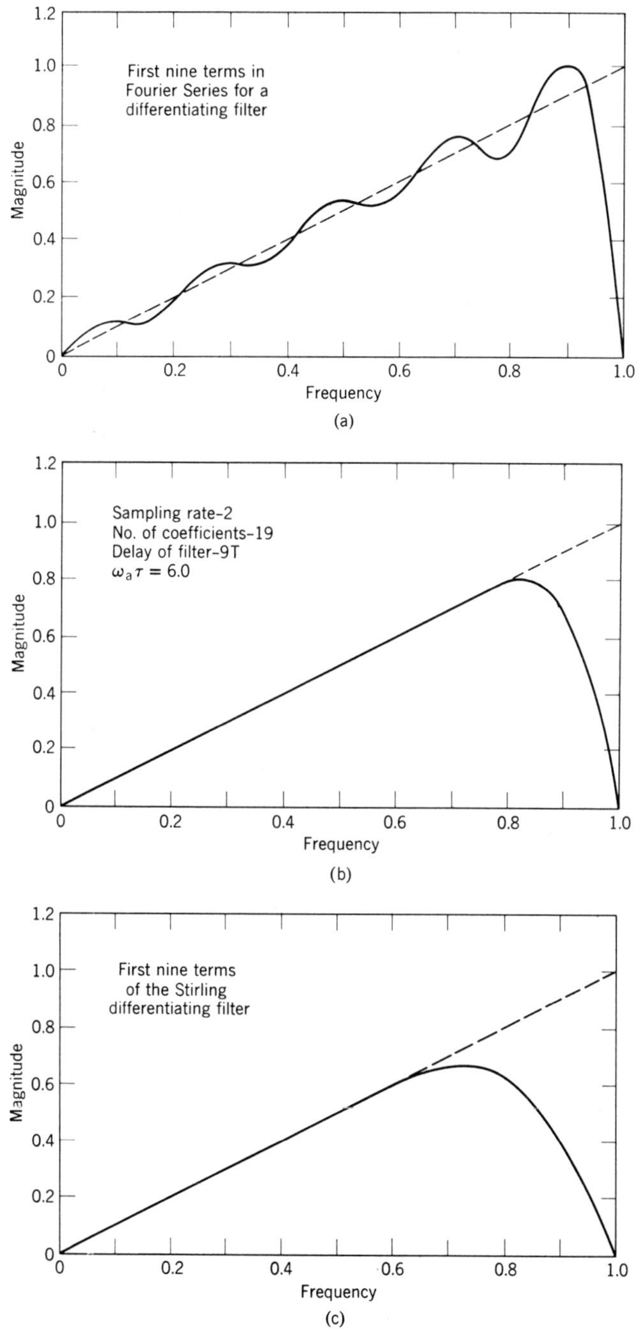

Fig. 7.10. The first-order differentiating filter as

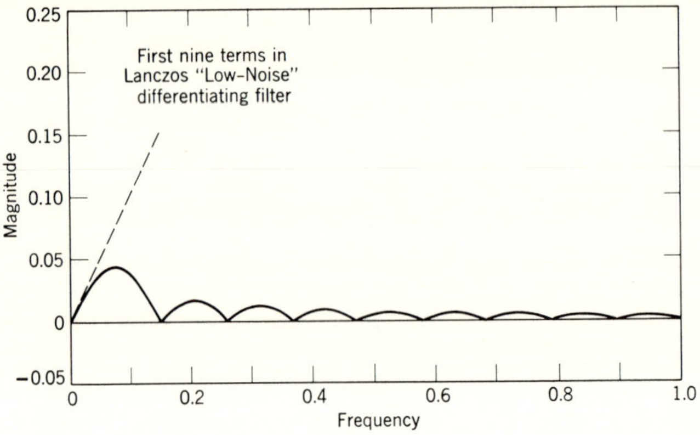

(d)

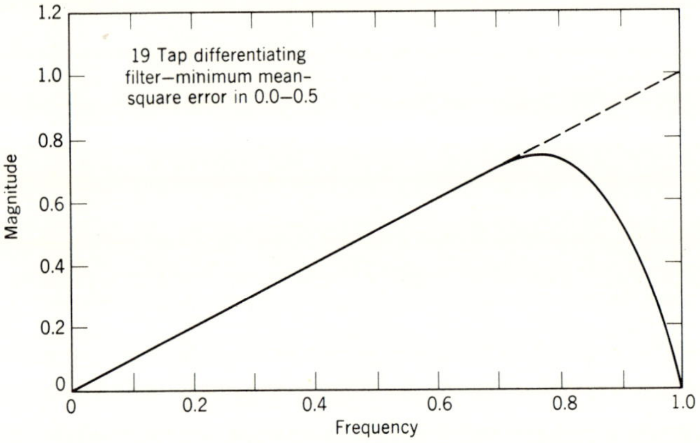

(e)

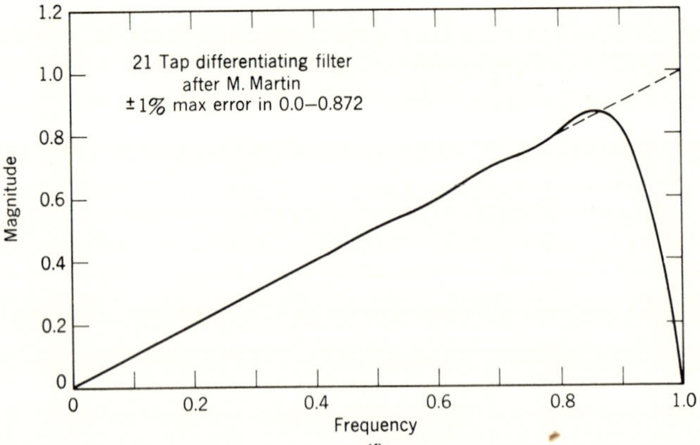

(f)

designed by six different methods yielding nonrecursive filters.

TABLE 7.2

DIFFERENTIATING FILTER CØEFFICIENTS

TAP	(A)	(B)	(C)	(D)	(E)	(F)
1	1.000000	0.967288	0.900000	0.001754	0.939722	0.983896
2	-0.500000	-0.437297	-0.327273	0.003509	-0.389068	-0.468363
3	0.333333	0.245513	0.127273	0.005263	0.187977	0.287290
4	-0.250000	-0.143478	-0.044056	0.007018	-0.088332	-0.191291
5	0.200000	0.081825	0.012587	0.008772	0.037527	0.130620
6	-0.166667	-0.043622	-0.002797	0.010526	-0.013638	-0.088958
7	0.142857	0.020715	0.000450	0.012281	0.003969	0.059197
8	-0.125000	-0.008042	-0.000046	0.014035	-0.000828	-0.038097
9	0.111111	0.001962	0.000002	0.015789	0.000095	0.025113
10	-	-	-	-	-	-0.009885

TABLE 7.3

DIFFERENTIATING FILTER FREQUENCY RESPØNSE MAGNITUDE

FREQ	(A)	(B)	(C)	(D)	(E)	(F)
0.00	0.000000	0.000000	0.000000	0.000000	0.000000	0.000000
0.05	0.083424	0.050055	0.050000	0.039774	0.050000	0.050291
0.10	0.105016	0.099960	0.100000	0.035258	0.100000	0.100912
0.15	0.116495	0.149931	0.150000	0.000461	0.150000	0.149672
0.20	0.189713	0.200065	0.200000	0.017187	0.200000	0.198072
0.25	0.283682	0.250056	0.250000	0.003678	0.250000	0.250262
0.30	0.316119	0.299859	0.300000	0.010960	0.300000	0.302910
0.35	0.316002	0.349907	0.349999	0.004562	0.350000	0.349783
0.40	0.377044	0.400182	0.399991	0.007686	0.400000	0.396110
0.45	0.484542	0.450055	0.449937	0.004922	0.450000	0.450285
0.50	0.531527	0.499649	0.499662	0.005584	0.500000	0.504930
0.55	0.514484	0.549930	0.548557	0.005101	0.550018	0.549490
0.60	0.556781	0.600516	0.594923	0.004057	0.600107	0.594200
0.65	0.687432	0.649795	0.634901	0.005200	0.649791	0.651518
0.70	0.761116	0.699105	0.661293	0.002845	0.695820	0.706458
0.75	0.708095	0.751297	0.663135	0.005257	0.728011	0.745921
0.80	0.706175	0.794590	0.627124	0.001814	0.726028	0.808830
0.85	0.906289	0.785264	0.541529	0.005289	0.662477	0.857662
0.90	1.077312	0.659812	0.401698	0.000884	0.515273	0.819302
0.95	0.790293	0.384161	0.214675	0.005303	0.284239	0.528082
1.00	0.000000	0.000000	0.000000	0.000000	0.000000	0.000000

coefficients yielded the filter whose response is shown in Fig. 7.10b; this filter has an 80% usable bandwidth for derivative action. If up to the ninth order differences are used in the Stirling differentiation formula the filter of Fig. 7.10c results. This filter maintains a slope error of less than 1% out to only 60% of the Nyquist rate (clearly the last two terms, a_8 and a_9, are so small has to have an almost negligible effect).

The filter of Fig. 7.10d is of a design postulated by Lanczos [12, p. 321-322] to have low effective noise transmission (minimum Σa_j^2) and first derivative action ($\Sigma j a_j = 1$). We immediately see the effect of the optimization technique; the filter is trying to approximate zero response everywhere on the real frequency axis except at the point $\omega = 0$ where the slope is to be unity. The usable frequency range for this derivative filter is less than 2%; hence it is usable only for very slowly changing signals.

If a least-squares design is used with the error (departure from true derivative action) being weighted only over the interval $0 < |\omega| < 0.5\ \omega_s/2$ then the response of the design is as shown in Fig. 7.10e. Note the almost perfect response to 60% and the effective (1% error) bandwidth of 70%, a marked improvement over the Stirling filter.

The last filter was taken from Martin [44, p. 77] where the design criterion was a min-max type relative error curve oscillating between ±1% error. The usable bandwidth is ≈87% and is the maximum obtainable within the specification of 1% relative error and 21 total taps.

Thus we see each of the nonrecursive design methods has its own particular advantages and disadvantages. The choice of method will ultimately depend on the desired filter characteristics.

From (7.20) we see that the number of taps required increases rapidly when sharp changes in amplitude are required. The recursive filters described in the next section offer a more economical solution to this problem than do the nonrecursive filters.

7.5 RECURSIVE FILTER DESIGN METHODS

The basic form of the recursive digital filter, $H^*(z)$, is given by

$$H^*(z) = \frac{\sum_{j=0}^{N} a_j z^{-1}}{1 + \sum_{j=1}^{N} b_j z^{-1}} \tag{7.22}$$

where z^{-1} is the unit delay operator. The general problem of recursive filter design is then the determination of the filter coefficients, the a_j and b_j, such that the filter specifications are satisfied.

There are two basic approaches to this design problem. The first, and most direct, approach is to determine the a_j and b_j coefficients of the digital filter by some computation procedure directly from the filter specifications. This may involve the application of the approximation theory of rational functions of trigonometric polynomials for which there is an extensive literature. See for example Walsh [48], Cheney and Loeb [49], and Stoer [50]. Specific iterative design procedures as well as fixed designs determined from a trial-and-error procedure are also forms of the direct approach.

The second basic approach to the determination of the digital filter coefficients is an indirect one consisting of two parts. The first part consists of the determination of a suitable continuous filter H(s) that satisfies essentially the same filter specification. The determination of a satisfactory digitalization of this continuous filter then comprises the second part of the method. Thus this method makes it possible to utilize effectively the extensive literature of continuous filter design built up so elaborately by the electrical engineers over the last thirty-five years. This indirect method is implicit in the digital simulation of continuous filters; here the continuous filter is already known and thus only the digitalization is required.

In this section three basically different methods are discussed for effecting the digitalization required by the indirect approach. These methods include (though not necessarily in the order of importance) the standard z-transform, the bilinear z-transform, and the Boxer and Thaler method. The descriptions of numerous other particular design techniques then follow.

7.5.1 The Standard z-Transform†

The indirect method requires the digitalization of a continuous filter, H(s), whose transfer characteristic has the form

$$H(s) = \frac{\sum_{m=0}^{M} c_m s_m}{\sum_{n=0}^{N} d_n s_n} \tag{7.23}$$

Since digital filters are necessarily effectively *bandlimited* and since the H(s) of (7.23) is not strictly bandlimited, some form of approximation must appear in the digitalization process.

The nature of the approximation of the standard z-transform can be seen from the defining equations for the transform. The standard z-transform [6] of this transfer characteristic is defined by

$$H^*(s) = \sum_{n=-\infty}^{\infty} H(s+jn\omega_s) + \frac{T}{2} h(0^+) \tag{7.24}$$

† The material in this section follows closely that of Kaiser [28] and Golden and Kaiser [51].

or equivalently by

$$H^*(z) = T \sum_{n=0}^{\infty} h(nT)z^{-n} \tag{7.25}$$

The asymptotic behavior for large s is assumed to be

$$\lim_{s \to j\infty} | H(s) | = \lim_{s \to j\infty} | 1/(s/\omega_c)^n | , \quad n > 0 \tag{7.26}$$

Thus, from (7.25), the z-transform is so defined that the impulse response of the sampled filter is identical to the sampled impulse response of the continuous filter. Further, $H^*(s)$ is periodic in ω of period ω_s. From (7.24) it follows that in the baseband $(-\omega_s/2 \leq \omega \leq \omega_s/2)$ the frequency response characteristics of the obtained sampled filter, H(s), differ from those of the continuous filter, H(s), the difference being the amount added or *folded* in through terms of the form $H(s+jn\omega_s)$ $n \neq 0$. If H(s) is bandlimited to the baseband, i.e., $| H(s) | = 0$ for $| \omega | > \omega_s/2$, then there is no folding error and the frequency response of the sampled-data filter is identical to that of the continuous filter. Unfortunately, when H(s) is a rational function of s, it is not bandlimited and therefore $H(s) \neq H^*(s)$ in the baseband.

The magnitude of the errors resulting from the folding is directly related to the high frequency asymptotic behavior of H(s) as defined in (7.26). If n is large and $\omega_c << \omega_s/2$, then the folding errors will be small and the standard z-transform generally will yield a satisfactory sampled-data filter design. However, in wideband simulations, ω_c is usually an appreciable fraction of $\omega_s/2$. Furthermore, many continuous filter designs, such as elliptic filters or Chebyshev Type II filters, yield transfer functions in which n is no greater than 1. These two conditions, namely $\omega_c \approx \omega_s/2$ and $n = 1$, can create large folding errors in the frequency response characteristics of the filter obtained by using the standard z-transform and thus render an *unusable* result.

To reduce the possibility of error due to folding, the H(s) can be modified by adding in cascade a wideband low-pass filter G(s), having an 'n' sufficiently large. This *guard* filter, G(s), is usually chosen to have flat magnitude and linear phase characteristics in the frequency range where it is desired to have the H*(s) faithfully approximate H(s). This guard filter is obtained by using one of the standard low-pass all-pole filter forms such as the Butterworth or Chebyshev Type I and if necessary followed by an all-pass filter for phase equalization.

The procedure is to form $H_m(s)$ as

$$H_m(s) = G(s)\, H(s) \tag{7.27}$$

and then to apply the standard z-transformation to this $H_m(s)$.

With H(s) the quotient of a numerator polynomial N(s) of order n_N and a denominator polynomial D(s) of order n_D where $n_D > n_N$, application of the standard z-transformation requires first determining the zeros of D(s), then obtaining a partial fraction expansion of H(s) in these poles, and finally z-transforming each of the individual terms. From the following transform pair all the necessary pairs can be derived.

$$\frac{1}{s+a} \longrightarrow \frac{T}{1-\varepsilon^{-aT}z^{-1}} \tag{7.28}$$

Thus it is seen that H*(z) will be rational in z^{-1} and will have at most $2n_D$ coefficients. If, in addition, a *guard* filter of order n_G is used then the obtained $H^*_m(z)$ will be more complex now having at most $2(n_D+n_G)$ coefficients.

Another limiting feature of the standard z-transformation is the fact that the moments of h(t) do not remain invariant, i.e., in general*

$$T \sum_{\ell=0}^{\infty} (\ell T)^n \, h(\ell T) \neq \int_0^{\infty} t^n h(t)dt \qquad (7.29)$$

However, the numerator coefficients of H*(z) may usually be varied slightly so as to satisfy an equality for the first few moments without seriously affecting the overall frequency characteristics of H*(z).

7.5.2 The Bilinear z-Transformation†

To circumvent the *folding* problem of the standard z-transform, a transformation can be employed which will map the entire complex s plane into the horizontal strip in the s_1 plane bounded by the lines $s_1 = -j\omega_S/2$ and $s_1 = j\omega_S/2$. Since a permissible H*(z) must also be periodic in ω of period ω_S this transformation must also cause H(s) to be mapped identically in each of the other horizontal strips bounded by the lines $s_1 = j(n - \frac{1}{2})\ \omega_S$ and $s_1 = j(n+\frac{1}{2})\ \omega_S$ where n is an integer.

A transformation having these properties is the bilinear z-transform or z-form which is defined as

$$s = \frac{2}{T} \tanh\left(\frac{s_1 T}{2}\right) \qquad (7.30)$$

which becomes upon substituting

---

*This noninvariance of moment values was the primary reason Blackman [11, p. 83] abandoned the standard z-transform for his particular data smoothing and prediction applications.

†One of the earliest applications of the bilinear transform for analyzing discrete systems was made by Tustin [19].

$$z^{-1} = \varepsilon^{-s_1 T}$$

$$s = \frac{2}{T}\frac{(1-z^{-1})}{(1+z^{-1})} \tag{7.31}$$

Thus in terms of the z^{-1} plane this algebraic transformation (7.31) has the property of uniquely mapping the left-half of the s-plane into the exterior of the unit circle in the z^{-1} plane. Folding errors are eliminated since no folding occurs.

The z-form is applied simply by making the substitution indicated by (7.31) in the transfer characteristic H(s). Hence

$$H^*(z) = H(s)\Big|_{s = \frac{2}{T}\frac{(1-z^{-1})}{(1+z^{-1})}} \tag{7.32}$$

The price paid for this feature is the nonlinear warping imparted to the frequency scale as can be seen by setting $s_1 = j\omega_1$ and $s = j\omega$ in (7.30) to yield

$$\frac{\omega T}{2} = \tan\left(\frac{\omega_1 T}{2}\right) \tag{7.33}$$

The deviation from linearity of this relation is shown in Fig. 7.11.

Because of this frequency warping aspect, the z-form is most useful in obtaining sampled-data filter approximations for continuous filters whose magnitude characteristics can be divided along the frequency scale into successive pass and stop bands where the gain or loss is essentially constant in each band. This type of frequency behavior is typical of many low-pass, bandpass and bandstop filters. Compensation can be made for the effect of warping by prewarping the continuous

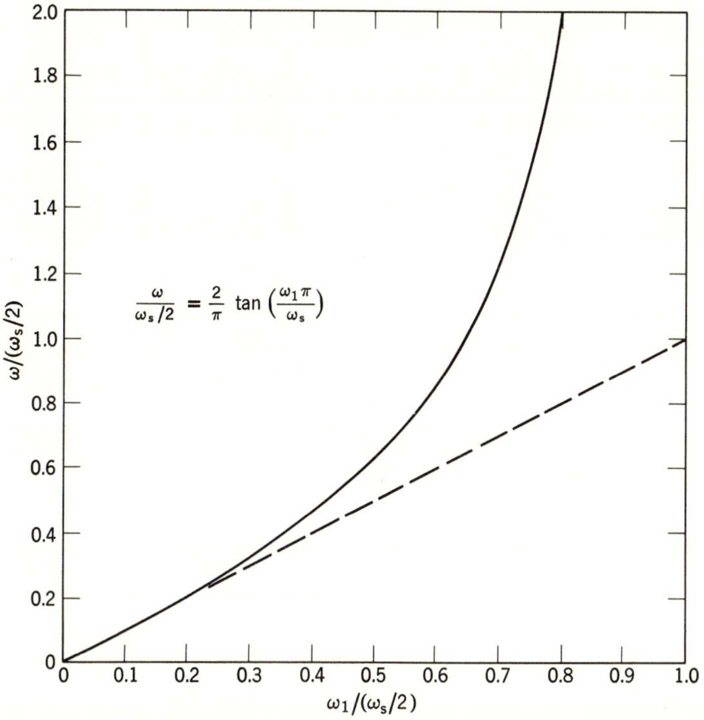

Fig. 7.11. Nonlinear warping of the frequency scale in the bilinear z-transformation.

filter design in the opposite way such that upon applying the z-form the critical frequencies will be shifted back to the desired values [51].

For a simple example of the application of the bilinear z-transform with compensation for frequency warping consider the digitalization of the class of Butterworth low-pass filters $B_n(s)$. The magnitude-squared characteristic of this class of filters is given by

$$B_n(s)B_n(-s) = \frac{1}{1 + (-)^n\left(\frac{s}{\omega_o}\right)^{2n}} \tag{7.34}$$

where ω_0 is the cutoff frequency. The digitalization consists simply of replacing ω_0 by its prewarped value determined from (7.33) as

$$\omega_0 \rightarrow \frac{2}{T}\tan(\omega_0 T/2) \tag{7.35}$$

Then using (7.30) and this (7.35) in (7.34) yields the magnitude squared characteristic of the obtained digital filter

$$B_n^*(s)B_n^*(-s) = \frac{1}{1 + (-)^n\left[\dfrac{\tanh(s_1 T/2)}{\tan(\omega_0 T/2)}\right]^{2n}} \tag{7.36}$$

which upon setting $s_1 = j\omega$ immediately gives

$$|B_n^*(j\omega)|^2 = \frac{1}{1 + \left[\dfrac{\tan(\omega T/2)}{\tan(\omega_0 T/2)}\right]^{2n}} \tag{7.37}$$

The transfer characteristic of the required digital filter is found by using (7.31) along with (7.35) in (7.34) to give

$$B_n^*(z^{-1}) = \frac{1}{\left\{1 + (-)^n\left[\dfrac{\left(\dfrac{1-z^{-1}}{1+z^{-1}}\right)}{\tan(\omega_0 T/2}\right]^{2n}\right\}^+} \tag{7.38}$$

where the (+) sign denotes keeping only those n roots of the denominator polynomial in z^{-1} that lie outside the unit circle. (The other n roots lie inside the unit circle with geometric symmetry about the unit circle.)

It is interesting to note that this design technique, especially with regard to the Butterworth filters is being continually *rediscovered* albeit different terminology. See for example Rader and Gold [53, p. 17] and Holtz and Leondes [54, p. 270]; both sources list filters of the form of (7.37) without realizing or pointing out the direct relation to the bilinear transformation. Both of these references also list five other forms related to (7.37) one of which corresponds to the reciprocal of (7.31) and is essentially the high-pass version of it.

Returning to the bilinear z-form we see that if the magnitude-frequency characteristics are not essentially piecewise constant over the major portion of the Nyquist frequency interval, the frequency warping inherent in the z-form method can render an unsatisfactory filter design. This is immediately obvious for the wideband differentiating filter, $H(s) = s$, if we refer to Fig. 7.11 and interprete it as the magnitude-frequency characteristics of this filter.

Thus the z-form has two primary advantages. First, the magnitude-frequency characteristics of the continuous filter are carried directly over to the obtained digital filter except for a warping of the frequency scale, i.e., continuous filters with equiripple magnitude characteristics transform into equiripple sampled filters with only the position of the maxima and minima being shifted in frequency a calcuable and compensatable amount. Secondly, application of the z-form is algebraic in form and can be applied equally well to the rational transfer characteristics of the continuous filter in either polynomial or factored form.

With respect to the moments of $h(t)$ it can be shown [11, p. 74] that only the first two non-zero moments remain invariant under application of the bilinear z-transformation. As in the case of the standard z-transform, agreement can be forced for higher order moments if slight perturbations are made in the numerator coefficients of $H^*(z)$. This modification scheme corresponds somewhat to Blackman's modified method of moments [11, p. 75].

In comparing the z-form method with the standard z-transform method it is seen that both methods yield sampled-data transfer functions which are rational in z^{-1} and of the same degree namely that of the degree of the continuous filter. Further when the ratio of the moduli of the poles of the continuous filter to the sampling frequency approaches zero the two methods give identical results as one would expect. For obtaining *wideband* designs by use of the standard z-transformation guard filters must usually be used thus raising the order of the digital filter. For obtaining *wideband* designs using the z-form the frequency warping must always be compensated for; although no increase in filter order is necessary, this method is restricted to filters with essentially piecewise constant magnitude characteristics.

Golden and Kaiser [52] have utilized the versatility of these two design methods in the implementation of a general digital filter design program. The program first generates continuous filter designs for the basic low-pass filter structures* (Bessel, Butterworth, Chebyshev, elliptic, etc), then transforms these by any of the four band-to-band transformations-low-pass to low-pass, bandpass, bandstop, or high-pass. Finally the program applies either the standard z- or bilinear z-transform method (prewarping included) to obtain the resulting digital filter. Complete graphical and tabular data are supplied for each digital filter design. This program has been a valuable aid in determing the digital filter coefficients required for simulation of complex systems via the block diagram compiler BLODI (see Chapter 8).

7.5.3 Two Examples of Recursive Filters

To illustrate the range of application and limitations of both the standard and the bilinear z-transforms two designs are briefly presented.

*These basic filter structures are discussed in detail in such network synthesis references as Weinberg [55], Calahan [56], and Storer [57].

The first consists of the design of a *wideband* differentiating filter whose magnitude characteristics are to be within 0.1 db out to 80% of the Nyquist limit. Because the magnitude characteristic is not piecewise constant, the bilinear z-form may not be used satisfactorily. Instead the standard z-transform is used. A guard filter is necessary to reduce folding error at all frequencies to within specifications. The guard used is a 0.1 db ripple Chebyshev low-pass filter of ninth-order chosen to give 50 db loss at 120% of the Nyquist limit. The magnitude response of the resulting sampled filter is shown in Fig. 7.12. The delay of 3T seconds can be compared with the delay of 9T seconds for the nonrecursive design as shown in Fig. 7.10b, p. 240. It is interesting to note that both designs require eighteen coefficients; the recursive filter has only one-third the delay of the nonrecursive design but at the expense of a phase characteristic that departs from linearity as the Nyquist frequency is approached.

As the second example, consider the design of a digital bandstop filter having equiripple passband magnitude characteristics and equiripple stopband loss. The bilinear z-form is used with prewarping of the band-edge frequencies being required. The result of the design is shown in Fig. 7.13. This result could not have been arrived at by either the standard z-transform method or any reasonable nonrecursive design scheme. In fact if these specifications were to be realized by a nonrecursive filter the number of taps would be, using (7.20), approximately

$$n_s \approx \frac{11.5}{8/5000} \approx 7200 \text{ terms!}$$

7.5.4 The Boxer-Thaler Technique

Another method, related to the bilinear z-transform, for digitalizing continuous filters is that of applying the technique of Boxer and Thaler [21, 22]. An excellent description of the

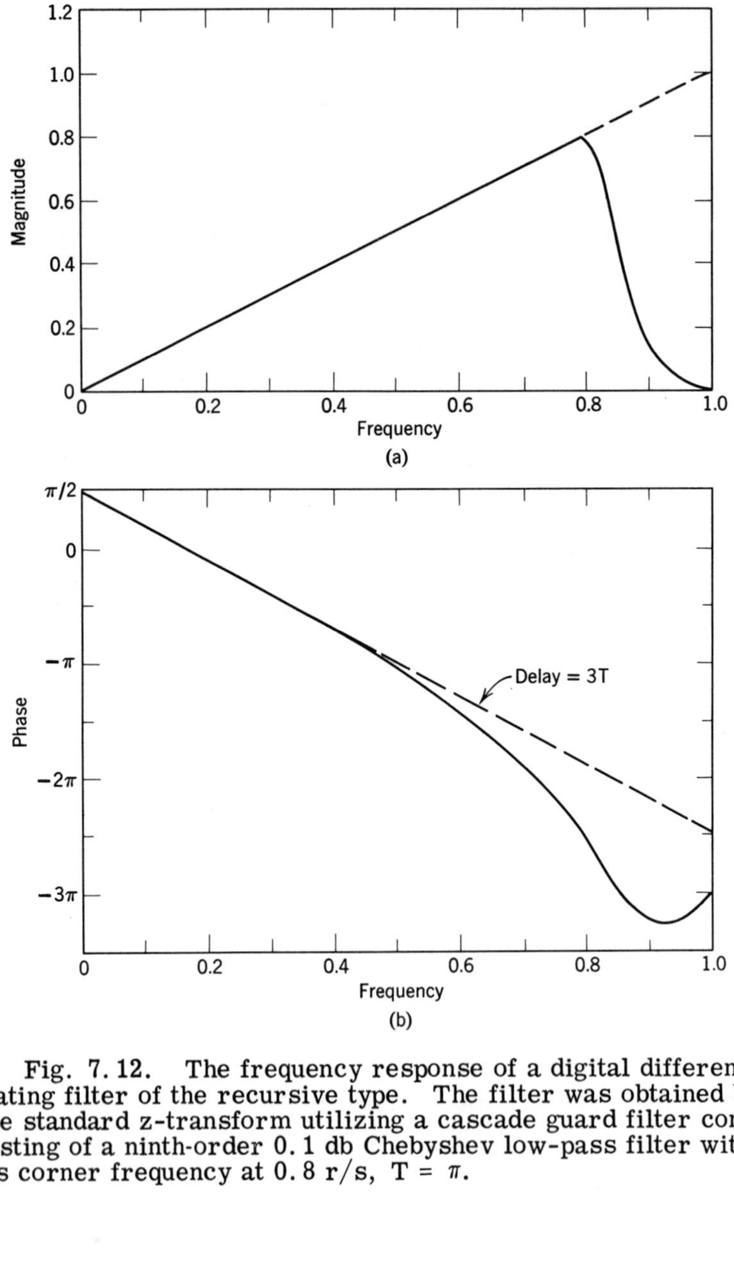

Fig. 7.12. The frequency response of a digital differentiating filter of the recursive type. The filter was obtained by the standard z-transform utilizing a cascade guard filter consisting of a ninth-order 0.1 db Chebyshev low-pass filter with its corner frequency at 0.8 r/s, $T = \pi$.

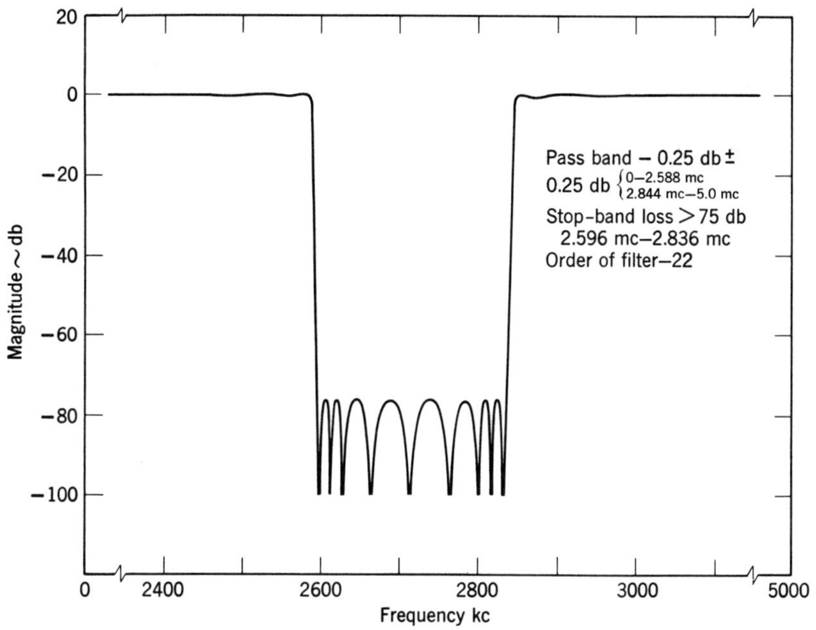

Fig. 7.13. Frequency response of a sampled data band-stop filter of the elliptic type obtained using the bilinear z-transformation with warping correction.

method as applied to determining numerically the response of continuous systems is given by Gibson [25].

The method as applied to the design of digital filters consists first of writing the continuous filter transfer characteristic in powers of s^{-1} rather than s. The transformation scheme then involves replacing each particular power of s^{-1} by a corresponding rational z-form expression.* Upon collecting terms the resulting rational expression in z^{-1} is the required digital filter. As with the standard and bilinear z-form

*Tables of the corresponding z-form expressions are given in [21, 22, 25] of which [25] is the most complete giving expressions up to and including eleventh order.

methods the order of the resulting filter in z^{-1} is identical to that of the original continuous filter, H(s); i.e., if H(s) is of order M/N in s, where $M \leq N$, then the resulting H*(z) will be of order N/N in z^{-1}. Further, the substitution is purely algebraic in nature and thus the original H(s) need not be in factored or partial-fraction form.

The physical interpretation of this digitalizing scheme from a frequency domain viewpoint is complicated by the fact that each higher power of s^{-1} is approximated by a different and more exact rational function of z^{-1}. This fact results also in the property that

$$\beta(G_1G_2) \neq \beta(G_1)\beta(G_2) \tag{7.39}$$

where β denotes the Boxer-Thaler transformation and where G_1 and G_2 are continuous filters of greater than zeroth order. Thus the precise nature of the approximation depends upon the Boxer-Thaler z-form of s^{-n} and upon the particular H(s). As a general rule equiripple continuous filters, H(s), do not transform via the Boxer-Thaler z-form to equiripple digital filters.

A further complicating factor is that the complexity of the z-form increases as the power of s^{-1} increases thus increasing both the amount and the required precision (whose effect we shall see in Section 7.6) of the algebra to be performed. That this fact has an effect on the error was noticed by Gibson [25, p. 155-159] who briefly discusses the sources and effects of error.

7.5.5 Some Other Recursive Design Methods

Thus far the schemes advanced for designing digital bandpass filters have been either to design first a continuous bandpass filter and then to digitalize the result or to cascade overlapping high-pass and low-pass filters. Broome [58] suggests obtaining the digital bandpass filter, $H_1^*(z)$, directly

from the digital low-pass filter, H*(z), by using the translation expression

$$H_1^*(z) = H^*\left(z\varepsilon^{j\omega_0 T}\right) + H^*\left(z\varepsilon^{-j\omega_0 T}\right) \tag{7.40}$$

where ω_0 is the new center frequency. If H*(z) is given in rational form by (7.22) then Broome shows also that $H_1^*(z)$ may be written down directly as

$$H_1^*(z) = \frac{2\sum_{j=0}^{N}\sum_{k=0}^{N} b_j a_k z^{-(j+k)} \cos[\omega_0 T(j-k)]}{\sum_{j=0}^{N}\sum_{k=0}^{N} b_j b_k z^{-(j+k)} \cos[\omega_0 T(j-k)]} \tag{7.41}$$

The passband characteristic of $H_1^*(z)$ is simply the low-pass characteristic of H*(z) centered at $+\omega_0$ and disturbed only by what Broome terms the *distortion error* which results from the tails of the image of H*(z) centered at $-\omega_0$. The use of a good low-pass filter design can reduce the distortion error to a very small level.

A second and more direct method of obtaining digital band-pass filters is to use a digitalizing transformation that will transform a continuous low-pass design directly into a digital bandpass design. Rader and Gold [53, p. 26-28] advance the second-order transformation

$$s \to \frac{z^2 - 2z\cos\omega_c T + 1}{z^2 - 1} \tag{7.42}$$

for this purpose. This transform maps the entire imaginary axis of the s-plane onto *both* the top and the bottom arcs of the unit circle with the origin of the s plane mapping into the two points $\epsilon^{\pm j\omega_c T}$. Thus applying (7.42) to a low-pass H(s) yields a digital band-pass filter with a center frequency of ω_c. In their report [53] Rader and Gold briefly outline a design procedure including warping correction and giving the necessary design equations.

Another recursive filter design method with considerable flexibility involves forming an $H^*(z)$ with its denominator fixed by applying either (7.28) or (7.31) to the denominator of the continuous filter, H(s), and with its numerator left as a polynomial in z. The unknown numerator coefficients can then be determined to satisfy moment constraints as Blackman does in his modified method of moments [11, p. 75] or to satisfy other time or frequency domain criteria. Mantey [59] has investigated in considerable detail the convergence problems associated with the automatic adjustment of these unknown numerator coefficients to satisfy different performance criteria.

Blackman and Tukey [13] and Blackman [11] have shown the computational advantages to be gained by building up a nonrecursive filter from a cascade of various simple sum, difference, and decimation operations. Whittlesey has extended these ideas by including an elementary recursive operation. In [60] Whittlesey tabulates a number of combinations of cascade designs with specific low-pass and bandpass properties.

Young [61] and later Broome [62] have designed a series of digital filters of essentially the all-pass type which generate a set of discrete orthonormal responses. The use of such filters in signal representation studies leads to a considerable simplification in the analysis work. Young [61], in addition, gives many of the details regarding application of the filters.

Recursive digital filters while computationally efficient do not generally possess linear phase characteristics. Broome shows [63] that recursive filters may be made both phaseless and recursive by filtering the input time series in both the forward and backward time sense and summing the two results. This process of course requires the *availability* of the input in both the forward and reverse time sense.

Golden describes [17] a novel way to very efficiently realize a set of many contiguous digital bandpass filters. The method employs a cascade of both second-order recursive filters and a simple *comb* or nonrecursive filter. The single comb filter is common to all the bandpass filters.

Having presented a wide variety of recursive digital filter design methods we now turn to the important questions associated with the actual implementation of these filters. The realization will be either by hardware or by a computational algorithm. In both cases our attention becomes focused on accuracy considerations as related to coefficient representation, to rounding effects, and to the choice of a canonical realization form. These considerations are taken up in the next section.

7.6 ACCURACY CONSIDERATIONS IN DIGITAL FILTER DESIGN*

The increased computational accuracy of high speed digital computers, the broader dynamic ranges in both amplitude and frequency of the simulated system variables, and the increasing order of complexity of the simulated dynamic systems themselves have made it necessary to take a much closer look at the computational and realization details of the designed recursive digital filters. Many of the problems now coming to light were not noticed before [3, 4, 5, 8, 9, 16] either because the filters being used were of low order with low (two or three decimal) accuracy requirements or because the sampling frequencies were comparable to the dynamic system and signal frequencies. An understanding of the computational problems and realization considerations is of vital interest to the users of the different recursive digital filter design methods as their presence may often spell the success or failure of a particular application.

*The material in this section was taken in a large part from Kaiser [29].

The problems treated in this section relate to the stability of the digital filters, the precision of the arithmetic necessary to carry out the desired filtering operation, the numerical determination of the digital filter coefficients for each basic filter design method and, the choice of filter realization form.

7.6.1 The Stability and Coefficient Accuracy Problem

The discussion is necessarily limited to digital filters of the recursive type as it is the inherent feedback action in this type filter that precipatates the accuracy problems. The non-recursive filters on the other hand have no feedback paths and hence no stability problem per se. The magnitude of their coefficient accuracy problem can be quickly estimated by simply looking at the relative magnitudes of the coefficients making up the weighting sequence.

For the recursive filters the problem is not as straightforward. We shall concentrate our attention on the two main recursive filter design procedures, the bilinear and the standard z-transformations, extending the results where applicable to the other recursive design schemes.

Let us first look at the stability problem.

We recall that the bilinear z-transform is algebraic in nature and consists simply of a substitution or change of variable in the continuous filter transfer characteristic H(s). Thus if the transformation indicated by (7.32) is carried out *exactly* then $H^*(z^{-1})$ will be stable if H(s) is stable. Further $H^*(z^{-1})$ will be of precisely the same order as H(s). The bilinear z-form can theoretically be applied directly to any rational transfer characteristic, H(s), in either polynomial or factored form. It will be shown later which form is to be preferred.

The standard z-transformation method consists first of obtaining a partial fraction expansion of H(s) in its poles and then z-transforming each partial fraction by making use of the transform pair

$$\frac{1}{s+a} \rightarrow \frac{T}{1-e^{-aT}z^{-1}} \tag{7.28}$$

or transform pairs derived therefrom. Thus the standard z-transformation when applied to a stable transfer characteristic H(s) always yields an $H^*(z^{-1})$ that is also stable *provided* the arithmetic is carried out with infinite precision.

For both transforms the resulting $H^*(z^{-1})$ for linear lumped continuous filters is of the general form

$$H^*(z^{-1}) = \frac{\sum_{k=0}^{n} a_k z^{-k}}{1 + \sum_{k=1}^{n} b_k z^{-k}} = \frac{N_d(z^{-1})}{D_d(z^{-1})} = \frac{N_d(z^{-1})}{\prod_{k=1}^{n} (1-z^{-1}/z_k)} \tag{7.43}$$

where b_0 has been set to unity with no loss in generality.

The first question we shall consider is to what accuracy must the coefficients b_k be known to insure that the zeros of $D_d(z^{-1})$ all lie external to the unit circle (the requirement for a stable digital filter). First a crude bound will be established, then to be followed by a more refined evaluation of the coefficient sensitivity.

The polynomial $D_d(z^{-1})$ can be written in factored form as follows

$$D_d(z^{-1}) = \prod_{k=1}^{n} (1-z^{-1}/z_k) \tag{7.44}$$

For ease of presentation only simple poles are assumed for the basically low-pass transfer characteristic H(s), there being no difficulty in extending the analysis to the multiple order pole and non-low-pass filter cases. If the standard z-transform is used then $D_d(z^{-1})$ becomes

$$D_d(z^{-1}) = \prod_{k=1}^{n} (1-\varepsilon^{p_k T} z^{-1}) \tag{7.45}$$

where p_k represents the kth pole of H(s) and may be complex. For the bilinear z-transform there results

$$D_d(z^{-1}) = \prod_{k=1}^{n} \left[1 - \frac{(1+p_k T/2)}{(1-p_k T/2)} z^{-1} \right] \tag{7.46}$$

Assuming now that the sampling rate, 1/T, has been chosen to be at least twice the highest critical frequency in the H(s) and in the signals to be processed by the H(s), it is of interest to see how $D_d(z^{-1})$ behaves as the sampling rate is increased further.

The Nyquist limit constraint dictates that for the standard z-transformation

$$|\mathcal{J}\,[p_k T]| \leq \pi \tag{7.47}$$

where $\mathcal{J}$ denotes the imaginary part of. The critical frequencies are normalized with respect to half the sampling frequency as

$$\mu_k = \frac{p_k T}{\pi} = p_k/\omega_n \tag{7.48}$$

where ω_n is the Nyquist frequency or one-half the sampling frequency. Normally

$$|\mu_k| < 1. \tag{7.49}$$

Thus as the sampling frequency is increased the μ_k decrease from unity and approach zero. Then one can write for the standard z-transform case

$$\left[1-\varepsilon^{p_k T} z^{-1}\right] \rightarrow \left[1-(1+\mu_k \pi)z^{-1}\right] \quad \text{as } T \rightarrow 0 \tag{7.50}$$

and for the bilinear case

$$\left[1 - \frac{(1+p_k T/2)}{(1-p_k T/2)} z^{-1}\right] \rightarrow \left[1-(1+\mu_k \pi)z^{-1}\right] \quad \text{as } T \rightarrow 0 \tag{7.51}$$

which illustrates that the two design methods yield essentially the same characteristic polynomials, $D_d(z^{-1})$, in the limit as T is made small.

Inspection of (7.50) and (7.51) show that the zeros of $D_d(z^{-1})$ tend to cluster about the point $z^{-1} = +1$ in the z^{-1} plane, i.e.,

$$z_k \approx \frac{1}{1+\mu_k \pi} \approx 1 - \mu_k \pi \tag{7.52}$$

where for a stable system the μ_k has a negative real part. Now the filter $H^*(z^{-1})$ will become unstable if any of its poles move across the unit circle to the interior as a result of some perturbation or change in the coefficients b_j. To estimate the order of this effect one computes the change necessary to cause a zero of $D_d(z^{-1})$ to occur at the point $z^{-1} = 1$. From (7.44) and (7.52) there results

$$D_d(z^{-1})\Big|_{z^{-1}=1} \approx \prod_{k=1}^{n} (\mu_k \pi) = \prod_{k=1}^{n} (p_k T) \tag{7.53}$$

But

$$D_d(z^{-1})\Big|_{z^{-1}=1} = 1 + \sum_{k=1}^{n} b_k z^{-k}\Big|_{z^{-1}=1} = 1 + \sum_{k=1}^{n} b_k \tag{7.54}$$

The right hand side of this expression is an important quantity and is therefore defined as

$$F_0 \triangleq 1 + \sum_{k=1}^{n} b_k \tag{7.55}$$

Thus by combining (7.53) and (7.54) it is immediately seen that if any of the b_k are changed by the amount given by (7.55) then the $D_d(z^{-1})$ will have a zero at $z^{-1} = 1$ and the filter $H^*(z^{-1})$ will thus have a singularity on the stability boundary. A zero of $D_d(z^{-1})$ at $z^{-1} = 1$ causes $H^*(z^{-1})$ to behave as if an integration were present in the H(s). Any further change in the magnitudes of any combination of the b_k in such a manner as to cause $D_d(z^{-1})\Big|_{z^{-1}=1}$ to change sign will result in an unstable filter, i.e., with some of the zeros of $D_d(z^{-1})$ lying inside the unit circle. Hence (7.53) is the desired crude bound on coefficient accuracy.

Equation (7.53) has a significant interpretation; it states that for small μ_k (large sampling rates) the bound on coefficient accuracy is dependent on both the order n of the filter and the sampling rate of normalized filter pole locations. Thus going from an nth order filter to a $(2n)^{th}$ order filter at the same normalized frequency will require approximately twice as many digits accuracy for the representation of the b_k. Similarly doubling the sampling rate for an nth order filter requires $n \times \log_{10} 2$ or $0.3 \times n$ additional decimal digits in the representation of each of the b_k.

Equation 7.54 has the interpretation that it represents the return difference [64] at zero frequency when the filter $H^*(z^{-1})$ is realized in the single feedback loop. This expression is also recognized as simply the reciprocal of Blackman's deadband factor [11, p. 75-76] λ. Thus for complex filters with fairly large sampling rates the quantity F_0 will usually be very small. For example a fifth-order Butterworth low-pass filter with its break frequency at 1/10 the sampling frequency yields an $H^*(z^{-1})$ having $F_0 = 7.9 \times 10^{-5}$.

The coefficient accuracy problem is somewhat further aggravated by the fact that as T is made smaller the b_k tend to approach in magnitude the binomial coefficients [65], $\binom{n}{k}$, and tend to alternate in sign. Thus the evaluation of $D_d^*(z^{-1})$ involves the perennial computational problems associated with the differencing of large numbers. A better bound on coefficient accuracy is obtained by dividing F_0 as obtained from (7.55) by the magnitude of the largest b_k. The largest b_k is given approximately by

$$\max b_k \approx \binom{n}{[n/2]} \approx \frac{2^n}{\sqrt{n}} \sqrt{\frac{2}{\pi}}$$

$$\therefore \quad \max b_k \approx \frac{4}{5} * \frac{2^n}{\sqrt{n}} \tag{7.56}$$

Hence from (7.53), (7.55), and (7.56) an absolute minimum bound on the number of decimal digits m_d required for representing the b_k is found as

$$m_d \geq 1 + \left[-\log_{10}\left(\frac{5\sqrt{n}}{2^{n+2}} \prod_{k=1}^{n} p_k T \right) \right] \tag{7.57}$$

where [x] denotes the "greatest integer in x".

While the foregoing analysis has yielded an easily computable absolute accuracy bound on the denominator coefficients of the recursive digital filter, the bound is not necessarily the best possible nor does it say anything about what happens to the zeros of $D_d(z^{-1})$ as small perturbations (less than F_0) are introduced in the values of the b_k. It is not enough to say that the digital filter $H^*(z^{-1})$ is simply stable; what is necessary is that the obtained digital filter have response characteristics close in some sense to those of the continuous filter it is approximating. This means that the sensitivity of the zeros of $D_d(z^{-1})$ to changes in the b_k must be determined. The most direct way to establish this relationship is to equate the two forms of the denominator of (7.43) and then to compute $\partial z_i/\partial b_k$. There results for filters with simple poles only

$$\left.\frac{\partial z_i}{\partial b_k}\right|_{z^{-1}=z_i} = \frac{z_i^{k+1}}{\prod_{\substack{\ell=1 \\ \ell \neq i}}^{n} \left(1 - \frac{z_i}{z_\ell}\right)} \tag{7.58}$$

from which the total differential change in any zero may be evaluated as

$$dz_i = \sum_{k=1}^{n} \left.\frac{\partial z_i}{\partial b_k}\right|_{z^{-1}=z_i} db_k \tag{7.59}$$

These results extend directly to the multiple-order pole case [66].

Utilizing the fact that a pole, p_k, of H(s) transforms to a zero of $D_d(z^{-1})$ at $\varepsilon^{-p_k T}$ for the standard z-transform and that for T such that $p_k T \ll 1$ the zero becomes approximately equal to $1 - p_k T$ as given previously by (7.52), the fractional change δ_i in a zero location z_i can be expressed in terms of the fractional change e_k in a coefficient b_k of the polynomial $D_d(z^{-1})$. Using (7.58) there results

$$e_k \approx \left[\prod_{\substack{\ell=1 \\ \ell \neq i}}^{n} (z_\ell - z_i) \times \frac{\delta_i}{b_k} \right] \tag{7.60}$$

where

$$e_k = \frac{\Delta b_k}{b_k}, \quad \delta_i = \frac{\Delta z_i}{z_i}$$

and $| 1 - z_i | << 1$, i.e., tightly clustered zeros have been assumed. For the purpose of quickly estimating the value required for e_k, the product of the (n-1) factors can be approximated coarsely by the product (7.53) which is simply F_0. Thus (7.60) illustrates that the precision required for the representation of the b_k is increased by the factor $\log_{10}(1/\delta_i)$ over that given by (7.57).

Returning to (7.58) and (7.59) it is seen that the detailed changes in the positions of the zeros resulting from changes in the b_k are in general complex functions as the z_i may be complex. The changes in the b_k can occur as a result of imprecise arithmetic used in their computation or as a result of truncating or rounding the obtained b_k coefficients to a smaller number of significant digits.

The qualitative evaluation of (7.58) can also be carried out by using the well developed ideas of the root locus [67]. For example the changes in location of all the zeros of $D_d(z^{-1})$ as a result of a change ρ_k in b_k are found from

$$\left(1 + \sum_{i=1}^{n} b_i z^{-1}\right) + \rho_k z^{-k} = 0 \tag{7.61}$$

or

$$\prod_{i=1}^{n}\left(1 - \frac{z^{-1}}{z_i}\right) + \rho_k z^{-k} = 0 \tag{7.62}$$

This has the appearance of the standard root-locus problem for a single feedback loop having the loop transmission poles of the z_i, a $k^{'th}$ order zero at the origin, and a loop gain factor of ρ_k. The parameter F_0 is simply the *gain* ρ_k required when the root-locus passes through the point $z^{-1} = 1$. Thus all the techniques of the root-locus method and the insight gained thereby can be brought to bear on the problem.

By viewing the coefficient sensitivity problem in terms of root loci the effects of both increasing filter order and especially increasing the sampling rate can be easily observed. Increasing the sampling rate tends to cluster the poles of $H^*(z^{-1})$ even more compactly about the point $z^{-1} = 1$ as Fig. 7.14 shows for a third-order filter. As filter order increases so does the possible order k of the zero at the origin of the z^{-1} plane. All n branches of the root loci begin at the roots z_i; as ρ_k increases, k branches converge on the kth order zero at the origin and n-k branches move off toward

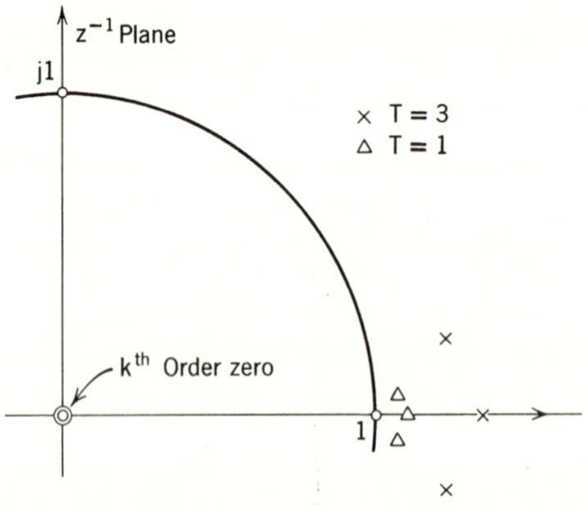

Fig. 7.14. The effect of a change in sampling rate on the pole-zero configuration.

infinity with eventually radial symmetry. The angles the loci make as they leave the z_i are simply the angles given by evaluating (7.58) at each z_i. The value of ρ_k at which a branch of the locus first crosses the unit circle (the stability boundary) gives the measure of total variation that can be made in b_k and still keep the filter stable. Clearly the closer the roots z_i are to the unit circle initially, the smaller will be the value of ρ_k necessary to move them to lie on the boundary. Thus by varying the ρ_k (the changes in b_k) the extent of the stability problem can be viewed.

The development up to this point assumed that the H(s) was basically low-pass with simple poles. Extension to filters of high-pass, bandpass, or bandstop types and with multiple-order poles presents no real problems when viewed using the root locus idea. For example a digital version of a narrow bandpass filter with center frequency at ω_c would have its poles and zeros located, as shown in Fig. 7.15, about the radial lines at $\pm\omega_c T$ radians. The coefficient sensitivity analysis proceeds in the same way as before except that now the points on the unit circle in the vicinity of $z^{-1} = \cos\ \omega_c T \pm j\ \sin\ \omega_c T$

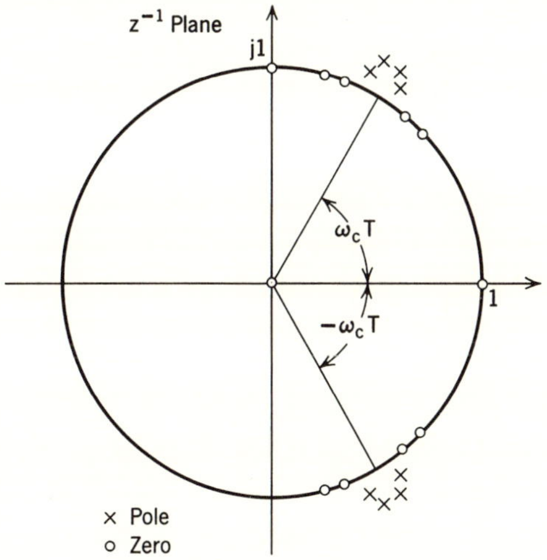

Fig. 7.15. The pole-zero configuration for a digital bandpass filter.

replace the point $z^{-1} = 1$ in the stability computations. This is easily seen for the standard z-transform where a continuous pole of H(s) at $p_k = \sigma_k + j\omega_k$ transforms to a zero of $D_d(z^{-1})$ at $\varepsilon^{-p_k T}$, i.e., at $\varepsilon^{-\sigma_k T} \angle \omega_k T$ when written in polar form.

Thus it follows that it is primarily the smallness of the real parts of the filter poles of H(s) that cause the z^{-1} plane poles to be very near the unit circle and as a result to contribute measurably to the coefficient accuracy and the related sensitivity problems. It can be shown that expressions quite similar to (7.53) and (7.57) can be developed for digital bandpass and bandstop filters. The expressions differ primarily in that for an nth order filter only n/2 terms in the product $\prod^{n} p_k T$ will in general be small and thus contribute to this measure of sensitivity.

In this section some of the relationships between filter order, pole location, sampling frequency and digital filter coefficient accuracy have been established. The questions then logically arise, how do these results affect the choice of a filter design method and the form chosen for realizing the digital filter ? These questions are discussed in the following two sections.

7.6.2 Computational Aspects of the Recursive Design Methods

The results of the preceeding section, especially (7.55) and (7.53) which state

$$F_o = 1 + \sum_{k=1}^{n} b_k \approx \prod_{k=1}^{n} p_k T \qquad (7.63)$$

have important implications in the choice of a filter design method. First, *if* the filter being digitalized is of low order, having no very lightly damped poles, and having poles of approximately the same modulus and *if* the sampling rate is not too much higher than five to ten times the average pole modulus *then* most any of the design procedures described in

Section 7.5 will work satisfactorily. However, when the order of the filter is high, i.e., n is large, and when T is such that $p_k T << \pi$ then care must be exercized in the choice of a design method. The cardinal rule to follow is to use the design method that involves (or at least permits) decomposing the high-order filter into a group of low-order subfilters. The digitalization is then performed on the low-order subfilters.

Thus extreme care must be taken in the direct digitilization of a transfer characteristic of the form

$$H(s) = \frac{\sum_{m=0}^{M} c_m s^m}{\sum_{n=0}^{N} d_n s^n} \tag{7.23}$$

where N is large. This direct procedure is characteristic of the straight-forward application of the Boxer-Thaler method and the bilinear method. Instead a far better policy, from accuracy considerations, is to break H(s) into either a cascade or a parallel connection of subfilters as

$$H(s) = \prod_{k=1}^{N} \frac{s - r_k}{s - p_k} = \sum_{j=1}^{N} \frac{k_j}{s - p_j} \tag{7.64}$$

and then to proceed with the digitalization of the subfilters with the results being left as a set of digital subfilters. The decomposition of H(s) into partial fractions is inherent in the standard z-transform method and straight-forward to carry out [68, 69].

Although the bilinear z-transform method does not require factoring the denominator of H(s), the method is much more accurate and hence potentially useful if the partial-fraction expansion of (7.64) is made. This procedure is followed in the program of Golden and Kaiser [52], where in fact the design shown in the example of Fig. 7.13, page 256, could not have otherwise been so accurately obtained if at all.

As a general rule it is unwise, where reasonable accuracy is desired, to generate or manipulate high-order polynomials in z. The accurate factoring of polynomials in z is usually quite difficult because of the tight clustering of the roots for moderate to high relative sampling rates. Thus all polynomial manipulation and factoring should be done in the s-domain before digitalization.

The application of iterative design or coefficient determination [11, 59] methods to general digital filters of the rational form of (7.22) should be approached numerically with caution. Again it is better to postulate the digital form as a product or sum of first- and second-order terms than as a quotient of two high-degree polynomials in z.

Assuming that the digitalization process has been satisfactorily completed what form should be chosen for the physical realization (hardware or computation algorithm) of the digital filter or subfilters? This question we now consider.

7.6.3 Digital Filter Realization Schemes

The three basic forms for realizing linear digital filters of the recursive type are the direct, the cascade and the parallel forms as shown in Fig. 7.16. As far as the stability question goes the two variations of the direct form, Fig. 7.16a and Fig. 7.16b, are entirely equivalent, with the configuration of Fig. 7.16a requiring fewer delay elements. The stability results developed in the Section 7.6.1 indicate clearly that the coefficient accuracy problem will be by far the most acute for the direct form realization. For any reasonably complex filter with steep transitions between pass and stop bands the use of the direct form should be avoided.

The choice between the utilization of either the cascade, Fig. 7.16c, or parallel, Fig. 7.16d, forms is not clear cut

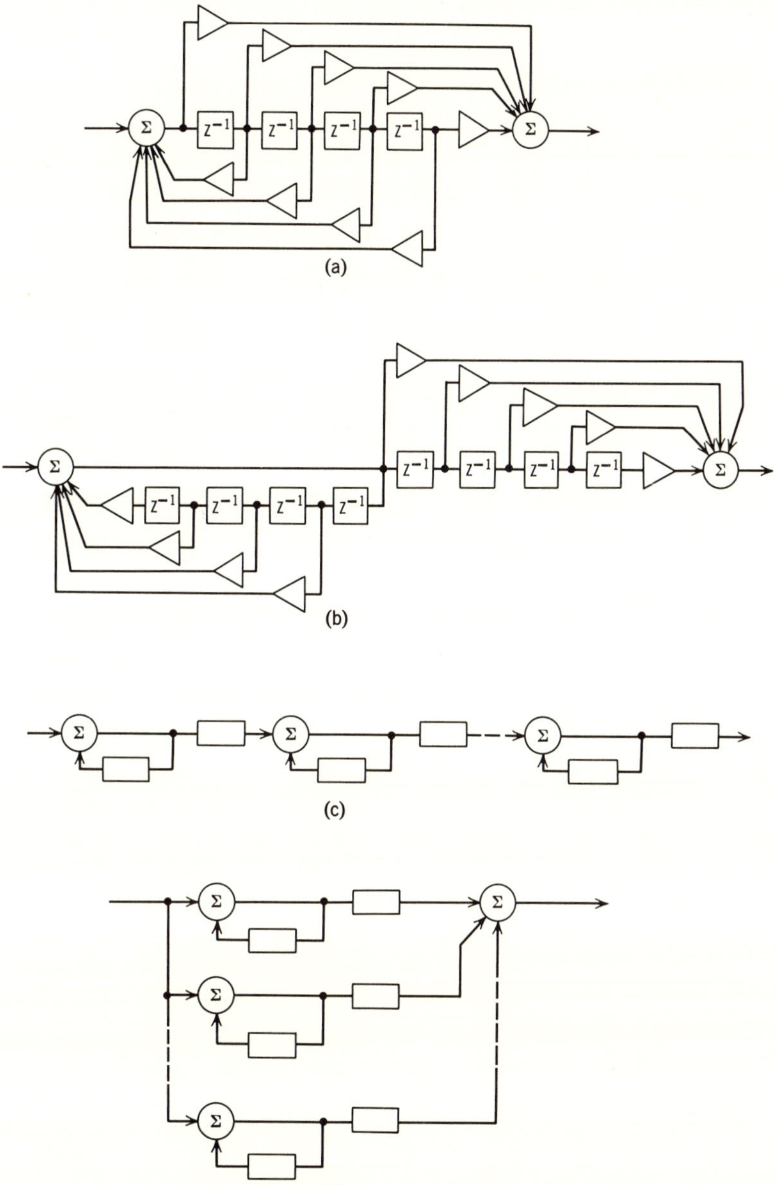

Fig. 7.16. The canonical realization forms for digital filters; (a) and (b) are the direct form, (c) the cascade form and (d) the parallel form.

but depends somewhat on the initial form of the continuous filter and on the digitalization scheme used. In any case the denominator of H(s) must be known in factored form. If the parallel form is desired then a partial fraction expansion of H(s) must first be made. This is followed by a direct application of either (7.32) or (7.28) if the bilinear or standard z-transforms are used respectively. For bandpass or bandstop structures the midfrequency gains of the individual parallel sections may vary considerably in magnitude thus introducing a small problem of the differencing of large numbers. This parallel form is perhaps the most widely used realization form.

For cascade realization the bilinear z-form requires that the numerator, in addition to the denominator, of H(s) be known in factored form. The splitting into simpler cascaded forms can then be done rather arbitrarily since the bilinear z operator has the property that

$$Z(G_1G_2) = Z(G_1)\ Z(G_2) \tag{7.65}$$

If the standard z-transform is utilized, a partial fraction expansion must first be made followed by z-transforming term by term. Then the fractions must be collapsed to yield an $N_d(z^{-1})$ which must then be factored to permit the cascade realization. This more involved procedure is necessary because the standard z-transform does not possess the transform property given by(7.65).This collapsing of fractions to yield the numerator polynomial $N_d\,(z^{-1})$ and the concomitant factoring must be done with great care for high-order filters in order to give a usable result.

The discussion up to this point has centered on satisfactory means for obtaining the digital filter coefficients required for the desired realization form. In actually using the digital filter to process data streams, the performance of the filter will also be affected by the quantization of the data and by roundoff in the required multiplication and addition operations. These factors are briefly considered in the next section.

7.6.4 Round-Off Errors and Quantization Effects in Digital Filters

In realizing a digital filter with either physical hardware or a digital computer program there will be errors introduced by the finite-word-length constraint. There are basically three types of error. The first is that caused by finite precision in the representation of the filter coefficients. This type of error we have already treated in considerable detail in Section 7.6.1. The second type of error results from the quantization of the input data to a specified number of bits. Provided the quantization is suitably fine and the signal being quantized is suitably well behaved with reference to the sampling rate then the effect of the quantization may be treated as an additive random noise source at the input of mean-square value $q^2/12$ where q is the basic quanta size [70, 71, 72]. The resulting output signal due to quantization noise alone can be either calculated analytically or determined by actual simulation of the digital filter using a random noise generator.

The third type of error is that due to the round-off in the multiplication and addition operations of the digital filter. Hills [73] looked into these errors briefly for first and second-order systems. Knowles and Edwards have treated the general problem in considerable detail [74, 75, 76, 77] both theoretically and experimentally with corroborating results. In essence their results show that the round-off errors resulting from the finite-arithmetic operations of the recursive filter can be accounted for statistically by placing a number of additive white noise sources as inputs to a noiseless filter. Their results tend to indicate also that the cascade form of realization exhibits somewhat less performance error than the parallel form and that the direct form is definitely inferior to both the cascade and parallel forms.

Gold and Rader [78] have taken a similar approach to the effects of round-off error. They have worked out in some detail the error analysis of a first- and a second-order system. They also present corroborating experimental evidence.

Wilkinson [79] has treated the effect of quantization errors in nonrecursive filters from a purely numeric point of view.

7.6.5 Fast Fourier Transformation Methods

One further technique for the *efficient* realization of complex digital filters should be included. The filtering of signals by complex nonrecursive filters having fifty or more terms in their weighting sequences involves a large amount of repetitive calculation. If the input signal has M samples and the nonrecursive filter has N taps (where $M > N$) then NM multiplications are required to evaluate the filter response. Using the scheme of Cooley and Tukey [80] as a starting point, Stockham [81] devised a method to reduce the number of multiplications required in the evaluation of the filter response. He showed that if N is chosen to be highly composite (such as of the form 2^n) then approximately only $M \log_2 N$ multiplications are required when the computations are properly organized. Helms [82] discusses the computational details of the method and shows the measure of improvement in efficiency of computation. Stockham cites $N = 28$ as being the approximate crossover point; i.e. for filters having greater than 28 taps the fast Fourier technique holds a computational advantage that increases with N approximately as $(\log_2 N)/N$.

7.7 SUMMARY

After a brief introduction to the early development of the field of digital filter design and utilization, a wide range of design procedures are described for digital filters of both recursive and nonrecursive types. Design examples showing the applicability of some of the more important methods are given. Some of the computation problems associated with physical realization of the digital filters are discussed in detail. Means for rapidly estimating the magnitude of the accuracy problem are developed.

Because of the vastness of the application of digital filtering techniques only a sampling of the more important work could be given. Many of the techniques described here have doubtless been independently discovered by many different workers in different disciplines as in the digital filter field one finds circuit theorists, sampled-data specialists, control engineers, numerical analysists, approximation theorists, statisticians, etc. all working to develop new and more usable methods.

REFERENCES

1. C. Jordan, Calculus of Finite Differences, Chelsea, 1960, (Reprint of 1939 Edition).
2. H. M. James, N. B. Nichols, R. S. Phillips, Theory of Servomechanisms, McGraw Hill, New York, 1947, pp. 231-261.
3. J. R. Ragazzini and G. F. Franklin, Sampled-Data Control Systems, McGraw Hill, 1958.
4. E. I. Jury, Sampled-Data Control Systems, John Wiley and Sons, Inc., 1958.
5. J. T. Tou, Digital and Sampled-Data Control Systems, McGraw Hill, 1959.
6. C. H. Wilts, Principles of Feedback Control, Addison-Wesley, Reading, Massachusetts, 1960, pp. 195-220.
7. E. Mishkin and L. Braun, Jr., Adaptive Control Systems, McGraw Hill, New York, 1961, pp. 119-183.
8. E. I. Jury, Theory and Application of the z-Transform Method, John Wiley, New York, 1964.
9. H. Freeman, Discrete-Time Systems, John Wiley, 1965.
10. P. M. DeRusso, R. J. Roy, C. M. Close, State Variables for Engineers, John Wiley, 1965, p. 158-186.
11. R. B. Blackman, Linear Data-Smoothing and Prediction in Theory and Practice, Addison-Wesley, Reading, Massachusetts, 1965.

12. C. Lanczos, Applied Analysis, Prentice Hall, Inc., Englewood Cliffs, New Jersey, 1956.

13. R. B. Blackman, J. W. Tukey, The Measurement of Power Spectra from the Point of View of Communication Engineering, Dover, 1959.

14. M. A. Martin, Frequency domain applications to data processing, IRE Transactions on Space Electronics and Telemetry, Vol. SET-5, No. 1, March 1959, pp. 33-41.

15. J. F. A. Ormsby, Design of numerical filters with applications to missle data processing, Jour. ACM, Vol. 8, No. 3, July 1961, pp. 440-466.

16. A. J. Monroe, Digital Processes for Sampled Data Systems, John Wiley and Sons, Inc., New York, 1962.

17. R. M. Golden, Digital computer simulation of a sampled-data voice-excited vocoder, Jour. Acous. Soc. Amer., Vol. 35, September 1963, pp. 1358-1366.

18. R. M. Golden, Digital computer simulation of communication systems using the block diagram compiler: BLODIB, Third Annual Allerton Conference on Circuit and System Theory, Monticello, Illinois, October 1965, pp. 690-707.

19. A. Tustin, A method of analyzing the behavior of linear systems in terms of time series, Jour. IEE, Vol. 94, Part II A, May 1947, pp. 130-142.

20. J. M. Salzer, Frequency analysis of digital computers operating in real time, Proc. IRE, Vol. 42, February 1954, pp. 457-466.

21. R. Boxer, S. Thaler, A simplified method of solving linear and nonlinear systems, Proc. IRE, Vol. 44, January 1956, pp. 89-101.

22. R. Boxer, A note on numerical transform calculus, Proc. IRE, Vol. 45, No. 10, October 1957, pp. 1401-1406.

23. D. C. Baxter, The Digital Simulation of Transfer Functions, National Research Laboratories, Ottawa, Canada, DME Report No. MK-13, April 1964.

24. D. C. Baxter, Digital Simulation Using Approximate Methods, National Research Council, Ottawa, Canada, Report MK-15, Division of Mechanical Engineering, July 1965.
25. J. E. Gibson, Nonlinear Automatic Control, McGraw Hill, New York, 1963, pp. 147-159.
26. K. Steiglitz, The General Theory of Digital Filters with Applications to Spectral Analysis, AFOSR Report No. 64-1664, New York University, New York, May 1963.
27. K. Steglitz, The equivalence of digital and analog signal processing, Information and Control, Vol. 8, No. 5, October 1965, pp. 455-467.
28. J. F. Kaiser, Design methods for sampled-data filters, Proceedings First Allerton Conference on Circuit and System Theory, November 1963, pp. 221-236.
29. J. F. Kaiser, Some practical considerations in the realization of linear digital filters, Proceedings Third Allerton Conference on Circuit and System Theory, Monticello, Illinois, October 1965, pp. 621-633.
30. G. M. Jenkins, A survey of spectral analysis, Applied Statistics, Vol. XIV, No. 1, 1965, pp. 2-32.
31. H. H. Robertson, Approximate design of digital filters, Technometrics, Vol. 7, No. 3, August 1965, pp. 387-403.
32. W. K. Linvill, Sampled-data control systems studied through comparison of sampling with amplitude modulation, Trans. AIEE, Vol. 70, Part II, 1951, pp. 1779-1788.
33. A. Susskind, Notes on Analog-Digital Conversion Techniques, John Wiley, New York, 1957. See especially Chapter II, Sampling and Quantization by D. Ross.
34. D.T. Ross, Improved Computational Techniques for Fourier Transformation, Servomechanisms Laboratory Report, No. 7138-R-5, Massachusetts Institute of Technology, Cambridge, Massachusetts, June 25, 1954.

35. G. E. Heyliger, The Scanning Function Approach to the Design of Numerical Filters, Report R-63-2, Martin Co., Denver, Colorado, April 1963.

36. R. J. Graham, Determination and Analysis of Numerical Smoothing Weights, NASA Technical Report No. TR-R-179, December 1963.

37. E. B. Anders et al., Digital Filters, NASA Contractor Report CR-136, December 1964.

38. D. G. Watts, Optimal Windows for Power Spectra Estimation, Mathematics Research Center, University of Wisconsin, MRC-TSR-506, September 1964.

39. E. Parzen, Notes on Fourier Analysis and Spectral Windows, Applied Mathematics and Statistical Laboratories, Technical Report No. 48, May 15, 1963, Stanford University, California.

40. G. A. Campbell and R. M. Foster, Fourier Integrals for Practical Applications, D. Van Nostrand, 1948, p. 113 pair 872.1.

41. J. F. Kaiser, A family of window functions having nearly ideal properties, November 1964, unpublished memorandum.

42. D. Slepian and H. O. Pollak, Prolate spheroidal wave functions, Fourier analysis and uncertainty - I and II, B.S.T.J. Vol. 40, No. 1, January 1961, pp. 43-84.

43. P. E. Fleischer, Digital realization of complex transfer functions, Simulation, Vol. 6, No. 3, March 1966, pp. 171-180.

44. M. A. Martin, Digital Filters for Data Processing, General Electric Co., Missle and Space Division, Tech. Info. Series Report No. 62-SD484, 1962.

45. S. A. Schelkunoff, A mathematical theory of linear arrays, B.S.T.J. Vol. 22, January 1943, pp. 80-107.

46. Mark Tsu-Han Ma, A New Mathematical Approach for Linear Array Analysis and Synthesis, Ph.D. thesis, Syracuse University, 1961, University Microfilms No. 62-3050.

47. C. J. Drane, Directivity and Beamwidth Approximations for Large Scanning Dolph-Chebyshev Arrays, AFCRL Physical Science Research Papers No. 117, AFCRL-65-472, June 1965.

48. H. L. Garabedian (Ed.), Approximation of Functions, Elsevier Publishing Co., 1965, see especially Walsh pp. 1-16 and Cheney pp. 101-110.

49. E. W. Cheney and H. L. Loeb, Generalized rational approximation, J. SIAM, Numerical Analysis, B, Vol. 1, 1964, pp. 11-25.

50. Josef Stoer, A direct method for Chebyshev approximation by rational functions, JACM, January 1964, Vol. 11, No.1, pp. 59-69.

51. R. M. Golden and J. F. Kaiser, Design of wideband sampled-data filters, BSTJ, Vol. 43, Part 2, July 1964, pp. 1533-1546.

52. R. M. Golden and J. F. Kaiser, A computer program for the design of continuous and sampled-data filters, to be published.

53. C. M. Rader and B. Gold, Digital Filter Design Techniques, Lincoln Laboratory Report, M.I.T., Preprint JA2612, September 1965.

54. H. Holtz and C. T. Leondes, The synthesis of recursive filters, Jour. ACM, Vol. 13, No. 2, April 1966, pp. 262-280.

55. L. Weinberg, Network Analysis and Synthesis, McGraw Hill, 1962.

56. D. A. Calahan, Modern Network Synthesis, Hayden, New York, 1964.

57. J. E. Storer, Passive Network Synthesis, McGraw Hill, New York, 1957, pp. 287-302.

58. P. Broome, A frequency transformation for numerical filters, Proc. IEEE, Vol. 52, No. 2, February 1966, pp. 326-7.

59. P. E. Mantey, Convergent Automatic-Synthesis Procedures for Sampled-Data Networks with Feedback, Stanford Electronics Laboratories, Technical Report No. 6773-1, SU-SEL-64-112, Stanford University, October 1964.

60. J. R. B. Whittlesey, A rapid method for digital filtering, Comm. ACM, Vol. 7, No. 9, September 1964, pp. 552-556.

61. T. Y. Young, Representation and Analysis of Signals, Part X. Signal Theory and Electrocardiography, Department of Electrical Engineering, Johns Hopkins University, May 1962.

62. P. W. Broome, Discrete orthonormal sequences, Jour. ACM, Vol. 12, No. 2, April 1965, pp. 151-168.

63. Archambeau et al., Data processing techniques for the detection and interpretation of teleseismic signals, Proc. IEEE, Vol. 53, No. 12, December 1965, pp. 1860-1884, see especially p. 1878.

64. H. W. Bode, Network Analysis and Feedback Amplifier Design, Van Nostrand, 1945, pp. 47-49.

65. M. Mansour, Instability criteria of linear discrete systems, Automatica, Vol. 2, No. 3, January 1965, pp. 167-178.

66. C. E. Maley, The effect of parameters on the roots of an equation system, Computer Journal Vol. 4, 1961-2, pp. 62-63.

67. J. G. Truxal, Automatic Feedback Control System Synthesis, McGraw Hill Book Co., Inc., New York, 1955, pp. 223-250.

68. F. F. Kuo, Network Analysis and Synthesis, John Wiley, New York, First Edition, 1962, pp. 136-137, (Second Edition, pp. 148-155).

69. C. Pottle, On the partial-fraction expansion of a rational function with multiple poles by a digital computer, IEEE Trans. Circuit Theory, Vol. CT-11, March 1964, pp. 161-162.

70. W. R. Bennett, Spectra of quantized signals, BSTJ, Vol. 27, July 1948, pp. 446 - 472.

71. B. Widrow, A study of rough amplitude quantization by means of Nyquist sampling theory, Trans. IRE on Circuit Theory, Vol. CT-3, No. 4, December 1956, pp. 266-276.

72. B. Widrow, Statistical analysis of amplitude quantized sampled-data systems, Trans. AIEE Applications and Industry, No. 52, January 1961, pp. 555-568.

73. F. B. Hills, A Study of Incremental Computation by Difference Equations, Servomechanisms Laboratory Report No. 7849-R-1, Massachusetts Institute of Technology, Cambridge, Massachusetts, May 1958.

74. J. B. Knowles and R. Edwards, Effect of a finite-word-length computer in a sampled-data feedback system, Proc. IEE Vol. 112, No. 6, June 1965, pp. 1197-1207.

75. J. B. Knowles and R. Edwards, Simplified analysis of computational errors in a feedback system incorporating a digital computer, S.I.T. Symposium on Direct Digital Control, April 22, 1965, London.

76. J. B. Knowles and R. Edwards, Complex cascade programming and associated computational errors, Electronics Letters, Vol. 1, No. 6, August 1965, pp. 160-161.

77. J. B. Knowles and R. Edwards, Finite word-length effects in multirate direct digital control systems, Proc. IEE, Vol. 112, No. 12, December 1965, pp. 2376-2384.

78. B. Gold and C. Rader, Effects of quantization noise in digital filters, Proceedings Spring Joint Computer Conference, 1966, Vol. 28, pp. 213-219.

79. J. H. Wilkinson, Rounding Errors in Algebraic Processes, Prentice Hall, Englewood Cliffs, New Jersey, 1963.

80. J. W. Cooley and J. W. Tukey, An algorithm for the machine calculation of complex Fourier series, Mathematics of Computation, Vol. 19, No. 90, April 1965, pp. 297-301.

81. T. G. Stockham, Jr., High speed convolution and correlation, Proceedings Spring Joint Computer Conference, 1966, Vol. 28, pp. 229 - 233.

82. H. D. Helms, Fast Fourier transform methods of computing difference equations arising from z-transforms and autoregressions, (to appear).

Supplement
A Summary of Some Basic Relationships

In the following, we assume that all systems are linear and time invariant. The range of summation and of integration is from $-\infty$ to $+\infty$. $\mathfrak{F}$ denotes Fourier transform; superscripts a and d denotes analog and digital, respectively. $\mathcal{Z}$ denotes Z-transform.

1. *Continuous-Time (Analog)*

$$y(t) = \int h(t-\tau)x(\tau)\,d\tau$$
$$Y(\omega) = H(\omega)X(\omega)$$

where $\mathfrak{F}[x(t)] = X(\omega)$, $\mathfrak{F}[y(t)] = Y(\omega)$, and $\mathfrak{F}[h(t)] = H(\omega)$.

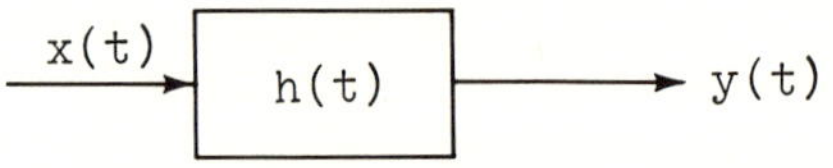

2. *Discrete-time (digital)*

$$y_k = \sum_n h_{k-n}x_n$$
$$Y(z) = H(z)X(z)$$

where $\mathcal{Z}\left[\{x_k\}\right] = X(z)$, $\mathcal{Z}\left[\{y_k\}\right] = Y(z)$, and $\mathcal{Z}\left[\{h_k\}\right] = H(z)$.

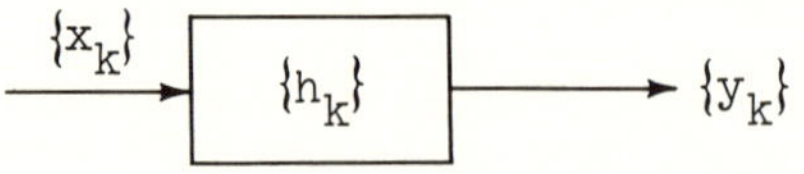

3. *Sampling of Analog Signal*

$$x_k = x(kT)$$
$$x^d(e^{j\omega}) = \frac{1}{T}\sum_k x^a\left(\frac{\omega + 2\pi k}{T}\right)$$

where $x^d(z) = \mathcal{Z}\left[\{x_k\}\right]$ and $x^a(\omega) = \mathfrak{F}[x(t)]$.

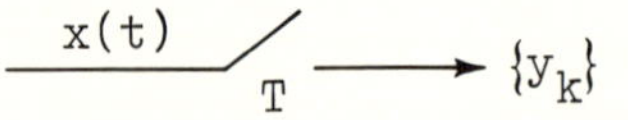

4. *Reconstruction of Analog Signal*

$$y(t) = \sum_k x_k h(t - kT)$$
$$Y(\omega) = H(\omega) X(e^{j\omega T})$$

where $Y(\omega) = \mathfrak{F}[y(t)]$, $X(z) = \mathcal{Z}\left[\{x_k\}\right]$, and $H(\omega) = \mathfrak{F}[h(t)]$.

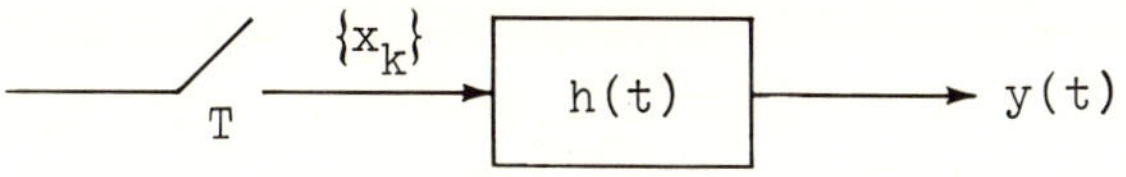

5. *Sampling, Digital Filtering, and Reconstruction*

$$x^a(\omega) = \mathfrak{F}\left\{x(t)\right] \qquad X^d(z) = \mathcal{Z}\left[\{x_k\}\right]$$
$$H(z) = \mathcal{Z}\left[\{h_k\}\right] \qquad Y^d(z) = \mathcal{Z}\left[\{y_k\}\right]$$
$$G(\omega) = \mathfrak{F}\left[g(t)\right] \qquad Y^a(\omega) = \mathfrak{F}\left[y(t)\right]$$

Then

$$X^d(e^{j\omega}) = \frac{1}{T}\sum_k X^a\left(\frac{\omega + 2\pi k}{T}\right)$$
$$Y^d(z) = H(z) X^a(z)$$
$$Y^a(\omega) = G(\omega) Y^d(e^{j\omega T})$$
$$= \frac{1}{T} G(\omega) \sum_k X^a\left(\omega + \frac{2\pi k}{T}\right) H(e^{j\omega T})$$

If

$$G(\omega) \approx \begin{cases} 1 & |\omega| < \pi/T \\ 0 & |\omega| > \pi/T \end{cases}$$

and $X^a(\omega) = 0$, $|\omega| > \pi/T$, then

$$Y^a(\omega) \approx \frac{1}{T} X^a(\omega) H(e^{j\omega T})$$

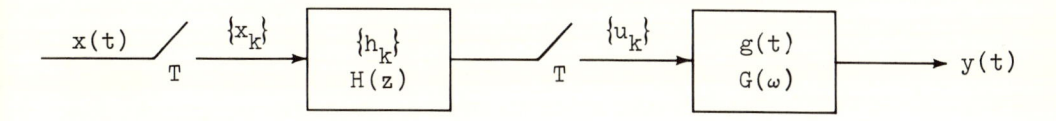

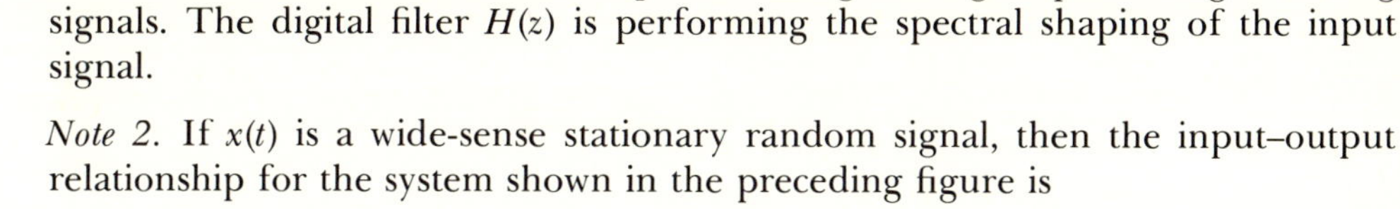

Note 1. This is the basic relationship concerning the digital processing of analog signals. The digital filter $H(z)$ is performing the spectral shaping of the input signal.

Note 2. If $x(t)$ is a wide-sense stationary random signal, then the input–output relationship for the system shown in the preceding figure is

$$\Phi^a_{yy}(\omega) = \frac{1}{T}|G(\omega)|^2 \sum_k \Phi^a_{xx}\left(\omega + \frac{2\pi k}{T}\right) |H(e^{j\omega T})|^2$$

where $\Phi^a_{yy}(\omega)$ and $\Phi^a_{xx}(\omega)$ are, respectively, the power spectral density of the analog signals $y(t)$ and $x(t)$. As in the deterministic signal case, if $G(\omega)$ is an ideal lowpass filter and the input $x(t)$ is properly bandlimited, then

$$\Phi^a_{yy}(\omega) \approx \frac{1}{T}\Phi^a_{xx}(\omega)\left|H(e^{j\omega T})\right|^2$$

II
Design and Implementation

Editor's Comments on Papers 2 Through 12

2 **Golden and Kaiser:** *Design of Wideband Sampled-Data Filters*

3 **Steiglitz:** *Computer-Aided Design of Recursive Digital Filters*

4 **Deczky:** *Equiripple and Minimax (Chebyshev) Approximations for Recursive Digital Filters*

5 **McClellan and Parks:** *A Unified Approach to the Design of Optimum FIR Linear-Phase Digital Filters*

6 **McClellan, Parks, and Rabiner:** *A Computer Program for Designing Optimum FIR Linear Phase Digital Filters*

7 **Fettweis:** *A Simple Design of Maximally Flat Delay Digital Filters*

8 **Thiran:** *Equal-Ripple Delay Recursive Digital Filters*

9 **Jackson, Kaiser, and McDonald:** *An Approach to the Implementation of Digital Filters*

10 **Gabel:** *A Parallel Arithmetic Hardware Structure for Recursive Digital Filtering*

11 **Peled and Liu:** *A New Approach to the Realization of Nonrecursive Digital Filters*

12 **Peled and Liu:** *Some New Realizations of Dedicated Hardware Digital Signal Processors*

The term "design" refers here to the determination of a transfer function from a given set of specifications. Typical design specifications include the desired characteristics and some tolerance requirements. The desired characteristics may be given either in the frequency domain, or in the time domain, or, for some applications, in both. Excellent methods have been developed for the design of both recursive filters and nonrecursive filters.

Consider first the design of recursive filters with a specified frequency response. Frequently, a filter of the conventional type is called for, such as a lowpass filter with a certain ripple requirement in the passband and a specified minimum attenuation in the stopband. In this case, the basic approach is to rely heavily on the well-developed theory of analog filter design and to use a bilinear transformation on the frequency variables to arrive at the digital filter transfer function. This approach is discussed in the paper by Golden and Kaiser [2],† which also considers the problem of frequency warping and its correction.

When the specifications are not in the standard form, analog filter design theory cannot be applied directly and numerical approaches must be used. The paper by Steiglitz [3] deals with this problem of determining a rational function in z^{-1} which approximates a specified amplitude response in the best mean-square sense. All-pass phase equalizers can also be designed using a similar approach (Deczky, 1970). The next paper [4], by Deczky, treats the problem using the minimax criterion. That is,

---

†Numbers in brackets refer to articles included in this volume. References in parentheses are listed at the end of the Comments.

the maximum of the deviation from the desired response is minimized. His approach uses the Remez exchange algorithm (Remez, 1962) to find the best solution. It is also possible to find an approximate solution to the minimax approach by using the L_p error criterion with a very large value of p (Deczky, 1970; Bandler et al., 1973). Recently, linear programming has been applied to recursive filter design (Rabiner et al., 1970).

A number of design methods for nonrecursive filters have been discussed in the article by Kaiser [1] in Part I. The transfer function of a nonrecursive filter is a polynomial in z^{-1}. These filters are also called finite impulse response (FIR) filters because their impulse responses have only a finite number of nonzero terms. A nonrecursive filter has a linear phase characteristic in the frequency domain if the impulse response is symmetric with respect to its midpoint.

The paper by McClellan and Parks [5] discusses the minimax design of linear-phase nonrecursive filters. A general-purpose computer program using this approach to design a large class of nonrecursive filters is presented in the next paper, by McClellan, Parks, and Rabiner [6]. Space does not allow us to include many excellent papers on digital filter design. In addition to the effective method using window functions discussed in Kaiser's paper [1] in Part I, we mention in particular the frequency-sampling approach (Rabiner et al., 1970) and the linear-programming approach (Rabiner, 1972). A technical report (Hankins, 1972) containing the tabulation of more than 500 nonrecursive lowpass filters can be a very useful reference.

Comparisons have been made between recursive and nonrecursive filters that meet the same frequency-response specifications (Rabiner et al., 1974b). For lowpass frequency response, recursive filters can usually be realized more efficiently than linear-phase nonrecursive filters, *if* the phase response can be ignored. On the other hand, if a linear phase or constant group delay is required, all-pass sections are needed to correct the phase of recursive filters designed on the basis of amplitude response only. The resulting combination of recursive filter and the phase equalizer is found to be less efficient than the linear-phase nonrecursive filter alone.

The order of the transfer function is a measure of the complexity of the filter. The designer would like to know the minimum order needed to meet the design specification. For lowpass filters, the orders can be estimated rather accurately (Rabiner et al., 1974a). For a nonrecursive filter with the frequency variable normalized such that the sampling frequency is 1.0, if the amplitude variation is limited to $1 \pm \delta_1$ in the passband $(0, F_p)$ and to $\pm \delta_2$ in the stopband $(F_s, 0.5)$, the minimum order N can be estimated by the following expression:

$$N = 1 + \frac{D}{F_s - F_p} - (F_s - F_p)f$$

where

$$D = [0.005309(\log_{10}\delta_1)^2 + 0.07114 \log_{10}\delta_1 - 0.4761] \log_{10}\delta_2 - [0.00266(\log_{10}\delta_1)^2 + 0.5941 \log_{10}\delta_1 + 0.4278]$$

and

$$f = 0.51244 \log_{10}\frac{\delta_1}{\delta_2} + 11.01$$

This estimate is usually accurate to within ±2.

For a recursive filter, if the magnitude (absolute value of amplitude) is limited to the range between $1-2\delta_1/(1 + \delta_1)$ and 1 in the passband $(0. F_p)$ and to the range between 0 and $\delta_2/(1 + \delta_1)$ in the stopband $(F_s, 0.5)$, the minimum order of the filter transfer function is given by

$$N = \frac{K(k)K(\sqrt{1 - k_1^2})}{K(k_1)K(\sqrt{1 - k^2})}$$

where $K(\cdot)$ is the complete elliptic integral of the first kind,

$$k = \frac{\tan(\pi F_p/2)}{\tan(\pi F_s/2)}$$

and

$$k_1 = \frac{2\delta_2\sqrt{\delta_1}}{(1 - \delta_1)\sqrt{(1 + \delta_1)^2 - \delta_2^2}}$$

The paper by Fettweis [7] deals with lowpass recursive filters with maximally flat group delay. These filters may be regarded as digital-equivalent to the familiar analog Bessel filters. The paper by Thiran [8] treats the same problem but with equal-ripple group delay.

A digital filter can be implanted either by programming a general-purpose computer or by constructing a special-purpose processor using electronic hardware. The programming of the algorithm of a digital filter is straightforward. The subject of hardware implementation is dealt with in the next four papers in this section. There are a number of practical considerations that must receive careful attention in the implementation; these are discussed in the papers of the next part.

The paper by Jackson, Kaiser, and McDonald [9] can be regarded as a classic in this field. It discusses a basic approach to the implementation of digital filters. It also treats a number of important problems that arise in the implementation, such as canonical forms, multiplexing, and analog-to-digital conversion. Two's-complement serial arithmetic is used. The paper by Gabel [10] describes a hardware structure for digital filter implementation using a parallel arithmetic. As noted in the beginning of Gabel's paper, the principal difference between these two types of arithmetic lies in the nature of tradeoff. In a serial processor, the tradeoff involves accuracy versus processing speed, and in a parallel processor, the tradeoff involves accuracy versus hardware.

Multipliers are a major factor in determining the speed of operation, power consumption, and hardware complexity of digital signal processors. The two papers by Peled and Liu [11 and 12] discuss two approaches to hardware realization of digital filters that eliminate the need for multipliers. The first paper exploits the relative simplicity of delta modulation for analog-to-digital conversion and for realization of nonrecursive filters. The second paper capitalizes on the recent advances in semiconductor technology to realize digital filters, as well as fast Fourier transform processors.

References

Bandler, J. W., and B. J. Bardkjian (1973). Least *p*th Optimization of Recursive Digital Filters, *IEEE Trans. Audio Electroacoustics,* **AU-21,** 460–470.

Deczky, A. G. (1970). Synthesis of Recursive Digital Filter Using the Minimum *p*-Error Criterion, *IEEE Trans. Audio Electroacoustics,* **AU-18**(2), 123–129.

Hankins, R. W. (1972). Design Procedure for Equiripple Nonrecursive Digital Filters, *MIT RLE Tech. Rep. 485.*

Rabiner, L. R. (1972). The Design of Finite Impulse Response Digital Filters Using Linear Programming Techniques, *Bell Syst. Tech. J.,* **51,** 1177–1198.

———, B. Gold, and C. A. McConegal (1970). An Approach to the Approximation Problem for Nonrecursive Digital Filters, *IEEE Trans. Audio Electroacoustics,* **AU-18**(2), 83–106.

———, N. Y. Graham, and H. D. Helms (1974a). Linear Programming Design of IIR Digital Filters with Arbitrary Magnitude Function, *IEEE Trans. Acoustics, Speech, Signal Processing,* **ASSP-22,** 117–123.

———, J. F. Kaiser, O. Hermann, and M. T. Dolan (1974b). Some Comparisons Between FIR and IIR Digital Filters, *Bell Syst. Tech. J.,* **53,** 305–331.

Remez, E. Y. (1962). *General Computational Methods for Chebyshev Approximation* (AEC Trans. 4491), U.S. Atomic Energy Commission, Washington, D.C.

2

Reprinted from *Bell Syst. Tech. J.*, **43,** Pt. 2, 1533–1546 (July 1964)

Design of Wideband Sampled-Data Filters

By R. M. GOLDEN and J. F. KAISER

(Manuscript received March 6, 1964)

A design procedure is presented for readily obtaining sampled-data filter representations of continuous filters. The procedure utilizes the bilinear z transformation and preserves the essential amplitude characteristics of the continuous filter over the frequency range between zero and one-half the sampling frequency. It is shown that the procedure can yield meaningful sampled-data filter designs for many of those filters where the standard z transform cannot be used directly.

I. INTRODUCTION

Sampled-data filter representations for continuous filters can be obtained using several different design procedures.[1] A particular design method utilizing the bilinear transformation is developed herein. The method is especially useful in designing wideband* sampled-data filters which exhibit relatively flat frequency-magnitude characteristics in successive pass and stop bands. Filters of this type are widely used in network simulation and data processing problems.[2] The design method possesses two chief advantages over the standard z transform.[3] The first is that the transformation used is purely algebraic in form. This means it can be applied easily to a continuous filter having a rational transfer characteristic expressed in either polynomial or factored form. The second advantage is the elimination of aliasing[4] errors inherent in the standard z transform. Thus, the sampled-data filter obtained by this design method exhibits the same frequency response characteristics as the continuous filter except for a nonlinear warping of the frequency scale. Compensation for this warping can be made by a suitable frequency scale modification. Some of the more common filter networks to which the design method can be applied effectively are the Butterworth, Bessel, Chebyshev, and elliptic filter structures.

The essential properties of the bilinear transformation are presented

* A sampled-data filter design will be termed "wideband" if the frequency range of useful approximation approaches half the sampling frequency.

in the next section. For comparison purposes, properties of the standard z transform are also given. This is followed by a detailed description of a filter design procedure using frequency transformations. Examples illustrating the design procedure are then presented. A short discussion concerning computer simulation of the obtained sampled-data filters is also included.

II. THE STANDARD AND BILINEAR z TRANSFORMATIONS

In this section it is assumed that a satisfactory rational expression is known for the transfer function of a continuous filter for which a sampled-data approximation is sought. What is then necessary is a transformation which will convert this transfer characteristic into a sampled-data transfer function rational in z^{-1}, the unit delay operator.

Two transformations applicable to this problem are the standard z transform and the bilinear z transformation.

The standard z transform applied to $H(s)$, the transfer function of the filter, is[3]

$$H^*(s) = \sum_{m=-\infty}^{\infty} H(s + jm\omega_s) \tag{1}$$

or equivalently in terms of the impulse response, $h(t)$, of the filter

$$\mathcal{H}^*(z) = T \sum_{l=0}^{\infty} h(lT) z^{-l} \tag{2}$$

where

$$\begin{aligned}
s &= \sigma + j\omega \\
H(s) &= \text{Laplace transform of } h(t) \\
\omega_s &= 2\pi/T = \text{radian sampling frequency} \\
H^*(s) &= \text{Laplace transform of the sampled filter impulse response} \\
z^{-1} &= \exp(-sT) = \text{the unit delay operator} \\
\mathcal{H}^*(z) &= H^*(s) \mid_{s = (\ln z)/T} = z \text{ transform of } h(t).
\end{aligned}$$

The behavior of $H(s)$ for s greater than some critical frequency $j\omega_c$ is assumed to be of the form

$$H(s) \mid_{\text{all } s > j\omega_c} = K/s^n, \qquad n > 0 \tag{3}$$

where K is a determined constant.

Equation (1) or (2) is the transfer function of a sampled-data filter which *approximates* the continuous filter. In the time domain, the impulse response of the sampled-data filter is the sampled impulse response of the continuous filter. This can be shown by taking the inverse transform of (2). Equation (1) shows that in the baseband

$$(-\omega_s/2 \leqq \omega \leqq \omega_s/2)$$

the frequency response characteristics of the sampled-data filter, $H^*(s)$, differ from those of the continuous filter, $H(s)$. The difference is the amount added or "aliased"[4] in through terms of the form

$$H(s + jm\omega_s),\ m \neq 0.$$

If $H(s)$ is bandlimited to the baseband, i.e., $| H(s) | = 0$ for $\omega > \omega_s/2$, then there is no aliasing error and the sampled-data filter frequency response is identical to that of the continuous filter. Unfortunately, when $H(s)$ is a rational function of s, it is not bandlimited and therefore $H(s) \neq H^*(s)$ in the baseband.

The magnitude of the errors resulting from aliasing is directly related to the high-frequency asymptotic behavior of $H(s)$ as defined in (3). If n is large and $\omega_c \ll \omega_s/2$, then the aliasing errors will be small and the standard z transform generally will yield a satisfactory sampled-data filter design. However, in wideband designs, ω_c is usually an appreciable fraction of $\omega_s/2$. Furthermore, many continuous filter designs result in transfer functions in which n is no greater than 1. These two conditions, namely $\omega_c \approx \omega_s/2$ and $n = 1$, can create large aliasing errors in the frequency response characteristics, thus yielding an unusable result.

Fortunately, even when $\omega_c \approx \omega_s/2$ and $n = 1$, a design method employing the bilinear z transformation* may provide satisfactory wideband designs. This z form is defined from the mapping transformation,

$$s = (2/T) \tanh (s_1 T/2) \tag{4}$$

where

$$s_1 = \sigma_1 + j\omega_1 .$$

The right-hand side of (4) is periodic in ω_1 with period $2\pi/T$. Considering only the principal values of ω_1, $-\pi/T < \omega_1 < \pi/T$, it is seen that the transformation given by (4) maps the *entire* complex s plane into the strip in the s_1 plane bounded by the lines $\omega_1 = -\pi/T$ and $\omega_1 = +\pi/T$. For this reason the bilinear transformation can be looked upon as a bandlimiting transformation. Therefore, when this transformation is applied to a transfer function $H(s)$, the entire s-plane frequency characteristics of $H(s)$ are uniquely carried over into the frequency characteristics of $H(s_1)$.

* This transformation will be referred to as the bilinear z form or z form.

With the substitution

$$z^{-1} = e^{-s_1 T},$$

(4) can be written immediately as

$$s = \frac{2}{T} \frac{(1 - z^{-1})}{(1 + z^{-1})}. \tag{5}$$

Thus,

$$\mathcal{H}(z) \equiv H(s) \bigg|_{s = \frac{2}{T} \frac{(1-z^{-1})}{(1+z^{-1})}} \tag{6}$$

where $\mathcal{H}(z)$ denotes a sampled-data transfer function obtained by the use of the bilinear z form.*

The transfer function $\mathcal{H}(z)$ obtained by means of the bilinear z form and the function $\mathcal{H}^*(z)$ obtained by means of the standard z transform are both rational in z^{-1} and of the same denominator order as the continuous filter. These two functions, $\mathcal{H}(z)$ and $\mathcal{H}^*(z)$, become essentially equal to each other as the sampling frequency becomes large compared to the moduli of each of the poles of the continuous filter function, $H(s)$. When the sampling frequency is not large, representation of some filters by the standard z transform can be quite unsatisfactory because of aliasing errors. However, the bilinear z form, with its absence of aliasing errors, may give a satisfactory representation for these filters. A particular set of filters to which the bilinear z form always can be applied successfully are those which exhibit relatively flat frequency-magnitude characteristics in successive pass and stop bands. This follows directly from (6).

Thus, sampled-data filters designed by using the bilinear z form preserve the essential amplitude characteristics of the continuous filter. In the baseband ($-\omega_s/2 \leqq \omega \leqq \omega_s/2$), the frequency characteristics of the sampled-data filter are identical to those of the continuous filter *except* for a nonlinear warping of the frequency scale.

This warping is found from (4) upon substituting $j\omega$ for s and is

$$\omega = (2/T) \tan (\omega_1 T/2). \tag{7}$$

For small values of ω_1, the relation is essentially linear, producing

* It should be noted that the bilinear transform is used here in a distinctly different way than it is commonly used in sampled-data control system design. In the control system literature it is used to transform the sampled-data function $\mathcal{H}^*(z)$ from the discrete domain back to the continuous domain for conventional stability and frequency response analysis. See for example J. T. Tou, *Digital and Sampled-Data Control Systems*, McGraw-Hill, New York, 1959, pp. 244–247 and pp. 466–470.

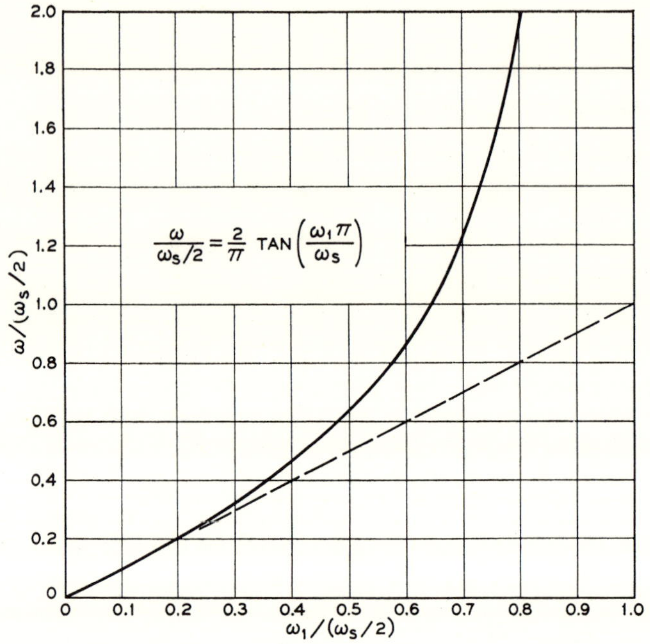

Fig. 1 — The frequency scale warping of the bilinear z transformation.

negligible warping at the lower end of the frequency scale. Fig. 1 shows the nature of this warping. Compensation for the effect of warping can be made by prewarping the band-edge frequencies of the continuous filter in such a way that application of the z-form transformation will shift the band-edge frequencies back to the desired values. The incorporation of this prewarping compensation into a sampled-data filter design procedure is discussed in the next section.

III. A SAMPLED-DATA FILTER DESIGN METHOD

A sampled-data filter design may be obtained by applying the z-form transformation of (6) to the rational transfer characteristic for a continuous filter. However, in order to compensate for the frequency warping imposed by the z form, the frequency characteristics of the continuous filter first must be altered or prewarped. Hence the transfer characteristic for the continuous filter must be redesigned such that the band-edge (cutoff) and maximum loss frequencies are computed according to,

$$\omega_c = (2/T) \tan (\omega_d T/2) \tag{8}$$

where:

ω_c = computed cutoff or loss frequency
ω_d = desired cutoff or loss frequency.

The redesign of the continuous transfer characteristic cannot be accomplished simply by applying (8) to each pole and zero of the original transfer characteristic. On the contrary a completely new transfer characteristic must be obtained for the continuous filter. It is then possible to obtain the desired sampled-data filter by applying the z-form of (6) directly to the redesigned transfer characteristic of the continuous filter. The sampled-data filter so obtained will then have the desired magnitude-frequency characteristics.

Compensation for frequency warping becomes especially simple to apply if the original continuous filter design was obtained by applying a frequency-band transformation to a suitable low-pass design such as Butterworth, Chebyshev, etc. Thus the extensive literature available on tabulated low-pass filter designs can be used to great advantage to simplify the filter design problem. The well-known frequency transformations which convert a normalized low-pass filter to a low-pass, a bandpass, a bandstop, or a high-pass design are

$$s_n = s/\omega_u \text{ low-pass to low-pass} \tag{9}$$

$$s_n = \frac{(s^2 + \omega_u \omega_l)}{s(\omega_u - \omega_l)} \text{ low-pass to bandpass} \tag{10}$$

$$s_n = \frac{s(\omega_u - \omega_l)}{(s^2 + \omega_u \omega_l)} \text{ low-pass to bandstop} \tag{11}$$

$$s_n = \omega_u/s \text{ low-pass to high-pass} \tag{12}$$

where:

s_n = the complex variable of the normalized low-pass filter transfer function
s = the complex variable of the desired filter transfer function
ω_u = the upper radian cutoff frequency
ω_l = the lower radian cutoff frequency.

When continuous filters are designed with the aid of these transformations, prewarping is accomplished by properly choosing the cutoff frequencies used in the frequency transformations. The choice of these cutoff frequencies is determined from the desired cutoff frequencies by means of (8). Using these values, the new prewarped transfer function is determined by applying (9), (10), (11) or (12) to the original low-pass function. Transformation is made to a sampled-data filter by applying (5) to the prewarped continuous function. This sampled-data filter will now have the correct cutoff frequencies. The transfer function thus obtained can be used directly in a digital computer simulation.

IV. SIMULATION OF SAMPLED-DATA FILTERS

Application of either the standard z transform or the bilinear z form to a rational transfer function yields a transfer function rational in z^{-1} for the sampled-data filter. The programming or simulation of this sampled-data filter on a digital computer can be accomplished by either the direct, the cascade or the parallel form. These forms, as commonly defined, are shown in Fig. 2. In this figure $\mathcal{G}(z)$ and $\mathcal{F}(z)$ represent finite polynomials in z^{-1} for feed-forward and feedback transmissions re-

(a) DIRECT FORM $\qquad \mathcal{H}(z) = \dfrac{\mathcal{G}(z)}{1 + \mathcal{F}(z)}$

(b) CASCADE FORM $\qquad \mathcal{H}(z) = \dfrac{\mathcal{G}(z)}{\prod\limits_{\iota=1}^{n} \left[1 + \mathcal{F}_\iota(z)\right]}$

(C) PARALLEL FORM $\qquad \mathcal{H}(z) = \mathcal{G}_T(z) \sum\limits_{\iota=1}^{n} \dfrac{\mathcal{G}_\iota(z)}{1 + \mathcal{F}_\iota(z)}$

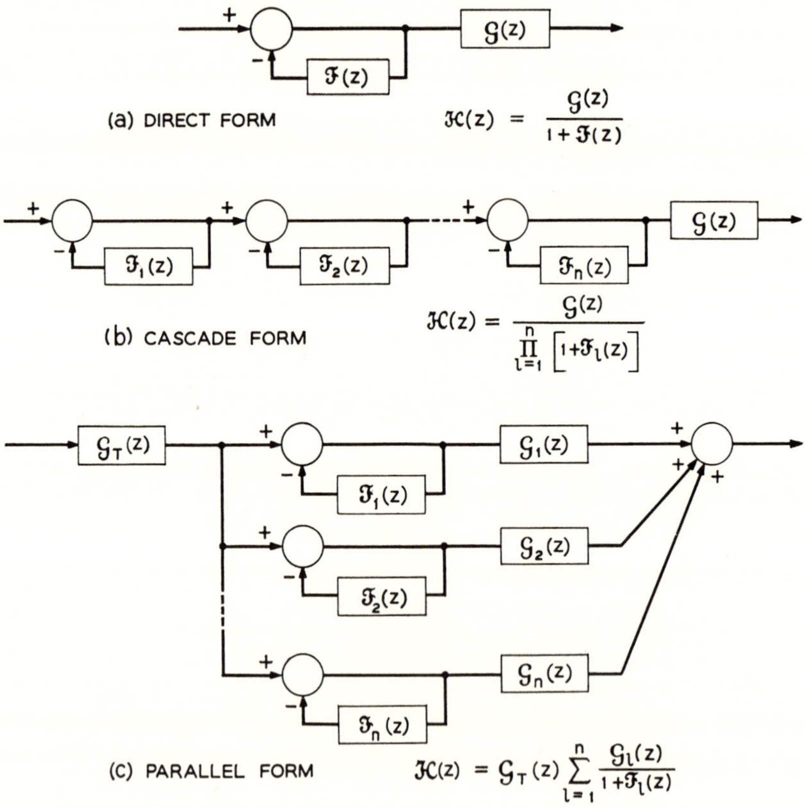

Fig. 2 — Some possible simulation forms for sampled-data filters.

spectively; where subscripted, the order of the polynomial is at most second.

The choice of which of the three forms to use for simulation of the sampled-data filter depends on the complexity of the filter function $H(s)$, on the form of $H(s)$, and on the particular z transformation used. Generally, simulation by the direct form requires considerably greater accuracy in the determination of the filter parameters than either of the other two forms. This is especially true when the order of $H(s)$ is large and when $H(s)$ has poles with real parts that are a very small fraction of the sampling frequency. For this reason either the cascade or parallel form may be preferred.

The choice between using the cascade or the parallel form depends largely on which z-transform method is used to obtain the sampled filter and how that particular method is applied. Realization in the cascade form requires calculation of the numerator polynomial, $\mathcal{G}(z)$, or its factors. This computation consists of a simple algebraic substitution when the bilinear z form is applied to a filter function $H(s)$ expressed in the form,

$$H(s) = \frac{G(s)}{\prod_{k=1}^{m} (s - \alpha_k)}. \tag{13}$$

Determination of $\mathcal{G}(z)$ in polynomial or product form respectively allows either of the following cascade realizations to be made:

$$\mathcal{H}(z) = \mathcal{G}(z) \prod_{k=1}^{n} \left(\frac{1}{1 + b_{1k}z^{-1} + b_{2k}z^{-2}} \right) \tag{14}$$

or

$$\mathcal{H}(z) = \prod_{k=1}^{n} \left(\frac{a_{0k} + a_{1k}z^{-1} + a_{2k}z^{-2}}{1 + b_{1k}z^{-1} + b_{2k}z^{-2}} \right) \tag{15}$$

If the numerator $G(s)$ is in polynomial form, considerable care must be taken in the calculation of the coefficients of the polynomial $\mathcal{G}(z)$, as this computation involves differencing nearly equal numbers.

For realization in the parallel form the partial fraction expansion of $H(s)$ must be known. Since in the standard z transform method obtaining the partial fraction expansion is a necessary step, simulation of filters designed by this method is most directly accomplished in the parallel form. Here the continuous filter transfer characteristic is represented by

$$H(s) = \sum_{k=1}^{N} \frac{P_{1k}s + P_{0k}}{Q_{2k}s^2 + Q_{1k}s + Q_{0k}}. \tag{16}$$

Transforming this expression by use of the standard z transformation yields

$$\mathcal{H}(z) = \sum_{k=1}^{N} \frac{A_{1k}z^{-1} + A_{0k}}{B_{2k}z^{-2} + B_{1k}z^{-1} + 1} \tag{17}$$

whereas transforming by use of the bilinear z form yields the similar expression

$$\mathcal{H}(z) = (1 + z^{-1}) \sum_{j=1}^{N} \frac{A_{1j}z^{-1} + A_{0j}}{B_{2j}z^{-2} + B_{1j}z^{-1} + 1}. \tag{18}$$

Each rational function in either of the above summations can be synthesized by the recursive structure shown in functional block diagram form in Fig. 3(a). This recursive structure uses only two delays, four multiplications, and five additions. The complete realization of (18) is shown in Fig. 3(b).

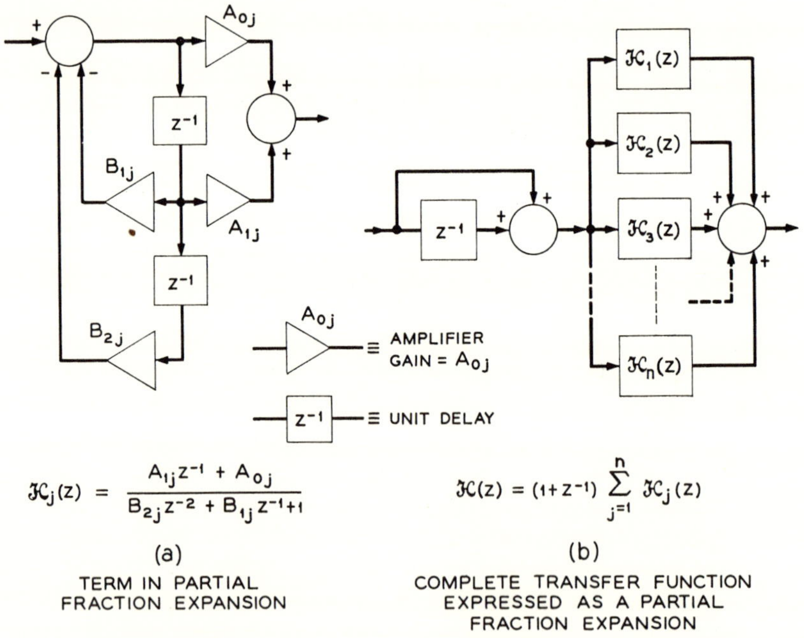

Fig. 3 — Simulation in parallel form of a sampled-data filter obtained by the bilinear z transformation.

The programming of these sampled-data filters for computer simulation can be greatly simplified if a compiler such as the Block Diagram (BLODI) compiler[5] developed at Bell Telephone Laboratories is used. The compiler permits specification of a sampled-data system in functional block diagram form.

In the following section an example is presented for a filter designed, synthesized, and simulated by the foregoing method.

V. DESIGN, SYNTHESIS, AND SIMULATION OF A WIDEBAND BANDSTOP SAMPLED-DATA FILTER

As an example of the application of the bilinear z-form to the simulation of a practical filter, consider the design of a particular bandstop filter. The filter is to exhibit at least 75 db loss in a rejection band which extends between 2596 cps and 2836 cps. Below 2588 cps and above 2844 cps, the loss is to be between 0 and +0.5 db. The sampling rate of the discrete filter is to be 10 kc. The complexity or order of the filter is to be held to a minimum. Fig. 4 shows a sketch of the amplitude response characteristics desired of the filter.

Minimum filter complexity and sharp transition between pass and stop bands suggest the use of an elliptic filter[6] (equiripple) as the basic low-pass type. However, before a suitable low-pass structure can be determined, the above specified critical frequencies must be prewarped by

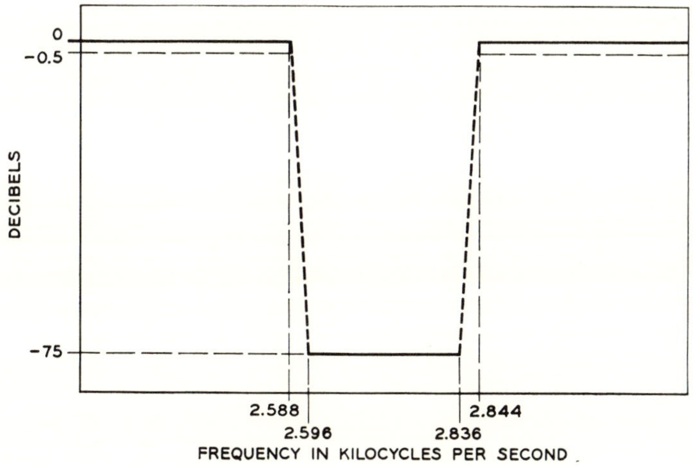

Fig. 4 — Desired amplitude response characteristic of a bandstop filter.

means of (8). The warped values are:

f_{lp} = lower cutoff frequency in passband (2588 cps) = 3364.15 cps
f_{lr} = lower cutoff frequency in rejection band (2596 cps)
= 3381.13 cps
f_{ur} = upper cutoff frequency in rejection band (2836 cps)
= 3937.54 cps
f_{up} = upper cutoff frequency in passband (2844 cps) = 3957.84 cps.

The warped values at the lower band edge require a low-pass filter with a transition ratio of 0.93792, while the values at the upper band edge require a transition ratio[6] of 0.93658. Therefore, to meet the original specifications, the larger of the two transition ratios must be chosen. Hence specifications required for the basic low-pass elliptic filter are:

in-band ripple = 0.5 db
out-of-band minimum attenuation = 75.0 db
transition ratio = 0.93792.

Application of elliptic filter design procedure with these specifications yields a basic low-pass structure of eleventh order. The poles and zeros for the transfer function of this low-pass filter are listed in Table I. The low-pass filter has been normalized to have a cutoff frequency of one radian per second and amplitude gain of unity at zero frequency.

TABLE I—POLES AND ZEROS OF NORMALIZED ELEVENTH-ORDER ELLIPTIC LOW-PASS FILTER

In-band ripple = 0.500 db
Minimum attenuation = 76.504 db
Transition ratio = 0.937917

Gain factor = 0.0011060

Poles
$-0.0069130 \pm j\ 1.0010752$
$-0.0257616 \pm j\ 0.9756431$
$-0.0615122 \pm j\ 0.9063786$
$-0.1269215 \pm j\ 0.7504391$
$-0.2142976 \pm j\ 0.4483675$
-0.2611853

Zeros
$\pm j\ 1.0695414$
$\pm j\ 1.1009005$
$\pm j\ 1.1946271$
$\pm j\ 1.4652816$
$\pm j\ 2.5031313$

The desired bandstop filter is obtained next by applying the low-pass-to-bandstop transformation, given in (11), to the normalized elliptic low-pass filter. The cutoff frequencies used are the warped values obtained above for f_{lp} (3364.15 cps) and f_{up} (3957.84 cps). The resulting bandstop filter is then transformed by the bilinear z form to yield the required sampled-data filter. Table II lists the coefficients of the resulting sampled-data filter needed for parallel realization of the form shown in Fig. 3. Fig. 5 shows the frequency response characteristics of this sampled-data filter. It is seen that the original filter specifications are met by the sampled-data filter. For comparison purposes, the standard z transform was applied directly to the twenty-second-order continuous filter. The frequency response characteristic of this filter is shown in Fig. 6. It is seen that the standard z transform has yielded an unusable result.

VI. SUMMARY

The need for sampled-data filters in wideband simulations of many processing systems has led to a synthesis method which overcomes the shortcomings of the standard z transform. The method presented consists of directly transforming a suitable continuous transfer function to a sampled-data filter by means of the bilinear z form. For wideband filters the method is particularly suited to those filters that exhibit relatively constant magnitude-frequency characteristics in successive pass and stop bands. Conventional design techniques of continuous filters are used

TABLE II—PARTIAL-FRACTION EXPANSION COEFFICIENTS FOR PARALLEL REALIZATION OF SAMPLED-DATA BANDSTOP FILTER

Term	Numerator Coefficients		Denominator Coefficients	
	A_1	A_0	B_2	B_1
1	0.0001628	0.0008827	0.9987854	0.1106416
2	−0.0009283	−0.0001764	0.9989898	0.4285348
3	−0.0024098	−0.0027894	0.9956089	0.1063723
4	0.0031774	0.0026966	0.9957459	0.4317548
5	0.0102446	0.0026026	0.9879911	0.0940731
6	−0.0037799	−0.0112135	0.9883051	0.4414974
7	−0.0277640	0.0127415	0.9651789	0.0616261
8	−0.0108027	0.0289421	0.9661438	0.4663508
9	0.0272223	−0.1163873	0.8694592	−0.0204564
10	0.1206914	−0.0054765	0.8742300	0.5186036
11	0.2973946	−0.2973227	0.5283651	0.2074591

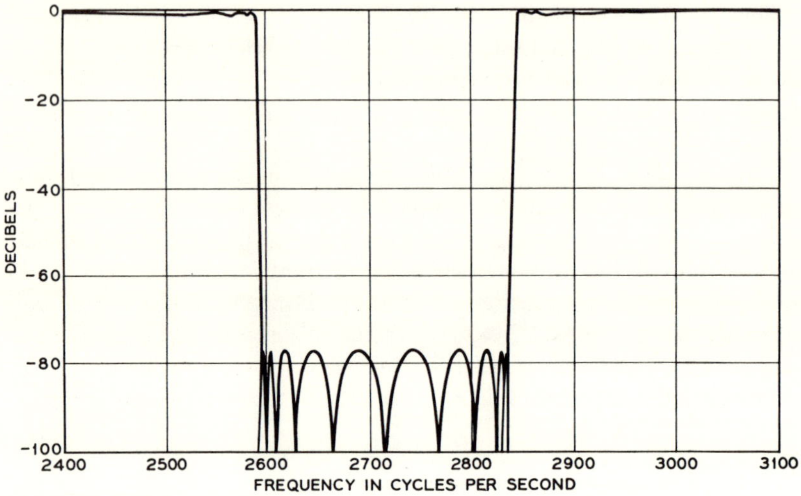

Fig. 5 — Frequency response characteristics of the sampled-data bandstop filter designed by the bilinear z transformation.

directly in the synthesis procedure of the sampled-data filters. (Thus the synthesized filters have frequency characteristics comparable to those of continuous filters.) An example has been presented of a filter function synthesized by this procedure and easily programmed for a simulation. Results obtained from this example demonstrate the usefulness and accuracy of the bilinear z-form method.

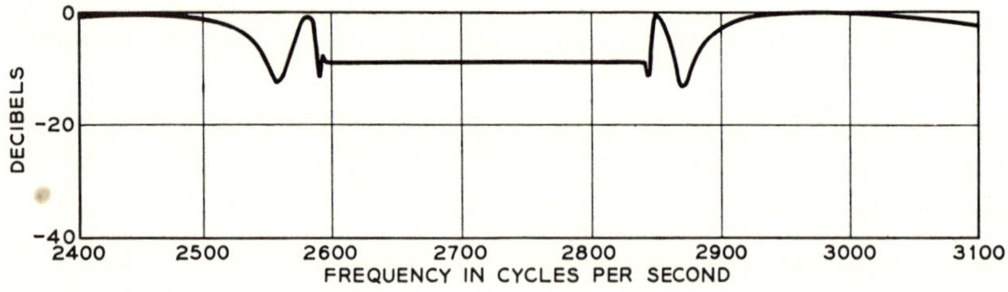

Fig. 6 — Frequency response characteristics of the sampled-data bandstop filter designed by the standard z transformation.

REFERENCES

1. Kaiser, J. F., Design Methods for Sampled-Data Filters, Proc. First Allerton Conference on Circuit and System Theory, Nov., 1963, Monticello, Illinois.

2. Golden, R. M., Digital Computer Simulation of a Sampled-Data Voice-Excited Vocoder, J. Acoust, Soc. Am., **35**, Sept., 1963, pp. 1358–1366.
3. Wilts, C. H., *Principles of Feedback Control,* Addison-Wesley, 1960, pp. 197–207.
4. Hamming, R. W., *Numerical Methods for Scientists and Engineers,* McGraw-Hill, New York, 1962, pp. 277–280.
5. Kelly, J. L., Jr., Lochbaum, C., and Vyssotsky, V. A., A Block Diagram Compiler, B. S. T. J., **40**, May, 1961, pp. 669–676.
6. Storer, J. E., *Passive Network Synthesis,* McGraw-Hill, New York, 1957, pp. 293–296.

3

Reprinted from *IEEE Trans. Audio Electroacoustics,* **AU-18**(2), 123–129 (1970)

Computer-Aided Design of Recursive Digital Filters

KENNETH STEIGLITZ, Member, IEEE
Department of Electrical Engineering
Princeton University
Princeton, N. J. 08540

Abstract

A practical method is described for designing recursive digital filters with arbitrary, prescribed magnitude characteristics. The method uses the Fletcher–Powell optimization algorithm to minimize a square-error criterion in the frequency domain. A strategy is described whereby stability and minimum-phase constraints are observed, while still using the unconstrained optimization algorithm. The cascade canonic form is used, so that the resultant filters can be realized accurately and simply. Design examples are given of low-pass, wide-band differentiator, linear discriminator, and vowel formant filters.

Manuscript received September 10, 1969.

This work was supported by the U.S. Army Research Office-Durham, under Contract DA HC04 69 C 0012.

I. Introduction

While the problem of choosing the coefficients of a nonrecursive digital filter to approximate a specified magnitude characteristic has been thoroughly explored, the corresponding problem for recursive digital filters remains open [1], [2]. Design procedures for recursive filters generally deal only with the piecewise constant case, and involve transformations of well known continuous-time filter designs, such as the Butterworth or Chebyshev. The purpose of this paper is to describe a practical method for choosing the coefficients of a recursive digital filter to meet arbitrary specifications of the magnitude characteristic.

The proposed method uses the optimization algorithm described by Fletcher and Powell [3] to minimize a square-error criterion in the frequency domain. This technique has been used to design continuous-time filters [4]. In order to deal with the realization problem in the continuous-time case, a network topology is usually fixed, and the optimization method must incorporate the constraints that the element values be nonnegative. These restrictions are not present for digital filters, since any coefficients can be used for realization. The resulting digital filter must be stable, however, and this imposes the constraint that the poles lie inside the unit circle in the z-plane. It will be shown how this constraint, and an additional minimum-phase constraint, can be observed, while still using the unconstrained minimization method of Fletcher and Powell.

II. Choice of Canonic Form

The first important question to be resolved is the choice of the canonic form of the digital filter. A general recursive filter can be assumed to have the transfer function

$$Y(z) = \frac{\sum_{k=1}^{K} a_k z^{-(k-1)}}{1 + \sum_{k=1}^{N} b_k z^{-k}} . \tag{1}$$

This so-called direct form suffers from the following difficulties. First, if control is to be exercised over the pole locations, the denominator must be factored at certain stages in the optimization process. Second, the pole locations may be extremely sensitive functions of the coefficients b_k for high-order filters [1]. This means that the b_k must be found to very high precision, and that the error surface may be badly skewed. The cascade form

$$Y(z) = A \prod_{k=1}^{K} \frac{1 + a_k z^{-1} + b_k z^{-2}}{1 + c_k z^{-1} + d_k z^{-2}} \tag{2}$$

avoids these difficulties, and has the additional advantage of yielding the realization shown in Fig. 1, which is known to be practical for high-order filters. This form also has the advantage of making the zeros easy to find, a feature not shared by a third possibility, the parallel form. For

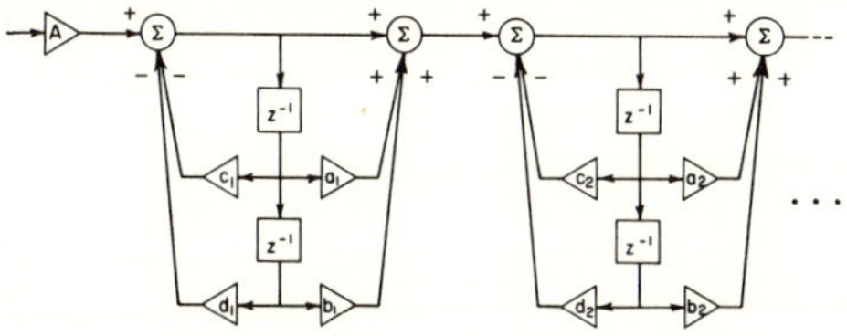

Fig. 1. Cascade realization corresponding to the canonic form.

these reaons the cascade canonic form will be assumed in what follows.

III. Statement of the Problem

Suppose now that the desired magnitude characteristic is prescribed at the discrete set of frequencies $W_1, \cdots, W_M$ where W_i is given in fractions of the Nyquist rate. These correspond to values of the variable z

$$z_i = e^{jW_i\pi} \qquad i = 1, \cdots, M. \tag{3}$$

Call the desired magnitude at these frequencies Y_i^d. Then the square-error in the frequency domain is

$$Q(\theta) = \sum_{i=1}^{M} (|Y(z_i)| - Y_i^d)^2 \tag{4}$$

where θ is the $(4K+1)$-vector of unknown coefficients

$$\theta = (a_1, b_1, c_1, d_1, a_2, b_2, c_2, d_2, \cdots, A)'. \tag{5}$$

The problem is to find a value of θ, say θ^*, such that for all θ

$$Q(\theta^*) \leq Q(\theta). \tag{6}$$

This square-error is a nonlinear function of the parameter vector θ, and an iterative method must be used to accomplish its minimization. Such numerical methods as are available seek a relative (local) minimum from a given starting point, and cannot in general be relied upon to find the global solution. Computational experience, gained by using different starting points for the same problem, often gives some indication of the likelihood that a given local solution is in fact global. In addition, a suboptimal value of θ can often provide a useful design.

IV. Elimination of *A* and Calculation of the Gradient

The method of Fletcher and Powell appears to be the most efficient and powerful nonlinear optimization method now available [3], [4]. It need not be described here, except to say that it performs a one-dimensional minimization at each cycle, along a direction determined by the gradient and an updated estimate of the Hessian. The double precision FORTRAN IV program DFMFP, supplied by IBM in the scientific subroutine package [5], was used without change. The Fletcher–Powell method requires the computation of the gradient of Q with respect to the parameter vector. This computation was performed using double precision complex arithmetic in FORTRAN IV.

The error function Q can be minimized analytically with respect to A for fixed a_i, b_i, c_i, d_i; and A need not be considered an unknown parameter. To eliminate A from Q, define the $4K$-dimensional parameter vector

$$\phi = (a_1, b_1, c_1, d_1, a_2, b_2, c_2, d_2, \cdots, d_K)' \tag{7}$$

and write

$$\begin{aligned} Y(z, A, \phi) &= A \prod_{k=1}^{K} \frac{1 + a_k z^{-1} + b_k z^{-2}}{1 + c_k z^{-1} + d_k z^{-2}} \\ &=: AH(z, \phi). \end{aligned} \tag{8}$$

Then

$$Q(A, \phi) = \sum_{i=1}^{M} (|AH(z_i, \phi)| - Y_i^d)^2. \tag{9}$$

Differentiating with respect to $|A|$ and setting the result to zero yields the following optimum value of $|A|$, say $|A^*|$:

$$|A^*| = \frac{\sum_{i=1}^{M} |H(z_i, \phi)| Y_i^d}{\sum_{i=1}^{M} |H(z_i, \phi)|^2} \tag{10}$$

The Fletcher–Powell method is then used to minimize the new error criterion

$$\hat{Q}(\phi) = Q(A^*, \phi). \tag{11}$$

Notice that the sign of A^* is immaterial, since it does not affect the magnitude characteristic. It will be taken positive. The gradient of $\hat{Q}$ with respect to ϕ can be computed as follows:

$$\frac{\partial \hat{Q}}{\partial \phi_n} = \frac{\partial Q(A^*, \phi)}{\partial \phi_n} + \frac{\partial Q(A^*, \phi)}{\partial A^*} \frac{\partial A^*}{\partial \phi_n} \qquad n = 1, \cdots, 4K. \tag{12}$$

The second term is zero, since A^* is chosen to minimize Q. Hence by (9),

$$\frac{\partial \hat{Q}}{\partial \phi_n} = 2A^* \sum_{i=1}^{M} (A^* |H(z_i, \phi)| - Y_i^d) \frac{\partial |H(z_i, \phi)|}{\partial \phi_n} \cdot \tag{13}$$

Writing

$$|H(z_i, \phi)| = [H(z_i, \phi) \overline{H(z_i, \phi)}]^{1/2} \tag{14}$$

we have

$$\frac{\partial |H(z_i, \phi)|}{\partial \phi_n} = \frac{1}{|H(z_i, \phi)|} \cdot \mathrm{Re}\left\{\overline{H(z_i, \phi)} \, \frac{\partial H(z_i, \phi)}{\partial \phi_n}\right\} \tag{15}$$

which can be computed directly from (8) using complex arithmetic.

The subroutine which calculates $\hat{Q}(\phi)$ and grad $\hat{Q}(\phi)$, given ϕ, is summarized below.

1) Calculate

$$H_i = \prod_{k=1}^{K} \frac{1 + a_k z_i^{-1} + b_k z_i^{-2}}{1 + c_k z_i^{-1} + d_k z_i^{-2}}, \quad i = 1, \cdots, M. \tag{16}$$

2) Calculate

$$A^* = \frac{\sum_{i=1}^{M} |H_i| \, Y_i^d}{\sum_{i=1}^{M} |H_i|^2}. \tag{17}$$

3) Calculate

$$E_i = A^* |H_i| - Y_i^d, \; i = 1, \cdots, M. \tag{18}$$

4) Calculate

$$Q = \sum_{i=1}^{M} E_i^2. \tag{19}$$

5) Calculate

$$\begin{aligned} \frac{\partial |H_i|}{\partial a_k} &= \frac{1}{|H_i|} \mathrm{Re}\left\{\overline{H_i} \, \frac{\partial H_i}{\partial a_k}\right\} \\ &= \frac{1}{|H_i|} \mathrm{Re}\left\{\overline{H_i} \, H_i \frac{z_i^{-1}}{1 + a_k z_i^{-1} + b_k z_i^{-2}}\right\} \\ &= |H_i| \; \mathrm{Re}\left\{\frac{z_i^{-1}}{1 + a_k z_i^{-1} + b_k z_i^{-2}}\right\} \\ & k = 1, \cdots, K; \quad i = 1, \cdots, M \end{aligned} \tag{20}$$

and similarly,

$$\frac{\partial |H_i|}{\partial b_k} = |H_i| \; \mathrm{Re}\left\{\frac{z_i^{-2}}{1 + a_k z_i^{-1} + b_k z_i^{-2}}\right\} \tag{21}$$

$$\frac{\partial |H_i|}{\partial c_k} = - |H_i| \; \mathrm{Re}\left\{\frac{z_i^{-1}}{1 + c_k z_i^{-1} + d_k z_i^{-2}}\right\} \tag{22}$$

$$\frac{\partial |H_i|}{\partial d_k} = - |H_i| \; \mathrm{Re}\left\{\frac{z_i^{-2}}{1 + c_k z_i^{-1} + d_k z_i^{-2}}\right\}. \tag{23}$$

6) Calculate

$$\frac{\partial \hat{Q}}{\partial \phi_n} = 2A^* \sum_{i=1}^{M} E_i \frac{\partial |H_i|}{\partial \phi_n}, \qquad n = 1, \cdots, 4K. \tag{24}$$

The elimination of A as an unknown parameter reduces by one the dimensionality of the search performed by the optimization program. An additional savings in computation time is achieved by computing the z_i once at the beginning of execution and storing them for later use.

V. Stability and Minimum-Phase Constraints

Suppose $Y(z)$ has a real pole at $z = \alpha$. Replacing this by a pole at $z = 1/\alpha$ is equivalent to multiplying by the function

$$\frac{z - \alpha}{z - 1/\alpha} \tag{25}$$

which has magnitude $|\alpha|$ when z is on the unit circle. Hence the inversion of a real pole with respect to the unit circle does not affect the shape of the magnitude characteristic. Since the gain constant A is chosen optimally, $\hat{Q}(\phi)$ is not affected at all by such inversion. Similarly, $\hat{Q}(\phi)$ is unaffected by inversions of complex pairs of poles, or real or complex pairs of zeros. At convergence of the optimization program, poles and zeros will appear randomly inside or outside the unit circle, depending on the starting point of the iteration, and upon the course of the iteration itself. It will be taken as a design criterion that all the poles and zeros lie within the unit circle. The poles must do so in order that the filter be stable. The zeros lying inside the unit circle ensure that there is no excess phase.

VI. Final Strategy and Example 1: A Low-Pass Filter

At first thought, it would appear that the following procedure would yield an optimum transfer function with all its poles and zeros inside the unit circle.

1) Use of the optimization program to minimize $\hat{Q}(\phi)$ without constraining pole or zero locations.
2) At convergence, invert all poles or zeros outside the unit circle.

If the optimization program is started anew from the result of step 2, however, it is found that further reduction in $\hat{Q}(\phi)$ is sometimes possible. The following example will show how this can happen. Consider the specification of an ideal low-pass filter with cutoff frequency at one-tenth the Nyquist frequency:

$$\begin{aligned} W &= 0.00, 0.09\ (0.01); & Y^d &= 1.0 \\ W &= 0.10; & Y^d &= 0.5 \\ W &= 0.11, 0.20\ (0.01); & Y^d &= 0.0 \\ W &= 0.2, \quad 1.0 \quad (0.1); & Y^d &= 0.0. \end{aligned} \tag{26}$$

This specification weights frequencies below $W = 0.2$ more heavily than those above. If the optimization for a one-section filter ($K = 1$) is started at

$$\phi = (0., 0., 0., -0.25)' \tag{27}$$

convergence is obtained after 93 function evaluations. The resulting zeros and poles are given below and are plotted in Fig. 2(A):

$$\begin{aligned} &\text{zeros: } 0.67834430 \pm j\ 0.73474418 \\ &\text{poles: } 0.75677793, \quad 1.3213916. \end{aligned} \tag{28}$$

The corresponding value of the error criterion is

$$\hat{Q} = 1.2611. \tag{29}$$

Notice that the two poles are very nearly inverses of each other. After inversion, 62 more function evaluations are required to produce convergence[1] to the following parameters (see Fig. 2(B)):

$$\begin{aligned} &\text{zeros: } 0.82191163 \pm j\ 0.56961501 \\ &\text{poles: } 0.89676390 \pm j\ 0.19181084 \\ &A = 0.11733978 \\ &\hat{Q} = 0.56731. \end{aligned} \tag{30}$$

The introduction of a complex pole pair is prevented in the first sequence of iterations by the fact that one pole is inside the unit circle and one is outside. After inversion, the two poles can split and become a complex pair, leading to the final minimum. The magnitude characteristic of the final filter is shown in Fig. 3.

The following algorithm was used to allow such convergence to take place.

1) Use the Fletcher–Powell optimization program until convergence takes place, or for a maximum of 25 cycles, and go to 2.
2) If any poles or zeros are outside the unit circle, invert them and go to 1. Otherwise, go to 3.
3) If convergence has not taken place, go to 1. Otherwise, go to 4.
4) Print out the final parameters and stop.

Fig. 3 also shows the resultant filter characteristic for $K=2$, corresponding to the following parameters at convergence:

$$\begin{aligned} &\text{zeros: } 0.92538461 \pm j\ 0.37902945 \\ &\qquad\quad 0.61137175 \pm j\ 0.79134343 \\ &\text{poles: } 0.93121838 \pm j\ 0.27718988 \\ &\qquad\quad 0.86454727 \pm j\ 0.13353860 \\ &A = 0.024867372 \\ &Q = 0.033959. \end{aligned} \tag{31}$$

The final parameters obtained from the $K=1$ design were used as a starting point for the $K=2$ optimization, and 376 further function evaluations were required for convergence.

[1] The convergence criterion for this and all succeeding examples corresponds to the parameter EPS = 10^{-5} in DFMFP.

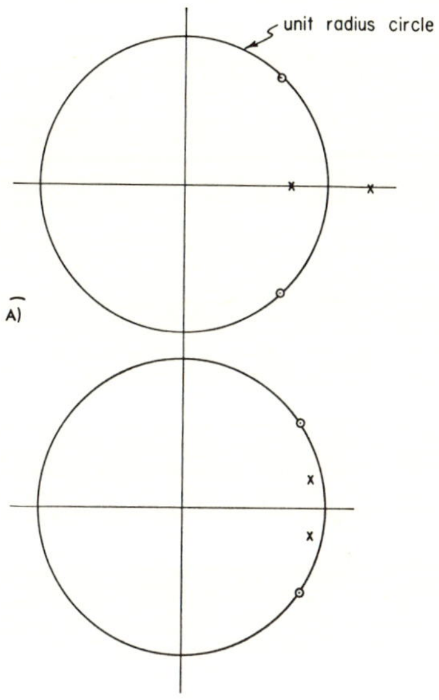

Fig. 2. Pole-zero configurations for the low-pass filter designs. (A) Intermediate local minimum. (B) Final minimum.

Fig. 3. Magnitude characteristic of the one- and two-section low-pass filters of Example 1.

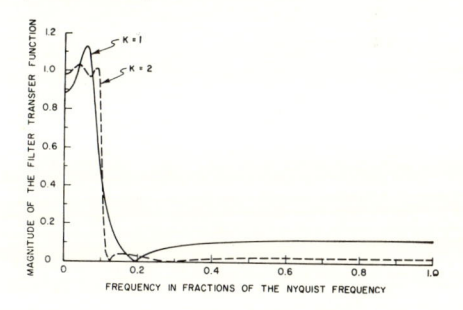

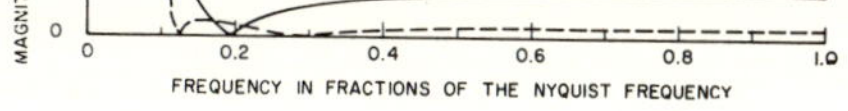

VII. Examples

Example 2: A Wide-Band Differentiator

Consider the following specification,

$$W = 0.0,\ 1.0\ (0.05); \qquad Y^d = W, \tag{32}$$

which represents a linear amplitude characteristic, and hence an ideal differentiating filter, ignoring consideration of the phase for the moment. The one-section design

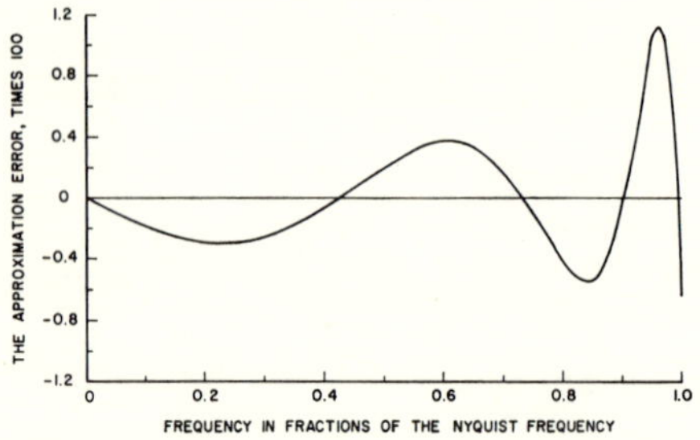

Fig. 4. The approximation error for the one-section wide-band differentiator filter.

Fig. 5. Phase characteristic of the one-section wide-band differentiator filter.

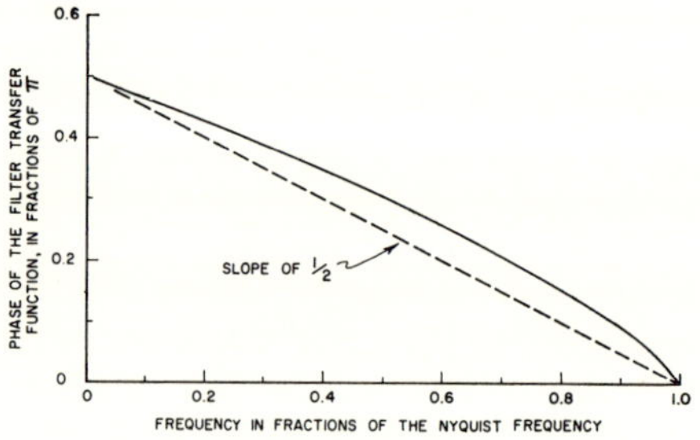

converged after 96 function evaluations to the following design:

$$\begin{aligned} \text{zeros:}\quad & 1.0000000, \quad -0.67082621 \\ \text{poles:}\quad & -0.14240300, \quad -0.71698670 \\ A &= 0.36637364 \\ \hat{Q} &= 2.7480 \times 10^{-4}. \end{aligned} \tag{33}$$

Fig. 4 shows the approximation error over the entire band of frequencies from zero to the Nyquist frequency. Of particular interest is the fact that the approximation is within about 1 percent of maximum over this entire range, in contrast with designs based on guard-band filters, which usually are accurate only up to about 80 percent of the Nyquist frequency (see [1]). Fig. 5 shows the phase characteristic, which approximates the phase of an ideal differentiator with an additional lag of one-half sampling period. Thus, this design introduces significantly less lag than designs reported in [1].

Starting with the one-section design above, 500 more function evaluations produced convergence to a more accurate two-section approximation, with $\hat{Q}=6.15\times 10^{-7}$. This two-section filter exhibited a similar characteristic ripple near the Nyquist frequency, and had almost the same phase characteristic. This points out the desirability of extending the method to include specifications on the phase characteristic.

Example 3: A Linear Discriminator

For the next example, consider the specification

$$W = 0.0,\ 1.0\ (0.05); \qquad Y^d = |1 - 2W| \tag{34}$$

which represents a linear discriminator with a zero at one-half the Nyquist frequency. After 40 function evaluations, the following one-section design was produced:

$$\begin{aligned} \text{zeros:}\quad & 0.00000000 \pm j\ 1.00000000 \\ \text{poles:}\quad & \pm 0.49614741 \\ A &= 0.35765018 \\ \hat{Q} &= 1.2299 \times 10^{-2}. \end{aligned} \tag{35}$$

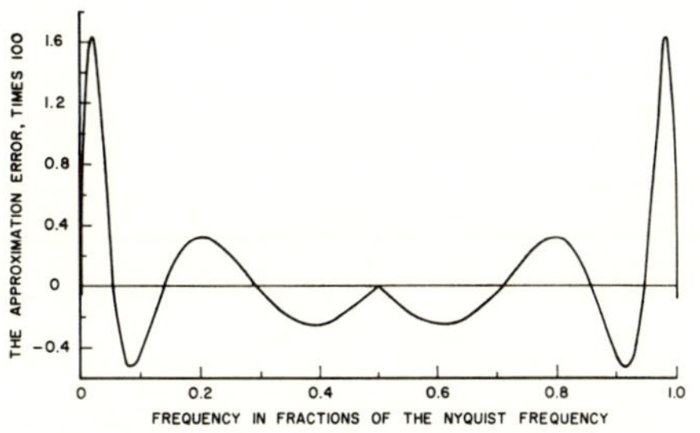

Fig. 6. The approximation error for the two-section linear discriminator filter.

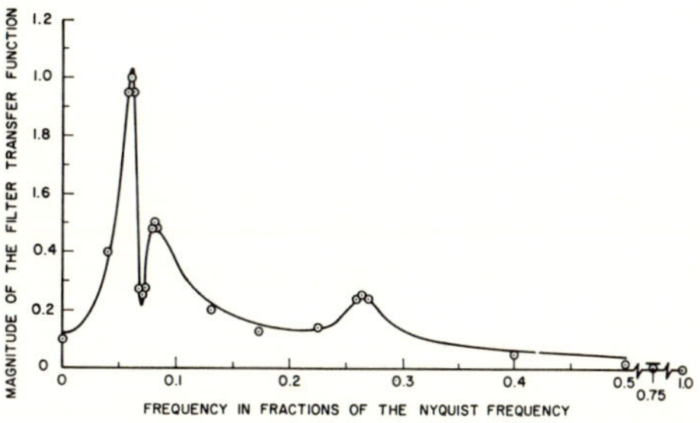

Fig. 8. Magnitude characteristic of the three-section vowel formant filter Circles indicate specification points.

Fig. 7. Phase characteristic of the two-section linear discriminator filter.

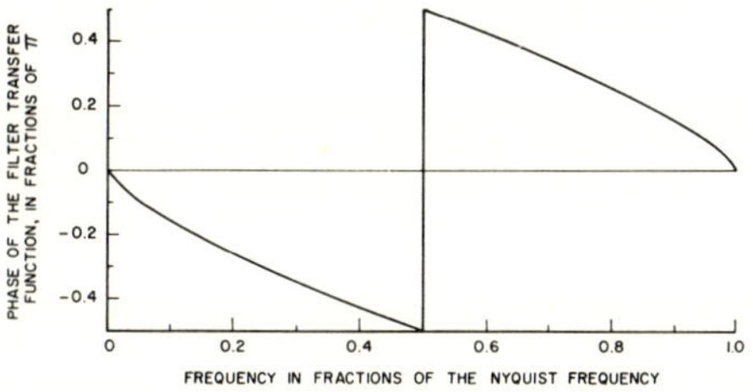

One-hundred and thirty more function evaluations produced convergence to the following two-section design:

$$\begin{aligned}
\text{zeros:}\ & 0.00000004 \ \pm j\ 0.99999931 \\
& 0.81492900, \ -\ 0.81492888 \\
\text{poles:}\ & 0.84492845, \ -\ 0.84492830 \\
& 0.37204922, \ -\ 0.37204934 \\
& A = 0.36676649 \\
& \hat{Q} = 1.0807 \times 10^{-4}.
\end{aligned} \tag{36}$$

As might be expected, the resulting pole-zero patterns are symmetric with respect to the imaginary axis, within the precision allowed by the convergence criterion. Figs. 6 and 7 show the approximation error and phase characteristic of the two-section filter.

Example 4: A Vowel Formant Filter

Fig. 8 shows the specification of a filter which is to have a magnitude characteristic corresponding to the formant for the vowel ⊃ (as in "law") [6]. The principal requirements are taken to be that peaks occur at $W=0.06$, 0.08, and 0.26; with values of 1.0, 0.5, and 0.25, respectively; and that troughs occur midway between these peaks, with values one-half the lower peak. This design problem is considerably more difficult than the previous ones, since it involves approximating a rather arbitrary and complex shape. An acceptably good design required three sections, 1809 function evaluations and 2 minutes 34 seconds of computation time on the IBM 360/65. The final design is also shown in Fig. 8 and corresponds to the following parameters:

$$\begin{aligned}
\text{zeros:}\ & 0.93470084, \ -\ 0.99966051 \\
& 0.62177384 \ \pm j\ 0.62464737 \\
& 0.97101465 \ \pm j\ 0.21383044 \\
\text{poles:}\ & 0.64343825 \ \pm j\ 0.70167849 \\
& 0.96361229 \ \pm j\ 0.19280318 \\
& 0.93515982 \ \pm j\ 0.20909432 \\
& A = 0.041075206 \\
& \hat{Q} = 9.4712 \times 10^{-3}.
\end{aligned} \tag{37}$$

VIII. Conclusions

A practical method has been described for designing recursive digital filters with arbitrary, prescribed magnitude characteristics. Examples have been given of such designs with 1, 2, and 3 cascade sections, corresponding to 5, 9, and 13 parameters. The most difficult of these designs takes about 2.5 minutes of computation time on the IBM 360/65 computer.

The important considerations in the development of this method have been: 1) a strategy for ensuring that the resulting filters are stable and minimum phase, 2) elimination of the gain factor *A* as an unknown parameter, 3) choice of the canonic form of the filter, and 4) choice of the unconstrained optimization algorithm.

Further work along these lines might take into account specifications on the phase characteristic, and arbitrary weighting of the errors at different specification points.

Acknowledgment

The author is indebted to the following for valuable discussions about this problem: Dr. G. C. Winham, Department of Music, Princeton University; C. M. Rader, Lincoln Laboratories; and Dr. J. F. Kaiser, Bell Telephone Laboratories.

References

[1] J. F. Kaiser, "Digital filters," in *System Analysis by Digital Filter*, F. F. Kuo and J. F. Kaiser, Eds. New York: Wiley, 1966, ch. 7.

[2] C. M. Rader and B. Gold, "Digital filter design techniques in the frequency domain," *Proc. IEEE*, vol. 55, pp. 149–171, February 1967.

[3] R. Fletcher and M. J. D. Powell, "A rapidly convergent descent method for minimization," *Computer J.*, vol. 6, no. 2, pp. 163–168, 1963.

[4] G. C. Temes and D. A. Calahan, "Computer-aided network design—the state of the art," *Proc. IEEE*, vol. 55, pp. 1832–1863, November 1967.

[5] "System/360 scientific subroutine package (360A-CM-03X), version III, programmer's manual," IBM Data Processing Division, White Plains, N. Y., Document H20-0205-3, 1968.

[6] R. K. Potter and J. C. Steinberg, "Toward the specification of speech," *J. Acoust. Soc. Am.*, vol. 22, pp. 807–820, November 1950.

4

Reprinted from *IEEE Trans. Acoustics, Speech, Signal Processing,* **ASSP-22**(2), 98–111 (1974)

Equiripple and Minimax (Chebyshev) Approximations for Recursive Digital Filters

ANDREW G. DECZKY, MEMBER, IEEE

Abstract—**The problem of designing recursive digital filters whose frequency response approximates an arbitrarily prescribed function in the Chebyshev sense on a single interval is considered. Certain degenerate cases where the best Chebyshev approximation is not equiripple are studied in detail, and an algorithm is given for determining the best Chebyshev as well as the best equiripple approximation. Finally, a number of examples illustrating applications of this algorithm are given.**

I. INTRODUCTION

THE PROBLEM we shall treat here is the design of a recursive digital filter whose frequency response approximates an arbitrarily prescribed function in the Chebyshev sense on a single interval. Mathematically, this problem reduces to the approximation of a continuous function by ratios of polynomials or simple (monotone) functions of such ratios and is completely equivalent to the corresponding analog filter synthesis problem. However, in certain cases (group delay synthesis) the polynomials involved are not independent, thus having nonlinear constraints among the coefficients. This effect often leads to degenerate situations where the best Chebyshev approximation is not equiripple, as would be expected from the well known theorem of alternation [1],[2]. Although this effect has been observed in practice [3]–[5], it is not well understood. Our aim here will thus be to clear up this question, and also to give an algorithm that may be used to calculate the best Chebyshev approximation for both the nondegenerate and the degenerate case.

II. STATEMENT OF THE PROBLEM

The Chebyshev approximation is precisely stated as follows [6].

The Chebyshev Approximation Problem

Let $f(x)$ be a given real-valued function defined on a compact set $X \subset E^1$, and let $R(C,x)$ be a real-valued approximating function depending continuously on $x \in X$ and on N parameters $C \in P \subset E^N$.[1]

Then we wish to determine the N parameters $C^* \in P$ such that

$$\max_{x \in X} | f(x) - R(C^*,x) | \leq \max_{x \in X} | f(x) - R(C,x) | \qquad (1)$$

for all $C \in P$.

We shall find it useful to define the L_∞ norm

$$\| f(x) - R(C,x) \| \equiv \max_{x \in X} | f(x) - R(C,x) |. \qquad (2)$$

On terms of this definition equation, (1) becomes simply

$$\| f(x) - R(C^*,x) \| \leq \| f(x) - R(C,x) \| \qquad (3)$$

for all $C \in P$.

For digital filter synthesis, we have the following identification. Let the transfer function $H(z)$ be a function of N parameters (e.g., the filter coefficients or the poles and zeros) and order these in a vector $C = [c_1, c_2, \cdots, c_N]$. The independent variable in this case is the digital frequency $\varphi = \omega T$ where $1/T$ is the sampling frequency, and the set X is an arc of the unit circle in the Z plane, $X = \{\varphi : \varphi_A \leq \varphi \leq \varphi_B\}$. Finally, the function $R(C,\varphi)$ is the frequency response of the filter (such as magnitude, loss, phase or group delay) and $f(\varphi)$ is the desired frequency response.

III. THE APPROXIMATING FUNCTION

As mentioned above, the approximating function $R(C,\varphi)$ is some frequency response characteristic of the filter under consideration. Its form depends on the structure of the filter, as well as on the choice of the parameter vector C. Here we shall consider the following transfer function

$$H(z) = k_0 \left[\prod_{i=1}^{n} (z - z_{oi}) / \prod_{i=1}^{m} (z - z_{pi}) \right] \qquad (4)$$

where z_{oi}, z_{pi} are real or occur in complex conjugate pairs. Next the parameter vector C is chosen as

$$C = [\varphi_{o1}, r_{o1}, \cdots, \varphi_{p1}, r_{p1}, \cdots, c_N]$$

with a total of $N = n + m + 1$ parameters where

$$z_{oi} = r_{oi} \exp[j\varphi_{oi}] \qquad z_{pi} = r_{pi} \exp[j\varphi_{pi}]$$

and some of the $\varphi_{oi}, \varphi_{pi}$ may be zero.

The frequency responses of interest will be taken as the loss (in dB) and the group delay, defined by

$$\alpha(\varphi) = -20 \log_{10} | H(\exp[j\varphi]) | \text{ dB}$$

$$\tau(\varphi) = -\frac{d}{d\varphi} \{ \measuredangle H(\exp[j\varphi]) \}. \qquad (5)$$

Manuscript received January 20, 1973; revised July 5, 1973. This work was supported by the Swiss National Science Foundation and formed part of a dissertation submitted to the Swiss Federal Institute of Technology, Zurich, Switzerland, in partial fulfillment of the requirements for the Ph.D. degree.

The author is with Bell Northern Research, Ottawa, Ont., Canada.

[1] P here should be a closed and bounded subset of E^N.

TABLE I

$\frac{\partial\alpha}{\partial\varphi_{pi}} = 20 \log_{10}(e) \frac{-r_{pi}\sin(\varphi-\varphi_{pi})}{1-2r_{pi}\cos(\varphi-\varphi_{pi})+r_{pi}^2}$
$\frac{\partial\alpha}{\partial r_{pi}} = 20 \log_{10}(e) \frac{r_{pi}-\cos(\varphi-\varphi_{pi})}{1-2r_{pi}\cos(\varphi-\varphi_{pi})+r_{pi}^2}$
$\frac{\partial\alpha}{\partial\varphi_{oi}} = 20 \log_{10}(e) \frac{r_{oi}\sin(\varphi-\varphi_{oi})}{1-2r_{oi}\cos(\varphi-\varphi_{oi})+r_{oi}^2}$
$\frac{\partial\alpha}{\partial r_{oi}} = 20 \log_{10}(e) \frac{-r_{oi}+\cos(\varphi-\varphi_{oi})}{1-2r_{oi}\cos(\varphi-\varphi_{oi})+r_{oi}^2}$
$\frac{\partial\tau}{\partial r_{pi}} = \frac{(1+r_{pi}^2)\cos(\varphi-\varphi_{pi})-2r_{pi}}{(1-2r_{pi}\cos(\varphi-\varphi_{pi})+r_{pi}^2)^2}$
$\frac{\partial\tau}{\partial\varphi_{pi}} = \frac{r_{pi}(1-r_{pi}^2)\sin(\varphi-\varphi_{pi})}{(1-2r_{pi}\cos(\varphi-\varphi_{pi})+r_{pi}^2)^2}$
$\frac{\partial\tau}{\partial r_{oi}} = \frac{-(1+r_{oi}^2)\cos(\varphi-\varphi_{oi})+2r_{oi}}{(1-2r_{oi}\cos(\varphi-\varphi_{oi})+r_{oi}^2)^2}$
$\frac{\partial\tau}{\partial\varphi_{oi}} = \frac{-r_{oi}(1-r_{oi}^2)\sin(\varphi-\varphi_{oi})}{(1-2r_{oi}\cos(\varphi-\varphi_{oi})+r_{oi}^2)^2}$

Therefore, for the transfer function of (4), $R(C,\varphi)$ becomes [5], for the loss

$$R(C,\varphi) = \alpha(\varphi) = \sum_{i=1}^{m} 10\log_{10}(1-2r_{pi}\cos(\varphi-\varphi_{pi})+r_{pi}^2) - \sum_{i=1}^{n} 10\log_{10}(1-2r_{oi}\cos(\varphi-\varphi_{oi})+r_{oi}^2) - c_N, \tag{6}$$

for the group delay

$$R(C,\varphi) = \tau(\varphi) = \sum_{i=1}^{m} \frac{1-r_{pi}\cos(\varphi-\varphi_{pi})}{1-2r_{pi}\cos(\varphi-\varphi_{pi})+r_{pi}^2} - \sum_{i=1}^{n} \frac{1-r_{oi}\cos(\varphi-\varphi_{oi})}{1-2r_{oi}\cos(\varphi-\varphi_{oi})+r_{oi}^2} - c_N \tag{7}$$

where

$$N = n + m + 1$$

$c_N = \log_{10} k_o$ for the loss

$c_N = \tau_o$ for the group delay

since the absolute value of the loss or group delay is generally a free parameter. In (6) and (7), we have to remember that the poles and zeros are either real or occur in complex conjugate pairs.

The advantage of this representation is twofold. The stability of the filter is readily tested at any stage of the calculations, and the effect of the parameters is made explicit, permitting easy estimation of the initial conditions in many cases. Furthermore, the partial derivatives of $R(C,\varphi)$ with respect to the parameters is readily obtained in explicit form. These are given in Table I for later reference.

IV. REDUCTION TO RATIONAL FORM

As mentioned above, the frequency response functions $R(C,\varphi)$ of a digital filter are all basically rational functions or simple (monotone) functions of rational functions. To prove this statement, we could use (6) and (7), together with the transformation $x = \cos\varphi$. However, this leads to a large amount of algebraic manipulation. Thus we shall find it convenient to use (4) directly.

For the magnitude squared function we have

$$|H(z)|^2_{z=\exp(j\varphi)} = H(z)H(z^{-1})\big|_{z=\exp(j\varphi)} = k_0^2 \left.\frac{\prod_{i=1}^{n}(z-z_{oi})(z^{-1}-z_{oi})}{\prod_{i=1}^{m}(z-z_{pi})(z^{-1}-z_{pi})}\right|_{z=\exp(j\varphi)} \tag{8}$$

which may be written as

$$|H(z)|^2_{z=\exp(j\varphi)} = k_0' \left.\frac{\prod_{i=1}^{n}[(z+z^{-1})/2-(z_{oi}+z_{oi}^{-1})/2]}{\prod_{i=1}^{m}[(z+z^{-1})/2-(z_{pi}+z_{pi}^{-1})/2]}\right|_{z=\exp(j\varphi)} \tag{9}$$

Now letting

$$x = \frac{z+z^{-1}}{2} \tag{10}$$

we get

$$R(C',x) = |H(x)|^2_{x=\cos\varphi} = k_0' \left.\frac{\prod_{i=1}^{n}(x-x_{oi})}{\prod_{i=1}^{m}(x-x_{pi})}\right|_{x=\cos\varphi} \tag{11}$$

where we have to remember that x_{oi} and x_{pi} are either real or occur in complex conjugate pairs such that this is a real (positive or zero) expression. Clearly (11) is a ratio of polynomials in x, as asserted.[2] Furthermore, taking the logarithm of $|H(x)|^2$ we have

$$\alpha(x) = -10\log_{10}|H(x)|^2$$

[2] Strictly speaking the rational function in (11) is constrained to be positive semidefinite, or by the fact that any real zeros in the interval $-1 \le x \le +1$ must occur with even multiplicity. For our purposes we shall assume that $R(C,x)$ has no real zeros in this interval (i.e., $f(x)$ is everywhere positive) to assure that the alternation theorem may be applied in its classical form. Clearly in the general case, some modification of the classical theory is necessary.

which is a monotone (decreasing) function of a ratio of polynomials.

For the group delay function, on the other hand, we have [8]

$$\tau(\varphi) = -\frac{d}{d\varphi}[\measuredangle H(z)]_{z=\exp(j\varphi)}$$

$$= \mathrm{Re}\left\{\frac{z}{H(z)}\frac{dH(z)}{dz}\right\}_{z=\exp(j\varphi)}. \quad (12)$$

Using (4), this becomes after some algebra [8],[9]

$$\tau(\varphi) = \frac{m-n}{2} + \tfrac{1}{2}\sum_{i=1}^{m}\frac{(z_{pi}-z_{pi}^{-1})/2}{(z+z^{-1})/2-(z_{pi}+z_{pi}^{-1})/2}$$

$$-\tfrac{1}{2}\sum_{i=1}^{n}\frac{(z_{oi}-z_{oi}^{-1})/2}{(z+z^{-1})/2-(z_{oi}+z_{oi}^{-1})/2}\Bigg|_{z=\exp(j\varphi)}. \quad (13)$$

Again using the transformation of (10), we get

$$R(C',x) = \tau(x) = \frac{m-n}{2} + \tfrac{1}{2}\sum_{i=1}^{m}\frac{(x_{pi}^2-1)^{1/2}}{x-x_{pi}}$$

$$-\tfrac{1}{2}\sum_{i=1}^{n}\frac{(x_{oi}^2-1)^{1/2}}{x-x_{oi}}\Bigg|_{x=\cos\varphi} \quad (14)$$

where again the x_{po},x_{oi} are either real or occur in complex conjugate pairs, such that this is a real expression. Bringing (14) to a common denominator, we thus get

$$R(C',x) = \sum_{i=0}^{m+n} a_i x^i \Big/ \sum_{i=0}^{m+n} b_i x^i \quad (15)$$

where in general

$$a_i = f(b_0,b_1,\cdots,b_{n+m})$$

i.e., the a_i's are functions of the b_i's. Thus, in this case, $R(C',x)$ is a ratio of polynomials with nonlinear constaints among the coefficients. This may lead to certain degenerate situations where the best Chebyshev approximation is not equiripple, as will be shown below, and hence forces us to look a little more closely at the theory of Chebyshev approximations.

V. THE CHARACTERIZATION OF BEST CHEBYSHEV APPROXIMATIONS

The best Chebyshev approximation is generally characterized by the well known alternating property of the error function. This result was first established for the case of linear approximating functions (e.g., polynomials) [1] and was later extended to certain nonlinear approximating functions (e.g., rational functions) [2],[6], [7]. Let us state this result in a precise form first. For a proof see [6] or [7].

Theorem I [6]

Let $f(x)$ be a continuous real-valued function on a closed interval $X = \{x:x_a \le x \le x_b\}$, and let $R(C,x)$ be a ratio of polynomials with no common factors,

$$R(C,x) = \sum_{i=\epsilon}^{n} a_i x^i \Big/ \sum_{i=0}^{m} b_i x^i \qquad C \in P,\ x \in X \quad (16)$$

$$P = \{a_i,b_i : a_n \ne 0, b_m \ne 0, \quad \sum_{i=0}^{m} b_i x^i \ne 0 \quad \text{for } x \in X,\ b_0 = 1\}.$$

A necessary and sufficient condition that $R(C^*,x)$ be the best Chebyshev approximation to $f(x)$ is that there exist $n+m+2$ points $x_a \le x_1 < \cdots < x_{n+m+2} \le x_b$ such that

$$f(x_i) - R(C^*,x_i) = -[f(x_{i-1}) - R(C^*,x_{i-1})]$$

$$= \|f(x) - R(C^*,x)\|$$

$$i = 2,\cdots,n+m+2. \quad (17)$$

A solution such as that of (17) where the error function $e(x) = (f-R)(x)$ alternates in sign from extremum to extremum with equal magnitude will be called equiripple. Further, a point x_i such that $e(x_i) = \|e(x)\|$ will be called an extremal point.

As we have seen above, in the case of filter synthesis, the approximating function is often some simple monotone function of a rational function. For these cases we have the following extension of the above theorem.

Corollary I

Let $f(x)$ and $R(C,x)$ be as in Theorem I above, and let $g(y)$ be a continuous, real-valued, monotone function for $y \in Y$, where Y is the set $Y = Y_1 \cup Y_2, Y_1 = \{y_1 : y_1 = f(x), x \in X\}$ and $Y_2 = \{y_2 : y_2 = R(C,x), x \in X, C \in P\}$. Then Theorem I applies to $g[R(C,x)]$.

Proof: The proof of Theorem I is based on the fact that $R(C,x)$ is a varisolvent function (Rice [6]), i.e., it has the following properties.

1) Given a set of $N = n+m+2$ distinct points $x_i \in [x_a,x_b]$ and a set of N arbitrary real numbers $y_i \in Y_2$, there is a $C \in P$ such that

$$R(C,x_i) = y_i \qquad i = 1,\cdots,N.$$

2) Given $C_1,C_2 \in P$, $C_1 \ne C_2$ then $R(C_1,x) - R(C_2,x)$ has at most $N-1$ zeros in $[x_a,x_b]$.

Since $g(y)$ is monotone and continuous, $g(y_1) = g(y_2)$ if and only if $y_1 = y_2$. Letting $y_1 = R(C_1,x)$, $y_2 = R(C_2,x)$ shows that $g[R(C_1,x)] - g[R(C_2,x)]$ has at most $N-1$ zeros in $[x_a,x_b]$ if $C_1 \ne C_2$.

Also, given a set of N distinct points $x_i \in [x_a,x_b]$ and a set of N arbitrary real numbers $z_i \in Z$, where $Z = \{z : z = g[R(C,x)], x \in X, C \in P\}$, then there exist a set of distinct real numbers

$$y_i = g^{-1}(z_i) \qquad y_i \in Y_2 \qquad i = 1,\cdots,N.$$

Therefore by 1) there exists a $C \in P$ such that

$$R(C,x_i) = y_i \qquad i = 1,\cdots,N$$

and therefore

$$g[R(C,x_i)] = g(y_i) = z_i \qquad i = 1,\cdots,N.$$

Therefore $g[R(C,x)]$ is a varisovent function, and Theorem I applies. Q.E.D.

As we have seen above, the group delay function does not fulfill the conditions of Theorem I. Nevertheless it has been widely assumed that the alternation of the error function must still hold in this case, and such equiripple approximations have been computed [3],[4],[11]–[14]. This suggests the existence of such an equiripple solution in most cases of practical interest. However, as will be shown below, such an equiripple solution may not be optimal in the Chebyshev sense. Thus, two important questions arise. 1) Having found an equiripple solution, how do we test it for optimality? 2) If the equiripple solution is not optimal, how do we find a better one? The first question is answered by the following important theorem, the proof of which is postponed to the Appendix. The second question is taken up in Section VII.

Theorem II

Let X be a compact subset of E^1, and $f(x)$ be a continuous real-valued function for $x \in X$. Let $R(C,x)$ be a continuous, real-valued function having no poles in the range of approximation and posessing continuous first order derivatives $\partial R(C,x)/\partial c_j$, $j = 1,\cdots,N$ for all $x \in X$ and $C \in P$, where P is a compact subset of E^N. Further suppose that $f(x) - R(C^e,x)$ has q extremal points $x \in X$, $i = 1,\cdots,q$.

A necessary and sufficient condition that $R(C^e,x)$ be the locally best Chebyshev[3] approximation to $f(x)$ is that the system of inequalities

$$\sigma_i dR(C^e,x_i)\cdot\delta C > 0 \qquad i = 1,\cdots,q \tag{18}$$

be inconsistent for all $\delta C \in P$, where

$$\sigma_i = \operatorname{sgn}\left(f(x_i) - R(C^e,x_i)\right)$$

$$dR(C^e,x_i) = \left[\frac{\partial R(C^e,x_i)}{\partial c_1},\cdots,\frac{\partial R(C^e,x_i)}{\partial c_N}\right]$$

$$\delta C = [\delta c_1,\cdots,\delta c_N]. \tag{19}$$

A geometrical interpretation of this theorem is provided by the following result from the theory of linear inequalities (see for example Cheney [7]).

Lemma [7]

Let U be a compact subset of E^N. A necessary and sufficient condition that the system of linear inequalities

$$u_i\cdot z > 0 \qquad i = 1,\cdots,q \tag{20}$$

where $u_i \in U$ and z is any vector in E^N, be inconsistent is that there exist constants $\lambda_i \geq 0$ such that

$$\sum_{i=1}^{q}\lambda_i u_i = 0 \quad \text{and} \quad \sum_{i=1}^{q}\lambda_i = 1. \tag{21}$$

[3] Locally best Chebyshev approximation here means that there is a neighborhood $W \subset P$ of C^e, such that

$$\| R(C^e,x) - f(x) \| \leq \| R(C,x) - f(x) \qquad \text{for all } C \in W.$$

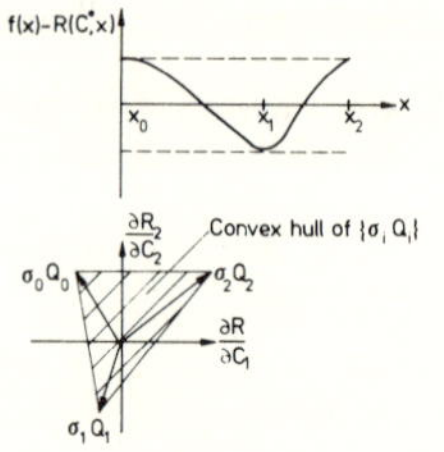

Fig. 1. Illustrating Theorem III.

In this case we say that 0 lies in the convex hull of the set $\{u_i\}$.

Combining this result with that of Theorem II, we get the following.

Theorem III

Under the conditions of Theorem II, if $f(x) - R(C^e,x)$ has q extremal points $x_i \in X$, then a necessary and sufficient condition that $R(C^e,x)$ be the locally best Chebyshev approximation to $f(x)$ is that 0 lies inside the donvex hull of the N tuples

$$\sigma_i Q_i \qquad i = 1,\cdots,q \tag{22}$$

where σ_i is the same as in Theorem II, and we have written for convenience

$$Q_i = dR(C^e,x_i) = \left[\frac{\partial R(C^e,x_i)}{\partial c_1},\cdots,\frac{\partial R(C^e,x_i)}{\partial c_N}\right].$$

(See Fig. 1.)

It often happens that the approximating function contains an additive constant (as in the case of group delay equalizers where the absolute value of the final delay is not important). In this case we may use the following result.

Corollary II

Under the conditions of Theorem II, if $R(C,x)$ is of the form

$$R(C,x) = R_0(C_0,x) + c_N \tag{23}$$

where

$$C = [c_1,\cdots,c_N],\ C_0 = [c_1,\cdots,c_{N-1}]$$

and if $f(x) - R(C^e,x)$ has q extremal points $x_i \in X$, a necessary and sufficient condition that $R(C^e,x)$ be the locally best Chebyshev approximation to $f(x)$ is that 0 lies inside the convex hull of the $(N-1)$ tuples

$$\sigma_i(Q_i - Q_j) \qquad \sigma_i\sigma_j = -1, \qquad i < j \tag{24}$$

where Q_i is the same as in Theorem III.

Proof: By the Lemma above and (24) this condition is equivalent to the one that the system of inequalities

$$(\sigma_i Q_i + \sigma_j Q_j)\cdot\delta C > 0, \sigma_i\sigma_j = -1, \quad i < j \tag{25}$$

be inconsistent for all $\delta C \in P$. Now if (25) is inconsistent, so is (18), for all i. Thus, suppose that (25) is consistent for some $\delta C_0 \in P$ and let

$$V_i = dR(C_0^e, x_i) \cdot \delta C_0.$$

$$V_{\max} = \max_{i \in I} V_i \qquad \text{where } I = \{i : \sigma_i = -1\}$$

$$V_{\min} = \min_{j \in J} V_j \qquad \text{where } J = \{j : \sigma_j = +1\}.$$

Then we have

$$\sigma_i V_i \geq \sigma_i V_{\max} > \sigma_i V_{\min} \qquad i \in I$$

$$\sigma_j V_j \geq \sigma_j V_{\min} > \sigma_j V_{\max} \qquad j \in J$$

and therefore

$$\sigma_i V_i > \sigma_i \frac{V_{\max} + V_{\min}}{2} \qquad \text{for all } i.$$

Therefore, for $\delta C = [\delta C_0, \delta c_N]$ with $\delta c_N = -(V_{\max} + V_{\min})/2$ we have

$$\begin{aligned} \sigma_i dR(C^e, x_i) \cdot \delta C &= \sigma_i (dR(C_0^e, x_i) \cdot \delta C_0 + \delta c_N) \\ &= \sigma_i V_i - \sigma_i (V_{\max} + V_{\min})/2 \\ &> 0 \end{aligned}$$

and hence (18) is consistent for $\delta C \in P$. Q.E.D.

Let us now summarize our results. We have seen that for rational functions and for continuous monotone functions of rational functions the best Chebyshev approximation is equiripple. For the case of the group delay function this is not generally true, as will be shown below. However, we may still attempt to find an equiripple approximation, and then test it by means of Theorem III. Although this theorem only guarantees local optimality, it does tell us when a solution is not even locally optimal. Further, if the best Chebyshev approximation is known to be unique, the local optimum is also global. In any case Theorem III suggests a method for finding the (locally) best Chebyshev approximation in those cases where the latter is not equiripple.

Before proceeding, however, let us give an algorithm that may be used to obtain the best equiripple approximation.

VI. THE REMEZ ALGORITHM

Remez' 2nd algorithm [1] was originally developed for the linear approximation problem, and was later extended to certain nonlinear problems [6], in particular to those cases where Theorem I applies.

The algorithm consists of two basic steps: "interpolation" and determination of the extrema. In the simple version presented here, we assume that a first approximation is available such that the error function alternates N times in sign (for N parameters in C). The $N + 1$ relative extrema of the error function are next determined, and a new approximating function is found such that the error function becomes $\pm\epsilon$ at these points (Fig. 2). This procedure is then repeated until all the extrema are equal in magnitude.

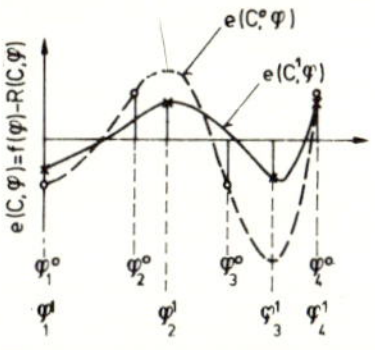

Fig. 2. One stage in the Remez algorithm.

Let us now state the algorithm in precise form. Let the approximating function be $R(C,\varphi)$, $C = [c_1, \cdots, c_N]$, the function to be approximated $f(\varphi)$, the error function $e(\varphi) = R(C,\varphi) - f(\varphi)$, and the range of approximation $[\varphi_A, \varphi_B]$. A point φ_i such that $e(\varphi_i) = \| e(\varphi) \|$ will further be called an extremal point.

Remez Algorithm

Step 1) Find an initial parameter vector C^0 such that the error function $e^0(\varphi)$ has $N + 1$ or more local extrema (not necessarily all equal) with alternating sign. Set $k = 0$, $\epsilon^0 = 0$ and go to Step 2.

Step 2) Find the $N + 1$ local extrema of the error function $e^k(\varphi)$ that are largest in absolute value with the condition that the alternation of sign is preserved. Call the abscissae at the extrema φ_i^k. If the extrema are equal in magnitude to within some tolerance, stop. Go to Step 3.

Step 3) Solve the system of (nonlinear) equations

$$R(C^{k+1}, \varphi_i^k) - f(\varphi_i^k) = (-1)^i \epsilon^{k+1} \qquad i = 1, \cdots, N + 1 \tag{26}$$

for the new parameters C^{K+1} and ϵ^{K+1}. Set $K = K + 1$ and go to Step 2.

The algorithm as formulated here presents three problems: the estimation of the initial parameters in Step 1, the determination of the extrema in Step 2, and the solution of the system (26) in Step 3.

The first problem is perhaps the most difficult. As an initial "guess" we may use a set of poles distributed equidistantly in the range of approximation (for all pole and all pass filters), with equal radii r chosen such that the individual poles are "resolved". Considering only two neighboring poles, this criterion is for the loss:

$$(1 - 2r\cos\Delta\varphi + r^2)^2 \geq (1 - r)^2(1 - 2r\cos 2\Delta\varphi + r^2),$$

for the group delay:

$$2\frac{1 - r\cos\Delta\varphi}{1 - 2r\cos\Delta\varphi + r^2} \leq \frac{1}{1 - r} + \frac{1 - r\cos 2\Delta\varphi}{1 - 2r\cos 2\Delta\varphi + r^2}.$$

The first of these may be solved explicitely, whereas for the second we drop the second factor on the right. This gives for the loss:

$$r \geq \frac{\cos^2\Delta\varphi/2}{\cos\Delta\varphi} - \sqrt{\left(\frac{\cos^2\Delta\varphi/2}{\cos\Delta\varphi}\right)^2 - 1}, \tag{27}$$

for the group delay:

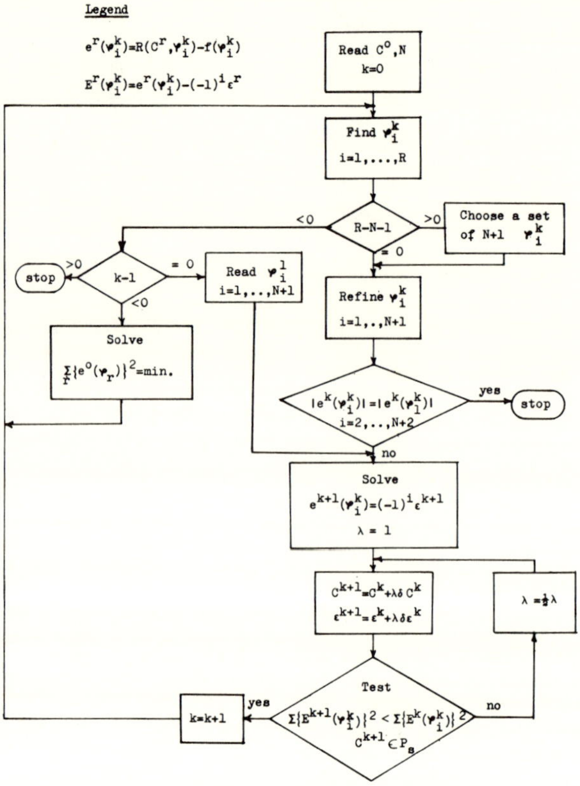

Fig. 3. General block diagram of the program.

$$r \geq \frac{1}{1 + 2\sin \Delta\varphi/2} \tag{28}$$

where

$$\Delta\varphi = (\varphi_B - \varphi_A)/(N + 1).$$

This pole distribution gives essentially a constant group delay and a Gaussian type of loss characteristic. For other types of characteristics the pole radii should be appropriately changed.

It is clear that the above method is limited to fairly simple functions $f(\varphi)$. In more complex cases this initial "guess" may be used as the initial parameter for the minimum square problem

$$\sum_{r=1}^{N_p} [R(C^0,\varphi_r) - f(\varphi_r)]^2 = \text{minimum}$$

which may be solved using the Fletcher–Powell algorithm [5]. The second problem is also solved in two steps. Thus, a preliminary set of extrema is found by evaluating the error function on a point set φ_r, $r = 1,\cdots N_p$, and detecting the points $\varphi_i^k = \varphi_r$ where $e(\varphi_{r+1}) - e(\varphi_r)$ changes sign. This method has the advantage of telling us the number and sign of the extrema. Next, these estimates are refined by using a linear search (such as Golden section [16]) between the points $\varphi_i^k + \Delta\varphi$ and $\varphi_i^k - \Delta\varphi$ where $\Delta\varphi = (\varphi_{r+1} - \varphi_r)$. The precision of this search depends on the tolerance to within which the extrema should be made equal. (We have found $N_p = 50$–100 and 5–7 steps of Golden section search satisfactory for cases up to $N = 30$, and a tolerance of 10^{-3}.)

The third problem may be solved using Newton's method. Thus linearizing the system (26) we get

$$\sum_{n=1}^{N} \frac{\partial R(C^k,\varphi_i)}{\partial c_n^k} \delta c_n^k - (-1)^i \delta\epsilon^k = -[e^k(\varphi_i^k) - (-1)^i \epsilon^k]$$

$$i = 1,\cdots,N+1 \tag{29}$$

where

$$e^k(\varphi_i^k) = R(C^k,\varphi_i^k) - f(\varphi_i^k)$$

which is a set of $N + 1$ linear equations in the $N + 1$ unknowns δc_j^k, $j = 1,\cdots,N$, $\delta\epsilon^k$, and may be solved using standard methods (e.g., Gauss–Jordan elimination).

Next we set

$$C^{k+1} = C^k + \lambda \delta C^k$$
$$\epsilon^{k+1} = \epsilon^k + \lambda \delta \epsilon k \tag{30}$$

where $0 < \lambda \leq 1$ is a damping parameter selected such that

$$\sum_{i=1}^{N+1} [e^{k+1}(\varphi_i^k) - (-1)^i \epsilon^{k+1}]^2 \leq \sum_{i=1}^{N+1} [e^k(\varphi_i^k) - (-1)^i \epsilon^k]^2 \tag{31}$$

$$C^{k+1} \in P_s \tag{32}$$

and P_s is a subset of P such that all the poles of $H(z)$ are inside the unit circle.

The condition (32) assures that the filter stays stable, and condition (31) that the error is reduced. The addition of these conditions usually assures the convergence of Newton's method provided that a stable solution of the system (26) exists, and the initial parameter C^0 is sufficiently close to this solution. (In practice this method always converged if the initial error function had $N + 1$ or more extrema). A general block diagram of the program is shown in Fig. 3.

VII. DEGENERATE CASES AND MODIFICATIONS OF THE REMEZ ALGORITHM

As mentioned in Section V, Theorem I does not apply to the case of the group delay function. Thus, although an equiripple solution based on the system (26) may exist, and may be found using the Remez algorithm above, this solution may not be the best minimax (Chebyshev) solution. At this point, it will be instructive to consider two simple examples.

Example 1

Consider the two parameter problems of approximating the delay function

$$f(\varphi) = -\gamma\varphi/\pi \qquad 0 \leq \varphi \leq 0.4\pi \tag{33}$$

where γ is a constant, using the approximating function[4]

$$R(C,\varphi) = \tau(\varphi) = \frac{1 - r_o \cos \varphi}{1 - 2r_o \cos \varphi + r_o^2} - \tau_o \tag{34}$$

where τ_o is an arbitrary constant delay.

Therefore, in this case we have $C = [r_o,\tau_o]$, $N = 2$. Ignoring any possible difficulties for the moment, we may solve for the equiripple solution (using the Remez algorithm)

$$R(C^e,\varphi_i) - f(\varphi_i) = \pm (-1)^i \epsilon \qquad i = 1,2,3 \tag{35}$$

where

$$\varphi_i = \{\varphi : | R(C^e,\varphi_i) - f(\varphi_i) | = || R(C^e,\varphi) - f(\varphi) ||\}$$

[4] This is essentially $\frac{1}{2}$ times the group delay of an all pass function.

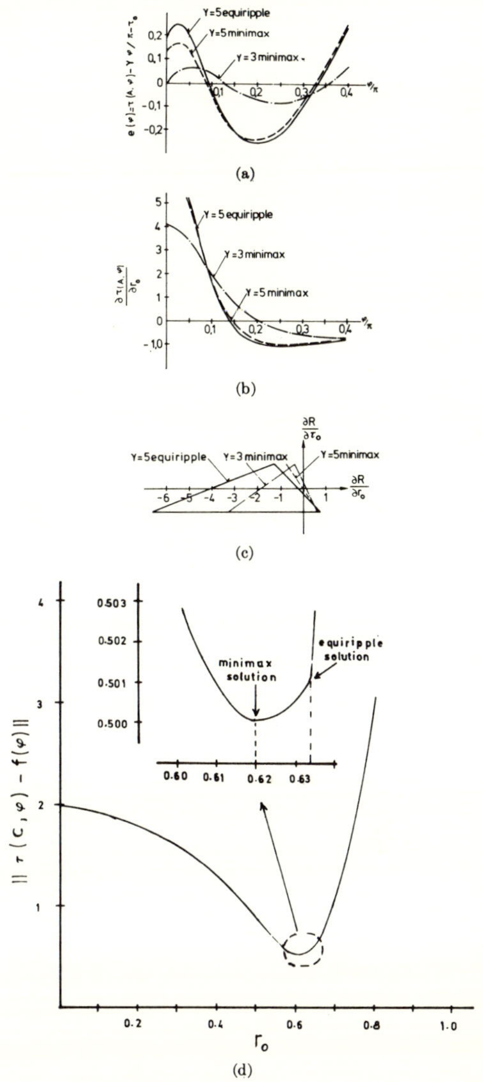

Fig. 4. (a)–(c) Equiripple and minimax solutions for the one parameter problem of Example 1. (d). Chebyshev norm as a function of the single parameter r_o in Example 1.

$$\epsilon = || R(C^e,\varphi) - f(\varphi) || .$$

This solution is shown in Fig. 4. For the two cases $\gamma = 3$ and $\gamma = 5$, together with the derivative $\partial R(C^e,\varphi)/\partial r_o$ for these cases. In order to check whether this is the best Chebyshev approximation, we may now apply Theorem III. Thus 0 must lie inside the convex hull of the 3 points $\sigma_i Q_i$, $i = 1,2,3$, which is a triangle in E^2 as shown in Fig. 4(c). From the figure we see that whereas for $\gamma = 3$, 0 is

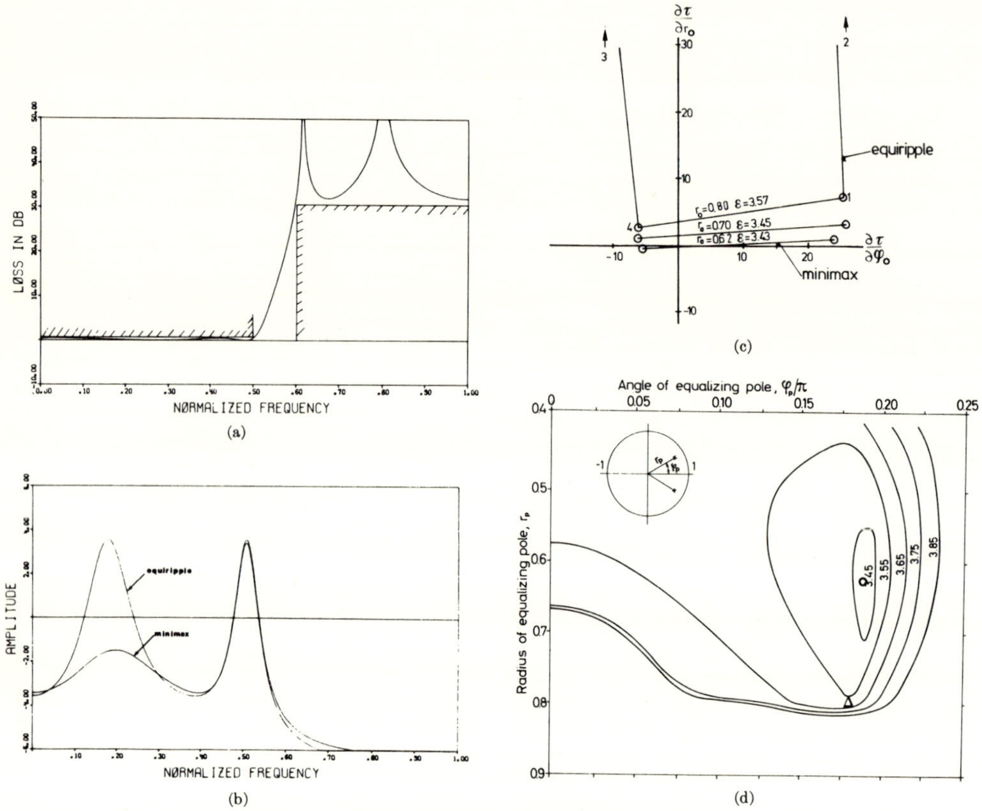

Fig. 5. (a) Loss characteristics of the elliptic digital low-pass filter of Example 2. (b) Equiripple and minimax solutions to the group delay problem of Example 2. (c) Illustrating Example 2. (d) Contours of the L_∞ norm for Example 2 as a function of r_p and φ_p/π: ○—minimax solution, △—equiripple solution.

inside this triangle, for $\gamma = 5$ it is not. Hence, we conclude that if $R(C^*,\varphi)$ is the best Chebyshev approximation then

$$\text{for } \gamma = 3 \qquad C^e = C^*$$

$$\text{for } \gamma = 5 \qquad C^e \neq C^*. \tag{36}$$

Indeed, in the latter case, by reducing r_o slightly (from 0.634 to 0.620) we obtain in the solution marked "minimax" in Fig. 4(a). Since we now have only two extremal points, Theorem III reduces to the fact that the line joining the points $\sigma_i Q_i$, $i = 1,2$ must pass through the origin [see Fig. 4(c)]. The fact that this is indeed the (unique) best Chebyshev approximation is verified in Fig. 4(d).

Example 2

As a second example we shall consider the group delay equalization of an elliptic digital low pass filter of order 4, having the following characteristics, [see Fig. 5(a)]

$$\text{passband ripple} = 1 \text{ dB}$$

$$\text{stopband loss} \geq 32 \text{ dB}$$

$$\text{group delay distortion} \simeq 10\ T \text{ s } (T = \text{sampling interval})$$

$$H(z) = k_o \prod_{i=1}^{2} \frac{z^2 + a_{1i}z + a_{2i}}{z^2 + b_{1i}z + b_{2i}} \tag{38}$$

and

$$k_o = 0.147295$$

where

$$a_{11} = 1.61178 \qquad b_{11} = -0.403133$$

$$a_{12} = 7.18956 \qquad b_{12} = 0.051401$$

$$a_{21} = 1.00000 \qquad b_{21} = 0.233280$$

$$a_{22} = 1.00000 \qquad b_{22} = 0.797295.$$

The group delay of this filter will be equalized using an all pass filter of the form

$$G(z) = \prod_{i=1}^{N} \frac{zz_{oi} - 1}{z - z_{oi}} \tag{39}$$

where the z_{oi} are real or occur in complex conjugate pairs. Taking $N = 2$, we thus have

$$f(\varphi) = -\tau_s(\varphi) \qquad 0 \leq \varphi \leq 0.51\pi$$

$$R(C,\varphi) = \tau(\varphi) = 2\left\{\frac{1 - r_o \cos(\varphi - \varphi_0)}{1 - 2r_o \cos(\varphi - \varphi_0) + r_o^2} + \frac{1 - r_o \cos(\varphi + \varphi_o)}{1 - 2r_o \cos(\varphi + \varphi_o) + r_o^2}\right\} - \tau_o \tag{40}$$

where

$$C = [\varphi_o, r_o, \tau_o].$$

τ_o is an additive constant delay and $\tau_s(\varphi)$ is the group delay of $H(z)$ in (38). Applying the Remez algorithm to this problem we get the solution [see Fig. 5(b)]

$$R(C^e,\varphi_i) - f(\varphi_i) = (-1)^i\epsilon \qquad i = 1,2,3,4 \tag{41}$$

where

$$\varphi_i = \{\varphi : | R(C^e,\varphi_i) - f(\varphi_i) | = || R(C^e,\varphi) - f(\varphi) ||\}$$

$$\epsilon = || R(C^e,\varphi) - f(\varphi) || = 3.57$$

and

$$C^e = [0.18, 0.80, 7.1].$$

Once more we shall test the optimality of this solution using Lemma III (in order to eliminate τ_o). Thus 0 must lie inside the convex hull of the four points $\sigma_1(Q_1 - Q_2)$, $\sigma_2(Q_2 - Q_3)$, $\sigma_3(Q_3 - Q_4)$, $\sigma_4(Q_4 - Q_1)$. These points are plotted in Fig. 5(c) where we see that 0 is not inside the polygon formed by these points. In fact by reducing r_o slightly (and keeping it fixed) we may solve the "constrained" problem

$$R(C,\varphi_i) - f(\varphi_i) = (-1)^i\epsilon \qquad i = 1,3,4$$

$$C = [r_o,\varphi_o,\tau_o], \qquad r_o \text{ fixed}$$

$$\varphi_i = \{\varphi : | R(C,\varphi_i) - f(\varphi_i) | = || R(C,\varphi) - f(\varphi) ||\} \tag{42}$$

$$\epsilon = || R(C,\varphi) - f(\varphi) || .$$

Such a solution for $r_o = 0.62$ is shown in Fig. 5(b), and it may be tested for optimality by plotting the two points $\sigma_3(Q_3 - Q_4)$, $\sigma_4(Q_4 - Q_1)$, as shown in Fig. 5(c). Since the line joining these points goes through the origin, clearly this is the best (local) Chebyshev (minimax) solution $R(C^*,\varphi)$, with

$$C^* = [0.19, 0.62, 7.8] \qquad \epsilon = 3.43.$$

The fact that this is indeed the (unique) best Chebyshev approximation in this case is verified in Fig. 5(d), where we have plotted the contours of $R(C,\varphi) - f(\varphi)$ as a function of r_o,φ_o.

Remarks

A solution of (42) with $r_o = 0.7$ is also shown in Fig. 5(c). We note that the line $\sigma_3(Q_3 - Q_4)$, $\sigma_4(Q_4 - Q_1)$ for this r_o is between those for $r_o = 0.80$ and $r_o = 0.62$. Thus this line moves gradually towards the origin as r_o is reduced, and the minimax solution may be found on the basis of a set of such solutions with different r_o, by interpolation.

Further we note that the difference (measured in terms of the norm $|| R(C,\varphi) - f(\varphi) ||$) between the minimax and equiripple solution is small. However, it turns out that the time response of the minimax solution is better, since the group delay is flat across most of the band. A similar effect has been reported in [4].

Finally, from the above examples, we see that the degeneracy occurs when the ripple is too large. This in turn means that the r_{oi} are too large (close to 1), and hence that the partial derivative(s) $\partial R(C^*,\varphi)/\partial r_o$ are not monotone in the interval of approximation. Still another way of looking at this is to compare the area under the group delay of the n poles of the approximating function with the area under the function $f(\varphi)$, since the ripple will in general be large if the former is not larger than the latter. This then gives us an estimate for the degree n as follows (note that the estimate will in general be too small).

Let

$$I_s = \int_{\varphi_A}^{\varphi_B} f(\varphi)\, d\varphi \tag{43}$$

$$\varphi_C = \varphi_B - \varphi_A$$

letting all the $r_{oi} = r$, we should have for n poles

$$I = \frac{n}{2}\int_0^{\varphi_C} \frac{1 - r\cos\varphi}{1 - 2r\cos\varphi + r^2}\, d\varphi \geq I_s. \tag{44}$$

Now

$$I = \frac{n}{2}\left\{\tan^{-1}\left[\frac{2r}{1 - r^2}\tan\frac{\varphi_C}{2}\right] + \frac{\varphi_C}{2}\right\} \tag{45}$$

which for r sufficiently close to 1 reduces to

$$I = \frac{n}{4}(\pi + \varphi_C) \tag{46}$$

and hence

$$n \geq \frac{4I_s}{\pi + \varphi_C}. \tag{47}$$

For our example 2 above, (remembering that here we used an all pass filter) (47) gives $n \geq (2 \times 3.84)/1.51 \simeq 5.1$, which is a little too small since in this case $n = 6$ is degenerate, but $n = 8$ is not.

Let us now recapitulate our results. We have seen that in certain cases of group delay approximation the best Chebyshev (minimax) approximation is not equiripple.

In these cases the N dimensional polyhedron formed by the points $\sigma_i Q_i$ does not contain the origin, and hence has a "face" contained in an $(N-1)$ dimensional hyperplane which is closest to the origin. The extremum corresponding to the vertex on the side of this hyperplane away from the origin is then not "essential", and may be eliminated by reducing the radius of the pole corresponding to this extremum. Thus solving the constrained problem at the N extrema (with one of the r_{oi} fixed at successively different, decreasing values) causes the $(N-1)$ dimensional hyperplane containing the N points $\sigma_i Q_i$ to move towards the origin until it passes through it. At this point the $(N-1)$ dimensional polyhedron formed by the points $\sigma_i Q_i$ either contains the origin (in which case we are finished) or it does not (in which case the above procedure is repeated by dropping another "nonessential" extremum). In general, we shall say that we have a degeneracy of order r if r extrema are dropped from the original $N+1$ of the equiripple solution.

In practice testing whether 0 is inside the polyhedron formed by the $\sigma_i Q_i$ is tedious for high orders. However, we may still apply the above procedure, and perturb the best equiripple approximation by reducing (the largest) r_{oi} slightly and finding the solution of the corresponding "constrained" problem. To make sure that we have the minimax solution, this would have to be repeated for each r_{oi} (i.e., for each extremum). In practice it is usually sufficient to do this for the largest r_{oi}, however.

Finally we shall give the necessary modification to the Remez algorithm for the degenerate case.

Remez Algorithm–Degenerate Case

1) Solve for the equiripple solution as for the normal case. Set $K = 1$.
2) Drop the equation corresponding to the "nonessential" extremum from the set (26). Fix the value of the r_{ok} corresponding to this extremum, and solve the Remez algorithm at the remaining $N-K$ extrema (using the $N-K$ parameters left).
3) Repeat for two more values of r_{ok} and pass a parabola through the three points $(\epsilon_i, r_{ok,i})$ $i = 1,2,3$, i.e., let $\epsilon = ar_{ok}^2 + br_{ok} + c$. Calculate $r_o^* = -b/2a$.
4) Solve the Remez algorithm with r_o^* at the $(N-K)$ extrema.
5) If 0 lies inside the polyhedron formed by the $\sigma_i Q_i$, $i = 1,\cdots,N-K$, stop. Otherwise set $K = K+1$ and go to 2.

VIII. RESULTS

In this section we shall give a few typical examples of the application of the Remez algorithm to the synthesis of recursive digital filters, for the normal (equiripple = minimax) case.

Example 3

Let us consider approximating the bandpass Gaussian magnitude characteristic

$$f(\varphi) = -20 \log_{10} |H(e^{j\varphi})| = -3\left(\frac{\varphi - \varphi_o}{\Delta\varphi}\right)^2 \qquad 0.3\pi \le \varphi \le 0.7\pi \quad (48)$$

where

$$\varphi_o = 0.5 \qquad \Delta\varphi = 0.05$$

which has a 3 dB bandwidth of 0.1π rad about the center frequency 0.5π rad. Using an all-pole filter of the form

$$H(z) = k_o \prod_{i=1}^{N} \frac{1}{z - r_{pi}\exp(j\varphi_{pi})} \quad (49)$$

with $N = 20$, and the initial parameters

φ_{pi}/π	r_{pi}
±0.32	0.94
±0.36	0.94
±0.40	0.94
±0.44	0.94
±0.48	0.94
±0.52	0.94
±0.56	0.94
±0.60	0.94
±0.64	0.94
±0.68	0.94

the Remez algorithm converged after 3 iterations to the solution shown in Fig. 6, with parameters

φ_{pi}/π	r_{pi}
±0.322084	0.886247
±0.365389	0.888391
±0.405404	0.892735
±0.443814	0.895749
±0.481363	0.897213
±0.518637	0.897213
±0.556186	0.895749
±0.594596	0.892735
±0.634611	0.888391
±0.677916	0.886247

$$k_o = 1.1371 \times 10^{-3}$$

$$\epsilon = 2.66 \times 10^{-2}.$$

Example 4

Here we shall consider the group delay equation of the elliptic low-pass filter considered in Example 2 above. The filter used in this case is an all pass of the form

$$H(z) = \prod_{i=1}^{N} \frac{zr_{pi}\exp(j\varphi_{pi}) - 1}{z - r_{pi}\exp(j\varphi_{pi})}. \quad (50)$$

The function $f(\varphi)$ is now

$$f(\varphi) = \tau_o - \tau_s(\varphi)$$

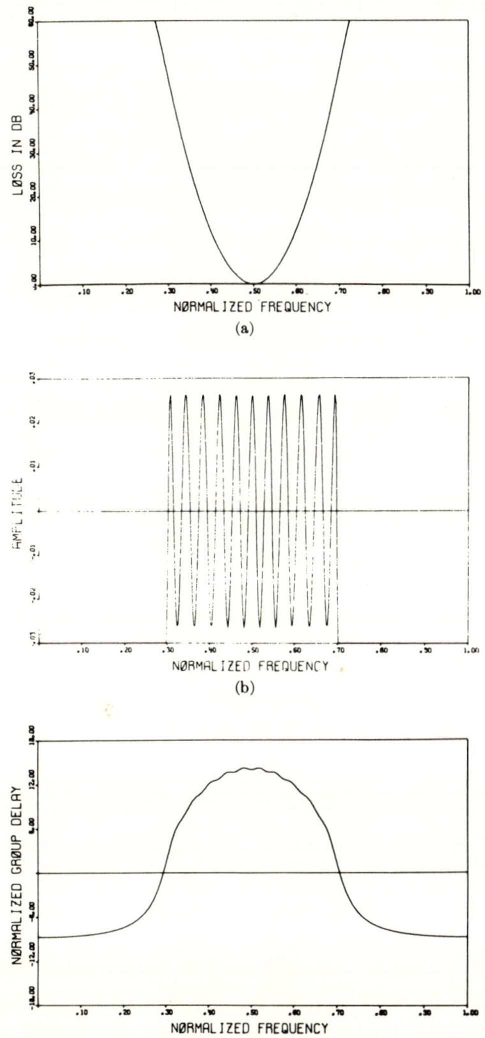

Fig. 6. (a) Loss characteristics for the Gaussian bandpass of Example 3. (b) Approximation error for the Gaussian bandpass of Example 3. (c) Group delay characteristics of the Gaussian bandpass of Example 3.

where $\tau_s(\varphi)$ is the group delay of the filter to be equalized (38), and τ_o is a constant additive delay.

Using $N = 11$ and the initial parameters

φ_{pi}/π	r_{pi}
0	0.85
±0.09	0.85
±0.18	0.85
±0.27	0.85
±0.35	0.85
±0.43	0.85

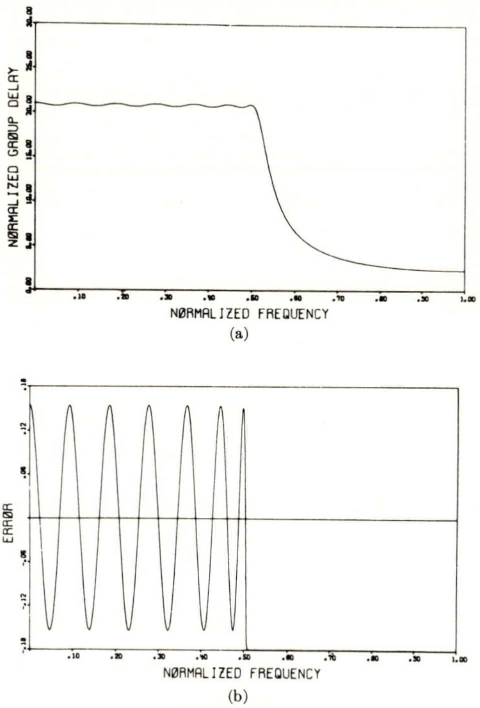

Fig. 7. (a) Group delay of the elliptic low-pass of Example 2 after equalization by an all-pass of order 11. (b) Approximation error for the group delay equalizer of Example 4.

the algorithm converged in 3 iterations to the solution shown in Fig. 7 with parameters

φ_{pi}/π	r_{pi}
0	0.770752
±0.092401	0.770745
±0.184837	0.770854
±0.277264	0.772255
±0.368346	0.780578
±0.452445	0.807923

$$\tau_o = 20.83$$

$$\epsilon = 0.154.$$

Example 5

Next we give an example of the approximation of an arbitrary group delay using all pass functions of the form given in (50). Let

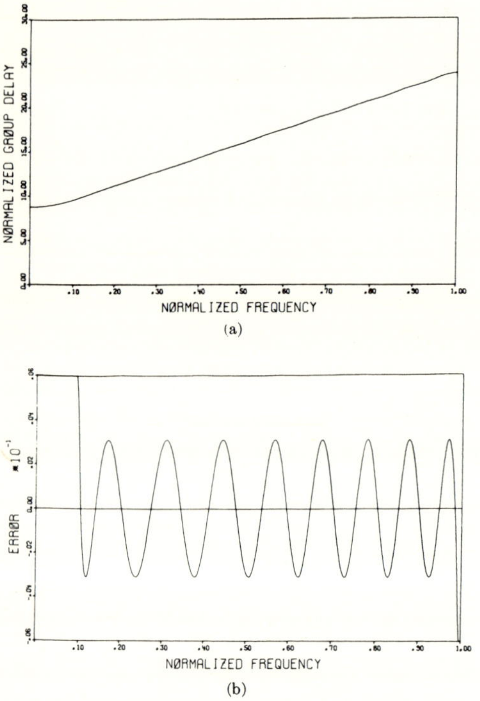

Fig. 8. (a) Group delay characteristics of the all-pass filter in Example 5. (b) Approximation error for the all-pass filter of Example 5.

$$f(\varphi) = 16\varphi/\pi \qquad 0.1\pi \leq \varphi \leq 0.99\pi$$

and further define the (relative) error as

$$e(\varphi) = (\tau(\varphi) - f(\varphi) - \tau_0)/f(\varphi).$$

Using $N = 16$ and the initial parameters (obtained from the minimum square solution)

φ_{pi}/π	r_{pi}
±0.1789	0.5469
±0.3312	0.5808
±0.4599	0.6176
±0.5772	0.6520
±0.6863	0.6783
±0.7872	0.6978
±0.8825	0.7134
±0.9841	0.6988

the algorithm converged after 6 iterations to the solution shown in Fig. 8, with parameters

φ_{pi}/π	r_{pi}
±0.173792	0.552191
±0.325451	0.591587
±0.456694	0.631520
±0.575408	0.664100
±0.684305	0.689599
±0.785469	0.709078
±0.880841	0.722128
±0.973011	0.716619

$$\tau_o = 7.974$$

$$\epsilon = 3.10 \times 10^{-3}.$$

Example 6

Finally we consider the approximation of a constant group delay using an all-pole filter of the form given in (49). We note that in this case the number of parameters is $N + 1$ as ϵ and τ_o cannot both be arbitrary since this yields the trivial solution $\epsilon = \tau_o = 0$. Another way of looking at this is to note that we have $N + 3$ "unknowns" $\varphi_{pi}, r_{pi}, i = 1, \cdots, \frac{1}{2}N, \epsilon, \tau_o$ and φ_C (the cutoff frequency) of which two must be fixed a priori. Here we shall treat the case ϵ and φ_C fixed, τ_o arbitrary. Thus we set

$$f(\varphi) = \tau_o \qquad 0.4\pi \leq \varphi \leq 0.6\pi.$$

$$\epsilon = 0.01$$

Using $N = 10$ and the initial parameters

φ_{pi}/π	r_{pi}
0.42	0.95
0.46	0.90
0.50	0.90
0.54	0.90
0.58	0.95

the Remez algorithm converged after 4 iterations to the solution shown in Fig. 9, with parameters

φ_{pi}/π	r_{pi}
±0.381944	0.847185
±0.439837	0.802570
±0.500000	0.791784
±0.560162	0.802570
±0.618056	0.847185

$$\tau_o = 6.81.$$

It is interesting to note the similarity of this solution to the Gaussian band pass of Example 3.

Finally, a remark is in order about the computer times needed for the above problems. On a CDC 6400 these problems took between 20 and 40 seconds, which may be compared to 100 to 200 seconds for the Fletcher–Powell algorithm [5] for similar problems.

APPENDIX

Proof of Theorem II

First we note that $R(C^e,x)$ is locally the best approximation to $f(x)$ if there exists a neighborhood W of C^e, $W =$

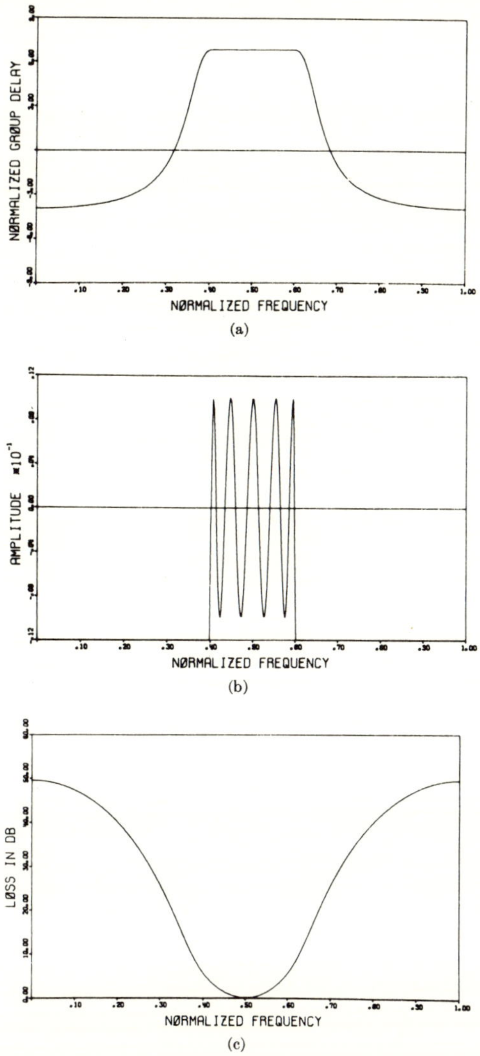

Fig. 9. (a) Group delay characteristics of the all-pole equiripple group delay bandpass of Example 6. (b) Approximation error for the equiripple group delay bandpass of Example 6. (c) Loss characteristics of the equiripple group delay bandpass of Example 6.

$\{C \in P: |C^e - C| < a_1\}$ such that

$$\|f(x) - R(C^e,x)\| \leq \|f(x) - R(C,x)\| \qquad \text{for all } C \in W. \quad \text{(A1)}$$

Now suppose that the system (18) is consistent for some $\delta C \in P$. Then by continuity and compactness there is an open set X_e, $x_i \in X_e$, and a neighborhood W_e of C^e, $W_e = \{C \in P: |C^e - C| < a_2\}$ such that

$$\sigma_i dR(C^e + \delta C,x) \cdot \delta C \geq d > 0 \qquad \text{for } x \in X_e, |\delta C| < a_2. \quad \text{(A2)}$$

Now consider $R(C^e + \lambda\delta C)$. We have as follows.

1) There exists a $\lambda_1 > 0$ such that $|\lambda_1 \delta C| < a_2$ and hence for $0 < \lambda < \lambda_1$

$$\sigma_i dR(C^e + \lambda\delta C,x) \cdot \delta C \geq d > 0 \qquad \text{for } x \in X_e. \quad \text{(A3)}$$

2) There exists a λ_2 so small that $0 < \lambda < \lambda_2$ implies

$$|f(x) - R(C^e + \lambda\delta C,x)| < \|f(x) - R(C^e + \lambda\delta C,x)\| \qquad \text{for } x \notin X_e. \quad \text{(A4)}$$

3) There exists a λ_3 so small that for $0 < \lambda < \lambda_3$

$$\operatorname{sgn}\{f(x) - R(C^e + \lambda\delta C,x)\} = \operatorname{sgn}\{f(x) - R(C^e,x)\} = \sigma(x) \quad \text{for } x \in X_e. \quad \text{(A5)}$$

4) Letting $\lambda_0 = \min[\lambda_1,\lambda_2,\lambda_3]$, then for $0 < \lambda < \lambda_o$

$$\|f(x) - R(C^e + \lambda\delta C,x)\| = \sup_{x \in X_e} \sigma(x)\{f(x) - R(C^e + \lambda\delta C,x)\}. \quad \text{(A6)}$$

Therefore for $x \in X_e$ and $0 < \lambda < \lambda_o$ consider

$$\begin{aligned} E &= \sigma(x)\{f(x) - R(C^e + \lambda\delta C,x)\} - \sigma(x)\{f(x) - R(C^e,x)\} \\ &= -\sigma(x)\{R(C^e + \lambda\delta C,x) - R(C^e,x)\} \qquad \text{(A7)} \\ &= -\sigma(x) dR(C^e + s\lambda\delta C,x) \cdot \lambda\delta C \end{aligned}$$

for some s, $0 < s < 1$, by the mean value theorem. But $|s\lambda\delta C| < a_2$ and $x_i \in X_e$, and therefore

$$E = -\sigma_i dR(C^e + s\lambda\delta C,x_i) \cdot \lambda\delta C \leq -\lambda d < 0 \qquad i = 1,\cdots,q \quad \text{(A8)}$$

and $R(C^e + \lambda\delta C,x)$ is a better approximation to $f(x)$ than $R(C^e,x)$.

For the converse suppose that for every neighborhood W of C^e, $W = \{C \in P: |C^e - C| < a_3\}$, $0 < a_3 < \min(a_2,\lambda_o a_2)$, there is a δC such that $|\delta C| < a_3$ and $R(C^e + \delta C,x)$ is a better approximation to $f(x)$ than $R(C^e,x)$. Then we must have from (A6) and (A7)

$$\sigma(x)\{R(C^e + \delta C,x) - R(C^e,x)\} > 0 \qquad \text{for } x \in X_e \quad \text{(A9)}$$

or by the mean value theorem

$$\sigma(x)\, dR(C^e + s\delta C,x) \cdot \delta C > 0 \qquad \text{for some } s, 0 < s < 1 \text{ and } x \in X_e. \quad \text{(A10)}$$

Therefore by continuity there is a neighborhood of $C^e + s\delta C$, $W' = \{C \in P: |C^e + s\delta C - C| < a_4\}$, such that

$$\sigma(x)\, dR(C,x) \cdot \delta C \geq h > 0 \qquad x \in X_e, C \in W'. \quad \text{(A11)}$$

Now let $a_3 < a_4/2$. Then it is readily verified that $W \subset W'$ and $C^e \in W \subset W'$. Therefore

$$\sigma_i\, dR(C^e,x_i) \cdot \delta C \geq h > 0 \qquad \text{since } x_i \in X_e\ i = 1,\cdots,q \quad C^e \in W'. \quad \text{(A12)}$$

Q.E.D.

ACKNOWLEDGMENT

The author wishes to thank Prof. E. Baumann for his encouragement and support during the course of this research.

REFERENCES

[1] E. Y. Remez, General computational methods for Chebyshev approximation, U. S. Atomic Energy Commission Translation, Washington, D. C. AEC-tr-4491, May 1962.
[2] E. P. Novordskii and I. S. Pinsker, "The process of equating maxima," *Usp. mat. Nauk.* (SSSR), vol. 6, pp. 174–181, 1951 (Transl. A. Shenitzer).
[3] R. L. Crane, "All-pass network synthesis," *IEEE Trans. Circuit Theory*, vol. CT-15, pp. 474–477, Dec. 1968.
[4] R. D. Rakovich and B. Djurich, "Chebyshev approximation of a constant group delay with constraints at the origin," *IEEE Trans. Circuit Theory*, vol. CT-19, pp. 466–475, Sept. 1972.
[5] A. G. Deczky, "Synthesis of recursive digital filters using the minimum *p*-error criterion," *IEEE Trans. Audio Electroacoust.*, vol. AU-20, pp. 257–263, Oct. 1972.
[6] J. R. Rice, *The Approximation of Functions.* Reading, Mass.: Addison-Wesley, 1964.
[7] E. W. Cheney, *Introduction to Approximation Theory.* New York: McGraw-Hill, 1966.
[8] A. G. Deczky, "General expression for the group delay of digital filters," *Electron. Lett.*, vol. 5, pp. 663–665, Dec. 1969.
[9] R. Unbehauen, "Zur synthese digitaler filter," *Arch. Elet. Ubertragung.*, vol. 24, pp. 305–313, 1970.
[10] A. G. Deczky, "Computer aided synthesis of digital filters in the frequency domain," Ph.D. dissertation, Swiss Federal Inst. Tech., Zurich, Switzerland, 1973.
[11] T. A. Abele, "Uebertragungsfaktoren mit tschebyscheffscher approximation konstanter gruppenlaufzeit," *Arch. Elek. Ubertragung.*, pp. 9–18, 1962.
[12] E. Ulbricht and H. Piloty, "Ueber den entwurf von allpässen und bandpässen im tschebyscheffschen sinne approximierten konstanten gruppenlaufzeit," *Arch. Elek. Ubertragung.*, vol. 14, pp. 451–467, 1960.
[13] J. P. Thiran, "Equal-ripple delay recursive digital filters," *IEEE Trans. Circuit Theory*, vol. CT-18, pp. 664–669, Nov. 1971.
[14] A. J. Gibbs, "The attainment of Tschebycheff approximations to prescribed phase delay and group delay characteristics with the aid of an electronic computer," *Proc. IEEE* (Australia), pp. 312–320, Nov. 1966.
[15] H. D. Helms, "Digital filters with equiripple and minimax responses," *IEEE Trans. Audio Electroacoust.*, vol. AU-19, pp. 87–93, Mar. 1971.
[16] D. A. Pierre, *Optimization Theory with Applications.* New York: Wiley, 1969.

5

Reprinted from *IEEE Trans. Circuit Theory,* **CT-20**(6), 697–701 (1973)

A Unified Approach to the Design of Optimum FIR Linear-Phase Digital Filters

JAMES H. McCLELLAN AND THOMAS W. PARKS

***Abstract*—A method for designing finite-duration impulse-response (FIR) linear-phase digital filters is presented in which the four possible cases for such filters are treated in a unified approach. It is shown how to reduce each case to the proper form so that the Remez exchange algorithm can be used to compute the best approximation to the desired frequency response. The result is that a very flexible and fast technique is available for FIR linear-phase filter design.**

I. Introduction

FINITE-DURATION impulse-response (FIR) linear-phase digital filters have several important properties which make them attractive for digital signal-processing applications. Among these features are the exact linear phase, the absence of any stability problems as are encountered in infinite impulse-response filters, and the availability of efficient iterative design methods. The design problem for such filters can be stated loosely as the approximation of an idealized magnitude shape in the weighted Chebyshev (min-max) sense. Several authors have considered this problem. The frequency-sampling method has been applied to the design of low-pass and bandpass filters [1], [2], as well as to wide-band differentiators [3]. However, the frequency-sampling technique does not yield optimal Chebyshev filters, nor does it allow for the exact specification of bandedge frequencies.

Herrmann [4] and Hofstetter *et al.* [5] employed a method which resulted in a restricted class of optimum Chebyshev filters, called extraripple filters [6]. The Chebyshev design problem with specified bandedge frequencies has been solved in the case of low-pass filters, and the Remez algorithm was shown to be a very efficient means for computing the best approximation [7]. Linear programming can design the same filters; however, the computation time required for linear programming is far greater than for the Remez algorithm [8].

Until now, the efficient design procedures have been restricted to special filter types and the general procedure (linear programming) is comparatively slow. The objective of this paper is to present a comprehensive summary of the linear-phase FIR filter design problem and to unify the design of the four types of filters in order to apply the Remez algorithm.

Manuscript received February 9, 1973; revised May 18, 1973. This work was supported by the National Science Foundation under Grant GK 23697.

J. H. McClellan (S'69) and T. W. Parks (S'66–M'67) are with the Department of Electrical Engineering, Rice University, Houston, Tex. 77001.

Thus it is now possible to approximate bandpass, bandstop, Hilbert-transform, and differentiator filters, as well as arbitrary filter specifications using the Remez algorithm.

II. Four Types of FIR Linear-Phase Filters

An FIR digital filter of length N with impulse response $\{h(k)\}, k=0,1,\cdots,N-1$, has a frequency response which is the z-transform evaluated on the unit circle ($z=e^{j2\pi F}$):

$$H(F)=\hat{H}(z)\Big|_{z=e^{j2\pi F}}=\sum_{k=0}^{N-1} h(k)e^{-j2\pi kF}.$$

By a linear-phase filter we mean that the frequency response can be written as

$$H(F)=G(F)e^{j(A+B2\pi F)}$$

where A and B are constants and $G(F)$ is a real-valued function. Notice that $G(F)$ is not the magnitude of the frequency response since it can be negative, but $|G(F)|$ is the magnitude.

First we show that $A=0$ or $\pi/2$. We can restrict A to lie between 0 and π because a minus sign can be incorporated into $G(F)$. Since $|G(F)|$ is the magnitude response, $|G(F)|$ is an even function. Thus $G(F)$ is either an even function or an odd function.

Recall that $H^*(F)=H(-F)$. If $G(F)$ is even, then

$$G(-F)e^{j(A-B2\pi F)}=H(-F)=H^*(F)=G(F)e^{-j(A+B2\pi F)}.$$

This implies that $e^{jA}=e^{-jA}$, and thus $A=0$. If $G(F)$ is odd, then $-e^{jA}=e^{-jA}$ and $A=\pi/2$. Note that when $A\neq 0$ we have a fixed delay, but not linear phase in the usual sense.

Now we show that $B=-(N-1)/2$ and that there are two types of symmetry for the impulse response, depending on whether $G(F)$ is even or odd. If $G(F)$ is even, then defining $\hat{H}_0(z)=z^{-B}\hat{H}(z)$, $H_0(F)=\hat{H}_0(2^{j2\pi F})$ is purely real because $H_0(F)=G(F)$. Thus $H_0(F)=H_0^*(F)=H_0(-F)$ and writing this out,

$$\sum_{k=0}^{N-1} h(k)z^{-B-k}=\sum_{k=0}^{N-1} h(k)z^{B+k}$$

$$=z^{2B+N-1}\sum_{k=0}^{N-1} h(N-1-k)z^{-B-k},$$

$$\text{for } z=e^{j2\pi F}.$$

Thus $B=-(N-1)/2$ and equating coefficients yields $h(0)=h(N-1)\cdots h(k)=h(N-1-k)$, for $k=0,1,\cdots,N-1$. This symmetry of the impulse response will be referred to as positive symmetry.

If $G(F)$ is odd, $A=\pi/2$ and with $\hat{H}_0(z)=z^{-B}\hat{H}(z)$, $H_0(F)$ is purely imaginary since $H_0(F)=jG(F)$. Thus $H_0(F)=-H_0^*(F)=-H_0(-F)$, which is written as

$$\sum_{k=0}^{N-1} h(k)z^{-B-k}=-\sum_{k=0}^{N-1} h(k)z^{B+k}, \quad \text{for } z=e^{j2\pi F}$$

so again $B=-(N-1)/2$, but the symmetry is different. $h(0)=-h(N-1)$, $h(1)=-h(N-2),\cdots$, etc. This symmetry will be called negative symmetry.

Now we examine the form of the function $G(F)$. Different functions result depending on whether N is even or odd and whether the symmetry required for linear phase is positive or negative.

A. Positive Symmetry–Odd Length

$$G(F)=\sum_{k=0}^{n} a(k)\cos 2\pi kF \tag{1}$$

where $n=(N-1)/2$, $a(0)=h(n)$, and $a(k)=2h(n-k)$, for $k=1,2,\cdots,n$. This is the case which was considered in an earlier paper by the authors [7].

B. Positive Symmetry–Even Length

$$G(F)=\sum_{k=1}^{n} b(k)\cos 2\pi(k-\tfrac{1}{2})F \tag{2}$$

where $n=N/2$ and $b(k)=2h(n-k)$, for $k=0,1,\cdots,n$. Recently, Rabiner and Herrmann have applied linear programming to this case [9].

C. Negative Symmetry–Odd Length

$$G(F)=\sum_{k=1}^{n} c(k)\sin 2\pi kF \tag{3}$$

where $n=(N-1)/2$ and $c(k)=2h(n-k)$, for $k=1,2,\cdots,n$. In order that $G(F)$ be odd, we must have $h(n)=0$. Herrmann [10] used this type in designing Hilbert transform filters.

D. Negative Symmetry–Even Length

$$G(F)=\sum_{k=1}^{n} d(k)\sin 2\pi(k-\tfrac{1}{2})F \tag{4}$$

where $n=N/2$ and $d(k)=2h(n-k)$, for $k=1,2,\cdots,n$. Differentiators have been designed for this case by using linear programming [8].

A unified approach to the approximation problem results when $G(F)$ is rewritten using (5)–(7) in the form $G(F)=Q(F)P(F)$, where $P(F)$ is a linear combination of cosine functions:

$$\sum_{k=1}^{n} b(k)\cos 2\pi(k-\tfrac{1}{2})F=\cos\pi F\sum_{k=0}^{n-1}\tilde{b}(k)\cos 2\pi kF \tag{5}$$

$$\sum_{k=1}^{n} c(k)\sin 2\pi kF=\sin 2\pi F\sum_{k=0}^{n-1}\tilde{c}(k)\cos 2\pi kF \tag{6}$$

$$\sum_{k=1}^{n} d(k)\sin 2\pi(k-\tfrac{1}{2})=\sin\pi F\sum_{k=0}^{n-1}\tilde{d}(k)\cos 2\pi kF. \tag{7}$$

Note that $G(F)$ is constrained to be zero at either $F = 0$ and/or $F = 0.5$ by (5)–(7).

E. Proof of (5)–(7)

Recall the trigonometric identity

$$\cos A \cos B = \tfrac{1}{2}[\cos(A+B) + \cos(A-B)].$$

Substituting this into the right-hand side of (5) we get

$$\sum_{k=0}^{n-1} \tilde{b}(k) \cos \pi F \cos 2\pi k F = \tfrac{1}{2} \sum_{k=0}^{n-1} \tilde{b}(k) [\cos 2\pi(k+\tfrac{1}{2})F + \cos 2\pi(k-\tfrac{1}{2})F] = \tfrac{1}{2} \sum_{k=1}^{n} \tilde{b}(k-1) \cos 2\pi(k-\tfrac{1}{2})F$$

$$+ \tfrac{1}{2} \sum_{k=0}^{n-1} \tilde{b}(k) \cos 2\pi(k-\tfrac{1}{2})F = \tfrac{1}{2} \tilde{b}(0) \cos 2\pi(-\tfrac{1}{2})F$$

$$+ \tfrac{1}{2} \sum_{k=1}^{n-1} [\tilde{b}(k) + \tilde{b}(k-1)] \cos 2\pi(k-\tfrac{1}{2})F + \tfrac{1}{2}\tilde{b}(n-1) \cos 2\pi(n-\tfrac{1}{2})F.$$

Thus

$$b(1) = \tilde{b}(0) + \tfrac{1}{2} b(1)$$

$$b(k) = \tfrac{1}{2}(\tilde{b}(k-1) + \tilde{b}(k)), \qquad k = 2, 3, \cdots, n-1$$

$$b(n) = \tfrac{1}{2}\tilde{b}(n-1).$$

F. Proof of (6)

Use the trigonometric identity

$$\cos A \sin B = \tfrac{1}{2}[\sin(A+B) - \sin(A-B)]$$

and proceed as in the proof of (5) to obtain

$$c(1) = \tilde{c}(0) - \tfrac{1}{2}\tilde{c}(2)$$

$$c(k) = \tfrac{1}{2}[\tilde{c}(k-1) - \tilde{c}(k+1)], \qquad k = 2, 3, \cdots, n-2$$

$$c(n-1) = \tfrac{1}{2}\tilde{c}(n-2)$$

$$c(n) = \tfrac{1}{2}\tilde{c}(n-1).$$

Equation (7) is derived in the same manner, resulting in the following relations between the coefficients:

$$d(1) = \tilde{d}(0) - \tfrac{1}{2}\tilde{d}(1)$$

$$d(k) = \tfrac{1}{2}[\tilde{d}(k-1) - \tilde{d}(k)], \qquad k = 2, 3, \cdots, n-1$$

$$d(n) = \tfrac{1}{2}\tilde{d}(n-1).$$

III. Formulation of the Approximation Problem and its Solution

In this section we will show why it is convenient to rewrite the function $G(F)$ in the general form:

$$G(F) = Q(F)P(F). \tag{8}$$

Equation (8) leads to a very clear and compact statement of the necessary and sufficient conditions for the optimality of the best Chebyshev approximation, as well as to a compact method for computing this best approximation.

First, we state the approximation problem for linear-phase filter design. Let the sampling frequency be normalized to one; then the magnitude response is completely specified in the frequency interval $[0, \frac{1}{2}]$. Given a compact subset $\mathcal{F}$ of $[0, \frac{1}{2}]$, a desired function $D(F)$ defined and continuous on $\mathcal{F}$, a positive weight function $W(F)$ defined and continuous on $\mathcal{F}$, and one of the four types of linear-phase filters which determines the form of $G(F)$, then one seeks to minimize $\|E(F)\| = \max_{F \in \mathcal{F}} W(F)|D(F) - G(F)|$ by choice of $G(F)$.

The formulation of the problem on a compact subset of $[0, \frac{1}{2}]$ allows the designer to do multiband filter design by specifying the desired response on subintervals of $[0, \frac{1}{2}]$. For example,

$$D(F) = \begin{cases} 0, & F \in [0, 0.10] \\ 1, & F \in [0.2, 0.35] \\ 0, & F \in [0.425, 0.5] \end{cases} \tag{9}$$

specifies a bandpass filter.

The set $\mathcal{F}$ is $\mathcal{F} = [0, 0.1] \cup [0.2, 0.35] \cup [0.425, 0.5]$ and $D(F)$ is continuous on $\mathcal{F}$ because transition regions have been inserted at the jumps of $D(F)$. When, as in (1), $G(F)$ is a linear combination of cosines, the alternation theorem [11] states the necessary and sufficient conditions which must be satisfied by the unique best approximation. However, in the other cases, the alternation theorem cannot be applied directly.

However, using (8) we can formulate an equivalent approximation problem to which the alternation theorem can be applied. Observe that

$$W(F)|D(F) - Q(F)P(F)| = W(F)Q(F)\left|\frac{D(F)}{Q(F)} - P(F)\right| \tag{10}$$

except possibly at the end points, where $Q(F) = 0$. So (10) is true on $\mathcal{F}' \subset \mathcal{F}$, where $\mathcal{F}'$ is a compact subset of $\mathcal{F}$ and $Q(F) > 0$ on $\mathcal{F}'$. Letting $\hat{D}(F) = D(F)/Q(F)$ and $\hat{W}(F) = W(F)/Q(F)$, the problem becomes that of minimizing

$$\|E(F)\| = \max_{F \in \mathcal{F}'} \hat{W}(F)|\hat{D}(F) - P(F)| \tag{11}$$

by choice of $P(F)$, where $P(F)$ is a linear combination of cosines.

The problem is not exactly the same because the set $\mathcal{F}$ has been replaced by $\mathcal{F}'$. For example, in (9), if the length of the approximating filter were even, then $Q(\frac{1}{2}) = 0$ and $\mathcal{F}$ could be modified to get $\mathcal{F}' = [0, 0.1] \cup [0.20, 0.35] \cup [0.425, (0.5 -$

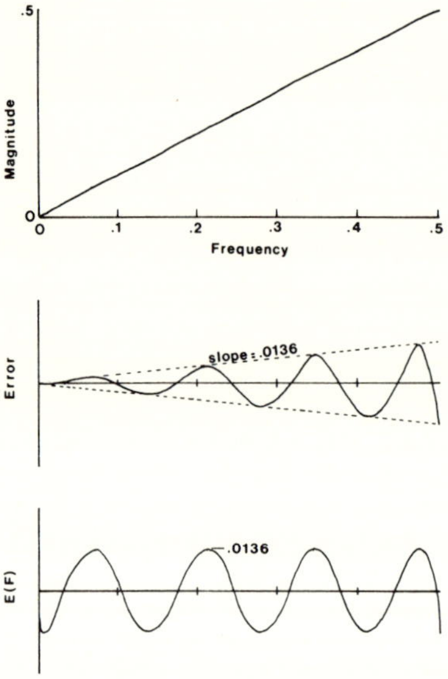

Fig. 1. Illustration of the alternation theorem for a length 16 differentiator. Slope error = 0.0136 with 9 extremal frequencies.

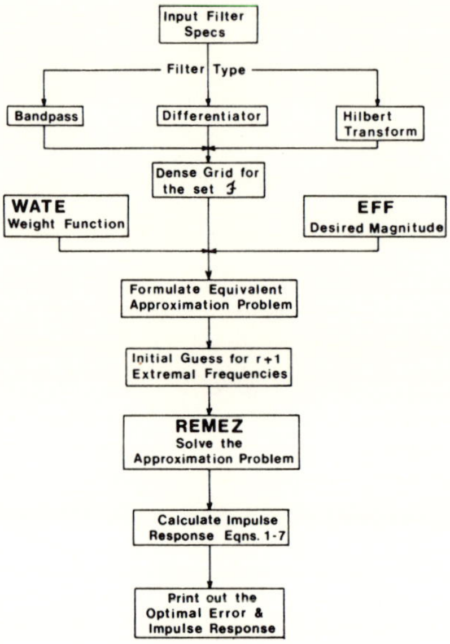

Fig. 2. Block diagram of the design program.

TABLE I

Symmetry / Length	POSITIVE	NEGATIVE
ODD	$G(F) = \sum_{k=0}^{n} a(k) \cos 2\pi kF$ $n = (N-1)/2$ At least $(N-1)/2 + 2$ extremal frequencies	$G(F) = \sin 2\pi F \sum_{k=0}^{n-1} \tilde{c}(k) \cos 2\pi kF$ $n = (N-1)/2$ $G(0) = G(\frac{1}{2}) = 0$ At least $(N-1)/2 + 1$ extremal frequencies
EVEN	$G(F) = \cos \pi F \sum_{k=0}^{n-1} \tilde{b}(k) \cos 2\pi kF$ $n = N/2$ $G(\frac{1}{2}) = 0$ At least $N/2 + 1$ extremal frequencies	$G(F) = \sin \pi F \sum_{k=0}^{n-1} \tilde{d}(k) \cos 2\pi kF$ $n = N/2$ $G(0) = 0$ At least $N/2 + 1$ extremal frequencies

$\epsilon)]$, where ϵ is a small positive number. As a matter of fact, $D(F)$ should be specified to be zero at any frequency, where $G(F) = 0$. Otherwise, there will be a fixed error in the Chebyshev approximation which cannot be reduced. This is the case when one tries to approximate a full-band differentiator with an even length filter because $D(\frac{1}{2}) = \frac{1}{2}$, while $G(\frac{1}{2}) = 0$.

The unity of our approach results from (11) because we can state one alternation theorem to cover all four cases.

A. Alternation Theorem

If $P(F)$ is a linear combination of r cosine functions (i.e., $P(F) = \sum_{k=0}^{r-1} \alpha(k) \cos 2\pi kF$), then a necessary and sufficient condition that $P(F)$ be the unique best-weighted Chebyshev approximation to a continuous function $\hat{D}(F)$ on $\mathcal{F}'$ is that the weighted error function $E(F) = \hat{W}(F)\,[\hat{D}(F) - P(F)]$ exhibit at least $r+1$ extremal frequencies in $\mathcal{F}'$. That is, points F_i, such that $F_1 < F_2 \cdots < F_n < F_{n+1}$, $E(F_i) = -E(F_{i+1})$, $i = 1, 2, \cdots, n$, and $|E(F_i)| = \max_{F \in \mathcal{F}'} E(F)$. Consider the application of the alternation theorem in the case of approximating a wide-band differentiator. The normalized frequency response of an ideal differentiator is $H(F) = jF$, so $D(F) = F$. Also, to get an error which is proportional to $D(F)$, let $W(F) = 1/F$. Thus $E(F) = 1/F|F - Q(F)P(F)|$. Fig. 1 illustrates the alternation of the error curve for $N = 16$ (i.e., $r = 8$ and 9 extremal frequencies). Table I summarizes the conclusion of this theorem for the four cases being considered.

IV. Calculation of the Best Approximation

The above development not only unifies the theory of linear-phase filter approximation, but it also motivates a unified technique for computing such approximations with the powerful Remez exchange algorithm [9]. It is easy to see that we only need a general algorithm which will do cosine approximations since the other three cases can be reduced to this case.

Hence, the structure of a general design program will consist of an input section, formulation of the appropriate equivalent approximation problem, solution of the approximation problem using the Remez algorithm, and calculation of the impulse response according to (1)-(7). Fig. 2 shows this structure in block-diagram form.

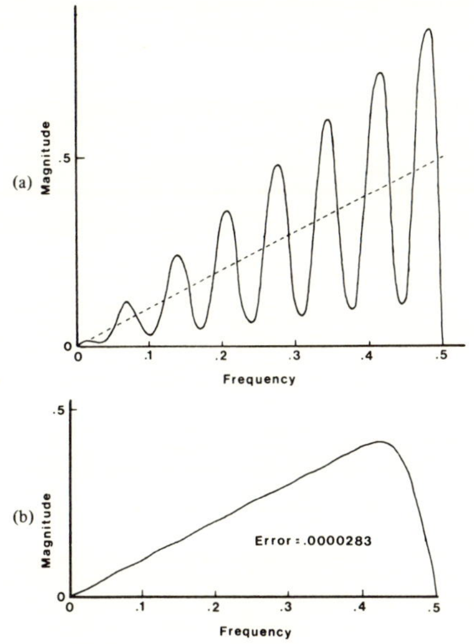

Fig. 3. (a) Length 31 wide-band differentiator. Slope error = 0.746. (b) Length 31 differentiator. Passband from $F = 0$ to $F = 0.4$; slope error = 0.0000283.

V. Examples

In order to display the flexibility of the present method and to point out some practical design considerations, the example filters of Figs. 3 and 4 have been designed.

Fig. 3(a) shows an attempt to design a length 31 wide-band differentiator. Since the desired response is $\frac{1}{2}$ at $F = \frac{1}{2}$ and $G(F)$ is constrained to be zero at $F = \frac{1}{2}$, the resulting Chebyshev error is very large. However, if the designer relaxes the specifications such that the desired filter is a differentiator on the interval [0. 0.4], then Fig. 3(b) shows that a reasonable filter can be obtained.

Finally, Fig. 4 shows a length 128 bandpass filter for which a special weight function was used. This weight was

$$W(F) = \begin{cases} \dfrac{10}{1\text{-}9F}, & 0 \leqslant F \leqslant 0.1 \\ 1, & 0.12 \leqslant F \leqslant 0.13 \\ \dfrac{10}{9F\text{-}1.25}, & 0.15 \leqslant F \leqslant 0.25 \\ 10, & 0.25 \leqslant F \leqslant 0.5 \end{cases}$$

which gives a tolerance scheme which is linear in the intervals [0, 0.1] and [0.15, 0.25]. At the edge of the stopbands the weight is 100, which makes the actual error small at this point. The error at the stopband edges is 0.0005 and the maximum error increases linearly to 0.005. This particular filter was designed in 36 s on an IBM 370/155 using a Fortran IV program.

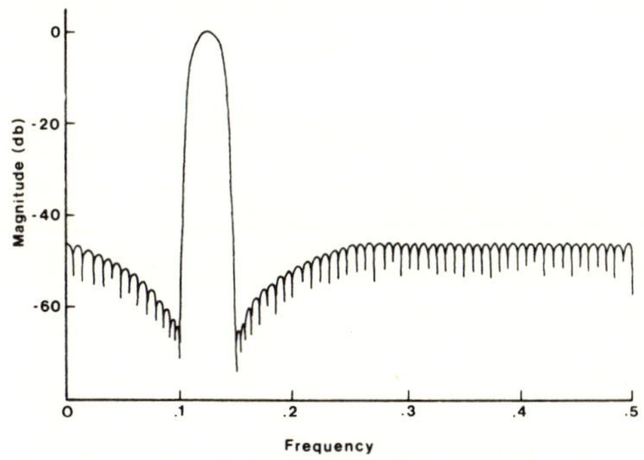

Fig. 4. Length 128 bandpass filter with a special weighting function. Stopband attenuation = 66 dB at the stopband edges increasing linearly to 46 dB; passband from 0.12 to 0.13; stopband edges at $F = 0.1$ and $F = 0.15$.

VI. Summary

An approach to the approximation of FIR linear-phase filters has been presented wherein the optimality conditions and the calculation of the best approximation are unified. Thus the four types of linear-phase filters can be dealt with by the same approach. This technique should provide a designer with an efficient tool for designing this class of filters. A Fortran program based on this technique is available from the authors.

Acknowledgment

The authors would like to acknowledge discussions with O. Herrmann and L. R. Rabiner which stimulated this work.

References

[1] B. Gold and K. L. Jordan, Jr., "A direct search procedure for designing finite duration impulse response filters," *IEEE Trans. Audio Electroacoust.*, vol. AU-17, pp. 33–36, Mar. 1969.

[2] L. R. Rabiner, B. Gold, and C. McGonegal, "An approach to the approximation problem for nonrecursive digital filters," *IEEE Trans. Audio Electroacoust. (Special Issue on Digital Filtering)*, vol. AU-18, pp. 83–106, June 1970.

[3] L. R. Rabiner and K. Steiglitz, "The design of wide-band recursive and nonrecursive digital differentiators," *IEEE Trans. Audio Electroacoust. (Special Issue on Digital Filtering)*, vol. AU-18, pp. 204–209, June 1970.

[4] O. Herrmann, "Design of nonrecursive digital filters with linear phase," *Electron. Lett.*, pp. 328–329, 1970.

[5] E. Hofstetter, A. V. Oppenheim, and J. Siegel, "A new technique for the design of non-recursive digital filters," in *Proc. 5th Annu. Princeton Conf. Information Sciences and Systems*, Mar. 1971, pp. 64–72.

[6] T. W. Parks, L. R. Rabiner, and J. H. McClellan, "On the transition width of finite impulse-response digital filters," *IEEE Trans. Audio Electroacoust.*, vol. AU-21, pp. 1–4, Feb. 1973.

[7] T. W. Parks and J. H. McClellan, "Chebyshev approximation for nonrecursive digital filters with linear phase," *IEEE Trans. Circuit Theory*, vol. CT-19, pp. 189–194, Mar. 1972.

[8] L. R. Rabiner, "The design of finite impulse response digital filters using linear programming techniques," *Bell Syst. Tech. J.*, vol. 51, pp. 1177–1198, July–Aug. 1972.

[9] L. R. Rabiner and O. Herrmann, "On the design of optimum FIR low-pass filters with even impulse response duration," *IEEE Trans. Audio Electroacoust.*, vol. AU-21, pp. 329–336, Aug. 1973.

[10] O. Herrmann, "Transversal filters for the Hilbert transformation," *AEÜ*, vol. 23, pp. 581–587.

[11] E. W. Cheney, *Introduction to Approximation Theory*. New York: McGraw-Hill, 1966, pp. 72–100.

[12] E. Ya. Remez, "General computational methods of Tchebycheff approximation," Kiev, U.S.S.R., Atomic Energy Transl. 4491, pp. 1–85, 1957.

6

Reprinted from *IEEE Trans. Audio Electroacoustics,* **AU-21**(6), 506–526 (1973)

A Computer Program for Designing Optimum FIR Linear Phase Digital Filters

JAMES H. McCLELLAN, Student Member, IEEE,
THOMAS W. PARKS, Member, IEEE, and
LAWRENCE R. RABINER, Member, IEEE

Abstract—This paper presents a general-purpose computer program which is capable of designing a large class of optimum (in the minimax sense) FIR linear phase digital filters. The program has options for designing such standard filters as low-pass, high-pass, bandpass, and bandstop filters, as well as multipassband-stopband filters, differentiators, and Hilbert transformers. The program can also be used to design filters which approximate arbitrary frequency specifications which are provided by the user. The program is written in Fortran, and is carefully documented both by comments and by detailed flowcharts. The filter design algorithm is shown to be exceedingly efficient, e.g., it is capable of designing a filter with a 100-point impulse response in about 20 s.

I. Introduction

This paper presents a general algorithm for the design of a large class of finite impulse response (FIR) linear phase digital filters. Emphasis is placed on a description of how the algorithm works, and several examples are included which illustrate specific applications. A unified treatment of the theory behind this approach is available in [1].

The algorithm uses the Remez exchange method [2], [3] to design filters with minimum weighted Chebyshev error in approximating a desired ideal frequency response $D(f)$. Several authors have studied the FIR design problem for special filter types using several different algorithms [4]-[13]. The advantage of the present approach is that it combines the speed of the Remez procedure with a capability for designing a large class of general filter types. While the algorithm to be described has a special section for the more common filter types (e.g., bandpass filters with multiple bands, Hilbert transform filters, and differentiators), an arbitrary frequency response can also be approximated.

Manuscript received August 6, 1973. The work of J. H. McClellan and T. W. Parks was supported by NSF Grant GK-23697.

J. H. McClellan and T. W. Parks are with Rice University, Houston, Tex.
L. R. Rabiner is with Bell Laboratories, Murray Hill, N.J. 07974.

II. Formulation of the Approximation Problem

The frequency response of an FIR digital filter with an N-point impulse response $\{h(k)\}$ is the z-transform of the sequence evaluated on the unit circle, i.e.,

$$H(f)^1 = H(z)\big|_{z=e^{j2\pi f}} = \sum_{k=0}^{N-1} h(k)\, e^{-j2\pi kf}. \quad (1)$$

The frequency response of a linear phase filter can be written as

$$H(f) = G(f)\, e^{j\left(\frac{L\pi}{2} - \left(\frac{N-1}{2}\right) 2\pi f\right)} \quad (2)$$

where $G(f)$ is a real valued function and $L = 0$ or 1. It is possible to show that there are exactly four cases of linear phase FIR filters to consider [1]. These four cases differ in the length of the impulse response (even or odd) and the symmetry of the impulse response [positive ($L = 0$) or negative ($L = 1$)]. By positive symmetry we mean $h(k) = h(N - 1 - k)$, and by negative symmetry $h(k) = -h(N - 1 - k)$.

In all cases, the real function $G(f)$ will be used to approximate the desired ideal magnitude specifications since the linear phase term in (2) has no effect on the magnitude response of the filter. The form of $G(f)$ depends on which of the four cases is being used. Using the appropriate symmetry relations, $G(f)$ can be expressed as follows.

Case 1: Positive symmetry, odd length:

$$G(f) = \sum_{k=0}^{n} a(k) \cos (2\pi kf) \quad (3)$$

where $n = (N - 1)/2$, $a(0) = h(n)$, and $a(k) = 2h(n - k)$ for $k = 1, 2, \cdots, n$.

Case 2: Positive symmetry, even length:

$$G(f) = \sum_{k=1}^{n} b(k) \cos [2\pi(k - \tfrac{1}{2}) f] \quad (4)$$

where $n = N/2$ and $b(k) = 2h(n - k)$ for $k = 1, \cdots, n$.

Case 3: Negative symmetry, odd length:

$$G(f) = \sum_{k=1}^{n} c(k) \sin (2\pi kf) \quad (5)$$

where $n = (N - 1)/2$ and $c(k) = 2h(n - k)$ for $k = 1, 2, \cdots, n$ and $h(n) = 0$.

Case 4: Negative symmetry, even length:

$$G(f) = \sum_{k=1}^{n} d(k) \sin [2\pi(k - \tfrac{1}{2}) f] \quad (6)$$

where $n = N/2$ and $d(k) = 2h(n - k)$ for $k = 1, \cdots, n$.

Earlier efforts at designing FIR filters concentrated on Case 1 designs, but it is now possible to combine

[1] For convenience, throughout this paper the notation $H(f)$ rather than $H(e^{j2\pi f})$ is used to denote the frequency response of the digital filter.

all four cases into one algorithm. This is accomplished by noting that $G(f)$ can be rewritten as $G(f) = Q(f)P(f)$ where $P(f)$ is a linear combination of cosine functions. Thus, results that have been worked out for Case 1 can be applied to the other three cases as well. For these purposes, it is convenient to express the summations in (4)-(6) as a sum of cosines directly. Simple manipulations of (4)-(6) yield the expressions.

Case 2:

$$\sum_{k=1}^{n} b(k) \cos [2\pi(k - \tfrac{1}{2})f] = \cos (\pi f) \sum_{k=0}^{n-1} \tilde{b}(k) \cos (2\pi kf). \quad (7)$$

Case 3:

$$\sum_{k=1}^{n} c(k) \sin (2\pi kf) = \sin (2\pi f) \sum_{k=0}^{n-1} \tilde{c}(k) \cos (2\pi kf). \quad (8)$$

Case 4:

$$\sum_{k=1}^{n} d(k) \sin [2\pi(k - \tfrac{1}{2})f] = \sin (\pi f) \sum_{k=0}^{n-1} \tilde{d}(k) \cos (2\pi kf) \quad (9)$$

where

$$\text{Case 2: } \begin{cases} b(1) = \tilde{b}(0) + \tfrac{1}{2}\tilde{b}(1) \\ b(k) = \tfrac{1}{2}[\tilde{b}(k-1) + \tilde{b}(k)], \quad k = 2, 3, \cdots, n-1 \\ b(n) = \tfrac{1}{2}\tilde{b}(n-1) \end{cases} \quad (10)$$

$$\text{Case 3: } \begin{cases} c(1) = \tilde{c}(0) - \tfrac{1}{2}\tilde{c}(2) \\ c(k) = \tfrac{1}{2}[\tilde{c}(k-1) - \tilde{c}(k+1)], \quad k = 2, 3, \cdots, n-2 \\ c(n-1) = \tfrac{1}{2}\tilde{c}(n-2) \\ c(n) = \tfrac{1}{2}\tilde{c}(n-1) \end{cases} \quad (11)$$

$$\text{Case 4: } \begin{cases} d(1) = \tilde{d}(0) - \tfrac{1}{2}\tilde{d}(1) \\ d(k) = \tfrac{1}{2}[\tilde{d}(k-1) - \tilde{d}(k)], \quad k = 2, 3, \cdots, n-1 \\ d(n) = \tfrac{1}{2}\tilde{d}(n-1). \end{cases} \quad (12)$$

The motivation for rewriting the four cases in a common form is that a single central computation routine (based on the Remez exchange method) can be used to calculate the best approximation in each of the four cases. This is accomplished by modifying both the desired magnitude function and the weighting function to formulate a new equivalent approximation problem.

The original approximation problem can be stated as follows: given a desired magnitude response $D(f)$ and a positive weight function $W(f)$, both continuous on a compact subset $F \subset [0, \tfrac{1}{2}]$ (note that the sampling rate is 1.0) and one of the four cases of linear phase filters [i.e., the forms of $G(f)$], then one wishes to minimize the maximum absolute weighted error, defined as

$$\|E(f)\| = \max_{f \in F} W(f) \,|D(f) - G(f)| \quad (13)$$

over the set of coefficients of $G(f)$.

The error function $E(f)$ can be rewritten in the form

$$E(f) = W(f)\,[D(f) - G(f)] = W(f)\,Q(f) \left[\frac{D(f)}{Q(f)} - P(f)\right] \quad (14)$$

if one is careful to omit those endpoint(s) where $Q(f) = 0$. Letting $\hat{D}(f) = D(f)/Q(f)$ and $\hat{W}(f) = W(f)\,Q(f)$, then an equivalent approximation problem would be to minimize the quantity

$$\|E(f)\| = \max_{f \in F'} \hat{W}(f) \,|\hat{D}(f) - P(f)| \quad (15)$$

by choice of the coefficients of $P(f)$. The set F is replaced by $F' = F - \{\text{endpoints where } Q(f) = 0\}$.

The net effect of this reformulation of the problem is a unification of the four cases of linear phase FIR filters from the point of view of the approximation problem. Furthermore, (15) provides a simplified viewpoint from which it is easy to see the necessary and sufficient conditions which are satisfied by the best approximation. Finally, (15) shows how to calculate this best approximation using an algorithm which can do only cosine approximations. The set of necessary and sufficient conditions for this best approximation is given in the following alternation theorem [2].

Alternation theorem: If $P(f)$ is a linear combination of r cosine functions i.e.,

$$P(f) = \sum_{k=0}^{r-1} \alpha(k) \cos 2\pi kf,$$

then a necessary and sufficient condition that $P(f)$ be the unique best weighted Chebyshev approximation to a continuous function $\hat{D}(f)$ on F' is that the weighted error function $E(f) = \hat{W}(f)\;[\hat{D}(f) - P(f)]$ exhibit *at least* $r + 1$ extremal frequencies in F'.

These extremal frequencies are a set of points $\{F_i\}$, $i = 1, 2, \cdots, r + 1$ such that $F_1 < F_2 < \cdots < F_r < F_{r+1}$, with $E(F_i) = -E(F_{i+1})$, $i = 1, 2, \cdots, r$ and $|E(F_i)| = \max_{f \in F'} E(f)$.

An algorithm can now be designed to make the

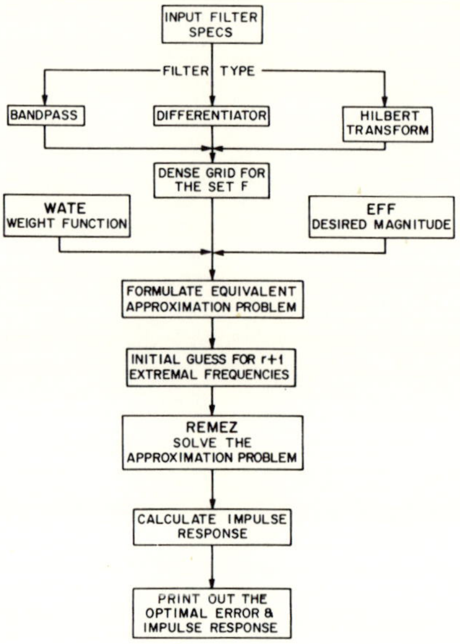

Fig. 1. Overall flowchart of filter design algorithm.

error function of the filter satisfy the set of necessary and sufficient conditions for optimality as stated in the alternation theorem. The next section describes such an algorithm along with details as to its implementation.

III. Description of the Design Algorithm

As seen in Fig. 1, the design algorithm consists of an input section, formulation of the appropriate equivalent approximation problem, solution of the approximation problem using the Remez exchange method, and calculation of the filter impulse response. The flowcharts of Figs. 2-5 give details of the exact structure of the computer program.

The input which describes the filter specifications consists of the following.

1) The filter length, $3 \leqslant$ NFILT $\leqslant$ NFMAX (the upper limit set by the programmer).

2) The type of filter (JTYPE):

a) Multiple passband/stopband (JTYPE=1)

b) Differentiator (JTYPE=2)

c) Hilbert transformer (JTYPE=3).

3) The frequency bands, specified by upper and lower cutoff frequencies (EDGE array) up to a maximum of 10 bands.

4) The desired frequency response (FX array) in each band.

5) A positive weight function (WTX array) in each band.

6) The grid density (LGRID), assumed to be 16 unless specified otherwise.

7) Impulse response punch option (JPUNCH).

Part 3) specifies the set F to be of the form $F = \cup B_i$ where each frequency band B_i is a closed subinterval of $[0, \frac{1}{2}]$. The inputs 4) and 5) are interpreted differently by the program for a differentiator than for the other two types of filters (see the EFF and WATE subroutines in Figs. 3 and 4). The weight specification in the case of a differentiator results in a *relative* error tolerance as is used in all other cases.

The set F must be replaced by a finite set of points for implementation on a computer. A dense grid of points is used with the spacing between points being $0.5/(\text{LGRID} \times r)$ where r is the number of cosine basis functions. Both $D(f)$ and $W(f)$ are evaluated on this grid by the subroutines EFF and WATE, respectively. Then the auxiliary approximation problem is set up by forming $\hat{D}(f)$ and $\hat{W}(f)$ as above, and an initial guess of the extremal frequencies is made by taking $r + 1$ equally spaced frequency values. The subroutine REMEZ (Fig. 5) is called to perform the calculation of the best approximation for the equivalent problem. The mechanics of the Remez algorithm will not be discussed here since they are treated elsewhere for the particular case of low-pass filters [9]. (The flowchart of Fig. 5 gives details about the mechanics of the Remez algorithm as implemented in this design program.)

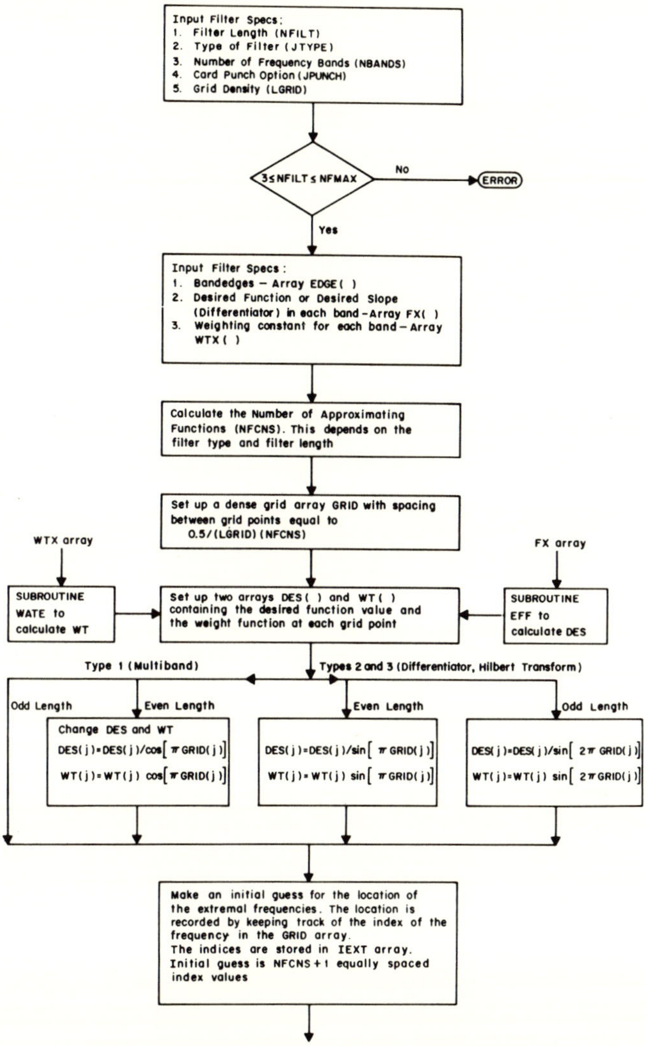

Fig. 2. Detailed flowchart for filter design algorithm.

The appropriate equations (3)-(12) are used to recover the impulse response from the coefficients of the best cosine approximation obtained in the REMEZ subroutine. The outputs of the program are the impulse response, the optimal error (min $\|E(f)\|$), and the $r+1$ extremal frequencies where $E(f) = \pm\|E(f)\|$.

It is possible that one might want to design a filter to approximate a magnitude specification which is not included in the scheme given above, or change the weight function to get a desired tolerance scheme. A flowchart of such a program is given in Fig. 6. In such cases, the user must code the subroutines EFF and WATE to calculate $D(f)$ and $W(f)$. The input is the same as before, except that there are only two types of filters, depending on whether the impulse symmetry is positive or negative.

A detailed program listing of the generalized design program is given in the Appendix. Representative

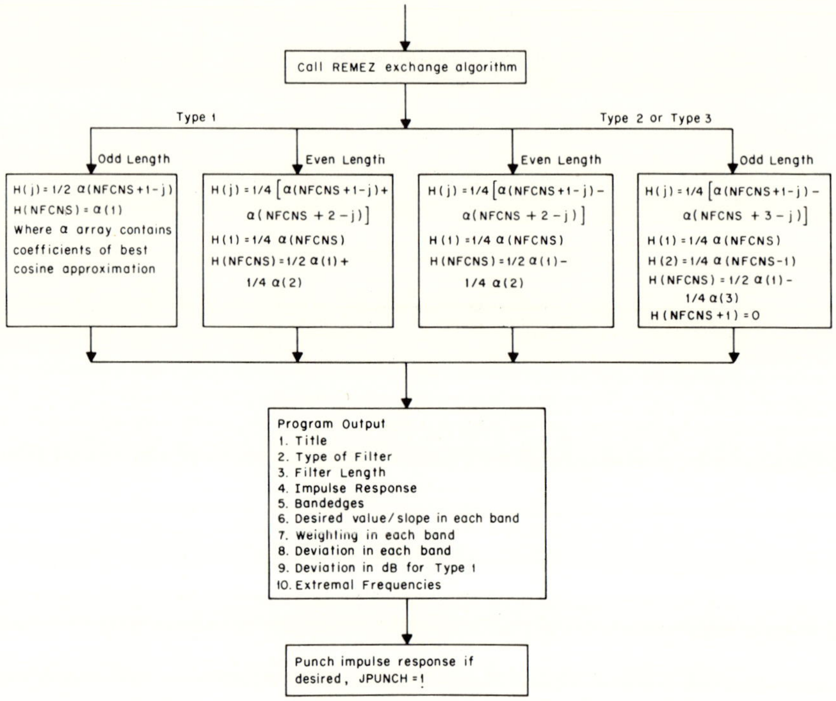

Fig. 2. (Continued.)

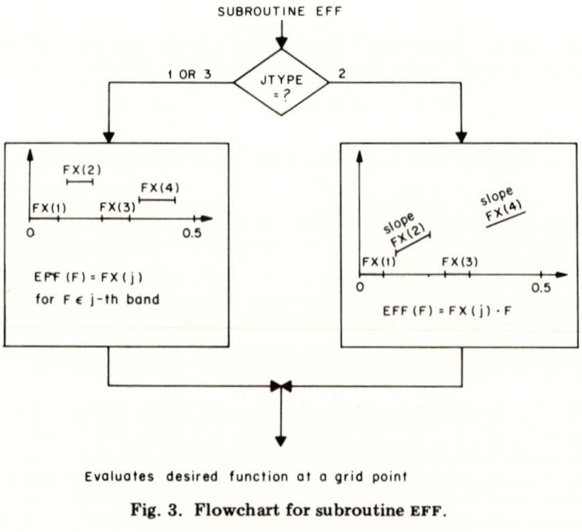

Fig. 3. Flowchart for subroutine EFF.

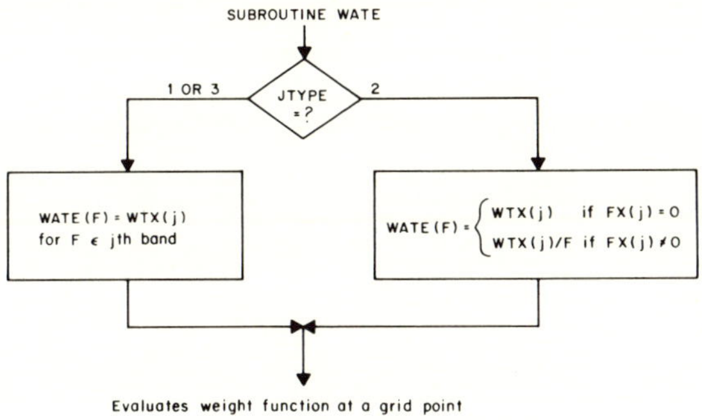

Fig. 4. Flowchart for subroutine WATE.

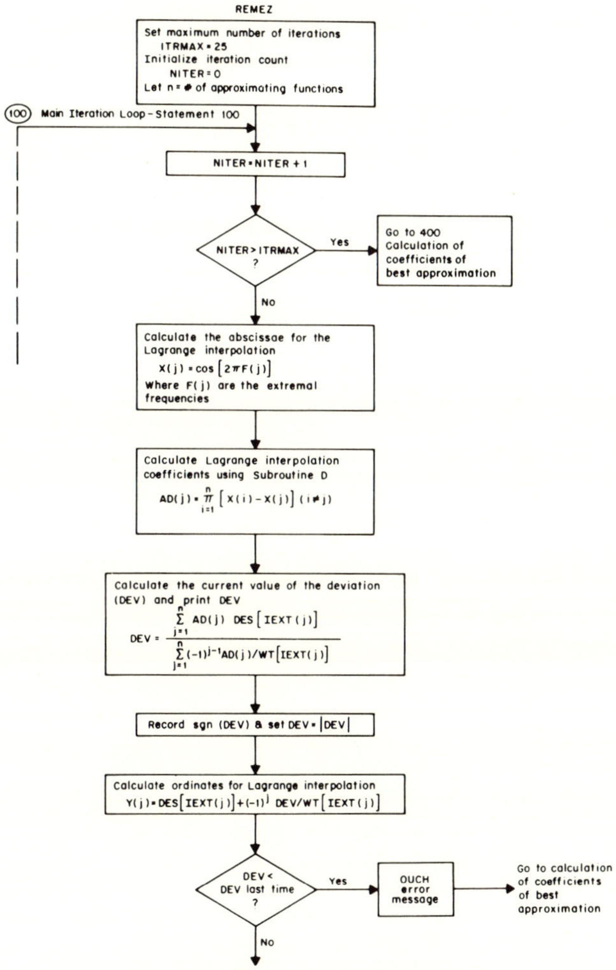

Fig. 5. Detailed flowchart for subroutine REMEZ.

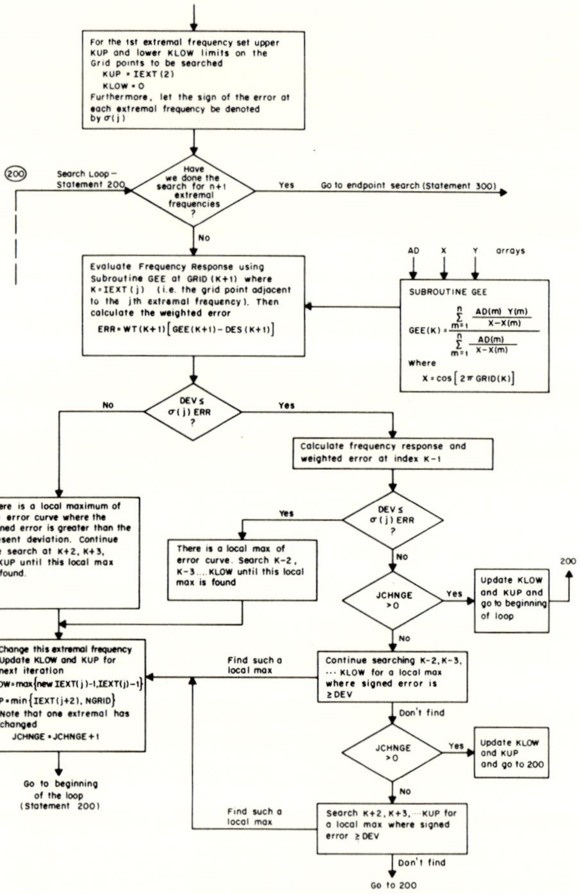

Fig. 5. (Continued.)

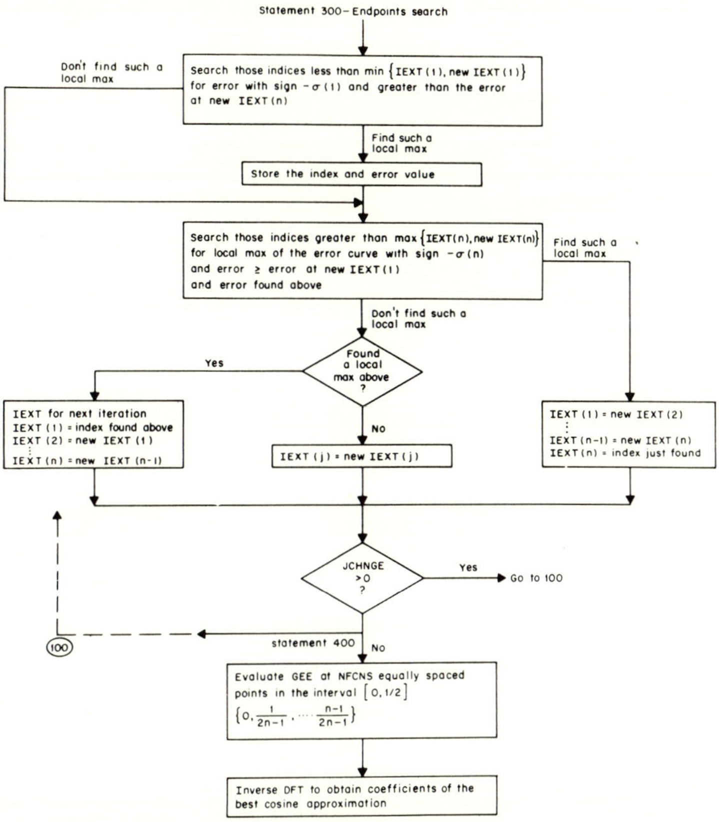

Fig. 5. (Continued.)

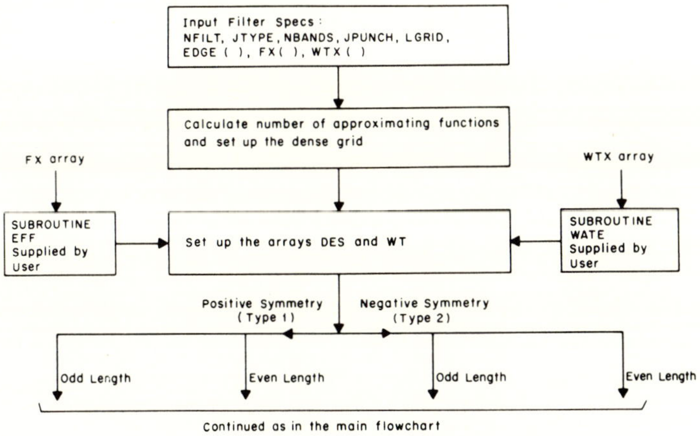

Fig. 6. Flowchart for arbitrary magnitude filter design algorithm.

```
**********************************************************************

                  FINITE IMPULSE RESPONSE (FIR)
                  LINEAR PHASE DIGITAL FILTER DESIGN
                  REMEZ EXCHANGE ALGORITHM
                  BANDPASS FILTER
          FILTER LENGTH =  24
          ***** IMPULSE RESPONSE *****
              H(  1) =  0.33740917E-02 = H(  24)
              H(  2) =  0.14938299E-01 = H(  23)
              H(  3) =  0.10569360E-01 = H(  22)
              H(  4) =  0.25415067E-02 = H(  21)
              H(  5) = -0.15929392E-01 = H(  20)
              H(  6) = -0.34085343E-01 = H(  19)
              H(  7) = -0.38112177E-01 = H(  18)
              H(  8) = -0.14629169E-01 = H(  17)
              H(  9) =  0.40089541E-01 = H(  16)
              H( 10) =  0.11540713E 00 = H(  15)
              H( 11) =  0.13850752E 00 = H(  14)
              H( 12) =  0.23354606E 00 = H(  13)

                       BAND 1          BAND 2          BAND
 LOWER BAND EDGE      0.              0.16000000
 UPPER BAND EDGE      0.08000000      0.50000000
 DESIRED VALUE        1.00000000      0.
 WEIGHTING            1.00000000      1.00000000
 DEVIATION            0.01243364      0.01243364
 DEVIATION IN DB    -38.10803413    -38.10803413

 EXTREMAL FREQUENCIES
    0.          0.0364583   0.0677083   0.0800000   0.1600000
    0.1730208   0.2068750   0.2459375   0.2876042   0.3318750
    0.3787500   0.4256251   0.4751043

**********************************************************************
TIME=        0.7694063 SECONDS
```

Fig. 7. Output listing for an $N = 24$ low-pass filter.

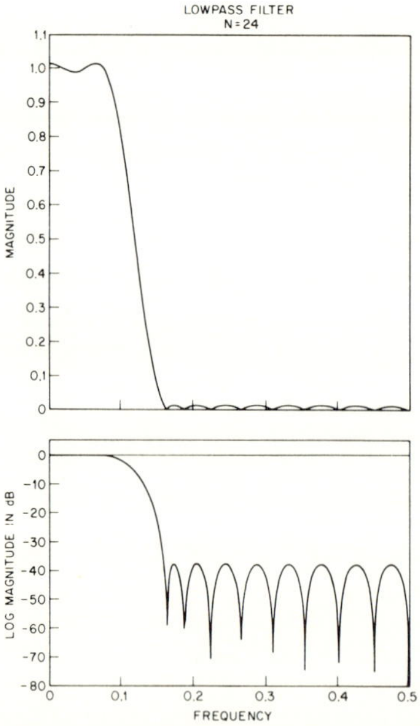

Fig. 8. Magnitude responses, on linear and log scales, for an $N = 24$ low-pass filter.

input card sequences are given for the design of a bandpass filter and a differentiator. To approximate an arbitrary magnitude response and/or an arbitrary weighting function, all the user has to do is change the subroutines EFF and WATE and use the program in the Appendix. In the next section, representative filters designed using these algorithms are presented.

IV. Design Examples

Figs. 7-22 show specific examples of use of the design program for several typical filters of interest. For each of these filters, one figure shows the computer output listing (including the run time on a Honeywell 6000 computer), and the other figure shows a plot of the filter frequency response on either a linear or a log magnitude scale (or sometimes both). Figs. 7 and 8 are for an $N = 24$ low-pass filter. For this example, the run time was 0.77 s. Figs. 9 and 10 are for an $N = 32$ bandpass filter. This example is the first example listed in the prologue to the program in the Appendix. The run time for this example was 0.82 s. Figs. 11 and 12 are for an $N = 50$ bandpass filter in which unequal weighting was used in the two stopbands. Thus the peak error in the upper stopband is ten times smaller than the peak error in the lower stopband. A total of 2.96 s was required to design this filter. Figs. 13 and 14 are for an $N = 31$ bandstop filter with equal weighting in both passbands. For the design of this filter 1.61 s were required.

To illustrate the multiband capability of the pro-

```
**********************************************************************

                 FINITE IMPULSE RESPONSE (FIR)
                 LINEAR PHASE DIGITAL FILTER DESIGN
                 REMEZ EXCHANGE ALGORITHM
                 BANDPASS FILTER
          FILTER LENGTH =  32
          ***** IMPULSE RESPONSE *****
              H(  1) = -0.57534121E-02 = H(  32)
              H(  2) =  0.39027198E-03 = H(  31)
              H(  3) =  0.75733545E-02 = H(  30)
              H(  4) = -0.65141192E-02 = H(  29)
              H(  5) =  0.13960525E-01 = H(  28)
              H(  6) =  0.22351469E-02 = H(  27)
              H(  7) = -0.13994067E-01 = H(  26)
              H(  8) =  0.71369560E-02 = H(  25)
              H(  9) = -0.39657363E-01 = H(  24)
              H( 10) =  0.11260114E-01 = H(  23)
              H( 11) =  0.66233643E-01 = H(  22)
              H( 12) = -0.10497223E-01 = H(  21)
              H( 13) =  0.85136133E-01 = H(  20)
              H( 14) = -0.12024393E 00 = H(  19)
              H( 15) = -0.29678577E 00 = H(  18)
              H( 16) =  0.30410317E 00 = H(  17)

                          BAND 1          BAND 2          BAND 3          BAND
 LOWER BAND EDGE        0.              0.20000000      0.42500000
 UPPER BAND EDGE        0.10000000      0.35000000      0.50000000
 DESIRED VALUE          0.              1.00000000      0.
 WEIGHTING             10.00000000      1.00000000     10.00000000
 DEVIATION              0.00151312      0.01513118      0.00151312
 DEVIATION IN DB      -56.40254641    -36.40254641    -56.40254641

 EXTREMAL FREQUENCIES
    0.           0.0273437    0.0527344    0.0761719    0.0937500
    0.1000000    0.2000000    0.2195312    0.2527344    0.2839844
    0.3132812    0.3386719    0.3500000    0.4250000    0.4328125
    0.4503906    0.4796875

**********************************************************************
TIME=        0.8245625 SECONDS
```

Fig. 9. Output listing for an N = 32 bandpass filter.

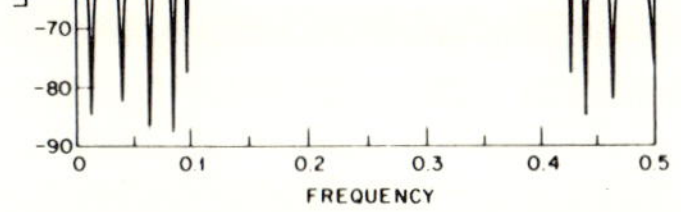

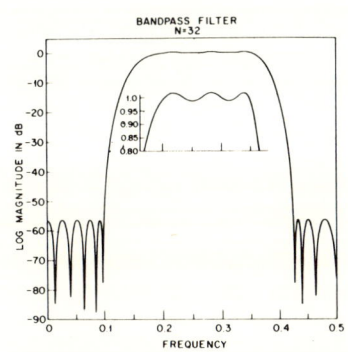

Fig. 10. Log magnitude response for an N = 32 bandpass filter.

```
**********************************************************************

                    FINITE IMPULSE RESPONSE (FIR)
                    LINEAR PHASE DIGITAL FILTER DESIGN
                    REMEZ EXCHANGE ALGORITHM
                    BANDPASS FILTER
          FILTER LENGTH =  50
          ***** IMPULSE RESPONSE *****
             H(  1) =  0.15648409E-02 = H(  50)
             H(  2) =  0.30816298E-02 = H(  49)
             H(  3) = -0.31745254E-02 = H(  48)
             H(  4) = -0.61980034E-02 = H(  47)
             H(  5) =  0.74350685E-02 = H(  46)
             H(  6) =  0.98368961E-02 = H(  45)
             H(  7) = -0.11103735E-01 = H(  44)
             H(  8) = -0.10101927E-01 = H(  43)
             H(  9) =  0.89949206E-02 = H(  42)
             H( 10) =  0.28980188E-02 = H(  41)
             H( 11) =  0.26632992E-02 = H(  40)
             H( 12) =  0.12021958E-01 = H(  39)
             H( 13) = -0.20657142E-01 = H(  38)
             H( 14) = -0.27189007E-01 = H(  37)
             H( 15) =  0.32337130E-01 = H(  36)
             H( 16) =  0.28305613E-01 = H(  35)
             H( 17) = -0.20922041E-01 = H(  34)
             H( 18) = -0.18761153E-02 = H(  33)
             H( 19) = -0.22823357E-01 = H(  32)
             H( 20) = -0.53926217E-01 = H(  31)
             H( 21) =  0.90472539E-01 = H(  30)
             H( 22) =  0.12315772E 00 = H(  29)
             H( 23) = -0.15639221E 00 = H(  28)
             H( 24) = -0.17733448E 00 = H(  27)
             H( 25) =  0.19078165E 00 = H(  26)

                          BAND 1          BAND 2          BAND 3          BAND
LOWER BAND EDGE        0.              0.20000000      0.35000000
UPPER BAND EDGE        0.15000000      0.30000000      0.50000000
DESIRED VALUE          0.              1.00000000      0.
WEIGHTING             10.00000000      1.00000000    100.00000000
DEVIATION              0.00370505      0.03705048      0.00037050
DEVIATION IN DB      -48.62412214    -28.62412214    -68.62412262

EXTREMAL FREQUENCIES
    0.          0.0200000   0.0400000   0.0612500   0.0812500
    0.1012500   0.1225000   0.1412500   0.1500000   0.2000000
    0.2100000   0.2287500   0.2500000   0.2712500   0.2900000
    0.3000000   0.3500000   0.3537500   0.3637500   0.3762500
    0.3925000   0.4099999   0.4287499   0.4487499   0.4687499
    0.4899999

**********************************************************************
TIME=       2.9562969 SECONDS
```

Fig. 11. Output listing for an $N = 50$ bandpass filter with unequal weighting in the stopbands.

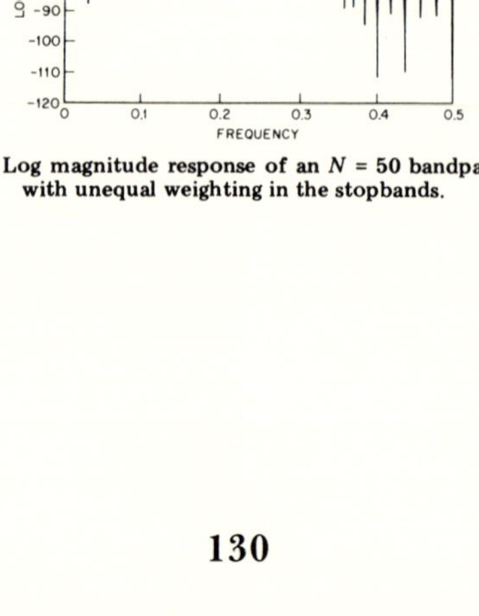

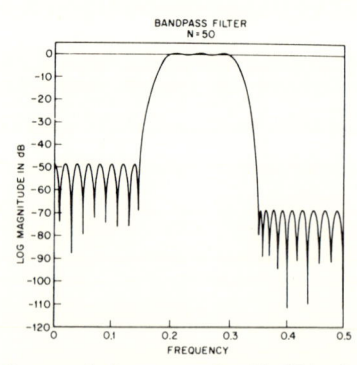

Fig. 12. Log magnitude response of an $N = 50$ bandpass filter with unequal weighting in the stopbands.

```
**********************************************************************
                    FINITE IMPULSE RESPONSE (FIR)
                    LINEAR PHASE DIGITAL FILTER DESIGN
                    REMEZ EXCHANGE ALGORITHM
                    BANDPASS FILTER
          FILTER LENGTH =  31
          ***** IMPULSE RESPONSE *****
               H(  1) = -0.43725800E-02 = H(  31)
               H(  2) =  0.19295933E-01 = H(  30)
               H(  3) = -0.56962301E-02 = H(  29)
               H(  4) =  0.52360280E-01 = H(  28)
               H(  5) =  0.31550241E-02 = H(  27)
               H(  6) =  0.43481227E-01 = H(  26)
               H(  7) =  0.11696224E-01 = H(  25)
               H(  8) = -0.37915416E-01 = H(  24)
               H(  9) =  0.34844146E-02 = H(  23)
               H( 10) = -0.87599027E-01 = H(  22)
               H( 11) = -0.10993063E-01 = H(  21)
               H( 12) =  0.44455166E-01 = H(  20)
               H( 13) = -0.69347169E-02 = H(  19)
               H( 14) =  0.31144524E 00 = H(  18)
               H( 15) =  0.90029834E-02 = H(  17)
               H( 16) =  0.45296733E 00 = H(  16)

                      BAND 1          BAND 2          BAND 3          BAND
LOWER BAND EDGE      0.              0.15000000      0.42000000
UPPER BAND EDGE      0.10000000      0.35000000      0.50000000
DESIRED VALUE        1.00000000      0.              1.00000000
WEIGHTING            1.00000000     50.00000000      1.00000000
DEVIATION            0.14402014      0.00288040      0.14402014
DEVIATION IN DB    -16.83153510    -50.81093502    -16.83153510

EXTREMAL FREQUENCIES
   0.           0.0390625    0.0781250    0.1000000    0.1500000
   0.1578125    0.1753906    0.2007813    0.2261719    0.2554688
   0.2828125    0.3062031    0.3335938    0.3500000    0.4258594
   0.4629688    0.5000000

**********************************************************************
TIME=        1.6110156 SECONDS
```

Fig. 13. Output listing for an N = 31 bandstop filter.

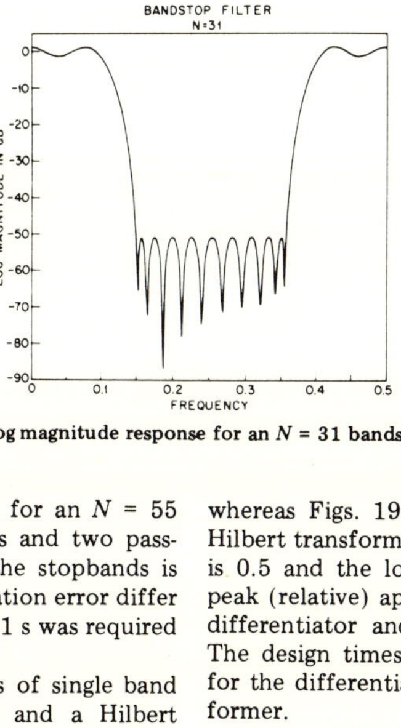

Fig. 14. Log magnitude response for an N = 31 bandstop filter.

gram, Figs. 15 and 16 show results for an N = 55 five-band filter with three stopbands and two passbands. The weighting in each of the stopbands is different, making the peak approximation error differ in each of these bands. A total of 3.81 s was required to design this filter.

Figs. 17-20 show typical examples of single band approximations to a differentiator and a Hilbert transformer. Figs. 17 and 18 show results for an N = 32 full band differentiator (this filter is the second example listed in the prologue to the Appendix), whereas Figs. 19 and 20 show results for an N = 20 Hilbert transformer where the upper cutoff frequency is 0.5 and the lower cutoff frequency is 0.05. The peak (relative) approximation error is 0.0062 for the differentiator and 0.02 for the Hilbert transformer. The design times for these two examples are 1.11 s for the differentiator and 0.48 s for the Hilbert transformer.

Finally, Figs. 21 and 22 show an example of an N = 128 bandpass filter with an arbitrary weighting function of the form

```
******************************************************************
                FINITE IMPULSE RESPONSE (FIR)
                LINEAR PHASE DIGITAL FILTER DESIGN
                REMEZ EXCHANGE ALGORITHM
                BANDPASS FILTER
     FILTER LENGTH =  55
     ***** IMPULSE RESPONSE *****
          H(  1) =  0.10062552E-02 = H(  55)
          H(  2) =  0.63777615E-02 = H(  54)
          H(  3) =  0.35755610E-02 = H(  53)
          H(  4) = -0.90677355E-02 = H(  52)
          H(  5) = -0.90906979E-02 = H(  51)
          H(  6) =  0.29155629E-02 = H(  50)
          H(  7) =  0.39637965E-02 = H(  49)
          H(  8) =  0.11172050E-01 = H(  48)
          H(  9) =  0.11646759E-01 = H(  47)
          H( 10) = -0.39630784E-02 = H(  46)
          H( 11) = -0.92384245E-02 = H(  45)
          H( 12) = -0.20406392E-01 = H(  44)
          H( 13) = -0.19460483E-01 = H(  43)
          H( 14) =  0.31243013E-01 = H(  42)
          H( 15) =  0.63045567E-02 = H(  41)
          H( 16) = -0.20482303E-01 = H(  40)
          H( 17) =  0.65740513E-02 = H(  39)
          H( 18) = -0.11202127E-02 = H(  38)
          H( 19) =  0.41956985E-01 = H(  37)
          H( 20) =  0.35784266E-01 = H(  36)
          H( 21) =  0.34744802E-01 = H(  35)
          H( 22) =  0.71496359E-01 = H(  34)
          H( 23) = -0.17138831E 00 = H(  33)
          H( 24) = -0.18255044E 00 = H(  32)
          H( 25) =  0.74059024E-01 = H(  31)
          H( 26) = -0.10317421E 00 = H(  30)
          H( 27) =  0.25716721E-01 = H(  29)
          H( 28) =  0.37813547E 00 = H(  28)

                     BAND 1          BAND 2          BAND 3          BAND 4
LOWER BAND EDGE     0.              0.10000000      0.18000000      0.30000000
UPPER BAND EDGE     0.05000000      0.15000000      0.25000000      0.36000000
DESIRED VALUE       0.              1.00000000      0.              1.00000000
WEIGHTING          10.00000000      1.00000000      3.00000000      1.00000000
DEVIATION           0.00344486      0.03444859      0.01148286      0.03444859
DEVIATION IN DB   -49.25657034    -29.25657034    -38.79839549    -29.25657034

                     BAND 5          BAND
LOWER BAND EDGE     0.41000000
UPPER BAND EDGE     0.50000000
DESIRED VALUE       0.
WEIGHTING          20.00000000
DEVIATION           0.00172243
DEVIATION IN DB   -55.27717018

EXTREMAL FREQUENCIES
    0.            0.0167411     0.0323661     0.0440429     0.0500000
    0.1000000     0.1089286     0.1267857     0.1424107     0.1500000
    0.1800000     0.1855804     0.1978571     0.2134821     0.2302232
    0.2436160     0.2500000     0.3000000     0.3122768     0.3323661
    0.3502232     0.3600000     0.4100000     0.4155804     0.4289732
    0.4457143     0.4635714     0.4814285     0.5000000

******************************************************************
TIME=      3.8164219 SECONDS
```

Fig. 15. Output listing for an N = 55 multiband filter.

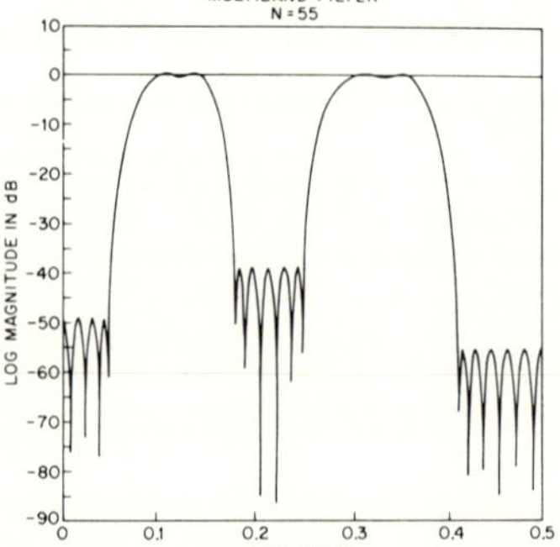

Fig. 16. Log magnitude response for an N = 55 multiband filter.

```
**********************************************************************
                    FINITE IMPULSE RESPONSE (FIR)
                    LINEAR PHASE DIGITAL FILTER DESIGN
                    REMEZ EXCHANGE ALGORITHM
                    DIFFERENTIATOR
          FILTER LENGTH =  32
          ***** IMPULSE RESPONSE *****
               H(  1) = -0.62713091E-03 = -H(  32)
               H(  2) =  0.85633433E-03 = -H(  31)
               H(  3) = -0.42418549E-03 = -H(  30)
               H(  4) =  0.39901518E-03 = -H(  29)
               H(  5) = -0.43437273E-03 = -H(  28)
               H(  6) =  0.49969450E-03 = -H(  27)
               H(  7) = -0.59634961E-03 = -H(  26)
               H(  8) =  0.73277031E-03 = -H(  25)
               H(  9) = -0.93002681E-03 = -H(  24)
               H( 10) =  0.12270042E-02 = -H(  23)
               H( 11) = -0.17012820E-02 = -H(  22)
               H( 12) =  0.25272341E-02 = -H(  21)
               H( 13) = -0.41601159E-02 = -H(  20)
               H( 14) =  0.81294555E-02 = -H(  19)
               H( 15) = -0.22539097E-01 = -H(  18)
               H( 16) =  0.20266535E 00 = -H(  17)

                    BAND 1          BAND
LOWER BAND EDGE     0.
UPPER BAND EDGE     0.50000000
DESIRED SLOPE       1.00000000
WEIGHTING           1.00000000
DEVIATION           0.00620231

EXTREMAL FREQUENCIES
   0.0019531   0.0332031   0.0664062   0.0996094   0.1328125
   0.1640625   0.1972656   0.2304687   0.2636719   0.2968750
   0.3300781   0.3632812   0.3945312   0.4277344   0.4589844
   0.4863281   0.5000000

**********************************************************************
TIME=       1.1072656 SECONDS
```

Fig. 17. Output listing for an $N = 32$ differentiator.

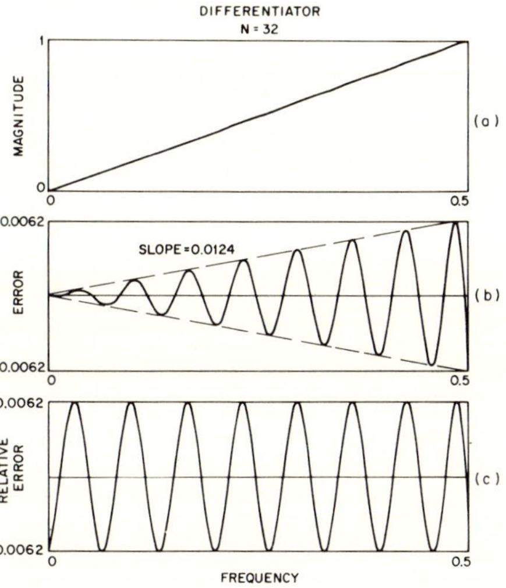

Fig. 18. Magnitude and error responses for an $N = 32$ differentiator.

```
**********************************************************************

                    FINITE IMPULSE RESPONSE (FIR)
                    LINEAR PHASE DIGITAL FILTER DESIGN
                    REMEZ EXCHANGE ALGORITHM
                    HILBERT TRANSFORMER
            FILTER LENGTH =  20
            ***** IMPULSE RESPONSE *****
                 H(  1) =  0.10026189E-01 = -H(  20)
                 H(  2) =  0.14173287E-01 = -H(  19)
                 H(  3) =  0.20452437E-01 = -H(  18)
                 H(  4) =  0.28736882E-01 = -H(  17)
                 H(  5) =  0.39852580E-01 = -H(  16)
                 H(  6) =  0.55333300E-01 = -H(  15)
                 H(  7) =  0.78542752E-01 = -H(  14)
                 H(  8) =  0.11823755E 00 = -H(  13)
                 H(  9) =  0.20664125E 00 = -H(  12)
                 H( 10) =  0.63475619E 00 = -H(  11)

                        BAND 1           BAND
 LOWER BAND EDGE      0.05000000
 UPPER BAND EDGE      0.50000000
 DESIRED VALUE        1.00000000
 WEIGHTING            1.00000000
 DEVIATION            0.02055604

 EXTREMAL FREQUENCIES
    0.0500000   0.0656250   0.1031250   0.1468750   0.1937500
    0.2437500   0.2937500   0.3468750   0.3968751   0.4500001
    0.5000000

**********************************************************************
TIME=       0.4825937 SECONDS
```

Fig. 19. Output listing for an N = 20 Hilbert transformer.

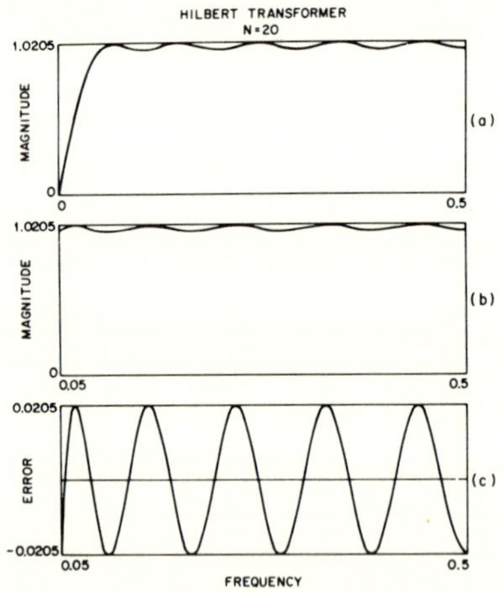

Fig. 20. Magnitude and error responses for an N = 20 Hilbert transformer.

```
**********************************************************************

                 FINITE IMPULSE RESPONSE (FIR)
                 LINEAR PHASE DIGITAL FILTER DESIGN
                 REMEZ EXCHANGE ALGORITHM
                 BANDPASS FILTER
 FILTER LENGTH = 128
 ***** IMPULSE RESPONSE *****
        H(  1) = -0.20662533E-02 = H( 128)
        H(  2) =  0.68616867E-03 = H( 127)
        H(  3) =  0.48620260E-03 = H( 126)
        H(  4) =  0.67255029E-03 = H( 125)
        H(  5) =  0.93764407E-03 = H( 124)
        H(  6) =  0.68975566E-03 = H( 123)
        H(  7) = -0.15411817E-03 = H( 122)
        H(  8) = -0.97912618E-03 = H( 121)
        H(  9) = -0.10910514E-02 = H( 120)
        H( 10) = -0.35942515E-03 = H( 119)
        H( 11) =  0.74040653E-03 = H( 118)
        H( 12) =  0.15280502E-02 = H( 117)
        H( 13) =  0.15296961E-02 = H( 116)
        H( 14) =  0.72595198E-03 = H( 115)
        H( 15) = -0.40411612E-03 = H( 114)
        H( 16) = -0.11471390E-02 = H( 113)
        H( 17) = -0.10636777E-02 = H( 112)
        H( 18) = -0.33855668E-03 = H( 111)
        H( 19) =  0.39066916E-03 = H( 110)
        H( 20) =  0.59175241E-03 = H( 109)
        H( 21) =  0.29497579E-03 = H( 108)
        H( 22) =  0.14988618E-04 = H( 107)
        H( 23) =  0.28683373E-03 = H( 106)
        H( 24) =  0.10519103E-02 = H( 105)
        H( 25) =  0.15913830E-02 = H( 104)
        H( 26) =  0.93133046E-03 = H( 103)
        H( 27) = -0.11501053E-02 = H( 102)
        H( 28) = -0.35816184E-02 = H( 101)
        H( 29) = -0.44120705E-02 = H( 100)
        H( 30) = -0.21995665E-02 = H(  99)
        H( 31) =  0.26225995E-02 = H(  98)
        H( 32) =  0.73509685E-02 = H(  97)
        H( 33) =  0.84703702E-02 = H(  96)
        H( 34) =  0.40213218E-02 = H(  95)
        H( 35) = -0.44717526E-02 = H(  94)
        H( 36) = -0.12166497E-01 = H(  93)
        H( 37) = -0.13554154E-01 = H(  92)
        H( 38) = -0.62183138E-02 = H(  91)
        H( 39) =  0.67914337E-02 = H(  90)
        H( 40) =  0.17947235E-01 = H(  89)
        H( 41) =  0.19518513E-01 = H(  88)
        H( 42) =  0.87717158E-02 = H(  87)
        H( 43) = -0.93473921E-02 = H(  86)
        H( 44) = -0.24259875E-01 = H(  85)
        H( 45) = -0.25905771E-01 = H(  84)
        H( 46) = -0.11425838E-01 = H(  83)
        H( 47) =  0.12021431E-01 = H(  82)
        H( 48) =  0.30564841E-01 = H(  81)
        H( 49) =  0.32239440E-01 = H(  80)
        H( 50) =  0.14010583E-01 = H(  79)
        H( 51) = -0.14528283E-01 = H(  78)
        H( 52) = -0.36550244E-01 = H(  77)
        H( 53) = -0.37904738E-01 = H(  76)
        H( 54) = -0.16241339E-01 = H(  75)

        H( 55) =  0.16669269E-01 = H(  74)
        H( 56) =  0.41368275E-01 = H(  73)
        H( 57) =  0.42361176E-01 = H(  72)
        H( 58) =  0.17925027E-01 = H(  71)
        H( 59) = -0.18188843E-01 = H(  70)
        H( 60) = -0.44592980E-01 = H(  69)
        H( 61) = -0.45118319E-01 = H(  68)
        H( 62) = -0.18854116E-01 = H(  67)
        H( 63) =  0.18954474E-01 = H(  66)
        H( 64) =  0.45897002E-01 = H(  65)
```

Fig. 21. Output listing for an $N = 128$ bandpass filter with arbitrary weighting characteristics.

```
                   BAND 1          BAND 2          BAND 3          BAND 4
LOWER BAND EDGE    0.              0.12000000      0.15000000      0.25000000
UPPER BAND EDGE    0.10000000      0.13000000      0.25000000      0.50000000
DESIRED VALUE      0.              1.00000000      0.              0.
WEIGHTING          10.00000000     1.00000000      10.00000000     10.00000000
DEVIATION          0.00500134      0.05001341      0.00500134      0.00500134
DEVIATION IN DB    -46.01827145    -26.01827145    -46.01827145    -46.01827145

EXTREMAL FREQUENCIES
    0.           0.0102539    0.0200195    0.0288086    0.0371094
    0.0454102    0.0532227    0.0610352    0.0688477    0.0761719
    0.0834961    0.0898437    0.0952148    0.0986328    0.1000000
    0.1200000    0.1248828    0.1300000    0.1500000    0.1514648
    0.1548828    0.1602539    0.1666016    0.1734375    0.1807617
    0.1880859    0.1958984    0.2037109    0.2115234    0.2193359
    0.2271484    0.2349609    0.2427734    0.2509766    0.2592773
    0.2670898    0.2749023    0.2827148    0.2910156    0.2988281
    0.3066406    0.3144531    0.3222656    0.3305664    0.3383789
    0.3461914    0.3540039    0.3618164    0.3696289    0.3779297
    0.3857422    0.3935547    0.4013672    0.4091797    0.4169922
    0.4252930    0.4331055    0.4409180    0.4487305    0.4565430
    0.4643555    0.4721680    0.4804687    0.4882812    0.4960937

**********************************************************************
TIME=      23.7952969 SECONDS
```

Fig. 21. (Continued.)

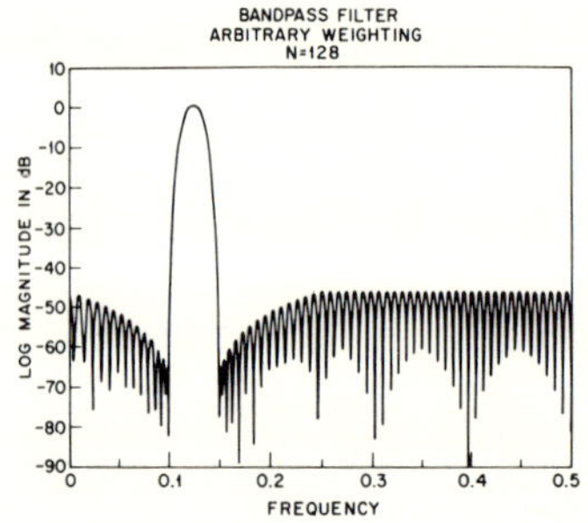

Fig. 22. Log magnitude response for an $N = 128$ bandpass filter with arbitrary weighting characteristics.

$$W(f) = \begin{cases} \dfrac{10}{1 - 9f} & 0 \leqslant f \leqslant 0.1 \\ 1 & 0.12 \leqslant f \leqslant 0.13 \\ \dfrac{10}{9f - 1.25} & 0.15 \leqslant f \leqslant 0.25 \\ 10 & 0.25 \leqslant f \leqslant 0.5. \end{cases}$$

Thus the tolerance scheme is linear in the intervals $0 \leqslant f \leqslant 0.1$ and $0.15 \leqslant f \leqslant 0.25$. The error at the stopband edges is 0.0005 (-66 dB), and the peak error increases linearly to 0.005 (-46 dB). The time required to design this filter was 23.8 s.

Summary

A general-purpose linear phase FIR filter design program is presented which is capable of designing a wide variety of standard filters as well as any desired magnitude response which can be specified by the user. The speed of the algorithm, as well as its generality, make this program an attractive one for a wide variety of design applications.

Appendix

```
C   PROGRAM FOR THE DESIGN OF LINEAR PHASE FINITE IMPULSE
C   RESPONSE (FIR) FILTERS USING THE REMEZ EXCHANGE ALGORITHM
C   JIM MCCLELLAN, RICE UNIVERSITY, APRIL 13, 1973
C   THREE TYPES OF FILTERS ARE INCLUDED--BANDPASS FILTERS
C   DIFFERENTIATORS, AND HILBERT TRANSFORM FILTERS
C
C   THE INPUT DATA CONSISTS OF 5 CARDS
C
C   CARD 1--FILTER LENGTH, TYPE OF FILTER.  1-MULTIPLE
C   PASSBAND/STOPBAND, 2-DIFFERENTIATOR, 3-HILBERT TRANSFORM
C   FILTER.  NUMBER OF BANDS, CARD PUNCH DESIRED, AND GRID
C   DENSITY.
C
C   CARD 2--BANDEDGES, LOWER AND UPPER EDGES FOR EACH BAND
C   WITH A MAXIMUM OF 10 BANDS.
C
C   CARD 3--DESIRED FUNCTION (OR DESIRED SLOPE IF A
C   DIFFERENTIATOR) FOR EACH BAND.
C
C   CARD 4--WEIGHT FUNCTION IN EACH BAND.  FOR A
C   DIFFERENTIATOR, THE WEIGHT FUNCTION IS INVERSELY
C   PROPORTIONAL TO F.
C
C   THE FOLLOWING INPUT DATA SPECIFIES A LENGTH 32 BANDPASS
C   FILTER WITH STOPBANDS 0 TO 0.1 AND 0.425 TO 0.5, AND
C   PASSBAND FROM 0.2 TO 0.35 WITH WEIGHTING OF 10 IN THE
C   STOPBANDS AND 1 IN THE PASSBAND.  THE IMPULSE RESPONSE
C   WILL BE PUNCHED AND THE GRID DENSITY IS 32.  THIS IS THE
C   FILTER IN FIGURES 9 AND 10 IN THE TEXT.
C   SAMPLE INPUT DATA SETUP
C   32,1,3,1,32
C   0,0.1,0.2,0.35,0.425,0.5
C   0,1,0
C   10,1,10
C
C   THE FOLLOWING INPUT DATA SPECIFIES A LENGTH 32 WIDEBAND
C   DIFFERENTIATOR WITH SLOPE 1 AND WEIGHTING OF 1/F.  THE
C   IMPULSE RESPONSE WILL NOT BE PUNCHED AND THE GRID
C   DENSITY IS ASSUMED TO BE 16.  THIS IS THE FILTER IN
C   FIGURES 17 AND 18 IN THE TEXT.
C   32,2,1,0,0
C   0,0.5
C   1.0
C   1.0
C
C
      COMMON PI2,AD,DEV,X,Y,GRID,DES,WT,ALPHA,IEXT,NFCNS,NGRID
      DIMENSION IEXT(66),AD(66),ALPHA(66),X(66),Y(66)
      DIMENSION H(66)
      DIMENSION DES(1045),GRID(1045),WT(1045)
      DIMENSION EDGE(20),FX(10),WTX(10),DEVIAT(10)
      DOUBLE PRECISION PI2,PI
      DOUBLE PRECISION AD,DEV,X,Y
      PI2=6.283185307179586
      PI=3.141592653589793
```

```
C
C  THE PROGRAM IS SET UP FOR A MAXIMUM LENGTH OF 128, BUT
C  THIS UPPER LIMIT CAN BE CHANGED BY REDIMENSIONING THE
C  ARRAYS IEXT, AD, ALPHA, X, Y, H TO BE NFMAX/2 + 2.
C  THE ARRAYS DES, GRID, AND WT MUST DIMENSIONED
C  16(NFMAX/2 + 2).
C
      NFMAX=128
  100 CONTINUE
      JTYPE=0
C
C  PROGRAM INPUT SECTION
C
      READ *,NFILT,JTYPE,NBANDS,JPUNCH,LGRID
      IF(NFILT.GT.NFMAX.OR.NFILT.LT.3) CALL ERROR
      IF(NBANDS.LE.0) NBANDS=1
C
C  GRID DENSITY IS ASSUMED TO BE 16 UNLESS SPECIFIED
C  OTHERWISE
C
      IF(LGRID.LE.0) LGRID=16
      JB=2*NBANDS
      READ *,(EDGE(J),J=1,JB)
      READ *,(FX(J),J=1,NBANDS)
      READ *,(WTX(J),J=1,NBANDS)
      IF(JTYPE.EQ.0) CALL ERROR
      NEG=1
      IF(JTYPE.EQ.1) NEG=0
      NODD=NFILT/2
      NODD=NFILT-2*NODD
      NFCNS=NFILT/2
      IF(NODD.EQ.1.AND.NEG.EQ.0) NFCNS=NFCNS+1
C
C  SET UP THE DENSE GRID.  THE NUMBER OF POINTS IN THE GRID
C  IS (FILTER LENGTH + 1)*GRID DENSITY/2
C
      GRID(1)=EDGE(1)
      DELF=LGRID*NFCNS
      DELF=0.5/DELF
      IF(NEG.EQ.0) GO TO 135
      IF(EDGE(1).LT.DELF) GRID(1)=DELF
  135 CONTINUE
      J=1
      L=1
      LBAND=1
  140 FUP=EDGE(L+1)
  145 TEMP=GRID(J)
C
C  CALCULATE THE DESIRED MAGNITUDE RESPONSE AND THE WEIGHT
C  FUNCTION ON THE GRID
C
      DES(J)=EFF(TEMP,FX,WTX,LBAND,JTYPE)
      WT(J)=WATE(TEMP,FX,WTX,LBAND,JTYPE)
      J=J+1
      GRID(J)=TEMP+DELF
      IF(GRID(J).GT.FUP) GO TO 150
      GO TO 145
  150 GRID(J-1)=FUP
      DES(J-1)=EFF(FUP,FX,WTX,LBAND,JTYPE)
      WT(J-1)=WATE(FUP,FX,WTX,LBAND,JTYPE)
      LBAND=LBAND+1
      L=L+2
      IF(LBAND.GT.NBANDS) GO TO 160
      GRID(J)=EDGE(L)
      GO TO 140
  160 NGRID=J-1
      IF(NEG.NE.NODD) GO TO 165
      IF(GRID(NGRID).GT.(0.5-DELF)) NGRID=NGRID-1
  165 CONTINUE
C
C  SET UP A NEW APPROXIMATION PROBLEM WHICH IS EQUIVALENT
C  TO THE ORIGINAL PROBLEM
C
      IF(NEG) 170,170,180
  170 IF(NODD.EQ.1) GO TO 200
      DO 175 J=1,NGRID
      CHANGE=DCOS(PI*GRID(J))
      DES(J)=DES(J)/CHANGE
  175 WT(J)=WT(J)*CHANGE
      GO TO 200
  180 IF(NODD.EQ.1) GO TO 190
      DO 185 J=1,NGRID
      CHANGE=DSIN(PI*GRID(J))
      DES(J)=DES(J)/CHANGE
  185 WT(J)=WT(J)*CHANGE
      GO TO 200
  190 DO 195 J=1,NGRID
      CHANGE=DSIN(PI2*GRID(J))
      DES(J)=DES(J)/CHANGE
  195 WT(J)=WT(J)*CHANGE
C
C  INITIAL GUESS FOR THE EXTREMAL FREQUENCIES--EQUALLY
C  SPACED ALONG THE GRID
C
  200 TEMP=FLOAT(NGRID-1)/FLOAT(NFCNS)
      DO 210 J=1,NFCNS
  210 IEXT(J)=(J-1)*TEMP+1
      IEXT(NFCNS+1)=NGRID
      NM1=NFCNS-1
      NZ=NFCNS+1
C
C  CALL THE REMEZ EXCHANGE ALGORITHM TO DO THE APPROXIMATION
C  PROBLEM
C
      CALL REMEZ(EDGE,NBANDS)
C
C  CALCULATE THE IMPULSE RESPONSE.
C
      IF(NEG) 300,300,320
  300 IF(NODD.EQ.0) GO TO 310
      DO 305 J=1,NM1
  305 H(J)=0.5*ALPHA(NZ-J)
      H(NFCNS)=ALPHA(1)
      GO TO 350
  310 H(1)=0.25*ALPHA(NFCNS)
      DO 315 J=2,NM1
  315 H(J)=0.25*(ALPHA(NZ-J)+ALPHA(NFCNS+2-J))
      H(NFCNS)=0.5*ALPHA(1)+0.25*ALPHA(2)
      GO TO 350
  320 IF(NODD.EQ.0) GO TO 330
      H(1)=0.25*ALPHA(NFCNS)
      H(2)=0.25*ALPHA(NM1)
      DO 325 J=3,NM1
  325 H(J)=0.25*(ALPHA(NZ-J)-ALPHA(NFCNS+3-J))
      H(NFCNS)=0.5*ALPHA(1)-0.25*ALPHA(3)
      H(NZ)=0.0
      GO TO 350
  330 H(1)=0.25*ALPHA(NFCNS)
      DO 335 J=2,NM1
  335 H(J)=0.25*(ALPHA(NZ-J)-ALPHA(NFCNS+2-J))
      H(NFCNS)=0.5*ALPHA(1)-0.25*ALPHA(2)
C
C  PROGRAM OUTPUT SECTION.
C
  350 PRINT 360
  360 FORMAT(1H1, 70(1H*)//25X,'FINITE IMPULSE RESPONSE (FIR)'/
     125X,'LINEAR PHASE DIGITAL FILTER DESIGN'/
     225X,'REMEZ EXCHANGE ALGORITHM'/)
      IF(JTYPE.EQ.1) PRINT 365
  365 FORMAT(25X,'BANDPASS FILTER'/)
      IF(JTYPE.EQ.2) PRINT 370
  370 FORMAT(25X,'DIFFERENTIATOR'/)
      IF(JTYPE.EQ.3) PRINT 375
  375 FORMAT(25X,'HILBERT TRANSFORMER'/)
      PRINT 378,NFILT
  378 FORMAT(15X,'FILTER LENGTH = ',I3/)
      PRINT 380
  380 FORMAT(15X,'***** IMPULSE RESPONSE *****')
      DO 381 J=1,NFCNS
      K=NFILT+1-J
      IF(NEG.EQ.0) PRINT 382,J,H(J),K
      IF(NEG.EQ.1) PRINT 383,J,H(J),K
  381 CONTINUE
  382 FORMAT(20X,'H(',I3,') = ',E15.8,' = H(',I4,')')
  383 FORMAT(20X,'H(',I3,') = ',E15.8,' = -H(',I4,')')
      IF(NEG.EQ.1.AND.NODD.EQ.1) PRINT 384,NZ
  384 FORMAT(20X,'H(',I3,') =  0.0')
      DO 450 K=1,NBANDS,4
      KUP=K+3
      IF(KUP.GT.NBANDS) KUP=NBANDS
      PRINT 385,(J,J=K,KUP)
  385 FORMAT(/24X,4('BAND',I3,8X))
      PRINT 390,(EDGE(2*J-1),J=K,KUP)
  390 FORMAT(2X,'LOWER BAND EDGE',5F15.9)
      PRINT 395,(EDGE(2*J),J=K,KUP)
  395 FORMAT(2X,'UPPER BAND EDGE',5F15.9)
      IF(JTYPE.NE.2) PRINT 400,(FX(J),J=K,KUP)
  400 FORMAT(2X,'DESIRED VALUE',2X,5F15.9)
      IF(JTYPE.EQ.2) PRINT 405,(FX(J),J=K,KUP)
  405 FORMAT(2X,'DESIRED SLOPE',2X,5F15.9)
      PRINT 410,(WTX(J),J=K,KUP)
  410 FORMAT(2X,'WEIGHTING',6X,5F15.9)
      DO 420 J=K,KUP
  420 DEVIAT(J)=DEV/WTX(J)
      PRINT 425,(DEVIAT(J),J=K,KUP)
  425 FORMAT(2X,'DEVIATION',6X,5F15.9)
      IF(JTYPE.NE.1) GO TO 450
      DO 430 J=K,KUP
  430 DEVIAT(J)=20.0*ALOG10(DEVIAT(J))
      PRINT 435,(DEVIAT(J),J=K,KUP)
  435 FORMAT(2X,'DEVIATION IN DB',5F15.9)
  450 CONTINUE
      PRINT 455,(GRID(IEXT(J)),J=1,NZ)
  455 FORMAT(/2X,'EXTREMAL FREQUENCIES'/(2X,5F12.7))
      PRINT 460
  460 FORMAT(/1X,70(1H*)/1H1)
      IF(JPUNCH.NE.0) PUNCH *,(H(J),J=1,NFCNS)
      IF(NFILT.NE.0) GO TO 100
      RETURN
      END

      FUNCTION EFF(TEMP,FX,WTX,LBAND,JTYPE)
C
C  FUNCTION TO CALCULATE THE DESIRED MAGNITUDE RESPONSE
C  AS A FUNCTION OF FREQUENCY.
C
      DIMENSION FX(5),WTX(5)
      IF(JTYPE.EQ.2) GO TO 1
      EFF=FX(LBAND)
      RETURN
```

```
    1 EFF=FX(LBAND)*TEMP
      RETURN
      END




      FUNCTION WATE(TEMP,FX,WTX,LBAND,JTYPE)
C
C  FUNCTION TO CALCULATE THE WEIGHT FUNCTION AS A FUNCTION
C  OF FREQUENCY.
C
      DIMENSION FX(5),WTX(5)
      IF(JTYPE.EQ.2) GO TO 1
      WATE=WTX(LBAND)
      RETURN
    1 IF(FX(LBAND).LT.0.0001) GO TO 2
      WATE=WTX(LBAND)/TEMP
      RETURN
    2 WATE=WTX(LBAND)
      RETURN
      END



      SUBROUTINE ERROR
      PRINT 1
    1 FORMAT(' ************ ERROR IN INPUT DATA **********')
      STOP
      END



      SUBROUTINE REMEZ(EDGE,NBANDS)
C
C  THIS SUBROUTINE IMPLEMENTS THE REMEZ EXCHANGE ALGORITHM
C  FOR THE WEIGHTED CHEBYCHEV APPROXIMATION OF A CONTINUOUS
C  FUNCTION WITH A SUM OF COSINES.  INPUTS TO THE SUBROUTINE
C  ARE A DENSE GRID WHICH REPLACES THE FREQUENCY AXIS, THE
C  DESIRED FUNCTION ON THIS GRID, THE WEIGHT FUNCTION ON THE
C  GRID, THE NUMBER OF COSINES, AND AN INITIAL GUESS OF THE
C  EXTREMAL FREQUENCIES.  THE PROGRAM MINIMIZES THE CHEBYCHEV
C  ERROR BY DETERMINING THE BEST LOCATION OF THE EXTREMAL
C  FREQUENCIES (POINTS OF MAXIMUM ERROR) AND THEN CALCULATES
C  THE COEFFICIENTS OF THE BEST APPROXIMATION.
C
      COMMON PI2,AD,DEV,X,Y,GRID,DES,WT,ALPHA,IEXT,NFCNS,NGRID
      DIMENSION EDGE(20)
      DIMENSION IEXT(66),AD(66),ALPHA(66),X(66),Y(66)
      DIMENSION DES(1045),GRID(1045),WT(1045)
      DIMENSION A(66),P(65),Q(65)
      DOUBLE PRECISION PI2,DNUM,DDEN,DTEMP,A,P,Q
      DOUBLE PRECISION AD,DEV,X,Y
C
C  THE PROGRAM ALLOWS A MAXIMUM NUMBER OF ITERATIONS OF 25
C
      ITRMAX=25
      DEVL=-1.0
      NZ=NFCNS+1
      NZZ=NFCNS+2
      NITER=0
  100 CONTINUE
      IEXT(NZZ)=NGRID+1
      NITER=NITER+1
      IF(NITER.GT.ITRMAX) GO TO 400
      DO 110 J=1,NZ
      DTEMP=GRID(IEXT(J))
      DTEMP=DCOS(DTEMP*PI2)
  110 X(J)=DTEMP
      JET=(NFCNS-1)/15+1
      DO 120 J=1,NZ
  120 AD(J)=D(J,NZ,JET)
      DNUM=0.0
      DDEN=0.0
      K=1
      DO 130 J=1,NZ
      L=IEXT(J)
      DTEMP=AD(J)*DES(L)
      DNUM=DNUM+DTEMP
      DTEMP=K*AD(J)/WT(L)
      DDEN=DDEN+DTEMP
  130 K=-K
      DEV=DNUM/DDEN
      NU=1
      IF(DEV.GT.0.0) NU=-1
      DEV=-NU*DEV
      K=NU
      DO 140 J=1,NZ
      L=IEXT(J)
      DTEMP=K*DEV/WT(L)
      Y(J)=DES(L)+DTEMP
  140 K=-K
      IF(DEV.GE.DEVL) GO TO 150
      CALL OUCH
      GO TO 400
  150 DEVL=DEV
      JCHNGE=0
      K1=IEXT(1)
      KNZ=IEXT(NZ)
      KLOW=0
      NUT=-NU
      J=1
C
C  SEARCH FOR THE EXTREMAL FREQUENCIES OF THE BEST
C  APPROXIMATION
C
  200 IF(J.EQ.NZZ) YNZ=COMP
      IF(J.GE.NZZ) GO TO 300
      KUP=IEXT(J+1)
      L=IEXT(J)+1
      NUT=-NUT
      IF(J.EQ.2) Y1=COMP
      COMP=DEV
      IF(L.GE.KUP) GO TO 220
      ERR=GEE(L,NZ)
      ERR=(ERR-DES(L))*WT(L)
      DTEMP=NUT*ERR-COMP
      IF(DTEMP.LE.0.0) GO TO 220
      COMP=NUT*ERR
  210 L=L+1
      IF(L.GE.KUP) GO TO 215
      ERR=GEE(L,NZ)
      ERR=(ERR-DES(L))*WT(L)
      DTEMP=NUT*ERR-COMP
      IF(DTEMP.LE.0.0) GO TO 215
      COMP=NUT*ERR
      GO TO 210
  215 IEXT(J)=L-1
      J=J+1
      KLOW=L-1
      JCHNGE=JCHNGE+1
      GO TO 200
  220 L=L-1
  225 L=L-1
      IF(L.LE.KLOW) GO TO 250
      ERR=GEE(L,NZ)
      ERR=(ERR-DES(L))*WT(L)
      DTEMP=NUT*ERR-COMP
      IF(DTEMP.GT.0.0) GO TO 230
      IF(JCHNGE.LE.0) GO TO 225
      GO TO 260
  230 COMP=NUT*ERR
  235 L=L-1
      IF(L.LE.KLOW) GO TO 240
      ERR=GEE(L,NZ)
      ERR=(ERR-DES(L))*WT(L)
      DTEMP=NUT*ERR-COMP
      IF(DTEMP.LE.0.0) GO TO 240
      COMP=NUT*ERR
      GO TO 235
  240 KLOW=IEXT(J)
      IEXT(J)=L+1
      J=J+1
      JCHNGE=JCHNGE+1
      GO TO 200
  250 L=IEXT(J)+1
      IF(JCHNGE.GT.0) GO TO 215
  255 L=L+1
      IF(L.GE.KUP) GO TO 260
      ERR=GEE(L,NZ)
      ERR=(ERR-DES(L))*WT(L)
      DTEMP=NUT*ERR-COMP
      IF(DTEMP.LE.0.0) GO TO 255
      COMP=NUT*ERR
      GO TO 210
  260 KLOW=IEXT(J)
      J=J+1
      GO TO 200
  300 IF(J.GT.NZZ) GO TO 320
      IF(K1.GT.IEXT(1)) K1=IEXT(1)
      IF(KNZ.LT.IEXT(NZ)) KNZ=IEXT(NZ)
      NUT1=NUT
      NUT=-NU
      L=0
      KUP=K1
      COMP=YNZ*(1.00001)
      LUCK=1
  310 L=L+1
      IF(L.GE.KUP) GO TO 315
      ERR=GEE(L,NZ)
      ERR=(ERR-DES(L))*WT(L)
      DTEMP=NUT*ERR-COMP
      IF(DTEMP.LE.0.0) GO TO 310
      COMP=NUT*ERR
      J=NZZ
      GO TO 210
  315 LUCK=6
      GO TO 325
  320 IF(LUCK.GT.9) GO TO 350
      IF(COMP.GT.Y1) Y1=COMP
      K1=IEXT(NZZ)
  325 L=NGRID+1
      KLOW=KNZ
      NUT=-NUT1
      COMP=Y1*(1.00001)
  330 L=L-1
      IF(L.LE.KLOW) GO TO 340
      ERR=GEE(L,NZ)
      ERR=(ERR-DES(L))*WT(L)
      DTEMP=NUT*ERR-COMP
      IF(DTEMP.LE.0.0) GO TO 330
      J=NZZ
      COMP=NUT*ERR
      LUCK=LUCK+10
      GO TO 235
```

```
340 IF(LUCK.EQ.6) GO TO 370
    DO 345 J=1,NFCNS
345 IEXT(NZZ-J)=IEXT(NZ-J)
    IEXT(1)=K1
    GO TO 100
350 KN=IEXT(NZZ)
    DO 360 J=1,NFCNS
360 IEXT(J)=IEXT(J+1)
    IEXT(NZ)=KN
    GO TO 100
370 IF(JCHNGE.GT.0) GO TO 100


C  CALCULATION OF THE COEFFICIENTS OF THE BEST APPROXIMATION
C  USING THE INVERSE DISCRETE FOURIER TRANSFORM
C
400 CONTINUE
    NM1=NFCNS-1
    FSH=1.0E-06
    GTEMP=GRID(1)
    X(NZZ)=-2.0
    CN=2*NFCNS-1
    DELF=1.0/CN
    L=1
    KKK=0
    IF(EDGE(1).EQ.0.0.AND.EDGE(2*NBANDS).EQ.0.5) KKK=1
    IF(NFCNS.LE.3) KKK=1
    IF(KKK.EQ.1) GO TO 405
    DTEMP=DCOS(PI2*GRID(1))
    DNUM=DCOS(PI2*GRID(NGRID))
    AA=2.0/(DTEMP-DNUM)
    BB=-(DTEMP+DNUM)/(DTEMP-DNUM)
405 CONTINUE
    DO 430 J=1,NFCNS
    FT=(J-1)*DELF
    XT=DCOS(PI2*FT)
    IF(KKK.EQ.1) GO TO 410
    XT=(XT-BB)/AA
    FT=ARCOS(XT)/PI2
410 XE=X(L)
    IF(XT.GT.XE) GO TO 420
    IF((XE-XT).LT.FSH) GO TO 415
    L=L+1
    GO TO 410
415 A(J)=Y(L)
    GO TO 425
420 IF((XT-XE).LT.FSH) GO TO 415
    GRID(1)=FT
    A(J)=GEE(1,NZ)
425 CONTINUE
    IF(L.GT.1) L=L-1
430 CONTINUE
    GRID(1)=GTEMP
    DDEN=PI2/CN
    DO 510 J=1,NFCNS
    DTEMP=0.0
    DNUM=(J-1)*DDEN
    IF(NM1.LT.1) GO TO 505
    DO 500 K=1,NM1
500 DTEMP=DTEMP+A(K+1)*DCOS(DNUM*K)
505 DTEMP=2.0*DTEMP+A(1)
510 ALPHA(J)=DTEMP
    DO 550 J=2,NFCNS
550 ALPHA(J)=2*ALPHA(J)/CN
    ALPHA(1)=ALPHA(1)/CN
    IF(KKK.EQ.1) GO TO 545
    P(1)=2.0*ALPHA(NFCNS)*BB+ALPHA(NM1)
    P(2)=2.0*AA*ALPHA(NFCNS)
    Q(1)=ALPHA(NFCNS-2)-ALPHA(NFCNS)
    DO 540 J=2,NM1
    IF(J.LT.NM1) GO TO 515
    AA=0.5*AA
    BB=0.5*BB
515 CONTINUE
    P(J+1)=0.0
    DO 520 K=1,J
    A(K)=P(K)
520 P(K)=2.0*BB*A(K)
    P(2)=P(2)+A(1)*2.0*AA
    JM1=J-1
    DO 525 K=1,JM1
525 P(K)=P(K)+Q(K)+AA*A(K+1)
    JP1=J+1
    DO 530 K=3,JP1
530 P(K)=P(K)+AA*A(K-1)
    IF(J.EQ.NM1) GO TO 540
    DO 535 K=1,J
535 Q(K)=-A(K)
    Q(1)=Q(1)+ALPHA(NFCNS-1-J)
540 CONTINUE
    DO 543 J=1,NFCNS
543 ALPHA(J)=P(J)
545 CONTINUE
    IF(NFCNS.GT.3) RETURN
    ALPHA(NFCNS+1)=0.0
    ALPHA(NFCNS+2)=0.0
    RETURN
    END


    DOUBLE PRECISION FUNCTION D(K,N,M)
C  FUNCTION TO CALCULATE THE LAGRANGE INTERPOLATION
C  COEFFICIENTS FOR USE IN THE FUNCTION GEE.
C
    COMMON PI2,AD,DEV,X,Y,GRID,DES,WT,ALPHA,IEXT,NFCNS,NGRID
    DIMENSION IEXT(66),AD(66),ALPHA(66),X(66),Y(66)
    DIMENSION DES(1045),GRID(1045),WT(1045)
    DOUBLE PRECISION AD,DEV,X,Y
    DOUBLE PRECISION Q
    DOUBLE PRECISION PI2
    D=1.0
    Q=X(K)
    DO 3 L=1,M
    DO 2 J=L,N,M
    IF(J-K)1,2,1
  1 D=2.0*D*(Q-X(J))
  2 CONTINUE
  3 CONTINUE
    D=1.0/D
    RETURN
    END


    DOUBLE PRECISION FUNCTION GEE(K,N)
C
C  FUNCTION TO EVALUATE THE FREQUENCY RESPONSE USING THE
C  LAGRANGE INTERPOLATION FORMULA IN THE BARYCENTRIC FORM
C
    COMMON PI2,AD,DEV,X,Y,GRID,DES,WT,ALPHA,IEXT,NFCNS,NGRID
    DIMENSION IEXT(66),AD(66),ALPHA(66),X(66),Y(66)
    DIMENSION DES(1045),GRID(1045),WT(1045)
    DOUBLE PRECISION P,C,D,XF
    DOUBLE PRECISION PI2
    DOUBLE PRECISION AD,DEV,X,Y
    P=0.0
    XF=GRID(K)
    XF=DCOS(PI2*XF)
    D=0.0
    DO 1 J=1,N
    C=XF-X(J)
    C=AD(J)/C
    D=D+C
  1 P=P+C*Y(J)
    GEE=P/D
    RETURN
    END


    SUBROUTINE OUCH
    PRINT 1
  1 FORMAT(' ************* FAILURE TO CONVERGE ***********'/
   1'0PROBABLE CAUSE IS MACHINE ROUNDING ERROR'/
   2'0THE IMPULSE RESPONSE MAY BE CORRECT'/
   3'0CHECK WITH A FREQUENCY RESPONSE')
    RETURN
    END
```

References

[1] J. H. McClellan and T. W. Parks, "A unified approach to the design of optimum FIR linear phase digital filters," *IEEE Trans. Circuit Theory*, vol. CT-20, pp. 697–701, Nov. 1973.

[2] E. W. Cheney, *Introduction to Approximation Theory*. New York: McGraw-Hill, 1966, pp. 72–100.

[3] E. Ya. Remez, "General computational methods of Tchebycheff approximation," Kiev, 1957 (Atomic Energy Translation 4491, pp. 1–85).

[4] B. Gold and K. L. Jordan, "A direct search procedure for designing finite duration impulse response filters," *IEEE Trans. Audio Electroacoust.*, vol. AU-17, pp. 33–36, Mar. 1969.

[5] L. R. Rabiner, B. Gold, and C. McGonegal, "An approach to the approximation problem for nonrecursive digital filters," *IEEE Trans. Audio Electroacoust.*, vol. AU-18, pp. 83–106, June 1970.

[6] L. R. Rabiner and K. Steiglitz, "The design of wideband recursive and nonrecursive digital differentiators," *IEEE Trans. Audio Electroacoust.*, vol. AU-18, pp. 204–209, June 1970.

[7] O. Herrmann, "Design of nonrecursive digital filters with linear phase," *Electron. Lett.*, pp. 328–329, 1970.

[8] E. Hofstetter, A. V. Oppenheim, and J. Siegel, "A new technique for the design of nonrecursive digital filters," in *Proc. 5th Annu. Princeton Conf. Inform. Sci. and Syst.*, Mar. 1971, pp. 64–72.

[9] T. W. Parks and J. H. McClellan, "Chebyshev approximation for nonrecursive digital filters with linear phase," *IEEE Trans. Circuit Theory*, vol. CT-19, pp. 189–194, Mar. 1972.

[10] L. R. Rabiner, "The design of finite impulse response

digital filters using linear programming techniques," *Bell Syst. Tech. J.*, vol. 51, pp. 1177–1198, July–Aug. 1972.

[11] T. W. Parks and J. H. McClellan, "A program for the design of linear phase finite impulse response digital filters," *IEEE Trans. Audio Electroacoust.*, vol. AU-20, pp. 195–199, Aug. 1972.

[12] O. Herrmann, "Transversal filters for the Hilbert Transformation," *Arch. Elek. Übertragung.*, vol. 23, pp. 581-587, 1969.

[13] J. H. McClellan, "On the design of one-dimensional and two-dimensional FIR digital filters," Ph.D. dissertation, Rice Univ., Houston, Tex., Apr. 1973.

7

Reprinted from *IEEE Trans. Audio Electroacoustics,* **AU-20**(2), 112–114 (1972)

A Simple Design of Maximally Flat Delay Digital Filters

ALFRED FETTWEIS, Senior Member, IEEE
Dep. Elec. Eng.
Univ. Bochum
Bochum, Germany

Abstract

The problem of maximally flat delay design of recursive digital filters has been solved by Thiran, using a direct *z*-domain approach. In this paper, the solution is obtained in a much simpler way by making use of the familiar bilinear transform *s* of the variable *z*. A suitable continued fraction expansion available in the mathematical literature is shown to lead immediately to the required solution. The approach corresponds to the one used by Abele in transmission line filter design.

I. Introduction

Thiran [1]–[3] has recently described an explicit solution for the problem of determining for recursive digital filters a transfer function that possesses a maximally flat delay characteristic. Thiran's method is based on a z-domain analysis. The purpose of this paper is to show that the same result can be obtained in a much simpler way by operating in the s domain, where s is the familiar bilinear transform of the variable z.

The present method is related to the one used by Storch [4] for solving the similar problem in classical LC filter design, and it corresponds to the solution described by Abele [5] for microwave unit element filters. It should be pointed out, however, that the lengthy proof given by Abele for verifying that the desired property has actually been obtained is not required. As we shall see, a simpler proof follows easily from known properties of continued fraction expansions of the type to be used.

Only the basic aspects of the method will be given, and the application to transfer functions of particular practical importance will be discussed. Further properties can be derived from the corresponding properties described by Abele [5]. In particular, the identity of the present result with the one obtained by Thiran can easily be established by comparing Abele's and Thiran's expressions in terms of hypergeometric functions.

Manuscript received October 15, 1971.

II. Direct Derivation of the Desired Transfer Function

Thiran bases his analysis on a transfer function having the general form

$$H = 1\Big/\left(\sum_{i=0}^{n} a_i z^{-i}\right), \qquad z = e^{pT} \tag{1}$$

where p is the complex frequency and T the sampling period. Using the definition

$$s = (z-1)/(z+1) = \tanh(pT/2), \tag{2}$$

we can thus write

$$H = \frac{(1+s)^n}{d} = \frac{(1+s)^n}{(1+F_n)d_e} \tag{3}$$

where $d = d(s)$ is a polynomial in s of degree n and where

$$F_n = d_o/d_e, \qquad d_o + d_e = d, \tag{4}$$

d_o and d_e being the odd and even part of d, respectively. Stability requires that the polynomial d is Hurwitz. According to (3), we henceforth consider H to be a function of s, i.e., $H = H(s)$.

We define a function $B_n = B_n(s)$ by means of

$$e^{j2B_n} = H_*/H \tag{5}$$

where H_* stands, in the usual notation, for $H(-s)$. The function $B_n(s)$ is analytic in s and thus also in p; for $p=j\omega$, B_n is equal to the phase corresponding to the transfer function H. Comparing (5) with (3), we can write

$$jB_n = -pnT/2 + \tanh^{-1} F_n. \quad (6)$$

In order to obtain the ideal phase characteristic, the function jB_n would have to be equal to the function

$$jB = p\tau \quad (7)$$

where τ is the delay to be approximated. Thus, replacing B_n in (6) by B and writing F instead of F_n, we see that the ideal function F to be realized is

$$F = \tanh(\mu p T/2) \quad (8)$$

where

$$\mu = n + 2\tau/T. \quad (9)$$

Due to (2), (8) can also be written

$$F = \tanh(\mu \tanh^{-1} s). \quad (10)$$

Our approximation problem now consists in determining B_n in such a way that it coincides with B up to the nth order in p. Due to the general properties of Taylor series expansions, this is equivalent to approximating F by F_n up to the nth order in p. Since $s=0$ for $p=0$, however, this latter problem is equivalent to approximating F up to the nth order in s; this must be done in such a way that the resulting function $F_n(s)$ is the ratio of an odd to an even polynomial in s whose sum d is Hurwitz.

The solution to the problem thus defined is easily obtained by making use of a suitable continued fraction expansion. By appropriately rewriting the expansion of tan ($\mu\tan^{-1}y$) given in the standard textbooks of Perron [6] and Wall [7], we obtain the following expression for the function F defined by (10), using Pringheim's notation for representing a continued fraction:

$$F = \frac{\mu|}{|1/s} + \frac{(\mu^2-1)|}{|3/s} + \frac{(\mu^2-4)|}{|5/s} + \cdots + \frac{(\mu^2-m^2)|}{|(2m+1)/s} + \cdots. \quad (11)$$

Let us choose for F_n the nth convergent of the continued fraction (11), i.e.,

$$F_n = \frac{\mu|}{|1/s} + \frac{(\mu^2-1)|}{|3/s} + \cdots + \frac{[\mu^2-(n-1)^2]|}{|(2n-1)/s}, \quad (12)$$

which, clearly, is a rational function of degree n in s and which consists of a ratio of an odd to an even polynomial in s. By well-known properties of continued fraction expansions of the type (11), the function $F_n(s)$ defined by (12) coincides with the function $F(s)$ defined by (10) precisely up to the nth order.

Even without regard to the Hurwitz property, still to be discussed, the function F_n thus obtained is certainly the only one that satisfies the approximation requirement. This can be seen by the fact that the approximation problem is also equivalent to finding a certain Padé approximant in s^2 of the even function $F(s)/s$, and this problem is known to have a unique solution [6].

The Routh–Hurwitz criterion for testing the Hurwitz character of a polynomial is known to amount to verifying that the ratio of its odd and even parts is a reactance function, which in turn amounts to requiring that the coefficients appearing in a continued fraction expansion of the type of (12) all are positive. Hence, due to (4) and (12), $d(s)$ is Hurwitz if $\mu > n-1$, i.e., because of (9) if

$$\tau > -T/2. \quad (13)$$

This condition is always fulfilled for positive values of the delay τ to be approximated.

III. Related Approximation Problems

For practical applications, transfer functions of the form

$$H = \frac{f}{d} = \frac{f}{(1+F_n)d_e}, \quad (14)$$

where d is still a Hurwitz polynomial in s of degree n, with F_n, d_e, and d_o defined as before, but where f is an even or odd polynomial in s, are of greater importance than those of the form (3). The degree m of the polynomial f then satisfies the condition $m \leq n$, since otherwise H would be infinite for points located on the imaginary p axis. In fact, the theory to be given hereafter remains valid even if f, although being either even or odd, contains, e.g., square root expressions, as may be the case for wave digital filters [8]–[11].

We now replace the definition (5) by

$$e^{j2B_n} = \pm H_*/H \quad (15)$$

where the plus sign has to be chosen for f even and the minus sign for f odd. The function B_n thus defined is, for $p=j\omega$, still equal to the phase corresponding to H, expect for an additive term equal to a multiple of $\pi/2$ appearing in case of f odd. This additional term does not affect the group delay and is thus of no importance to our problem. Due to (4) and (14), we derive from (15) the relation

$$jB_n = \tanh^{-1} F_n. \quad (16)$$

The ideal value for B_n is again the value of B given by (7). As can be seen from (16), the ideal value for F_n is thus again the value of F given by (8), and thus by (10), but with μ now defined by

$$\mu = 2\tau/T. \quad (17)$$

Using this definition rather than (9), the function F_n defined by (12) is still the desired approximation to be used, together with (4), in (14). The condition for positiveness of all coefficients in (12), i.e., $\mu > n-1$, however, is now given by

$$\tau > (n-1)T/2. \quad (18)$$

It is thus no longer sufficient to require τ to be positive in order to satisfy the stability requirement.

The transfer functions studied by Abele [5] are special cases of those given by (14), and the results described by this author coincide with those obtained in this section. Furthermore, the results of Section II can be derived easily from the present ones by writing (3) in the form

$$H = \frac{f}{d} e^{npT/2}, \tag{19}$$

which can easily be obtained by making use of (2) and by defining f by means of

$$f = (1 - s^2)^{n/2}.$$

Thus, the problem of choosing the function H defined by (19) in such a way that the corresponding delay approximates τ is equivalent to choosing f/d such that the corresponding delay approximates $\tau + nT/2$. This shows in particular why (18) reduces to (13) if τ is replaced by $\tau + nT/2$.

It may be surprising to find that condition (13) actually is compatible with a certain range of negative values of τ. This paradox can be explained by noticing that even for the extreme value $\tau = -T/2$, the value of $|\tau|$ is only equal to half a sampling period, which is too small a value to allow proper interpretation as the precise value of the time delay of a signal. If a discrete-time system approaches a continuous-time system, we have $T \to 0$, in which case (13) approaches the inequality $\tau > 0$.

References

[1] J.-P. Thiran, "Recursive digital filters with maximally flat group delay," in *1970 IEEE Int. Symp. Circuit Theory, Dig. Tech. Papers*, pp. 139–140.

[2] ——, "Recursive digital filters with maximally flat group delay," *IEEE Trans. Circuit Theory*, vol. CT-18, pp. 659–664, Nov. 1971.

[3] ——, "L'approximation des caractéristiques idéales par les filtres digitaux récursifs," Ph.D. dissertation, Catholic Univ. Louvain, Louvain, Belgium, 1971.

[4] L. Storch, "Synthesis of constant-time-delay networks using Bessel polynomials," *Proc. IRE*, vol. 42, pp. 1666–1675, Nov. 1954.

[5] T. A. Abele, "Transmission line filters approximating a constant delay in a maximally flat sense," *IEEE Trans. Circuit Theory*, vol. CT-14, pp. 298–306, Sept. 1967.

[6] O. Perron, *Die Lehre von den Kettenbrüchen*, vol. II. Stuttgart: Teubner, 1957, p. 154.

[7] H. S. Wall, *Analytic Theory of Continued Fractions*. New York: Van Nostrand, 1958.

[8] A. Fettweis, "Entwurf von Digitalfiltern in Anlehnung an Verfahren der klassischen Netzwerktheorie," NTZ Report in *NTG-Fachtagung "Analyse und Synthese von Netzwerken"* (Stuttgart), Oct. 1970, pp. 17–20.

[9] J. A. C. Bingham, "A new type of digital filter with a reciprocal ladder configuration," in *1970 IEEE Int. Symp. Circuit Theory, Dig. Tech. Papers*, pp. 129–130.

[10] A. Fettweis, "Digital filter structures related to classical filter networks," *Arch. Elek. Übertragung*, vol. 25, pp. 79–89, Feb. 1971.

[11] ——, "Some principles of designing digital filters imitating classical filter structures," *IEEE Trans. Circuit Theory* (Corresp.), vol. CT-18, pp. 314–316, Mar. 1971.

8

Reprinted from *IEEE Trans. Circuit Theory,* **CT-18**(6), 664–669 (1971)

Equal-Ripple Delay Recursive Digital Filters

JEAN-PIERRE THIRAN, MEMBER, IEEE

Abstract—All-pole recursive digital low-pass filters approximating a constant group delay in an equal-ripple sense are described. The equal-ripple conditions are given by a system of nonlinear equations in the poles of the transfer function. The closed-form solution obtained for a particular value of the cutoff frequency yields good first approximations for the iterative procedure required in the general case. The set of approximated delays and ripples leading to realizable filters is discussed. Numerical examples illustrating the procedure are included.

I. Introduction

Among the problems which cannot be satisfactorily solved by the classical method of designing recursive digital filters, i.e., digitalization of continuous transmittances, one finds the approximation of a constant group delay. The choice of the maximally flat criterion for the direct procedure thus required leads to the use of a Gauss hypergeometric function [1]. Obviously, this error norm is not the best one from the approximation standpoint and the present paper is intended to give the equal-ripple solution to the same problem.

For the continuous case, two methods have been proposed, one by Ulbrich and Piloty [2] and the other by Abele [3], to get the equal-ripple delay filters. Both methods are characterized by a system of nonlinear equations to be solved by iterative techniques. While, for the first solution, the only unknowns of the system are the poles of the transfer function, the second one computes, in addition, the abscissas of the extrema of the delay response: the order of the system equals approximately twice the degree of the filter. However, contrary to the Ulbrich–Piloty method, the Bessel filter is explicitly the solution of the Abele system and can be used as an excellent first guess for the iterative process. To take advantage of the lower order of the system, the method developed in this paper for digital filters parallels the Ulbrich–Piloty procedure, and as an explicit solution is obtained in a particular case, the lack of first approximations can be overcome. The allowable approximated delays and ripples for a given cutoff frequency are discussed and some amplitude and delay characteristics are shown.

II. Statement of the Problem

In an approximation problem for prescribed group delay, it is natural to determine only the poles of the transfer function. Indeed, it is well known that a "mirror-image polynomial" used for the numerator makes it possible to meet additional requirements on attenuation [4] without affecting

Manuscript received March 8, 1971.

The author is with the MBLE Research Laboratory, Brussels, Belgium.

group delay. Therefore the investigated recursive transmittances are of the all-pole type

$$H(z) = \frac{K}{p(z^{-1})}.$$

The constant K is a scaling factor in amplitude which will be fixed to have zero loss at dc, and the polynomial $p(z^{-1})$ can be defined either by its coefficients:

$$p(z^{-1}) = \sum_{i=0}^{n} a_i z^{-i}, \qquad \text{with } a_n = 1 \tag{1}$$

or by its roots:

$$p(z^{-1}) = \prod_{i=1}^{n} (z^{-1} - z_i^{-1}) \tag{2}$$

with $|z_i| < 1$ to guarantee stability.

The phase shift of $H(z)$ is obviously half that of the all-pass function $G(z) = p(z)/p(z^{-1})$ and is thus given by ln $[G(e^{j\omega})]/2j$ if the sampling frequency is normalized to unity. Hence the analytic continuation of the group delay is

$$t(z) = -\frac{1}{2}\frac{d \ln G(z)}{d \ln z}$$

or

$$t(z) = -\frac{1}{2}\left[\frac{zp'(z)}{p(z)} + \frac{z^{-1}p'(z^{-1})}{p(z^{-1})}\right]$$

where p' denotes the derivative of p.

Using, respectively, (1) and (2), one gets

$$t(z) = -\frac{1}{2}\left\{\frac{\sum_{i=0}^{n} a_i i z^i}{\sum_{i=0}^{n} a_i z^i} + \frac{\sum_{i=0}^{n} a_i i z^{-i}}{\sum_{i=0}^{n} a_i z^{-i}}\right\} \tag{3}$$

and

$$t(z) = -\frac{1}{2}\sum_{i=1}^{n}\left(\frac{z}{z - z_i^{-1}} + \frac{z^{-1}}{z^{-1} - z_i^{-1}}\right).$$

This last expression can be rewritten in the form

$$t(z) = -\frac{n}{2} + \frac{1}{2}\sum_{i=1}^{n}\frac{(z_i - z_i^{-1})/2}{(z + z^{-1})/2 - (z_i + z_i^{-1})/2} \tag{4}$$

which suggests [5] the use of the transformation of variable $w = (z + z^{-1})/2$ mapping the unit circle of the z plane onto the interval $[-1, +1]$ of the real axis of the w plane. If one denotes by w_i the transformed value of z_i, the group delay becomes in terms of the new variable

$$T(w) = -\frac{n}{2} + \frac{1}{2}\sum_{i=1}^{n}\frac{\sqrt{w_i^2 - 1}}{w - w_i}$$

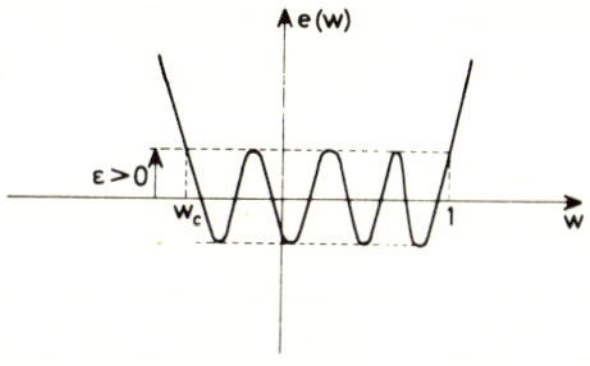

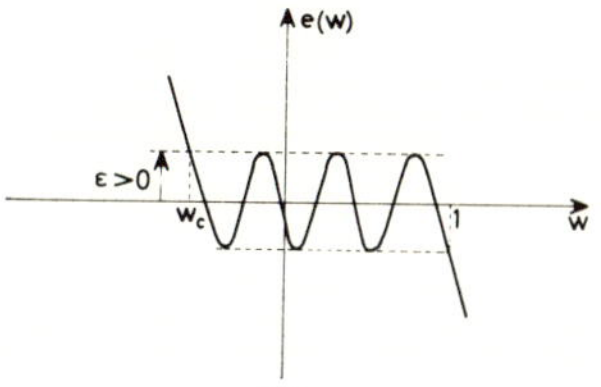

Fig. 1. Optimal error characteristics. (a) n even. (b) n odd.

where the sign of the square root is uniquely determined by the stability condition

$$|z_i| = |w_i + \sqrt{w_i^2 - 1}| < 1. \tag{5}$$

The problem can now be stated in the following terms: in the low-pass interval $[0, \omega_c]$, a constant delay τ must be approximated by $T(w)$ according to the equal-ripple norm or, in other words, so that the error

$$e(w) = \tau + \frac{n}{2} - \frac{1}{2}\sum_{i=1}^{n}\frac{\sqrt{w_i^2 - 1}}{w - w_i}$$

alternates [6, p. 54] in $[1, w_c = \cos \omega_c]$ the maximum number of times, i.e., n since $e(w)$ is a rational fraction of degree n in w. The optimal error is sketched on Fig. 1 for positive ripples, if it is agreed that the ripple ϵ is positive with the deviation at w_c.

III. Equal-Ripple Conditions

As is well known [7], [8], given $g(p)$ any strictly Hurwitzian polynomial of degree n, the rational fraction of degree n in p^2

$$R(p^2) = \frac{1}{2}\left[\frac{g(p)}{g(-p)} + \frac{g(-p)}{g(p)}\right]$$

alternates n times with a unit ripple on the whole negative real axis. In fact, the same result holds for $g(p)$ anti-Hurwitzian which amounts to a mere permutation between $g(p)$ and $g(-p)$, but if $g(p)$ had roots in both left and right half-planes, the number of alternations would be reduced.

Now, by using the transformation [9]

$$p^2 = \frac{w - w_c}{w - 1} \tag{6}$$

which maps the whole negative real axis of p^2 onto the low-pass interval $[w_c, 1]$, one gets the desired equal-ripple error in the w plane by

$$e(w) \equiv \epsilon R\left(\frac{w - w_c}{w - 1}\right). \tag{7}$$

Since $R(0)$ is equal to $+1$, $e(w_c)$ is positive with ϵ in accordance with Fig. 1.

Identifying the residues of both members of (7) at the poles w_i gives

$$-\frac{1}{2}\sqrt{w_i^2 - 1} = \epsilon \operatorname*{res}_{w=w_i}\left[R\left(\frac{w - w_c}{w - 1}\right)\right],$$

$$i = 1, 2, \cdots, n. \tag{8}$$

As is easily seen, the roots of $g(p)$, denoted by p_i $(i=1, 2, \cdots, n)$, correspond to the poles w_i so that one has

$$\sqrt{w_i^2 - 1} = \sqrt{2(1 - w_c)}\,\frac{\sqrt{p_i^2 - (w_c + 1)/2}}{p_i^2 - 1} \tag{9}$$

and, from the expression $\{d[1/f(t)]/dt\}_{t=t_0}^{-1}$ of the residue of a function $f(t)$ at a simple pole t_0,

$$\operatorname*{res}_{w=w_i}\left[R\left(\frac{w - w_c}{w - 1}\right)\right] = \operatorname*{res}_{p^2=p_i^2}[R(p^2)]\left(\frac{dw}{dp^2}\right)_{p^2=p_i^2}. \tag{10}$$

By means of (9) and (10) together with

$$\operatorname*{res}_{p^2=p_i^2}[R(p^2)] = -2p_i^2\prod_{\substack{j=1\\ \neq i}}^{n}\frac{p_j + p_i}{p_j - p_i}$$

$$\left(\frac{dw}{dp^2}\right)_{p^2=p_i^2} = -\frac{1 - w_c}{(p_i^2 - 1)^2}$$

(8) can be rewritten in the form

$$\sqrt{p_i^2-(w_c+1)/2}+2\epsilon\sqrt{2(1-w_c)}\,\frac{p_i^2}{p_i^2-1}\prod_{\substack{j=1\\ \neq 1}}^{n}\frac{p_j+p_i}{p_j-p_i}=0,$$

$$i = 1, 2, \cdots, n. \tag{11}$$

The sign of $[p_i^2-(\omega_c+1)/2]^{1/2}$ is fixed according to (5) transformed in the new plane:

$$\left|p_i^2 - w_c + \sqrt{2(1 - w_c)}\sqrt{p_i^2 - (w_c + 1)/2}\right| < \left|p_i^2 - 1\right|. \tag{12}$$

Finally, the identification of both members of (7) is completed by equating their values at $w=\infty$:

$$\tau + \frac{n}{2} = \frac{\epsilon}{2}\left(\prod_{i=1}^{n}\frac{p_i - 1}{p_i + 1} + \prod_{i=1}^{n}\frac{p_i + 1}{p_i - 1}\right). \tag{13}$$

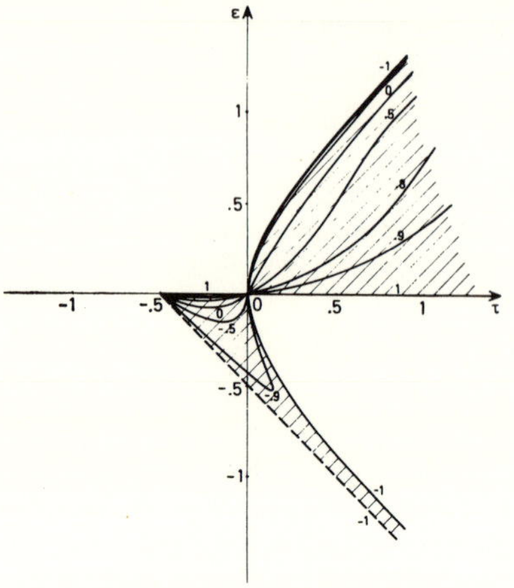

Fig. 2. Admissible values of ϵ and τ for $n=1$ and w_c ranging from -1 to $+1$.

Equations (11) and (13) together with conditions (12) form a system of $(n+1)$ equations in the roots p_i $(i=1, 2, \cdots, n)$ and one of the three parameters (ϵ, τ, w_c), the two others being freely chosen. For $|w_c| \leq 1$, any strictly Hurwitzian solution $g(p)$, whose roots are simple and do not belong to the transformed stopband $[-\sqrt{(1+w_c)/2}, 0]$, will yield a realizable equal-ripple delay filter.

IV. Explicit Solution for $n=1$

For $n=1$, the equal-ripple conditions are readily deduced from (4)

$$\tau + \epsilon = -\frac{1}{2} + \frac{1}{2}\,\frac{(z_1 - z_1^{-1})/2}{1 - (z_1 + z_1^{-1})/2}$$

$$\tau - \epsilon = -\frac{1}{2} + \frac{1}{2}\,\frac{(z_1 - z_1^{-1})/2}{w_c - (z_1 + z_1^{-1})/2}.$$

From the former condition it follows that $z_1=(\tau+\epsilon)/(\tau+\epsilon+1)$, and inserting this value into the latter one gets

$$w_c = 1 - \frac{\epsilon}{(\tau + \epsilon)(\tau + \epsilon + 1)(\tau - \epsilon + 1/2)}.$$

Obviously, the case $\epsilon=0$ corresponds to the maximally flat delay filter; i.e., $w_c=1$ or $\omega_c=0$ and $z_1=\tau/(\tau+1)$ [1]. The stability condition $|z_1|<1$ implies that $\tau+\epsilon+1/2$ must be strictly positive. On the other hand, the locus $\omega_c=\pi$ or $w_c=-1$ given by

$$(\tau + \epsilon)(\tau + \epsilon + 1)(\tau - \epsilon + 1/2) - \epsilon/2 = 0$$

can be factorized into

$$(\tau + \epsilon + 1/2)(\tau^2 + \tau - \epsilon^2) = 0.$$

Therefore, one easily verifies that the values of τ and ϵ yielding stable solutions ($|z_1|<1$) and admissible cutoff frequencies ($|w_c| \leq 1$) must lie in the shaded areas of Fig. 2.

V. Admissible Values of ϵ and τ for Arbitrary n

In order to generalize Fig. 2 for arbitrary degrees, one needs the locus corresponding to the extreme values of ω_c, i.e., 0 and π.

On the one hand, for $\omega_c=0$ or $\epsilon=0$, it has been proved in [1] that for any n the range of values of τ leading to stable maximally flat delay filters is the interval $]-1/2, +\infty[$.

On the other hand, for $\omega_c=\pi$, (11) reduces to

$$p_i\left(1 + 4\epsilon \frac{p_i}{p_i^2 - 1} \prod_{\substack{j=1 \\ \neq i}}^{n} \frac{p_j + p_i}{p_j - p_i}\right) = 0, \qquad i = 1, 2, \cdots, n$$

where the left member is the product of two factors. If $\omega_c=\pi$, $p_i=0$ corresponds to a root z_i at $1-$, and for a stable solution one has to choose the second factor which can be rewritten as

$$-\frac{1}{\epsilon} = \frac{4p_i}{p_i^2 - 1} \prod_{\substack{j=1 \\ \neq i}}^{n} \frac{p_j + p_i}{p_j - p_i}, \qquad i = 1, 2, \cdots, n \tag{14}$$

and after summation over all i becomes

$$-\frac{n}{\epsilon} = \sum_{i=1}^{n} \frac{4p_i}{p_i^2 - 1} \prod_{\substack{j=1 \\ \neq i}}^{n} \frac{p_j + p_i}{p_j - p_i}. \tag{15}$$

From (13), one easily deduces

$$2\tau + n + 2\epsilon = \epsilon \frac{\left[\prod_{i=1}^{n} (p_i + 1) + \prod_{i=1}^{n} (p_i - 1)\right]^2}{\prod_{i=1}^{n} (p_i^2 - 1)}$$

$$2\tau + n - 2\epsilon = \epsilon \frac{\left[\prod_{i=1}^{n} (p_i + 1) - \prod_{i=1}^{n} (p_i - 1)\right]^2}{\prod_{i=1}^{n} (p_i^2 - 1)}$$

so that one has

$$(2\tau + n + 2\epsilon)(2\tau + n - 2\epsilon) = \left[\epsilon\left(\prod_{i=1}^{n} \frac{p_i + 1}{p_i - 1} - \prod_{i=1}^{n} \frac{p_i - 1}{p_i + 1}\right)\right]^2. \tag{16}$$

Consider now the function

$$f(p) = \prod_{i=1}^{n} \frac{p + p_i}{p - p_i} - \prod_{i=1}^{n} \frac{p - p_i}{p + p_i}.$$

The partial fraction expansion of each term gives

$$f(p) = \sum_{i=1}^{n} 2p_i \prod_{\substack{j=1 \\ \neq i}}^{n} \frac{p_i + p_j}{p_i - p_j}\left(\frac{1}{p - p_i} + \frac{1}{p + p_i}\right)$$

and by evaluating both expressions of $f(p)$ at $p=1$, one gets the identity

$$\prod_{i=1}^{n} \frac{1 + p_i}{1 - p_i} - \prod_{i=1}^{n} \frac{1 - p_i}{1 + p_i} \equiv \sum_{i=1}^{n} \frac{4p_i}{1 - p_i^2} \prod_{\substack{j=1 \\ \neq i}}^{n} \frac{p_i + p_j}{p_i - p_j}. \tag{17}$$

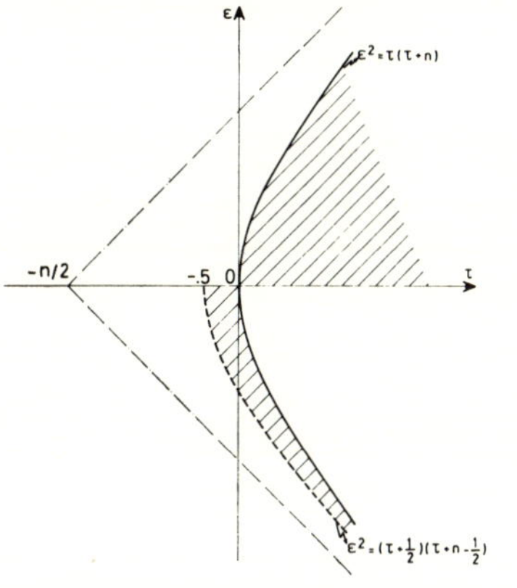

Fig. 3. Admissible values of ϵ and τ for arbitrary n.

Hence by means of (15)–(17) the locus is

$$(2\tau + n + 2\epsilon)(2\tau + n - 2\epsilon) = n^2$$

which, after simplification, turns out to be the hyperbola $\epsilon^2=\tau(\tau+n)$.

The second locus, which corresponds to the stability boundary drawn in the dashed line in Fig. 2, is obtained by taking one root, e.g., p_n, equal to zero while keeping the other $(n-1)$ equations (14). The equal-ripple conditions then become

$$\frac{1}{\epsilon} = \frac{4p_i}{p_i^2 - 1} \prod_{\substack{j=1 \\ \neq i}}^{n-1} \frac{p_j + p_i}{p_j - p_i}, \qquad i = 1, 2, \cdots, n - 1$$

$$\tau + \frac{n}{2} = -\frac{\epsilon}{2}\left(\prod_{i=1}^{n-1} \frac{p_i - 1}{p_i + 1} + \prod_{i=1}^{n-1} \frac{p_i + 1}{p_i - 1}\right).$$

Consequently, the second locus is readily deduced from the first one by replacing n, τ, and ϵ, respectively, by $n-1$, $\tau+1/2$, and $-\epsilon$, thereby yielding $\epsilon^2=(\tau+1/2)(\tau+n-1/2)$. To summarize, the allowable values of ϵ and τ lie in the shaded regions indicated in Fig. 3.

VI. Explicit Solution for $\omega_c=\pi$

For $\omega_c=\pi$, the passband interval corresponds to the whole unit circle of the z plane and, as could be expected, the z and p variables are related by the bilinear transformation $p=(1+z^{-1})/(1-z^{-1})$. Therefore, from the expression (3) of the group delay, the equal-ripple condition (7) becomes

$$\tau + \frac{1}{2}\left\{\frac{\sum_{i=0}^{n} a_i i z^i}{\sum_{i=0}^{n} a_i z^i} + \frac{\sum_{i=0}^{n} a_i i z^{-i}}{\sum_{i=0}^{n} a_i z^{-i}}\right\} \equiv \frac{\epsilon}{2}\left[\frac{g(p)}{g(-p)} + \frac{g(-p)}{g(p)}\right]_{p=(1+z^{-1})/(1-z^{-1})}$$

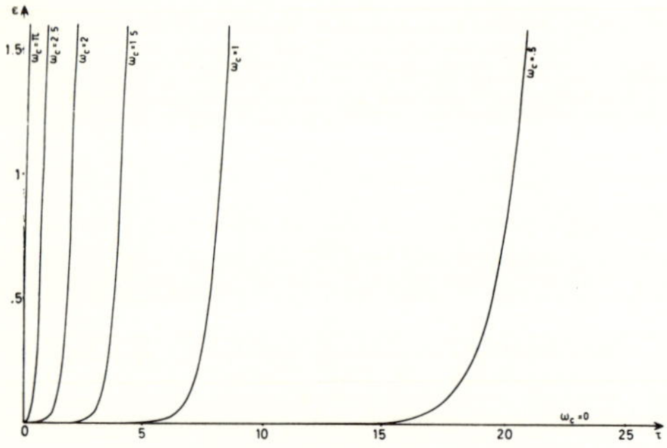

Fig. 4. Values of τ versus positive ϵ for $n=10$ and ω_c ranging from 0 to π.

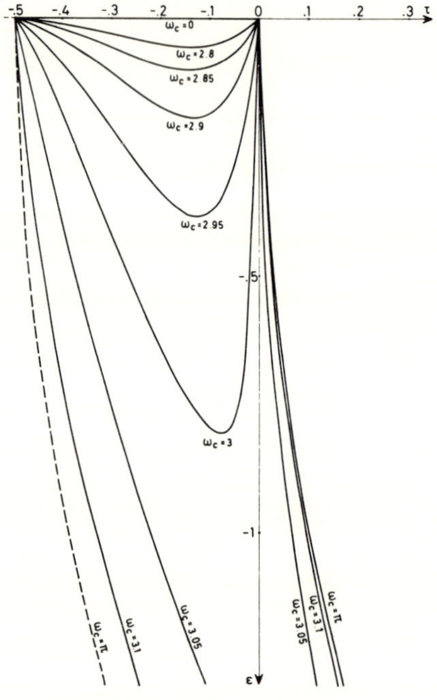

Fig. 5. Values of τ versus negative ϵ for $n=10$ and ω_c ranging from 0 to π.

which can be rewritten in the form

$$\frac{\sum_{i=0}^{n}[a_i(\tau+i)/\epsilon]z^{i}}{\sum_{i=0}^{n}a_i z^{i}}+\frac{\sum_{i=0}^{n}[a_i(\tau+i)/\epsilon]z^{-i}}{\sum_{i=0}^{n}a_i z^{-i}}$$

$$\equiv\frac{(-z)^{n}G(z^{-1})}{G(z)}+\frac{(-z^{-1})^{n}G(z)}{G(z^{-1})}$$

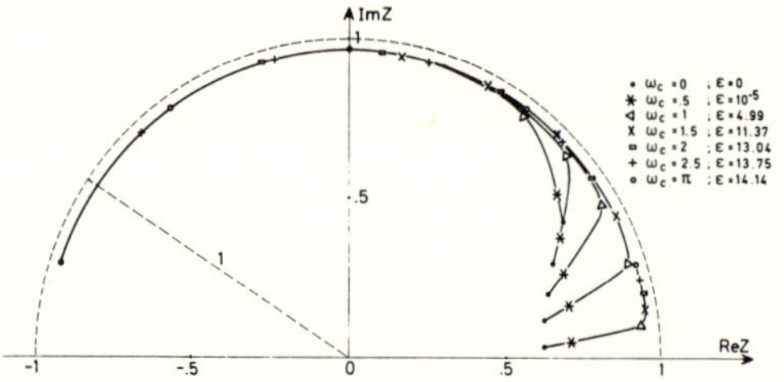

Fig. 6. Pole locus of the tenth-degree filter for $\tau=10$ and ω_c ranging from 0 to π.

where $G(z^{-1})=(1-z^{-1})^{n}g[(1+z^{-1})/(1-z^{-1})]$ has all its roots outside the unit circle if $g(p)$ is strictly Hurwitzian. The only way of obtaining a stable solution is thus to identify $G(z^{-1})$ with $\sum_{i=0}^{n} a_i z^{-i}$ so that one has

$$\sum_{i=0}^{n}[a_i(\tau+i)/\epsilon]z^{-i}\equiv(-z^{-1})^{n}G(z)$$

or

$$a_i(\tau+i)/\epsilon=(-1)^{n}a_{n-i}, \qquad i=0,1,\cdots,n.$$

Since a_n is normalized to unity, one gets $a_0=(-1)^{n}\epsilon/\tau$ and $(-1)^{n}(\tau+n)/\epsilon$ for $i=0$ and n, respectively, and the locus of stable solutions for $\omega_c=\pi$ is accordingly the hyperbola $\epsilon^2=\tau(\tau+n)$ as shown in Section V. All other coefficients are obviously zero, and the denominator of the equal-ripple delay transmittance turns out to be

$$p(z^{-1})=z^{-n}+(-1)^{n}\epsilon/\tau.$$

VII. Iterative Solution in the General Case and Numerical Results

For $n\neq1$ and $\omega_c\neq0$ or π, a closed-form expression of the equal-ripple delay filters could not be obtained. Therefore, a generalized Newton–Raphson procedure was implemented for solving the nonlinear equations (11) and (13). In order to reduce the system to only equation (11), it is preferable to choose ϵ and w_c as free parameters so that one can deduce τ from (13) after solving system (11). Obviously, the maximally flat delay filter is not an explicit solution of (11) since transformation (6) becomes meaningless for $\omega_c=0$ or $w_c=1$. Yet, the solution given in Section VI for $\omega_c=\pi$ is a starting point for generating the whole equal-ripple family when ω_c is decreased.

As an illustration for $n=10$, Figs. 4 and 5 give some details of the general Fig. 3 for positive and negative ripples, respectively. In particular, one notices in Fig. 4 that for any cutoff frequency it is possible to achieve a small error tolerance provided the approximated delay is less than some upper limit: the narrower the bandwidth, the larger this upper bound. On the other hand, Fig. 5 shows how the range $]-1/2, 0[$ of negative delays yielding stable maximally flat delay filters [1] is transformed when ω_c is increased from 0 to π. The pole locus of the tenth-degree filter is plotted in Fig. 6 for $\tau=10$ and ω_c ranging from 0 to π: the

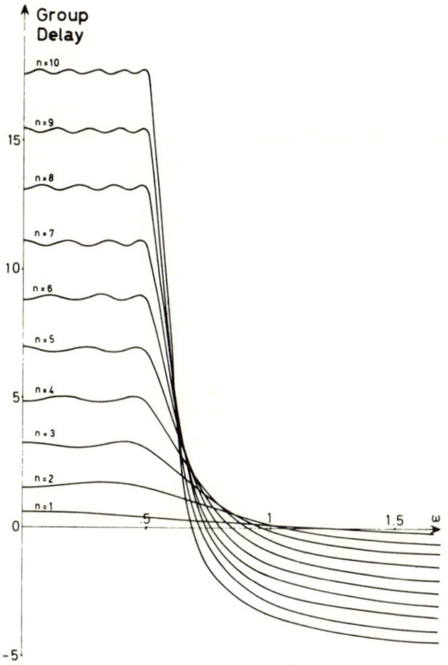

Fig. 7. Delay characteristics for $\omega_c = 0.5$, $\epsilon = 0.1$ and n ranging from 1 to n.

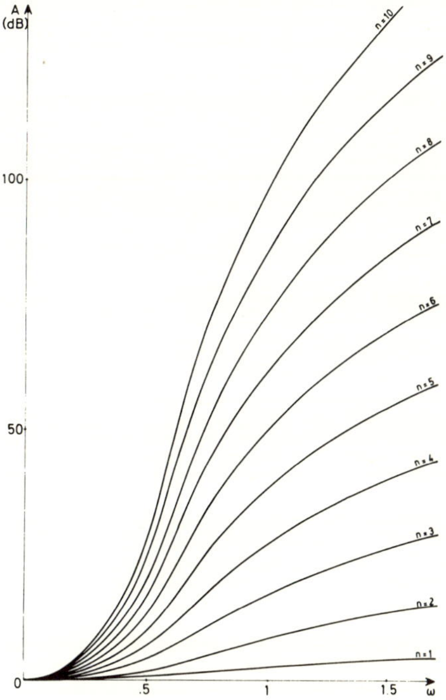

Fig. 8. Attenuation characteristics for $\omega_c = 0.5$, $\epsilon = 0.1$ and n ranging from 1 to n.

boundaries of the locus correspond to the solution given in [1] for $\omega_c = 0$ and in Section VI for $\omega_c = \pi$. Finally, in the case of $\omega_c = 0.5$ and $\epsilon = 0.1$, Figs. 7 and 8 give the delay and attenuation characteristics of filters of degrees 1 to 10.

Acknowledgment

The author wishes to thank M. J. Praet for carrying out the numerical computations.

References

[1] J. P. Thiran, "Recursive digital filters with maximally flat group delay," this issue, pp. 659–664.

[2] E. Ulbrich and H. Piloty, "Über den Entwurf von Allpässen, Tiefpässen, und Bandpässen mit einer in Tschebyscheffschen Sinne approximierten konstanten Gruppenlaufzeit," *Arch. Elek. Übertragung*, vol. 14, Oct. 1960, pp. 451–467.

[3] T. A. Abele, "Übertragungsfaktoren mit Tschebyscheffscher Approximation konstanter Gruppenlaufzeit," *Arch. Elek. Übertragung*, vol. 16, Jan. 1962, pp. 9–18.

[4] M. Sablatash, "Approximation theory for digital filters," in *Dig. IEEE Int. Symp. Circuit Theory* (Atlanta, Ga.), Dec. 1970, pp. 141–142.

[5] R. Unbehauen, "Zur Synthese digitaler Filter," *Arch. Elek. Übertragung*, vol. 24, July–Aug. 1970, pp. 305–313.

[6] J. R. Rice, *The Approximation of Functions*, vol. 1. Reading, Mass.: Addison-Wesley, 1964.

[7] B. J. Bennett, "Synthesis of electric filters with arbitrary phase characteristics," *IRE Conv. Rec.*, pt. 5, 1953, pp. 19–26.

[8] G. Szentirmai, "A filter synthesis program," in *System Analysis by Digital Computer*, F. F. Kuo and J. F. Kaiser, Eds. New York: Wiley, 1966.

[9] A. G. Constantinides, "Digital filters with equiripple passbands," *IEEE Trans. Circuit Theory* (Corresp.), vol. CT-16, Nov. 1969, pp. 535–538.

9

Reprinted from *IEEE Trans. Audio Electroacoustics,* **AU-16**(3), 413–421 (1968)

An Approach to the Implementation of Digital Filters

LELAND B. JACKSON, Member, IEEE
JAMES F. KAISER, Associate Member, IEEE
HENRY S. McDONALD, Member, IEEE
Bell Telephone Laboratories, Inc.
Murray Hill, N. J.

Abstract

An approach to the implementation of digital filters is presented that employs a small set of relatively simple digital circuits in a highly regular and modular configuration, well suited to LSI construction. Using parallel processing and serial, two's-complement arithmetic, the required arithmetic circuits (adders and multipliers) are quite simple, as are the remaining circuits, which consist of shift registers for delay and small read-only memories for coefficient storage. The arithmetic circuits are readily multiplexed to process multiple data inputs or to effect multiple, but different, filters (or both), thus providing for efficient hardware utilization. Up to 100 filter sections can be multiplexed in audio-frequency applications using presently available digital circuits in the medium-speed range. The filters are also easily modified to realize a wide range of filter forms, transfer functions, multiplexing schemes, and round-off noise levels by changing only the contents of the read-only memory and/or the timing signals and the length of the shift-register delays. A simple analog-to-digital converter, which uses delta modulation as an intermediate encoding process is also presented for audio-frequency applications.

Manuscript received May 6, 1968.

Introduction

The basic theory underlying the analysis and design of digital filters is well advanced (although by no means complete) and quite a few summaries of theoretical results are now available in the engineering literature [1]–[4]. However, the impact of digital filtering theory has not yet been felt by most of the engineers and technicians who design and use the wide variety of filters presently constructed from RLC or crystal circuits. This has been due, in part, to a general unawareness of the possibilities of digital filtering and also, until recently, to the prohibitive complexity and cost of constructing most digital filters. Hence, digital filter implementation has been confined primarily to computer programs for simulation work or for processing relatively small amounts of data, usually not in real time. However, the rapid development of the integrated-circuit technology and especially the potential for large-scale integration (LSI) of digital circuits now promise to reverse this situation in many instances and to make many digital filters more attractive than their analog counterparts, from the standpoints of cost, size, and reliability.

In this paper, we present an approach to the physical implementation of digital filters which has the following features.

1) The filters are constructed from a small set of relatively simple digital circuits, primarily shift registers and adders.

2) The configuration of the digital circuits is highly modular in form and thus well suited to LSI construction.

3) The configuration of the digital circuits has the flexibility to realize a wide range of filter forms, coefficient accuraices, and round-off noise levels (i.e., data accuracies).

4) The digital filter may be easily multiplexed to process multiple data inputs or to effect multiple, but different, filters with the same digital circuits, thus providing for efficient hardware utilization.

After a brief review of general digital filter forms, the advantages of serial, two's-complement, binary arithmetic in the implementation of digital filters are discussed. The required arithmetic circuits (adder/subtractor, complementer, and multiplier) are then described, followed by the techniques for multiplexing. Finally, several examples are presented of multiplexed digital filters that have been constructed and tested. A description of a simple analog-to-digital converter for relatively low-frequency applications is also included.

Canonical Forms

The transfer characteristics of a digital filter are commonly described in terms of its z-domain transfer function [1],

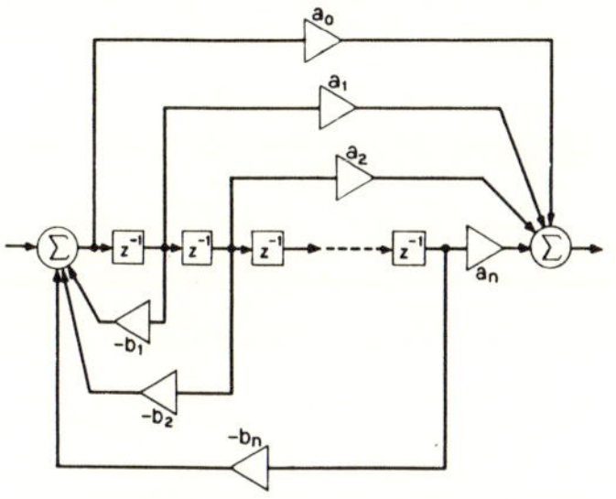

Fig. 1. The direct form for a digital filter.

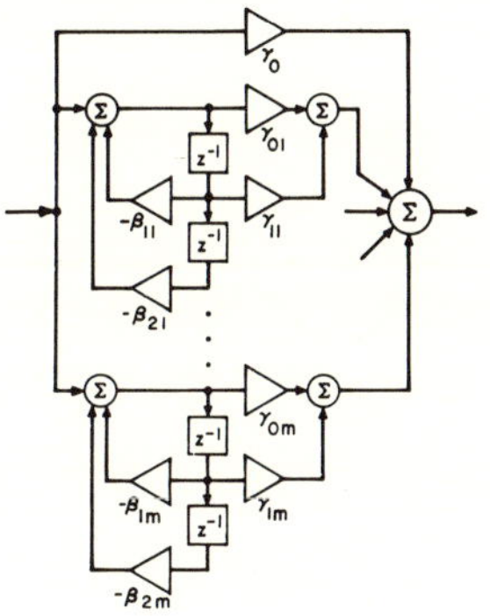

Fig. 3. The parallel form for a digital filter

Fig. 2. The cascade form for a digital filter.

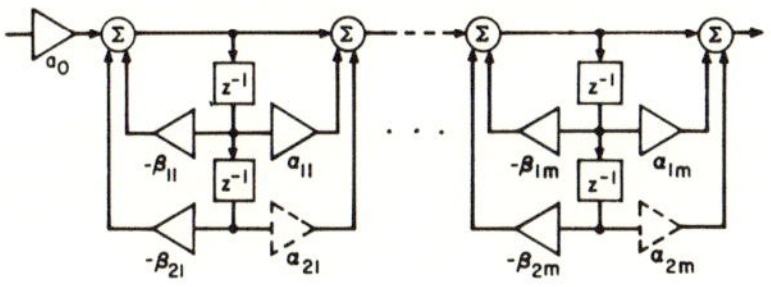

Fig. 4. (A) Second-order section for digital all-pass filter in cascade form. (B) Alternate configuration for digital all-pass filter in cascade form.

(A)

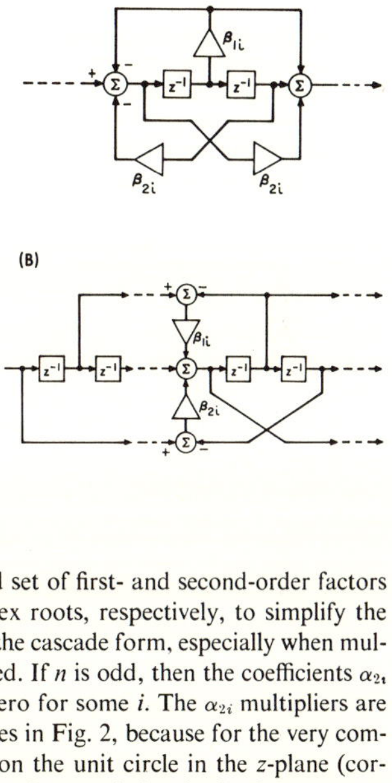

$$H(z) = \frac{\sum_{i=0}^{n} a_i z^{-i}}{1 + \sum_{i=1}^{n} b_i z^{-i}}, \qquad (1)$$

where z^{-1} is the unit delay operator. There are a multitude of equivalent digital circuit forms in which (1) may be realized, but three canonical forms, or variations thereof, are most often employed. These forms are canonical in the sense that a minimum number of adders, multipliers, and delays are required to realize (1) in the general case. The first of these forms, shown in Fig. 1, is a direct realization of (1) and as such is called the *direct form*. It has been pointed out by Kaiser [5] that use of the direct form is usually to be avoided because the accuracy requirements on the coefficients $\{a_i\}$ and $\{b_i\}$ are often severe. Therefore, although the implementation techniques presented here are applicable to any filter form, we will not specifically consider the direct form.

The second canonical form corresponds to a factorization of the numerator and denominator polynomials of (1) to produce an $H(z)$ of the form

$$H(z) = a_0 \prod_{i=1}^{m} \frac{\alpha_{2i} z^{-2} + \alpha_{1i} z^{-1} + 1}{\beta_{2i} z^{-2} + \beta_{1i} z^{-1} + 1}, \qquad (2)$$

where m is the integer part of $(n+1)/2$. This is the *cascade form* for a digital filter, depicted in Fig. 2. Second-order actors (with real coefficients) have been chosen for (2) rather than a mixed set of first- and second-order factors for real and complex roots, respectively, to simplify the implementation of the cascade form, especially when multiplexing is employed. If n is odd, then the coefficients α_{2i} and β_{2i} will equal zero for some i. The α_{2i} multipliers are shown in dotted lines in Fig. 2, because for the very common case of zeros on the unit circle in the z-plane (corresponding to zeros of transmission in the frequency response of the filter) the associated α_{2i} coefficients are unity. Thus, for these α_{2i} coefficients, no multiplications are actually required.

The third canonical form is the *parallel form*, shown in Fig. 3, which results from a partial faction expansion of

(1) to produce

$$H(z) = \gamma_0 + \sum_{i=1}^{m} \frac{\gamma_{1i}z^{-1} + \gamma_{0i}}{\beta_{2i}z^{-2} + \beta_{1i}z^{-1} + 1}, \tag{3}$$

where $\gamma_0 = a_n/b_n$ and we have again chosen to use all second-order (denominator) factors. Note that all three canonical forms are entirely equivalent with regard to the amount of storage required (n unit delays) and the number of arithmetic operations required ($2n+1$ multiplications and $2n$ additions per sampling period). As previously noted, however, the cascade form requires significantly fewer multiplications for zeros on the unit circle and is thus especially appropriate for filters of the band-pass and the band-stop variety (including low-pass and high-pass filters).

Another interesting filter form may be derived for the special case of an all-pass filter (APF), i.e., a filter or "equalizer" with unity gain at all frequencies. The transfer function for a discrete APF has the general form [6]

$$H_A(z) = \frac{\sum_{i=0}^{n} b_{n-i}z^{-i}}{\sum_{i=0}^{n} b_i z^{-i}}. \tag{4}$$

Thus, with $a_i = b_{n-i}$ and $b_0 = 1$, the direct form can be used to implement (4). However, to reduce the accuracy requirements on the filter coefficients, a modified cascade form can be derived for the APF, corresponding to an $H_A(z)$ of the form

$$H_A(z) = \prod_{i=1}^{m} \frac{z^{-2} + \beta_{1i}z^{-1} + \beta_{2i}}{\beta_{2i}z^{-2} + \beta_{1i}z^{-1} + 1}. \tag{5}$$

Second-order sections for the cascade form of the APF are shown in Fig. 4. Fig. 4(A) is a straightforward modification of the standard cascade form in Fig. 2. Note that because the β_{1i} multiplier may be shared by both the feed-forward and feedback paths, only three multiplications are required per second-order section rather than four. The number of multiplications may be further reduced by using the form of Fig. 4(B), which requires only two multiplications per second-order section. But now, two additional delays are required preceding the first second-order section to supply appropriately delayed inputs to the first section. Therefore, the cascade form of Fig. 4(B) requires a total of n multiplications and $n+2$ delays for an nth-order APF.

Serial Arithmetic

Using any of the canonical forms described in the preceding section, all of the coefficient multiplications and many of the additions during a given Nyquist interval may be performed simultaneously. Therefore, a high degree of parallel processing is possible in the implementation of a digital filter and this may be achieved by providing multiple adders and multipliers with appropriate interconnections. Economy is then realized by using serial arithmetic, and by sharing the adders and multipliers (using the multiplexed circuit configurations to be described) insofar as circuit speed will allow.

In addition to a significant simplification of the hardware, serial arithmetic provides for an increased modularity and flexibility in the digital circuit configurations. Also, the processing rate is limited only by the speed of the basic digital circuits and not by carry-propagation times in the adders and multipliers. Finally, with serial arithmetic, sample delays are realized simply as single-input single-output shift registers.

The two's-complement representation [7] of binary numbers is most appropriate for digital filter implementation using serial arithmetic because additions may proceed (starting with the least significant bits) with no advance knowledge of the signs or relative magnitudes of the numbers being added (and with no later corrections of the obtained sums as with one's complement). We will assume a two's-complement representation of the form

$$\delta_0 \cdot \delta_1 \delta_2 \cdots \delta_{N-1}, \tag{6}$$

which represents a number (δ) having a value of

$$\delta = -\delta_0 + \sum_{i=1}^{N-1} \delta_i 2^{-i}, \tag{7}$$

where each δ_i is either 0 or 1. Thus, the data is assumed to lie in the interval

$$-1 \le \delta < 1, \tag{8}$$

with the sign of the number δ being given by the last bit (in time) δ_0.

An extremely useful property of two's-complement representation is that in the addition of more than two numbers, if the magnitude of the correct total sum is small enough to allow its representation by the N available bits, then the correct total sum will be obtained regardless of the order in which the numbers are added, even if an overflow occurs in some of the partial sums. This property is illustrated in Fig. 5, which depicts numbers in two's-complement representation as being arrayed in a circle, with positive full scale ($1-2^{-N+1}$) and negative full scale (-1) being adjacent. The addition of positive addends produces counterclockwise rotation about the circle, whereas negative addends produce clockwise rotation. Thus, if the correct total sum satisfies (8), no information is lost with positive or negative overflows and the correct total sum will be obtained.

This overflow property is important for digital filter implementation because the summation points in the filters often contain more than two inputs (see Fig. 3);

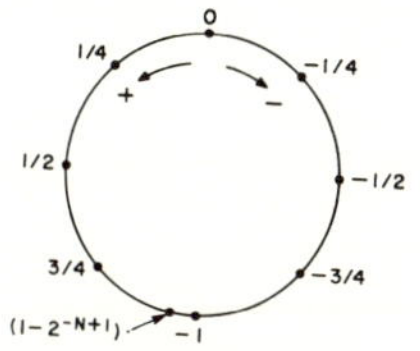

Fig. 5. Illustration of overflow property in two's-complement binary representation.

Fig. 6. Serial two's-complement adder.

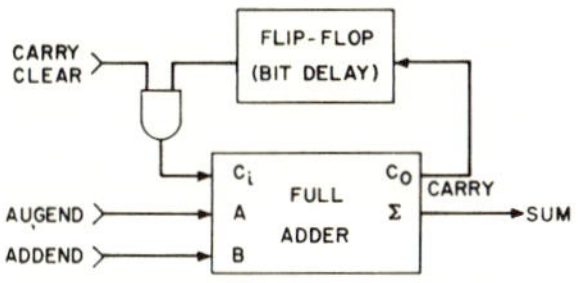

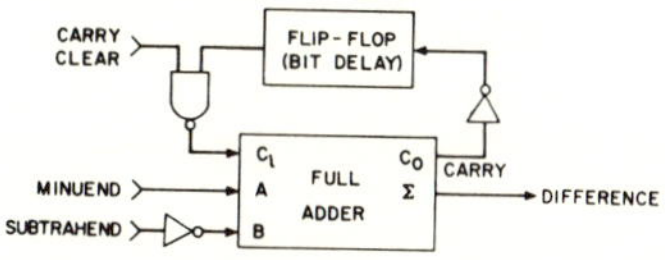

Fig. 7. Serial two's-complement subtractor.

Fig. 8. Serial two's complementer.

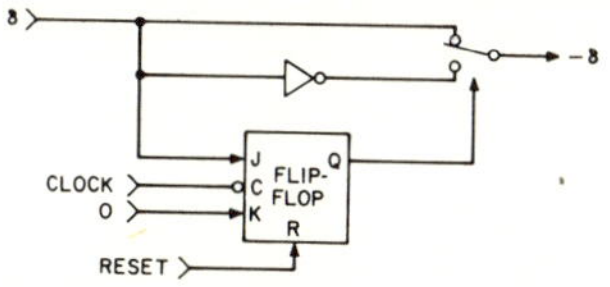

although it may be possible to argue that because of gain considerations the output of the summation point cannot overflow, there is no assurance that an overflow will not occur in the process of performing the summation. Note that this property also applies when one of the inputs to the summation has itself overflowed as a result of a multiplication by a coefficient of magnitude greater than one.

Arithmetic Unit

The three basic operations to be realized in the implementation of a digital filter are delay, addition (or subtraction), and multiplication. As previously mentioned, serial delays (z^{-1}) are realized simply as single-input single-output shift registers. Realizations for a serial adder (subtractor) and a serial multiplier are described in this section. The adders and multipliers, including their interconnections, will be said to comprise the arithmetic unit of the digital filter.

A serial adder for two's-complement addition is extremely simple to construct [7]. As shown in Fig. 6, it consists of a full binary adder with a single-bit delay (flip-flop) to transfer the carry output back to the carry input. A gate is also required in the carry feedback path to clear the carry to zero at the beginning of each sample. Accordingly, the carry-clear input is a timing signal, which is zero during the least significant bit of each sample and is one otherwise.

A serial two's-complement subtractor is implemented by first complementing (negating) the subtrahend input and then adding the complemented subtrahend to the minuend. To complement a number in two's-complement representation, each bit of the representation is inverted and a one is then added to the least significant bit of the inverted representation (i.e., 2^{-N+1} is added to the inverted number). The corresponding serial subtractor circuit is shown in Fig. 7. The subtrahend is inverted and a one is added to the least significant bit by clearing the initial carry bit to one, rather than to zero as in the adder. This is accomplished by means of two inverters in the carry feedback path, as shown.

A separate two's complementer (apart from a subtractor) may also be constructed; such circuits are required in the multiplier to be described. This operation is implemented with a simple sequential circuit which, for each sample, passes unchanged all initial (least significant) bits up to and including the first "1" and then inverts all succeeding bits. A corresponding circuit is depicted in Fig. 8.

A serial multiplier may be realized in a variety of configurations, but a special restriction imposed by this implementation approach makes one configuration most appropriate. This restriction is that no more than N bits per sample may be generated at any point in the digital network because successive samples are immediately adjacent in time and there are no "time slots" available for more than N bits per sample. Hence, the full ($N+K$)-bit product of the multiplication of an N-bit sample by a K-bit (fractional) coefficient may not be accumulated before rounding is performed. However, using the multiplication scheme described below, it is possible to obtain the same rounded N-bit product without ever generating more than N bits per sample. Rounding is usually preferable, rather than truncation, to limit the introduction of extraneous low-frequency components (dc drift) into the filter.

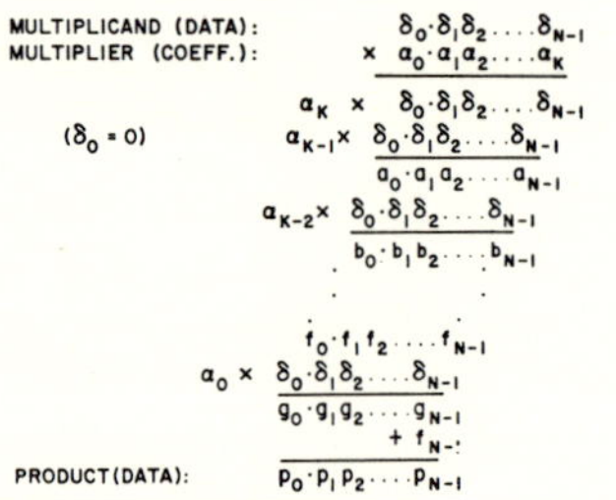

Fig. 9. Serial multiplication using no more than N bits per data word.

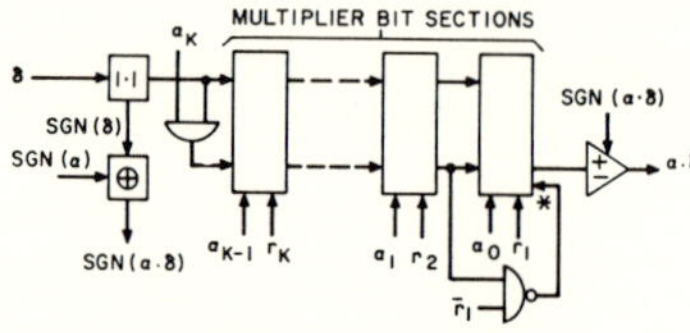

Fig. 10. Serial multiplier, showing modularity.

Fig. 11. A multiplier bit section.

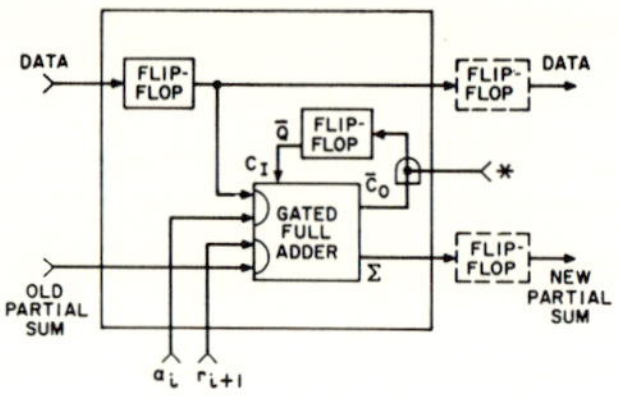

The serial multiplication scheme is depicted in Fig. 9. To simplify the required hardware, both the multiplicand (data) and the multiplier (coefficient) are constrained to be positive, with appropriate sign changes being made before and after the multiplication. Thus $\delta_0=0$ in Fig. 9 and the sign of the multiplicand (δ) is stored separately as SGN δ. For convenience, the multiplier (α) will be assumed to lie in the interval.

$$-2 < \alpha < 2. \tag{9}$$

Although (9) is not necessarily applicable in the general case, it does hold for the denominator coefficients of the cascade and parallel forms, and usually for the numerator coefficients of the cascade form as well. The magnitude of the multiplier is thus represented in Fig. 9 as

$$\alpha_0 \cdot \alpha_1\alpha_2 \cdots \alpha_K, \tag{10}$$

which represents a value of

$$|\alpha| = \sum_{i=0}^{K} \alpha_i 2^{-i}. \tag{11}$$

The restriction in (9) and the resulting representation in (10) and (11) are in no way essential to the serial multipliers to be described, but are meant only to be representative of the multiplication scheme. The sign of the multiplier is also stored separately as SGN α.

The multiplication scheme in Fig. 9 proceeds as follows. The multiplicand is successively shifted (delayed) and multiplied (gated) by the appropriate bit (α_i) of the multiplier. These delayed gated instances of the multiplicand are then added in the order indicated. After each addition (including the "addition" of $\alpha_K\delta$ to 0), the least significant bit of the resulting partial sum (i.e., $\delta_{N-1}, a_{N-1}, b_{N-1}, \cdots, f_{N-1}$) is truncated to prevent the succeeding partial sum from exceeding N bits in length. Note that these bits may be truncated because the full unrounded product would be

$$g_0 \cdot g_1 g_2 \cdots g_{N-1} f_{N-1} \cdots b_{N-1} a_{N-1} \delta_{N-1} \tag{12}$$

and to round (12) to N bits, only the value of the bit f_{N-1} is required. Thus, before truncating f_{N-1}, its value is stored elsewhere to be added in the final step to g, as shown in Fig. 9, to obtain the rounded product (p).

The serial multiplier corresponding to the scheme described above is shown in Fig. 10. The absolute value of each incoming datum (δ) is taken and its sign (SGN δ) is added modulo-2 to the coefficient sign (SGN α) to determine the product sign (SGN $\alpha\cdot\delta$). The (positive) multiplicand is then successively delayed and gated by the appropriate multiplier bits (α_i) and the partial sums are accumulated in the multiplier bit sections. A single multiplier bit section is shown in Fig. 11. The least significant bit of each partial sum is truncated (gated to zero) by the appropriate timing signal r_{i+1}. Rounding is accomplished by adding in the last truncated bit (f_{N-1}) via the * input to the last bit section. Finally, the sign of the product is inserted using a two's-complementer such as that in Fig. 8. At high data rates, it may be necessary to insert extra flip-flops between some or all of the multiplier bit sections, as shown in dotted lines in Fig. 11, to keep the propagation delay through the adder circuits from becoming excessive.

Several observations concerning the serial multiplier should be made at this point. First, there is a delay of K bits in going through the multiplier and this delay must be deducted from a delay (z^{-1}) that precedes the multiplier. (If the extra flip-flops in Fig. 11 are required, then the multiplier will yield a delay of up to $2K$ bits.) In addition, the absolute value operation at the first of the multiplier requires a delay of N bits (to determine the sign of each incoming datum) and this must be deducted from a preceding delay as well. Thus, to use this serial multiplier, the z^{-1} delays of the digital filter must be at least $N+K$ (or up to $N+2K$) bits in length. This in turn implies, as we shall see in the next section, that some form of multiplexing is required if the multipliers are to be implemented in this manner.

Another observation is that the adders in the multiplier bit sections do not require carry-clear inputs because only positive numbers are being added. However, output product overflows (in the sense of Fig. 5) are possible with coefficients (α) of magnitude greater than one. It may thus be necessary to restrict the amplitude of the data into certain multipliers to prevent output overflows; while in certain other multipliers, these overflows may be perfectly allowable as discussed in the preceding section. In general, however, the inputs to a summation must be scaled so that an overflow will not occur in the *final* output of the summation. Such overflows represent a severe nonlinearity in the system, and very undesirable effects can result in the output of the filter.

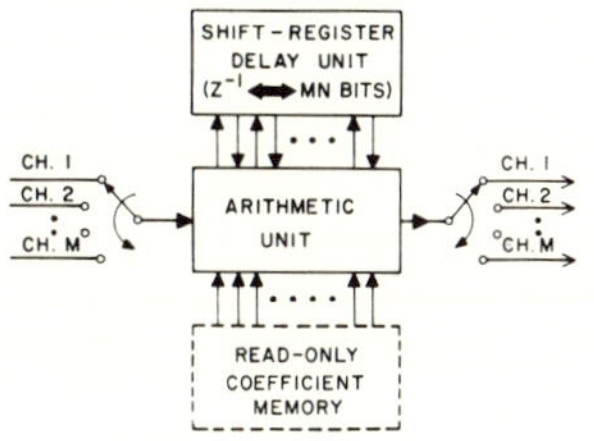

Fig. 12. Type-1 multiplexing for M input channels.

Multiplexing

Having realized the three basic digital filter components (delays, adders, and multipliers), the filter itself may be implemented by simply interconnecting these components in a configuration corresponding to one of the digital forms, canonical or otherwise, for the filter. However, if the input rate bit (sampling rate times bits per sample) is significantly below the capability of the digital circuits, the digital filter can be multiplexed to utilize the circuits more efficiently. The various multiplexing schemes are of two main types: 1) the multiplexed filter may operate upon a number of input signals simultaneously or 2) the multiplexed filter may effect a number of (different) filters for a single input signal. A combination of these two types is also possible.

To multiplex the filter to process M simultaneous inputs (type 1), the input samples from the M sources are interleaved sample by sample and fed (serially) into the filter. The bit rate in the filter is thus increased by a factor of M. The shift-register delays must also be increased by a factor of M to a length of MN bits. Otherwise, the filter is identical in its construction to the single-input case. In particular, the arithmetic unit containing the adders and multipliers is the same; it just operates M times faster. The output samples emerge in the same interleaved order as the input and are thus easily separated. Type-1 multiplexing is depicted in Fig. 12.

If the M channels in Fig. 12 are to be filtered differently or if type-2 multiplexing is also employed, the filter coefficients are stored in a separate read-only coefficient memory and are read-out as required by the multiplexed filter. A diode matrix provides a very fast and inexpensive form of read-only memory (ROM) for this purpose. If, however, all M channels are to be filtered identically and no type-2 multiplexing is employed, the coefficients may be wired into the multipliers and no ROM is then required. This is indicated by the dotted lines enclosing the ROM in Fig. 12. In this case, adders must be included only in those multiplier bit sections of the arithmetic unit for which the corresponding multiplier bits (α_i) equal one.

In many cases, a number of different, but similar, filters or subfilters are required for the same input signal. For example, all of the second-order subfilters comprising the cascade or parallel forms are similar in form, differing only in the values for the multiplying coefficients (see Figs. 2 and 3). Type-2 multiplexing refers to the implementation of these different (sub)filters with a signal multiplexed filter. An example of a multiplexed second-order filter is shown in Fig. 13. As with type-1 multiplexing, the combining of M separate filters into one multiplexed filter requires that the bit rate in the filter be increased by a factor of M and that the shift-register delays (z^{-1}) also be increased by a factor of M to MN bits in length. The coefficients are supplied from the read-only coefficient memory, which cycles through M values for each coefficient during every Nyquist interval. Data are routed in, around, and out of the filter by external routing switches, which are also controlled from the ROM.

As an example of type-2 multiplexing, consider the implementation of a 12th-order filter in cascade form, using the multiplexed second-order filter in Fig. 13. Here $M=6$, so the bit rate in the filter must be (at least) $6N$ bits per Nyquist interval. During the first N bits of each Nyquist interval, the input sample is introduced into and is processed by the arithmetic unit with the multiplying coefficients (α_1, α_2, β_1, β_2) of the first subfilter in the cascade form. The resulting output is delayed by N bits ($z^{-1/M}$) and fed back via the input routing switch to become the input to the filter during the second N-bit portion of the Nyquist interval. This feedback process is repeated four

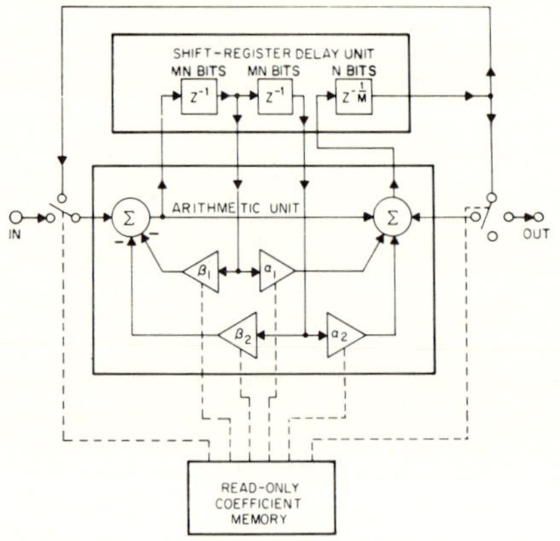

Fig. 13. General second-order filter for type-1 and type-2 multiplexing.

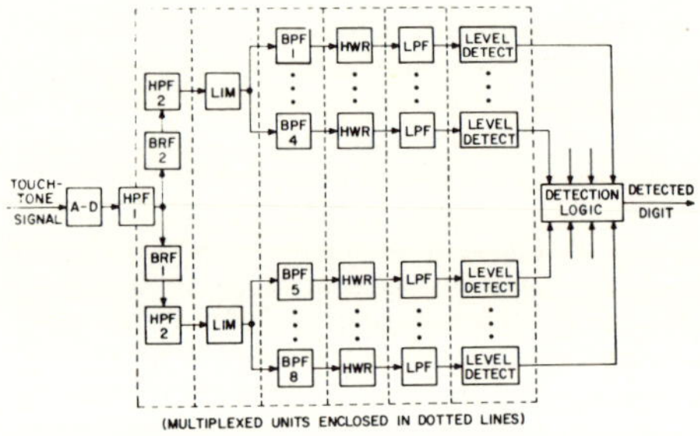

Fig. 14. Digital touch-tone receiver, showing multiplexed filters and nonlinear units.

more times, with the filter coefficients from the ROM being changed each time to correspond to the appropriate subfilter in the cascade form. The sixth (last) filter output during each Nyquist interval is the desired 12th-order filter output. The parallel form, or a combination of cascade and parallel filters, may be realized using the filter in Fig. 13 by simply changing the bits in the ROM which control the switching sequences of the input and output routing switches.

Sample System

As an example of this approach to the implementation of digital filters, we will take an experimental, all-digital touch-tone receiver (TTR) which has been designed and constructed at Bell Telephone Laboratories, Inc. The digital TTR is depicted in block-diagram form in Fig. 14. This is a straightforward digital version of the standard analog TTR described elsewhere [8]. Without going into the detailed operation of the system, we simply note that the combined high-pass filters (HPF's) are third order, the band-rejection filters (BRF's) are each sixth order, the bandpass filters (BPF's) are each second order, and the low-pass filters (LPF's) are each first order. The other signal-processing units required are the limiters (LIM's), half-wave rectifiers (HWR's), and level detectors. These nonlinear operations are, of course, easily implemented in digital form.

A multiplexing factor of $M=8$ is employed in the experimental TTR to combine all of the units enclosed in dotted lines into single multiplexed units. In particular, all of the HPF's and BRF's are multiplexed into one second-order filter (combined type-1 and type-2 multiplexing), the eight BPF's are multiplexed into another second-order filter (type-1 multiplexing with ROM coefficients), and the eight LPF's are multiplexed into one first-order section (type-1 multiplexing with wired-in coefficients). The nonlinear units are readily multiplexed as well and operate directly upon the interleaved output samples from the filters.

Some of the parameters of the experimental TTR design are as follows: the sampling rate is 10 K samples/second with an initial quantization (A/D conversion) of 7 bits/sample; the data word length (N) within the filter is 10 bits/sample; the filter coefficients have 6-bit fractional parts (K); and, as previously stated, the multiplexing factor (M) is eight. Thus, the bit rate within the filter (sampling rate × bits/sample × M) is 800 K bits/second. The number of bits required to represent the data and the coefficients of the TTR were determined through computer simulation of the system. The hardware required to implement this design consists primarily of about 40 serial adders and 400 bits of shift-register storage.

Analog-to-Digital Converter

In most applications of digital filters, the initial input signal is in analog form and must be converted to digital form for processing. It may or may not be necessary to reconvert the digital output signal to analog form, depending upon the application. Digital-to-analog (D/A) conversion is a relatively straightforward and inexpensive process, but the initial analog-to-digital (A/D) conversion is often quite a different situation. For audio-frequency applications, however, a simple and very accurate A/D converter may be implemented using delta modulation (Δ-mod) as an intermediate encoding process. This A/D converter will now be described.

The A/D converter is depicted in Fig. 15, with a Δ-mod encoder being shown in Fig. 16. The Δ-mod encoder produces a series of bivalued pulses (0's and 1's) which, when integrated, constitute an approximation to the input analog signal. The number of 1's (or 0's) occurring during each (eventual) sampling interval is accumulated in the counter as a measure of the change in signal amplitude during that interval. At the end of the interval, this number is transferred to the storage register and the counter is then reset to its initial value to begin counting during the next interval. The appropriate initial value for the counter is minus one-half the number of Δ-mod pulses per sampling interval.

The number stored in the storage register for each sampling interval is the difference between the desired sample value and the preceding sample value. Thus, if these difference samples are accumulated in a simple first-order accumulator, as shown, the full digital sample values result. A small "leak" is introduced into the accumulator by making the feedback gain slightly less than one ($1-2^{-K}$) to keep the dc gain of the accumulator from being infinite. This prevents a small dc bias in the Δ-mod output from generating an unbounded accumulator output. The accumulator leak should be matched by a similar leak in the integrator of the Δ-mod encoder.

Since the accumulator is itself a first-order digital filter, it can be implemented and multiplexed using the same circuits and techniques as previously described. The multiplexing may be either with other A/D conversion channels or with other filters, or both. Note, however, that if the digital filter following the A/D converter has, or can have, a zero at $z=1$ (corresponding to zero of transmission at dc), this zero would cancel the pole supplied by the accumulator (with no leak). Therefore, in this case, the accumulator may be eliminated from the A/D converter (along with the zero at $z=1$ from the following filter), making the A/D conversion even simpler.

The accuracy of A/D conversion implemented in this manner is a function of the following factors: 1) the ratio of the Δ-mod rate to the sampling rate, 2) the sensitivity of the input comparison amplifier in the Δ-mod encoder, and 3) the match between the accumulator leak (if an accumulator is required) and the integrator leak in the Δ-mod encoder. There is also a maximum-slope limitation with delta modulation and an accompanying slope overload noise results if this slope limitation is exceeded [9]. Assuming that the error resulting from 2) and 3) and from slope overload is negligible, a useful rule of thumb for the A/D conversion accuracy is that the number of quantization levels effected equals approximately the ratio of the Δ-mod rate to the sampling rate. Thus, for example, to effect 10-bit A/D conversion using this scheme, the Δ-mod rate must be approximately 1000 times the sampling rate.

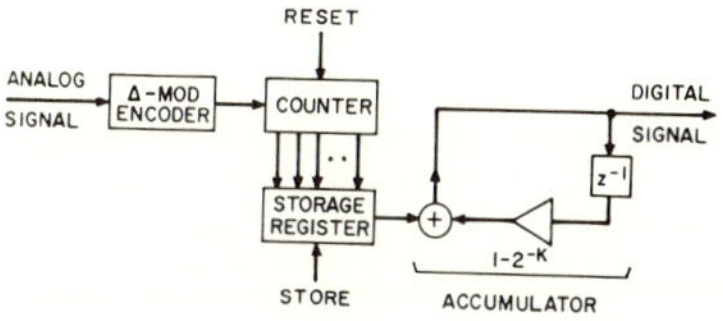

Fig. 15. Simple A/D converter using delta modulation.

Fig. 16. Delta-modulation encoder.

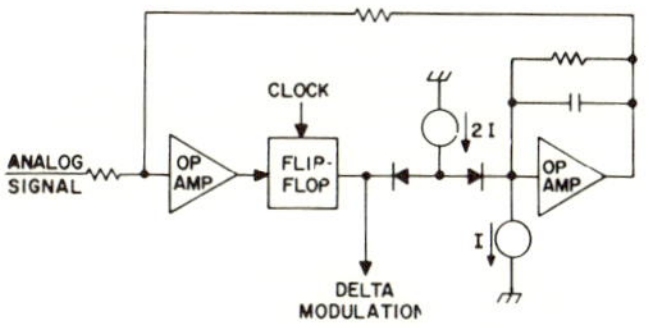

Conclusions

An approach to the implementation of digital filters has been described that employs a small set of relatively simple digital circuits in a highly regular and modular configuration, well suited to LSI construction. By using parallel processing and serial, two's-complement arithmetic, the required arithmetic circuits (adders and multipliers) are greatly simplified, and the processing rate is limited only by the speed of the basic digital circuits and not by carry-propagation times. The resulting filters are readily multiplexed to process multiple data inputs or to effect multiple, but different, filters (or both) using the same arithmetic circuits, thus providing for efficient hardware utilization. A multiplexing factor of 100 or so is possible in audio-frequency applications, using presently available digital logic in the medium-speed range. The filters are also easily modified to realize a wide range of filter forms, transfer functions, multiplexing schemes, and round-off noise levels (i.e., data accuracies) by changing only the contents of the read-only coefficient memory and/or the timing signals and the length of the shift-register delays. For audio-frequency applications, a simple A/D converter may be implemented using delta modulation as an intermediate encoding process.

REFERENCES

[1] J. F. Kaiser, "Digital filters," in *System Analysis by Digital Computer*, J. F. Kaiser and F. F. Kuo, Eds. New York: Wiley, 1966, pp. 218–85.
[2] C. M. Rader and B. Gold, "Digital filter design techniques in the frequency domain," *Proc. IEEE*, vol. 55, pp. 149–171, February 1967.
[3] R. M. Golden, "Digital filter synthesis by sampled-data transformation," this issue, pp. 321–329.
[4] B. Gold and C. M. Rader, *Digital Processing of Signals*. New York: McGraw-Hill, 1969.
[5] J. F. Kaiser, "Some practical considerations in the realization of linear digital filters," *1965 Proc. 3rd Allerton Conf. on Circuit and System Theory*, pp. 621–633.
[6] R. B. Blackman, unpublished memorandum.
[7] Y. Chu, *Digital Computer Design Fundamentals*. New York: McGraw-Hill, 1962.
[8] R. N. Battista, C. G. Morrison, and D. H. Nash: "Signaling system and receiver for touch-tone calling," *IEEE Trans. Communications and Electronics*, vol. 82, pp. 9–17, March 1963.
[9] E. N. Protonotarios, "Slope overload noise in differential pulse code modulation systems," *Bell Sys. Tech. J.*, vol. 46, pp. 2119–2161, November 1967.

10

Reprinted from *IEEE Trans. Acoustics, Speech, Signal Processing,* **ASSP-22**(4), 255–258 (1974)

A Parallel Arithmetic Hardware Structure for Recursive Digital Filtering

ROBERT A. GABEL, MEMBER, IEEE

***Abstract*—A flexible, low-cost hardware structure is described for implementing recursive digital filters. Either a parallel or cascade arrangement of any number of second-order filter sections may be implemented for each of the input channels. Although the internal arithmetic is fixed-point, filter coefficients may be represented in floating-point form to reduce coefficient roundoff error.**

INTRODUCTION

THIS PAPER describes a hardware structure for digital filter implementation which combines high-speed operation with low hardware cost and simple control and timing circuitry. In contrast with serial arithmetic units reported elsewhere [1]–[3], a parallel multiplier/adder unit is employed here.

The principal difference between serial arithmetic structures and this parallel arithmetic structure is the nature of the tradeoff available to the designer. Serial arithmetic units process signals one bit at a time; hence an increase in internal word length (i.e., increased precision) necessitates a longer processing time. The design of a serial processor thus involves a tradeoff of processing speed versus word length. With a parallel arithmetic unit, processing time is essentially independent of word length, since all bits are treated simultaneously. The parallel processor design thus involves a tradeoff of hardware reduction versus word length. Another difference is apparent in the timing and control circuitry, which is considerably simpler for the parallel structure.

CASCADE STRUCTURE

We begin with a cascade structure comprised of M second-order sections as shown in Fig. 1. The overall transfer function is given by

$$H(z) = \alpha_0 \prod_{i=1}^{M} \frac{1 + \alpha_{1i}z^{-1} + \alpha_{2i}z^{-2}}{1 + \beta_{1i}z^{-1} + \beta_{2i}z^{-2}} \quad (1)$$

where Z-transforms are defined in negative powers of z. Notice that all multipliers have either $w(k-1)$ or $w(k-2)$ as inputs, and that all multiplier outputs go to adders. This suggests that a single multiplier/adder unit could be time-shared among the four operations, as shown in Fig. 2. In this and the succeeding figures, a double line will be used to indicate a digital word of information, two or more bits wide.

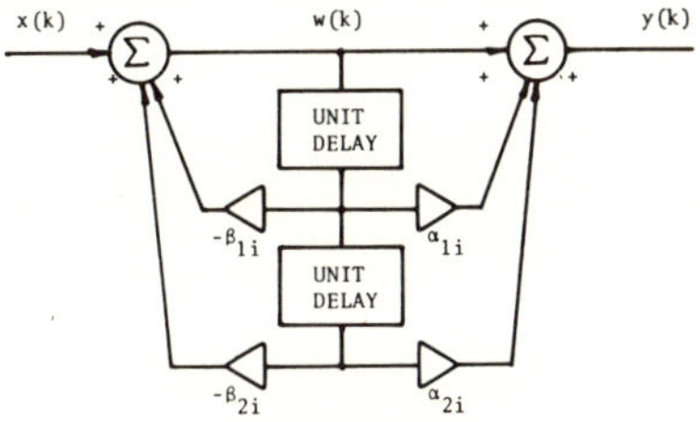

Fig. 1. Typical second-order section.

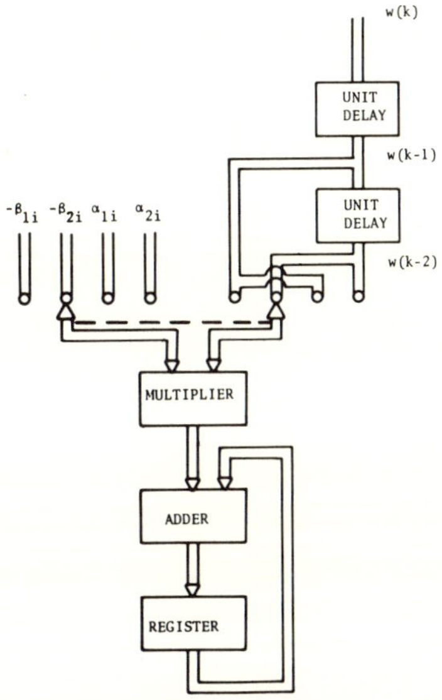

Fig. 2. Time multiplexing of multiplier and adder.

The complete implementation of this scheme, including input/output, feedback of $w(k)$ values, and interstage control, is shown in Fig. 3. In this figure, the input multiplexer MUX 1 is used to control selection of the filter section inputs (either from the A/D converter or from the previous section output) and accumulation of the partial sum in the AC register. Signal scaling by a power of 2 between sections (i.e., sign-extended binary shifting) is also implemented with this multiplexer, as illustrated in Fig. 4 for scaling of 1/4, 1/2, 1, and 2. The range of shift is

Manuscript received May 16, 1973; revised February 15, 1974.
The author is with the Department of Electrical Engineering, University of Colorado, Denver, Colo. 80202.

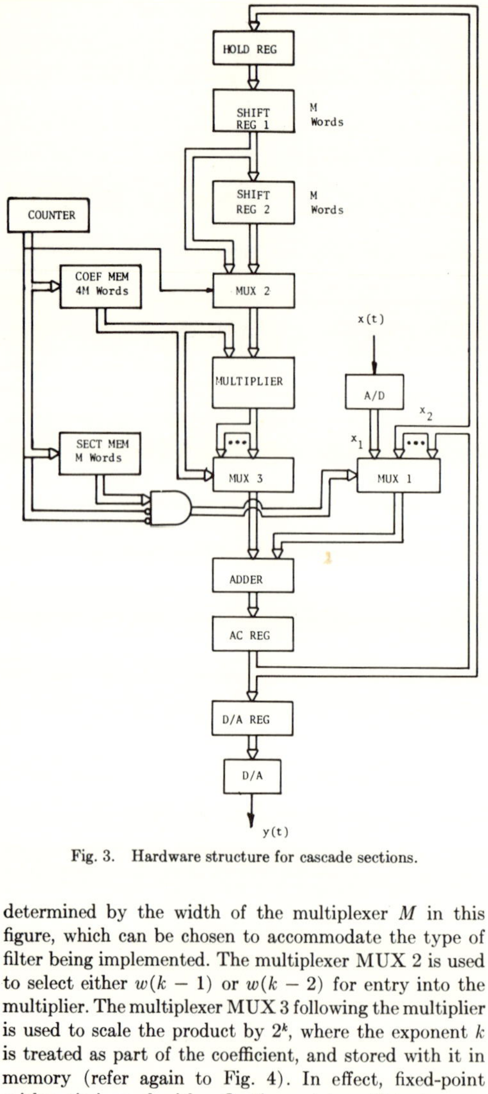

Fig. 3. Hardware structure for cascade sections.

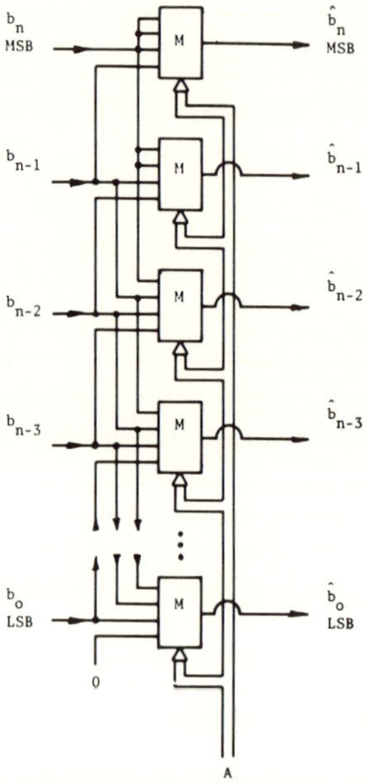

Fig. 4. Hardware implementation of MUX 1 and MUX 3 showing scaling by $\frac{1}{4}$, $\frac{1}{2}$, 1, and 2 (top position to bottom position). M = 1-of-4 data selectors with common address A.

determined by the width of the multiplexer M in this figure, which can be chosen to accommodate the type of filter being implemented. The multiplexer MUX 2 is used to select either $w(k-1)$ or $w(k-2)$ for entry into the multiplier. The multiplexer MUX 3 following the multiplier is used to scale the product by 2^k, where the exponent k is treated as part of the coefficient, and stored with it in memory (refer again to Fig. 4). In effect, fixed-point arithmetic is used with a floating-point coefficient representation. This feature allows the storage of coefficients and construction of the parallel multiplier to be carried out with a minimum number of bits, while still retaining significance in the representation. After the partial sum $w(k) = x(k) - \beta_{1i}w(k-1) - \beta_{2i}w(k-2)$ has been accumulated, it is held in a temporary storage register for subsequent entry into the shift register memory after $y(k)$ has been generated. To minimize the effects of roundoff error, more bits can be used in the adder and AC register than are required in the remainder of the circuit.

As discussed by Jackson *et al.* [2], 2's complement data representation can be used to help avoid problems caused by overflow of intermediate results. However, overflow in $w(k)$ must be avoided, since this variable is subsequently multiplied by the filter coefficients β_1 and β_2; and of course overflow in the final result would give rise to an erroneous output. These potential overflows are not detected by the system shown in Fig. 3 but can be partially controlled by intersection scaling. Complete control would require explicit overflow sensing and use of saturation arithmetic.

The control circuitry is exceptionally simple. A single counter controls all information flow, with the two least significant bits cycling through four states per filter section. Table I defines the events occurring during each of the counter states. One memory (COEF MEM) is used to hold coefficients for all sections; the counter state is used as its address register. A second memory (SECT MEM) holds data which directs the data input and output for each filter section. It is the information in this memory which governs the number of sections comprising each filter channel, controls the analog input multiplexer (not shown), and distributes output data to the appropriate digital-to-analog converter. This arrangement of control through stored data allows for an extremely flexible filter structure which can be easily programmed and modified.

The filter shown in Fig. 3 can be constructed from commercially available TTL integrated circuits for a cost of only a few hundred dollars. The maximum clock rate is on the order of 1–5 MHz, depending on the logic family chosen and the number of bits in the parallel multiplier.

TABLE I
TIMING EVENTS FOR THE FILTER OF FIG. 3 (SHOWN FOR $M = 4$ SECTIONS)

Count	Event
0000	Pass previous output to D/A Coef. Memory selects $-\beta_{11}$ MUX 2 selects $w(k-1)$ MUX 1 selects x_1 AC enters sum
0001	Coef. Memory selects $-\beta_{21}$ MUX 2 selects $w(k-2)$ MUX 1 selects x_2 AC accumulates sum
0010	Coef. Memory selects α_{11} MUX 2 selects $w(k-1)$ MUX 1 selects x_2 AC contents $= w(k)$ passed to Hold Reg. AC accumulates sum
0011	Coef. Memory selects α_{21} MUX 2 selects $w(k-2)$ MUX 1 selects x_2 SR Shifts right, entering contents of Hold Reg. AC accumulates sum
0100	Coef. Memory selects $-\beta_{12}$ MUX 2 selects $w(k-1)$ MUX 1 selects $2^s x_2$ (2^s is the intersection scaling factor controlled by the section memory; x_2 is the output of the previous section) AC enters sum $\cdots$ etc. for a total of 4 M counts

This multiplier, fabricated from individual 2 bit × 4 bit multiplier IC's, accounts for much of the processing time—typically 130 ns for an 8 bit × 8 bit multiplication to 300 ns for 16 bits × 16 bits. At even a 2 MHz clock rate, however, and a sampling rate of 8 kHz per analog signal, one unit can implement 60 2 pole–2 zero filter sections ($\frac{1}{4} \times f_{\text{clock}} \div f_{\text{samp}}$). Thus, for example, 15 analog signals could be accommodated, each processed with an 8 pole–8 zero transfer function. The hardware cost, exclusive of conversion equipment, is only tens of dollars per channel.

Multiple-rate processing of the various input signals can be achieved with additional taps incorporated in the delay line. For example, by including a tap at the midpoint of the first shift register section and an additional multiplexer input controlled by the word memory, rates of R and $2R$ can be accommodated in different channels, where R is the nominal sampling rate. With another tap at one-fourth the first section length, rates of R, $2R$, and $4R$ can be handled simultaneously, etc.

PARALLEL STRUCTURE

We now have a hardware structure to implement the cascade form transfer function of (1). To generate the equivalent parallel combination of second-order sections, we form the standard partial fraction expansion of (1) to obtain

$$H(z) = \gamma_0 + \sum_{i=1}^{M} \frac{\gamma_{0i} + \gamma_{1i}z^{-1}}{1 + \beta_{1i}z^{-1} + \beta_{2i}z^{-2}}. \tag{2}$$

We wish, however, to use only the 2-pole, 2-zero section of Fig. 1. To this end, we set

$$\gamma_0 = \sum_{i=1}^{M} c_i \tag{3}$$

to obtain

$$H(z) = \sum_{i=1}^{M} \left(c_i + \frac{\gamma_{0i} + \gamma_{1i}z^{-1}}{1 + \beta_{1i}z^{-1} + \beta_{2i}z^{-2}}\right)$$

$$= \sum_{i=1}^{M} \frac{(c_i + \gamma_{0i}) + (c_i\beta_{1i} + \gamma_{1i})z^{-1} + c_i\beta_{2i}z^{-2}}{1 + \beta_{1i}z^{-1} + \beta_{2i}z^{-2}} \tag{4}$$

where c_i is as yet undetermined. We now equate (2) and (4) to the desired form

$$H(z) = G\sum_{i=1}^{M} \frac{1 + a_{1i}z^{-1} + a_{2i}z^{-2}}{1 + \beta_{1i}z^{-1} + \beta_{2i}z^{-2}} \tag{5}$$

and set $z^{-1} = 0$ to obtain

$$\gamma_0 + \sum_{i=1}^{M} \gamma_{0i} = \sum_{i=1}^{M} (c_i + \gamma_{0i}) = MG$$

from which

$$G = \frac{1}{M}\left[\gamma_0 + \sum_{i=1}^{M} \gamma_{0i}\right]. \tag{6a}$$

Now equating powers of z^{-1} in the numerators of (4) and (5), we obtain expressions for a_{1i} and a_{2i}

$$c_i = G - \gamma_{0i}$$

$$a_{1i} = (c_i\beta_{1i} + \gamma_{1i})/G$$

$$= \beta_{1i} + \frac{\gamma_{1i} - \beta_{1i}\gamma_{0i}}{G} \tag{6b}$$

$$a_{2i} = c_i\beta_{2i}/G$$

$$= \beta_{2i}(1 - \gamma_{0i}/G). \tag{6c}$$

We note from (1) that G equals α_0/M, and hence will be nonzero.

The individual sections now have the same form of transfer function as that of the cascaded sections; thus, we can use the identical hardware to implement them as was used previously. For the parallel section filter, however, all sections have the same input, $x(k)$, and we must accumulate the section outputs, rather than passing them on as the inputs to succeeding stages. Thus the input multiplexer must be expanded to allow for scaling of the input x_1, and a second adder and accumulator must be incorporated to form the M-fold sum above. The section memory is used to control these latter functions. Fig. 5 shows the resulting hardware structure. In this form, either a cascade or parallel configuration of arbitrary order can be implemented. With a signal path from the second accumulator to the input multiplexer included, a combined parallel/cascade configuration can also be realized. In any of these structures, if the absolute filter gain is to be controlled more closely than ±3 dB, then the binary signal scaling through the input multiplexer must be augmented

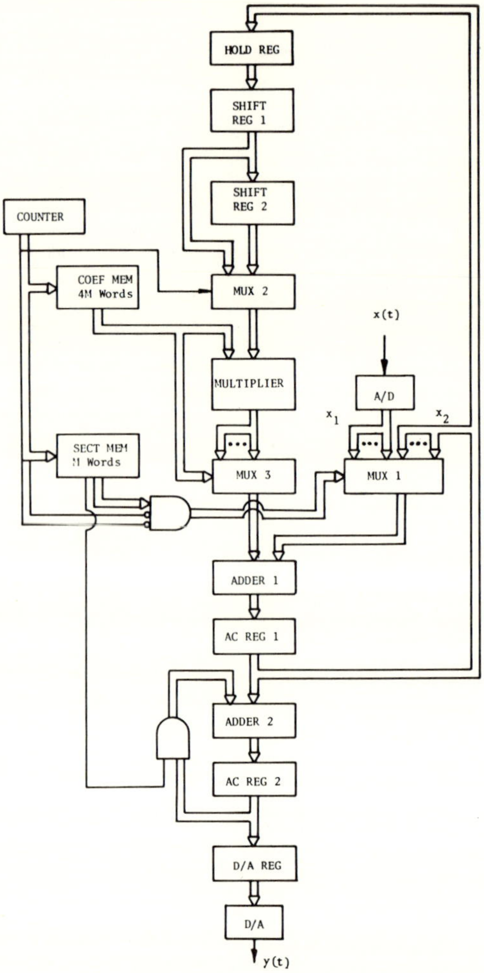

Fig. 5. Hardware structure for parallel or cascade sections.

with a final pass of the filtered signal through the multiplier and on to the output.

SUMMARY

The hardware structure described above offers the advantages of low cost, fast operation, simple control circuitry, and flexible programming. No restriction is placed on the number of bits used to represent external or internal signal variables, and filter coefficients are readily stored in floating point form. The structure appears to be quite suitable for large-scale integrated (LSI) construction. This filter has not yet been physically constructed, but it has been designed in detail and simulated at the bit level on a general purpose computer.

ACKNOWLEDGMENT

The author wishes to thank the reviewers of this paper for their helpful comments and suggestions.

REFERENCES

[1] S. L. Freeny, R. B. Kieburtz, K. V. Mina, and S. K. Tewksbury, "Design of digital filters for an all digital frequency division multiplex-time division multiplex translator," *IEEE Trans. Circuit Theory* (*Special Issue on Active and Digital Networks*), vol. CT-18, pp. 702–711, Nov. 1971.
[2] L. B. Jackson, J. F. Kaiser, and H. S. McDonald, "An approach to the implementation of digital filters," *IEEE Trans. Audio Electroacoust.* (*Special Issue on Digital Filters: The Promise of LSI Applied to Signal Processing*), vol. AU-16, pp. 413–421, Sept. 1968.
[3] S. Zohar, "The counting recursive digital filter," *IEEE Trans. Comput.*, vol. C-22, pp. 338–347, Apr. 1973.

11

Reprinted from *IEEE Trans. Audio Electroacoustics,* **AU-21**(6), 477–484 (1973)

A New Approach to the Realization of Nonrecursive Digital Filters

ABRAHAM PELED and BEDE LIU, Fellow, IEEE

Abstract—A new realization of nonrecursive digital filters that are used to operate on analog signals is proposed. This realization requires no multiplications, and exploits the relative simplicity of delta modulation as a means for analog to digital conversion. This realization also permits a mechanization as a "very fast" digital filter, using read-only memory (ROM). An evaluation of this realization in terms of computation time storage requirements and mean-squared error is presented. These characteristics are compared with their counterparts for existing realization methods of nonrecursive digital filters. Computer simulation results that tend to confirm the theoretical results of the error analysis are included.

I. Introduction

Nonrecursive digital filters possess many distinct features that make them attractive for a variety of applications. Their excellent phase characteristic, inherent stability, and conceptual simplicity are some of the causes for the extensive research efforts that have been invested in their design and analysis. Efficient methods are known for the design of nonrecursive digital filters using various approximation methods [1], and some common designs have been tabulated [2].

Nonrecursive filters require relatively longer computation time per output sample. Although this can be improved by using the fast convolution [3], [4], such an approach is only suitable when the filter is implemented on a computer or when a fast Fourier transform (FFT) box is available.

In this paper we present a new approach for nonrecursive digital filtering of analog signals that requires very simple hardware and does not require multiplication. We also suggest a particular hardware realization that permits very fast digital filtering, using read-only memory (ROM) so that each output sample requires only one addition.

Manuscript received March 26, 1973; revised June 20, July 17, and August 1, 1973. This work was supported by the Air Force Office of Scientific Research, USAF, under Grant AF-AFOSR 71-2101, and by the National Science Foundation under Grant GK-24187.

The authors are with the Department of Electrical Engineering, Princeton University, Princeton, N. J. 08540.

The proposed scheme uses a delta modulator [5] for analog-to-digital conversion at the input. The one-bit stream at the output of the delta modulator is then fed into a digital processor that involves no multiplication, but only additions and sign changes. We proceed to compare the efficiency and to study the various tradeoffs involved in the proposed realization versus existing realization schemes. An analysis is performed to show that the error for the proposed realization is due to two sources; one introduced by the delta modulation of the input and the other resulting from the use of finite word length to represent the filter coefficients. A simple expression is derived to predict the performance of the proposed realization. The influence of various design parameters on the resulting error is discussed and some guidelines are given to select the parameters. Finally, to confirm the validity of the error analysis computer simulation results are presented for various input signals.

Delta modulation has been used by Goodman [6] to convert an analog signal into a pulse-code modulated (PCM) signal. In his scheme one of the structures discussed in the present paper was used for minimum mean squared error estimation of the pulse-amplitude modulation (PAM) sample. Lockhart [7] also used delta modulation for filtering an analog signal. However, his realization is hybrid, employing analog multiplications and analog additions.

II. A New Approach for Implementing Nonrecursive Digital Filters

A nonrecursive digital filter is characterized by an input-output relationship of the form

$$y_n = \sum_{k=0}^{N-1} a_k \, x_{n-k} \tag{1}$$

where $\{x_n\}$ is the input sequence, $\{y_n\}$ the output sequence, and

$$\{a_k\}_{k=0}^{N-1}$$

are the filter coefficients. Consider the processing of an analog signal by such a filter realized with b-bit word hardware. The conventional implementation of this filter is illustrated in Fig. 1. The analog signal is sampled and each sample is quantized and converted to a b-bit word by a PCM. The output sequence of the PCM is fed to a digital processor that computes the output samples $\{y_n\}$ according to (1). The hardware and computational requirements can be easily determined from the block diagram. We need a b-bit PCM, storage for N filter coefficients, and a digital processor capable of performing addition and multiplication. The processor performs N multiplications and $N - 1$ additions per output sample.

The approach proposed in this paper is illustrated

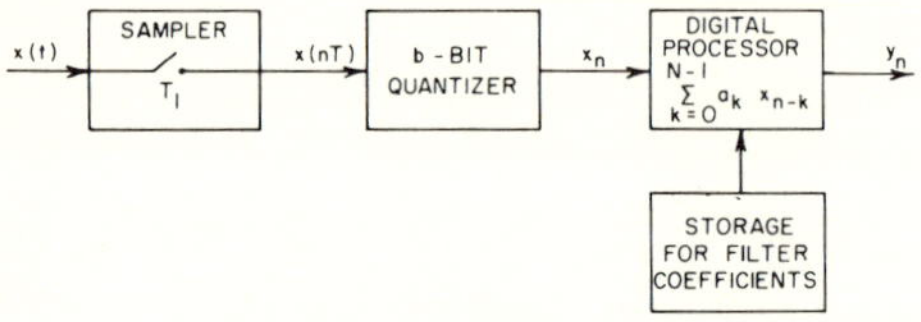

Fig. 1. Realization using PCM.

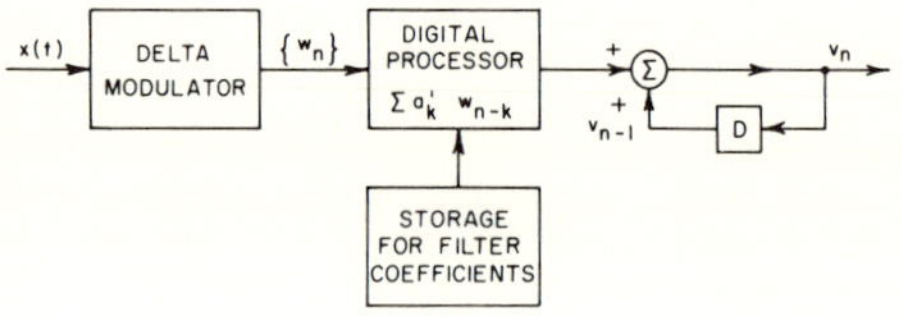

Fig. 2. Realization using delta modulation.

in Fig. 2. The input signal enters a delta modulator whose output is the sequence $\{w_n\}$ with each w_n taking on the values +1 or −1. This sequence enters a digital processor that computes

$$\sum_{k=0}^{N-1} a_k' w_{n-k}$$

and adds this sum to the content of an accumulator to obtain the output sample v_n. This operation is described by

$$v_n = v_{n-1} + \sum_{k=0}^{N-1} a_k' w_{n-k}. \tag{2}$$

We first show that if the error introduced by the delta modulator is negligible, the coefficients $\{a_k'\}$ can be chosen so that the output $\{v_n\}$ agrees with $\{y_n\}$. An error free delta modulator is characterized by the following relation between w_n and samples of the input signal $x(t)$ [5],

$$x(nT) = x[(n-1)T] + \Delta_x w_n \tag{3}$$

where T is the sampling interval, and Δ_x the step size of the delta modulation. Substituting (3) into (2), we obtain

$$v_n = \sum_{k=0}^{N-1} \frac{a_k'}{\Delta_x} x[(n-k)T] + C \tag{4}$$

where C is a constant, which can easily shown to be zero if we require that the output be zero when no signal is present at the input. With that assumption we immediately see that v_n equals the y_n given by (1) provided that

$$a_k' = a_k \Delta_x. \tag{5}$$

In the preceding discussion, we have assumed that no roundoff or quantization error occurs in the arithmetic operations. Thus we see that at least in an idealized situation, the new implementation scheme has the same capability as the conventional approach of Fig. 1.

Let us pursue this idea for a moment, since we know that the error introduced by the delta modulator and by the arithmetic operation can be made very small. A comparison of the realizations proposed in Figs. 1 and 2 reveals immediately two advantages of the delta modulation realization over the PCM realization. The first is the relative simplicity of a delta modulator as compared to a PCM. The second is a simpler processor due to the absence of multiplications. However, it is well known that in order to obtain a good fidelity with delta modulation, we have to sample much faster than the Nyquist rate, which is sufficient for the PCM realization. In the remainder of the paper, we will discuss the performance of the proposed realization and compare it to an equivalent PCM realization.

III. Possible Hardware Realizations

Two possible mechanizations of (2) are presented in Figs. 3 and 4. Both configurations will yield exactly the same output, but differ in storage requirements and speed of operation.

In Fig. 3, the input analog signal $x(t)$ is passed through a single integration delta modulator [8] whose output is $\{w_n\}$ with each w_n taking on +1 or −1 only. The sequence $\{w_n\}$ enters the arithmetic unit that computes in a straightforward manner the increment of the output according to

$$\Delta v_n = \sum_{j=0}^{N-1} a_j' w_{n-j}. \tag{6}$$

This increment is added to the previous value of the output to form the current output,

$$v_n = v_{n-1} + \Delta v_n. \tag{7}$$

The configuration of Fig. 4 has no arithmetic unit. Since the increment Δv_n can take on only a finite number (2^N) of values, depending on the array $\{w_{n-j}\}_{j=0}^{N-1}$, these 2^N possible values of Δv_n can be stored in a ROM at addresses that correspond to the binary word obtained from $\{w_{n-j}\}_{j=0}^{N-1}$ by substituting 0 for −1. The value of the increment Δv_n can be obtained once w_n is in the shift register. So to obtain the output v_n, only one addition is needed. The increase in the speed of operation is significant. However, there is a serious shortcoming in this configuration, since the exponential dependence of the storage capacity needed on the number of filter coefficients makes the storage requirement considerable even for a modest number of filter coefficients. It should be pointed out, that for linear phase filters, the coefficients $\{a_n\}_{n=0}^{N-1}$ are symmetric [9]. Therefore, only $2^{(N+1)/2}$ distinct values of the increment

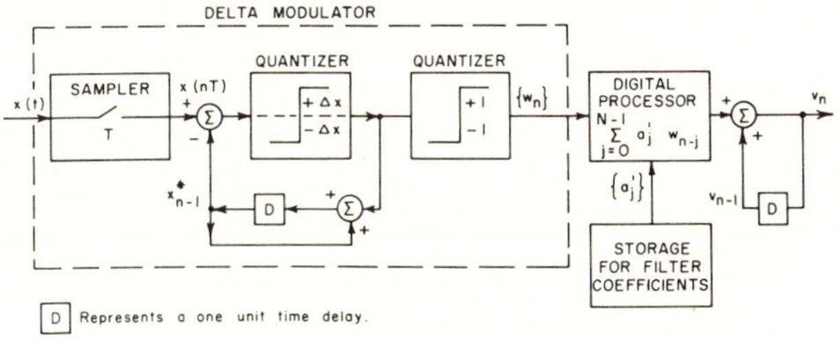

Fig. 3. Hardware realization of nonrecursive digital filter using delta modulation.

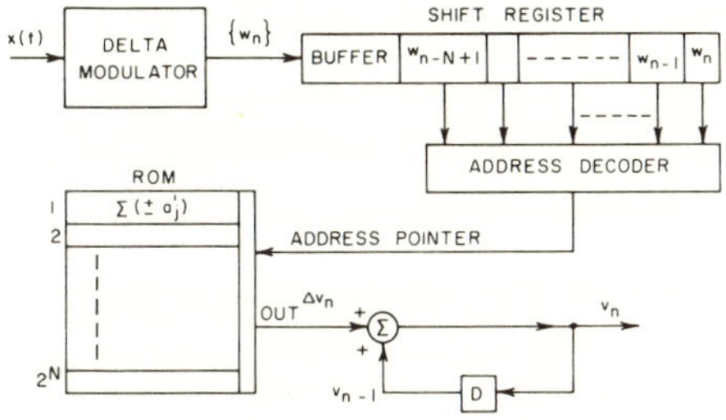

Fig. 4. A very fast realization of a nonrecursive digital filter.

Δv_n need to be stored. The storage requirements will be discussed further in the next section.

IV. Computation Time and Storage Considerations

In this section we attempt to compare the two equivalent realizations of a nonrecursive digital filter, the one using PCM as shown in Fig. 1 and the one using delta modulation as shown in Fig. 3. As mentioned previously, we have to sample much faster than the Nyquist rate in the case of delta modulation. Let K be the ratio of the sampling frequency in the delta modulation to the Nyquist frequency used in PCM.

In the case of the PCM realization the computation time per output sample is given by

$$C_{\mathrm{PCM}} = MP\alpha + (M - 1)\alpha \tag{8}$$

where M is the number of filter coefficients of the PCM realization, α is the time required for one full word addition, and $P\alpha$ is the time required for multiplication.

In the case of the delta modulation (DM) realization the computation time per output sample is

$$C_{\mathrm{DM}} = N\alpha \tag{9}$$

where N is the number of filter coefficients, and is in general different from M, because of the different sampling rates. In fact, in order for the two realizations to achieve the same or approximately the same frequency characteristic, it is reasonable to assume that $N = KM$. If we use the Fourier series method to design the filter, then it can be shown [10] that the number of coefficients needed to achieve a given transition region $\Delta\omega$ in the frequency characteristic is $N = 2\pi/T\Delta\omega$ in the case of Hamming weights or $N = 2\sqrt{R^2 + \pi^2}/T\Delta\omega$ in the case of Kaiser weights, where R is related to the overshoot. In both cases decreasing the sampling interval K times will require a K fold increase in the number of coefficients needed to achieve the same frequency characteristic. For other design methods $N = KM$ is a very good approximation.

We need to compare KC_{DM} with C_{PCM} because of the different sampling rates. From the previous discussion, the comparison to be made is, assuming $M >> 1$,

$$MK^2\alpha \overset{?}{\lessgtr} M(P+1)\alpha \tag{10}$$

or equivalently

$$K^2 \overset{?}{\lessgtr} P + 1. \tag{11}$$

It is therefore obvious that in order to save computation time through the delta modulation realization, one must have $K^2 < P + 1$.

If special purpose hardware is used to perform multiplication then the value of P would be approximately[1] 3, which requires $K < 2$. This results in an oversampling factor insufficient for good delta modulation for most practical signals. If the multiplication is realized by software the value of P is approximately[1] 50, which implies $K < 7$, a value that will allow good quality delta modulation for most practical

[1] According to a survey article by Butler [11] on 45 minicomputers by 27 manufacturers, the ratio P for fixed point hardware ranges between 2.0 and 12.5 with an average of 4.78. If the multiplication is done by software only, the ratio P was between 7.8 and 281 with an average of 91.7. Of the minicomputers surveyed, 22.2 percent had standard hardware multipliers, 35.5 percent had such multipliers as optional equipment, and 33.3 percent had no such option.

signals [8]. Thus, the implementation of Fig. 3 offers a saving in computation time only if the equivalent PCM realization uses software to perform multiplication. Thus the main advantage of this new form of realization is the simplicity in hardware, and the price paid is a slower speed of operation.

For the configuration of Fig. 4, the computation of each output sample requires one shift, one readout, and one actual addition. It is safe to assume that the total time needed is less than 2α, and the comparison with the implementation of Fig. 1 is

$$2K\alpha \overset{?}{\lessgtr} M(P+1)\alpha. \tag{12}$$

Thus this form of delta modulation realization would always offer a significant saving in computation time.

The realization proposed in Figs. 3 and 4 are really two extreme solutions to the mechanization of (2), one requires the performing of all computations each time, and the other requires the storing of all possible outcomes of the computation and using them when needed. We have seen that the later requires a large amount of storage even for a modest number of coefficients. However a very good compromise between the two is possible. To illustrate the idea, let us consider a 15 tap filter. The realization using ROM as described by Fig. 4 requires a storage of 2^{15} words. Now, suppose we rewrite

$$\sum_{j=0}^{14} a_j' w_{n-j}$$

as

$$\sum_{j=0}^{4} a_j' w_{n-j} + \sum_{j=5}^{9} a_j' w_{n-j} + \sum_{j=10}^{14} a_j' w_{n-j}$$

and compute separately each partial sum by using a ROM to store all its possible outcomes. For each partial sum, we need 2^5 words, so the total storage needed will be $3 \times 2^5 = 96$ words as compared to $2^{15} = 32\,768$ words in the straightforward ROM realization. However, two more additions are needed for each output sample. The tradeoff of an exponential decrease in storage for a linear increase in the number of additions needed is obviously attractive.

In general, let N be the number of filter coefficients and L be the number of coefficients in a partial sum; the last partial sum may contain less than L terms if L does not divide N. The computation time needed is $3\alpha + \left[\frac{N}{L}\right]_I \alpha$ and the storage requirement is $\left(\left[\frac{N}{L}\right]_I + 1\right)2^L$. Further reduction in storage is possible by taking advantage of the symmetry of the filter coefficients for linear phase filters. It is possible in most cases to use modest storage and achieve a saving in computation time over the PCM realization.

V. Error Analysis

The desired output of a nonrecursive digital filter applied to an analog signal is

$$y_n = \sum_{j=0}^{N-1} a_j x[(n-j)T] \tag{13}$$

where a_j are the filter coefficients, and $x[kT]$ are samples of the input signal. The proposed realization through delta modulation introduces two errors. One is due to the representation of each filter coefficient by a finite number of bits, and the second is due to delta modulation. The actual computed output is

$$v_n = v_{n-1} + \sum_{j=0}^{N-1} (a_j')^Q w_{n-j} \tag{14}$$

where $(a_j')^Q$ is a_j' quantized to b-bits and w_k is given by

$$w_k = (x_n^* - x_{n-1}^*)/\Delta_x \tag{15}$$

and x_n^* is the nth sample value of the reconstructed staircase signal approximating $x(t)$. Let e_{D_n} be the error

$$e_{D_n} = x(nT) - x_n^*. \tag{16}$$

Let ϵ_n be the error between the desired output and the actual output at time nT,

$$\epsilon_n = y_n - v_n. \tag{17}$$

One possible fidelity criterion of the approximation of y_n by v_n is the mean-squared error $E\{\epsilon_n^2\}$ where E denotes expectation.

In Appendix A an expression is derived for the mean-squared error

$$E\{\epsilon_n^2\} = \sum_{k=0}^{N-1} \sum_{j=0}^{N-1} a_k a_j E\{e_{D_{n-k}} e_{D_{n-j}}\} + \frac{q}{\Delta_x^2} \sum_{j=0}^{N-1} \cdot [E\{x_{n-j}^{*2}\} + E\{x_{n-j}^* x_{n-N+j+1}\}] \tag{18}$$

where q is the variance of the quantization error given by

$$q = \frac{2^{-2(b-1)}}{3}. \tag{19}$$

b is the number of bits including the sign bit used in fixed point representation of the filter coefficient. The conversion is done using rounding. The first term of (18) is due to the delta modulation and the second term is due to roundoff.

The delta modulation involves a nonlinear feedback system that makes the analysis impossible for arbitrary signal statistics. For Gaussian input signals with rational spectra, the statistics of the error introduced by delta modulation are well understood and theoretical and experimental results are available [12]. The statistics of the signal $\{x_n^*\}$ are known only for Gaussian input signal and when the step size Δ_x is chosen to prevent overload, say by the so called 4σ rule [13]. Under these assumptions an expression has been found by Goodman [13] for the autocorrelation function of $\{x_n^*\}$. However this results in rather complicated equations and no insight into the behavior of the error can be gained.

In order to obtain a simple expression for the mean-squared error, some simplifying assumptions are made. It is assumed that the delta modulation error $\{e_{D_n}\}$ is approximately white and has a power N_D. Thus the first term on the right side of (18) can easily be shown to be $N_D \sum_{j=0}^{N-1} a_j^2$. As for the second term, we have from (16)

$$E\{x_k^* x_j^*\} = E\{x(kT)x(jT)\} + E\{e_{D_k} e_{D_j}\} \quad (20)$$

when it is substituted into (18), the term $E\{e_{D_k} e_{D_j}\}$ can be dropped since it produces a second-order effect.

Thus assuming a wide-sense stationary input we can rewrite (18) as

$$E\{\epsilon_n^2\} = N_D A + \frac{q}{\Delta_x^2}\left[NR_{xx}(0) + \sum_{j=0}^{N-1} \cdot R_{xx}(N - 2j + 1)\right] \quad (21)$$

where

$$A = \sum_{j=0}^{N-1} a_j^2 .$$

Furthermore since $R_{xx}(0) \geqslant |R_{xx}(k)|$ for all k we can write the output error as

$$E\{\epsilon_n^2\} = N_D A + \frac{\beta q}{\Delta_x^2} NR_{xx}(0) \quad (22)$$

where β is some number, $1 \leqslant \beta \leqslant 2$. Or equivalently the signal to noise ratio in decibels at the output is

$$(S/N)_{0\,\mathrm{dB}} = 10 \log\left(\frac{Y}{E\{\epsilon_n^2\}}\right) = 10 \log \left(\frac{Y}{N_D A + \frac{\beta q}{\Delta_x^2} NR_{xx}(0)}\right) \quad (23)$$

where $Y = E\{y_i^2\}$ is the output signal power. Fig. 5 illustrates the qualitative behavior of the output signal to noise ratio as a function of the word length used to represent the filter coefficients.

We can see from the figure that there are two regions. One where the error is dominated by the first term in (22), that is the error due to delta modulation, and the second region corresponding to short word length b, where the second term of (22) dominates and the performance deteriorates rapidly.

From (22) we can derive a good estimate of the number of bits b needed to prevent deterioration of the performance, i.e., operation to the left of the knee in the signal to noise ratio curve given in Fig. 5.

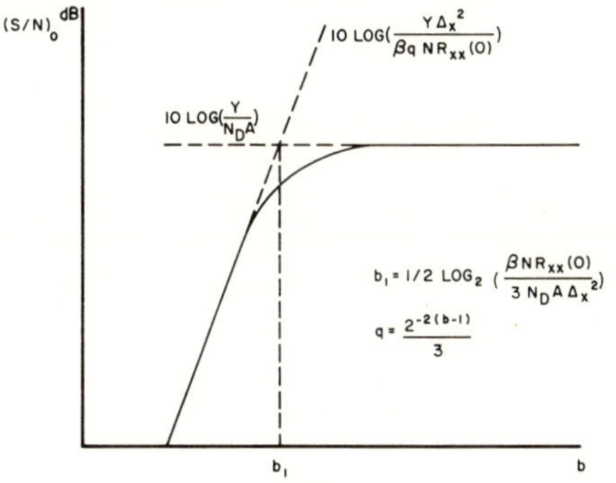

Fig. 5. Behavior of the output signal to noise ratio (23).

Let b be the number of bits for which the two terms on the right side of (22) equal (taking into account that b can be only an integer)

$$b_1 = \frac{1}{2}\left[\log_2\left(\frac{\beta NR_{xx}(0)}{3N_D A \Delta_x^2}\right)\right]_I + 1 \quad (24)$$

where $[\ ]_I$ denotes the integer part of the quantity inside the brackets. (Thus for $b > b_1$, the error is determined solely by the delta modulation, and for $b < b_1$ the performance deteriorates rapidly.) The b_1 given in (24) can be regarded as the minimum number of bits needed for satisfactory operation of the filter. It is worth mentioning that for most common filters the value of A is less than 1, since it represents the area under the filter transfer function relative to an ideal low pass up to the sampling frequency. This means that, for a sufficient word length, the output error will be smaller than the error caused by the delta modulation. The justification for the simplifying assumptions made earlier and the validity of the error expression (21) lies in the experimental verification, and that is the subject of the next section.

VI. Computer Simulation Results

The realization for nonrecursive digital filters through delta modulation of the input can be conveniently simulated on a digital computer. The delta modulator used was exactly the one in Fig. 3, that is a single integration, ideal integrator delta modulator. In practice it would be preferable to use a leaky integrator [8]. We used three types of input signals, for which theoretical and experimental results are known regarding their performance under delta modulation. These are sine wave with random phase, band-limited white Gaussian noise, and band-limited white Gaussian noise passed through an RC filter. The delta modulation step used was the optimum one as given in [8], [14], and [15]. We used the method described by

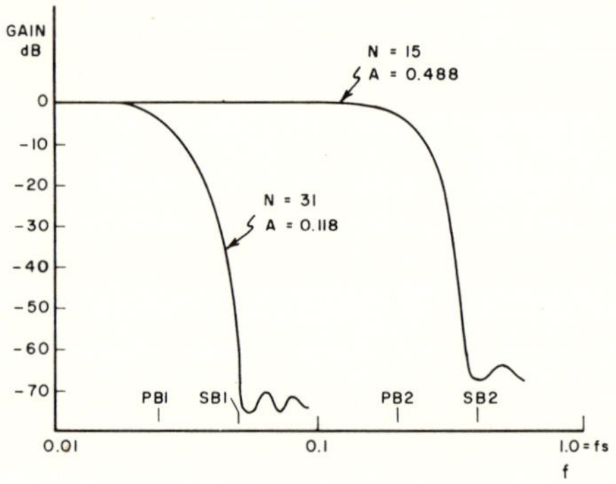

Fig. 6. Frequency response of the two filters used.

O'Neal [8] to obtain white band-limited Gaussian noise sampled K times the Nyquist rate, and we used the digital equivalent of an RC filter to obtain the RC-Gaussian input signal. Two digital filters were tested, one with 15 and the other with 31 coefficients, both have unity gain in the passband and at least 60-dB attenuation in the stopband. Their frequency response is plotted in Fig. 6. In order to examine the validity of the analysis presented in Section V, we computed the mean-squared error between the error-free output as given by (13), and the output with error as given by (14), using various word lengths for the filter coefficients. The input was scaled between -1 and 1. No overflow occurred because of the unity gain of the filter. We used fixed point with rounding to represent the filter coefficients. The signal to noise ratio was computed rather than the mean-squared error itself:

$$(S/N) = \frac{\sum_{n=1}^{L} y_n^2}{\sum_{n=1}^{L} (y_n - v_n)^2} \tag{25}$$

L is the number of points used. In Figs. 7-9 the results of the simulations are plotted for the different input signals and filters used. The ordinate represents the signal to noise ratio in decibels and the abscissa the number of bits used for the representation of the filter coefficients. To get the theoretical curve corresponding to (23) we computed the actual mean-squared error introduced by the delta modulation in our simulation, that is we constructed the staircase signal x_n^* and set

$$N_D = \frac{1}{L} \sum_{n=1}^{L} [x(nT) - x_n^*]^2. \tag{26}$$

The second term in (22) was computed using (19),

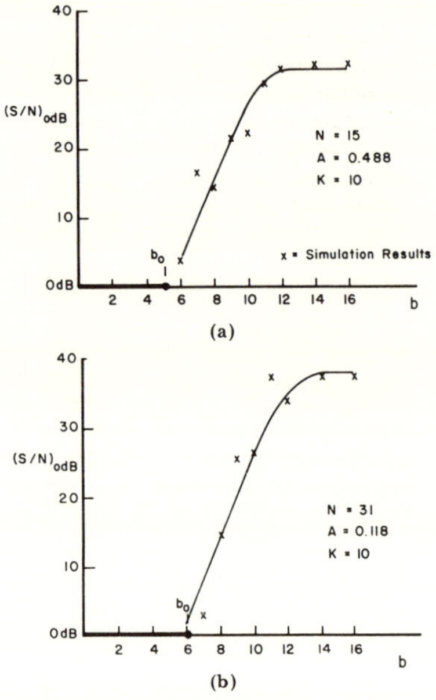

Fig. 7. Sine wave input. (a) $N = 15$. (b) $N = 31$.

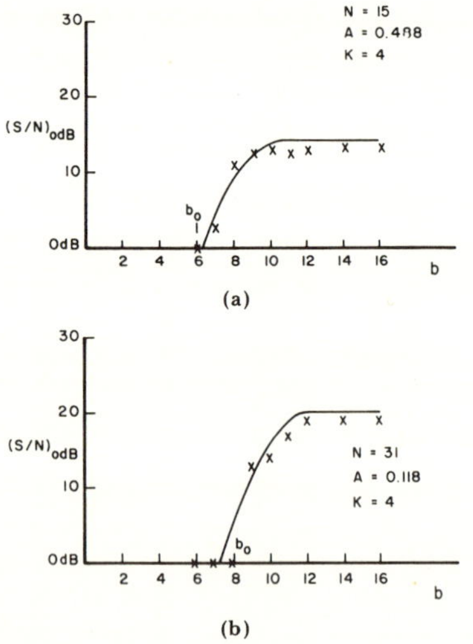

Fig. 8. Band-limited white Gaussian noise input, unity bandwith, zero mean, 0.25 standard deviation. (a) $N = 15$. (b) $N = 31$. *Note:* X = simulation results.

the autocorrelation values of the signals used, and the delta modulation step size.

It is worth mentioning a practical consideration that is the reason for the straight line corresponding

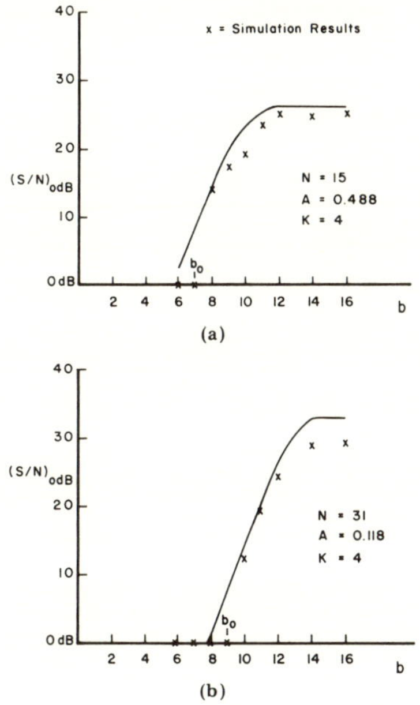

Fig. 9. Band-limited white Gaussian noise passed through RC filter, zero mean, $1/RC = 0.25$. (a) $N = 15$. (b) $N = 31$.

to a signal to noise ratio of 0 dB for very short word length. The reason for this is the fact that $\{a_j'\}_{j=0}^{N-1}$ are numbers between -1 and $+1$, and usually much smaller than 1 in absolute value. If the word length used b_0 is such that

$$\max |a_j'| < 2^{-b_0} \tag{27}$$

then all filter coefficients will be quantized to zero and the output v_n will be identically zero that corresponds to a signal to noise ratio of 0 dB.

It is evident from the experimental results that (23) is indeed a useful expression that will allow the designer to predict the approximate output signal to noise ratio for the realization through delta modulation. Furthermore, (24) gives a good estimate of the minimum word length needed to achieve the best possible signal to noise ratio. As we mentioned earlier theoretical and experimental results for the delta modulation error are known for a variety of input signals and can be directly applied to evaluate the expected performance of this realization of nonrecursive digital filters.

VII. Conclusion

We have proposed a new realization of nonrecursive digital filters using delta modulation. The main advantage of the scheme is the simplicity of the hardware, as a result of the elimination of multiplications in the processor, and of a simpler analog to digital conversion method. We have also suggested a realization permitting very fast filtering operation using ROM. The efficiency of the proposed realization was analyzed and compared to existing realization methods and some of the tradeoffs involved were discussed. An error analysis was carried out and a useful expression, which allows the system designer to choose the design parameters and predict the expected performance, was derived. Experimental results that confirm the error analysis were presented. In summary, the proposed realization requires very simple hardware and with a sufficient word length for the filter coefficients the output signal to noise ratio is determined by that of the delta modulation.

Appendix A

We derive here an expression for the mean-squared error between the desired output y_n and the computed output v_n of the new implementation. Equations (13)–(17) are repeated here:

$$y_n = \sum_{j=0}^{N-1} a_j x[(n-j)T] \tag{A1}$$

$$v_n = v_{n-1} + \sum_{j=0}^{N-1} (a_j')^Q w_{n-j} \tag{A2}$$

$$x_n^* = x_{n-1}^* + \Delta_x w_n \tag{A3}$$

$$e_{D_n} = x(nT) - x_n^* \tag{A4}$$

$$\epsilon_n = y_n - v_n. \tag{A5}$$

If the actual filter coefficients $(a_j')^Q$ are obtained from $a_j\Delta x$ by rounding to b bits we have a_j'.

$$(a_j')^Q = a_j\Delta_x - e_{Q_j} \tag{A6}$$

where rounding noise e_{Q_j} is known to have the following properties [16]:

$$E\{e_{Q_j}\} = 0 \tag{A7}$$

$$E\{e_{Q_j}e_{Q_k}\} = q\delta_{jk} + q\delta_{j,N-k-1} \tag{A8}$$

where $q = 2^{-2(b-1)}/3$. The second term in (A8) is due to the fact that for linear phase filters the coefficients are in pairs $a_j = a_{N-j-1}$. Using the previous we can rewrite (A2) as

$$v_n = v_{n-1} + \sum_{j=0}^{N-1} a_j x_{n-j}^* - \sum_{j=0}^{N-1} a_j x_{n-j-1}^* - \sum_{j=0}^{N-1} e_{Q_j} w_{n-j}. \tag{A9}$$

Combining (A1), (A4), (A5), and (A9) gives

$$\epsilon_n = \epsilon_{n-1} + \sum_{j=0}^{N-1} a_j e_{D_{n-j}} - \sum_{j=0}^{N-1} a_j e_{D_{n-1-j}} + \sum_{j=0}^{N-1} e_{Q_j} w_{n-j}. \quad \text{(A10)}$$

Equation (A10) provides a difference equation for the error in terms of the delta modulation error and quantization error. We assume the following initial conditions.

$$w_o = 0; \quad \epsilon_o = 0; \quad e_{D_o} = 0; \quad \text{(A11)}$$

with these assumptions the solution is

$$\epsilon_n = \sum_{j=0}^{N-1} a_j e_{D_{n-j}} + \sum_{j=0}^{N-1} e_{Q_j} \left(\sum_{i=1}^{n-j} w_i \right). \quad \text{(A12)}$$

Using the fact that summation over the past of the sequence $\{w_i\}$ gives us the value of x_n^*, the staircase function approximating the input, that is

$$x_n^* = \Delta_x \left(\sum_{i=1}^{n} w_i \right). \quad \text{(A13)}$$

On substituting (A13) into (A12) we finally get

$$\epsilon_n = \sum_{j=0}^{N-1} a_j e_{D_{n-j}} + \frac{1}{\Delta_x} \sum_{j=0}^{N-1} e_{Q_j} x_{n-j}^*. \quad \text{(A14)}$$

The mean-squared error $E\{\epsilon_n^2\}$ is given by

$$E\{\epsilon_n^2\} = E\left\{ \left[\sum_{j=0}^{N-1} a_j e_{D_{n-j}} + \frac{1}{\Delta_x} \sum_{j=0}^{N-1} e_{Q_j} x_{n-j}^* \right]^2 \right\}. \quad \text{(A15)}$$

Using (A7) and (A8) we can see that the cross term is zero if we assume independence of e_{Q_k} and e_{D_j}, which is obviously true. Thus,

$$E\{\epsilon_n^2\} = \sum_{k=0}^{N-1} \sum_{j=0}^{N-1} a_k a_j E\{e_{D_{n-k}} e_{D_{n-j}}\} + \frac{q}{\Delta_x^2} \sum_{j=0}^{N-1} \cdot [E\{x_{n-j}^{*2}\} + E\{x_{n-j}^* x_{n-N+j+1}^*\}] \quad \text{(A16)}$$

which is (18) in Section V.

References

[1] L. R. Rabiner, "Techniques for designing finite-duration impulse response digital filters," *IEEE Trans. Commun. Technol.*, vol. COM-19, pp. 188–195, Apr. 1971.

[2] R. W. Hankins, "Design procedures for equiripple nonrecursive digital filters," Tech. Rep. 485, Mass. Inst. Tech., Cambridge, May 1972.

[3] T. G. Stockham, "High speed convolution and correlation," in *1966 Spring Joint Computer Conf., AFIPS Conf. Proc.*, vol. 28, pp. 229–233.

[4] H. D. Helms, "Fast Fourier transform method of computing difference equations and simulating filters," *IEEE Trans. Audio Electroacoust.*, vol. AU-15, pp. 85–90, June 1967.

[5] K. W. Cattermole, *Principles of Pulse Code Modulation*. New York: American Elsevier, 1969, pp. 198–218.

[6] D. J. Goodman, "The application of delta modulation to analog-to-PCM encoding," *Bell Syst. Tech. J.*, vol. 40, pp. 321–343, Feb. 1969.

[7] G. B. Lockhart, "Digital encoding and filtering using delta modulation," in *Proc. Conf. Digital Processing of Signals in Communications*, University of Technology, Loughborough, England, Apr. 1972, pp. 43–52.

[8] J. B. O'Neal, "Delta modulation quantizing noise, analytical and computer simulation results for Gaussian and television input signals," *Bell Syst. Tech. J.*, vol. 45, pp. 117–141, Jan. 1966.

[9] J. F. Kaiser, "Digital filters," in *System Analysis by Digital Computer*, F. F. Kuo and J. F. Kaiser, Eds. New York: Wiley, 1966, ch. 7.

[10] J. V. Wait, "Digital filters," in *Active Filters: Lumped, Distributed, Integrated, Digital, and Parametric*, L. P. Huelsman, Ed. New York: McGraw-Hill, 1970, p. 243.

[11] J. L. Butler, "Comparative criteria for minicomputers," *Instrum. Technol.*, vol. 17, pp. 67–82, Oct. 1970.

[12] D. Slepian, "On delta modulation," *Bell Syst. Tech. J.*, vol. 51, pp. 2101–2137, Dec. 1972.

[13] D. J. Goodman, "Delta modulation granular quantizing noise," *Bell Syst. Tech. J.*, vol. 48, pp. 1197–1217, June 1969.

[14] A. A. Reqvicha and A. Voelcker, "Design of nonrecursive digital filters," *IEEE Trans. Audio Electroacoust.*, vol. AU-18, pp. 464–470, Dec. 1970.

[15] F. B. Johnson, "Calculating delta modulator performance," *IEEE Trans. Audio Electroacoust.*, vol. AU-16, pp. 121–129, Mar. 1968.

[16] B. Liu, "Effect of finite word length on accuracy of digital filters—A review," *IEEE Trans. Circuit Theory*, vol. CT-18, pp. 670–677, Nov. 1971.

12

Reprinted from *Proc. 1974 IEEE EASCON,* Institute of Electrical and Electronics Engineers, Inc., 1974, pp. 464–468

Some New Realizations of Dedicated Hardware Digital Signal Processors

ABRAHAM PELED and BEDE LIU

IBM Thomas J. Watson Research Center
Yorktown Heights, New York 10598

Princeton, University
Princeton, New Jersey 08540

ABSTRACT

The significant breakthroughs in the area of semiconductor technology have opened up new options for the implementation of digital signal processors. We suggest some new hardware realizations of such dedicated processors that capitalize on the advances in semiconductor memory technology to produce realizations that have a significantly lower package count and power consumption, and also make possible higher speeds of operation. We specifically discuss the realization of digital filters and high speed Fast Fourier Transformers.

I. INTRODUCTION

Digital signal processors are becoming an increasingly attractive replacement for analog signal processors due to the continuing advances in semiconductor technology and the general growth of the digital environment in the systems where such processors are needed. Such processors are usually required to operate in real time, and, depending on the nature of the system they serve, they will be called upon to perform a finite number of functions.

In this paper we propose a new hardware implementation of such dedicated digital signal processors that requires no multiplications, with specific applications shown for digital filters and Fast Fourier Transformers. The new realization calls for the storing of the finite number of possible outcomes of an intermediate arithmetic operation, and using them to obtain the output through repeated addition and shifting operations. This hardware implementation is highly flexible and modular and uses only standard available IC's. We show that the proposed realization produces significant savings in terms of hardware and power consumption over existing realizations, and also makes possible very high speeds of operation.

II. A NEW APPROACH TO IMPLEMENTING DIGITAL SIGNAL PROCESSORS

A digital signal processor is usually called upon to perform an operation of the type

$$y_n = \sum_{j=1}^{L} a_j \cdot v_{n,j} \tag{1}$$

where $\{y_n\}$ is the desired output sequence, $\{a_j\}_{j=1}^{L}$ are a set of fixed coefficients determined by the processing function to be performed, and $\{v_{n,j}\}$ is the data sequence and may include previous output values. Assume that the sequence $\{v_{n,j}\}$ is scaled to $|v_{n,j}| < 1$ and is represented in 2's complement form [1] with B bits accuracy, that is

$$v_{n,j} = \sum_{k=0}^{B-1} v_{n,j}^{k} 2^{-k} r_k$$
$$r_k = \begin{cases} -1 & k=0 \\ 1 & k>0 \end{cases} \tag{2}$$

where $v_{n,j}^{k}$ are 0 or 1. Then we can rewrite Eq. (1) as

$$y_n = \sum_{j=1}^{L} a_j \left(\sum_{k=0}^{B-1} v_{n,j}^{k} 2^{-k} r_k \right) \tag{3}$$

and interchanging the order of summation, we obtain

$$y_n = \sum_{k=0}^{B-1} r_k \cdot 2^{-k} \left(\sum_{j=1}^{L} v_{n,j}^{k} a_j \right) \tag{4}$$

Let

$$F(v_{n,1}^{k}, v_{n,2}^{k}, \ldots v_{n,L}^{k}) = \sum_{j=1}^{L} v_{n,j}^{k} \cdot a_j \tag{5}$$

then, since $v_{n,j}^{k}$ can take on only 0 or 1, the function F(...) can take on only 2^L possible values, which can be precomputed for any given set $\{a_j\}_{j=1}^{L}$ and stored in a ROM (read only memory). Given this function F(...), the output can be computed from the following relation

$$y_n = \sum_{k=0}^{B-1} r_k \cdot 2^{-k} F(v_{n,1}^{k}, \ldots v_{n,L}^{k}) \tag{6}$$

using addition (subtraction for k=0) operations

only with a hard-wired right-skewed accumulator. This is illustrated in Fig. 1. The data sequence $\{v_{n,j}\}$ shifts serially into shift registers SR1 to SRL with the least significant bit leading. The $v^k_{n,j}$ bits address a ROM of $2^L \times B$ bits in which the function F(...) resides. The output of the ROM is added (subtracted for k=0) to the value in register R2, which is simply the previous output of the adder shifted right one bit. After B additions, y_n is obtained and R2 is cleared.

Obviously the main shortcomings of this approach are the lack of flexibility inherent in using a multiplier which permits the changing of the coefficients $\{a_j\}$ with relative ease, and the exponential dependence of the storage requirements on the number of processing coefficients.

In the following two paragraphs we discuss the application of the new approach to the implementation of digital filters and of Fast Fourier Transformers and show that for fixed systems requiring a moderate number of processing coefficients the proposed approach will result in more economic realizations. Finally we should mention that if we are willing to trade off storage for speed, Eq. (6) permits computing the expression in Eq. (1) with only B additions regardless of the number of processing coefficients L.

III. IMPLEMENTING DIGITAL FILTERS

It is well known that out of the numerous equivalent configurations that realize digital filters, the most advantageous are the cascade and parallel forms [2]. These realizations use a second order section digital filter as a basic building block to realize higher order filters. A second order section of a digital filter is characterized by an input-output relationship of the form

$$y_n = a_0x_n + a_1x_{n-1} + a_2x_{n-2} - b_1y_{n-1} - b_2y_{n-2} \quad (7)$$

with a_2=0 in the case of parallel realization, $\{y_n\}$ is the output sequence, and $\{x_n\}$ the input sequence. Thus the function F(...), as defined from Eq. (5) by the appropriate correspondence between Eq. (7) and Eq. (1), in this case has five binary arguments and therefore takes on at most 32 possible values. Thus for each second order section to be realized we will need 32 words of memory.

Assuming that standard TTL IC's and bipolar ROM's are used and assuming B=12 bits data words, such a second order section can operate at a word rate of about 1.6 MHz, will have a package count of approximately 20 IC's, and consume 9.6 W. This word rate implies that up to 200 channels of 4 KHz bandwidth can be processed in a time-shared fashion. (That is, if the same filtering is performed on all channels, otherwise additional memory is needed.) If ECL IC's are used to implement the second order section, the word rate could be 4 MHz, the package count 33 IC's, and the power consumption 26.4 W.

Fig. 2 illustrates the mechanization of an 8th order digital filter (four sections) realized in cascade form, using the proposed approach. This filter could operate at a word rate of 400 KHz (12 bit words) and will require only 30 IC's consuming roughly 12 W.

It is worth mentioning that the availability of high density ROM's of 512 x 8 bits such as Signetics 8205 implies that the arithmetic unit (excluding data registers and multiplexers) of the filter in Fig. 2 could be shared between 16 different second order sections without increasing the package count or power consumption.

For comparison we mention that assuming that parallel TTL fast multipliers of 4 bit by 4 bit, such as Texas Instruments' SN74284, SN74285, which multiply 8 bit by 12 bit numbers in 90 nsec. and require 24 IC's consuming 10 W, are used to perform the minimum of 12 multiplications needed in the 8th order filter, then to achieve the same operating speed at least six such multipliers would be needed, thus requiring in excess of 140 IC's for the multipliers only versus the 30 IC's needed by the new approach. More details on the digital filter implementations using this approach can be found in [3].

IV. IMPLEMENTING PIPELINE FAST FOURIER TRANSFORMERS (FFT)

The pipeline FFT system is a hardware architecture designed to speed up the data rate throughput for real time applications [4]. Essentially such a system is characterized by its $K=\log_2 N$ independent arithmetic units operating concurrently to increase the throughput rate K times. (N is the number of points to be transformed.) These arithmetic units may be called basic computation units (BCU).

For a radix 2 decimation-in-time [5] algorithm, these BCU's repeatedly perform a computation of the form:

$$A' = A + C \cdot W^k$$
$$C' = A - C \cdot W^k \qquad 0 \le k < N/2 \quad (8)$$

where A and C are two complex numbers, $W = \exp(-i2\pi/N)$ with $i=\sqrt{-1}$, and k an integer that is determined by the control depending on how many steps of the algorithm have been executed. Thus the BCU accepts an input triple [A,C,k] and produces an output pair [A',C'] in accordance with Eq. (2). If T is the time required for the BCU to compute the output [A',C'], then 1/T is the complex data throughput rate of such a pipeline FFT processor.

Let us examine in more detail the operations to be performed by the BCU. In the FFT, because of dynamic range considerations, it is desirable to use floating point representations for the data. Let the complex numbers A and C be represented in the following floating point format:

$$A = 2^{m_a}(a_r + ia_i)$$
$$C = 2^{m_c}(c_r + ic_i) \tag{9}$$

where the subscripts r and i denote respectively the real and imaginary parts, and the integers m_a and m_c are common exponents such that $-1 \leq a_r$, a_i, c_r, $c_i < 1$. Suppose 2's complement code is used, with B bits, to represent the mantissa (see Eq. (2)).

Eq. (8) can be rewritten as

$$A' = 2^{m_a}(a_r + ia_i) + Z$$
$$C' = 2^{m_a}(a_r + ia_i) - Z \tag{10}$$

where Z is given by

$$Z = 2^{m_c}(c_r + ic_i)\left(\cos\frac{2\pi k}{N} - i\sin\frac{2\pi k}{N}\right) \tag{11}$$

To compute A' and C' we first compute Z, which involves four real multiplications and two real additions. Then A and Z are scaled jointly and added (subtracted) to give A' and C'. To achieve maximum speed, four real multipliers, a scaler and six adders are needed.

Now let us consider an alternate way of computing Z, based on Eq. (6), i.e. without multiplication. The real and imaginary parts of Z can be written in terms of the function F(...) as follows.

$$z_r = \sum_{j=0}^{B-1} r_j 2^{-j} F_{k,r}(c_r^j, c_i^j)$$
$$z_i = \sum_{j=0}^{B-1} r_j 2^{-j} F_{k,i}(c_r^j, c_i^j) \tag{12}$$
$$c_r^j, c_i^j = 0 \text{ or } 1$$

and

$$F_{k,r}(u,v) = u\cos\frac{2\pi k}{N} + v\sin\frac{2\pi k}{N}$$
$$F_{k,i}(u,v) = -u\sin\frac{2\pi k}{N} + v\cos\frac{2\pi k}{N} \tag{13}$$
$$u,v = 0 \text{ or } 1$$

Since $F_{k,r}(.,.)$ and $F_{k,i}(.,.)$ can take on only four values each, the memory requirement for each is 4 × B bits. Since k can take on any value between 0 and N/2 - 1, and each stage needs its own memory for parallel operation, the total number of memory bits needed is

$$M_a = 8(N-1)B \text{ bits} \tag{14}$$

A detailed discussion of the implementation of such a BCU can be found in [6]. Here we only mention that with standard TTL IC's and bipolar memories the BCU of a 1024 point FFT is capable of operating at a throughput rate of 2.5 MHz complex data points and its implementation requires about 45 IC's consuming 21 W.

Finally, to demonstrate the flexibility of the new approach, we present another possible design of the BCU which is capable of a 25 MHz data throughput rate.

Eq. (12), through which z is computed, treats one data bit at a time. It is possible to examine and to operate upon several data bits at once. Let us illustrate this for the case that 2 bits are used at one time. Eq. (12) can be rewritten as

$$z_r = \sum_{j=0,2,4} 4^{-j/2}\Phi_k(c_r^j, c_r^{j+1}, c_i^j, c_i^{j+1})$$
$$z_i = \sum_{j=0,2,4} 4^{-j/2}\theta_k(c_r^j, c_r^{j+1}, c_i^j, c_i^{j+1}) \tag{15}$$

and Φ_k and θ_k are defined by

$$\Phi_k(c_r^j, c_k^{j+1}, c_i^j, c_i^{j+1}) = \left(r_j c_r^j + \frac{c_r^{j+1}}{2}\right)\cos\frac{2\pi k}{N} + \left(r_j c_i^j + \frac{c_i^{j+1}}{2}\right)\sin\frac{2\pi k}{N}$$
$$\theta_k(c_r^j, c_r^{j+1}, c_i^j, c_i^{j+1}) = -\left(r_j c_r^j + \frac{c_r^{j+1}}{2}\right)\sin\frac{2\pi k}{N} + \left(r_j c_i^j + \frac{c_i^{j+1}}{2}\right)\cos\frac{2\pi k}{N} \tag{16}$$

Since Φ_k and θ_k have four binary arguments, each can take on only $2^4 = 16$ possible values.

Let us consider a concrete example of a 1024 point FFT using B=8 bits for the mantissa and 4 bits for the exponent. Fig. 3 shows the block diagram of a possible hardware implementation of Eq. (15). As we see, four ROM's are used, and their outputs are hard wired into the adders in a skewed fashion to provide for the appropriate shifting needed. (Note that the contents of ROM 1 are different because of the sign bit.) The outputs of the four ROM's needed to permit parallel operation on the data bits are added in a tree-like structure, and the adders are provided with storage to allow concurrent operation of all stages. Thus, if we assume that 40 nsec is the ROM access time and also the addition time, each 40 nsec a new data word C can be inputted to the BCU. The block diagram of the complete arithmetic unit is shown in Fig. 4.

Again assuming TTL IC's, and bipolar memories only, 95 IC's consuming about 50 W will be needed for each BCU, including all the memory required by that BCU. To achieve a comparable throughput using multipliers requires some 270 IC's consuming 130 W. Since 10 such BCU's are needed for the 1024 point pipeline FFT the total savings for the system of 1700 less IC's and 800 W less power are indeed significant.

V. CONCLUSION

We have presented in this paper a new approach to the hardware implementation of digital signal processors that eliminates the use of multipliers. Consequently, significant savings in hardware and power consumption are possible. We illustrated the application of the new approach to the implementation of digital filters and high speed pipeline FFT's.

Basically, we are trading memory bits for logic gates. Current developments that made available very high density memories with low power dissipation have made our approach attractive. New technologies that are emerging, such as silicon on sapphire (SOS), point to an even more favorable exchange between memory bits and logic gates, a fact that will make the proposed approach even more economical.

REFERENCES

1. L. B. Jackson, J. F. Kaiser, H. S. McDonald, "An Approach to the Implementation of Digital Filters", IEEE Trans. Audio Electroacoust. Vol. AU-16, pp. 413-421, Sept. 1968.

2. B. Liu, "Effect of Finite Word Length on the Accuracy of Digital Filters - A Review", IEEE Trans. Circuit Theory, Vol. CT-18, pp. 670-677, Nov. 1971.

3. A. Peled and B. Liu, "A New Hardware Realization of Digital Filters", to appear in IEEE Trans. Acoustics, Speech, and Signal Processing.

4. Herbert L. Groginsky and George A. Works, "A Pipeline Fast Fourier Transform", IEEE Trans. Comput. Vol. C-19, pp. 1015-1019, Nov. 1970.

5. B. Gold and C. Rader, Digital Signal Processing, McGraw-Hill, New York, 1969.

6. B. Liu and A. Peled, "A New Hardware Realization of High Speed Fast Fourier Transformers", to appear.

ACKNOWLEDGMENT:

This research at Princeton University was supported by the Air Force Office of Scientific Research under Grant AF-AFOSR 71-2101, and by the National Science Foundation under Grant GK-24187.

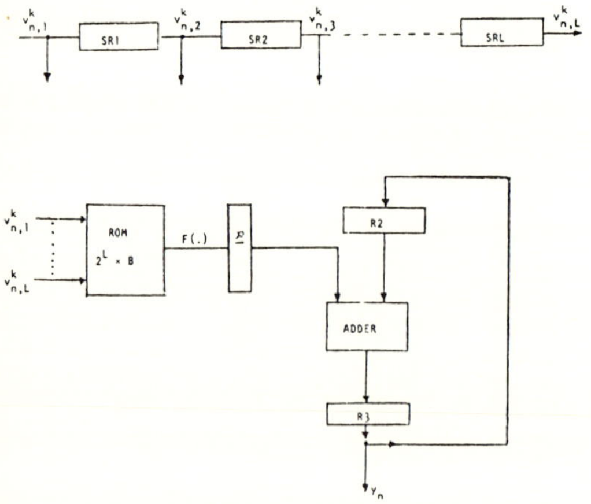

Fig. 1 - A possible mechanization of Eq. (6)

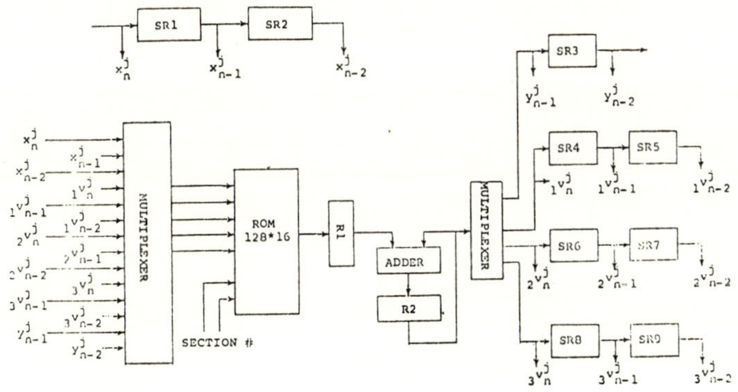

Fig. 2 An 8th Order Digital Filter Realized in Cascade Form

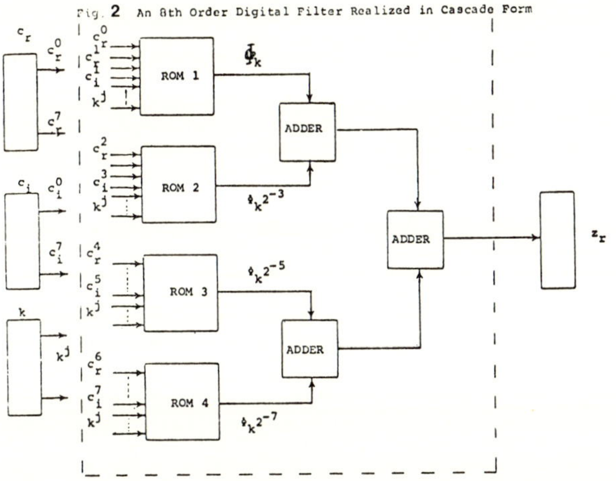

Figure 3 Half of MODULE 3, that performs a parallel computation of Z. (The other half of MODULE 3 computes z_i with an identical configuration.)

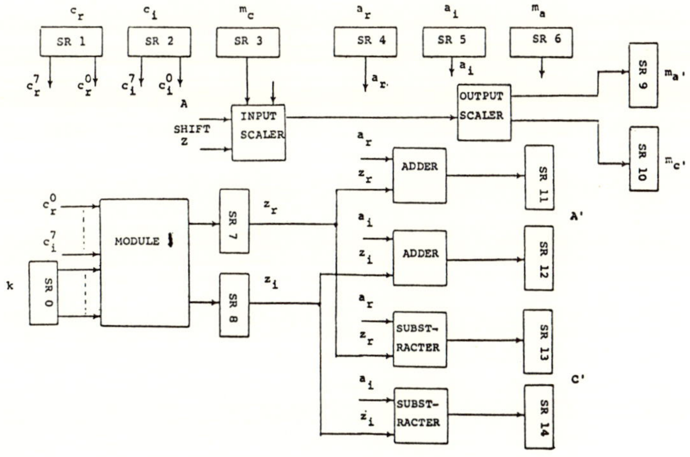

Figure 4 A BCU for pipeline FFT using MODULE 3 for parallel operation to achieve very high throughput rates.

III
Effects of Finite Word Length

Editor's Comments on Papers 13 Through 23

13 Oppenheim and Weinstein: *Effects of Finite Register Length in Digital Filtering and the Fast Fourier Transform*

14 Knowles and Edwards: *Effect of a Finite-Word-Length Computer in a Sampled-Data Feedback System*

15 Jackson: *Roundoff-Noise Analysis for Fixed-Point Digital Filters Realized in Cascade or Parallel Form*

16 Sandberg: *Floating-Point-Roundoff Accumulation in Digital-Filter Realizations*

17 Liu and Kaneko: *Error Analysis of Digital Filters Realized with Floating-Point Arithmetic*

18 Avenhaus: *On the Design of Digital Filters with Coefficients of Limited Word Length*

19 Otnes and McNamee: *Instability Thresholds in Digital Filters due to Coefficient Rounding*

20 Long and Trick: *An Absolute Bound on Limit Cycles due to Roundoff Errors in Digital Filters*

21 Sandberg and Kaiser: *A Bound on Limit Cycles in Fixed-Point Implementations of Digital Filters*

22 Ebert, Mazo, and Taylor: *Overflow Oscillations in Digital Filters*

23 Kaneko: *Limit-Cycle Oscillations in Floating-Point Digital Filters*

There is an inherent limitation in the accuracy of digital filters owing to the fact that all digital networks operate with a finite number of bits. The papers collected in this part deal with this class of problems. In a linear time-invariant digital filter, three sources of errors arise from the use of finite word length:

1. The quantization of the input signal to a finite number of levels, so that each sample can be coded into a word consisting of a fixed number of bits.
2. The representation of the filter coefficients by a finite number of bits.
3. The accumulation of roundoff errors committed at the arithmetic operations.

In addition to the word length, the accuracy of a digital filter is affected by the form of realization and the type of arithmetic used. If a digital filter is used to process analog signals, there may be other sources of error as a result of sampling of the input analog signal and the reconstruction of output analog signals (Thomas et al., 1964; Liu, et al., 1967; Liu, 1973).

Although problems that arise from the use of finite word length are basically nonlinear in nature, many of them can be treated as essentially linear problems with the error sources handled as independent additive or multiplicative noises. In other problems, the nonlinear nature must be dealt with directly.

The first paper, by Oppenheim and Weinstein [13], gives a thorough overall view of this subject. In addition, the discussion at the beginning of the paper on the binary representation of numbers and on the error caused by arithmetic roundoff are fundamental to the understanding of most of the articles in this part.

Although the paper by Knowles and Edwards [14] deals with a feedback system, its method of analysis can be applied directly to the calculation of roundoff errors in fixed-point digital filters. Their paper also demonstrates that a high-order filter should be realized as a cascade or parallel connection of first- and second-order filter sections from the error point of view. In a paper appearing at about the same time, Gold and Rader carry out a detailed analysis of the roundoff error for first- and second-order filters (Gold et al., 1966). The approach used in these analyses is based on two assumptions: (1) the roundoff error due to multiplication can be modeled statistically by an additive independent white-noise component at the output of each multiplier, and (2) the addition operation generates no roundoff error.

From simulation studies, it appears that assumption (1) holds very well if the word length is eight bits or more. The second asumption is valid only if there is no overflow of the results of addition. Since overflow also causes other problems, scaling must be introduced so that the signals stay in the proper dynamic range. Constraints on the dynamic range have been derived (Jackson, 1970) in terms of the L_p norms of the input signal spectrum and the transfer response to selected "nodes" within the filter. In the next paper, by Jackson [15], roundoff error is analyzed under this dynamic range constraint for series and parallel realizations. The paper also proposes a rule for arriving at a good ordering of the second-order sections and the pairing of poles and zeros in each section for cascade realization. This is an important problem, since the output noise depends on the ordering and pairing, and an exact solution for finding the optimum configuration would require an excessive amount of computing even for moderate-order filters.

While the roundoff noise for fixed-point arithmetic enters into the system additively, it is a multiplicative component in the case of floating-point arithmetic. This problem is analyzed in Sandberg's paper [16], where a bound on the time average of the squared error at the output is derived. The next paper [17], by Liu and Kaneko, calculates the statistical mean-squared value of the roundoff error for direct, parallel, and cascade realizations. The analysis is somewhat involved, as the problem cannot be reduced to one of a linear time-invariant system subject to white-noise input. The approach used by Liu and Kaneko is justified using a result in Sandberg's paper.

The roundoff error for digital filters using block floating point has been calculated (Oppenheim, 1970). Comparisons of the roundoff noise in filters using different types of arithmetics have also been reported (Oppenheim, 1970; Weinstein et al., 1969).

A further comment on the significance of the papers analyzing the roundoff error is in order. If a digital filter is implemented by programming a general-purpose computer, it is more likely that floating-point arithmetic will be used because of its better dynamic range. Since the word length is usually fixed in this case, floating-point roundoff error analysis is useful in that it enables one to calculate the error committed. If the error should turn out to be unsatisfactory, one can resort to double-precision arithmetic. If, on the other hand, a special-purpose hardware filter is constructed, it is more likely that fixed-point arithmetic is used because of its relative simplicity. In this case, one is more interested in determining the minimum word length that can keep the error within a prescribed tolerance.

Of the three sources of error mentioned previously, the input quantization can be analyzed straightforwardly by injecting an independent additive noise at the filter input. The coefficient quantization problem can be treated in a number of ways. The change in the filter coefficients results in a change in the filter characteristics. In addition, the filter poles and zeros are moved and the movement in the poles may result in instability. This problem was discussed in Kaiser's article [1] in Part I, and some results are extended in the paper on Otnes and McNamee [19]. For other works on the pole movements, see the References (Rader et al., 1967; Mitra et al., 1974; Kaneko et al., 1971).

Usually the transfer function of a filter is designed first without regard to the use of finite length word, the effects of which are then considered when the filter is implemented. The paper by Avenhaus [18] takes a different point of view; it attempts to design the transfer function with a fixed word length for the coefficients at the beginning by means of some search techniques in the (discrete parameter) coefficient space. An example shows a saving of three bits over simple coefficient rounding when using his approach. The design problem using limited word length is also discussed by Steiglitz (1971) and Suk et al. (1972).

Theoretically, a digital filter is stable if all its poles are inside the unit circle. However, when a stable filter is implemented with finite-precision arithmetics, oscillations are often observed at the output in the absence of input. The statistical approach to the study of roundoff error in fixed-point filters assumes that the error sequence generated at a multiplier is uncorrelated with the signal sequence. The results obtained using such an approach have been confirmed by experimental observation when the signal levels are high compared with the least-significant bit. The assumption of uncorrelatedness is no longer valid when oscillation occurs with zero input. In this case, the oscillation is caused by error sequences that are highly correlated with the signal sequence. Jackson used a linearized effective-value model to derive an approximate bound on the amplitude of such oscillation for first- and second-order filters (Jackson, 1969). Parker and Hess performed a detailed analysis of these limit cycles and obtained bounds on the amplitude (Parker et al., 1971). These bounds reduce to those of Jackson's for oscillations of periods 1 and 2. For arbitrary period oscillations, the bounds are expressed in determinant form. In the paper by Long and Trick [20], an amplitude bound is derived for second-order filters in terms of the filter coefficients. A bound on the root-mean-square value of limit cycles is derived in the paper by Sandberg and Kaiser [21].

In addition to the limit cycles due to roundoff multiplication, digital filters implemented in two's-complement or one's-complement arithmetics have been observed to produce large oscillations. This type of oscillation is caused by overflow at the adders. The paper by Ebert, Mazo, and Taylor [22] presents a thorough analysis of this problem. It is shown there that such oscillations can be eliminated by a modification of the adder so that it saturates when overflow occurs. Roundoff errors, however, have been neglected. Taking the roundoff errors into account, Sandberg showed that, when saturation arithmetic is used, the amplitude of these overflow oscillations can be made arbitrarily small by increasing the word length (Sandberg,

1969). Wilson considered a generalized saturation arithmetic and a "zeroing" arithmetic for the adder and derived conditions under which the amplitude of these limit cycles can be made arbitrarily small when sufficient numbers of bits are used to represent the data (Wilson, 1972).

For filters using floating-point arithmetic, it was once thought that limit-cycle oscillations would be negligibly small if and when they exist. However, this is not always true, and the paper by Kaneko [23] deals with this problem.

References

Gold, B., and C. M. Rader (1966). Effects of Quantization Noise in Digital Filters, *Proc. AFIPS Spring Joint Computer Conf.*, **28,** 213–219.

Jackson, L. B. (1969). An Analysis of Limit Cycles Due to Multiplication Rounding in Recursive Digital Filters, *Proc. 7th Allerton Conf. Circuit Syst. Theory,* pp. 69–79.

——— (1970). On the Interaction of Roundoff Noise and Dynamic Range in Digital Filters, *Bell Syst. Tech. J.*, **49,** 159–184.

Kaneko, T., and B. Liu (1971). Effect of Coefficient Rounding in Floating Point Digital Filters, *IEEE Trans. Aerospace Electron. Syst.*, **AES-7,** 995–1003.

Liu, B. (1973). Timing Jitter in Digital Filtering, *Proc. 16th Midwest Symp. Circuit Theory,* **1,** II.4.1–II.410.

———, and J. B. Thomas (1967). Error Problems in the Reconstruction of Signals from Sampled Data, *Proc. Natl. Electron. Conf.*, **23,** 803–807.

Mitra, S. K., and J. J. Sherwood (1974). Estimation of Pole–Zero Displacements of a Digital Filter Due to Coefficient Rounding, *IEEE Trans. Circuits Syst.* **CAS121,** 116–124.

Oppenheim, A. V. (1970). Realization of Digital Filters Using Block-Floating-Point Arithmetic, *IEEE Trans. Audio Electroacoust.*, **AU-18,** 130–136.

Parker, S. R., and S. F. Hess (1971). Limit-Cycle Oscillations in Digital Filters, *IEEE Trans. Circuit Theory,* **CT-8,** 687–697.

Rader, C. M., and B. Gold (1967). Effects of Parameter Quantization on the Poles of a Digital Filter, *Proc. IEEE (Lett.),* **55,** 688–689.

Sandberg, I. W. (1969). A Theorem Concerning Limit Cycles in Digital Filters, *Proc. 7th Allerton Conf. Circuit Syst. Theory,* pp. 63–38.

Steiglitz, K. (1971). Designing Short-Word Recursive Digital Filters, *Proc. 9th Allerton Conf. Circuit Syst. Theory,* pp. 778–788.

Suk, M., and S. K. Mitra (1972). Computer Aided Design of Digital Filters with Finite Word Length, *IEEE Trans. Audio Electroacoustics,* **AU-20,** 356–363.

Thomas, J. B., and B. Liu (1964). Error Problems in Sampling Representations, *IEEE Int. Conv. Rec.*, Pt. 5, pp. 269–277.

Weinstein, C., and A. V. Oppenheim (1969). A Comparison of Round-off Noise in Floating Point and Fixed Point Digital Filter Realizations, *Proc. IEEE (Corresp.),* **57,** 1181–1183.

Wilson, A. N. (1972). Limit Cycle Due to Overflow in Digital Filters, *IEEE Trans. Circuit Theory,* **CT-19,** 342–346.

13

Reprinted from *Proc. IEEE,* **60**(8), 957–976 (1972)

Effects of Finite Register Length in Digital Filtering and the Fast Fourier Transform

ALAN V. OPPENHEIM, SENIOR MEMBER, IEEE, AND
CLIFFORD J. WEINSTEIN, MEMBER, IEEE

Invited Paper

***Abstract*—When digital signal processing operations are implemented on a computer or with special-purpose hardware, errors and constraints due to finite word length are unavoidable. The main categories of finite register length effects are errors due to A/D conversion, errors due to roundoffs in the arithmetic, constraints on signal levels imposed by the need to prevent overflow, and quantization of system coefficients. The effects of finite register length on implementations of linear recursive difference equation digital filters, and the fast Fourier transform (FFT), are discussed in some detail. For these algorithms, the differing quantization effects of fixed point, floating point, and block floating point arithmetic are examined and compared.**

The paper is intended primarily as a tutorial review of a subject which has received considerable attention over the past few years. The groundwork is set through a discussion of the relationship between the binary representation of numbers and truncation or rounding, and a formulation of a statistical model for arithmetic roundoff. The analyses presented here are intended to illustrate techniques of working with particular models. Results of previous work are discussed and summarized when appropriate. Some examples are presented to indicate how the results developed for simple digital filters and the FFT can be applied to the analysis of more complicated systems which use these algorithms as building blocks.

I. Introduction

In PRACTICE, digital signal processing requires the representation of sequence values in a binary format with a finite register length. The effect of the finite word-length constraint manifests itself in several different ways. If a sequence to be processed is derived by sampling an analog waveform, then the finite word-length constraint requires that the analog-to-digital conversion produce only a finite number of values. This represents quantization of the input waveform. Even when we start with data representable with a finite word length, the result of processing will naturally lead to values requiring additional bits for their representation. For example, a b-bit data sample multiplied by a b-bit coefficient results in a product which is $2b$ bits long. If in a recursive digital filter we do not quantize the result of arithmetic operations, the number of bits required will increase indefinitely, since after the first iteration $2b$ bits are required, after the second iteration $3b$ bits are required, etc. The effect of quantization in such a context depends on such factors as whether we are considering fixed-point or floating-point arithmetic, and whether for fixed-point arithmetic we are using a representation of numbers in terms of fractions or integers, or perhaps a mixture. We will be treating the case of fixed-point arithmetic and floating-point arithmetic separately. For fixed-point arithmetic, it is natural in a signal processing context to consider a register as representing a fixed-point fraction. In this way the product of two numbers remains a fraction and the limited register length can be maintained by truncating or rounding the least significant bits. With this type of representation the result of addition on fixed-point fractions need not be truncated or rounded but it can increase in magnitude so that the sum eventually is not a fraction. This effect is commonly referred to as overflow, and can be handled by requiring that the input data be sufficiently small so that the possibility of overflow is avoided. In considering floating-point arithmetic, dynamic range considerations generally can be neglected due to the large range of representable numbers, but quantization is introduced both for multiplication and for addition.

A third effect of finite word length is inaccuracies in parameter values. While generally signal processing parameters are initially specified with unlimited accuracy, they can only be utilized with finite word length. This effect is similar to the effect which arises in implementing analog processing using inaccurate circuit elements. There are two possible approaches to handling the inaccuracies in parameter values. One possibility is to develop design procedures which inherently are insensitive to parameter inaccuracies. An alternate is to choose specifications which are consistent with the limited register length. There is a certain amount that is understood about the effect of inaccuracies in parameter values, but for the most part present results lead to guidelines rather than hard design or analytical strategies.

In the following discussion the relationship between the binary representation of numbers and truncation or rounding is discussed and a statistical model for arithmetic roundoff is presented. This statistical model is then applied to the analysis of fixed-point and floating-point rounding errors in digital filters. The analysis includes a consideration of the effect of dynamic range in developing and comparing signal-to-noise ratios for fixed-point and floating-point filters. It is not always possible to treat the effects of arithmetic roundoff in terms of a simple statistical model. Some approaches and results are available in the literature on the limit cycle behavior of digital filters due to arithmetic roundoff, and a discussion of some of these results is included.

For the analysis of arithmetic roundoff in computation of the discrete Fourier transform using the fast Fourier trans-

Manuscript received May 11, 1972. A. V. Oppenheim was supported in part by the National Science Foundation under Grant GK-31353 and in part by the Advanced Research Project Agency of the Department of Defense, monitored by ONR under Contract N00014-67-A-0204-0064; C. J. Weinstein was supported in part by the U. S. Air Force. *This invited paper is one of a series planned on topics of general interest—The Editor.*

A. V. Oppenheim is with the Department of Electrical Engineering and the Research Laboratory of Electronics, Massachusetts Institute of Technology, Cambridge, Mass. 02139.

C. J. Weinstein is with Lincoln Laboratory, Massachusetts Institute of Technology, Lexington, Mass. 02173.

form (FFT) algorithm a statistical model is used. With this model the signal-to-noise ratio is developed and compared for fixed-point and floating-point arithmetic.

While for any given filter configuration or spectral analysis problem it can be difficult to carry out a detailed analysis of the effects of finite register length there are a number of general guidelines that can be distilled from the results presented here. In Section IV some examples and guidelines are presented for filters implemented with fixed-point arithmetic and with floating-point arithmetic as well as for filters implemented with the FFT.

This paper is intended primarily as a tutorial review of a subject which has received considerable attention over the past few years. The analyses which are presented here are selected to illustrate techniques of working with particular models. Previous work is freely referenced, discussed, and borrowed from.

II. Number Representation and Its Effect on Quantization

A. Fixed-Point and Floating-Point Numbers

The manner in which finite word-length effects are manifested is closely tied to the way in which numbers are represented.

Digital computers and special purpose digital hardware for the most part use a number representation with a radix of 2, i.e., a binary representation. Therefore, a number is represented by a sequence of binary digits which are either zero or unity. Just as a decimal number is represented as a string of decimal digits with a decimal point dividing the integer part from the fractional part, the sequence of binary digits is divided by a binary point into those representing the integer part of the number and those representing the fractional part. Thus if Δ denotes the location of the binary point, the binary number $1001_\Delta 0110$ has the decimal value of $(1\times 2^3+0\times 2^2+0\times 2^1+1\times 2^0)+(0\times 2^{-1}+1\times 2^{-2}+1\times 2^{-3}+0\times 2^{-4})$. This representation always corresponds to a positive number.

The manner in which arithmetic is implemented in a digital computer or in a special purpose hardware depends on where in the register the binary point is located. For fixed-point arithmetic, the implementation is based on the assumption that the location of the binary point is fixed. The manner in which addition is carried out will not depend on the location of the binary point for fixed-point arithmetic as long as it is the same for every register. For multiplication, however, the location of the binary point must be known. For example, consider the product of the two 4-bit numbers 1001_Δ and 0011_Δ. In general, of course, the product of two b-bit numbers will be $2b$ bits long. The 8-bit product of the above number is 00011011_Δ. If, on the other hand, we consider the 4-bit fractions $_\Delta 1001$ and $_\Delta 0011$, then the 8-bit product is $_\Delta 00011011$. In digital filtering applications, it is usually necessary to approximate the $2b$-bit product of two b-bit numbers by a b-bit result. In integer arithmetic this is difficult. With fractional arithmetic, on the other hand, this can be accomplished by truncating or rounding to the most significant b bits. For multiplication with fractions, overflow can never occur since the product of two fractions is a fraction. Thus for the 4-bit example previously mentioned, the product $_\Delta 00011011$ can be approximated by $_\Delta 0001$ (truncation) or $_\Delta 0010$ (rounding).

An alternative to fixed-point arithmetic is a floating-point representation. In this case, a positive number F is represented as $F=2^c M$, where M, the mantissa, is a fraction between 1/2 and 1, and c, the characteristic, can be either positive or negative. The product of two floating-point numbers is carried out by multiplying the mantissa as fixed-point fractions and adding the characteristics. Since the product of the mantissas will be between 1/4 and 1, a normalization of the mantissa and corresponding adjustment of the characteristic may be necessary. The sum of two floating-point numbers is carried out by scaling the mantissas of the smaller number to the right until the characteristics of the two numbers are equal and then adding the mantissas. For example, consider the sum of F_1 and F_2 with $F_1=4$ and $F_2=5/4$. Then in floating-point notation, $F_1=2^{c_1}M_1$, and $F_2=2^{c_2}M_2$ with

$$
\begin{aligned}
c_1 &= 11_\Delta \quad (=3 \text{ decimal})\\
M_1 &= {}_\Delta 1000 \quad (=0.5 \text{ decimal})\\
c_2 &= 1_\Delta \quad (=1 \text{ decimal})\\
M_2 &= {}_\Delta 1010 \quad (=5/8 \text{ decimal}).
\end{aligned}
$$

In order to carry out the addition, c_2 must be changed to equal c_1 and M_2 must be adjusted accordingly. Thus first the representation of F_2 is changed to $F_2=2^{\hat{c}_2}\hat{M}_2$ with

$$
\begin{aligned}
\hat{c}_2 &= 11_\Delta\\
\hat{M}_2 &= {}_\Delta 00101
\end{aligned}
$$

in which case the mantissas can now be added. The resulting sum is $F=2^c M$ with $c=11_\Delta$ and $M={}_\Delta 10101$. In this case the sum of M_1 and $\hat{M}_2$ is a fraction between 1/2 and 1 and therefore no further adjustment of c has to be carried out. In a more general case, the sum may not be in that range, and consequently, c would be adjusted to bring the mantissa into the proper range. From this example it should be clear that in general with floating-point arithmetic, the mantissa can exceed the register length and must therefore be truncated or rounded for both addition and multiplication whereas this is only necessary for multiplication in the fixed-point case. On the other hand, if the result of addition in the fixed-point case exceeds the register length, truncation or rounding will not help, i.e., the dynamic range has been exceeded. Thus while floating point introduces error due to arithmetic roundoff, it provides much greater dynamic range than fixed point. As we will see later, both of these effects must be considered when comparing fixed-point and floating-point realizations of digital filters.

B. Representation of Negative Numbers

There are three common means used for representing fixed-point negative numbers. The first, and most familiar, is sign and magnitude, i.e., the magnitude (which is of course positive) is represented as a binary number and the sign is represented by the leading binary digit which, if 0 corresponds to a + and if 1 corresponds to a − (or vice versa). Thus for example, in sign and magnitude $0_\Delta 0011$ represents 3/16 and $1_\Delta 0011$ represents −3/16. Two other related representations of negative numbers are often referred to as one's-complement and two's-complement representations. Considering all numbers to be fractions, a positive number is represented as before. For two's complement representation a negative number is represented by 2.0 minus its magnitude. For example $-(0_\Delta 0110)$ in sign and magnitude is represented as $1_\Delta 1010$ in two's-complement since $10_\Delta 000-0_\Delta 0100=1_\Delta 1010$. For one's-complement, the negative number is represented by subtracting the magnitude from the largest number representable in the register. Thus $-(0_\Delta 0110)$ is represented by

$(1_\Delta 1111)-(0_\Delta 0110)=1_\Delta 1001$. One's complement representation is equivalent to representing a negative number by the bit-by-bit complement of its magnitude. The choice of representation for negative numbers in a particular system is usually based almost entirely on hardware considerations.

For the representation of negative floating-point numbers there are a variety of conventions that have been used. In this paper we will consider the sign of the number to be associated with the mantissa so that the mantissa is a signed fraction. The representation of this signed fraction can of course be in sign and magnitude, one's-complement of two's-complement notation.

C. A Model for Arithmetic Roundoff

In formulating a model for arithmetic roundoff, we shall consider both fixed-point numbers and mantissas of floating-point numbers to be represented as $b+1$-bit binary fractions, with the binary point just to the right of the highest order bit (or sign bit). This convention represents no loss of generality, and its convenience has been alluded to above. The numerical value (for positive numbers) of a one in the least significant bit is 2^{-b}, and this quantity can be referred to as the width of quantization.

As indicated previously, the effect of finite register length on the result of arithmetic operations depends on whether fixed-point or floating-point arithmetic is used, and how negative numbers are represented. Let us consider first the effect of truncation and rounding in the fixed-point case. For sign and magnitude, one's-complement and two's-complement, the representation of positive numbers is identical and, consequently, so is the effect of truncation and rounding. If E_T denotes the error due to truncation, i.e., the value after truncation minus the value before truncation, this error will always be negative for positive numbers. That is, the effect of truncation is to reduce the value of the numbers. More specifically, if b_2 denotes the number of bits (exclusive of sign) after truncation, and b_1 denotes the number of bits before truncation, then the result satisfies $0\geq E_T\geq -(2^{-b_2}-2^{-b_1})$.

With sign and magnitude representation of negative numbers, truncation reduces the magnitude of the number and the error E_T satisfies $0\leq E_T\leq(2^{-b_2}-2^{-b_1})$. For a two's-complement negative number represented by the bit string $1_\Delta, a_1, a_2, \cdots, a_{b_1}$, the magnitude is given by

$$M_1 = 2.0 - x_1$$

where

$$x_1 = 1 + \sum_{i=1}^{b_1} a_i 2^{-i}.$$

Truncation to b_2 bits ($b_2<b_1$) produces the bit string $1_\Delta, a_1, a_2, \cdots, a_{b_2}$, where now the magnitude is

$$M_2 = 2.0 - x_2$$

with

$$x_2 = 1 + \sum_{i=1}^{b_2} a_i 2^{-i}.$$

The change in magnitude is

$$\Delta M = M_2 - M_1 = \sum_{i=b_2+1}^{b_1} a_j 2^{-j}$$

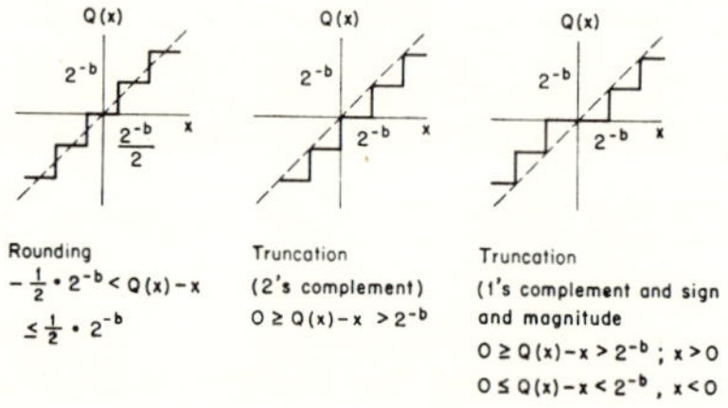

Fig. 1. Transfer characteristics for rounding and truncation.

and it is easily seen that

$$0 \leq \Delta M \leq 2^{-b_2} - 2^{-b_1}.$$

Hence the effect of truncation for two's-complement negative numbers is to *increase* the magnitude of the negative number; the truncation error is negative, and satisfies $0\geq E_T\geq -(2^{-b_2}-2^{-b_1})$.

For a one's-complement negative number represented by the bit string $1_\Delta, a_1, a_2, \cdots, a_{b_1}$, the magnitude is given by

$$M_1 = 2.0 - 2^{-b_1} - x_1$$

and truncation to b_2 bits yields a magnitude

$$M_2 = 2.0 - 2^{-b_2} - x_2$$

where x_1 and x_2 are as defined above. The change in magnitude is

$$\Delta M = M_2 - M_1 = \sum_{i=b_2+1}^{b_1} a_i 2^{-i} - (2^{-b_2} - 2^{-b_1})$$

and now

$$0 \geq \Delta M \geq -(2^{-b_2} - 2^{-b_1}).$$

Hence the effect of truncation for one's complement negative numbers is to *decrease* the magnitude of the negative number; the truncation error is positive, and satisfies $0\leq E_T\leq 2^{-b_2}-2^{-b_1}$.

The effect of rounding, of course, will be the same independent of how negative numbers are represented and the rounding error will always be greater than or equal to $(-1/2)2^{-b}$ and less than or equal to $(+1/2)2^{-b}$. The effect of truncation and rounding for the fixed-point case is summarized in Fig. 1, where x represents the value before truncation or rounding and $Q(x)$ represents the value after. In the figure it is assumed that x can take on a continuous range of values, corresponding to $b_1=\infty$ in the discussion above, and that the quantized word length is b bits plus sign.

For the case of floating-point arithmetic, the effect of truncation or rounding is reflected only in the mantissa. It is convenient in the floating-point case to describe the error in a multiplicative sense rather than in an additive sense as is done in fixed-point arithmetic. In other words, for a floating-point word, if x represents the value before truncation or rounding and $Q(x)$ represents the value after, then we express $Q(x)$ as equal to $x(1+\epsilon)$. For the case of rounding, for example, the error in the mantissa is between $\pm 2^{-b}/2$, and consequently the error in the value of the floating-point word is

$$-2^c \cdot \frac{2^{-b}}{2} \leq Q(x) - x \leq 2^c \cdot \frac{2^{-b}}{2}$$

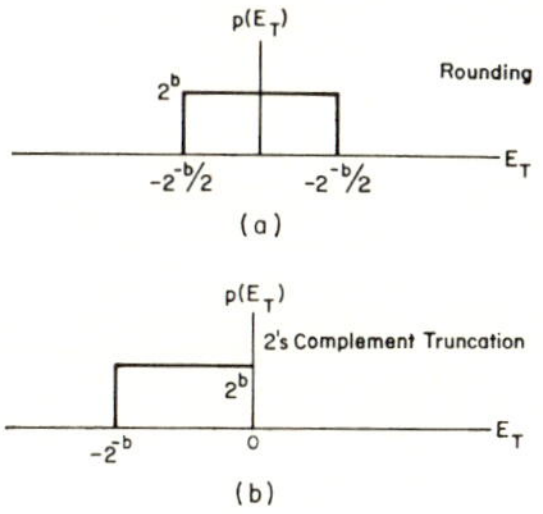

Fig. 2. (a) Probability density function for rounding noise. (b) Probability density function for noise due to two's-complement truncation.

or, since $Q(x)-x=\epsilon x$

$$-2^c\cdot\frac{2^{-b}}{2}\leq \epsilon x\leq 2^c\cdot\frac{2^{-b}}{2}$$

and since $2^{c-1}\leq x<2^c$, we can write that for the case of rounding $-2^{-b}\leq\epsilon\leq 2^{-b}$. In a similar manner we can show that for one's-complement and for sign and magnitude truncation $0\geq\epsilon\geq -2\cdot 2^{-b}$. For two's-complement truncation

$$0\geq\epsilon\geq -2\cdot 2^{-b},\qquad x>0$$

$$0\leq\epsilon\leq 2\cdot 2^{-b},\qquad x<0.$$

D. Statistical Model of Arithmetic Roundoff

A convenient means for analyzing the effect of quantization is to represent the error statistically [1], [2]. In particular, for the case of fixed-point arithmetic and rounding E_T is represented as a random variable with a probability density shown in Fig. 2(a). For the case of two's-complement truncation, the probability density is shown in Fig. 2(b).

In each of these cases, the assumption is that the random variable E_T is independent of x. For one's complement and sign magnitude truncation, this assumption cannot be made since the mean value of the error is directly correlated with the sign of x. In the analysis that follows for fixed-point arithmetic, the discussion is phrased in terms of rounding. The results are easily modified for two's-complement truncation. In particular the variance of the noise is identical for both cases. However, for rounding the noise is zero mean and for two's-complement truncation it is not zero mean.

For the floating-point case, the parameter ϵ is considered to be a random variable which is independent of x. In that case the assumption of independence is reasonable for rounding, sign and magnitude truncation, and one's-complement truncation, but not for two's-complement truncation. The random variable ϵ is bounded by $-2^{-b}\leq\epsilon\leq 2^{-b}$. We will generally assume ϵ to be uniformly distributed in this range with a variance $\sigma_\epsilon^2=(1/3)2^{-2b}$. Empirical work has shown that the distribution is not quite uniform so that while σ_ϵ^2 is proportional to 2^{-2b}, the constant of proportionality is slightly less than 1/3. However, the interpretation of the results depends primarily on the proportionality to 2^{-2b}.

III. Finite Register Length Effects for Digital Filters [3]

A. Introduction

The basic arithmetic operations involved in implementation of a digital filter are multiplication by a constant and addition. For fixed-point arithmetic, roundoff is introduced only after the multiplication. Because of the possibility of overflow due to addition, there is a dynamic range limitation in fixed-point filters. In contrast, floating-point filter implementation has a much less severe dynamic range constraint, although arithmetic roundoff is introduced due to both multiplication and addition. In the next sections we will first develop the statistical analysis of arithmetic roundoff for fixed-point filters including dynamic range considerations. This is followed by a statistical analysis for floating-point arithmetic and a discussion of zero input limit cycle behavior for fixed-point arithmetic.

B. Statistical Analysis of Fixed-Point Errors in a Digital Filter [4], [5]

In many situations it is reasonable to model the effect of rounding in a digital filter by a simple statistical model. The approach is to model the effect of the rounding at each multiplier by a white-noise source uniformly distributed in amplitude between plus and minus $(1/2)2^{-b}$. Each of the noise sources is assumed to be linearly independent of each other and of the input. Experimentally these assumptions have been justified for a broad class of inputs including random signals, speech, etc. The model is clearly not valid for certain inputs, such as constant inputs. If the impulse response from the kth noise source to the output is $h_k(n)$ then the steady-state output noise variance due to the kth noise source is

$$\sigma_{ok}^2=\sigma_\epsilon^2\sum_{n=0}^{\infty}h_k^2(n)\tag{1}$$

where $\sigma_\epsilon^2=(1/12)2^{-2b}$. Since all the noise sources are assumed to be uncorrelated, the total output noise is

$$\sigma_o^2=\sum_k\sigma_{ok}^2.\tag{2}$$

For example, if we consider the first-order filter in Fig. 3 one noise source is introduced. In this case, the impulse response from the noise source input to the output is $h(n)=a^n u_{-1}(n)$ where u_{-1} denotes a unit step sequence, so that

$$\sigma_o^2=\frac{1}{12}2^{-2b}\frac{1}{1-a^2}.\tag{3}$$

For a second-order filter with one complex pole pair there are two noise sources as indicated in Fig. 4. The resulting output noise is

$$\sigma_o^2=\frac{2}{12}2^{-2b}\left(\frac{1+r^2}{1-r^2}\frac{1}{r^4+1-2r^2\cos 2\theta}\right).\tag{4}$$

C. Dynamic Range Considerations for Fixed-Point Filters

As indicated previously, the possibility of overflow must be considered in the implementation of digital filters with fixed-point arithmetic. With the convention that each fixed-point register represents a signed fraction, each node in the filter must be constrained to maintain a magnitude less than unity in order to avoid overflow. Letting $x(n)$ denote the filter input and $y_k(n)$ and $h_k(n)$ denote the output and unit sample response for the kth node in the filter, then

$$y_k(n)=\sum_{r=0}^{\infty}h_k(r)x(n-r).\tag{5}$$

If x_{max} denotes the maximum of the absolute value of the in-

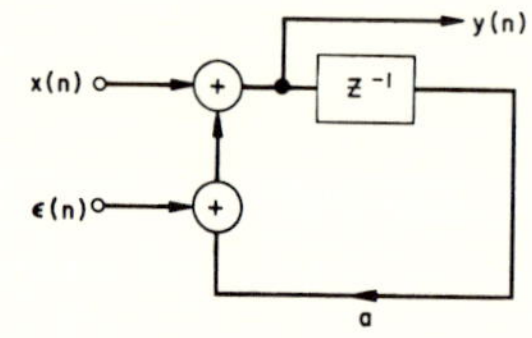

Fig. 3. Noisy first-order filter (fixed point).

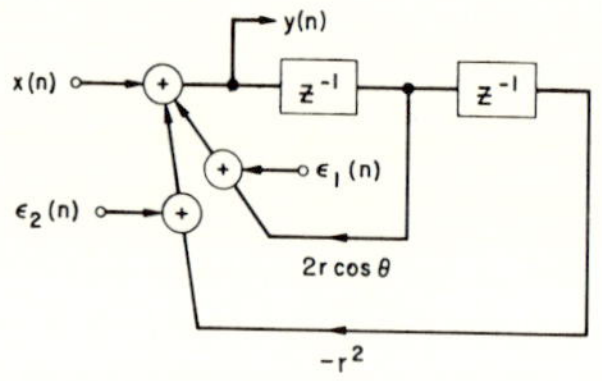

Fig. 4. Noisy second-order filter (fixed point).

put then

$$|y_k(n)| \leq x_{\max} \sum_{r=0}^{\infty} |h_k(r)|. \tag{6}$$

Thus, since we require that $|y_k(n)| < 1$, (6) requires that

$$x_{\max} < 1 \Big/ \sum_{r=0}^{\infty} |h_k(r)|, \qquad \text{for all } k. \tag{7}$$

Equation (7) thus provides an upper bound on the maximum value of the input to insure that no overflow occurs in the kth node. For a general input (7) in fact provides a least upper bound, i.e., if the maximum value of the input exceeds the bound, overflow can occur. This is a consequence of the fact that equality can be achieved in (6) with a sequence $x(n)$ for which at $n=n_o$, $x(n_o-r)=[\text{sgn } h_k(r)]$ for $r=0$ to ∞. (Where sgn $(x)=1$ for $x\geq 0$ and sgn $(x)=-1$ for $x<0$.) Thus in the most general case, (7) is required to guarantee that no overflow occurs. The condition in (7) would generally be satisfied by applying attenuation to the signal at the filter input.

If we assume, for example, that the input $x(n)$ is a white-noise sequence with a uniform amplitude distribution, we would choose for the case of the first-order filter a maximal input amplitude of $(1-a)$. For this case, if σ_x^2 denotes the variance of the input signal, and σ_y^2 denotes the variance of the output signal, then

$$\sigma_x^2 = \left(\frac{1}{3}\right)(1-a)^2 \tag{8a}$$

$$\sigma_y^2 = \left(\frac{1}{3}\right)\left(\frac{(1-a)^2}{1-a^2}\right). \tag{8b}$$

For this example, we can then compute a *noise-to-signal* ratio as the ratio σ_o^2/σ_y^2 with the result

$$\frac{\sigma_o^2}{\sigma_y^2} = \frac{1}{4}2^{-2b}\frac{1}{(1-a)^2}. \tag{9}$$

In a similar manner we can derive a noise-to-signal ratio for the second-order filter shown in Fig. 4. As in the first-order case, we restrict the maximum input in order to guarantee that the dynamic range of the registers is not exceeded. If we consider the input sequence to be uniformly distributed white noise, the resulting output noise-to-signal ratio will be

$$\frac{\sigma_o^2}{\sigma_y^2} = \frac{1}{2}2^{-2b}\left(\sum_{n=0}^{\infty}|h_n|\right)^2$$

$$= \frac{1}{2}2^{-2b}\left(\frac{1}{\sin\theta}\sum_{n=0}^{\infty} r^n \left|\sin[(n+1)\theta]\right|\right)^2. \tag{10}$$

While it is difficult to evaluate this expression exactly, it is possible to obtain an upper and lower bound. Since $\sum_{n=0}^{\infty}|h_n|$ is the largest possible output obtainable with an input that never exceeds unity, it must be larger than the response of the second-order filter to a sinusoid of unity amplitude at the resonant frequency. With this consideration, we can write that

$$\left(\sum_{n=0}^{\infty}|h_n|\right)^2 \geq 1/(1-r)^2(1+r^2-2r\cos 2\theta) \tag{11}$$

since the right-hand side of this inequality is the gain at resonance. Furthermore,

$$\left(\frac{1}{\sin\theta}\sum_{n=0}^{\infty} r^n\left|\sin[(n+1)\theta]\right|\right)^2 \leq \left(\frac{1}{\sin\theta}\sum_{n=0}^{\infty} r^n\right)^2. \tag{12}$$

Therefore, for the second-order case

$$\frac{1}{2}2^{-2b}\frac{1}{(1-r)^2(1+r^2-2r\cos 2\theta)} \leq \frac{\sigma_o^2}{\sigma_y^2} \leq \frac{1}{2}2^{-2b}\frac{1}{\sin^2\theta(1-r)^2}. \tag{13}$$

For both the first- and second-order filter an expression for the noise-to-signal ratio can be obtained which provides some insight into the behavior of the noise-to-signal ratio as the poles approach the unit circle. For the first-order filter let $\delta = 1-a$ so that as $\delta\to 0$, the pole approaches the unit circle. Then in terms of δ, the noise-to-signal ratio for the first-order filter is

$$\frac{\sigma_o^2}{\sigma_y^2} = \frac{1}{4}2^{-2b}\frac{1}{\delta^2}. \tag{14}$$

For the second-order filter, let $\delta = 1-r$ so that, again, as $\delta\to 0$ the poles approach the unit circle. Then if we assume that $\delta \ll 1$, we can approximate $(1+r^2-2r\cos 2\theta)$ as

$$(1+r^2-2r\cos 2\theta) \cong 4\sin^2\theta + \delta^2 \tag{15}$$

which for $4\sin^2\theta$ large compared with δ^2 we will approximate as $4\sin^2\theta$. Consequently, incorporating this approximation,

$$\left(\frac{1}{2}\right)2^{-2b}\frac{1}{4\delta^2\sin^2\theta} \leq \frac{\sigma_o^2}{\sigma_y^2} \leq \frac{1}{2}2^{-2b}\frac{1}{\delta^2\sin^2\theta}. \tag{16}$$

Thus we observe that the noise-to-signal ratio as considered thus far can be considered to be proportional to $2^{-2b}/\delta^2$. We note from this dependence that if δ is halved, then to maintain the same noise-to-signal ratio b must be increased by 1, i.e., one bit must be added to the register length. This dependence provides a convenient basis for comparison of different overflow strategies and different kinds of arithmetic.

In the above analysis, the filter input was assumed to be uniformly distributed white noise. As δ approaches zero the frequency response of both the first- and second-order filter becomes more selective so that more and more of the input energy is out of band. An alternative basis for determining the noise-to-signal ratio is for an input which is sinusoidal. For this choice of inputs, of course, we would not use the general condition of (7) to avoid overflow since we can determine exactly the maximum allowable input amplitude as a function of the filter parameters.

In particular, if the input is of the form $x(n)=x_{\max}\cos n\phi$ then the steady-state output is of the form $y(n)=y_{\max}\cos(n\phi+\psi)$. To prevent overflow, $y_{\max}$ must be less than unity and to maximize the output signal energy, $y_{\max}$ is chosen to be as large as possible. Thus the maximum noise-to-signal ratio is obtained when $x_{\max}$ is chosen so that $y(n)=\cos(n\phi+\psi)$. Note that in order to choose $x_{\max}$ in this way, the frequency of the input signal must be known. For an input sinusoid of unknown frequency $x_{\max}$ must be attenuated so that overflow will not occur even in the worst case, where the frequency of the input coincides with the peak gain in the filter's transfer function.

For fixed-point filters, within the validity of the statistical model for roundoff error, the output noise is independent of the form and amplitude of the input signal. Thus for this choice of inputs, the noise-to-signal ratio obtained for the first-order filter is

$$\frac{\sigma_o^2}{\sigma_y^2}=\frac{1}{24}2^{-2b}\frac{1}{1-a^2}. \tag{17}$$

If, as before, we let $a=1-\delta$, then for $\delta \ll 1$

$$\frac{\sigma_o^2}{\sigma_y^2}=\frac{1}{48}\frac{2^{-2b}}{\delta}. \tag{18}$$

Thus in this case, the noise-to-signal ratio is proportional to $1/\delta$ rather than $1/\delta^2$ so that if δ is multiplied by 1/4 and the register length is increased by one bit, the noise-to-signal ratio will remain constant. We can consider the second-order case in a similar manner. Again for a sinusoidal input, the output with maximum amplitude has the form $y(n)=\cos(n\phi+\psi)$ so that the noise-to-signal ratio in this case is

$$\frac{\sigma_o^2}{\sigma_y^2}=\frac{1}{12}2^{-2b}\left(\frac{1+r^2}{1-r^2}\frac{1}{1+r^4-2r^2\cos 2\theta}\right). \tag{19}$$

Again, choosing $r=1-\delta$ with $\delta \ll 1$,

$$\frac{\sigma_o^2}{\sigma_y^2}\cong\frac{2^{-2b}}{4\delta\sin^2\theta} \tag{20}$$

so that, as with the first-order filter, the noise-to-signal ratio is proportional to $1/\delta$ rather than $1/\delta^2$. The comparison in the noise-to-signal ratio for a white-noise input and a sinusoidal input serves to illustrate the dependence of the effect of dynamic range considerations on the particular form of the input. In some sense, the two cases considered represent extremes. As the input becomes more confined to a known narrow band of frequencies the above analysis with a sinusoidal input would be more representative, and as the input becomes more wide-band the above analysis with a white-noise input is more representative.

In the above discussion, the noise-to-signal ratio for the case of white-noise input was derived on the basis that overflow must be avoided. In a practical case, a scaling of the input on the basis of (7) can be considered to be somewhat pessimistic since the probability of equality being attained in (7) is extremely small. Furthermore, for many filters it is difficult to compute the sum in (7). Jackson [7] has formulated the dynamic range constraints on fixed-point digital filters in terms of L_p norms. In particular, let $Y(\omega)$, $X(\omega)$, and $H(\omega)$ denote the Fourier transforms of the filter output, input, and system impulse response, respectively. Then it can be shown in general that

$$|y(n)| \le \|H\|_p\|X\|_q \quad 1/p+1/q=1 \tag{21}$$

where $\|H\|_p$ and $\|X\|_q$ are the L_p norm and L_q norm of $H(\omega)$ and $X(\omega)$, respectively, where these norms are defined as

$$\|H\|_p=\left[\frac{1}{2\pi}\int_{-\pi}^{\pi}|H(\omega)|^p\,d\omega\right]^{1/p}$$

and

$$\|X\|_q=\left[\frac{1}{2\pi}\int_{-\pi}^{\pi}|X(\omega)|^q\,d\omega\right]^{1/q}$$

For example, with $H(\omega)$ chosen as unity, a consequence of (21) is that

$$|x(n)| \le \|X\|_q, \qquad \text{all } q \ge 1.$$

As another consequence, if we choose $p=1$, $q=\infty$, and use the fact that the L_∞ norm of $|X(\omega)|$ is the maximum value of $|X(\omega)|$ then we obtain the statement that

$$|y(n)| \le \max\left[|X(\omega)|\right]\frac{1}{2\pi}\int_{-\pi}^{\pi}|H(\omega)|\,d\omega.$$

As an alternative, with $p=2$, $q=2$,

$$|y(n)| \le \left[\frac{1}{2\pi}\int_{-\pi}^{\pi}|H(\omega)|^2\,d\omega\right]^{1/2}\left[\frac{1}{2\pi}\int_{-\pi}^{\pi}|X(\omega)|^2\,d\omega\right]^{1/2}$$

To prevent overflow in the output we require that $|y(n)|<1$ and to insure this from (21) we will require that $\|H\|_p\|X\|_q<1$. Consequently, the input must be scaled in such a way that

$$\|X\|_q < 1/\|H\|_p. \tag{22}$$

This condition is somewhat less general than (7) but in many cases is easier to apply. According to (22) with $p=2$, $q=2$, the condition is in terms of the energy in the input signal and the energy in the system impulse response. For $q=1$, $p=\infty$, (22) provides a bound in terms of the peak value of the magnitude of the transfer function, which is perhaps most appropriate for a sinusoidal input.

For the case of a random input (21) cannot be applied since the input and output do not have Fourier transforms. In this case the corresponding condition is phrased in terms of $\phi_{yy}(n)$ the autocorrelation function of the output, $\Phi_{xx}(\omega)$ the power density spectrum of the input, and $H(\omega)$ the magnitude of the system function. In particular, the inequality corresponding to (21) is

$$\phi_{yy}(n) \le \|H^2\|_p\|\Phi_{xx}\|_q \tag{23a}$$

or equivalently

$$\phi_{yy}(n) \le \|H\|_{2p}^2\|\Phi_{xx}\|_q. \tag{23b}$$

Since, if the input is zero mean, $\phi_{yy}(0)=\sigma_y^2$ it follows that

$$\sigma_y^2 \leq \|H\|_{2p}^2 \|\Phi_{xx}\|_q. \qquad (24)$$

Two particular cases of interest are $p=1$, $q=\infty$ and $p=\infty$, $q=1$ so that

$$\sigma_y^2 \leq \|H\|_2^2 \|\Phi_{xx}\|_\infty \qquad (25)$$

and

$$\sigma_y^2 \leq \|H\|_\infty^2 \|\Phi_{xx}\|_1. \qquad (26)$$

As Jackson points out, (25) implies the most stringent condition on the input spectrum $\Phi_{xx}(\omega)$ whereas (26) implies the most stringent condition on the transfer function. From (25), if the input spectrum is white so that $\Phi_{xx}(\omega)=\sigma_x^2$ for all ω, then

$$\sigma_y^2 \leq \sigma_x^2 \|H\|_2^2 \qquad (27)$$

with the input sequence Gaussian, then, the output will overflow no more often than the input overflows if

$$\|H\|_2 \leq 1. \qquad (28)$$

More generally, (27) provides a basis for choosing the input variance to control the maximum percentage of time that the output can overflow.

D. Statistical Analysis of Roundoff Errors with Floating-Point Arithmetic

For the case of floating-point arithmetic, noise is introduced due both to the adds and the multiplies. In analyzing the effect of floating-point roundoff the effect of rounding will be represented multiplicatively so that if $[x]$ denotes rounding of the mantissa in a floating-point number, then

$$[x] = x(1+\epsilon). \qquad (29)$$

To illustrate the analysis of roundoff errors with floating-point arithmetic let us consider a first-order filter. Let $w(n)$ denote the ideal response of the filter, that is, the response with no roundoff noise and let $y(n)$ denote the response of the filter in the presence of roundoff noise. Then following Liu and Kaneko [8] we can write that

$$w(n) = aw(n-1) + x(n) \qquad (30)$$

$$y(n) = [ay(n-1)(1+\epsilon_n) + x(n)](1+\xi_n). \qquad (31)$$

We assume that ϵ_n and ξ_n are uniformly distributed between -2^{-b} and 2^{-b}, are uncorrelated from iteration to iteration, are independent of each other, and also are independent of the signal. Letting $E(n)$ represent the error in the output, so that $E(n)=y(n)-w(n)$, we can write from the above two equations that

$$E(n) - aE(n-1) = aw(n-1)(\epsilon_n + \xi_n) + x(n)\xi_n = u(n) \qquad (32)$$

where we have neglected second-order terms in ϵ, ξ, and E. Since ϵ and ξ are statistically independent of x, and of $w(n-1)$, the term $u(n)$ is easily shown to be a white-noise sequence. Its variance, of course, depends on the excitation $x(n)$. The derivation of (32) with the second-order terms neglected corresponds to representing the roundoff noise as an additive noise source that is statistically independent of the signal but whose variance depends on the signal variance. Specifically, consider the first-order network drawn in Fig. 5 with the two noise sources $e_1(n)$ and $e_2(n)$. From the model for multiplier

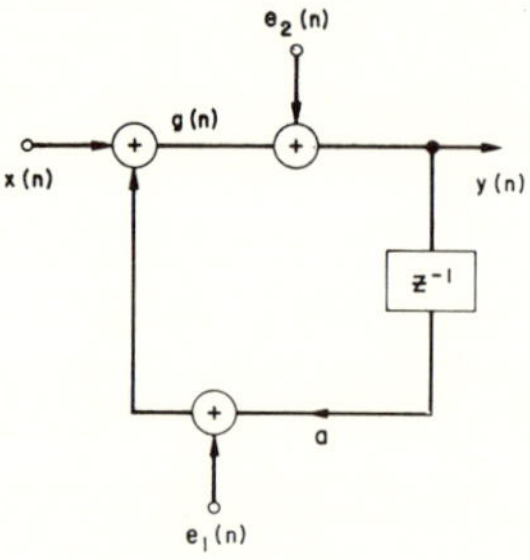

Fig. 5. Noisy first-order filter (floating point).

roundoff noise, the noise source $e_1(n)$ is given by

$$e_1(n) = ay(n-1)\epsilon_n \qquad (33)$$

and the noise source $e_2(n)$ is given by

$$e_2(n) = g(n)\xi_n. \qquad (34)$$

The analysis above in which we neglected second-order terms corresponds in this case to evaluating the variance of $e_1(n)$ and $e_2(n)$ by using the mean-square values for $y(n-1)$ and $g(n)$ that would result if no roundoff noise were present. Therefore, if we assume that $x(n)$ is a zero-mean white-noise input, with variance σ_x^2, then the variances of $e_1(n)$ and $e_2(n)$ are, respectively,

$$\sigma_{e_1}^2 = a^2\sigma_\epsilon^2\overline{y^2(n-1)} = a^2\sigma_\epsilon^2\sigma_x^2 \frac{1}{1-a^2} \qquad (35)$$

$$\sigma_{e_2}^2 = \sigma_\xi^2\overline{g^2(n)} = \sigma_\xi^2\sigma_x^2 \frac{1}{1-a^2} \qquad (36)$$

where the bar denotes expected value. Then, since $e_1(n)$ and $e_2(n)$ are independent, because ϵ_n and ξ_n are independent, the output noise variance is

$$\sigma_o^2 = \sigma_\epsilon^2\sigma_x^2 \frac{1+a^2}{(1-a^2)^2} = \sigma_\epsilon^2\sigma_y^2 \frac{1+a^2}{1-a^2} \qquad (37a)$$

where we have assumed again that σ_ϵ^2 and σ_ξ^2 are equal. The output noise-to-signal ratio is

$$\frac{\sigma_o^2}{\sigma_y^2} = \sigma_\epsilon^2 \frac{1+a^2}{1-a^2} \qquad (37b)$$

We can analyze the effect of roundoff noise in the second-order filter in a similar manner. In Fig. 6 is shown the network for a second-order filter with roundoff noise sources included. Note that since noise sources must be included due to addition, two summers are included to add the three variables in the feedback loop. The noise sources $e_3(n)$ and $e_4(n)$ represent the noise due to the multiplies and the noise sources $e_1(n)$ and $e_2(n)$ represent the noise due to the additions. With assumptions similar to those above in which we neglected second-order terms, we write that

$$\begin{aligned} e_1(n) &= y(n)\epsilon_1(n) \\ e_2(n) &= [y(n) - x(n)]\epsilon_2(n) \\ e_3(n) &= 2r\cos\theta\, y(n-1)\epsilon_3(n) \\ e_4(n) &= -r^2 y(n-2)\epsilon_4(n) \end{aligned} \qquad (38)$$

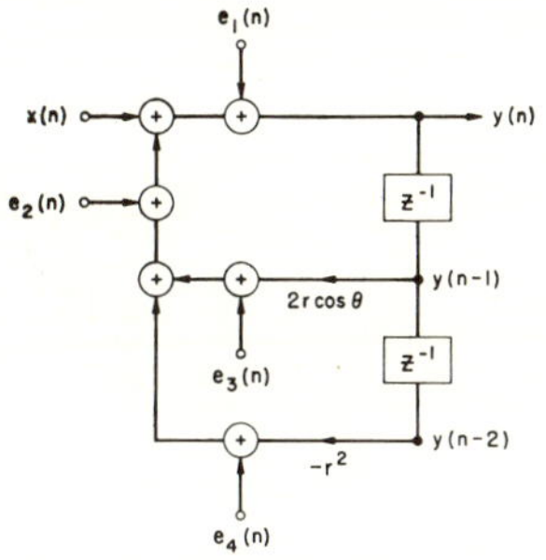

Fig. 6. Noisy second-order filter (floating point).

where ϵ_1, ϵ_2, ϵ_3, and ϵ_4 are independent random variables with equal variance σ_ϵ^2. If as before, $x(n)$ is assumed to be a white random process with variance σ_x^2, then the output noise-to-signal ratio for the second-order case is

$$\frac{\sigma_o^2}{\sigma_y^2} = \sigma_\epsilon^2\left[1 + G\left(3r^4 + 12r^2\cos^2\theta - 16\frac{r^4\cos^2\theta}{1+r^2}\right)\right] \quad (39)$$

where

$$G = \frac{1+r^2}{1-r^2}\left(\frac{1}{r^4+1-4r^2\cos^2\theta+2r^2}\right). \quad (40)$$

For the high gain case, it is possible to compare fixed-point and floating-point arithmetic by approximating the expressions for the noise-to-signal ratio. For the first-order case, with $a=1-\delta$, and $\delta \ll 1$, the result (37b) for the first-order filter, can be approximated as

$$\frac{\sigma_o^2}{\sigma_y^2} \simeq \frac{1}{3}2^{-2b}\frac{1}{\delta}. \quad (41)$$

Similarly for the second-order filter

$$\frac{\sigma_o^2}{\sigma_y^2} \simeq \frac{1}{3}2^{-2b}\left(\frac{3+4\cos^2\theta}{4\delta\sin^2\theta}\right) \quad (42)$$

where in (41) and (42) we have taken $\sigma_\epsilon^2=(1/3)2^{-2b}$.

For fixed-point arithmetic we recall that for a white-noise input the noise-to-signal ratio behaved as $1/\delta^2$ and for a sinusoidal input as $1/\delta$. Comparison of (41) and (42) with (14) and (16) and (18) and (20) indicates a slightly larger noise-to-signal ratio for floating-point arithmetic as compared with fixed-point arithmetic with a sinusoidal input of known frequency but a significantly smaller noise-to-signal ratio for floating-point arithmetic as compared with fixed-point arithmetic with a white-noise input. It is important to keep in mind, however, that the noise-to-signal ratios for the fixed-point filters were computed on the basis that the input signal was as large as possible. If the input signal level decreases, the noise-to-signal ratio will increase since the output noise variance is independent of the input signal level. For floating-point arithmetic, on the other hand, the output noise variance is proportional to the output signal variance and as the input level is scaled up or down so is the roundoff noise. It is also important to note that the comparison just discussed assumes that the floating-point mantissa is equal in length to the entire fixed point word, and does not account for the extra bits needed for the characteristic. The authors [6] have previously compared fixed- and floating-point filters on the basis of equal total word length. However, in completing such a comparison one must take account of the large difference in hardware complexity between implementing floating-point arithmetic, and adding a few bits to a fixed-point arithmetic element.

Oppenheim [9] has proposed a realization of recursive digital filters using block floating-point arithmetic. Here the input and filter states (i.e., the inputs to the delay registers) are jointly normalized before the multiplications and additions are performed in fixed-point arithmetic. The scale factor (or exponent) obtained during the normalization is then applied to the final output to obtain a fixed-point output. The roundoff noise properties of such a realization were studied, and the noise-to-signal ratio was found to lie between that for fixed and floating point.

E. Zero-Input Limit Cycle Behavior of Digital Filters for Fixed-Point Arithmetic

In the preceding discussion the effect of arithmetic roundoff was modelled as an additive white-noise source, uncorrelated with the data. Justification of this model assumes that from iteration to iteration, the input can be expected to pass through several quantization levels. Consequently, this model is applied primarily when the input signal has a complicated behavior and cannot be expected to be valid in general. For example, consider a first-order filter for which the difference equation is

$$y_n = \alpha y_{n-1} + x_n \quad (43)$$

and for which the register length for the data is 4 bits and the coefficient α is 0.5. If the input x_n is 7/8 and if rounding is applied after the arithmetic then on successive iterations of the filter, the output will be:

$$y_0 = 7/8$$
$$y_1 = 1/2$$
$$y_2 = 1/4$$
$$y_3 = 1/8$$
$$y_n = 1/8, \quad \text{for } n \geq 4.$$

Thus due to rounding, the output reaches a steady-state nonzero value and since the ideal steady-state output is zero, this nonzero value represents roundoff error. Clearly this kind of roundoff error cannot be modelled as white noise, but in fact represents a limit cycle due to the nonlinearity corresponding to the quantizer which implements the rounding. Limit cycle behavior of this type was first noted by Blackman [10] who referred to the amplitude intervals within which these limit cycles are confined as "deadbands." Blackman considered only first-order limit cycles corresponding to a dc behavior in the deadband. More generally, Jackson [11] has considered limit cycle behavior in first- and second-order filter sections with an analysis based on the location of the "effective" poles in the filter due to roundoff. Following the approach presented by Jackson, consider a first-order filter with a difference equation of the form of (43). Due to the register length constraint, the product αy_{n-1} must be rounded. Let $(\cdot)'$ denote the operation of rounding. If the register length is $(b+1)$ bits and if data are represented as fractions then

$$|(\alpha y_{n-1})' - \alpha y_{n-1}| \leq (\tfrac{1}{2})2^{-b}. \quad (44)$$

If y_{n-1} is such that $|(\alpha y_{n-1})'| = |y_{n-1}|$ then the magnitude of the effective value of the coefficient is unity corresponding to the pole of the filter being on the unit circle. The range of values for which this condition is met is

$$|y_{n-1}| - |\alpha y_{n-1}| \leq (\tfrac{1}{2})2^{-b} \tag{45}$$

or

$$|y_{n-1}| \leq \frac{(0.5)2^{-b}}{1 - |\alpha|}. \tag{46}$$

This range of values is referred to as the deadband. Due to rounding, of course, values within the deadband must be in steps of 2^{-b}. For the first-order filter, when the filter state falls within this range and the input is zero, the effective pole is on the unit circle and the filter will support a limit cycle behavior. If the coefficient α is positive, as in the above example, the limit cycle response is dc, i.e., has constant magnitude and sign. For α negative the limit cycle behavior has constant magnitude but alternating sign.

For a second-order filter there is a larger variety of modes of limit cycle behavior. In particular, consider the second-order difference equation

$$y_n = x_n - \beta_1 y_{n-1} - \beta_2 y_{n-2}. \tag{47}$$

With $\beta_1^2 < +4\beta_2$ the filter poles occur as a complex conjugate pair and with $\beta_2 = 1$ the poles occur on the unit circle. The approach proposed by Jackson for examining the limit cycle behavior of the second-order filter corresponds to considering the filter behavior when the effect of rounding places the effective poles of the filter on the unit circle. With zero input the effective poles will be on the unit circle if

$$|y_{n-2}| - |(\beta_2 y_{n-2})| \leq \tfrac{1}{2}\, 2^{-b} \tag{48}$$

or

$$|y_{n-2}| \leq \frac{(0.5)2^{-b}}{1 - |\beta_2|}. \tag{49}$$

Thus if the output falls within this range the effective value of β_2 is unity so that the effective poles are on the unit circle. With the effective value of β_2 as unity, the effective value of β_1 controls the oscillation frequency.

A second mode of limit cycle behavior occurs in second-order filters when the effect of rounding is to place an effective pole at $z = \pm 1$. As shown by Jackson, the deadband corresponding to this mode is for values less than or equal to $1/(1 - |\beta_1| + \beta_2)$ in steps of integer multiples of 2^{-b}.

While this approach is somewhat heuristic, Jackson has found that these bounds are consistent with experimental results and hence he has hypothesized that they represent necessary and sufficient conditions. These bounds for second-order filters are summarized in Fig. 7, showing different deadband subregions in the β_1, β_2 plane. The number within an area in the β_1, β_2 plane represents the maximum magnitude of the limit cycle in multiples of 2^{-b} and the cross hatched region represents the region for which no limit cycles can occur.

Recently, Parker and Hess [12] have studied the limit cycle problem further, and found that these bounds are approximately correct and sufficient, but not necessary. In other words, there exist some limit cycles outside the regions specified by Fig. 7.

In addition to the above classes of limit cycles, a more severe type of limit cycle can occur due to overflow in filters implemented using one's-complement or two's-complement

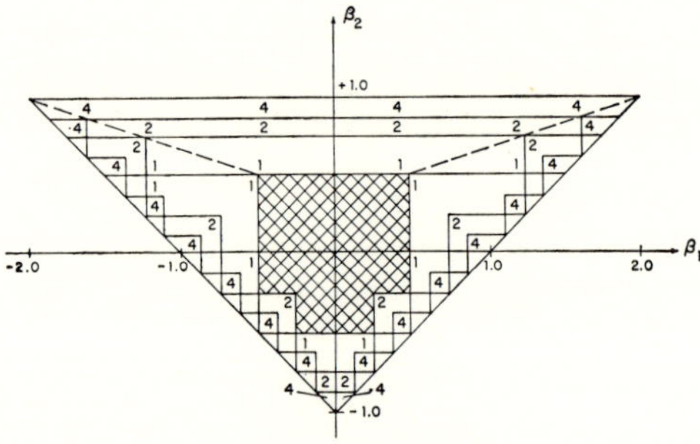

Fig. 7. Deadband subregions.

arithmetic. These limit cycles have been referred to as overflow oscillations [13] and can be avoided by using saturation arithmetic.

F. *Effects of Parameter Quantization in Digital Filters*

In the preceding sections we focussed on the effects of arithmetic roundoff in digital filters. Another consequence of the requirement of finite register length is that the filter coefficients cannot be specified exactly. Classical design procedures generally lead to filter coefficients with arbitrary accuracy and the implementation of the filter then requires that the coefficients be modified to fit the available register length. For hardware realizations of digital filters it is, of course, desirable to keep the register length as small as possible.

One common approach to the problem of parameter quantization is the use of filter configurations or structures which in some sense are least sensitive to inaccuracies in the parameters. One of the difficulties in evaluating the sensitivity of filter structures is the choice of a meaningful measure of the sensitivity. Most commonly, the sensitivity of the filter is tied to the movement of the poles of the filter. For this choice Kaiser [14] has shown that for a filter with clustered poles a cascade or parallel combination of first- and second-order sections provides more accuracy in the pole positions than a direct form realization. This is basically a consequence of the fact that for a polynomial whose roots are clustered, the sensitivity of the roots to changes in the polynomial coefficients increases as the order of the polynomial increases. Thus the roots can be more accurately controlled if the polynomial is factored into first- and second-order factors.

Even within the choice of first- and second-order sections some flexibility remains. For a direct form implementation of a pole pair as shown in Fig. 8(a) the coefficients are $-r^2$ and $2r\cos\theta$. For a given quantization on the coefficients the poles must lie on a grid in the z plane defined by the intersection of concentric circles, corresponding to quantization of r^2 and vertical lines, corresponding to quantization of $2r\cos\theta$. Such a grid is illustrated in Fig. 8(b). An alternative realization of a pole pair is the coupled form proposed by Rader and Gold [15], as shown in Fig. 9(a). In this case the coefficients are $r\cos\theta$ and $r\sin\theta$ and consequently the poles must lie on a rectangular grid as illustrated in Fig. 9(b). We note, for example, that for a given coefficient word length the direct form permits more accurate placement of poles with r close to unity and θ large while the coupled form is more advantageous for θ small. There are, in theory, many other structures in addition to the direct and coupled forms for implementing pole

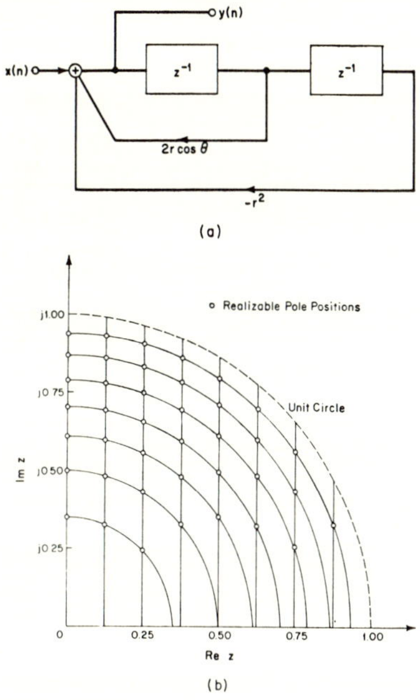

Fig. 8. (a) Direct form implementation of a pole pair. (b) Grid of allowable pole positions—direct form.

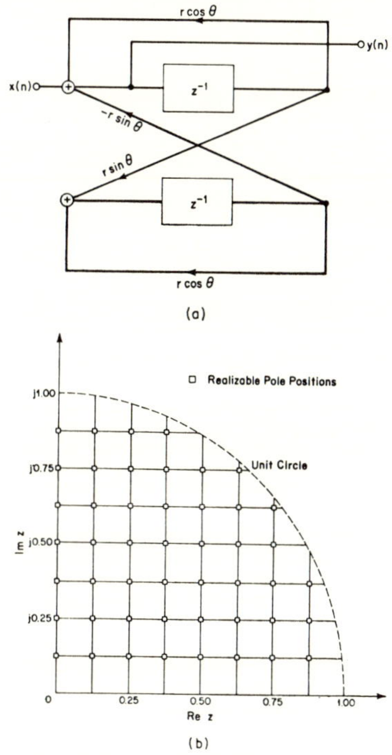

Fig. 9. (a) Coupled form implementation of a pole pair. (b) Grid of allowable pole positions—coupled form.

pairs although they are the most commonly considered [16]. Different structures, of course, imply different grids in the z plane and generally it is advantageous to choose a structure for which the grid is dense in the region of the z plane where the poles are to be located.

With a given choice of structure there remains the question as to how the pole locations on the grid should be chosen. A common procedure is to truncate or round the ideal coefficients. An alternative used by Avenhaus and Schussler [17] and also by Steiglitz [18] is to search over the grid in the vicinity of the ideal pole locations to select a grid point which locally minimizes the maximum error in the filter frequency response. As an alternative to the use of cascade or parallel connections of first- and second-order sections, more general filter structures can be considered. Digital wave filters, as proposed by Fettweis [19] and investigated by Bingham [20] and by Crochiere [21], appear to have much less sensitivity to parameter inaccuracies than the cascade form.

It would, of course, be desirable to incorporate the constraint of quantized coefficients into the design of digital filters. For nonrecursive filters, algorithmic design falls within the framework of integer linear programming. For recursive filters, however, the equations become nonlinear. In general, the development of design procedures with quantized coefficients remains an important area of research.

IV. Effects of Arithmetic Roundoff in the FFT

A. *Introduction*

The FFT algorithm [22] for computing the discrete Fourier transform (DFT) plays a central role in many signal processing applications [23]. As with the implementation of digital filters, it is important to understand the effect of finite register length arithmetic on the performance of the algorithm.

There are many forms of the FFT algorithm and the detailed effects of quantization will differ depending on the form used. The most commonly used forms of the algorithm are the radix-2 forms for which the size of the transform computed is an integer power of two. For the most part, the discussion below is phrased in terms of a particular form of the radix-2 FFT, commonly referred to as the decimation in time form of the algorithm; the results however are applicable with only minor modification to the decimation in frequency form. We feel that most of the ideas employed in the error analysis of the radix-2 forms of the algorithm can be utilized in other forms such as mixed radix, etc.

Our approach in analyzing noise in the FFT is basically statistical. In most cases, the predictions of the models are supported with experimental data (from Weinstein [24], unless otherwise stated). For floating and block floating point arithmetic, in order to simplify the analysis and obtain concrete results, it is convenient to assume a simple, white-noise model for the signal being transformed. Discussion of how the results might be expected to change for other types of signals is included, as are experimental noise measurements on FFT's of nonwhite signals.

B. *The FFT Algorithm*

The FFT algorithm is directed toward computing the DFT of a finite duration sequence $f(n)$, defined as

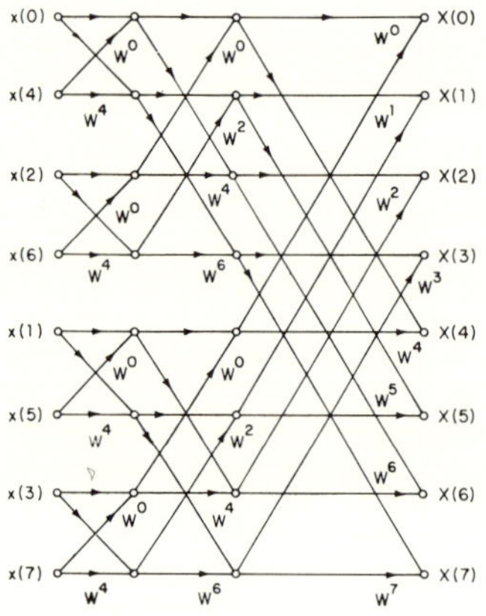

Fig. 10. FFT flow graph. $N=8$.

$$F(k) = \sum_{n=0}^{N-1} f(n)W^{nk}, \qquad W = e^{-j(2\pi/N)}. \tag{50}$$

A flow chart depicting the FFT algorithm for $N=8=2^3$ is shown in Fig. 10. A specific decimation in time algorithm is depicted. (An implementation of this particular form of the algorithm was used for the reported experimental work.) Some key aspects of this diagram, which are common to all standard radix-2 algorithms, are as follows. The DFT is computed in $\nu=\log_2 N$ stages. At each stage, the algorithm passes through the entire array of N complex numbers, two at a time, generating a new N number array. The νth array contains the desired DFT. The basic numerical computation operates on a pair of numbers in the $(m+1)$th array. This computation, referred to as a "butterfly" is

$$X_{m+1}(i) = X_m(i) + \tilde{W}X_m(j)$$
$$X_{m+1}(j) = X_m(i) - \tilde{W}X_m(j). \tag{51}$$

Here, $X_m(i)$ and $X_m(j)$ represent a pair of numbers in the mth array, and $\tilde{W}$ is some appropriate integer power of W, that is

$$\tilde{W} = W^p = e^{-j2\pi p/N}. \tag{52}$$

The form of the butterfly computation is actually somewhat different for a decimation in frequency algorithm, where the computation is

$$X_{m+1}(i) = X_m(i) + X_m(j)$$
$$X_{m+1}(j) = [X_m(i) - X_m(j)]\tilde{W}. \tag{53}$$

At each stage, $N/2$ separate butterfly computations are carried out to produce the next array. The integer p varies with i, j, and m in a manner which depends on the specific form of the FFT algorithm that is used. Fortunately, our analysis is not tied to the specific way in which p varies. Also, the specific relationship between i, j, and m, which determines how we index through the mth array, is not important for the analysis. The details of the analysis for decimation in time and decimation in frequency differ somewhat due to the different butterfly forms, but the basic results for the dependence of noise-to-signal ratio on N do not change significantly. In our analysis we will assume a butterfly of the form (51), corresponding to decimation in time.

C. FFT Roundoff Noise with Fixed-Point Arithmetic

We will model the roundoff noise by associating an independent white-noise generator with each multiplier. This means that a noise source feeds into each node of the signal flow graph of Fig. 10 (excluding the initial array of nodes, since we are not considering A/D noise here). Since we are dealing with complex multiplications, these elemental noise sources are complex. Defining the complex variance σ_B^2 as the expected squared magnitude of such a noise source, we have

$$\sigma_B^2 = 4\,\frac{2^{-2b}}{12} \tag{54}$$

where it is assumed that each of the four real multiplications used to perform the complex multiplication is rounded separately. In Fig. 10, $3\times 8=24$ such noise sources must be inserted. To add the effects of each of the noise sources in evaluating the total roundoff noise in the output, we note that the transmission function from any node in the flow graph to any other connected node is multiplication by a complex constant of unity magnitude. Since we assume that all noise sources are uncorrelated, the noise variance at any output node is equal to σ_B^2 times the number of noise sources that propagate to that node. The general result which is easily verified for the case $N=8$ by inspection of Fig. 10 is that $(N-1)$ noise sources propagate to each output node so that the output noise variance σ_E^2 is given by

$$\sigma_E^2 = (N-1)\sigma_B^2$$

which for large N we take as

$$\sigma_E^2 \simeq N\sigma_B^2. \tag{55}$$

According to this result, the variance of the output noise is proportional to N, the number of points transformed. The effect of doubling N, or adding another stage in the FFT, is to double the output noise variance. Using the assumptions we have made thus far about the noise generators in the FFT (all uncorrelated, with equal variances), the output noise is white, i.e., the N noise samples $E(k)$ are mutually uncorrelated, with independent real and imaginary parts. This follows from the fact that the output of any butterfly is white (two outputs uncorrelated with equal variance, real and imaginary parts uncorrelated) if the input is white. Since the noise sources in our system are white, and all connected to the output via some combination of butterfly computations, the output noise must also be white.

In order to simplify the analysis leading to (55), we have neglected some details. First, we have associated equal variance noise sources with all multipliers, including where $W=1$ and j. In many programmed FFT's these multiplications are performed noiselessly. If we assume in the analysis that these multiplications are noiseless, the output noise variance will no longer be uniform over the output array. For example, the zeroth output point would be noiseless. The average variance over the output array will be somewhat lower than the result in (55), but will retain a linear dependence on N. Second, the assumption that all noise sources are uncorrelated is contradicted by the fact that the two noise sources associated with a given butterfly are negatives of each other, and therefore completely correlated. This does not affect the result for output noise variance, since the two outputs of a butterfly connect to a disjoint set of output points. However, it implies that the output noise samples $E(k)$ are somewhat correlated. These details are worth mentioning, but not worth analyzing here at length, because they cloud the essential ideas of the analysis, are quite program-dependent, and do not change the essential character of the dependence of mean-squared output noise on N.

In implementing the FFT with fixed-point arithmetic we must insure against overflow. From (51) it follows that

$$\max [|X_m(i)|, |X_m(j)|] \leq \max [|X_{m+1}(i)|, |X_{m+1}(j)|] \quad (56)$$

and also that

$$\max [|X_{m+1}(i)|, |X_{m+1}(j)|] \leq 2 \max [|X_m(i)|, |X_m(j)|]. \quad (57)$$

Equation (56) implies that the maximum modulus is nondecreasing from stage to stage so that, if the magnitude of the output of the FFT is less than unity then the magnitude of the points in each array must be less than unity,[1] i.e., there will be no overflow in any of the arrays.

In order to express this constraint as a bound on the input sequence, we note that the maximum possible output can be expressed in terms of the maximum input as

$$|X(k)|_{\max} \leq |x(n)|_{\max} \sum_{n=0}^{N-1} |W^{nk}| = N|x(n)|_{\max}. \quad (58)$$

Thus bounding the input sequence so that

$$|x(n)| < 1/N \quad (59)$$

will prevent overflow. To obtain an explicit expression for output signal variance, we assume $x(n)$ white, with real and imaginary parts each uniformly distributed in $(-1/\sqrt{2}N, 1/\sqrt{2}N)$. Then we have

$$\sigma_X^2 = \overline{|X(k)|^2} = N\sigma_x^2 = N\overline{|x(n)|^2} = \frac{1}{3N}. \quad (60)$$

Combining this with (55) yields

$$\frac{\sigma_E^2}{\sigma_X^2} = 3N^2\sigma_B^2. \quad (61)$$

The assumption of white input signal is not critical here. For example, if a complex sinusoid $x(n) = (1/N) \exp j(2\pi k_0 n/N + \phi)$ had been selected σ_E^2/σ_X^2 would still be proportional to N^2, which is the essential point of (61).

Equation (57) suggests an alternative procedure for preventing overflow. Since the maximum modulus increases by no more than a factor of two from stage to stage we can prevent overflow by requiring that $|x(n)| < 1$ and incorporating an attenuation factor of 1/2 at each stage. Using this step-by-step scaling, the attainable output signal level (for white input) is the same as in (60) since the output signal level does not depend on where the scaling is done, but only on how much overall scaling is done. However, the output noise level will be much less than in (55) since the noise introduced at early stages of the FFT will be attenuated by the scaling which takes place at the later array. Quantitatively for $N = 2^\nu$

$$\sigma_E^2 = \sigma_{B'}^2 \sum_{k=1}^{N/2} \frac{1}{k} \quad (62)$$

where $\sigma_{B'}^2$ represents the roundoff noise introduced due to multiplication by W and scaling and will consequently be slightly higher than σ_B^2. In particular, if we assume that the scaling is accomplished with rounding, it can be shown

$$\sigma_{B'}^2 = \frac{5}{6} 2^{-2b}. \quad (63)$$

For large N, (62) is approximately

$$\sigma_E^2 = 2\sigma_{B'}^2 \quad (64)$$

and thus is much less than the noise variance resulting when all of the scaling is carried out on the input data.

Now, we can combine (64) with (60) to obtain the output noise-to-signal ratio for the case of step-by-step scaling and white input. We obtain

$$\frac{\sigma_E^2}{\sigma_X^2} = 6N\sigma_{B'}^2 = (5N)2^{-2b} \quad (65)$$

a result proportional to N, rather than to N^2. An interpretation of (65) is that the rms output noise-to-signal ratio increases as N, or by half a bit per stage. This result was first obtained by Welch [25]. It is important to note that the assumption of white signal is not essential in the analysis. The basic result of half-a-bit-per-stage increase holds for a broad class of signals, with only the constant multiplier in (65) being signal-dependent. In particular, for a general input with scaling at each array, the output variance is related to the variance of the input array by

$$\sigma_X^2 = \frac{1}{N} \sigma_x^2 = \frac{1}{N} \overline{|x(n)|^2} \quad (66)$$

so that

$$\frac{\sigma_E^2}{\sigma_X^2} = \frac{\frac{5}{3} N 2^{-2b}}{\sigma_x^2} \quad (67)$$

where, to reduce noise-to-signal ratio, we would like to make σ_x^2 as large as possible but are limited by the constraint $|x(n)| < 1$. The result (67) has been verified experimentally for both wide-band and narrow-band signals [24], [25].

We should also note that the dominant factor causing the increase of σ_E^2/σ_X^2 with N is the decrease in signal level (required by the overflow constraint) as we pass from stage to stage. According to (63) and (64), very little noise (only a bit or two) is present in the final array. Most of the noise has been shifted off by the scalings. However, the mean-squared signal level has decreased by a factor of $1/N$ from its initial value, due to the scalings. Our output consists not of the DFT defined by (50) but of $1/N$ times this DFT.

We have assumed straight fixed point computation in this section, i.e., only preset attenuations were allowed, and we were not permitted to rescale on the basis of an overflow test. Clearly, if the hardware or programming facility are such that straight fixed point must be used, we should, if possible, incorporate attenuators of 1/2 at each array rather than using a large attenuation of the input array.

[1] Actually one should discuss overflow in terms of the real and imaginary parts of the data, rather than the magnitude. However, $|x| < 1$ implies that $|\mathrm{Re}\,(x)| < 1$ and $|\mathrm{Im}\,(x)| < 1$, and only a slight increase in allowable signal level is achieved by scaling on the basis of Re and Im parts.

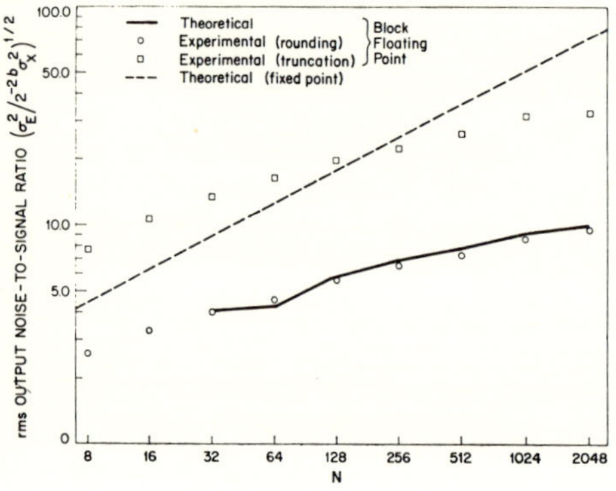

Fig. 11. Experimental and theoretical noise-to-signal ratios for block floating-point FFT.

A third approach to avoiding overflow is the use of block floating point. In this procedure the original array is normalized to the far left of the computer word, with the restriction that $|x(n)|<1$; the computation proceeds in a fixed point manner, except that after every addition there is an overflow test; if overflow is detected, the entire array is shifted right 1 bit and the computation continues. The number of necessary shifts are counted to determine a scale factor or exponent for the entire final array. The output noise-to-signal ratio depends strongly on how many overflows occur, and at what stages of the FFT they occur. The positions and timing of overflows are determined by the signal being transformed, and thus, in order to analyze noise-to-signal ratio in block floating FFT, one needs to know the signal statistics. This is in contrast to the fixed point analysis above, where it was not necessary to assume specific signal statistics.

The necessary number of right shifts of the array is related to the peakiness of the DFT of the signal being transformed. If the constant signal $x(n)=1$ or the single-frequency input $x(n)=\exp j(2\pi/N)k_o n$ is transformed, the output (with k_o an integer) will consist of a single nonzero point and (for $N=2^\nu$) ν scalings of the array will be necessary, one for each stage.

A reasonable case to examine is the case of a white input signal; the DFT of a white signal is white, and one might expect (since the spectral energy is spread) that scalings at all stages would not be necessary, and a noise-to-signal ratio advantage over fixed point would be gained. This problem can be analyzed theoretically [24] but the analysis is quite involved and will be omitted. Instead, we will present some experimental results.

In Fig. 11 experimentally measured values of output noise-to-signal ratio are presented for block floating FFT's of white inputs, using rounded arithmetic. The quantity plotted is $(\sigma_E^2/2^{-2b}\sigma_x^2)^{1/2}$, the rms noise-to-signal ratio. For comparison, a theoretical curve representing fixed point noise-to-signal ratio (for rounded arithmetic) is also shown. We see that for white input block floating point provides some advantages over fixed point, especially for the larger transforms. For $N=2048$, the rms noise-to-signal ratio for block floating point is about 1/8 that of fixed point, representing a 3-bit improvement.

An experimental investigation was used to examine how

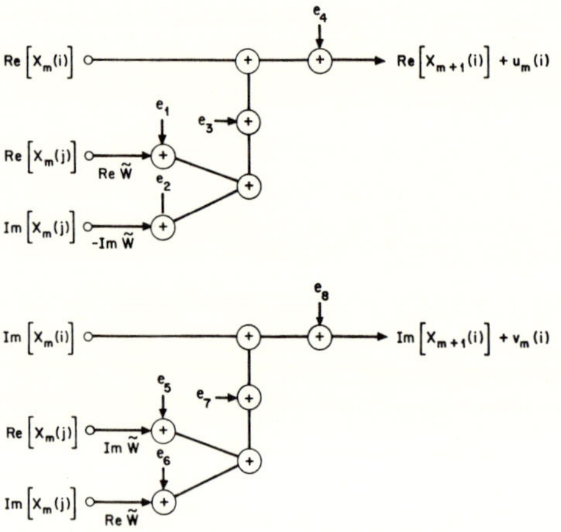

Fig. 12. Noisy butterfly computation (floating point).

the results for block floating point change, when truncation rather than rounding is used. The results of this experiment are also shown in Fig. 11. Noise-to-signal ratios are generally a bit or two worse than for rounding. The rate of increase of noise-to-signal ratio with N seems to be about the same as for rounding.

D. FFT Roundoff Noise with Floating-Point Arithmetic

The effect of arithmetic roundoff with floating-point arithmetic has been analyzed theoretically and experimentally by Gentleman and Sande [26], by Weinstein [27], and by Kaneko and Liu [28]. As with the statistical analysis of roundoff errors with fixed-point arithmetic, noise is introduced due to each butterfly computation. As with floating-point errors in digital filters, we neglect second-order error terms so that noise sources are introduced after each multiplication and addition that are assumed to be white but for which the variance is proportional to the variance of the signal at that node. Unless the input signal is assumed to be white, the analysis becomes quite complicated due to the variation of the variance of the signal and therefore of the noise sources within each array. Kaneko and Liu have obtained detailed formulas for a general stochastic model of the input signal. We will confine attention here to the case of white input signal, where the signal at any array in the FFT is also white, with constant variance across the array.

In Fig. 12 a typical butterfly computation (only top half) is indicated, including the noise sources due to multiplication and addition. The assumption of white input signal implies that

$$\overline{[\mathrm{Re}\,(X_m)]^2} = \overline{[\mathrm{Im}\,(X_m)]^2} = \tfrac{1}{2}\,\overline{|X_m|^2} \tag{68}$$

and application of our floating-point noise model as in Section III-D yields the noise source variances

$$\sigma_{e_1}^2 + \sigma_{e_2}^2 = \sigma_{e_5}^2 + \sigma_{e_6}^2 = \sigma_{e_3}^2 = \sigma_{e_7}^2 = \tfrac{1}{2}\,\sigma_\epsilon^2 \overline{|X_m|^2} \tag{69}$$

$$\sigma_{e_4}^2 = \sigma_{e_8}^2 = \sigma_\epsilon^2 \overline{|X_m|^2}. \tag{70}$$

The variance of the complex noise source $U_m = u_m + jv_m$ is then

$$\overline{|U_m|^2} = 4\sigma_\epsilon^2 \overline{|X_m|^2} \tag{71}$$

so that the variance of the noise generated in computing the $(m+1)$th array is $4\sigma_\epsilon^2$ times the variance of the signal in the mth array. If the input (zeroth) array is white noise with variance σ_x^2 then the noise generated in the $(m+1)$th array is $2^m\sigma_x^2(4\sigma_\epsilon^2)$. If σ_{om}^2 is the output noise due to the noise generated in the $(m+1)$th array, then

$$\sigma_{om}^2 = 2^{\nu-(m+1)}2^m\sigma_x^2(4\sigma_\epsilon^2) = 2N\sigma_\epsilon^2\sigma_x^2. \tag{72}$$

Since the noise generated in each array is assumed to be independent, the total output noise variance σ_E^2 is

$$\sigma_E^2 = 2\nu N\sigma_\epsilon^2\sigma_x^2. \tag{73}$$

By noting that the output signal variance is related to the input signal variance by

$$\sigma_X^2 = N\sigma_x^2 \tag{74}$$

the result follows:

$$\frac{\sigma_E^2}{\sigma_X^2} = 2\sigma_\epsilon^2\nu. \tag{75}$$

A further result, which can be derived from our model, is an expression for the final expected output noise-to-signal ratio which results after performing an FFT and an inverse FFT on a white signal $x(n)$. The inverse FFT introduces just as much roundoff noise as the FFT itself, and thus the resulting output noise-to-signal ratio is

$$\frac{\sigma_E^2}{\sigma_x^2} = 4\sigma_\epsilon^2\nu \tag{76}$$

or just double the result in (75).

In order to see the implications of (75) or (76) in terms of register length requirements, it is useful to express these results in units of bits. We use

$$(\sigma_E^2/\sigma_X^2\sigma_\epsilon^2) \text{ bits} = \tfrac{1}{2}\log_2(2\nu) \tag{77}$$

to represent the number of bits by which the rms noise-to-signal ratio increases in passing through a floating point FFT. For example, for $\nu=8$ this represents 2 bits and for $\nu=11$ it represents 2.23 bits. The number of bits of rms noise-to-signal ratio increases as $\log_2(\log_2 N)$, so that doubling the number of points in the FFT produces a very mild increase in output noise, significantly less than the half-bit-per-stage increase for fixed-point computation. In fact, to obtain a half-bit increase in the result above, we would have to double ν, or square N.

In the analysis leading to (75), we have not considered the fact that multiplications by 1 can be performed noiselessly. For a specified radix-2 algorithm, such as the decimation in time algorithm shown in Fig. 10, these reduced variances for $\tilde{W}=1$ and j can be included in the model to obtain a slightly reduced prediction for output noise-to-signal ratio. However, for reasonably large N, this modified noise analysis yields only slightly better predictions of output noise than does the simplified analysis above.

A consequence of our analysis leading to (75) is that the output noise is white. This follows from the fact that each array of noise sources is white. The reduced noise source variance for $\tilde{W}=1$ and j implies that for some arrays there will be a variation of noise source variance over the array. This implies a slight variation of output noise variance over the output array, and thus our modified noise analysis will only predict an average noise variance over the output array.

The results discussed above have been verified with excellent agreement as shown in Fig. 13(a) and (b). To obtain this agreement, however, it was necessary to use randomized rounding, i.e., randomly rounding up or down when the value of mantissa was exactly $(1/2)2^{-b}$. The modified theoretical

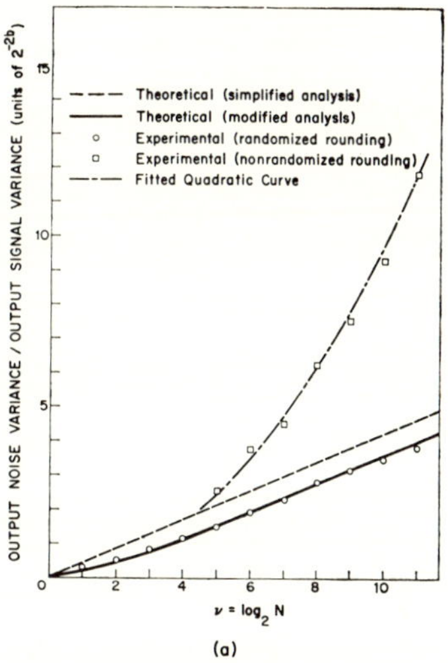

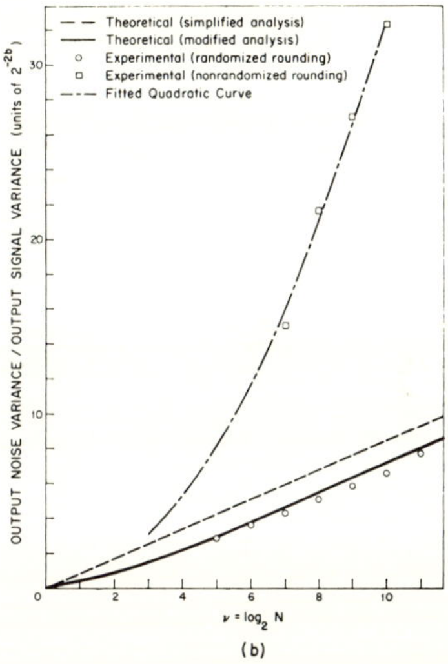

Fig. 13. (a) Experimental and theoretical noise-to-signal ratios for floating point FFT. (b) Experimental and theoretical noise-to-signal ratios for floating-point FFT and inverse.

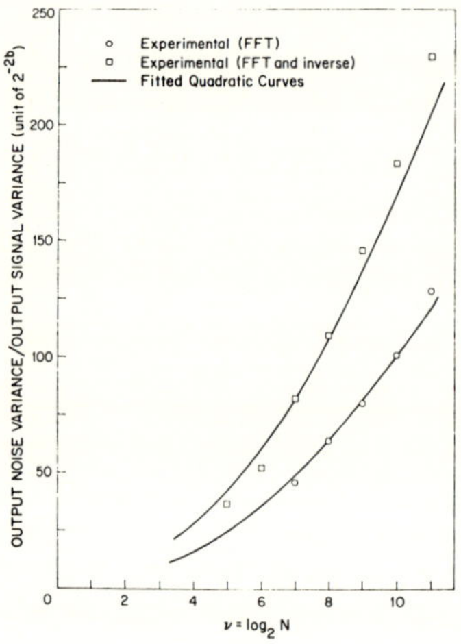

Fig. 14. Experimental noise-to-signal ratios for floating point FFT and FFT-inverse FFT; truncation used instead of rounding.

curve shown was obtained by taking into account reduced noise source variances for $\tilde{W}=1$ and $\tilde{W}=j$. Also shown are experimental results for nonrandomized rounding. These results were fitted empirically with a curve of the form $a\nu^2$, but this quadratic dependence was not established theoretically. Noise-to-signal ratios were also measured for the case where truncation rather than rounding was used in the arithmetic; the results, with empirically fitted quadratic curves, are shown in Fig. 14.

Our analysis, and all the above experiments, applied to the case of white signal. Some experimental investigation has been carried out as to whether the predictions are valid when the signal is nonwhite. Specifically, the noise introduced in computing an FFT was measured for sinusoidal signals of several frequencies, for $\nu=8$, 9, 10, and 11. The results, averaged over the input frequencies used, were within 15 percent of those predicted by (75). In these experiments, the "randomized" rounding procedure was used.

E. Effects of Coefficient Quantization in the FFT

As with the implementation of digital filters, the implementation of the FFT algorithm requires the use of quantized coefficients. While a completely definitive study of the effects of coefficient quantization in the FFT remains to be done, two approaches have been pursued for which some results have been obtained.

Although the nature of coefficient quantization is inherently nonstatistical, Weinstein [24] has obtained some useful results by means of a rough statistical analysis. This statistical analysis corresponds to introducing random jitter in the coefficients and determining the output noise-to-signal ratio due to this noise. While the detailed effect due to coefficient error due to quantization is different than that due to jitter, it is reasonable to expect that in a gross sense the magnitude of the errors is comparable.

To develop this statistical analysis, we let $F(k)$ denote the DFT of a sequence $f(n)$ and $\hat{F}(k)$ the result of transforming $f(n)$ with a radix-2 FFT algorithm with jittered coefficients. Then

$$F(k) = \sum_{n=0}^{N-1} f(n) W^{nk} \tag{78}$$

and

$$\hat{F}(k) = \sum_{n=0}^{N-1} f(n) \Omega_{nk}. \tag{79}$$

Because of the form of the FFT algorithm each element Ω_{nk} will be a product of $\nu=\log_2 N$ quantized coefficients. Thus

$$\Omega_{nk} = \prod_{i=1}^{\nu} (W^{a_i} + \delta_i) \tag{80}$$

where

$$\prod_{i=1}^{\nu} W^{a_i} = W^{nk} \tag{81}$$

with b bits for the real and imaginary parts of each of the coefficients, excluding sign, $|\delta_i|$ is less than or equal to $(\sqrt{2})2^{-b}$. If we assume that the real and imaginary parts of the jitter in the coefficients are uncorrelated and uniformly distributed between plus and minus $(1/2)2^{-b}$ then σ_δ^2, the variance of δ_i is $\sigma_\delta^2 = 2^{-2b}/6$. The error in the computation of the DFT can be expressed as

$$E(k) = \hat{F}(k) - F(k) = \sum_{n=0}^{N-1} f(n)(\Omega_{nk} - W^{nk}). \tag{82}$$

From (80) and (81) we can express the factor $(\Omega_{nk} - W^{nk})$ as

$$(\Omega_{nk} - W^{nk}) = \sum_{i=1}^{\nu} \delta_i \prod_{\substack{j=1 \\ j\neq i}}^{\nu} W^{a_i} + \text{higher order terms.} \tag{83}$$

If we neglect higher order error terms, and assume that δ_i are mutually uncorrelated then the variance of $(\Omega_{nk} - W^{nk})$ is equal to $\nu(2^{-2b}/6)$. Finally, assuming that all elements Ω_{nk} are uncorrelated with each other and with the input signal, the output error variance σ_E^2 is

$$\sigma_E^2 = \nu \frac{2^{-2b}}{6} \sum_{n=0}^{N-1} |f(n)|^2. \tag{84}$$

Since from Parseval's relation

$$\sum_{n=0}^{N-1} |f(n)|^2 = \frac{1}{N} \sum_{n=0}^{N-1} |F(k)|^2$$
$$= \text{mean-squared output signal} \tag{85}$$

the ratio of mean-squared output error to mean-squared output signal is thus

$$\sigma_E^2 \Big/ \left[\left(\frac{1}{N}\right)\left(\sum_{n=0}^{N-1} |F(k)|^2\right)\right] = \left(\frac{\nu}{6}\right) 2^{-2b}. \tag{86}$$

Although we would not expect (86) to predict with great accuracy the error in an FFT due to coefficient quantization, it is helpful as a rough estimate of the error. The key result of (86), which we would like to test experimentally, is that the error-to-signal ratio increases very mildly with N, being proportional to $\nu=\log_2 N$, so that doubling N produces only a slight increase in the error-to-signal ratio.

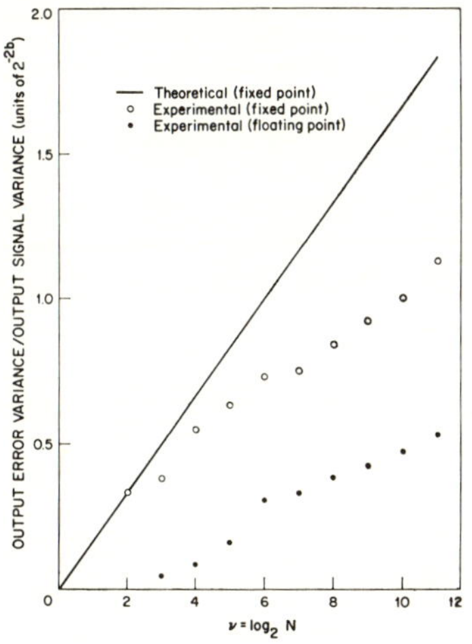

Fig. 15. Errors due to coefficient quantization in FFT.

To test this result, experimental measurements on errors due to coefficient quantization were made. In each run, a sequence $f(n)$—white in the sense that all $2N$ real numbers making up the N-point complex sequence were mutually uncorrelated, with zero means, and equal variances—was obtained using a random number generator. This sequence was transformed twice, once using a 36-bit coefficient table, and once using a coefficient table rounded to much shorter word length (e.g., 12 bits). For each transform, 36-bit accuracy was used in the arithmetic to make the effect of roundoff error negligible. The results were subtracted, squared, averaged over the output array, and divided by the output signal variance (N times the input signal variance) to obtain an experimental output error-to-signal ratio. For each value of N, several random sequences were transformed and the results were averaged to obtain statistically convergent estimates.

These results are displayed in Fig. 15; the quantity plotted is $\sigma_E^2/2^{-2b}\sigma_F^2$ where σ_F^2 is the mean-squared output signal as defined in (85). The theoretical curve corresponding to (86) is shown, and the circles represent measured output error-to-signal ratio for the fixed-point case. We note that the experimental results generally lie below the theoretical curve. No experimental result differs by as much as a factor of two from the theoretical result, and since a factor of two in σ_E^2/σ_F^2 corresponds to only half-a-bit difference in the rms output error, it seems that (86) is a reasonably accurate estimate of the effect of coefficient errors. The experimental results do seem to increase essentially linearly with ν, but with smaller slope than given in (86).

In the above, fixed-point arithmetic has been assumed. However, since a block floating-point FFT will generally use fixed-point coefficients, our results are valid for the block floating-point case also. With some slight modifications, it is possible to obtain similar results for the floating-point case. Except for a constant factor, the floating- and fixed-point results are the same. Experimental results for the floating-point case are represented by the solid dots in Fig. 15, and are observed to be slightly lower than the results for the fixed-point case.

A different approach to the characterization of FFT coefficient quantization has been taken by Tufts, Hersey, and Mosier [29]. In their analysis the effect of coefficient quantization is represented in terms of the level of spurious sidelobes introduced. In particular, a sequence $f(n)=u_o(n-n_o)$ where $u_o(n)$ denotes a unit sample, has a DFT with a purely sinusoidal real and imaginary part, i.e.,

$$F(k) = W^{n_o k} \tag{87}$$

and the inverse DFT of $F(k)$ should, of course, have only a single nonzero component. Due to coefficient quantization, however, the DFT obtained is

$$\hat{F}(k) = \Omega_{n_o k} \tag{88}$$

and since the real and imaginary parts are not exactly sinusoidal, the inverse DFT, with exact coefficients, of $F(k)$ will have spurious components. For each of the set of N sequences

$$f_k(n) = u_o(n - n_k), \qquad k = 0, 1, \cdots, N - 1. \tag{89}$$

Tufts *et al.* compute the DFT with quantized coefficients followed by the inverse DFT with accurate coefficients. At the output of the inverse DFT, the size and frequency locations of the spurious sidelobe components produced due to the quantized coefficients are observed. Since any function $f(n)$ can be constructed as a weighted sum

$$f(n) = \sum_{k=0}^{N-1} a_k f_k(n) \tag{90}$$

the spurious sidelobes produced for any $f(n)$ can in principle be determined by combining the responses due to a set of impulses. But carrying out such a combination is not practical for arbitrary $f(n)$. Tufts *et al.* have, however, tabulated the worse sidelobe levels encountered for any $f_k(n)$ as a function of the number of bits retained in the coefficients, for the case of a 64-point FFT and sign-magnitude representation of coefficients.

V. Examples

A. *Introduction*

In the preceding sections the effects of arithmetic roundoff have been analyzed for simple (first- and second-order) digital filters and the FFT. These algorithms are the basic building blocks in more complicated digital processing such as a higher order digital filter or a convolutional filter realized via the FFT. Examples will be presented in this section to indicate how some of the ideas developed above can be applied to analyze and to choose the most advantageous configuration for such systems. The first two examples concern the realization of higher order recursive filters and have borrowed from the work of other authors. The third example deals with an FFT filter.

B. *Fixed-Point Digital Filter in Cascade and Parallel Form*

After a digital filter has been specified in terms of its poles and zeroes, and the type of arithmetic has been selected, a choice must still be made among the various possible configurations of the filter which will differ with respect to the effects of roundoff noise. An exhaustive study of the selection

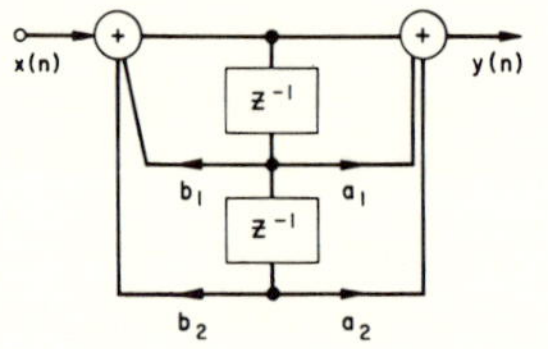

Fig. 16. Second-order section with poles preceding zeros. Used in Jackson's $1P$ parallel form with $a_2=0$. Also used in $1D$ direct form.

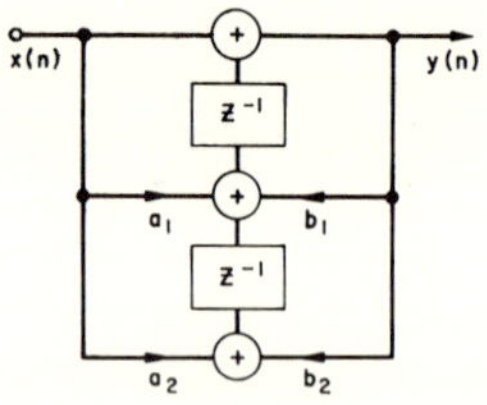

Fig. 17. Second-order section with zeros preceding poles. Used in $2P$ parallel form with $a_2=0$. Also used in $2D$ direct form.

of filter form is beyond the scope of this paper, but an excellent example of the necessary considerations is given by Jackson [30] in his analysis of roundoff noise for fixed-point digital filters realized in cascade and parallel form

Jackson considers two parallel form realizations: the $1P$ form where the individual second-order sections are realized, as shown in Fig. 16, with the poles preceding the zeros, and the $2P$ form where zeros precede poles in the individual sections, as shown in Fig. 17. (Figs. 16 and 17 do not show all the scaling coefficients needed to prevent overflow.) His analysis indicates that for a variety of scaling criteria (based on different L_p norms[2] of the input signal) and for various measures of the output noise (such as its total power, or its peak spectral value), the output signal-to-noise ratios of the two forms are very close. Generally, a very slight advantage will be gained with the $1P$ form.

Comparison of the two parallel forms basically reduces to a comparison of the noise properties of the two forms of second-order sections, since the noises from the second-order sections are simply additive in the output of the parallel form. Hence the discussion above applies to a comparison of second-order section realizations. Another form of second-order section which could be considered is a coupled form as shown in Fig. 18. For the case $a_1=r\cos\theta$, $a_2=r\sin\theta$, this filter has poles at $z=re^{\pm j\theta}$ and a zero at $z=r\cos\theta$. The coupled form noise-to-signal ratio has been compared [24] to ratios for forms essentially the same as those in Figs. 16 and 17 for the

[2] If

$$F(\omega)=\sum_{n=-\infty}^{\infty} f(n)\exp\left(-j2\pi\frac{\omega}{\omega_s}n\right)$$

represents the Fourier transform of a signal or of a filter impulse response $f(n)$, then the corresponding L_p norm is

$$\|F\|_p=\left(\frac{1}{\omega_s}\int_0^{\omega_s}|F(\omega)|^p\,d\omega\right)^{1/p}$$

where ω_s denotes sampling frequency.

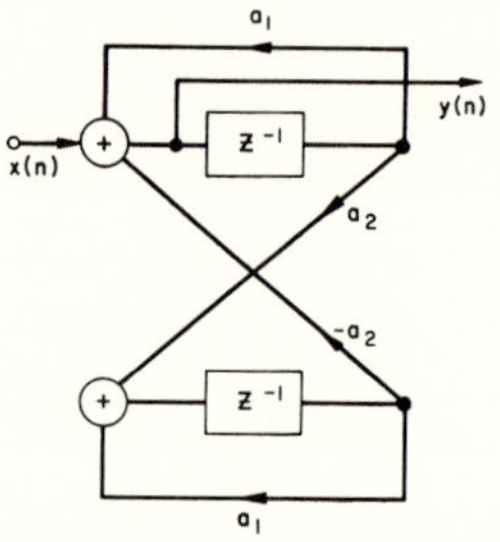

Fig. 18. Coupled form for second-order section.

case of white-noise input and an absolute overflow constraint (through the type of analysis given in Section III-C). The coupled form was found to have substantially lower noise-to-signal ratio for filters with high gain and low resonant frequencies. For $\delta=1-r$, and $\delta\ll 1$, the results vary as

$$\frac{\sigma_e^2}{\sigma_y^2}\sim 1/\delta^2\sin^2\theta \qquad \text{(forms of Figs. 16 and 17)}$$

$$\frac{\sigma_e^2}{\sigma_y^2}\sim 1/\delta^2 \qquad \text{(coupled form).} \qquad (91)$$

The implication of this result, together with the somewhat reduced coefficient sensitivity for the coupled form, is that this form may be a good choice in some situations, despite the fact that its implementation requires four multiplications instead of three for a pole pair and a single zero.

As stated above, Jackson found not much difference between the noise properties of the $1P$ and $2P$ parallel forms. However, the situation is more interesting in the case of the cascade form. Here he finds that large differences are possible between the roundoff noise outputs of the $1D$ (poles before zeros in individual sections) and $2D$ (zeros before poles in each section) forms. Also the ordering of the sections and the pairing of poles and zeros have important effect on the output signal-to-noise ratio. Jackson's analyses lead to several rules of thumb for selection of $1D$ or $2D$, for ordering of sections, and for pairing of poles and zeros.

In general, the choice of configuration depends on which L_p norm of the scaled transfer functions is constrained to prevent overflow and on which L_r norm of the output noise spectrum is used as a measure of performance. Two L_p constraints on the filter are of particular interest: the $p=\infty$ case, where the peak value of the transfer function to each possible overflow node is constrained; and the $p=2$ case where the rms transfer function to each node is constrained. The choice $p=\infty$ is just slightly less stringent than the absolute overflow constraint (7), and prevents overflow even when the input is a narrow-band signal at resonance of the relevant transfer function. The $p=2$ constraint is more appropriate for preventing overflow when the input is wide-band in nature. Two L_r norms on the output noise spectrum are of particular interest: the $r=1$ norm which measures the total output noise power, and the $r=\infty$ norm which measures the peak value of the spectrum of the output noise.

With regard to selection of $1D$ or $2D$ forms, Jackson's rule of thumb says to select $1D$ when $p=2$, $r=\infty$ and $2D$ when

$p=\infty$, $r=1$; for $p=2$, $r=1$ and for $p=\infty$, $r=\infty$, either form may be selected.

The choice of ordering of sections also depends on the norms which are selected. For $p=2$, $r=\infty$, the sections should be ordered in decreasing peakedness, where peakedness is defined as the peak gain of a section divided by its rms gain. For $p=\infty$, $r=1$, the sections should be ordered in increasing peakedness. For $p=2$, $r=1$ and for $p=\infty$, $r=\infty$, the choice of ordering depends on whether form $2D$ or $1D$ is chosen for the individual sections. Decreasing peakedness should be chosen with form $2D$, and increasing peakedness with form $1D$.

The rule for pairing of poles and zeros is as follows: Let $|H_n(\omega)|$ denote the magnitude of the frequency response of the nth section, and M_n denote the maximum over ω of $|H_n(\omega)|$. Then the pairing should be chosen such that the maximum over n of M_n is minimized.

The above rules are illustrated by Jackson with a specific filter example—a sixth-order Chebyshev band rejection filter, and the results are in accord with his rules. He analyzes the output noise of this filter for parallel forms $1P$ and $2P$ and for all orderings of cascade forms $1D$ and $2D$ (with proper pole–zero pairing). Little difference is seen between the two parallel forms. For $p=2$, $r=\infty$ the peak output noise spectrum is 7–12 dB worse for the $2D$ cascade forms than for the $1D$ forms; while for $p=\infty$, $r=1$, the output noise power is 7–12 dB worse for $1D$ than for $2D$. The effects of pole–zero pairing and of ordering of sections also follow quite well the rules previously stated. The parallel forms turn out to be slightly superior to the best cascade forms with respect to roundoff noise.

As Jackson indicates, these rules of thumb have certain qualifications and are not always valid. However, they have been shown to be helpful in a variety of types of examples [31].

C. Choice of Form for Floating-Point Digital Filter

By means of an example, Liu and Kaneko [8] have compared the direct, cascade, and parallel form realizations of a floating-point digital filter. The filter selected was an eighth-order low-pass elliptic filter. The noise-to-signal ratio for the parallel form was about 20 dB worse than for the direct form, while the cascade form was comparable to (about 1.5 dB worse than) the parallel form.

Various orderings of cascade form floating-point filters have not been studied in detail. Probably floating-point cascade filters are not too sensitive to ordering since large variations in signal level from stage to stage can be accommodated by the floating-point exponent.

A comparison of the noise-to-signal ratio properties of floating-point second-order sections where poles precede zeros (Fig. 16) and where zeros precede poles (Fig. 17) indicates that at least for white-noise inputs the behavior of the two forms is essentially identical. For a high-gain second-order section of low resonant frequency, a coupled form realization yields some noise-to-signal ratio advantage over both of these two forms.

D. FFT Filter

The results of our roundoff noise analysis for fixed-point FFT will now be applied to obtain an expression for the output noise-to-signal ratio of a finite impulse response digital filter, implemented by means of the FFT. The overflow constraints of this type of filter will be accounted for in the analysis. Attention will be focussed on a prototype low-pass filter with 256-point impulse response and a cutoff frequency of 1/4 the half sampling frequency. Rounded arithmetic will be assumed.

Let us examine the basic steps in the filtering computation, tracing the buildup of noise variance as we proceed. First the FFT is used to compute the DFT of a section of input. In the implementation of a filter with 256-point impulse response, it is reasonable to compute a 512-point FFT, where the input consists of 256 data samples and 256 zeros. Actually 512 real input samples would be treated simultaneously, by placing sections of 256 real samples in both the real and imaginary parts of the input to the FFT. To guarantee against overflow, a scaling of 1/2 is needed at each stage of the FFT yielding an overall attenuation of 1/512. The samples of the input sequence must be less than unity in magnitude. The noise $E_1(k)$ at the output of this first DFT has variance

$$\sigma_{E_1}^2 = \overline{|E_1(k)|^2} = \frac{5}{3} 2^{-2b}. \tag{92}$$

This noise variance is small, because most of the roundoff noise has been shifted off by the attenuations. However, the scalings have also caused the mean-squared signal to decrease by a factor of 1/512.

Next this computed transform is multiplied by a sequence $H(k)$ representing the DFT of a 512-point sequence $\bar{h}(n)$ consisting of the filter impulse response plus 256 zeros. This complex multiplication introduces roundoff noise of variance $2^{-2b}/3$. Assuming that we have chosen $|H(k)|<1$, the mean square of the noise E_1 becomes reduced by

$$\frac{1}{512}\sum_{k=0}^{511} |H(k)|^2 = B \tag{93}$$

a ratio of the filter bandwidth to the sampling frequency. Thus after the multiplication, the variance of the total noise $E_2(k)$ is

$$\sigma_{E_2}^2 = \frac{2^{-2b}}{3} + B\frac{5}{3}2^{-2b}. \tag{94}$$

This noise is not white, but has a component whose spectrum has been shaped by the filter.

For the example under consideration $B \approx 1/4$, so

$$\sigma_{E_2}^2 \approx \frac{3}{4}2^{-2b}. \tag{95}$$

Note that $\sigma_{E_2}^2$ is slightly less than $\sigma_{E_1}^2$ and represents only about a bit of noise. However, if the signal spectrum is flat, the mean-squared signal will also be reduced somewhat due to the multiplication by $H(k)$.

Now an inverse transform is computed to obtain a section of output. The noise variance at the output of this transform depends on how many scalings are necessary in the inverse FFT. In order to determine how many scalings are necessary, a bound on the output of the circular convolution is required [32]. For a particular filter, such a bound can be stated as

$$|y(n)| < \sum_{l=0}^{M-1} |h(l)| \tag{96}$$

where $y(n)$ is the output and M is the length of the impulse response. The prototype filter has an impulse response

$$h(n) = \frac{2}{256}\left[\frac{1}{2} + \sum_{k=0}^{31} (-1)^k \cos\frac{2\pi kn}{256} + 0.7\cos\frac{2\pi(32)n}{256} - 0.225\cos\frac{2\pi(33)n}{256} + 0.01995\cos\frac{2\pi(33)n}{256}\right] \quad (97)$$

and

$$\sum_{n=0}^{255} |h(n)| = 3.12. \quad (98)$$

Hence, only two scalings (at the first two arrays) are necessary in the inverse transform. Then, in propagating through the IFFT (inverse FFT), the variance of the noise $E_2(k)$ increases by a factor of 512/16. (The 512 represents the gain of the inverse DFT, and the 1/16 is due to the scalings.) The variance of the additional output noise $E_3(k)$ caused by roundoff in the IFFT can be estimated easily via the method of Section IV-C. The result is

$$\sigma_{E_3}^2 = \sigma_{B'}^2\left(\frac{512}{8} + \frac{512}{4}\right) + \sigma_{B'}^2 \sum_{k=1}^{6} 2^k = (202)2^{-2b}. \quad (99)$$

The total mean-squared output noise is

$$\sigma_E^2 = \frac{512}{16}\sigma_{E_2}^2 + \sigma_{E_3}^2 = (226)2^{-2b} \quad (100)$$

or in units of bits of rms output noise

$$\tfrac{1}{2}\log_2 (2^{2b}\sigma_E^2) = 3.91 \text{ bits}. \quad (101)$$

The mean-squared output signal can be estimated if specific statistics are assumed for the input signal $x(n)$. As an example, assume that $x(n)$ is white with variance $\sigma_x^2 = 2/3$. This variance goes through an attenuation of 1/512 in the first FFT, an attenuation of $B = 1/4$ due to multiplication by $H(k)$, and a gain of 512/16 in the inverse transform. The mean-squared output signal is then

$$\sigma_y^2 = \left(\frac{1}{512}\right)\left(\frac{1}{4}\right)\left(\frac{512}{16}\right)\left(\frac{2}{3}\right) = \frac{1}{96} \quad (102)$$

and the output noise-to-signal ratio is

$$\frac{\sigma_E^2}{\sigma_y^2} \approx (22\,000)2^{-2b}. \quad (103)$$

Assuming an input noise-to-signal ratio (due to A/D noise) of $(1/4)2^{-2b}$, the noise-to-signal ratio has worsened by a factor of about 5500, or

$$\tfrac{1}{2}\log_2 5500 = 6.15 \text{ bits} \quad (104)$$

in passing through the FFT filter.

References

[1] W. R. Bennett, "Spectra of quantized signals," *Bell Syst. Tech. J.*, vol. 27, pp. 446–472, 1948.

[2] B. Widrow, "Statistical analysis of amplitude-quantized sampled-data systems," *AIEE Trans. (Appl. Indust.)*, vol. 81, pp. 555–568, Jan. 1961.

[3] B. Liu, "Effect of finite word length on the accuracy of digital filters—A review," *IEEE Trans. Circuit Theory*, vol. CT-18, pp. 670–677, Nov. 1971.

[4] J. B. Knowles and R. Edwards, "Effect of a finite-word-length computer in a sampled-data feedback system," *Proc. Inst. Elec. Eng.*, vol. 112, pp. 1197–1207, 1965.

[5] B. Gold and C. M. Rader, "Effect of quantization noise in digital filters," in *Proc. Spring Joint Comput. Conf., AFIPS Conf. Proc.*, vol. 28, pp. 213–219, 1966.

[6] C. Weinstein and A. V. Oppenheim, "A comparison of roundoff noise in floating point and fixed point digital filter realizations," *Proc. IEEE* (Lett.), vol. 57, pp. 1181–1183, June 1969.

[7] L. B. Jackson, "On the interaction of roundoff noise and dynamic range in digital filters," *Bell Syst. Tech. J.*, vol. 49, pp. 159–184, 1970.

[8] B. Liu and T. Kaneko, "Error analysis of digital filters realized with floating-point arithmetic," *Proc. IEEE*, vol. 57, pp. 1735–1747, Oct. 1969.

[9] A. V. Oppenheim, "Realization of digital filters using block-floating-point arithmetic," *IEEE Trans. Audio Electroacoust.*, vol. AU-18, pp. 130–136, Jan. 1970.

[10] R. B. Blackman, *Linear Data-Smoothing and Prediction in Theory and Practice*. Reading, Mass.: Addison-Wesley, 1965.

[11] L. Jackson, "An analysis of limit cycles due to multiplication rounding in recursive digital filters," in *Proc. 7th Allerton Conf. Circuit and System Theory*, pp. 69–78, 1969.

[12] S. R. Parker and S. F. Hess, "Limit-cycle oscillations in digital filters," *IEEE Trans. Circuit Theory*, vol. CT-8, pp. 687–697, Nov. 1971.

[13] P. M. Ebert, J. E. Mazo, and M. C. Taylor, "Overflow oscillations in digital filters," *Bell Syst. Tech. J.*, vol. 48, pp. 2999–3020, 1969.

[14] J. F. Kaiser, "Some practical considerations in the realization of linear digital filters," in *Proc. 3rd Allerton Conf. Circuit and Systems Theory*, pp. 621–633, 1965.

[15] C. M. Rader and B. Gold, "Effects of parameter quantization on the poles of a digital filter," *Proc. IEEE* (Lett.), vol. 55, pp. 688–689, May 1967.

[16] E. Avenhaus, "An optimization procedure to minimize the word length of digital filter coefficients," presented at the London Conf. on Digital Filtering, Aug. 31, 1971.

[17] E. Avenhaus and W. Schuessler, "On the approximation problem in the design of digital filters with limited wordlength," *Arch. Elek. Übertragung*, vol. 24, pt. 12, pp. 571–572, 1970.

[18] K. Steiglitz, "Designing short-word recursive digital filters," in *Proc. 9th Allerton Conf. Circuit and System Theory* (Monticello, Ill.), pp. 778–788, Oct. 1971.

[19] A. Fettweis, "Some principles of designing digital filters imitating classical filter structures," *IEEE Trans. Circuit Theory* (Corresp.), vol. CT-18, pp. 314–316, Mar. 1971.

[20] J. Bingham, "A new type of digital filter with a reciprocal ladder configuration," in *Proc. 1970 IEEE Int. Symp. Circuit Theory, Dig. of Tech. Papers* pp. 129–130.

[21] R. Crochiere, "Digital ladder filter structures and coefficient sensitivity," M.I.T. Res. Lab. of Electronics, Quart. Progr. Rep. 103, Oct. 15, 1971.

[22] J. W. Cooley and J. W. Tukey, "An algorithm for the machine calculation of complex Fourier series," *Math. Comput.*, vol. 19, pp. 297–301, 1965.

[23] B. Gold and C. M. Rader, *Digital Processing of Signals*. New York: McGraw-Hill, 1969.

[24] C. J. Weinstein, "Quantization effects in digital filters," M.I.T. Lincoln Lab. Tech. Rep. 468, ASTIA Doc. DDC AD-706862, Nov. 21, 1969.

[25] P. D. Welch, "A fixed-point fast Fourier transform error analysis," *IEEE Trans. Audio Electroacoust.*, vol. AU-17, pp. 153–157, June 1969.

[26] W. M. Gentleman and G. Sande, "Fast Fourier transforms—For fun and profit," in *Proc. Fall Joint Computer Conf., AFIPS Conf. Proc.*, pp. 563–578, 1966.

[27] C. J. Weinstein, "Roundoff noise in floating point fast Fourier transform computation," *IEEE Trans. Audio Electroacoust.*, vol. AU-17, pp. 209–215, Sept. 1969.

[28] T. Kaneko and B. Liu, "Accumulation of roundoff error in fast Fourier transforms," *J. Ass. Comput. Mach.*, vol. 17, pp. 637–654, Oct. 1970.

[29] D. W. Tufts, H. S. Hersey, and W. E. Mosier, "Effects of FFT coefficient quantization on bin frequency response," *Proc. IEEE* (Lett.), vol. 60, pp. 146–147, Jan. 1972.

[30] L. B. Jackson, "Roundoff-noise analysis for fixed-point digital filters realized in cascade or parallel form," *IEEE Trans. Audio Electroacoust.*, vol. AU-18, pp. 107–122, June 1970.

[31] L. B. Jackson, "An analysis of roundoff noise in digital filters," Sc.D. dissertation, Stevens Inst. Technol., Dep. Elec. Eng., Castle Point, Hoboken, N. J., 1969.

[32] A. V. Oppenheim and C. Weinstein, "A bound on the output of a circular convolution with application to digital filtering," *IEEE Trans. Audio Electroacoust.*, vol. AU-17, pp. 120–124, June 1969.

Bibliography

[33] J. E. Bertram, "The effect of quantization in sampled-feedback systems," *AIEE Trans. (Appl. Industry)*, vol. 77, pp. 177–182, July 1958.

[34] F. Bonzanigo, "Constant-input behavior of recursive digital filters," presented at IEEE Arden House Workshop in Digital Filtering, Harriman, N. Y., Jan. 1970.

[35] R. B. Blackman, *Linear Data-Smoothing and Prediction in Theory and Practice.* Reading, Mass.: Addison-Wesley, 1965, pp. 75–79.

[36] E. E. Curry, "The analysis of round-off and truncation errors in a hybrid control system," *IEEE Trans. Automat. Contr.* (Short Papers), vol. AC-12, pp. 601–604, Oct. 1967.

[37] L. D. Divieti, C. M. Rossi, R. M. Schmid, and A. E. Vereschkin, "A note on computing quantization errors in digital control systems," *IEEE Trans. Automat. Contr.* (Corresp.), vol. AC-12, pp. 622–623, Oct. 1967.

[38] R. Edwards, J. Bradley, and J. Knowles, "Comparison of noise performances of programming methods in the realization of digital filters," in *Proc. 1969 Polytech. Inst. Brooklyn Symp. on Computer Processing in Communications.*

[39] A. Fettweis, "A general theorem for signal-flow networks, with applications," *Arch. Elek. Übertragung*, vol. 25, pp. 557–561, Dec. 1969.

[40] W. A. Gardner, "Reduction of sensitivities in sampled-data filters," *IEEE Trans. Circuit Theory* (Corresp.), vol. CT-17, pp. 660–663, Nov. 1970.

[41] B. Gold and L. Rabiner, "Analysis of digital and analog formant synthesizers," *IEEE Trans. Audio Electroacoust.*, vol. AU-16, pp. 81–94, Mar. 1968.

[42] B. Gold and C. M. Rader, *Digital Processing of Signals.* New York: McGraw-Hill, 1969.

[43] R. M. Golden and S. A. White, "A holding technique to reduce number of bits in digital transfer functions," *IEEE Trans. Audio Electroacoust.*, vol. AU-16, pp. 433–437, Sept. 1968.

[44] D. Herrmann and W. Schuessler, "On the accuracy problem in the design of nonrecursive digital filters," *Arch. Elek. Übertragung*, vol. 24, pt. 11, pp. 525–526, 1970.

[45] J. F. Kaiser, "Digital filters," in F. F. Kuo and J. F. Kaiser, Eds., *System Analysis by Digital Computer.* New York: Wiley, 1966, ch. 7.

[46] T. Kaneko and B. Liu, "Roundoff error of floating-point digital filters," in *Proc. 6th Allerton Conf. on Circuit and System Theory*, pp. 219–227, Oct. 1968.

[47] J. Katzenelson, "On errors introduced by combined sampling and quantization," *IRE Trans. Automat. Contr.*, vol. AC-7, pp. 58–68, Apr. 1962.

[48] W. C. Kellogg, "Information rates in sampling and quantization," *IEEE Trans. Informat. Theory*, vol. IT-13, pp. 506–511, July 1967.

[49] J. B. Knowles and R. Edwards, "Simplified analysis of computational errors in a feedback system incorporating a digital computer," presented at S.I.T. Symp. on Direct Digital Control, London, England, Apr. 22, 1965.

[50] ——, "Complex cascade programming and associated computational errors," *Electron. Lett.*, vol. 1, pp. 160–161, Aug. 1965.

[51] ——, "Finite word-length effects in a multirate direct digital control system," *Proc. Inst. Elec. Eng.*, vol. 112, pp. 2376–2384, Dec. 1965.

[52] J. B. Knowles and E. M. Olcayto, "Coefficient accuracy and digital filter response," *IEEE Trans. Circuit Theory*, vol. CT-15, pp. 31–41, Mar. 1968.

[53] A. A. Kosyakin, "The statistical theory of amplitude quantization," *Avtomat. Telemekh.*, vol. 22, p. 722, 1969.

[54] I. M. Langenthal, "Coefficient sensitivity and generalized digital filter synthesis," in *EASCON 1968 Rec.*, pp. 386–392, 1968.

[55] C. E. Maley, "The effect of parameters on the roots of an equation system," *Comput. J.*, vol. 4, pp. 62–63, 1961–1962.

[56] P. E. Mantey, "Eigenvalue sensitivity and state-variable selection," *IEEE Trans. Automat. Contr.*, vol. AC-13, pp. 263–269, June 1968.

[57] R. K. Otnes and L. P. McNamee, "Instability thresholds in digital filters due to coefficient rounding," *IEEE Trans. Audio Electroacoust.*, vol. AU-18, pp. 456–463, Dec. 1970.

[58] C. Rader and B. Gold, "Digital filter design techniques in the frequency domain," *Proc. IEEE*, vol. 55, pp. 149–171, Feb. 1967.

[59] Q. I. Rahman, "The influence of coefficients on the zeros of polynomials," *J. London Math Soc.*, vol. 36, pt. 1, pp. 57–64, Jan. 1961.

[60] I. W. Sandberg, "Floating-point-roundoff accumulation in digital filter realization," *Bell Syst. Tech. J.*, vol. 46, pp. 1775–1791, Oct. 1967.

[61] ——, "A theorem concerning limit cycles in digital filters," in *Proc. 7th Ann. Allerton Conf. Circuit and System Theory*, pp. 63–68, 1968.

[62] J. B. Slaughter, "Quantization errors in digital control systems," *IEEE Trans. Automat. Contr.*, vol. AC-9, pp. 70–74, Jan. 1964.

[63] O. Sornmoonpin, "Investigation of quantization errors," M.Sc. thesis, Univ. of Manchester, England, 1966.

[64] T. G. Stockham, Jr., "A-D and D-A converters: Their effect on digital audio fidelity," presented at 41st Meeting of the Audio Engineering Society, New York, Oct. 5–8, 1971.

[65] D. W. Tufts, W. Knight, and D. W. Rorabacher, "Effects of quantization and sampling in digital correlators and in power spectral estimation," *Proc. IEEE* (Lett.), vol. 56, pp. 79–82, Jan. 1969.

[66] D. W. Tufts, D. W. Rorabacher, and W. E. Mosier, "Designing simple, effective digital filters," *IEEE Trans. Audio Electroacoust.*, vol. AU-18, pp. 142–158, June 1970.

[67] C. Weinstein, "Quantization effects in frequency sampling fiilters," in *NEREM Rec.*, p. 222, 1968.

[68] B. Widrow, "A study of rough amplitude quantization by means of Nyquist sampling theory," *IRE Trans. Circuit Theory*, vol. CT-3, pp. 266–276, Dec. 1956.

[69] J. H. Wilkinson, "Error analysis of floating-point comparison," *Numerisch. Math.*, vol. 2, pp. 319–340, 1960.

[70] ——, *Rounding Errors in Algebraic Processes.* Englewood Cliffs, N. J.: Prentice-Hall, 1963.

[71] E. P. F. Kan and J. K. Aggarwal, "Error analysis of digital filter employing floating-point arithmetic," *IEEE Trans. Circuit Theory*, vol. CT-18, pp 678–686, Nov. 1971.

Reprinted from *Proc. IEE,* **112**(6), 1197–1207 (1965)

Effect of a finite-word-length computer in a sampled-data feedback system

J. B. Knowles, M.Sc.Tech., Ph.D., Graduate I.E.E., and R. Edwards, M.Sc.Tech., Graduate I.E.E.

Synopsis

When a digital computer acts as the compensating element of a sampled-data feedback system, the system performance is degenerated by the computational errors inherent in the finite-word-length machine.

The statistical analysis presented in this paper enables the increase in the mean-square system error due to quantisation and computer roundoff errors to be assessed for a closed-loop linear sampled-data system. In fact, the loss in overall performance can be determined by slide-rule calculation for all computer word lengths. A further advantage of the present approach over previous work is that the necessary computations are performed on the conventional block diagram of the system, rather than its matrix equivalent. One application of the analysis is to specify the precision required in the digital compensator and peripheral equipment for consistency with the overall accuracy demanded of the closed-loop system. It also enables the selection of the programming technique which requires the shortest computer word length for this accuracy.

As a means of verifying the analysis, the performance of two different digital compensators for a practical system are each compared with a counterpart having negligible quantisation and roundoff errors.

List of principal symbols

γ = binary digit
M = computer word length
q = width of quantisation
$x(k)$ = input sequence to a sampled-data system
$y(k)$ = output sequence of a sampled-data system
z^{-k} = ordering variable of the kth term in a sequence
$H(z)$ = discrete shaping filter
$h(k)$ = weighting sequence of the filter $H(z)$
$S(z)$ = discrete shaping filter
$e(k)$ = ideal error-channel sequence
$e'(k)$ = actual error-channel sequence
$r(k)$ = roundoff- or quantisation-error sequence
$R(z)$ = Ztransformation of the sequence $r(k)$
$\epsilon(k)$ = computational-error sequence
$\mathcal{E}(z)$ = Ztransformation of the sequence $\epsilon(k)$
$f(k)$ = ideal plant-forcing sequence
$f'(k)$ = actual plant-forcing sequence
$D(z)$ = pulsed transfer function of a discrete compensator
μ = number of multiplications
t = time
T = sampling period
σ^2 = variance of a random process
$U(t)$ = unit step function $U(t) = \begin{cases} 0 \text{ for } t < 0 \\ 1 \text{ for } t \geqslant 0 \end{cases}$
$\phi^*_{xx}(k)$ = autocorrelation sequence of the process $x(n)$
s = complex variable of the Laplace transformation
Re = real part of a complex quantity
Im = imaginary part of a complex quantity

1 Introduction

An electronic digital computer processes information in binary-number format. When the so-called 'scientific-programming' convention is employed in a fixed-point machine, a binary number $\gamma_0\gamma_1 \ldots \gamma_{M-1}$ in the machine has the significance

$$\gamma_0\gamma_1 \ldots \gamma_{M-1} = -\gamma_0 + \sum_{i=1}^{M-1} \gamma_i 2^{-i} \quad . \quad . \quad . \quad . \quad (1)$$

Paper 4760 S, first received 9th November 1964 and in revised form 23rd February 1965

Dr. Knowles and Mr. Edwards are with the Department of Electrical Engineering, Faculty of Technology, University of Manchester

where $\gamma_j = 0$ or 1 for all j . . . . (2

In connection with eqn. 1, the width of quantisation is defined as

$$q = 2^{-(M-1)} \quad . \quad . \quad . \quad . \quad . \quad . \quad . \quad . \quad . \quad . \quad . \quad (3)$$

It is evident, from eqns. 1, 2 and 3, that all numbers in the machine lie in the interval $(-1, 1 - q)$.

By virtue of the finite computer word length, errors are inherently present in computer calculations. Firstly, when continuous data are read into the computer, a quantisation error is incurred. Secondly, a roundoff error arises in the evaluation of each arithmetic product in the computation. Depending on whether a multiplication is rounded or unrounded, the roundoff error lies in the interval $(-q/2, +q/2)$ or $(-q, 0)$, respectively. It should be noted that the processes of arithmetic addition and subtraction are not subject to further error.

When a digital computer acts as the compensating element of a sampled-data feedback system, its inherent computational errors may cause the system performance to deviate considerably from the ideal. It would appear that Bertram[1] was the first to examine the degenerative effect of roundoff errors in a closed-loop linear, sampled-data system incorporating a digital computer as the compensating element. The approach adopted involves the use of the state-vector concept and the evaluation of the upper error bound on a matrix product. Evidently, the computation of the upper error bound involves a simulator model which is normally quite different from that usually employed in the investigation of the system's conventional performance. Moreover, an upper error bound would seem to be a pessimistic measure of the loss of system performance, because roundoff errors are intrinsically statistical. A later paper, by Slaughter,[2] applies Bertram's analysis to the evaluation of the steady-state upper error bound for the case of direct programming only. He confirms that the upper bounds derived from Bertram's analysis are pessimistic. Tsypkin[3] investigates a closed-loop sampled-data system in which, effectively, only the input and output data are quantised and multiplicative roundoff errors in the computer are not considered.

This paper shows that quantisation and roundoff errors in a digital computer correspond, in the steady state, to a number of additive noise processes at the input of an errorless system.

Using the mean squares of these noise processes, a measure of the degeneration of system performance is obtained. The necessary calculations are simple and involve the conventional block diagram of the system rather than its matrix equivalent. For a given programming technique, it is possible to specify the minimum computer word length consistent with the overall accuracy demanded of the system. By selecting the programming technique which requires the shortest computer word length to give this accuracy, it may be possible to achieve an economy in computer storage by packing more than one system variable into a single word.

2 A statistical bound

The total steady-state output of the linear sampled-data system shown in Fig. 1 is evidently

$$y(k) = \sum_{p=1}^{N} \left[\sum_{n=0}^{\infty} h_p(n) r_p(k-n) \right] \quad . \quad . \quad . \quad . \quad . \quad (4)$$

where $\{r_p(k)\}$ are uncorrelated stationary processes and $\{h_p(k)\}$ are the weighting sequences of the systems $\{H_p(z)\}$. Thus the mean-square steady-state output sequence is given by

$$\overline{y^2} = \sum_{q=1}^{N} \sum_{p=1}^{N} \left[\sum_{m=0}^{\infty} \sum_{n=0}^{\infty} h_p(n) h_q(m) \overline{r_p(k-n) r_q(k-m)} \right] \quad . \quad . \quad . \quad . \quad . \quad (5)$$

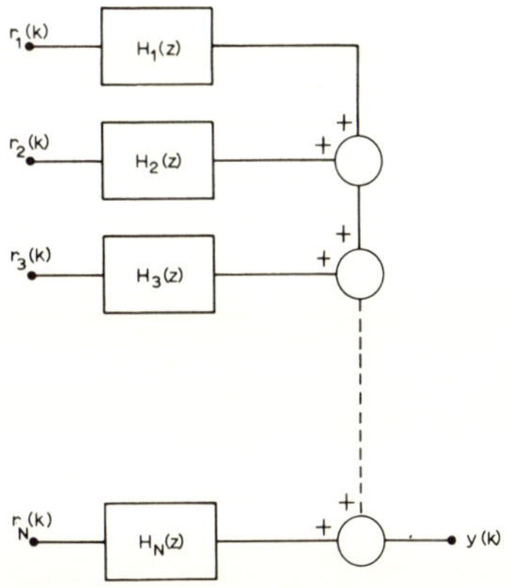

Fig. 1
Generalised, multi-input sampled-data system

Assuming the input sequences to be bounded and each system to be asymptotically stable in the large, the summations in eqn. 5 are absolutely convergent, and their order can be changed to

$$\overline{y^2} = \sum_{p=1}^{N} \sum_{m=0}^{\infty} \sum_{n=0}^{\infty} \left[\sum_{q=1}^{N} h_p(n) h_q(m) \overline{r_p(k-n) r_q(k-m)} \right] \quad . \quad . \quad . \quad . \quad (6)$$

As the individual input processes are uncorrelated, eqn. 6 reduces to

$$\overline{y^2} = \sum_{p=1}^{N} \sum_{m=0}^{\infty} \sum_{n=0}^{\infty} [h_q(m) h_q(n) \phi_{pp}^*(n-m)] \quad . \quad . \quad (7)$$

or in terms of frequency spectra,[7,8]

$$\overline{y^2} = \sum_{p=1}^{N} \frac{T}{2\pi} \int_0^{2\pi/T} |H^*(j\omega)|^2 \Phi_{pp}^*(\omega) d\omega \quad . \quad . \quad . \quad . \quad . \quad (8)$$

Denoting the maximum gain of the system $H_p(z)$ in Fig. 1 by $|H_p^*(j\omega_m)|$, so that

$$|H_p^*(j\omega_m)| \geqslant |H^*(j\omega)| \text{ for all } \omega \quad . \quad . \quad . \quad . \quad (9)$$

then, from the integral mean value theorem,

$$\int_0^{2\pi/T} |H^*(j\omega)|^2 \Phi_{pp}^*(\omega) d\omega \leqslant \int_0^{2\pi/T} |H^*(j\omega_m)|^2 \Phi_{pp}^*(\omega) d\omega \quad (10)$$

Hence

$$\overline{y^2} \leqslant \sum_{p=1}^{N} |H_p^*(j\omega_m)|^2 \frac{T}{2\pi} \int_0^{2\pi/T} \Phi_{pp}^*(\omega) d\omega \quad . \quad . \quad . \quad (11)$$

or

$$\overline{y^2} \leqslant \sum_{p=1}^{N} |H_p^*(j\omega_m)|^2 \phi_{pp}^*(0) \quad . \quad . \quad . \quad . \quad . \quad . \quad . \quad (12)$$

3 Data-quantisation error

An ideally linear, sampled-data system having a digital compensator is shown in Fig. 2. In the absence of data quantisation, the true error sequence is given by

$$e(k) = x(k) - y(k) \quad . \quad . \quad . \quad . \quad . \quad . \quad . \quad . \quad (13)$$

However, owing to the limited accuracy of the peripheral equipment and the machine, the computed error sequence is

$$e'(k) = x'(k) - y'(k) \quad . \quad . \quad . \quad . \quad . \quad . \quad . \quad . \quad (14)$$

$$\text{where } \left.\begin{aligned} x'(k) &= x(k) + r_x(k) \\ y'(k) &= y(k) + r_y(k) \end{aligned}\right\} \quad . \quad . \quad . \quad . \quad . \quad . \quad . \quad (15)$$

and $r_x(k)$ and $r_y(k)$ represent the error incurred in quantising the actual input and output of the system. From eqns. 13, 14 and 15, we obtain

$$e'(k) = e(k) + [r_x(k) - r_y(k)] \quad . \quad . \quad . \quad . \quad . \quad (16)$$

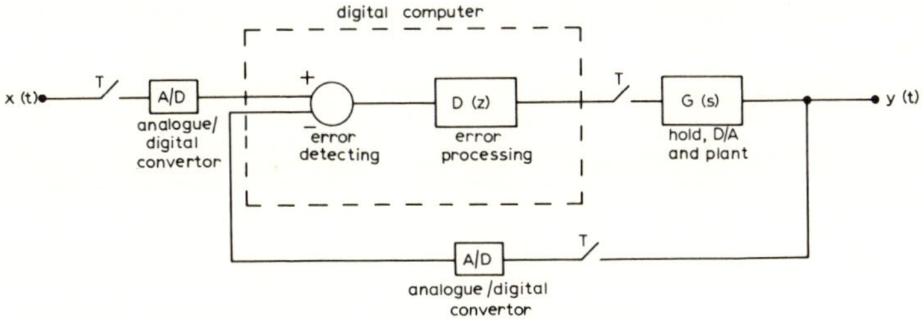

Fig. 2
Ideally linear sampled-data feedback system

Thus the computed error sequence $e'(k)$ may be regarded as the true error sequence additively contaminated by the error processes $r_x(k)$ and $-r_y(k)$. That is to say, quantisation of the input and output information in an ideally linear, sampled-data feedback system may be represented by the injection of noise processes into its error channel. By elementary block-diagram manipulation, the noise processes may be represented as inputs to the ideal system in the manner shown in Fig. 4. In practice, the accuracy demanded of a closed-loop computer-controlled system generally implies that q is small enough for the quantisation and multiplicative roundoff errors to have a uniform probability density function.[4,5,6] Further, it is assumed throughout the following analysis that both quantisation and roundoff processes are mutually uncorrelated and wide-sense stationary in the steady state. Thus, with rounded operations, the mean and mean-square values of each steady-state error process are given by

$$\bar{r} = 0; \; \overline{r^2} = q^2/12 \quad \ldots \ldots \ldots \quad (17)$$

and, for unrounded operations,

$$\bar{r} = -q/2; \; \overline{r^2} = q^2/3 \quad \ldots \ldots \quad (18)$$

4 Computational errors for direct programming

The digital compensator of the system shown in Fig. 2 possesses a pulsed transfer function

$$D(z) = \frac{B(z)}{A(z)} = \frac{\sum_{n=1}^{N} b_n z^{-n}}{\sum_{n=0}^{N} a_n z^{-n}} \quad \ldots \ldots \quad (19)$$

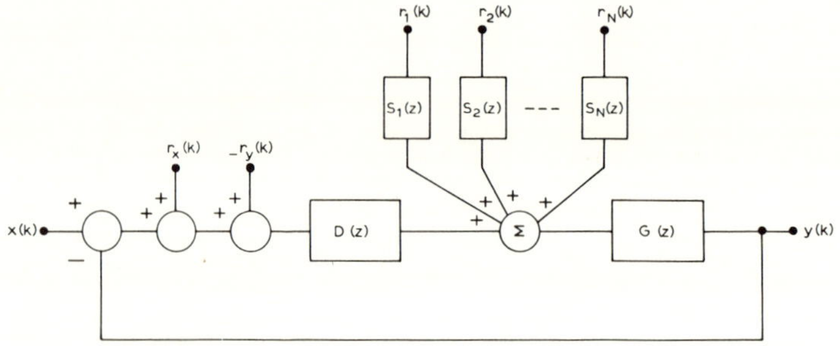

Fig. 3
Block diagram of actual system

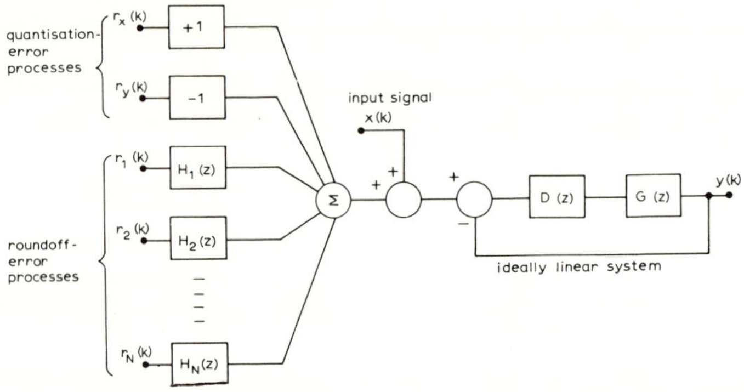

Fig. 4
Equivalent system

For physical realisability, a_0 is nonzero and, without loss of generality, may be taken as unity. In the direct method of programming,[7,8] the realisation of eqn. 19 is attempted using the recursion formula

$$f(k) = \sum_{n=1}^{N} b_n e'(k-n) - \sum_{n=1}^{N} a_n f(k-n) \quad \ldots \quad (20)$$

However, owing to the finite word length of the computer, the actual output is

$$f'(k) = \sum_{n=1}^{N} b_n e'(k-n) - \sum_{n=1}^{N} a_n f'(k-n) + r_D(k) \quad (21)$$

where $r_D(k)$ is the total error incurred in evaluating

$$\sum_{n=1}^{N} b_n e'(k-n) - \sum_{n=1}^{N} a_n f'(k-n) \quad \ldots \ldots \quad (22)$$

Substituting eqn. 20 into eqn. 21 so as to eliminate the input terms yields

$$f'(k) = f(k) - \sum_{n=1}^{N} a_n[f'(k-n) - f(k-n)] + r_D(k) \quad (23)$$

Defining a computational error quantity $\epsilon_d(k)$ by

$$f'(k) = f(k) + \epsilon_d(k) \quad . \quad . \quad . \quad . \quad . \quad . \quad . \quad . \quad (24)$$

$$\epsilon_d(k) = r_D(k) - \sum_{n=1}^{N} a_n \epsilon_d(k-n) \quad . \quad . \quad . \quad . \quad . \quad (25)$$

The intrinsic statistical nature of $r(k)$ and eqn. 24 enable the actual plant-forcing function to be regarded as the true forcing sequence contaminated with an additive noise sequence, $\epsilon_d(k)$. It is now proposed to determine the relationship existing between the computational-error sequence $\epsilon_d(k)$ and the error sequence $r_D(k)$.

Because $a_0 = 1$, it follows that eqn. 25 may be written as

$$\sum_{n=0}^{N} a_n \epsilon_d(k-n) = r_D(k) \quad . \quad . \quad . \quad . \quad . \quad . \quad . \quad (26)$$

Taking the Z transformation of eqn. 26, one obtains

$$\epsilon_d(z) = \frac{1}{A(z)} R_D(z) \quad . \quad . \quad . \quad . \quad . \quad . \quad . \quad . \quad . \quad (27)$$

Eqn. 27 specifies the computational-error sequence to be added to the plant-forcing function of the ideal system to account for the finite word length of the computer. By elementary block-diagram manipulation, the error sequence $r_D(k)$ can be represented as a spurious input to the ideal closed-loop system in the manner of Fig. 4. In order to compute the degeneration in overall system performance by eqn. 12, it is necessary to evaluate the steady-state mean square of the sequence $r_D(k)$. Since all the variables in eqn. 22 are quantised, the error quantity $r_D(k)$ is composed entirely of roundoff errors incurred in the multiplications. If there are μ multiplications,

$$r_D(k) = \sum_{p=1}^{\mu} r_p(k) \quad . \quad . \quad . \quad . \quad . \quad . \quad . \quad . \quad . \quad (28)$$

where $r_p(k)$ is the error from a single multiplication at time kT. Assuming the roundoff error sequences to be mutually uncorrelated and wide-sense stationary, it follows that the steady-state mean-square value of $r_D(k)$ is

$$\overline{r_D^2} = \mu \overline{r^2} + (\mu^2 - \mu)(\bar{r})^2 \quad . \quad . \quad . \quad . \quad . \quad . \quad . \quad (29)$$

The result for rounded operations is derived from eqn. 17 as

$$\overline{r_D^2} = \frac{\mu q^2}{12} \quad . \quad . \quad . \quad . \quad . \quad . \quad . \quad . \quad . \quad . \quad . \quad (30)$$

and the corresponding result for unrounded operations is obtained from eqn. 18 as

$$\overline{r_D^2} = \frac{\mu q^2}{12}(1 + 3\mu) \quad . \quad . \quad . \quad . \quad . \quad . \quad . \quad . \quad (31)$$

If the roundoff-error sequence $r_D(k)$ is assumed to be linearly independent of the actual input signal to the system, the maximum increase in mean-square error of the system due to the finite word length of the compensator can be calculated independently of the input.[9] The necessary calculation is performed on the conventional block diagram of the system preceded by the appropriate shaping filters according to Fig. 4. In practice, the magnitude of the increase in system error due to quantisation and roundoff errors is essentially determined by the form of compensation employed, because the d.c. gain of the ideal closed-loop system is exactly, or very nearly, unity.

If $G_p(z)$ denotes the transfer function which relates the output of the system shown in Fig. 4 to the samples of the pth spurious input, the total response of the overall system is given by

$$y(t) = \sum_{p=1}^{N} \sum_{k=-\infty}^{t} g_p(t - kT) r_p(kT) \quad . \quad . \quad . \quad . \quad (32)$$

The inherent character of the noise processes $r_p(k)$ and the form of eqn. 32 make the steady-state output of the system due to quantisation and roundoff errors a normal process.† This conclusion is not contradicted by the experimental evidence.

5 Computational errors for parallel programming

For parallel programming, the pulsed transfer function of the discrete compensator is expressed in the form

$$D(z) = \frac{A_0}{z} + \sum_i \frac{A_i}{z - z_i} + \sum_j \frac{B_j}{z - C_j} \quad . \quad . \quad . \quad (33)$$

where $\{z_i\}$ and $\{C_j\}$ represent the complex and real poles, respectively, of $D(z)$. As is shown in References 7, 8 and 12, the parallel programming of $D(z)$ consists of solving by direct programming each difference equation in the three sets:

$$\{f_0(k) = A_0 e'(k-1)\} \quad . \quad . \quad . \quad . \quad . \quad . \quad . \quad (34)$$

$$\left.\begin{aligned} \mathrm{Re}\,[f_i(k)] &= \mathrm{Re}\,[A_i e'(k-1)] \\ &\quad + \mathrm{Re}\,(z_i)\,\mathrm{Re}\,[f_i(k-1)] \\ &\quad - \mathrm{Im}\,(z_i)\,\mathrm{Im}\,[f_i(k-1)] \\ \mathrm{Im}\,[f_i(k)] &= \mathrm{Im}\,[A_i e'(k-1)] \\ &\quad + \mathrm{Re}\,(z_i)\,\mathrm{Im}[f_i(k-1)] \\ &\quad + \mathrm{Im}\,(z_i)\,\mathrm{Re}\,[f_i(k-1)] \end{aligned}\right\} \quad (35)$$

$$\{f_j(k) = B_j e'(k-1) + C_j f_j(k-1)\} \quad . \quad . \quad (36)$$

and then forming an output according to

$$f(k) = f_0(k) + \sum_i \{\mathrm{Re}[f_i(k)] + j\,\mathrm{Im}\,[f_i(k)]\} + \sum_j f_j(k) \quad . \quad . \quad . \quad . \quad (37)$$

It should be observed that in a realisable, discrete compensator any complex poles occur in conjugate pairs. Thus the summation $\sum_i \{\mathrm{Re}\,[f_i(k)] + j\,\mathrm{Im}\,[f_i(k)]\}$ is always real because pairs of imaginary components are opposite in sign and equal in magnitude. Hence, eqn. 37 reduces to

$$f(k) = f_0(k) + \sum_i \mathrm{Re}\,[f_i(k)] + \sum_j f_j(k) \quad . \quad . \quad . \quad (38)$$

which involves only real arithmetic operations.[12]

Owing to the finite word length of the computer, the actual outputs of the compensator elements are

$$\{f'_0(k) = A_0 e'(k-1) + r_0(k)\} \quad . \quad . \quad . \quad . \quad (39)$$

† Central-limit theorem[10, 11]

$$\left.\begin{aligned}\operatorname{Re}[f'_i(k)] &= \operatorname{Re}[A_i e'(k-1)] \\ &\quad + \operatorname{Re}(z_i)\operatorname{Re}[f'_i(k-1)] \\ &\quad - \operatorname{Im}(z_i)\operatorname{Im}[f'_i(k-1)] + r_i(k) \\ \operatorname{Im}[f'_i(k)] &= \operatorname{Im}[A_i e'(k-1)] \\ &\quad + \operatorname{Re}(z_i)\operatorname{Im}[f'_i(k-1)] \\ &\quad + \operatorname{Im}(z_i)\operatorname{Re}[f'_i(k-1)] + r_i^*(k) \end{aligned}\right\} \quad (40)$$

$$\{f'_j(k) = B_j e'(k-1) + C_j f'_j(k-1) + r_j(k)\} \quad (41)$$

It is shown in Appendix 12.1 that the actual plant-forcing sequence may be regarded as the true forcing sequence contaminated with an additive noise sequence $\epsilon_p(k)$, where

$$\epsilon_p(z) = R_0(z) + \sum_i \left(\frac{1 - z^{-1}|z_i|\cos\angle z_i}{1 - 2z^{-1}|z_i|\cos\angle z_i + z^{-2}|z_i|^2}\right) R_i(z) - \left(\frac{z^{-1}|z_i|\sin\angle z_i}{1 - 2z^{-1}|z_i|\cos\angle z_i + z^{-2}|z_i|^2}\right) R_i^*(z)$$
$$+ \sum_j \left(\frac{1}{1 - z^{-1}C_j}\right) R_j(z) \quad \ldots \quad (42)$$

Eqn. 42 specifies the computational-error sequence to be added to the plant-forcing function of the ideal system to account for the finite word length of the computer. By elementary block-diagram manipulation, the error sequences $r_0(k)$, $\{r_i(k)\}$, $\{r_i^*(k)\}$ and $\{r_j(k)\}$ can be represented as spurious inputs to the ideal closed-loop system in the manner of Fig. 4. Assuming these input sequences to be wide-sense stationary, their mean-square values for rounded operations are derived from eqn. 30 as

$$\left.\begin{aligned}\overline{r_0^2} &= \frac{q^2}{12}; \quad \overline{r_i^2} = \frac{\mu_C q^2}{12} \\ \overline{r_i^{*2}} &= \frac{\mu_C q^2}{12}; \quad \overline{r_j^2} = \frac{\mu_R q^2}{12}\end{aligned}\right\} \quad \ldots \quad (43)$$

where $\mu_C(=3)$ and $\mu_R(=2)$ represent the total number of multiplications for a single complex pole and a single real pole, respectively. Corresponding results for unrounded operations are obtained from eqn. 31 as

$$\left.\begin{aligned}\overline{r_0^2} &= \frac{q^2}{3}; \quad \overline{r_i^2} = \frac{\mu_C q^2}{12}(1 + 3\mu_C) \\ \overline{r_i^{*2}} &= \frac{\mu_C q^2}{12}(1 + 3\mu_C); \quad \overline{r_j^2} = \frac{\mu_R q^2}{12}(1 + 3\mu_R)\end{aligned}\right\} \quad (44)$$

If the error sequences $r_0(k)$, $\{r_i(k)\}$, $\{r_i^*(k)\}$ and $\{r_j(k)\}$ are mutually uncorrelated amongst themselves and with the actual input signal to the system, the maximum increase in mean-square system error due to the finite word length of the compensator can be calculated, independently of the input signal, using eqn. 12.

6 Computational errors for cascade programming

For cascade programming, the pulsed transfer function of the digital compensator, having real poles and zeros only, is expressed in the form

$$D(z) = B_1 z^{-1} \prod_{i=1}^{N} \frac{(1 - \alpha_i z^{-1})}{(1 - z_i z^{-1})} \quad \ldots \quad (45)$$

The cascade programming[7,8] of $D(z)$ consists of successively solving by direct programming the following difference equations to determine $f(k)$:

$$e_1(k) = B_1 e'(k) \quad \ldots \quad (46)$$

$$e_{i+1}(k) = e_i(k) - \alpha_i e_i(k-1) + z_i e_{i+1}(k-1) \quad (47)$$
$$\text{for } i = 1, 2, \ldots N$$

$$f(k) = e_{N+1}(k-1) \quad \ldots \quad (48)$$

However, owing to the finite word length of the computer, the actual outputs of the compensator elements are

$$e'_1(k) = B_1 e'(k) + r_1(k) \quad \ldots \quad (49)$$

$$\begin{aligned}e'_{i+1}(k) &= e'_i(k) - \alpha_i e'_i(k-1) \\ &\quad + z_i e'_{i+1}(k-1) + r_{i+1}(k)\end{aligned} \quad \ldots \quad (50)$$
$$\text{for } i = 1, 2, \ldots N$$

$$f'(k) = e'_{N+1}(k-1) \quad \ldots \quad (51)$$

Defining the computational error quantities,

$$e'_1(k) = e_1(k) + \epsilon_1(k) \quad \ldots \quad (52)$$

and $\quad e'_{i+1}(k) = e_{i+1}(k) + \epsilon_{i+1}(k) \quad \ldots \quad (53)$
$$\text{for } i = 1, 2, \ldots N$$

then, from eqn. 53, the computational error $\epsilon_c(k)$ on the plant-forcing sequence for cascade programming is evidently $\epsilon_{N+1}(k-1)$, so that

$$f'(k) = f(k) + \epsilon_c(k) \quad \ldots \quad (54)$$

It is apparent from eqns. 49, 50 and 51 that the error sequence $\epsilon_c(k)$ is determined by the random roundoff variables $\{r_i(k)\}$. Thus the actual plant-forcing sequence may again be regarded as the true forcing sequence contaminated with an additive noise sequence $\epsilon_c(k)$. It is now proposed to determine the relationship existing between the computational error sequence $\epsilon_c(k)$ and the error sequences $\{r_i(k)\}$.

Taking the Z transformation of eqns. 46–51 inclusive and applying the principle of mathematical induction, it is shown in Appendix 12.2 that

$$\epsilon_C(z) = z^{-1} \epsilon_{N+1}(z) = \sum_{m=1}^{N+1} H_m(z) R_m(z) \quad \ldots \quad (55)$$

where $\quad H_m(z) = z^{-1} \dfrac{\prod_{i=m}^{N}(1 - \alpha_i z^{-1})}{\prod_{j=m-1}^{N}(1 - z_j z^{-1})} \quad \ldots \quad (56)$
$$\text{for } m \leq N$$

and $\quad H_{N+1}(z) = \dfrac{z^{-1}}{1 - z_N z^{-1}} \quad \ldots \quad (57)$

Eqn. 55 specifies the computational-error sequence to be added to the plant-forcing function of the ideal system to account for the finite word length of the computer. By slight block-diagram manipulation, the error sequences $r_1(k)$ and $\{r_i(k)\}$ can again be represented as spurious inputs to the ideal closed-loop system in the manner of Fig. 4. Assuming these input sequences to be wide-sense stationary, their mean-square values for rounded operations are derived from eqn. 30 as

$$\overline{r_1^2} = \frac{q^2}{12}; \quad \overline{r_i^2} = \frac{q^2}{6} \quad \ldots \quad (58)$$
$$\text{for } i = 2, 3, \ldots N$$

The corresponding results for unrounded operations are obtained from eqn. 31 as

$$\overline{r_1^2} = \frac{q^2}{3}; \overline{r_i^2} = \frac{7q^2}{12} \quad \text{for } i = 2, 3, \ldots N \qquad (59)$$

If the error sequences $r_1(k)$ and $\{r_i(k)\}$ are mutually uncorrelated amongst themselves and with the actual input signal to the system, the maximum increase in mean-square error of the system can again be calculated, independently of the input signal, by means of eqn. 12.

7 Stability and drift

The analysis presented in Sections 3, 4, 5 and 6 indicates that, if the linear system is stable without quantisation, the increase in the mean-square system error due to the finite word length of the machine is statistically bounded. In addition, this error process is shown to be normal, and thus, for practical purposes, its individual terms are also bounded.† These observations concur with those of Bertram.[1]

Assuming that a system is stable if, and only if, its response to a bounded input is also bounded, Bertram concludes that a system is not made unstable by the introduction of quantisation. However, while this concept of stability is satisfactory for linear systems, it is generally inadequate for nonlinear systems, and a system including a finite-word-length computer is inherently nonlinear.[8, 17] In fact, it is only by considering the ensemble properties of the quantisation and roundoff errors, rather than the term-by-term properties of a single realisation, that the present analysis gives rise to solvable linear equations.

It follows from the analysis given in the previous Sections that if the input to the system shown in Fig. 2 is given by

$$x(k) = x_0 \text{ for } k = 0 \qquad (60)$$
$$= 0 \text{ for } k \neq 0$$

for rounded operations, quantisation and computational error processes are initiated, provided that

$$|x_0| > q/2 \qquad (61)$$

and provided that

$$|b_{max}\, x_0| > q/2 \qquad (62)$$

where $|b_{max}|$ is the largest coefficient in the numerator of $D(z)$. In practice, the scalar gain of the compensator is generally greater than unity, so that the initiation of the computational-error processes is determined by eqn. 61 rather than eqn. 62. If the statistical bound of the steady-state output noise process is less than the larger of $q/2$ and $q/(2|b_{max}|)$, no information is processed by the computer. Under these circumstances, the output-noise process is evidently not self-maintaining and the output of the free system decays to zero. In this case, a linear system which is generally asymptotically stable[13, 14] retains its stability after the introduction of quantisation. It is evident that similar conclusions may be drawn for the case of unrounded operations.

In general, however, the d.c. gain of the overall noise transfer function in Fig. 4 is significantly greater than unity, and this causes the steady-state output noise process to be self-maintained. A typical record of this process is shown in Fig. 5, and the bound of its mean-square value is calculable from the previous analysis. This output noise process of the free system is sometimes referred to as drift, which must be constrained within tolerable limits. Thus the previous analysis is of importance even when a system is designed from a nonstatistical approach such as step-function or frequency-domain criteria.

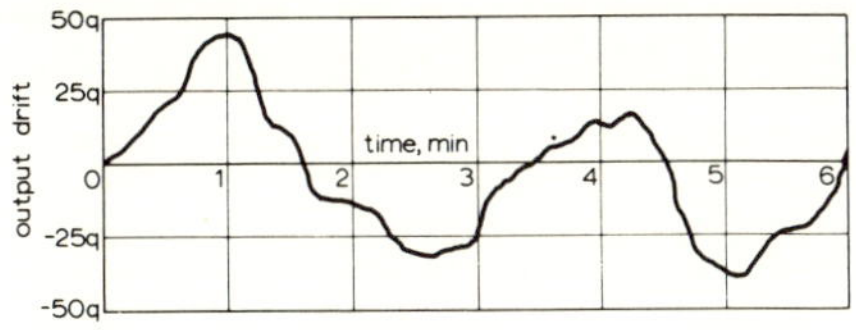

Fig. 5
Typical drift record

These conclusions regarding the general asymptotic stability of a computer-controlled system support the doubt expressed by Tsypkin[3] of the general validity of Bertram's conclusions. From an engineering viewpoint, it is the authors' experience that the requirement of asymptotic stability in the large for a nonlinear control system can be too stringent. Frequently, a nonlinear engineering system may be considered as 'stable' provided that the amplitude of any inherent oscillations is always consistent with the steady-state accuracy of the system. In addition, the amplitude of the oscillations should not produce any undue wear in the system components. It is evident that the previous analysis specifies the width of quantisation for the system to be considered stable in this sense. Such a situation is analogous to that found in 'bang-bang' control systems, where the steady-state amplitude of the inherent limit-cycle oscillations must also be contained within tolerable limits.

8 Experimental verification

If the system shown in Fig. 2 has a plant

$$G(s) = \frac{1 - e^{-sT}}{s^3(s+1)} \qquad (63)$$

and a sampling period of 0·1 s, it may be satisfactorily compensated using either of the discrete compensators

$$D_1(z) = \frac{4\cdot 8z^{-1}(1 - 0\cdot 975z^{-1})^2}{(1 - 0\cdot 800z^{-1})^2} \qquad (64)$$

or

$$D_2(z) = \frac{12z^{-1}(1 - 0\cdot 900z^{-1})(1 - 0\cdot 975z^{-1})}{[1 - (0\cdot 8 + j.0\cdot 3)z^{-1}][1 - (0\cdot 8 - j0\cdot 3)z^{-1}]} \qquad (65)$$

The shaping filters for the systems specified in eqns. 63, 64 and 65 were evaluated, for the cases of direct, parallel and cascade programming, according to eqns. 27, 42 and 55, respectively. Slight manipulation of the resulting block diagram of the system produced an equivalent system of the form of Fig. 4. The maximum increase in the mean-square error of the system was evaluated by means of eqn. 12, and the results are presented as the curves in Figs. 7–13. The maximum gain $|H_p^*(j\omega_m)|$ for a roundoff-error shaping filter cascaded with the ideal closed-loop system was determined very easily for the examples considered. Cursory

† A practical bound of individual terms in a discrete normal process is taken to be $\pm 4\sigma$, where σ is the r.m.s. value of the process.

examination of the Nyquist diagram for the control system and the gain curves for poles and zeros in the z plane[18] indicates that the real frequency response of such a series combination is determined by the shaping filter. Furthermore, the maximum gain of the shaping filter is readily seen to occur with d.c. Because the gain of the ideal control system is unity at d.c., $|H_p^*(j\omega_m)|$ is simply calculated by setting $z = 1$ in the appropriate shaping filter. In the case of the quantisation error, whose shaping filter is 2, the maximum gain through it and the control system is evidently obtained as the peak magnification $[1{\cdot}2$ with $D_1(z)$ and $1{\cdot}26$ with $D_2(z)]$ of the latter.

As a means of verifying these statistical upper bounds and the Gaussian nature of the computational error process, the systems under consideration were each compared with an 'ideal' counterpart having negligible quantisation and round-off errors. In the 'ideal' counterpart, all arithmetic operations were performed using the full word length of the Atlas machine. The recursion formulas for the discrete compensator in the actual system were evaluated using a significantly smaller word length, but the recursion formula for the plant was again computed employing the full machine accuracy. By computing the mean-square value of the steady-state difference in the output of the two systems, the mean-square value of the steady-state degeneration in performance due to the finite word length of the compensator was calculated. In addition, the first, third and fourth moments were evaluated so as to assess the normality of the noise process produced by the finite word-length compensator in terms of skew and Kurtosis factors.[11]

For obvious practical reasons, measurements can be made only on a finite sample of the difference process. On this account, the experimental results are inherently subject to a statistical estimation error. A measure of the error on an estimate of the mean square of a continuous process is given by[15]

$$\sigma_m^2 = \int_{-\infty}^{\infty}\int_{-\infty}^{\infty}[U(v) - U(v - T)][U(v + \eta) - U(v + \eta - T)][\phi_{zz}(\eta) - (\bar{z})^2]\,dv\,d\eta \quad (66)$$

where $U(v)$ = unit step function

$\phi_{zz}(n)$ = autocorrelation function of the process being measured

$\bar{z}$ = its mean value.

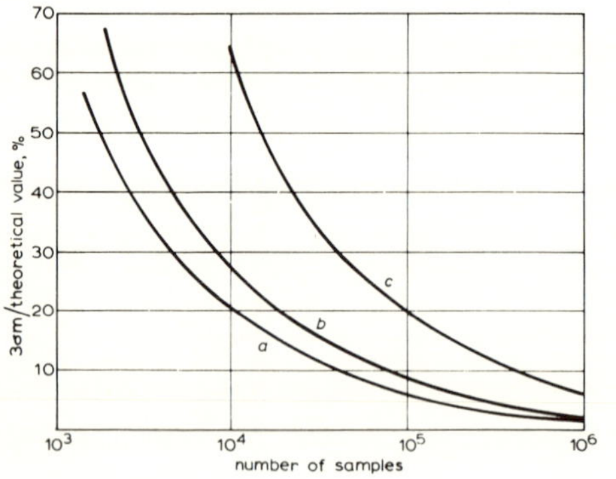

Fig. 6

Sample accuracy

a Parallel programming $D_2(z)$
b Direct programming $D_2(z)$
c Cascade programming $D_1(z)$

The corresponding result for sampled data is readily derived as[16]

$$\sigma_m^2 = \frac{1}{N^2}\sum_{p=-\infty}^{\infty}\sum_{m=-\infty}^{\infty}[U(m-1) - U(m-N)][U(m+p-1) - U(m-N)][\phi_{zz}^*(p) - (\bar{z})^2] \quad (67)$$

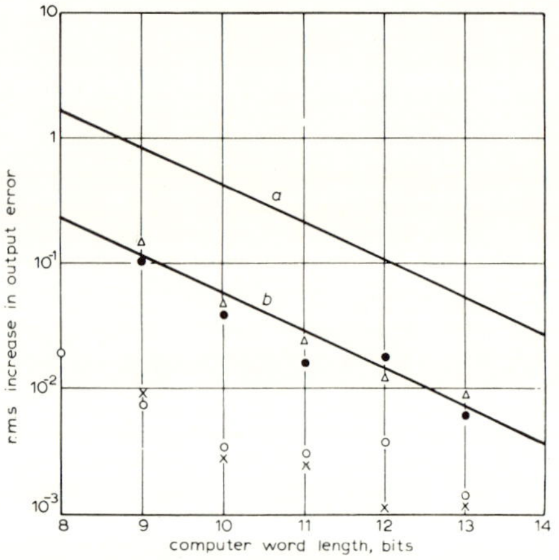

Fig. 7

Compensation with $D_1(z)$, rounded operations

R.M.S. output-error bound
a Direct-programming. Experimental results:
● ● ● 0·01 c/s input
0·10 c/s input
b Cascade-programming. Experimental results:
0·01 c/s input
0·01 c/s input

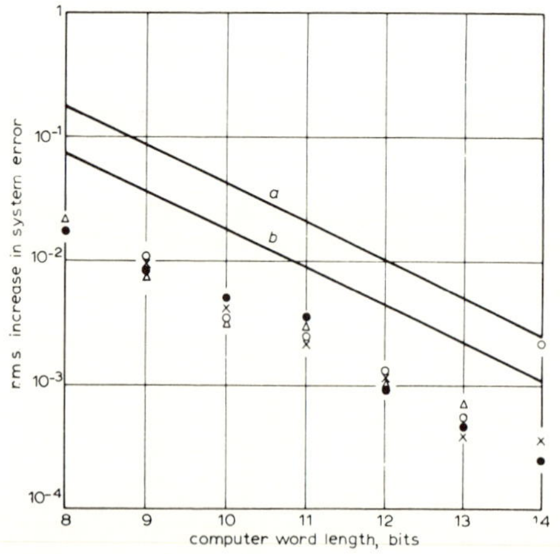

Fig. 8

Compensation with $D_2(z)$, rounded operations

R.M.S. output-error bound
a Direct-programming. Experimental results:
× × × 0·01 c/s input
○ ○ ○ 0·10 c/s input
b Parallel-programming. Experimental results:
△ △ △ 0·01 c/s input
● ● ● 0·10 c/s input

where N is the number of samples. Using eqn. 67, the dependence of σ_m on the number of samples was computed for the examples considered. A graphical presentation of these results is shown in Fig. 6. In the experimental investigation the number of samples was selected so that, in each case,

$$\frac{3\sigma_m}{\text{Theoretical value}} < 7\% \quad \ldots \ldots \quad (68)$$

Thus, with reasonable certainty, the estimates of the increase in mean-square system error due to quantisation and roundoff errors can be considered correct to within $\pm 7\%$.

In the cases of direct- and cascade-programming techniques, the scalar gain λ may be arbitrarily distributed in the compensator. It is evident that, if the entire scalar gain is after the frequency-dependent routine, the mean squares of both quantisation and computational noise are increased by a

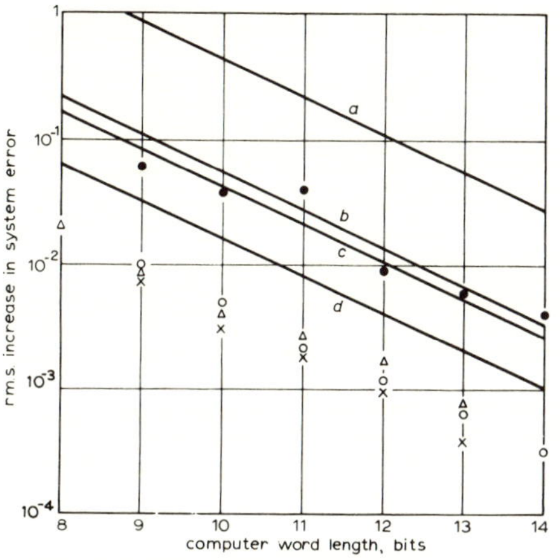

Fig. 9

Output drift with rounded operations

R.M.S. error bound
a Direct programming of $D_1(z)$: ● ● ● experimental results
b Direct programming of $D_2(z)$: ○ ○ ○ experimental results
c Cascade programming of $D_1(z)$: △ △ △ experimental results
d Parallel programming of $D_2(z)$: × × × experimental results

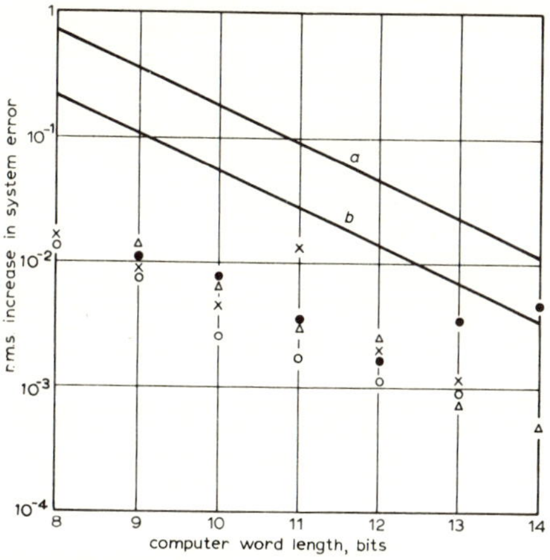

Fig. 11

Compensation with D(z), unrounded operations

R.M.S. output-error bound
a Direct-programming. Experimental results:
△ △ △ 0·01 c/s input
● ● ● 0·10 c/s input
b Parallel-programming. Experimental results:
○ ○ ○ 0·01 c/s input
× × × 0·10 c/s input

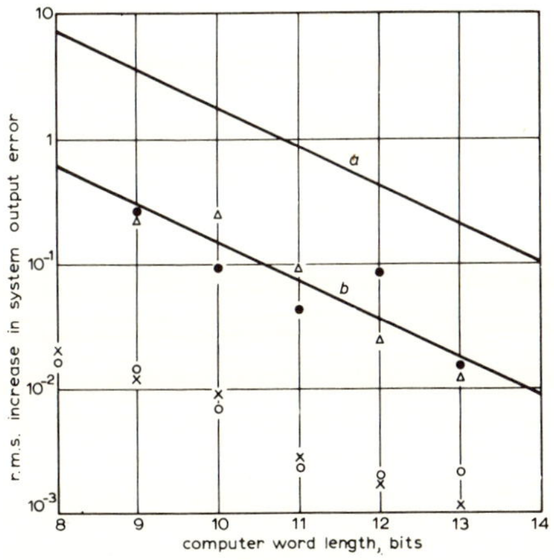

Fig. 10

Compensation with $D_1(z)$, unrounded operations

R.M.S. output error
a Direct-programming. Experimental results:
● ● ● 0·01 c/s input
△ △ △ 0·10 c/s input
b Cascade-programming. Experimental results:
× × × 0·01 c/s input
○ ○ ○ 0·10 c/s input

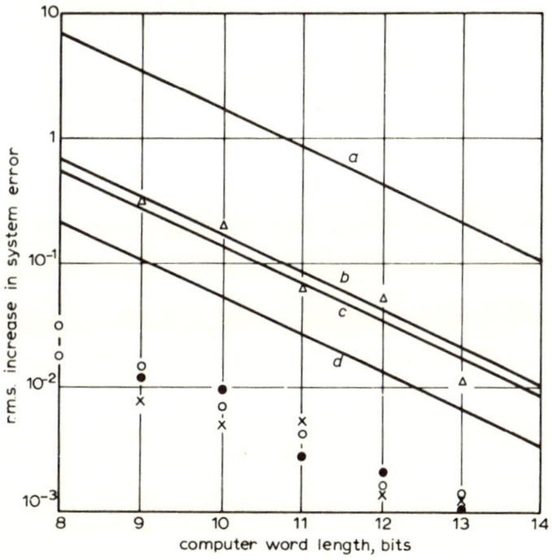

Fig. 12

Output drift with unrounded operations

R.M.S. error bound
a Direct programming of $D_1(z)$: △ △ △ experimental results
b Direct programming of $D_2(z)$: ○ ○ ○ experimental results
c Cascade programming of $D_1(z)$: ● ● ● experimental results
d Parallel programming of $D_2(z)$: × × × experimental results,

factor of λ^2. However, if the entire scalar gain is located before the frequency-dependent routine, only the mean-square value of quantisation noise is accentuated by λ^2. This is confirmed by the theoretical and experimental results in Fig. 13, which correspond to the direct programming of the compensator $D_2(z)$ and an input sequence specified by

$$x(k)\begin{Bmatrix} = 1\ ,\ k = 0 \\ = 0\ ,\ k \neq 0 \end{Bmatrix} \quad . \quad . \quad . \quad . \quad . \quad . \quad . \quad . \quad (69)$$

Thus, from the viewpoint of minimising the degeneration in system performance, it would seem desirable to precede the frequency-dependent routine by the entire scalar gain of the compensator. However, under these circumstances, saturation of the error channel of the system is more likely because all numbers in the machine must lie in the interval $(-1, 1 - q)$. Within the authors' experience, a saturation unit before a phase-advance-type routine tends to reduce the effectiveness of compensation. On this account, some compromise may have to be considered in practice.

In order to demonstrate the invariance of the mean-square increase of system error with respect to system input, the simulation was performed for two different frequencies and for an input specified by eqn. 69. The simulations of direct and cascade programming were, in fact, performed with the entire scalar gain preceding the frequency-dependent part of the compensator. The results obtained are shown in Figs. 7–13 and are observed to be within the predicted bounds.

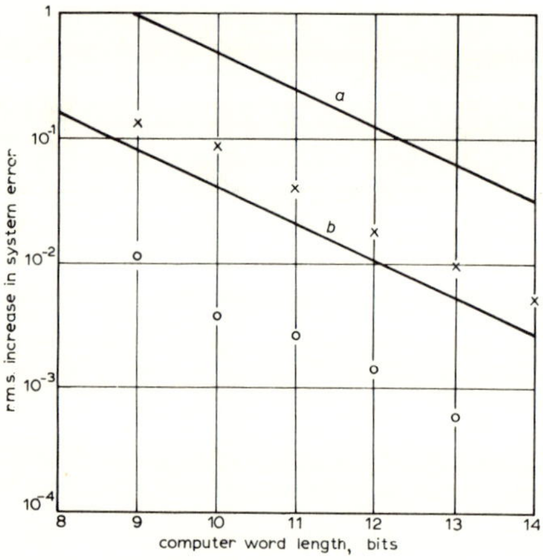

Fig. 13

Compensation with $D_2(z)$ for direct-programming rounded operations

R.M.S. output-error bound

a Compensating routine preceding entire scalar gain. Experimental results: × × × 0·10c/s input

b Entire scalar gain preceding the compensating routine. Experimental results: ○ ○ ○ 0·10c/s input

Table 1

CHARACTERISTICS OF THE OUTPUT-NOISE PROCESS DUE TO QUANTISATION AND ROUNDOFF ERRORS

Parameter	Mean value	Standard deviation
Mean of process	−0·00128	0·00187
Kurtosis factor	3·03	0·43
Skew factor	0·055	0·272

The average values of the measured Kurtosis and skew factors, together with the mean of the difference in output processes, are given in Table 1. These results present no compelling evidence to suggest that the output noise due to the finite word length of the compensator is other than a normal process with zero mean.

9 Conclusions

The increase in the mean-square error of a closed-loop sampled-data system due to the finite word length of the digital compensator has been analysed in this paper. The calculations may be performed on a slide rule using the conventional block diagram of the system preceded by a series of easily calculable shaping filters. The experimental results confirm the analysis and the assumptions made in its derivation.

When the shaping filters for roundoff errors in Fig. 3 are manipulated to form the equivalent system shown in Fig. 4, their denominators become the numerator $B(z)$ of the discrete compensator. Therefore, as the zeros of the compensator are brought closer to the point $z = 1\angle 0$, the d.c. gains of the roundoff shaping filters in Fig. 4 become larger. It is therefore to be expected that an increase in the derivative action of a digital compensator will aggravate quantisation and roundoff effects. This conclusion is confirmed by comparing the direct programming of the compensators $D_1(z)$ and $D_2(z)$ in Figs. 7–12.

Although a statistical measure is used to assess the degradation in system performance, the analysis is applicable to both conventional and statistical design procedures in that it permits a determination of the amplitude of the output 'drift' process. A further application of the analysis is to the selection of the computing technique requiring the minimum word length consistent with the required system accuracy. Such a calculation may indicate the possibility of 'packing' several system variables into a single computer word. Alternatively, it may enable the purchase of the shortest-possible-word-length computer consistent with the system accuracy. Both these considerations effectively save computer storage space and, thereby, capital outlay.

10 Acknowledgment

The authors record their gratitude to the members of the Computing Laboratory at the University of Manchester. In particular, they would like to thank Miss C. M. Popplewell for her assistance in obtaining the computing time necessary for the satisfactory completion of this project.

11 References

1 BERTRAM, J. E.: 'The effect of quantization in sampled-feedback systems', *Trans. Amer. Inst. Elect. Engrs.*, 1958, **77**, Pt. II, p. 177
2 SLAUGHTER, J. B.: 'Quantization errors in digital control systems', *IEEE Trans.*, 1964, **AC-1**, p. 70
3 TSYPKIN, YA .Z.: 'An estimate of the influence of amplitude quantization on processes in digital automatic control systems', *Avtomat. i Telemekh.*, 1960, **21**, (3), p. 195
4 WIDROW, B.: 'Statistical analysis of amplitude-quantized sampled-data systems', *Trans. Amer. Inst. Elect. Engrs.*, 1960, **79**, Pt. II, p. 555
5 WATTS, D. G.: 'A general theory of amplitude quantization with applications to correlation determination', *Proc. IEE*, 1962, **109** C, p. 209
6 KOSYAKIN, A. A.: 'The statistical theory of amplitude quantization', *Avtomat. i Telemekh.*, 1961, **22**, (6), p. 722
7 JURY, E. I.: 'Sampled data control systems' (Wiley, 1958)
8 RAGAZZINI, J. R., and FRANKLIN, G.: 'Sampled data control systems' (McGraw-Hill, 1958)
9 LANING, J. H., and BATTIN, R.: 'Random processes in automatic control' (McGraw-Hill, 1956)

10 DOOB, J. L.: 'Stochastic processes' (Wiley, 1953)
11 CRAMER, H.: 'Mathematical methods of statistics' (Princetown Press, 1946)
12 MURTHY, N.: 'Quantization and instability', M.Sc.Tech. Thesis, University of Manchester, 1963
13 KALMAN, R. E., and BERTRAM, J. E.: 'Control system analysis and design via the "second method" of Lyapunov', *Trans. Amer. Soc. Mech. Engrs.*, 1960, **82**, p. 371
14 PARKS, P. C.: 'A survey of analytical methods for investigating stability—linear and non-linear systems'. Discussion on the stability of systems by the Automatic Control Group of the Institution of Mechanical Engineers, May 1964
15 DAVENPORT, W. B., JOHNSON, R. A., and MIDDLETON, D.: 'Statistical errors in measurements on random time functions', *J. Appl. Phys.*, 1952, **23**, p. 377
16 EDWARDS, R.: Ph.D. dissertation to be presented to Manchester University, 1966
17 TSYPKIN, YA. Z.: 'Sampling systems theory and its application' (Pergamon Press, 1964)
18 KNOWLES, J. B.: 'A contribution to computer control', Ph.D. dissertation, University of Manchester, 1962

Therefore, the actual output of the discrete compensator when realised by the parallel-programming technique is

$$f'(k) = f(k) + \epsilon_p(k) \qquad (77)$$

where $$\epsilon_p(k) = \epsilon_0(k) + \sum_i \epsilon_i(k) + \sum_j \epsilon_j(k) \qquad (78)$$

The intrinsic statistical nature of the roundoff-error sequences $r_0(k)$, $\{r_i(k)\}$, $\{r_i^*(k)\}$ and $\{r_j(k)\}$, together with the form of eqn. 77, enables the actual plant-forcing sequence to be regarded as the true forcing sequence contaminated with an additive noise sequence $\epsilon_p(k)$. It is now proposed to determine the relationship existing between the computational-error sequence $\epsilon_p(k)$ and the error sequences $r_0(k)$, $\{r_i(k)\}$, $\{r_i^*(k)\}$ and $\{r_j(k)\}$.

Taking the Z transformation of eqns. 74, 75 and 76 yields

$$\epsilon_0(z) = R_0(z) \qquad (79)$$

$$\left\{\begin{aligned} \epsilon_i(z) &= \left(\frac{1 - z^{-1}|z_i| \cos \angle z_i}{1 - 2z^{-1}|z_i| \cos \angle z_i + z^{-2}|z_i|^2}\right) R_i(z) - \left(\frac{z^{-1}|z_i| \sin \angle z_i}{1 - 2z^{-1}|z_i| \cos \angle z_i + z^{-2}|z_i|^2}\right) R_i^*(z) \\ \epsilon_i^*(z) &= \left(\frac{1 - z^{-1}|z_i| \cos \angle z_i}{1 - 2z^{-1}|z_i| \cos \angle z_i + z^{-2}|z_i|^2}\right) R_i^*(z) + \left(\frac{z^{-1}|z_i| \sin \angle z_i}{1 - 2z^{-1}|z_i| \cos \angle z_i + z^{-2}|z_i|^2}\right) R_i(z) \end{aligned}\right\} \qquad (80)$$

$$\left\{\epsilon_j(z) = \left(\frac{1}{1 - z^{-1}C_j}\right) R_j(z)\right\} \qquad (81)$$

so that from eqn. 78,

$$\epsilon_p(z) = R_0(z) + \sum_i \left\{\left(\frac{1 - z^{-1}|z_i| \cos \angle z_i}{1 - 2z^{-1}|z_i| \cos \angle z_i + z^{-2}|z_i|^2}\right) R_i(z) - \left(\frac{z^{-1}|z_i| \sin \angle z_i}{1 - 2z^{-1}|z_i| \cos \angle z_i + z^{-2}|z_i|^2}\right) R_i^*(z)\right\} + \sum_j \left(\frac{1}{1 - z^{-1}C_j}\right) R_j(z) \qquad (82)$$

12 Appendixes

12.1 Derivation of eqn. 42

Substituting eqns. 34, 35 and 36 into eqns. 39, 40 and 41, one obtains

$$\{f_0'(k) - f_0(k) = r_0(k)\} \qquad (70)$$

$$\left\{\begin{aligned} \mathrm{Re}[f_i'(k)] - \mathrm{Re}[f_i(k)] &= \{\mathrm{Re}[f_i'(k-1)] - \mathrm{Re}[f_i(k-1)]\}\,\mathrm{Re}(z_i) - \{\mathrm{Im}[f_i'(k-1)] - \mathrm{Im}[f_i(k-1)]\}\,\mathrm{Im}(z_i) + r_i(k) \\ \mathrm{Im}[f_i'(k)] - \mathrm{Im}[f_i(k)] &= \{\mathrm{Im}[f_i'(k-1)] - \mathrm{Im}[f_i(k-1)]\}\,\mathrm{Re}(z_i) + \{\mathrm{Re}[f_i'(k-1)] - \mathrm{Re}[f_i(k-1)]\}\,\mathrm{Im}(z_i) + r_i^*(k) \end{aligned}\right\} \qquad (71)$$

$$\{f_j'(k) - f_j(k) = C_j[f_j'(k-1) - f_j(k-1)] + r_j(k)\} \qquad (72)$$

Defining the computational-error quantities

$$\left.\begin{aligned} f_0'(k) &= f_0(k) + \epsilon_0(k) \\ \mathrm{Re}[f_i'(k)] &= \mathrm{Re}[f_i(k)] + \epsilon_i(k) \\ \mathrm{Im}[f_i'(k)] &= \mathrm{Im}[f_i(k)] + \epsilon_i^*(k) \\ f_j'(k) &= f_j(k) + \epsilon_j(k) \end{aligned}\right\} \qquad (73)$$

where

$$\{\epsilon_0(k) = r_0(k)\} \qquad (74)$$

$$\left\{\begin{aligned} \epsilon_i(k) &= \epsilon_i(k-1)\,\mathrm{Re}(z_i) - \epsilon_i^*(k-1)\,\mathrm{Im}(z_i) + r_i(k) \\ \epsilon_i^*(k) &= \epsilon_i^*(k-1)\,\mathrm{Re}(z_i) + \epsilon_i(k-1)\,\mathrm{Im}(z_i) + r_i^*(k) \end{aligned}\right\} \qquad (75)$$

$$\{\epsilon_j(k) = \epsilon_j(k-1)C_j + r_j(k)\} \qquad (76)$$

which is the result quoted as eqn. 42 in Section 5.

12.2 Derivation of eqns. 55, 56 and 57

The deviation is by mathematical induction; i.e. eqns. 55, 56 and 57 will be assumed to hold true for a compensator with N poles and will be then proved to hold true for a compensator with $N+1$ poles. The proof will be completed by demonstrating the validity of eqns. 55, 56 and 57 for the cases of one pole and two poles.

Taking the Z transformation of eqns. 46 to 51 gives

$$E_1(z) = B_1 E'(z) \qquad (83)$$

$$\left\{E_{i+1}(z) = \frac{(1 - \alpha_i z^{-1})}{(1 - z_i z^{-1})} E_i(z)\right\} \text{ for } i = 1, 2 \ldots N \qquad (84)$$

$$F(z) = z^{-1} E_{N+1}(z) \qquad (85)$$

$$E_1'(z) = B_1 E'(z) + R_1(z) \qquad (86)$$

$$\left\{E_{i+1}'(z) = \left[\frac{(1 - \alpha_i z^{-1})E_i'(z) + R_{i+1}(z)}{(1 - z_i z^{-1})}\right]\right\} \text{ for } i = 1, 2 \ldots N \qquad (87)$$

$$F'(z) = z^{-1}E'_{N+1}(z) \quad . \quad . \quad . \quad . \quad . \quad . \quad . \quad (88)$$

Assuming eqns. 55, 56 and 57 to be true for the cascade programming of an N-pole compensator,

$$F'(z) = F(z) + \sum_{m=1}^{N+1} H_m(z)R_m(z) \quad . \quad . \quad . \quad . \quad (89)$$

where $$H_m(z) = z^{-1}\frac{\prod_{i=m}^{N}(1-\alpha_i z^{-1})}{\prod_{j=m-1}^{N}(1-z_j z^{-1})} \text{ for } m \leqslant N \quad . \quad (90)$$

and $$H_{N+1}(z) = \frac{z^{-1}}{(1-z_N z^{-1})}$$

For an $(N+1)$-pole compensator, the equations corresponding to eqns. 83–88 are

$$E_1(z) = B_1 E'(z) \quad . \quad . \quad . \quad . \quad . \quad . \quad . \quad . \quad (91)$$

$$\left\{E_{i+1}(z) = \frac{(1-\alpha_i z^{-1})}{(1-z_i z^{-1})}E_i(z)\right\}; \text{ for } i = 1, 2, \ldots N \quad . \quad . \quad . \quad . \quad (92)$$

$$\left.\begin{aligned} E_{N+2}(z) &= \frac{(1-\alpha_{N+1}z^{-1})}{(1-z_{N+1}z^{-1})}E_{N+1}(z) \\ F(z) &= z^{-1}E_{N+2}(z) \end{aligned}\right\} \quad . \quad . \quad . \quad (93)$$

$$E'_1(z) = B_1 E'(z) + R_1(z) \quad . \quad . \quad . \quad . \quad . \quad . \quad (94)$$

$$\left\{E'_{i+1}(z) = \left[\frac{(1-\alpha_i z^{-1})E'_i(z) + R_{i+1}(z)}{(1-z_i z^{-1})}\right]\right\} \text{ for } i = 1, 2, \ldots N \quad (95)$$

$$\left.\begin{aligned} E'_{N+2}(z) &= \frac{(1-\alpha_{N+1}z^{-1})E'_{N+1}(z) + R_{N+2}(z)}{(1-z_{N+1}z^{-1})} \\ F'(z) &= z^{-1}E'_{N+2}(z) \end{aligned}\right\} \quad (96)$$

Combining eqns. 93 gives

$$F(z) = z^{-1}\frac{(1-\alpha_{N+1}z^{-1})}{(1-z_{N+1}z^{-1})}E_{N+1}(z) \quad . \quad . \quad . \quad . \quad (97)$$

and eqns. 96 become

$$F'(z) = z^{-1}\frac{(1-\alpha_{N+1}z^{-1})E'_{N+1}(z) + R_{N+2}(z)}{(1-z_{N+1}z^{-1})} \quad (98)$$

Defining the quantities

$$X(z) = \frac{(1-z_{N+1}z^{-1})}{(1-\alpha_{N+1}z^{-1})}F(z) \quad . \quad . \quad . \quad . \quad . \quad . \quad (99)$$

and

$$X'(z) = \frac{(1-z_{N+1}z^{-1})F'(z) - z^{-1}R_{N+2}(z)}{(1-\alpha_{N+1}z^{-1})} \quad (100)$$

then eqns. 91–96 reduce to

$$E_1(z) = B_1 E'(z) \quad . \quad . \quad . \quad . \quad . \quad . \quad . \quad . \quad (101)$$

$$\left\{E_{i+1}(z) = \frac{(1-\alpha_i z^{-1})}{(1-z_i z^{-1})}E_i(z)\right\} \text{ for } i = 1, 2, \ldots N \quad (102)$$

$$X(z) = z^{-1}E_{N+1}(z) \quad . \quad . \quad . \quad . \quad . \quad . \quad . \quad . \quad (103)$$

$$E_1(z) = B_1 E'(z) + R_1(z) \quad . \quad . \quad . \quad . \quad . \quad . \quad (104)$$

$$\left\{E'_{i+1}(z) = \frac{(1-\alpha_i z^{-1})E'_1(z) + R_{i+1}(z)}{(1-z_i z^{-1})}\right\} \text{ for } i = 1, 2, \ldots N \quad (105)$$

$$X'(z) = z^{-1}E'_{N+1}(z) \quad . \quad . \quad . \quad . \quad . \quad . \quad . \quad . \quad (106)$$

It is evident that eqns. 101–106 are identical to eqns. 83–88 except that $F(z)$ has been replaced by $X(z)$ and $F'(z)$ by $X'(z)$. Hence, it follows from eqn. 101 that

$$X'(z) = X(z) + \sum_{m=1}^{N+1} H_m(z)R_m(z) \quad . \quad . \quad . \quad . \quad . \quad (107)$$

where $H_m(z)$ is given by eqn. 90. Substituting eqns. 99 and 100 into eqn. 107 yields

$$F'(z) = F(z) + \frac{(1-\alpha_{N+1}z^{-1})}{(1-z_{N+1}z^{-1})}\sum_{m=1}^{N+1} H_m(z)R_m(z) + \frac{z^{-1}}{(1-z_{N+1}z^{-1})}R_{N+2}(z) \quad . \quad . \quad (108)$$

and combining the error terms of this equation gives

$$F'(z) = F(z) + \sum_{m=1}^{N+2} W_m(z)R_m(z) \quad . \quad . \quad . \quad . \quad (109)$$

The expressions for $W_m(z)$ in the above equation are given by eqns. 90 and 108 as

$$W_m(z) = z^{-1}\frac{\prod_{i=m}^{N+1}(1-\alpha_i z^{-1})}{\prod_{j=m-1}^{N+1}(1-z_i z^{-1})}; \text{ for } m \leqslant N+1 \quad . \quad . \quad . \quad . \quad (110)$$

and $$W_{N+2}(z) = \frac{z^{-1}}{(1-z_{N+1}z^{-1})}$$

Evidently, eqns. 109 and 110 are identical to eqns. 89 and 90 except that N has been replaced by $N+1$. Thus, if eqns. 89 and 90 are true for order N, they are true for order $N+1$.

By writing eqns. 84 and 87 in full for the cases of one compensator pole and two compensator poles, it is easily shown that

$$F'(z) = F(z) + z^{-1}\frac{(1-\alpha_1 z^{-1})}{(1-z_1 z^{-1})}R_1(z) + \frac{z^{-1}}{(1-z_1 z^{-1})}R_2(z) \quad . \quad . \quad . \quad . \quad (111)$$

and $$F'(z) = F(z) + z^{-1}\frac{(1-\alpha_1 z^{-1})(1-\alpha_2 z^{-1})}{(1-z_1 z^{-1})(1-z_2 z^{-1})}R_1(z) + z^{-1}\frac{(1-\alpha_2 z^{-1})R_2(z)}{(1-z_1 z^{-1})(1-z_2 z^{-1})} + \frac{z^{-1}R_2(z)}{(1-z_2 z^{-1})} \quad . \quad . \quad . \quad . \quad . \quad . \quad (112)$$

respectively. Hence, by the process of mathematical induction, the proof of eqns. 55, 56 and 57 quoted in Section 6 is complete.

15

Reprinted from *IEEE Trans. Audio Electroacoustics,* **AU-18**(2), 107–122 (1970)

Roundoff-Noise Analysis for Fixed-Point Digital Filters Realized in Cascade or Parallel Form

LELAND B. JACKSON, Member, IEEE
Bell Telephone Laboratories, Inc.
Murray Hill, N. J.

Abstract

The roundoff-noise outputs from two transpose configurations, each for the cascade and parallel forms of a digital filter, are analyzed for the case of uncorrelated roundoff noise and fixed dynamic range. Corresponding transpose configurations are compared on the basis of the variance, or total average power, and the peak spectral density of the output roundoff noise. In addition to providing general computational techniques to be employed in choosing an appropriate configuration for the digital filter, these results also indicate useful "rules of thumb" relating to this choice of configuration. Included are indications of good (although not necessarily optimum) sequential orderings and pole-zero pairings for the second-order sections comprising the cascade form. Computational results are presented which indicate that the analysis is quite accurate and useful.

Manuscript received January 12, 1970.

This paper is taken in part from a thesis submitted by the author in partial fulfillment of the requirements for the degree of Doctor of Science to the Department of Electrical Engineering, Stevens Institute of Technology, Hoboken, N. J. [1].

The author is now with Rockland Systems Corp., Blauvelt, N. Y. 10913.

Introduction

Several analyses of the roundoff-noise output from a digital filter with fixed dynamic range (i.e., implemented using fixed-point arithmetic) have recently appeared in the literature [1]–[4]. Comparisons between the roundoff-noise outputs of different circuit configurations for a digital filter have been of particular interest because of desire to maximize some measure of the output signal-noise ratio. A general framework for these comparisons has been provided by the author [2], and two configurations for the direct form of a digital filter have been analyzed and compared using this approach. In this paper the techniques in [2] are applied to two configurations each for the cascade and parallel forms, and the sequential ordering of the second-order sections comprising the cascade form is also investigated.

Background

The interaction between the roundoff-noise output from a digital filter and the associated dynamic-range limitations is investigated in [2] for the case of uncorrelated rounding errors from sample to sample and from one error source to another. The spectrum of the output roundoff noise from fixed-point implementations is readily shown to be of the form

$$N_y(\omega) = \sigma_0^2 \sum_j k_j' \, | G_j'(\omega) |^2 \tag{1}$$

where the $G_j'(\omega)$ are scaled transfer responses from certain "summation nodes" in the digital circuit to the filter output. σ_0^2 is the variance of the rounding errors from each multiplier (or other rounding point), and the k_j' are integers indicating the number of error inputs to the respective summation nodes.

Defining $F_i'(\omega)$ to be the scaled transfer response from the input to the ith "branch node" at which a dynamic-range constraint is required, constraints of the form

$$\|F_i'\|_p \leq 1 \tag{2}$$

for $p \geq 1$ are then derived, where $\|F_i'\|_p$ is the L_p norm of the response $F_i'(\omega)$ defined by

$$\|F_i'\|_p = \left[\frac{1}{\omega_s}\int_0^{\omega_s} | F_i'(\omega) |^p d\omega\right]^{1/p}. \tag{3}$$

It can be shown [5] that for $F_i'(\omega)$ continuous (as is always the case for digital filters), the limit of (3) as $p \rightarrow \infty$ exists and is given by

$$\|F_i'\|_\infty = \max_{0 \leq \omega \leq \omega_s} | F_i'(\omega) | \tag{4}$$

where ω_s is the radian sampling frequency $2\pi/T$. The appropriate value of p in (2) is determined by assumed conditions on the spectra of the input signals to the filter. The effect of (2) is to bound the maximum signal amplitude (for deterministic inputs) or the maximum average power (for random inputs) at the ith branch node.

As implied above, the prime is used to indicate that the filter has been scaled to satisfy (2). It is intuitively clear that to preserve the greatest possible signal-to-noise ratio, this scaling should reduce the magnitude of $F_i'(\omega)$ no more than is required (or increase it as much as possible, as the case may be). In other words, $F_i'(\omega)$ should actually satisfy

$$\|F_i'\|_p = 1. \tag{5}$$

This condition will be satisfied if the scaling factors s_i, defined by

$$F_i'(\omega) = s_i F_i(\omega), \tag{6a}$$

are given by

$$s_i = 1/\|F_i\|_p. \tag{6b}$$

The unscaled responses $F_i(\omega)$ and $G_j(\omega)$, as well as the overall filter response $H(\omega)$, are depicted in Fig. 1 where

$$\begin{aligned} H(\omega) &= H^*(e^{j\omega T}) \\ F_i(\omega) &= F_i^*(e^{j\omega T}) \\ G_k(\omega) &= G_k^*(e^{j\omega T}). \end{aligned} \tag{7}$$

This notation will be used throughout this paper. That is, for any z transform $A^*(z)$ which converges for $|z| = 1$, the corresponding Fourier transform is given by

$$A(\omega) = A^*(e^{j\omega T}).$$

The sequences $\{e_j(n)\}$ in Fig. 1 represent the roundoff-noise "inputs" at the summation nodes; while the sequences $\{v_i(n)\}$ are the total "outputs" from those branch nodes at which dynamic-range restraints are required.

A state-variable description is then employed in [2] to formulate the general concept of "transpose configurations" for a digital network and to illustrate the usefulness of this concept in digital-filter synthesis. A particularly important result is that for a given unscaled configuration with transpose responses $F_i(\omega)$ and $G_j(\omega)$, as described above, the responses $F_i^t(\omega)$ and $G_j^t(\omega)$ for the corresponding transpose configuration are given by

$$F_i^t(\omega) = G_i(\omega),$$

and

$$G_j^t(\omega) = F_j(\omega). \tag{8}$$

Hence, although the overall transfer responses $H(\omega)$ for these two configurations are the same, their roundoff-noise outputs and/or dynamic-range limitations can be quite different, in general. The transpose configuration is obtained by simply reversing the direction of all branches in the given network configuration, and the poles and zeros of the network are thus realized in reverse order in the transpose configuration.

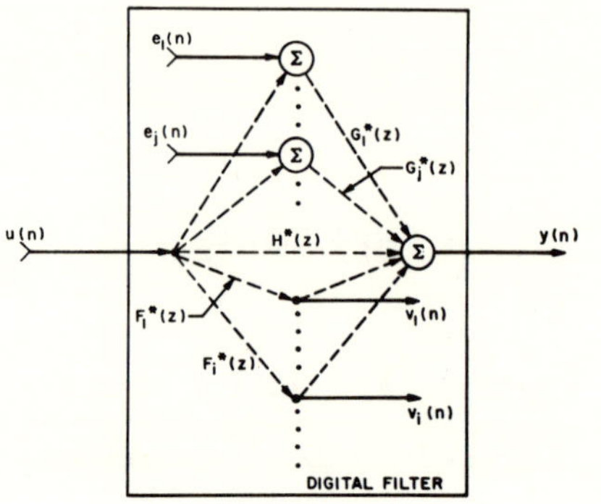

Fig. 1. General digital-filter model.

The Parallel Form

Two transpose configurations for the parallel form of a digital filter are shown in Figs. 2 and 3. The parallel form corresponds to an $H^*(z)$ of the general form [6], [7]

$$H^*(z) = \gamma_0 + \sum_{i=1}^{M} \frac{\gamma_{1i}z^{-1} + \gamma_{0i}}{\beta_{2i}z^{-2} + \beta_{1i}z^{-1} + 1}. \tag{9}$$

However, in order to satisfy (5) in the general case, additional scaling multipliers ρ_i have been added to the configurations in Figs. 2 and 3 (called forms $1P$ and $2P$, respectively). Hence, in form $1P$, $H^*(z)$ is actually realized as

$$H^*(z) = \gamma_0 + \sum_{i=1}^{M} \rho_{1i} \frac{{}_1\gamma_{1i}'z^{-1} + {}_1\gamma_{0i}'}{\beta_{2i}z^{-2} + \beta_{1i}z^{-1} + 1} \tag{10}$$

where the scaling multipliers ρ_{1i} are given by

$$\rho_{1i} = {}_{1P}s_i = 1/\|{}_{1P}F_i\|_p$$

and

$${}_1\gamma_{ki}' = {}_1\gamma_{ki}/\rho_{1i} \qquad k = 0, 1. \tag{11}$$

The transfer function in (10) obviously equals that in (9), but now the transfer responses $F_i'(\omega)$ [from the filter input to the nodes marked by (*)] also satisfy (5). The unscaled transfer functions ${}_{1P}F_i^*(z)$ are given by

$${}_{1P}F_i^*(z) = \frac{1}{\beta_i^*(z)} \qquad i = 1, 2, \cdots, M \tag{12a}$$

where

$$\beta_i^*(z) = \beta_{2i}z^{-2} + \beta_{1i}z^{-1} + 1,$$

and thus

$${}_{1P}F_i(\omega) = \frac{1}{\beta_i(\omega)} \qquad i = 1, 2, \cdots, M. \tag{12b}$$

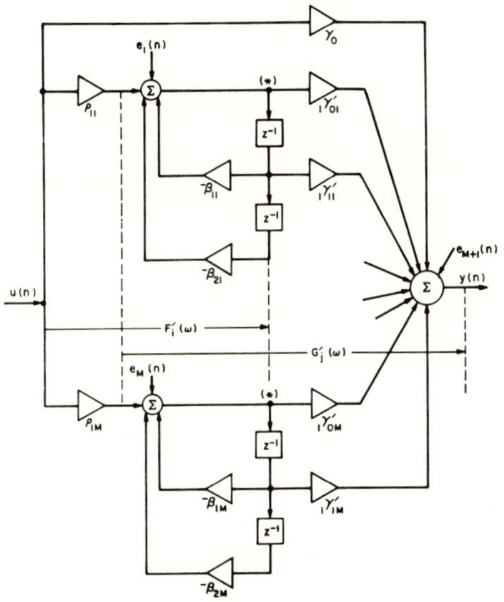

F g. 2. Parallel form 1*P* with scaling.

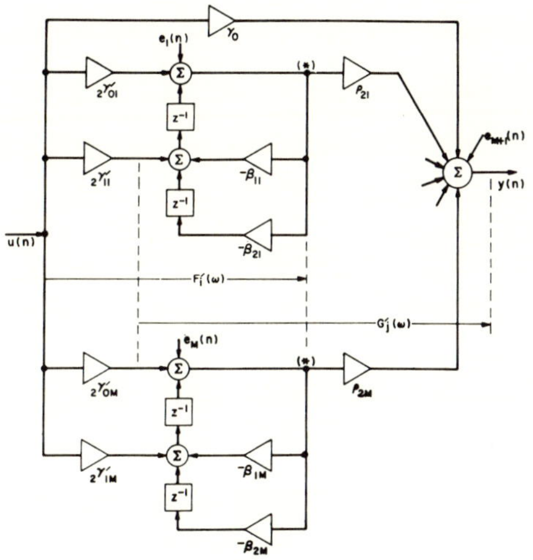

Fig. 3. Parallel form 2*P* with scaling.

The sealing for parallel from 2*P* is shown in Fig. 3. The transfer function $H^*(z)$ for this form is also given by (10), except that the scaling multipliers ρ_{1i} are replaced by

$$\rho_{2i} = 1/{}_{2P}s_i = \|{}_{2P}F_i\|_p \tag{13a}$$

and now, of course,

$${}_2\gamma_{ki}' = {}_2\gamma_{ki}/\rho_{2i} \qquad k = 0, 1. \tag{13b}$$

The transfer responses to the form-2*P* branch nodes marked by (*) are

$${}_{2P}F_i'(\omega) = \frac{{}_2\gamma_i'(\omega)}{\beta_i(\omega)} = ({}_{2P}s_i)\,{}_{2P}F_i(\omega) \tag{14}$$

where

$$\gamma_i^{*\prime}(z) = \gamma_{1i}'z^{-1} + \gamma_{0i}'$$

and thus ${}_{2P}F_i'(\omega)$ indeed satisfies (5), as required.

Note that in form 1*P* the scaling is accomplished via the ρ_{1i} multipliers and is then compensated for in the γ_{ki}' multipliers; whereas in form 2*P*, the scaling is performed in the γ_{ki}' multipliers and is then compensated for in the ρ_{2i} mutlipliers. Note also that scaling has increased the required number of constant multipliers in these parallel forms to $2N+M+1$ from $2N+1$, in the general case. We will find that this applies to the cascade forms as well. Practically speaking, however, it is assumed that in most cases scaling will be performed with only a few (extra) bits in the multipliers. In particular, scaling may be restricted to powers of 2 in a binary system, in which case the overall transfer function can be maintained exactly and no extra multiplication hardware is required (just shifting operations). In this case, (5) is not satisfied generally, but we can ensure that

$$\tfrac{1}{2} < \|F_i'\|_p \leq 1. \tag{15}$$

Since such finite-accuracy considerations are not expected to influence significantly the noise comparisons to follow, we will assume for the purpose of analysis that ideal scaling is performed as indicated in (11) and (13), and thus that (5) is indeed satisfied.

The roundoff-noise output from each of the two parallel forms is readily analyzed, as one would expect, because this output noise is simply the sum of the noise produced by the individual second-order sections comprising the filter. Referring to Fig. 2, the transfer responses $G_j'(\omega)$ for form 1*P*, corresponding to the roundoff-noise inputs $e_j(n)$, are easily seen to be

$$\begin{aligned} {}_{1P}G_j'(\omega) &= \frac{{}_1\gamma_j'(\omega)}{\beta_j(\omega)} \qquad i = 1, 2, \cdots, M \\ {}_{1P}G_{M+1}'(\omega) &= 1. \end{aligned} \tag{16}$$

Assuming that the coefficients β_{kj}, ${}_1\gamma_{kj}'$, and ρ_{1j} for all k and j are nonintegers, the weights k_j' corresponding to the $G_j'(\omega)$ are simply

$$\begin{aligned} {}_{1P}k_j' &= 3 \quad j = 1, 2, \cdots, M-1 \\ {}_{1P}k_M' &= 3 - [2M - N] \\ {}_{1P}k_{M+1}' &= N + 1 \end{aligned} \tag{17}$$

where [ ] denotes the "integer part of," and we have assumed that if the order (N) of the filter is odd, the Mth section is the one (degenerate) first-order section. As indicated previously, with all noninteger coefficients the k_j' for each of the configurations to be considered satisfy

$$\sum_j k_j' = 2N + M + 1. \tag{18}$$

Using the fact that $M = [(N+1)/2]$, (17) is readily shown to satisfy (18).

The effect of scaling on the transfer responses ${}_{1P}G_j'(\omega)$ is seen by substituting (11) into (16) to yield

$${}_{1P}G_j'(\omega) = \|{}_{1P}F_j\|_p \frac{\gamma_j(\omega)}{\beta_j(\omega)} \quad j = 1, 2, \cdots, M \tag{19a}$$

or, from (12),

$${}_{1P}G_j'(\omega) = \left\|\frac{1}{\beta_j}\right\|_p \frac{\gamma_j(\omega)}{\beta_j(\omega)} \quad j = 1, 2, \cdots, M. \tag{19b}$$

The total roundoff noise in the output of parallel form 1P is then, from (1),

$${}_{1P}N_y(\omega) = \sigma_0^2 \left\{(N+1) + \sum_{j=1}^{M} {}_{1P}k_j' \left\|\frac{1}{\beta_j}\right\|_p^2 \left|\frac{\gamma_j(\omega)}{\beta_j(\omega)}\right|^2\right\}. \tag{20}$$

Turning to parallel form 2P, shown in Fig. 3, we note that only one noise input $e_j(n)$ has been shown for each second-order section, rather than two as would be indicated by the number of summation nodes per section. This is done for convenience, and it is possible because the two summation nodes are separated only by a delay and, therefore, actually comprise a single summation. The corresponding transfer responses $G_j'(\omega)$ are then given by

$${}_{2P}G_j'(\omega) = \frac{\rho_{2j}}{\beta_j(\omega)} \quad j = 1, 2, \cdots, M. \tag{21}$$

But from (13) through (15) this is just

$${}_{2P}G_j'(\omega) = \left\|\frac{\gamma_j}{\beta_j}\right\|_p \frac{1}{\beta_j(\omega)} \quad j = 1, 2, \cdots, M. \tag{22a}$$

Note the difference between (19b) and (22a). The transfer response for the input $e_{M+1}(n)$ is, of course, just

$${}_{2P}G_{M+1}'(\omega) = 1 \tag{22b}$$

as in (16).

The noise weights k_j' for form 2P are

$$\begin{aligned} {}_{2P}k_j' &= 4 \quad j = 1, 2, \cdots, M-1 \\ {}_{2P}k_M' &= 4 - 2[2M - N] \\ {}_{2P}k_{M+1}' &= M + 1 \end{aligned} \tag{23}$$

and again (23) is readily shown to satisfy (18). The total roundoff noise in the output of form 2P is then

$${}_{2P}N_y(\omega) = \sigma_0^2 \left\{(M+1) + \sum_{j=1}^{M} {}_{2P}k_j' \left\|\frac{\gamma_j}{\beta_j}\right\|_p^2 \left|\frac{1}{\beta_j(\omega)}\right|^2\right\}. \tag{24}$$

We now compare the output roundoff noise for form 1P given by (20) with that for form 2P given in (24). First note that if N is odd, the noise contribution of the one degenerate first-order section (the Mth section) is the same for both forms. This is seen by noting that, in this case, ${}_{1P}k_M' = {}_{2P}k_M' = 2$, while

$$\begin{aligned} ({}_{1P}G_M'(\omega))_{N \text{ odd}} &= ({}_{2P}G_M'(\omega))_{N \text{ odd}} \\ &= \gamma_{0M} \left\|\frac{1}{\beta_M}\right\|_p \frac{1}{\beta_M(\omega)}. \end{aligned} \tag{25}$$

We next note that $N = M + [N/2]$ and hence that

$${}_{1P}k_{M+1}' = {}_{2P}k_{M+1}' + \left[\frac{N}{2}\right]. \tag{26}$$

Therefore, defining

$$N_1(\omega) = \sigma_0^2 \left\{(M+1) + 2[2M-N]\gamma_{0M}^2 \left\|\frac{1}{\beta_M}\right\|_p^2 \left|\frac{1}{\beta_M(\omega)}\right|^2\right\}, \tag{27}$$

we may rewrite the equations for $N_y(\omega)$ in (20) and (24) as follows:

$$\begin{aligned} {}_{1P}N_y(\omega) &= N_1(\omega) + \sigma_0^2 \sum_{j=1}^{[N/2]} \left(1 + 3\left\|\frac{1}{\beta_j}\right\|_p^2 \left|\frac{\gamma_j(\omega)}{\beta_j(\omega)}\right|^2\right) \\ {}_{2P}N_y(\omega) &= N_1(\omega) + \sigma_0^2 \sum_{j=1}^{[N/2]} 4\left\|\frac{\gamma_j}{\beta_j}\right\|_p^2 \left|\frac{1}{\beta_j(\omega)}\right|^2. \end{aligned} \tag{28}$$

Consider now the variance, or total average power, of the output roundoff noise, given by

$$\sigma_y^2 = \|N_y\|_1 = \frac{1}{\omega_s}\int_0^{\omega_s} N_y(\omega)\,d\omega \tag{29}$$

(since $N_y(\omega)$ is a nonnegative real function). From (28) and the definition of $\|\ \|_2$ in (3), it follows that

$$\|{}_{1P}N_y\|_1 = \|N_1\|_1 + \sigma_0^2 \sum_{j=1}^{[N/2]} \left(1 + 3\left\|\frac{1}{\beta_j}\right\|_p^2 \left\|\frac{\gamma_j}{\beta_j}\right\|_2^2\right); \tag{30a}$$

whereas

$$\|{}_{2P}N_y\|_1 = \|N_1\|_1 + \sigma_0^2 \sum_{j=1}^{[N/2]} 4\left\|\frac{\gamma_j}{\beta_j}\right\|_p^2 \left\|\frac{1}{\beta_j}\right\|_2^2. \tag{30b}$$

Hence, letting

$${}_{pq}\theta_j = \left\|\frac{1}{\beta_j}\right\|_p^2 \left\|\frac{\gamma_j}{\beta_j}\right\|_q^2, \tag{31}$$

we have the following simple result for the case $p = q = 2$:

$$\sum_{j=1}^{[N/2]} {}_{pq}\theta_j \begin{cases} < [N/2] \Rightarrow \text{form } 2P \\ > [N/2] \Rightarrow \text{form } 1P. \end{cases} \tag{32}$$

Note, however, that for $p = 2$, the output noise variances $\|N_y\|_1$ for these two forms can differ at most by a factor of

4/3, or 1.25 dB, and thus it really makes little difference in this case whether form $1P$ or form $2P$ is chosen.

The comparison of (30a) and (30b) in the case of $p=\infty$ is not so obvious as for $p=2$, but it is basically the same. Let ω_{0j} be a frequency at which the maximum of $|1/\beta_j(\omega)|$ is achieved; that is,

$$\left|\frac{1}{\beta_j(\omega_{0j})}\right| = \left\|\frac{1}{\beta_j}\right\|_\infty. \tag{33}$$

Because of the very peaked nature of $|1/\beta_j(\omega)|$ relative to $|\gamma_j(\omega)|$, we may then approximate the L_p norm of $\gamma_j(\omega)/\beta_j(\omega)$ by

$$\left\|\frac{\gamma_j}{\beta_j}\right\|_p \approx \gamma_j(\omega_{0j})\left\|\frac{1}{\beta_j}\right\|_p, \tag{34}$$

and with this approximation (31) becomes

$${}_{pq}\theta_j \approx \gamma_j(\omega_{0j})\left\|\frac{1}{\beta_j}\right\|_p\left\|\frac{1}{\beta_j}\right\|_q. \tag{35a}$$

Hence, in this case

$${}_{pq}\theta_j \approx {}_{qp}\theta_j \tag{35b}$$

and (32) again applies. As before, it really makes little difference which parallel form is chosen. However, experience has shown that usually ${}_{\infty 2}\theta_j>1$, and hence form $1P$ is generally to be preferred over form $2P$ for $p=\infty$.

The other norm of $N_y(\omega)$ to be considered is the L_∞ norm, i.e., the maximum of $N_y(\omega)$ over all ω. Here we actually compare bounds on $\|N_y\|_\infty$ for forms $1P$ and $2P$, rather than expressions or approximations for $\|N_y\|_\infty$ itself. Although comparisons based on bounds for $\|N_y\|_\infty$ do not, of course, necessarily hold for $\|N_y\|_\infty$ itself, experimental results supporting the application of these comparisons to $\|N_y\|_\infty$ are presented in a later section.

Again employing $N_1(\omega)$ as defined by (27) in (20) and (24), the following bounds on the L_∞ norms of $N_y(\omega)$ for forms $1P$ and $2P$ are readily obtained:

$$\begin{aligned}\|{}_{1P}N_y\|_\infty &\le \|N_1\|_\infty + \sigma_0^2 \sum_{j=1}^{[N/2]} (1 + 3{}_{p\infty}\theta_j)\\ \|{}_{2P}N_y\|_\infty &\le \|N_1\|_\infty + \sigma_0^2 \sum_{j=1}^{[N/2]} 4{}_{\infty p}\theta_j\end{aligned} \tag{36}$$

where ${}_{pq}\theta_j$ is given by (31). Utilizing the approximation for ${}_{pq}\theta_j$ in (35), the conclusions contained in (32) apply in this case as well. As indicated previously, we usually find that ${}_{2\infty}\theta_j>1$ and almost always that ${}_{\infty\infty}\theta_j>1$.

Therefore, for all four cases considered (i.e., $\|N_y\|_r$ for $r=1,\infty$ and $p=2,\infty$) there is really little advantage to be gained by using form $1P$ over form $2P$ (or vice versa); but as a general "rule of thumb," form $1P$ is somewhat to be preferred. The situation is much more interesting in the case of the cascade form, where large differences are possible between the roundoff-noise outputs of the form-1 and form-2 configurations. In addition, there is the question of how to sequentially order the cascade-form sections.

The Cascade Form

The two most commonly employed (transpose) configurations for the cascade form of a digital filter are shown in Figs. 4 and 5 and are designated as forms $1D$ and $2D$, respectively. (The letter D indicates that each second-order section is realized in direct form, as opposed to other possible configurations [1].) The cascade form corresponds to an $H^*(z)$ of the general form [6], [7]

$$H^*(z) = a_0 \prod_{i=1}^{M} \frac{\alpha_{2i}z^{-2} + \alpha_{1i}z^{-1} + 1}{\beta_{2i}z^{-2} + \beta_{1i}z^{-1} + 1}. \tag{37}$$

However, in the scaled versions of forms lD, $l=1, 2$, $H^*(z)$ is realized as

$$H^*(z) = {}_l\alpha_0' \prod_{i=1}^{M} \frac{{}_l\alpha_i^{*\prime}(z)}{\beta_i^*(z)} \tag{38}$$

where for form $1D$ (letting ${}_{1D}s_{M+1}=a_0$)

$${}_1\alpha_0' = {}_{1D}s_1$$

$${}_1\alpha_i^{*\prime}(z) = \frac{{}_{1D}s_{i+1}}{{}_{1D}s_i}\alpha_i^*(z) \qquad i = 1, 2, \cdots, M;$$

while for form $2D$ (letting ${}_{2D}s_0=1$)

$${}_2\alpha_i^{*\prime}(z) = \frac{{}_{2D}s_i}{{}_{2D}s_{i-1}}\alpha_i^*(z) \qquad i = 1, 2, \cdots, M$$

$${}_2\alpha_0' = \frac{a_0}{{}_{2D}s_M}.$$

The corresponding transfer responses to the branch nodes marked by (*) are then

$$\begin{aligned}{}_{1D}F_i'(\omega) &= {}_{1D}s_i \frac{1}{\beta_i(\omega)} \prod_{j=1}^{i-1} \frac{\alpha_j(\omega)}{\beta_j(\omega)} \qquad i = 1, 2, \cdots, M\\ {}_{2D}F_i'(\omega) &= {}_{2D}s_i \prod_{j=1}^{i} \frac{\alpha_j(\omega)}{\beta_j(\omega)} \qquad i = 1, 2, \cdots, M\end{aligned} \tag{39}$$

where we define $\prod_{j=1}^{0}(\cdot)=1$. Hence, from the definition of s_i in (6), the ${}_{lD}F_i'(\omega)$ in (39) satisfy (5), as required.

Scaling is accomplished in cascade forms $1D$ and $2D$ via the altered multipliers ${}_l\alpha_{ki}'$ *within* the second-order sections, rather than by inserting multipliers *between* the second-order sections, for the following reasons. In form $1D$ the insertion of additional multipliers between the second-order sections would require that an additional overflow constraint be added at the output of each section. In a large number of cases (although not in every case), these additional overflow constraints would significantly reduce the potential signal-to-noise ratio which could otherwise be realized in the filter. In form $2D$, on the other hand, it is readily apparent that even if the additional overflow constraints (now required at the inputs to the second-order sections) do not actually change the required scaling, the total rounding error in the outputs of the α_{1i} and α_{2i} multipliers must be greater with scaling (and rounding) between sections, rather than within them.

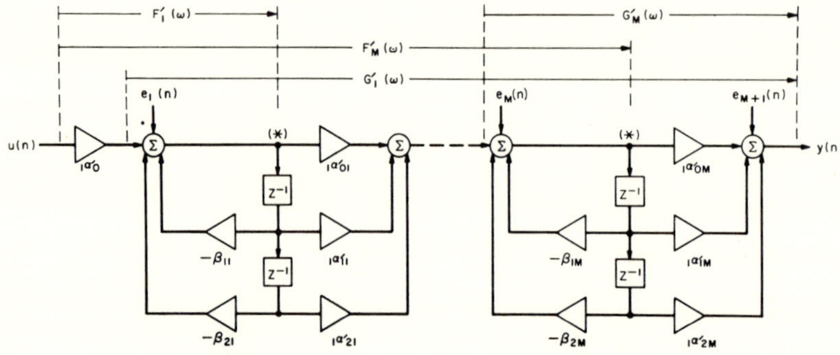

Fig. 4. Cascade form 1D with scaling.

Fig. 5. Cascade form 2D with scaling.

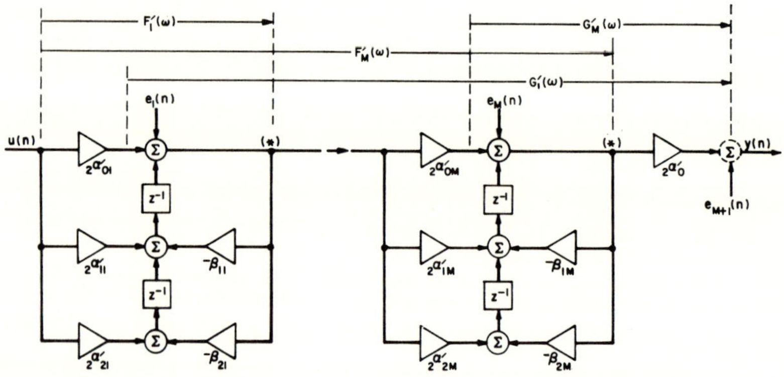

The transfer responses $G_j'(\omega)$ for form $1D$ are given by

$$
\begin{aligned}
{}_{1D}G_i'(\omega) &= \prod_{i=j}^{M} \frac{{}_1\alpha_i'(\omega)}{\beta_i(\omega)} \qquad i = 1, 2, \cdots . M \\
{}_{1D}G_{M+1}'(\omega) &= 1.
\end{aligned}
\tag{40}
$$

Assuming that the coefficients ${}_1\alpha_{ki}'$ and β_{ki} for all k and i are nonintegers, the corresponding weights k_j' for form $1D$ are

$$
\begin{aligned}
{}_{1D}k_1' &= 3 \\
{}_{1D}k_j' &= 5 \qquad j = 2, 3, \cdots, M-1 \\
{}_{1D}k_M' &= 5 - [2M - N] \\
{}_{1D}k_{M+1}' &= 3 - [2M - N]
\end{aligned}
\tag{41}
$$

where we have assumed, for the present, that if the order (N) of the filter is odd, the Mth section is the one first-order section. Equation (41) is readily shown to satisfy (18). The effect of scaling on the transfer responses ${}_{1D}G_j'(\omega)$ is seen by substituting the expressions for ${}_1\alpha_i'(\omega)$ in (38) into (40), which yields

$$
{}_{1D}G_j'(\omega) = \frac{a_0}{{}_{1D}s_j} \prod_{i=j}^{M} \frac{\alpha_i(\omega)}{\beta_i(\omega)} \qquad j = 1, 2, \cdots, M \tag{42}
$$

or, from (39) and (6),

$$
{}_{1D}G_j'(\omega) = a_0 \left\| \frac{1}{\beta_j} \prod_{i=1}^{j-1} \frac{\alpha_i}{\beta_i} \right\|_p \prod_{i=j}^{M} \frac{\alpha_i(\omega)}{\beta_i(\omega)} \qquad j = 1, 2, \cdots, M. \tag{43}
$$

In cascade form $2D$, shown in Fig. 5, we have again chosen to represent the roundoff-noise inputs to each second-order section by a single error source $e_j(n)$. The corresponding $G_j'(\omega)$ are then

$$
{}_{2D}G_j'(\omega) = {}_2\alpha_0' \frac{1}{\beta_j(\omega)} \prod_{i=j+1}^{M} \frac{{}_2\alpha_i'(\omega)}{\beta_i(\omega)} \qquad j = 1, 2, \cdots, M. \tag{44}
$$

But from (38) and (39) this becomes

$$ {}_{2D}G_j'(\omega) = a_0 \left\| \prod_{i=1}^{j} \frac{\alpha_i}{\beta_i} \right\|_p \frac{1}{\beta_j(\omega)} \prod_{i=j+1}^{M} \frac{\alpha_i(\omega)}{\beta_i(\omega)} \qquad j = 1, 2, \cdots, M. \tag{45a} $$

The transfer response for the input $e_{M+1}(n)$ is, of course, just

$$ {}_{2D}G_{M+1}'(\omega) = 1. \tag{45b} $$

The noise weights k_j' for form $2D$ are

$$ \begin{aligned} {}_{2D}k_j' &= 5 \qquad j = 1, 2, \cdots, M-1 \\ {}_{2D}k_M' &= 5 - 2[2M - N] \\ {}_{2D}k_{M+1}' &= 1. \end{aligned} \tag{46} $$

The variance, or total average power, of the output roundoff noise from forms $1D$ and $2D$ is derived from (1) and (42) through (46) to be

$$ \|{}_{1D}N_y\|_1 = \sigma_0^2 \left\{ {}_{1D}k_{M+1}' + a_0 \sum_{j=1}^{M} {}_{1D}k_j' \cdot \left\| \frac{1}{\beta_j} \prod_{i=1}^{j-1} \frac{\alpha_i}{\beta_i} \right\|_p^2 \left\| \prod_{i=j}^{M} \frac{\alpha_i}{\beta_i} \right\|_2^2 \right\} \tag{47a} $$

$$ \|{}_{2D}N_y\|_1 = \sigma_0^2 \left\{ 1 + a_0 \sum_{j=1}^{M} {}_{2D}k_j' \cdot \left\| \prod_{i=1}^{j} \frac{\alpha_i}{\beta_i} \right\|_p^2 \left\| \frac{1}{\beta_j} \prod_{i=j+1}^{M} \frac{\alpha_i}{\beta_i} \right\|_2^2 \right\}. \tag{47b} $$

There are two separate questions to be considered in the minimization of (47): first, which of these two forms is best; and second, what should be the sequential ordering of the M sections. We investigate the first question in this section, and the second in the next.

Cascade forms $1D$ and $2D$ are related via the concept of transpose configurations, as previously described. In particular, assuming that the $\alpha_i(\omega)$ and $\beta_i(\omega)$ appearing in (47) are the same for both forms, form $2D$ corresponds to the *section-by-section* transpose of form $1D$. Therefore, the sequential ordering of the individual sections in form $1D$ is not reversed in form $2D$. There is, of course, another form-2 configuration (denoted as form $2D_t$) which corresponds to the *overall* transpose of the given form-$1D$ configuration, and likewise, a form-1 configuration which corresponds to the overall transpose of the given form-$2D$ configuration. The sequential ordering of the individual sections of forms $1D$ and $2D$ is thus reversed in forms $1D_t$ and $2D_t$.

Since we will separately investigate the ordering of the sections of the cascade forms, we may here compare the noise output from form $1D$ with that from form $2D$ or $2D_t$. We will find it most convenient to consider form $2D_t$ in the case of $p=2$. Denoting the numerator and denominator factors of the D_t forms by $\alpha_i^t(\omega)$ and $\beta_i^t(\omega)$, respectively, we then note that

$$ \beta_i^t(\omega) = \beta_{M-i+1}(\omega) \qquad i = 1, 2, \cdots, M $$

and hence that

$$ \prod_{i=r}^{s} \frac{\alpha_i^t(\omega)}{\beta_i^t(\omega)} = \prod_{i=M-s+1}^{M-r+1} \frac{\alpha_i(\omega)}{\beta_i(\omega)} \qquad 1 \le r \le s \le M. \tag{48} $$

Therefore, letting $l = M-j+1$, we may rewrite (47b) for form $2D_t$ as follows:

$$ \|{}_{2D_t}N_y\|_1 = \sigma_0^2 \left\{ 1 + a_0 \sum_{l=1}^{M} {}_{2D}k_l' \left\| \prod_{i=l}^{M} \frac{\alpha_i}{\beta_i} \right\|_p^2 \cdot \left\| \frac{1}{\beta_l} \prod_{i=1}^{l-1} \frac{\alpha_i}{\beta_i} \right\|_2^2 \right\} \tag{49} $$

where we have employed the fact that

$$ {}_{2D_t}k_{M-l+1}' = {}_{2D}k_l' \qquad l = 1, 2, \cdots, M. $$

Comparing $\|N_y\|_1$ for form $1D$ as given in (47a) with that for form $2D_t$ in (49), we see that these two expressions differ in only two respects: first, the functions appearing in the L_p and L_2 norms are reversed; and second, the k_j' correspond to forms $1D$ and $2D$, respectively. From (41) and (46), the k_j' for these two forms are related in the following way:

$$ \begin{aligned} {}_{1D}k_1' &= {}_{2D}k_1' - 2 \\ {}_{1D}k_j' &= {}_{2D}k_j' \qquad j = 2, 3, \cdots, M-1 \\ {}_{1D}k_M' &= {}_{2D}k_M' + [2M - N] \\ {}_{1D}k_{M+1}' &= {}_{2D}k_{M+1}' + 2 - [2M - N]. \end{aligned} \tag{50} $$

Therefore, for $p=2$, $\|N_y\|_1$ differs for forms $1D$ and $2D_t$ only by virtue of the k_j' for $j=1$, M, and $M+1$. And for N even, only k_1' and k_{M+1}' are different for these two forms. Hence, as was the case for the parallel form, it really makes little difference which form is used when $p=2$ (and we are considering $\|N_y\|_1$).

In the case of $p=\infty$, however, there can be a significant difference between $\|{}_{1D}N_y\|_1$ and $\|{}_{2D_t}N_y\|_1$ due to the reversal of the L_∞ and L_2 norms in (47a) and (49). Neglecting the small difference due to the different k_j' for $j=1$, $M+1$, and (for N odd) M, the choice between forms $1D$ and $2D_t$ is determined by the relative magnitudes of ${}_{pq}\theta_j$ and ${}_{qp}\theta_j$ for $p=\infty$ and $q=2$, where

$$ {}_{pq}\theta_j = \left\| \frac{1}{\beta_j} \prod_{i=1}^{j-1} \frac{\alpha_i}{\beta_i} \right\|_p^2 \left\| \prod_{i=j}^{M} \frac{\alpha_i}{\beta_i} \right\|_q^2. \tag{51} $$

Note that this definition for ${}_{pq}\theta_j$ in the case of the D forms is consistent with that given in (31) for the P forms because in both cases

$$ {}_{pq}\theta_j = \|{}_1F_j\|_p^2 \|{}_1G_j\|_q^2 = \|{}_{2t}G_j\|_p^2 \|{}_{2t}F_j\|_q^2. \tag{52} $$

In particular, then,

$$ \sum_{j=1}^{M} ({}_{pq}\theta_j - {}_{qp}\theta_j) \begin{cases} < 0 \Rightarrow \text{form } 1D \\ > 0 \Rightarrow \text{form } 2D_t \end{cases} \tag{53} $$

for $p=\infty$ and $q=2$.

Although it was especially convenient to compare form $1D$ with form $2D_t$ in the case of $p=2$, another useful comparison can be made using form $2D$ when $p=\infty$. Noting that

$$\begin{aligned} {}_{2D}F_j(\omega) &= \alpha_j(\omega)\,{}_{1D}F_j(\omega) \\ {}_{1D}G_j(\omega) &= \alpha_j(\omega)\,{}_{2D}G_j(\omega) \end{aligned} \qquad j = 1, 2, \cdots, M$$

we may rewrite (47) in the following form:

$$\|{}_{1D}N_y\|_1 = \sigma_0^2 \Big\{ {}_{1D}k_{M+1}' + \sum_{j=1}^{M} {}_{1D}k_j' \cdot \|{}_{1D}F_j\|_p^2 \, \|(\alpha_j)\,{}_{2D}G_j\|_2^2 \Big\} \tag{54a}$$

$$\|{}_{2D}N_y\|_1 = \sigma_0^2 \Big\{ 1 + \sum_{j=1}^{M} {}_{2D}k_j' \cdot \|(\alpha_j)\,{}_{1D}F_j\|_p^2 \, \|{}_{2D}G_j\|_2^2 \Big\} . \tag{54b}$$

Hence, neglecting the difference in (54a) and (54b) due to the different k_j' for $j=1$, $M+1$, and (for N odd) M, the primary difference between these two expressions results from the occurrence of the extra factors $\alpha_j(\omega)$ multiplying ${}_{2D}G_j(\omega)$ in the former and ${}_{1D}F_j(\omega)$ in the latter for all j.

The implication of the extra factors $\alpha_j(\omega)$ in (54) for the case of $p=\infty$ is as follows. In (54a) the factors $\alpha_j(\omega)$ affect the L_2 norms, while in (54b) they affect the L_∞ norms. But the L_∞ norm of a function "concentrates" exclusively on the maxima of that function, whereas the L_2 norm of a function reflects the rms value of that function over all argument values. Therefore, the effect of the extra $\alpha_j(\omega)$ in (54b) results from the alteration of the maxima of ${}_{1D}F_j(\omega)$ in $\alpha_j(\omega)\,{}_{1D}F_j(\omega)$; while in (54a), the effect concerns the difference in ${}_{2D}G_j(\omega)$ and $\alpha_j(\omega)\,{}_{2D}G_j(\omega)$ over all ω.

Intuitively, one expects that the former effect is potentially much greater; that is, in some cases the extra $\alpha_j(\omega)$ should affect the L_∞ norms in (54b) much more than the L_2 norms in (54a). The experimental evidence presented later will support this conclusion. In particular, we will find that when the $\alpha_j(\omega)$ provide significant attenuation in the neighborhood of the maxima of the corresponding responses ${}_{1D}F_j(\omega)$, $\|N_y\|_1$ can be significantly less for form $2D$ than for form $1D$ (when $p=\infty$). On the other hand, when the $\alpha_j(\omega)$ do not provide such attenuation, there is not a great difference between the $\|N_y\|_1$ for these two forms.

Turning now to the consideration of $\|N_y\|_\infty$ (i.e., the maximum of $N_y(\omega)$ over all ω), the analysis goes through exactly as above, but with the L_2 norms being replaced by L_∞ norms. As in the case of the parallel forms, the comparisons are actually based on bounds for $\|N_y\|_\infty$, rather than on $\|N_y\|_\infty$ itself, with intuition and experimental evidence supporting this approach. By analogy with (54) and (49), these bounds on $\|N_y\|_\infty$ for forms $1D$, $2D$, and $2D_t$ are readily derived to be

$$\|{}_{1D}N_y\|_\infty \le \sigma_0^2 \Big\{ {}_{1D}k_{M+1}' + \sum_{j=1}^{M} {}_{1D}k_j' \cdot \|{}_{1D}F_j\|_p^2 \, \|(\alpha_j)\,{}_{2D}G_j\|_\infty^2 \Big\} \tag{55a}$$

$$\|{}_{2D}N_y\|_\infty \le \sigma_0^2 \Big\{ 1 + \sum_{j=1}^{M} {}_{2D}k_j' \cdot \|(\alpha_j)\,{}_{1D}F_j\|_p^2 \, \|{}_{2D}G_j\|_\infty^2 \Big\} \tag{55b}$$

$$\|{}_{2D_t}N_y\|_\infty \le \sigma_0^2 \Big\{ 1 + \sum_{j=1}^{M} {}_{2D}k_j' \cdot \|(\alpha_j)\,{}_{2D}G_j\|_p^2 \, \|{}_{1D}F_j\|_\infty^2 \Big\} \tag{55c}$$

where, as before,

$${}_{1D}F_j(\omega) = \frac{1}{\beta_j(\omega)} \prod_{i=1}^{j-1} \frac{\alpha_i(\omega)}{\beta_i(\omega)}$$

and

$${}_{2D}G_j(\omega) = \frac{1}{\beta_j(\omega)} \prod_{i=j+1}^{M} \frac{\alpha_i(\omega)}{\beta_i(\omega)} .$$

Therefore, it is now the case of $p=\infty$ for which the only difference between these bounds for forms $1D$ and $2D_t$ lies in the k_j' for $j=1$, $M+1$, and (for N odd) M. And for $p=2$, the attenuation provided by the extra factors $\alpha_j(\omega)$ in (55a) and (55b) should, when significant, benefit from $1D$ over form $2D$.

The general results of this section are thus summarized as follows. Significant differences in $\|N_y\|_r$ for cascade forms $1D$ and $2D$ (or $2D_t$) can occur in the cases of $r=1$, $p=\infty$ and $r=\infty$, $p=2$. In the first case, form $2D$ tends to be superior; whereas in the latter case, form $1D$ tends to be superior. For $p=r+1$, $r=1$, ∞; however, there seems to be little difference between these two forms. A chart incorporating these conclusions is given in Fig. 6.

$\|F_j'\|_p$ \ $\|N_y\|_r$	$r = 1$	$r = \infty$
$p = 2$	EITHER	FORM 1D
$p = \infty$	FORM 2D	EITHER

Fig. 6. General rules for selection of cascade-form configuration.

Sequential Ordering of Cacsade Sections

In this section we present a heuristic discussion of the sequential ordering of the second-order sections comprising the cascade D forms, based upon the analytical results of the preceding section. We must not only consider the ordering of given second-order factors $\alpha_i^*(z)$ $\beta_i^*(z)$, but also the pairing of each numerator factor $\alpha_i^*(z)$ with a denominator factor $\beta_i^*(z)$. An analytical technique to determine the best sequential ordering, other than a complete enumeration and evaluation of all possible permutations, has not yet been devised, but it is possible to state some general rules which can be quite helpful in the design of a given digital filter. In view of the fact that there are up to $M!$ possible pairings of the $\alpha_i^*(z)$ and $\beta_i^*(z)$ and $M!$ possible permutations of each set of $\alpha_i^*(z)/\beta_i^*(z)$, it is important to determine such general rules, where possible.

Consider the relations in (54) and (55) for $\|N_y\|_r$, $r=1$ and $r=\infty$, respectively. Each term of these expressions corresponds to the noise contribution from one section of the filter (i.e., the jth section). In each of these terms, the transfer response ${}_{1D}F_j(\omega)$ [containing the factors $\alpha_i(\omega)/\beta_i(\omega)$ for the $j-1$ sections preceding that section] appears as the argument of an L_p norm, while the response ${}_{2D}G_j(\omega)$ [containing the factors $\alpha_i(\omega)/\beta_i(\omega)$ for the $M-j$ sections following that section] appears as the argument of an L_{r+1} norm. This, of course, just reflects the fact that it is the sections preceding a given section which determine the overflow scaling for that section, while the succeeding sections filter the roundoff noise produced by that section.

Therefore, for $p\neq r+1$ the primary difference between different sequential orderings of the M sections most likely results from the different characteristics of the L_∞ and L_2 norms, as discussed in the preceding section. In particular, since the L_∞ norm is much more sensitive to the maxima of its argument function than the L_2 norm, one would expect that the preferred sequential ordering should minimize (in some sense) the peaked nature of those functions which appear as arguments of the L_∞ norms of (54) and (55). This would, in turn, indicate that the $\alpha_i(\omega)/\beta_i(\omega)$ should be ordered from "most peaked" to "least peaked", or vice versa, depending upon whether $r=\infty$ or $p=\infty$, respectively.

Since it is the L_2 and L_∞ norms with which we are concerned, a reasonable measure of the peakedness of the $\alpha_i(\omega)/\beta_i(\omega)$ for our purposes is simply

$$\mathcal{P}_i = \left\|\frac{\alpha_i}{\beta_i}\right\|_\infty \Big/ \left\|\frac{\alpha_i}{\beta_i}\right\|_2. \tag{56}$$

We would then expect from the above discussion that for $p\neq r=\infty$; a good (but not necessarily optimum) ordering for the M sections of the cascade form is provided by having $\mathcal{P}_i$ decrease (i.e., decreasing peakedness) with increasing i; while for $r\neq p=\infty$, $\mathcal{P}_i$ should increase with increasing i. Our experimental results support this general rule. Note that because of the numerator factors $\alpha_i(\omega)$, $\mathcal{P}_i$ is not necessarily proportional to the Q of the section, although this is often the case.

For $p=r+1$, however, the factors $\alpha_i(\omega)$ $\beta_i(\omega)$ for all $i\neq j$ appear in one L_p norm or another, and hence the output roundoff noise in this case should be less sensitive to the ordering of the M sections of the cascade form than for $p\neq r+1$. There will still, of course, be some effect due to the grouping of the $\alpha_i(\omega)/\beta_i(\omega)$ within these norms; but the primary effect is probably due, as before, to the attenuation provided by the extra factors $\alpha_j(\omega)$ which occur in the L_p norms for form $1D$ and the L_{r+1} norms for form $2D$. In particular, when the $\alpha_j(\omega)$ provide significant attenuation in the neighborhood of the peaks of the responses ${}_{1D}F_j(\omega)$ and ${}_{2D}G_j(\omega)$, this attenuation should be applied to the most peaked of these two sets of responses. This, in turn, implies (from the preceding discussion for $p\neq r+1$) that for $p=r+1$, a good ordering for the M sections of cascade form $1D$ should result from having $\mathcal{P}_i$ increase with increasing i; while for cascade form $2D$, $\mathcal{P}_i$ should decrease with increasing i. This and the preceding rules for sequential ordering are contained in the chart of Fig. 7.

It is instructive to note that something like the above rule for $p=r+1$ was to be expected from the results of the preceding section, where each form-$1D$ configuration was found to be comparable to the corresponding transpose configuration ($2D_t$) in this case. Note from Figs. 6 and 7, however, that this rule is *opposite* to that for $p\neq r+1$ since, for $p\neq r=\infty$, form $1D$ with decreasing $\mathcal{P}_i$ is indicated; while for $r\neq p=\infty$, form $2D$ with increasing $\mathcal{P}_i$ is indicated. An alternate, and perhaps clearer, presentation of these results is given in the charts of Fig. 8, which show the recommended combination(s) of form and ordering for each p, r pair.

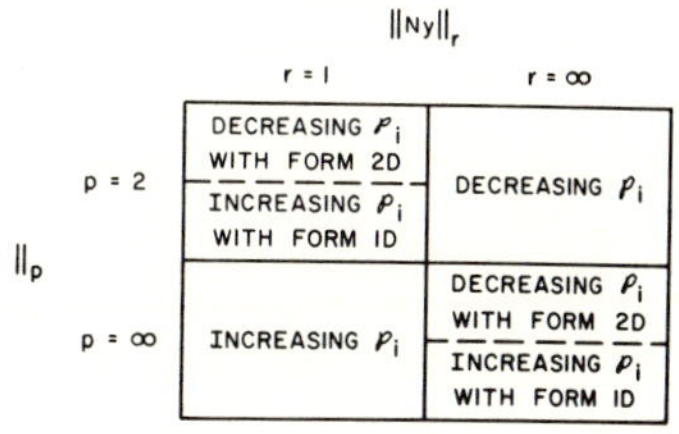

Fig. 7. General rules for sequential ordering of cascade-form sections.

Fig. 8. Alternative description of general rules in Figs. 6 and 7 [use either (A) or (B)].

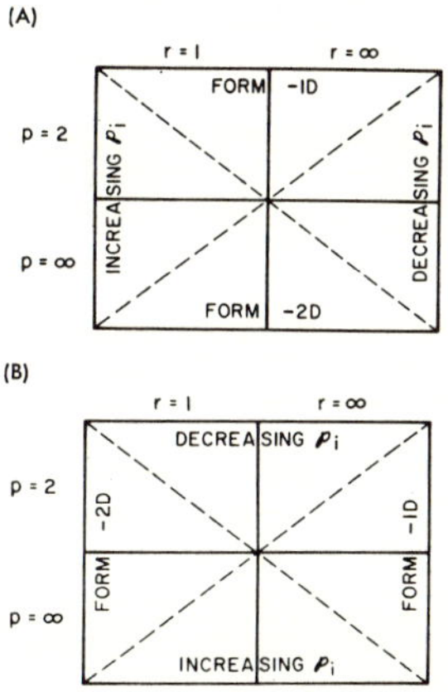

Having considered the sequential ordering of the section responses $\alpha_i(\omega)/\beta_i(\omega)$, we now address the question of how best to pair the numerator factors $\alpha_i(\omega)$ with the denominator factors $\beta_i(\omega)$ to produce these responses. There are, of course, many cases where this question does not arise including, for example, the cases of Butterworth or Chebyshev (type-1) low-pass or high-pass filters, where all zeros occur at $z=1$ or $z=-1$ (assuming that the bilinear z transform has been employed). In these cases the $\alpha_i(\omega)$ are equal for all i (except for one first-order factor when N is odd), and there is no choice to be made in pairing the $\alpha_i(\omega)$ and $\beta_i(\omega)$ (assuming that the first-order factors, if present, are paired together).

When the second-order factors $\alpha_i(\omega)$ are not equal, however, there is almost always a significant effect on the roundoff-noise output due to the pairing of these factors with the $\beta_i(\omega)$. The reason is apparent from (54) and (55), where the $\alpha_i(\omega)$ occur only in ratio with the corresponding $\beta_i(\omega)$, and the ratios $\alpha_i(\omega)/\beta_i(\omega)$ for all i appear in each term of the summations. Since the norms of $2M-1$ different combinations of the $\alpha_i(\omega)/\beta_i(\omega)$ occur in each relation for $\|N_y\|_r$ in (54) and (55), it is most reasonable to assume that the minimum $\|N_y\|_r$ will result when the individual $\alpha_i(\omega)/\beta_i(\omega)$ are minimized (in some sense). And since

$$\left\|\frac{\alpha_i}{\beta_i}\right\|_p \leq \left\|\frac{\alpha_i}{\beta_i}\right\|_\infty$$

for all $p\geq 1$, it is reasonable to minimize the L_∞ norms of the individual $\alpha_i(\omega)/\beta_i(\omega)$, insofar as this is possible.

Although the above discussion may seem inconclusive as it stands, the proper pairing of the $\alpha_i(\omega)$ and $\beta_i(\omega)$ is really quite evident from such considerations in most cases of practical interest. Consider, for example, the z-plane diagram of Fig. 9 for a sixth-order band-rejection filter. The dotted lines indicate the best pairing of the zeros corresponding to the $\alpha_i(\omega)$ with the poles corresponding to the $\beta_i(\omega)$. This pairing obviously minimizes the L_∞ norms of the $\alpha_i(\omega)/\beta_i(\omega)$ *for all i* and thus, most likely, the values of $\|\alpha_i/\beta_i\|_p$ for all i and p, as well. Almost as obvious are the pairings of Fig. 10 for a sixth-order elliptic bandpass filter. The two "higher-Q" pole pairs are combined with the nearest zero pairs, leaving the real zeros to be combined with the "lower-Q" pole pair. This should minimize the maximum $\|\alpha_i/\beta_i\|_\infty$ over i. Experimental evidence of the effect of proper pole-zero combination is presented in the next section.

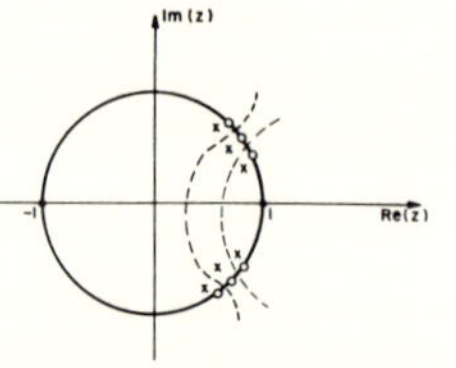

Fig. 9. Proper pairing of poles and zeros in cascade form for sixth-order band-rejection filter.

Fig. 10. Proper pairing of poles and zeros in cascade form for sixth-order bandpass filter.

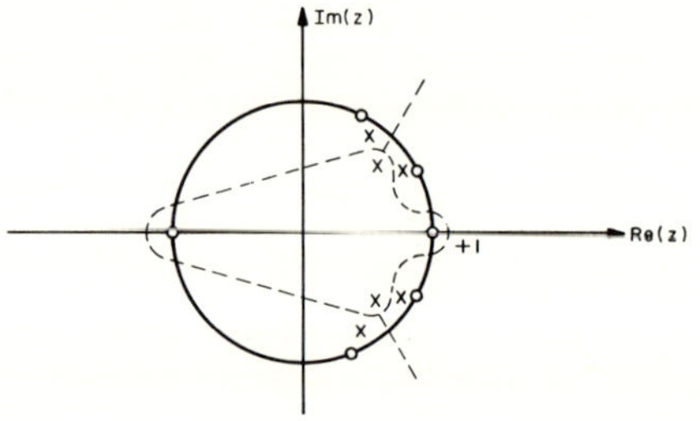

Computer Implementation: An Example

Two digital computer programs have been written to implement and test the analytical results of the preceding sections. These programs are written in FORTRAN IV and have been run successfully on the GE635 digital computer at Bell Telephone Laboratories, Inc., Murray Hill, N. J. The first is the Noise Analysis Program (NAP), which, given the transfer function $H^*(z)$ for a digital filter in either cascade or parallel form, computes the scaling required for that form to satisfy the overflow constraints in (5) (for $p=2$ or for $p=\infty$) and then predicts the resulting roundoff-noise spectrum $N_y(\omega)$ from (1). The other is the Filter Simulation Program (FSP), which simulates the (scaled) digital filter in the appropriate form using sinusoidal or white-noise input signals and estimates the power-density spectrum $N_y(\omega)$ of the output roundoff noise. In other words, NAP implements the synthesis and analysis procedures developed in this paper, and FSP tests the validity of these results for specific filters and input signals.

In both NAP and FSP the output noise density $N_y(\omega)$ is normalized with respect to $N_0=\sigma_0^2$ (i.e., the density or variance of the white noise from a single rounding operation). That is, instead of $N_y(\omega)$ we actually compute

$$\frac{N_y(\omega)}{N_0} = \sum_j k_j' \mid G_j'(\omega) \mid^2. \tag{57}$$

Hence $N_y(\omega)$ and norms thereof are always given in decibels relative to N_0. This normalization is very helpful in relating the results of NAP and FSP, for it eliminates any (direct) reference to the input signal level. In [1] NAP and FSP are described in greater detail, and several representative examples of their operation are given.

One of these examples is the sixth-order Chebyshev (type-2) band-rejection filter (BRF) described in Table I. The unscaled coefficients for both the cascade and parallel

TABLE

Sixth-Order Chebyshev-2 Band-Rejection Filter

$\alpha_0=0.76091619$; $\gamma_0=1.3142206$

	Section 1	2	3	1′	2′
β_{2i}	0.90352914	0.84506679	0.75829007	0.90352914	0.84506679
β_{1i}	−1.7636952	−1.4427789	−1.5334490	−1.7636952	−1.4427789
α_{2i}	1.0	1.0	1.0	1.0	1.0
α_{1i}	−1.8118373	−1.6545862	−1.7442502	−1.6545862	−1.8118373
$\|\alpha_i/\beta_i\|_\infty$	1.65	1.37	1.137	5.717	2.694
$\|\alpha_i/\beta_i\|_2$	1.056	1.076	1.066	1.602	1.280
ρ_i	1.56	1.27	1.07	3.57	2.11
γ_{1i}	0.09494903	0.17123073	0.24439853	—	—
γ_{0i}	−0.10898306	−0.16408810	−0.28023324	—	—

Fig. 11. Overall frequency response of sixth-order Chebyhev-2 BRF in parallel form.

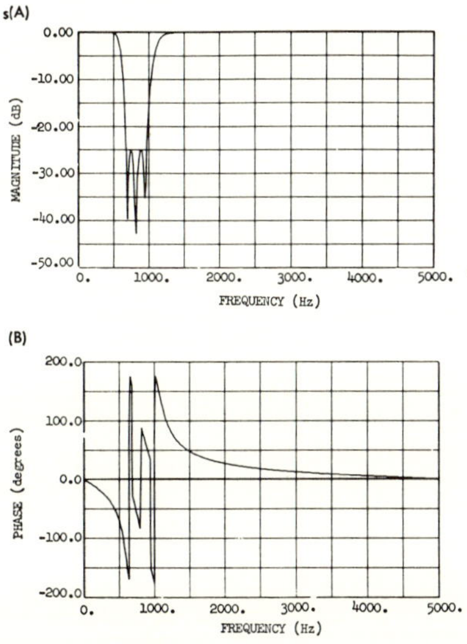

forms of this filter are given in Table I, as well as the L_2 and L_∞ norms of the cascade-form responses $\alpha_i(\omega)/\beta_i(\omega)$ and the corresponding ratios $\mathcal{P}_i$. The effect of proper pole–zero pairing is also illustrated by reversing the pairings of Sections 1 and 2 to yield Sections 1′ and 2′, as described in Table I. The pairings in Sections 1, 2, and 3 are in accordance with the discussion contained in the preceding section, and illustrated in Fig. 9.

The frequency response (both magnitude and phase) of the BRF is shown in Fig. 11. The specifications for the filter are 2.26-dB passband ripple, 25-dB stopband attenuation, and a transition ratio of 0.53. The filter was

TABLE II

Sixth-Order Chebyshev-2 Band-Rejection Filter

	Variance ($r=1$)				Peak Noise ($r=\infty$)			
	$p=\infty$		$p=2$		$p=\infty$		$p=2$	
	F1	F2	F1	F2	F1	F2	F1	F2
Parallel (*P*) Forms	24.1*	25.2	13.8**	14.3	35.6*	37.0	23.0*	24.4
Cascade (*D*) Forms								
(123)	34.2	23.5**	22.1	21.0	34.8	34.8	22.7**	31.2
(321)	31.8	24.2	20.4*	23.6	35.0	36.0	23.0	35.7
(213)	33.6	23.5**	21.3	22.4	36.0	35.0	23.3	33.7
(312)	34.0	23.8	22.0	21.7	35.0	35.5	23.1	32.5
(231)	31.6	24.5	20.3*	23.8	35.0	37.0	23.3	36.0
(132)	34.4	23.7	22.3	20.8	35.0	34.5**	23.1	30.7

Note: $\|N_y\|_r/N_0$ in decibels.

designed from these specifications using a FORTRAN-IV computer program developed and written by R. M. Golden,[1] J. F. Kaiser, and E. J. Sitar of Bell Telephone Laboratories, Inc. The plots were generated on a Stromberg–Carlsen 4060 microfilm plotter using another subroutine due to Kaiser and Sitar.

Predictions of the output roundoff-noise spectra $N_y(\omega)$ were made by NAP for parallel forms 1*P* and 2*P* and all orderings of cascade forms 1*D* and 2*D* with $p=2$ and $p=\infty$. The results are summarized in Table II, where the predicted $\|N_y\|_r$ for $r=1$, ∞ are given in decibels relative to N_0 for all cases. The form-1 results are indicated by "F1" column headings, and form-2 by "F2."

The ordering (123) corresponds to decreasing $\mathcal{P}_i$, i.e., decreasing peakedness. Note from Table II that there is not much difference between the different cascade-form orderings, although for $p=r+1$ form 2*D* is somewhat better when Section 1 is first and form 1*D* is better when

[1] Now with Technology Service Corporation, Santa Monica, Calif.

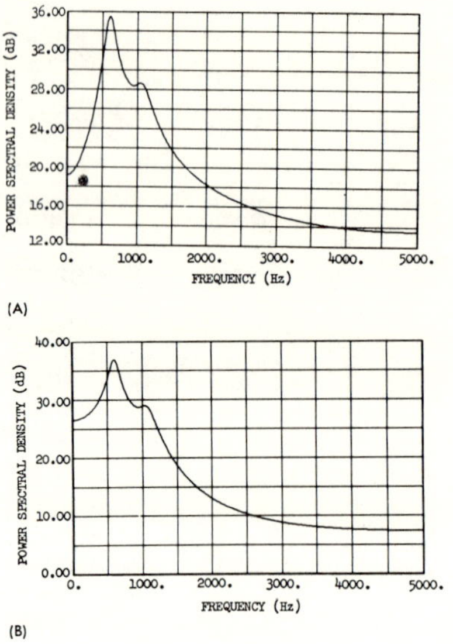

Fig. 12. Predicted roundoff-noise densities for sixth-order BRF in parallel forms 1*P* (A) and 2*P* (B) with $p = \infty$.

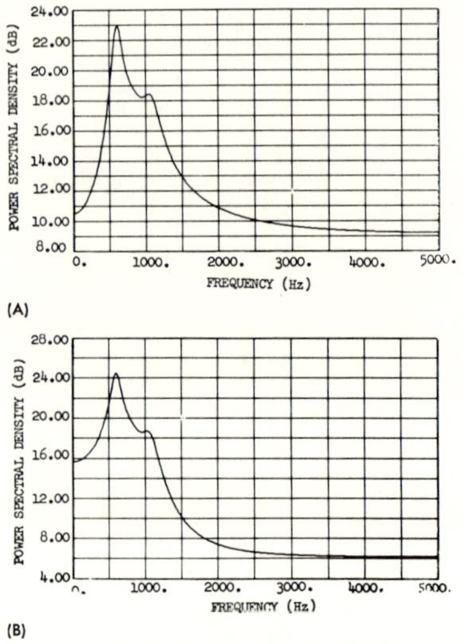

Fig. 13. Predicted roundoff-noise densities for sixth-order BRF in parallel forms 1*P* (A) and 2*P* (B) with $p = 2$.

Fig. 14. Predicted roundoff-noise densities for sixth-order BRF in cascade forms 1*D* (A) and 2*D* (B) with $p = \infty$ (123 ordering).

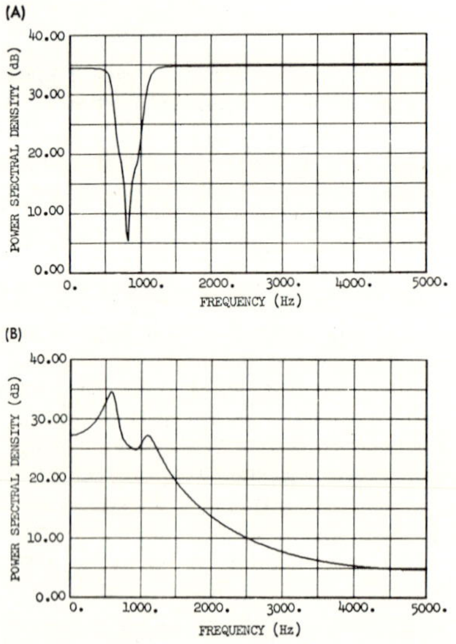

Fig. 15. Predicted roundoff-noise densities for sixth-order BRF in cascade forms 1*D* (A) and 2*D* (B) with $p = 2$ (123 ordering).

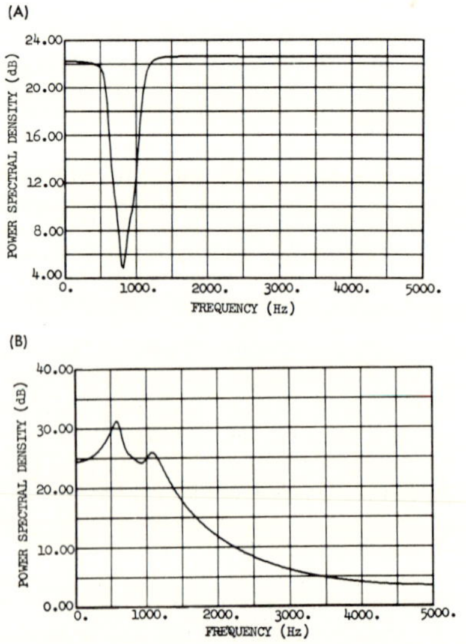

TABLE III
Sixth-Order Chebyshev-2 Band-Rejection Filter (Pairing Reversed in Sections 1 and 2)

Cascade (*D*) Forms	Variance ($r=1$)				Peak Noise ($r=\infty$)			
	$p=\infty$		$p=2$		$p=\infty$		$p=2$	
	F1	F2	F1	F2	F1	F2	F1	F2
(1′2′3)	36.5	28.3	23.9	20.6	39.4	37.7*	26.7*	29.0
(32′1′)	29.6	27.7*	19.9*	24.9	37.7*	40.4	28.1	37.6
(2′1′3)	30.7	27.6*	20.4	23.8	38.2	39.8	28.0	35.8
(31′2′)	35.8	28.2	23.5	20.9	39.5	38.4	27.1	30.6
(2′31′)	32.5	32.9	21.6	28.4	42.1	46.0	31.2	41.6
(1′32′)	41.1	31.3	27.9	22.0	45.8	42.1	32.6	32.9

Note: $||N_y||_r/N_0$ in decibels.

Section 1 is last, in agreement with the general rules in Fig. 7. This relative insensitivity to ordering is explained by the relatively small variation in $\mathcal{P}_i$ for Sections 1, 2, and 3, as given in Table I.

The largest differences in the predicted $||N_y||_r$ for the BRF occur between forms 1D and 2D when $p\neq r+1$ and between the P and D forms when $p=r+1=2$. Note from Table II that these differences are of the order of 7 to 12 dB. In agreement with Fig. 6, form 2D is superior for $r\neq p=\infty$, while form 1D is best when $p\neq r=\infty$. Parallel form 1P is comparable to the best cascade form for each p, r pair except $p=r+1=2$, where it (1P) is definitely superior.

The above differences in $||N_y||_r$ are readily apparent from the predicted densities in Figs. 12 through 15. Figs. 12 and 13 show the predicted $N_y(\omega)$ for the parallel forms, and Figs. 14 and 15 give the predicted $N_y(\omega)$ for the (123)-ordered cascade forms. Note that the shapes of the form-2D densities are very similar to those for the P forms. In particular, the densities for form 1D resemble the wide-band response of the BRF itself, whereas those for forms 2D, 1P, and 2P are narrow-band in character and achieve their peak values in the BRF stopband.

With LI scaling ($p=\infty$) the peak values $||N_y||_\infty$ are comparable in every case, but because of the wide-band character of ${}_{1D}N_y(\omega)$, the total power $||N_y||_1$ for form 1D is much greater than for the other forms. For L2 scaling ($p=2$), however, the peak density for form 1D is much less than for form 2D although their total noise powers are very comparable. The peak density for parallel form 1P is almost the same as for form 1D when $p=2$, but because of its narrow-band character the total noise power for the parallel form is much less in this case.

Another implication of the wide-band response of the BRF is that the form-2D noise densities are not much less for $p=2$ than for $p=\infty$, in contrast with other examples presented in [1]. The reason for this is that the transfer responses $F_i(\omega)$ are all wide-band in this case, and hence the scaling for wide-band inputs ($p=2$) is not much different from that for narrow-band inputs ($p=\infty$). For form 1D, however, the $F_i(\omega)$ are not wide-band, and the noise densities are, therefore, much less for $p=2$ than for $p=\infty$ due to scaling.

As an example of the effect of proper pole-zero pairing, we now consider the corresponding results for the BRF when the pairings of the $\alpha_i(\omega)$ and $\beta_i(\omega)$ for Sections 1 and 2 are reversed to form Sections 1′ and 2′. The resulting $||N_y||_r$ are given in Table III for all orderings of the cascade forms. As before, the ordering (1′2′3) corresponds to decreasing $\mathcal{P}_i$, but now the variation in $\mathcal{P}_i$ is much greater. Note that the results in Table III are in complete agreement with the general rules in Figs. 6 and 7 concerning form and ordering.

Our main point is, however, that the lowest $||N_y||_r$ for each p, r pair in Table III is significantly greater (3 to 4 dB) than that for the cascade forms in Table II in every case except for $p=r+1=2$. And in the latter case, the results are comparable. Hence, proper pole-zero pairing is indeed important in the synthesis of this digital filter, as expected.

The results of four representative FSP simulations of the sixth-order BRF are presented in Table IV and Figs. 16 through 19. The cascade-form results correspond to the (123) ordering (with proper pole-zero pairing). The scaling appropriate to the input signal is employed (i.e., $p=\infty$ for sinusoidal inputs and $p=2$ for white noise). The input signal amplitudes, peak output roundoff errors, estimated error means, and peak signal levels at overflow-constrained branch nodes are listed in Table IV in quantization-step units (assuming rounding to the nearest integer). The estimated and predicted error variances are given in decibels relative to N_0. Parzen lag windows have been employed in each case.

Note first of all the excellent agreement between the estimated noise densities in Figs. 16 through 19 and the corresponding predictions in Figs. 12(A), 13(A), 14(B), and 15(A), respectively. For example, note that the peak densities and those at dc and 5000 Hz agree quite closely with their predicted values. The estimated error variances, or average noise powers, in these cases are all within 0.9

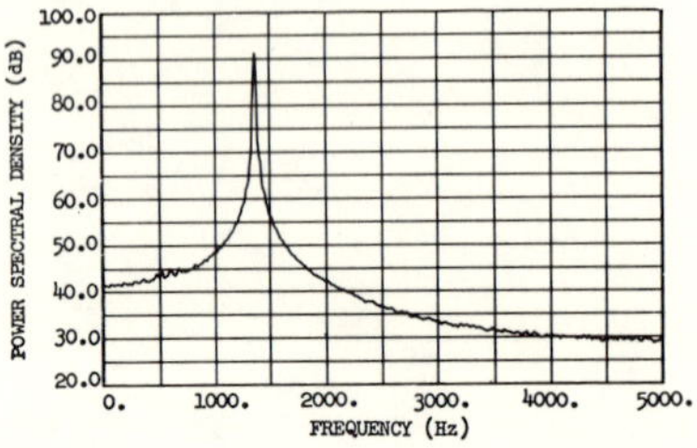

(A)

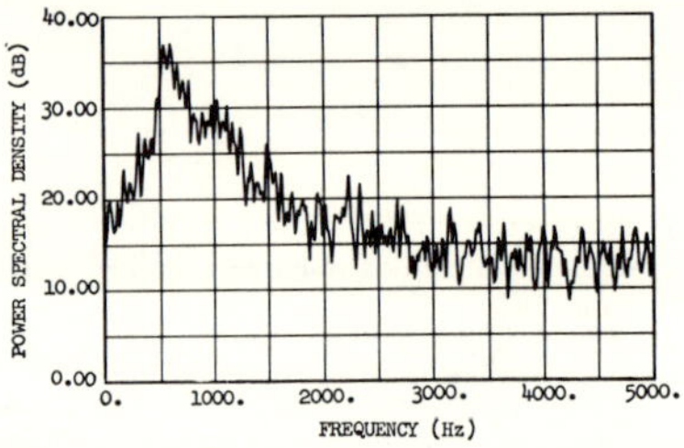

(B)

Fig. 16. Estimated densities of filter output and roundoff noise from simulation of sixth-order BRF in parallel from 1*P* with 1369-Hz input, $p=\infty$. (A) Actual filter output including roundoff error. (B) Roundoff error in actual filter output.

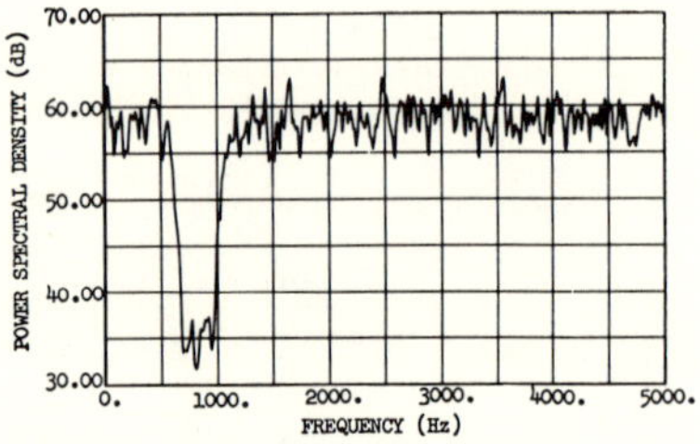

(A)

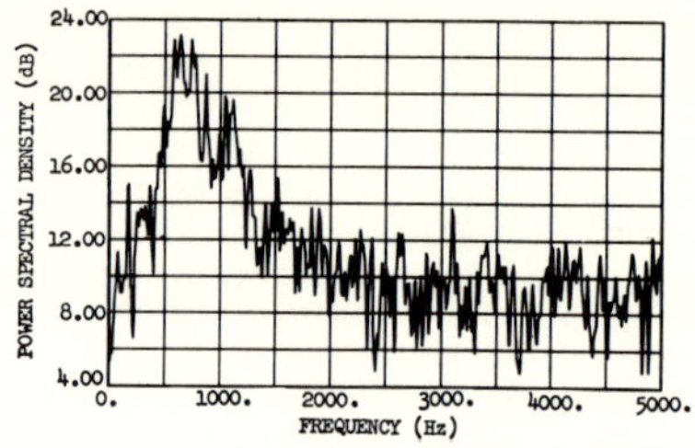

(B)

Fig. 17. Estimated densities of filter output and roundoff noise from simulation of sixth-order BRF in parallel form 1*P* with white-noise input, $p=2$. (A) Actual filter output including roundoff error. (B) Roundoff error in actual filter output.

TABLE IV

Simulation of Sixth-Order Chebyshev-2 Band-Rejection Filter (2048 Samples)

Scaling	Form	Input	Amplitude	Frequency or PF (p)	Peak Error	Error Mean	Error Variance (dB)	Predicted Variance (dB)	Peak Data
LI	1*P*	sine	1024	1369 Hz	15.54	−0.0253	24.16	24.1	372
L2	1*P*	noise	1024	4.0	4.545	0.0218	13.45	13.8	1076
LI	2*D*	sine	1024	1369 Hz	17.99	−0.284	24.42	23.5	1025
L2	1*D*	noise	1024	4.0	11.88	0.0707	22.05	22.1	909

dB of their predicted values. This agreement is perhaps surprising in the first and third cases because the input signal is a single sinusoid (at 1369 Hz) and significant correlation might be expected in the output roundoff error as a result. However, the correlation in the individual roundoff-error inputs seems to average out in the total output from the filter. Cases can, of course, be generated with significant correlation being apparent in the form of harmonics of the sinusoid [1], but these cases are the exception, not the rule.

There is also good agreement between our analytical results and other measured data. The estimated error means are very close to zero in all cases, and the peak data at overflow-constrained branch nodes are close to the input signal amplitude (1024). The one case (form 2*D*, LI) where the peak data exceeded 1024 (plus the peak error) resulted from the residual transient remaining in the data after the 200-sample initialization period. Otherwise, the overflow constraints appear to have been satisfied in all cases.

Summary

The roundoff-noise outputs from two transpose configurations each for the cascade and parallel forms of a digital filter have been analyzed and compared using the

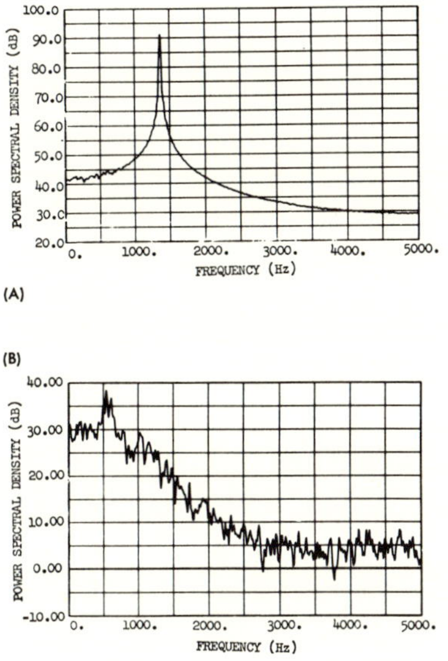

Fig. 18. Estimated densities of filter output and roundoff noise from simulation of sixth-order BRF in cascade form 2D with 1369-Hz input, $p=\infty$. (A) Actual filter output including roundoff error. (B) Roundoff errror in actual filter output.

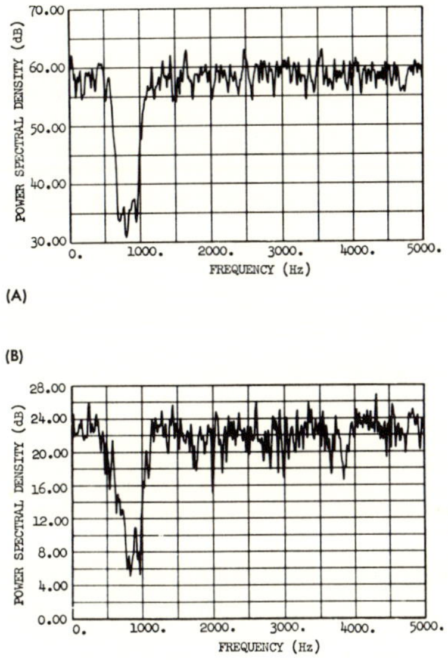

Fig. 19. Estimated densities of filter output and roundoff noise from simulation of sixth-order BRF in cascade form 1D with white-noise input, $p=2$. (A) Actual filter output including roundoff error. (B) Roundoff error in actual filter output.

techniques developed in [2]. The spectrum of the output roundoff noise from these four configurations has been shown to be of the form

$$N_y(\omega) = \sigma_0^2 \left\{ k_{M+1}' + \sum_{j=1}^{M} k_j' \|F_j\|_p^2 \, | G_j(\omega)|^2 \right\} \qquad (58)$$

where $F_i(\omega)$ is the (unscaled) transfer response from the input to the ith branch node, $G_j(\omega)$ is the (unscaled) transfer response from the jth summation node to the output, k_j' is the number of noise sources inputting (directly) to the jth summation node, M is the number of second-order sections comprising the cascade or parallel form, and σ_0^2 is the variance of the noise from each rounding operation. Hence, the variance, or total average power, of the output roundoff noise is simply

$$\sigma_y^2 = \sigma_0^2 \left\{ k_{M+1}' + \sum_{j=1}^{M} k_j' \|F_j\|_p^2 \|G_j\|_2^2 \right\}, \qquad (59)$$

while the peak spectral density $\|N_y\|_\infty$ is bounded by

$$\|N_y\|_\infty \leq \sigma_0^2 \left\{ k_{M+1}' + \sum_{j=1}^{M} k_j' \|F_j\|_p^2 \|G_j\|_\infty^2 \right\}. \qquad (60)$$

Corresponding transpose configurations have been compared on the basis of (59) and (60), and the comparisons based on the bounds in (60), rather than on $\|N_y\|_\infty$ itself, have been justified by the excellent experimental agreement reported here and in [1]. A heuristic discussion of the sequential ordering and pole–zero pairing of the cascade-form sections has also been presented on the basis of (59) and (60). From these results, the following general "rules of thumb" may be stated.

1) Parallel form $1P$ is generally preferable to form $2P$, although the maximum difference in $\|N_y\|_r$ for these two forms (for a given p) is only of the order of 4/3, or 1.25 dB.

2) For $p=r+1$ there is little difference in $\|N_y\|_r$ between any given form-$1D$ configuration and the corresponding transpose configuration (form $2D_t$).

3) For $p \neq r+1$, however, cascade form $2D$ is generally superior to form $1D$ for $p=\infty$, and vice versa for $r=\infty$; and the difference can be quite significant.

4) Good (but not necessarily optimum) sequential orderings for the M sections of the cascade form are indicated in the charts of Figs. 7 and 8, where they are stated in terms of the variation in

$$\mathcal{P}_i = \left\| \frac{\alpha_i}{\beta_i} \right\|_\infty \Big/ \left\| \frac{\alpha_i}{\beta_i} \right\|_2$$

with increasing i.

5) The numerator factors $\alpha_i(\omega)$ and denominator factors $\beta_i(\omega)$ of the cascade D forms should be paired so

as to minimize the individual $\|\alpha_i/\beta_i\|_\infty$, insofar as this is possible.

The analytical results obtained do not provide a general rule relating to the choice between parallel form 1*P* and one of the cascade *D* forms. The experimental results do indicate, however, that form 1*P* is usually comparable to or somewhat better than the best *D* form(s). Therefore, from the viewpoint of roundoff noise alone, the simplest and most reliable choice of a digital-filter configuration would seem to be parallel form 1*P*. However, the cascade *D* forms often have other advantages which make them most desirable [8], especially when the zeros of $H^*(z)$ lie on the unit circle ($|z| = 1$). In these cases, the *D*-form options of form, sequential ordering, and pole–zero pairing contained in (2) through (5) must be considered.

Finally, two computer programs written to implement and test the analytical results have been briefly described. A representative example has been included to illustrate the operation of these programs and, more importantly, to demonstrate the effectiveness of the analysis and synthesis procedures upon which they are based. These experimental results indicate that the uncorrelated-error analysis is quite accurate, even for sinusoidal inputs, unless the input is periodic in a multiple of the sampling period *T* and/or is of insufficient magnitude [1]. The general rules concerning circuit configuration have also been found to be very helpful and effective.

References

[1] L. B. Jackson, "An analysis of roundoff noise in digital filters," Sc.D. dissertation, Dept. of Elec. Engrg., Stevens Institute of Technology, Hoboken, N. J., 1969.

[2] ——, "On the interaction of roundoff noise and dynamic range in digital filters," *Bell Sys. Tech. J.*, vol. 49, no. 2, February 1970.

[3] R. Edwards, J. Bradley, and J. B. Knowles, "Comparison of noise performance of programming methods in the realization of digital filters," *Proc. Symp. on Computer Processing in Communications*, PIB-MRI Symposia Ser., vol. 19, 1969.

[4] C. Weinstein and A. V. Oppenheim, "A comparison of roundoff noise in floating point and fixed point digital filter realizations," *Proc. IEEE* (Letters), vol. 57, pp. 1181–1183, June 1969.

[5] J. R. Rice, *The Approximation of Functions.* Reading, Mass.: Addison-Wesley, 1964, pp. 4–10.

[6] C. M. Rader and B. Gold, *Digital Processing of Signals.* New York: McGraw-Hill, 1969.

[7] J. F. Kaiser, "Digital filters," in *System Analysis by Digital Computer*, F. F. Kuo and J. F. Kaiser, Eds. New York: Wiley, 1966, ch. 7, pp. 218–285.

[8] L. B. Jackson, J. F. Kaiser, and H. S. McDonald, "An approach to the implementation of digital filters," *IEEE Trans. Audio and Electroacoustics*, vol. AU-16, pp. 413–421, September 1968.

16

Reprinted from *Bell Syst. Tech. J.*, **46**(8), 1775–1791 (1967)

Floating-Point-Roundoff Accumulation in Digital-Filter Realizations

By I. W. SANDBERG

(Manuscript received June 20, 1967)

In this paper, several results are presented concerning the effects of roundoff in the floating-point realization of a general discrete filter governed ideally by a stable difference equation of the form

$$w_n = \sum_{k=0}^{M} b_k x_{n-k} - \sum_{k=1}^{N} a_k w_{n-k}, \qquad n \geqq N \tag{1}$$

in which $\{w_n\}$ *and* $\{x_n\}$ *are output and input sequences, respectively.*

In particular, for a large class of filters it is proved that there is a function $f(K)$ *with* $f(K) \to 0$ *as* $K \to \infty$ *and a constant* c*, both dependent on the* b_k*, the* a_k*, the order in which the products on the right side of (1) are summed in the machine, and* t*, the number of bits allotted to the mantissa, such that*

$$\langle e \rangle_K \leqq c \langle y \rangle_K + f(K)$$

for all $K \geqq N$*, in which, with* $\{y_n\}$ *the computed output sequence of the realized filter,*

$$\langle y \rangle_K = \left(\frac{1}{K+1} \sum_{n=0}^{K} | y_n |^2 \right)^{\frac{1}{2}}$$

and

$$\langle e \rangle_K = \left(\frac{1}{K+1} \sum_{n=0}^{K} | w_n - y_n |^2 \right)^{\frac{1}{2}}.$$

Bounds on $f(K)$ *and* c *are given that are not difficult to evaluate, and which, in many realistic cases, are informative. For example, for the second-order bandpass filter:*

$$w_n = x_n - a_1 w_{n-1} - a_2 w_{n-2}, \qquad n \geqq 2 \tag{2}$$

with a_1 *and* a_2 *chosen so that its poles are at approximately* $\pm 45°$ *and at distance approximately (but not less than) 0.001 from the unit circle, we find that c, an upper bound on the "asymptotic output error-to-signal ratio", is not greater than* 0.58×10^{-4}*, assuming that* $t = 27$*, that the terms on the right side of (2) are summed in the machine in the order indicated (from right to left), and that the* x_n *in (2) are machine numbers. If the* x_n *are not machine numbers, and hence must be quantized before processing, then* $c \leqq 0.76 \times 10^{-4}$*.*

In addition to error bounds, an inequality is derived which, if satisfied, rules out certain types of generally undesirable behavior such as self-sustained output limit cycles due to roundoff effects. This inequality is satisfied for the example described above.

I. INTRODUCTION

The difference equation

$$w_n = \sum_{k=0}^{M} b_k x_{n-k} - \sum_{k=1}^{N} a_k w_{n-k}, \qquad n \geqq N \tag{1}$$

with $M \leqq N$ defines the behavior of a general time-invariant discrete filter which acts on an input sequence $x_0, x_1, x_2, \cdots$ to produce an output sequence $w_N, w_{N+1}, w_{N+2}, \cdots$ that depends on the starting values $w_0, w_1, \cdots, w_{N-1}$.

There is a vast literature concerned with techniques for designing discrete filters [i.e., for determining the a_k and the b_k in (1)] to meet specifications of various types (see, for example, Refs. 1, 2, and 3), and a good deal of material is available on the subject of roundoff effects in fixed-point realizations of discrete filters (see, for instance, Refs. 4 and 5). In this paper, we derive some bounds on a meaningful measure of the overall effect of roundoff errors for discrete filters realized as digital filters on a machine employing floating-point arithmetic operations. This type of realization, as opposed to the fixed-point kind, is of particular importance in connection with, for example, digital computer simulations of systems, as a result of the large dynamic range afforded by the floating-point mode.

There are basic differences concerning fixed-point and floating-point error estimation problems which stem from the fact that the modulus of every individual arithmetic error in the fixed-point mode is bounded by a constant determined by the machine, whereas the maximum modulus of the error in forming, for example, the floating-point sum of two floating-point numbers is proportional to the magnitude of the true sum. For this reason, the approach* presented here, as well as the

* The approach can be extended in several different directions. For example, it can be used to obtain statistical error estimates based on the assumption that each roundoff error is an independent random variable.

character of the results, are quite different from those of earlier writers concerned with fixed-point realizations.

In addition to error bounds, an inequality is derived which, if satisfied, rules out certain types of generally undesirable behavior such as self-sustained output limit cycles due to roundoff effects.

II. ASSUMPTIONS AND RESULTS

2.1 *Assumptions*

It is assumed that:

(*i*) each machine number q is equal to sgn (q) a 2^b in which the exponent b is an integer, and a, the mantissa, is a t-bit number contained in $[\frac{1}{2}, 1]$ or $[\frac{1}{2}, 1] \cup \{0\}$;

(*ii*) the range of values of b is adequate to ensure that all computed numbers lie within the permissible range;

(*iii*) the machine operations of addition and multiplication are performed in accordance with standard rounding conventions* (described, for example, by Wilkinson[6]); and

(*iv*) the coefficients a_k and b_k in (1) are machine numbers.†

2.2 *Results: x_n Machine Numbers*

It is assumed throughout Section 2.2 that the x_n of (1) are floating-point machine numbers.

If the discrete filter (1) is realized on a floating-point machine, then

$$y_n = fl\left(\sum_{k=0}^{M} b_k x_{n-k} - \sum_{k=1}^{N} a_k y_{n-k} \right), \qquad n \geqq N \tag{2}$$

in which the y_n are approximations to the infinite precision numbers w_n, and $fl(\Sigma - \Sigma)$ denotes the machine number corresponding to $(\Sigma - \Sigma)$ with the understanding that the floating-point numbers corresponding to the products $b_k x_{n-k}$ and $a_k y_{n-k}$ are to be machine-added in some specified order.

Let

$$D(z) \triangleq 1 + \sum_{k=1}^{N} a_k z^{-k}, \tag{3}$$

---

* That is, conventions for which the first two equations of Section III are satisfied.

† It is certainly true that *preliminary* design considerations may lead to coefficients that are not machine numbers, and one may then be interested also in the overall effect of approximating the coefficients by machine numbers. That problem also can be treated with the approach used here.

let

$$\langle q \rangle_K \triangleq \left(\frac{1}{K+1} \sum_{k=0}^{K} | q_k |^2 \right)^{\frac{1}{2}}$$

for every sequence $\{q_k\}$ and all $K \geqq 0$, and let e_n denote the nth error $(y_n - w_n)$ for $n \geqq 0$.

Our first result (all proofs are given in Section III) is as follows. If $D(z) \neq 0$ for $| z | \geqq 1$ [i.e., if the discrete filter (1) is stable], then

$$\begin{aligned} \langle e \rangle_K \leqq & \max_{0 \leqq \omega \leqq 2\pi} | D(e^{i\omega})^{-1} | \left(\frac{1}{K+1} \sum_{n=0}^{N-1} | \eta_n |^2 \right)^{\frac{1}{2}} \\ & + 2^{-t} \left(\sum_{k=0}^{M} | b_k | \beta_k \right) \max_{0 \leqq \omega \leqq 2\pi} | D(e^{i\omega})^{-1} | \left(\frac{1}{K+1} \sum_{n=N-M}^{K} | x_n |^2 \right)^{\frac{1}{2}} \\ & + 2^{-t} \left(\sum_{k=1}^{N} | a_k | \alpha_k \right) \max_{0 \leqq \omega \leqq 2\pi} | D(e^{i\omega})^{-1} | \langle y \rangle_K \end{aligned} \tag{4}$$

for all $K \geqq N$, in which, with $y_n = w_n = 0$ for $n < 0$,

$$\eta_n = \sum_{k=0}^{N} a_k (y_{n-k} - w_{n-k}) \qquad n = 0, 1, 2, \cdots, (N-1)$$

and the α_k and β_k are easily evaluated nonnegative numbers which depend on the order in which the products in (2) are summed.

Since the first term on the right side of (4), which arises as a result of the possibility of differences in the starting values, approaches zero as $K \to \infty$, we see that, after a reasonable number of evaluations of the successive y_n, $\langle e \rangle_K$ is bounded essentially by a constant times the root-mean-squared value of the input sequence, plus another constant times the root-mean-squared value of the output sequence.

In order to determine the α_k and β_k, we draw a signal-flow graph that indicates the ordering of the operations that would be used to compute

$$fl\left(\sum_{k=0}^{M} b_k x_{n-k} - \sum_{k=1}^{N} a_k y_{n-k} \right) \tag{5}$$

if x_n and y_n were unity for all n. This graph is to contain an input node with input b_k' for each $b_k \neq 0$, an input node with input a_k' for each $a_k \neq 0$, no other input nodes, and a single output node θ which is associated with

$$\sum_{k=0}^{M} b_k' - \sum_{k=1}^{N} a_k' .$$

All other nodes represent an addition or subtraction of two signals to produce a third signal. Exactly one branch is connected to each of the input nodes and to the output node. We assign the value ρ to all of the branch transmissions with the exception of those branches, if any, which terminate on an input b_k' or a_k' for which b_k or a_k, respectively, is equal to unity. These branches are assigned unity transmission. Then, by inspection, we evaluate the signal at θ, which must clearly be of the form

$$\sum_{k=0}^{M} b_k' \rho^{\varphi_\beta(k)} + \sum_{k=1}^{N} a_k' \rho^{\varphi_\alpha(k)} \tag{6}$$

in which $\varphi_\beta(k)$ and $\varphi_\alpha(k)$ are positive-integer valued functions. In terms of these functions*

$$\beta_k = (1.06)\varphi_\beta(k)$$

$$\alpha_k = (1.06)\varphi_\alpha(k).$$

For example, if the right side of (2) is computed as the floating-point difference of the machine sums

$$fl(b_0x_n + b_1x_{n-1} + \cdots + b_Mx_{n-M})$$

and

$$fl(a_1y_{n-1} + a_2y_{n-2} + \cdots + a_Ny_{n-N}),$$

each obtained by performing machine summations in the order indicated (from left to right), if all of the b_k and a_k are nonzero and not unity, and if $M \geqq 1$ and $N \geqq 2$, then the relevant flow graph is shown in Fig. 1, from which it follows that

$$\beta_0 = (1.06)(M + 2)$$

$$\beta_1 = (1.06)(M + 2)$$

$$\beta_k = (1.06)(3 + M - k); \qquad k = 2, 3, \cdots, M$$

$$\alpha_1 = (1.06)(N + 1)$$

$$\alpha_2 = (1.06)(N + 1)$$

$$\alpha_k = (1.06)(3 + N - k); \qquad k = 3, 4, \cdots, N.$$

The bound (4), although revealing, requires a knowledge of both $\langle x \rangle_K$ and $\langle y \rangle_K$ and is, therefore, not as explicit as we would like.

* We are assuming here only that $\max_k |\varphi_\beta(k)| 2^{-t} < 0.1$ and $\max_k |\varphi_\alpha(k)| 2^{-t} < 0.1$. Also if $\varphi_\beta(k) = 1$, then we can take $\beta_k = 1$, and similarly for $\varphi_\alpha(k)$.

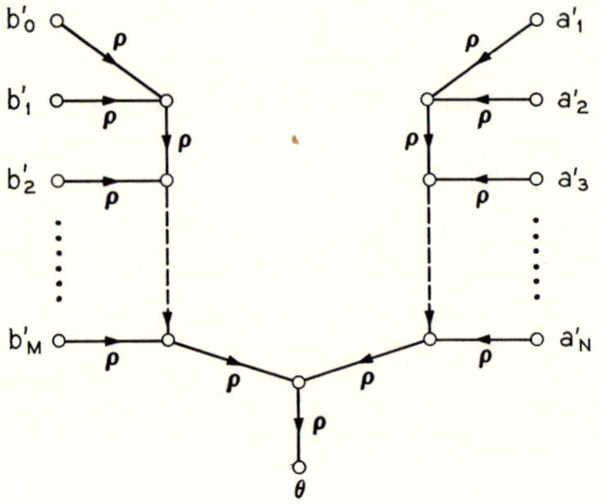

Fig. 1 — Flow graph for the example.

For the important case in which $b_0 \neq 0$ and $N(z) \triangleq \sum_{k=0}^{M} b_k z^{-k} \neq 0$ for $|z| \geqq 1$ (i.e., for the minimum-phase filter case) we prove that if the filter (1) is stable and if

$$\min_{0 \leqq \omega \leqq 2\pi} | N(e^{i\omega}) | > 2^{-t} \sum_{k=0}^{M} | b_k | \beta_k , \tag{7}$$

then there exists a constant c, independent of K, and a function $f(K)$ with the property that $f(K) \to 0$ as $K \to \infty$ such that

$$\langle e \rangle_K \leqq c \langle y \rangle_K + f(K) \tag{8}$$

for all $K \geqq N$. Moreover, it is proved that

$$c \leqq 2^{-t} \max_{0 \leqq \omega \leqq 2\pi} | D(e^{i\omega})^{-1} | \left\{ \sum_{k=1}^{N} | a_k | \alpha_k + \sum_{k=0}^{M} | b_k | \beta_k \right.$$
$$\left. \cdot \frac{\max_{\omega} | D(e^{i\omega})/N(e^{i\omega}) | + \max_{\omega} | N(e^{i\omega})^{-1} | 2^{-t} \sum_{k=1}^{N} | a_k | \alpha_k}{1 - 2^{-t} \sum_{k=0}^{M} | b_k | \beta_k \max_{\omega} | N(e^{i\omega})^{-1} |} \right\} \tag{9}$$

and

$$f(K) \leqq \max_{\omega} | D(e^{i\omega})^{-1} | \left(\frac{1}{K+1} \sum_{n=0}^{N-1} | \eta_n |^2 \right)^{\frac{1}{2}} + \max_{\omega} | D(e^{i\omega})^{-1} | 2^{-t}$$

$$\cdot\left(\sum_{k=0}^{M} | b_k | \beta_k\right) \frac{\max | N(e^{i\omega})^{-1} | \left(\frac{1}{K+1} \sum_{n=0}^{N-1} | q_n |^2\right)^{\frac{1}{2}}}{1 - 2^{-t} \sum_{k=0}^{M} | b_k | \beta_k \max_{\omega} | N(e^{i\omega})^{-1} |} \tag{10}$$

for all $K \geqq N$, in which, with $a_0 = 1$ and $x_n = y_n = 0$ for $n < 0$,

$$q_n = \sum_{k=0}^{N} a_k y_{n-k} - \sum_{k=0}^{M} b_k x_{n-k}$$

for $n = 0, 1, 2, \cdots, (N - 1)$.

Since $\langle y \rangle_K$ is the root-mean-squared value of the *computed output*, and since $f(K) \to 0$ fairly rapidly as $K \to \infty$, we may interpret the smallest value of c for which (8) is satisfied (for all input sequences) as an "output error-to-signal ratio" of the realized digital filter. Note that the bound (9) on c is not difficult to evaluate.

2.2.1. *Stability in the Presence of Roundoff*

If roundoff effects are ignored, it is well known that the discrete filter is stable in several different senses of the word if $D(z) \neq 0$ for $| z | \geqq 1$. In Section III it is proved that, with roundoff effects taken into account, the digital filter is stable in the sense that there is a constant c_1 and a function $f_1(K)$, with $f_1(K)$ independent of the values of x_n for $n \geqq N$ and $f_1(K) \to 0$ as $K \to \infty$, such that

$$\langle y \rangle_K \leqq c_1 \langle x \rangle_K + f_1(K) \tag{11}$$

for all $K \geqq N$, provided that $D(z) \neq 0$ for $| z | \geqq 1$, and

$$\min_{\omega} | D(e^{i\omega}) | > 2^{-t} \sum_{k=1}^{N} | a_k | \alpha_k . \tag{12}$$

Roughly speaking, inequality (12) is satisfied if the damping of the infinite precision counterpart of the digital filter is sufficiently large relative to the number of bits allotted to the mantissa. Stability in the sense of (11) rules out, for example, the possibility, due to roundoff effects, of a limit-cycle response to a zero input sequence or to an input sequence $\{x_n\}$ that approaches zero as $n \to \infty$.*

* There are simple examples which illustrate that instability may result with $D(z) \neq 0$ for $| z | \geqq 1$ if (12) is not satisfied. For instance, suppose that each machine number is represented in the form $(-m_0 2^0 + m_1 2^{-1} + m_2 2^{-2} + \cdots + m_t 2^{-t}) 2^b$ with the m_j zeros or ones, and $t > 1$. Let

$$w_n = (1 - 2^{-t}) w_{n-1} + (1 - 2^{-t}) 2^{-t} w_{n-2} \text{ for } n \geqq 2, \text{ with } w_0 = w_1 = -1.$$

Then $fl[(1 - 2^{-t}) w_1] = -(1 - 2^{-t})$, $fl[(1 - 2^{-t}) 2^{-t} w_0] = -(1 - 2^{-t}) 2^{-t}$, and $fl[-(1 - 2^{-t}) - (1 - 2^{-t}) 2^{-t}] = -1$, which shows that the computed approximation y_n to w_n satisfies $y_n = -1$ for *all* $n \geqq 0$. This example is a slight modification of one suggested by S. Darlington.

2.3 *A Result Concerning the Overall Effect of Input Quantization Errors*

In many applications the sequence $\{x_n\}$ of (1) is obtained by quantizing an input sequence $\{\bar{x}_n\}$ [i.e., by replacing each $\bar{x}_n$ with the machine number (or one of the possibly two machine numbers) of closest value]. The infinite precision response $\bar{w}_N, \bar{w}_{N+1}, \cdots$ to the sequence $\{\bar{x}_N\}$ satisfies

$$\bar{w}_n = \sum_{k=0}^{M} b_k \bar{x}_{n-k} - \sum_{k=1}^{N} a_k \bar{w}_{n-k}, \qquad n \geqq N \tag{13}$$

with $\bar{w}_0, \bar{w}_1, \cdots, \bar{w}_{N-1}$ some set of starting values. Let $w_N, w_{N+1}, \cdots$ be defined by (1) with $w_n = \bar{w}_n$ for $n = 0, 1, 2, \cdots, (N-1)$. It is clear that $\langle y - \bar{w}\rangle_K$, the root-mean-squared value of the difference of the computed output and the infinite precision response to $\{\bar{x}_n\}$, satisfies

$$\langle y - \bar{w}\rangle_K \leqq \langle y - w\rangle_K + \langle w - \bar{w}\rangle_K . \tag{14}$$

Bounds on the first term on the right side of (14) are given in Section 2.2. In Section III it is proved that if both $N(z)$ and $D(z)$ have no zeros on or outside the unit circle, $b_0 \neq 0$, and

$$\min_{\omega} | N(e^{i\omega}) | > 2^{-t} \sum_{k=0}^{M} | b_k | \beta_k ,$$

then* there is a constant c_2 and a function $f_2(K)$ such that $f_2(K) \to 0$ as $K \to \infty$, and

$$\langle w - \bar{w}\rangle_K \leqq c_2 \langle y\rangle_K + f_2(K) \tag{15}$$

for all $K \geqq N$. It is proved also that

$$c_2 \leqq 2^{-t} \sum_{k=0}^{M} | b_k | \max_{\omega} | D(e^{i\omega})^{-1} |$$

$$\cdot \frac{\max_{\omega} | D(e^{i\omega})/N(e^{i\omega}) | + \max_{\omega} | N(e^{i\omega})^{-1} | \, 2^{-t} \sum_{k=1}^{N} | a_k | \alpha_k}{1 - 2^{-t} \sum_{k=0}^{M} | b_k | \beta_k \max_{\omega} | N(e^{i\omega})^{-1} |} . \tag{16}$$

2.4 *A Realistic Example*

For the ideally stable second-order bandpass filter

$$w_n = x_n - a_1 w_{n-1} - a_2 w_{n-2}, \qquad n \geqq 2$$

---

* It is assumed here that the range of values assigned to the mantissa includes the number zero.

with poles in the z-plane at angles $\approx \pm 45°$ and at distance ≈ 0.001 (but not less than 0.001) from the unit circle, we have $a_1 \approx -1.41$, $a_2 \approx 1$, and $\min_\omega |D(e^{i\omega})^{-1}| \approx (0.00141)^{-1}$. We assume that the operations are performed as indicated in Fig. 2, so that $\beta_0 = 1$, $\alpha_1 = 3(1.06)$, and $\alpha_2 = 3(1.06)$. Assuming that $t = 27$, we find that c our bound on the "asymptotic output error-to-signal ratio," ignoring input quantization effects, is approximately 0.584×10^{-4}. For this problem, our bound on c_2 is approximately 0.18×10^{-4}. Thus, even taking into account input quantization effects, the error-to-signal ratio is not more than 0.764×10^{-4}. Finally, a simple calculation shows that this filter is stable in the presence of roundoff, in the sense of inequality (11).

III. PROOFS

3.1 *Derivation of Inequality (4)*

If a and b are floating-point machine numbers, then the floating-point product and sum $fl(ab)$ and $fl(a+b)$, respectively, satisfy[6]

$$fl(ab) = ab(1+\epsilon)$$

$$fl(a+b) = (a+b)(1+\delta)$$

with $|\epsilon| \leqq 2^{-t}$ and $|\delta| \leqq 2^{-t}$. Thus,

$$fl\left(\sum_{k=0}^{M} b_k x_{n-k} - \sum_{k=1}^{N} a_k y_{n-k}\right)$$

is equal to the value of the output signal θ of the flow graph described in Section II with

(*i*) $b_k' = b_k x_{n-k}$

$a_k' = a_k y_{n-k}$

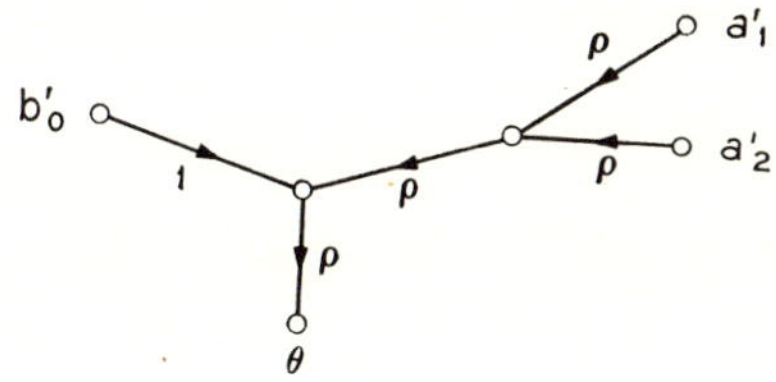

Fig. 2 — Flow graph for the second-order band-pass filter.

and

(*ii*) each of the branch transmissions of the form: $(1 + \epsilon)$ with $|\epsilon| \leqq 2^{-t}$ (recall that in certain special cases ϵ is *taken* to be zero), or $-(1 + \epsilon)$ with $|\epsilon| \leqq 2^{-t}$. Therefore,

$$fl\left(\sum_{k=0}^{M} b_k x_{n-k} - \sum_{k=1}^{N} a_k y_{n-k}\right)$$

is equal to

$$\sum_{k=0}^{M} b_k x_{n-k} q_k - \sum_{k=1}^{N} a_k y_{n-k} r_k$$

in which

$$(1 - 2^{-t})^{\varphi\beta(k)} \leqq q_k \leqq (1 + 2^{-t})^{\varphi\beta(k)} \tag{17}$$

and

$$(1 - 2^{-t})^{\varphi\alpha(k)} \leqq r_k \leqq (1 + 2^{-t})^{\varphi\alpha(k)}. \tag{18}$$

Inequalities (17) and (18) imply[6]

$$1 - (1.06)\varphi_\beta(k)2^{-t} \leqq q_k \leqq 1 + (1.06)\varphi_\beta(k)2^{-t}$$

$$1 - (1.06)\varphi_\alpha(k)2^{-t} \leqq r_k \leqq 1 + (1.06)\varphi_\alpha(k)2^{-t}$$

provided that $2^{-t} \max_k \varphi_\beta(k) < 0.1$ and $2^{-t} \max_k \varphi_\alpha(k) < 0.1$.

Thus, for $n \geqq N$

$$\begin{aligned} y_n &= fl\left(\sum_{k=0}^{M} b_k x_{n-k} - \sum_{k=1}^{N} a_k y_{n-k}\right) \\ &= \sum_{k=0}^{M} b_k x_{n-k} - \sum_{k=1}^{N} a_k y_{n-k} + \eta_n \end{aligned} \tag{19}$$

with

$$|\eta_n| \leqq 2^{-t} \sum_{k=0}^{M} |b_k| \cdot |x_{n-k}| \beta_k + 2^{-t} \sum_{k=1}^{N} |a_k| \cdot |y_{n-k}| \alpha_k \tag{20}$$

and

$$\beta_k = (1.06)\varphi_\beta(k), \qquad \alpha_k = (1.06)\varphi_\alpha(k).$$

Using (1) and (19),

$$\sum_{k=0}^{N} a_k e_{n-k} = \eta_n, \qquad n \geqq 0$$

in which, with $y_n = w_n = 0$ for $n < 0$,

$$\eta_n = \sum_{k=0}^{N} a_k(y_{n-k} - w_{n-k})$$

for $n = 0, 1, \cdots, (N-1)$. By Propositions 1 and 2 (see Sections 3.5 and 3.6)

$$\langle e \rangle_K \leqq \max_{0 \leqq \omega \leqq 2\pi} | D(e^{i\omega})^{-1} | \langle \eta \rangle_K , \qquad K \geqq 0. \tag{21}$$

By Proposition 3 (Section 3.7), inequality (20), and Minkowski's inequality

$$\langle \eta \rangle_K \leqq \left(\frac{1}{K+1} \sum_{n=0}^{N-1} | \eta_n |^2 \right)^{\frac{1}{2}} + 2^{-t} \sum_{k=0}^{M} | b_k | \beta_k \left(\frac{1}{K+1} \sum_{n=N-M}^{K} | x_n |^2 \right)^{\frac{1}{2}} + 2^{-t} \sum_{k=1}^{N} | a_k | \alpha_k \langle y \rangle_K \tag{22}$$

for all $K \geqq N$. This proves inequality (4).

3.2 *Inequality (8)*

Here we assume that both $D(z)$ and $N(z)$ are zero free for $| z | \geqq 1$, that $b_0 \neq 0$, and that

$$\min_{0 \leqq \omega \leqq 2\pi} | N(e^{i\omega}) | > 2^{-t} \sum_{k=0}^{M} | b_k | \beta_k . \tag{23}$$

From (19), we have, with $a_0 \triangleq 1$,

$$\sum_{k=0}^{N} a_k y_{n-k} = \sum_{k=0}^{M} b_k x_{n-k} + q_n , \qquad n \geqq 0, \tag{24}$$

where

$$\begin{aligned} q_n &= \eta_n , \qquad n \geqq N \\ &= \sum_{k=0}^{N} a_k y_{n-k} - \sum_{k=0}^{M} b_k x_{n-k} , \qquad n = 0, 1, 2, \cdots, (N-1) \end{aligned}$$

with $x_n = y_n = 0$ for $n < 0$. Therefore, by Propositions 1 and 2,

$$\langle x \rangle_K \leqq \max_{\omega} | D(e^{i\omega})/N(e^{i\omega}) | \langle y \rangle_K + \max_{\omega} | N(e^{i\omega})^{-1} | \langle q \rangle_K , \quad K \geqq 0. \tag{25}$$

Using Proposition 3, Minkowski's inequality, and (20),

$$\langle q \rangle_K \leqq \left(\frac{1}{K+1} \sum_{n=0}^{N-1} | q_n |^2 \right)^{\frac{1}{2}} + 2^{-t} \sum_{k=0}^{M} | b_k | \beta_k \langle x \rangle_K + 2^{-t} \sum_{k=1}^{N} | a_k | \alpha_k \langle y \rangle_K , \qquad K \geqq N. \tag{26}$$

Therefore,

$$\langle x\rangle_K \leqq \frac{\max\limits_{\omega} | D(e^{i\omega})/N(e^{i\omega}) | + \max\limits_{\omega} | N(e^{i\omega})^{-1} | 2^{-t} \sum\limits_{k=1}^{N} | a_k | \alpha_k}{1 - 2^{-t} \sum\limits_{k=0}^{M} | b_k | \beta_k \max\limits_{\omega} | N(e^{i\omega})^{-1} |} \langle y\rangle_K$$

$$+ \frac{\max\limits_{\omega} | N(e^{i\omega})^{-1} | \left(\frac{1}{K+1} \sum\limits_{n=0}^{N-1} | q_n |^2\right)^{\frac{1}{2}}}{1 - 2^{-t} \sum\limits_{k=0}^{M} | b_k | \beta_k \max\limits_{\omega} | N(e^{i\omega})^{-1} |} \tag{27}$$

for all $K \geqq N$, which together with (21) and (22) yields

$$\langle e\rangle_K \leqq 2^{-t} \max_{0 \leqq \omega \leqq 2\pi} | D(e^{i\omega})^{-1} | \left\{ \sum_{k=1}^{N} | a_k | \alpha_k + \sum_{k=0}^{M} | b_k | \beta_k \right.$$

$$\left. \cdot \frac{\max\limits_{\omega} | D(e^{i\omega})/N(e^{i\omega}) | + \max\limits_{\omega} | N(e^{i\omega})^{-1} | 2^{-t} \sum\limits_{k=1}^{N} | a_k | \alpha_k}{1 - 2^{-t} \sum\limits_{k=0}^{M} | b_k | \beta_k \max\limits_{\omega} | N(e^{i\omega})^{-1} |} \right\} \langle y\rangle_K$$

$$+ \max_{\omega} | D(e^{i\omega})^{-1} | \left(\frac{1}{K+1} \sum_{n=0}^{N-1} | \eta_n |^2\right)^{\frac{1}{2}} + \max_{\omega} | D(e^{i\omega})^{-1} | 2^{-t}$$

$$\cdot \sum_{k=0}^{M} | b_k | \beta_k \frac{\max\limits_{\omega} | N(e^{i\omega})^{-1} | \left(\frac{1}{K+1} \sum\limits_{n=0}^{N-1} | q_n |^2\right)^{\frac{1}{2}}}{1 - 2^{-t} \sum\limits_{k=0}^{M} | b_k | \beta_k \max\limits_{\omega} | N(e^{i\omega})^{-1} |}. \tag{28}$$

This proves that there exists a constant c and a function $f(K)$ with the property that $f(K) \to 0$ as $K \to \infty$ such that (8) is satisfied for all $K \geqq N$, and of course it also proves that c and $f(K)$ are bounded as stated in Section 2.2.

3.3 *Proof of (11) Under the Conditions Stated*

From (24) and Propositions 1 and 2,

$$\langle y\rangle_K \leqq \max_{\omega} | N(e^{i\omega})/D(e^{i\omega}) | \langle x\rangle_K + \max_{\omega} | D(e^{i\omega})^{-1} | \langle q\rangle_K ,$$

and using (26)

$$\langle y\rangle_K \leqq \frac{\max_\omega | N(e^{i\omega})/D(e^{i\omega}) | + \max_\omega | D(e^{i\omega})^{-1} | 2^{-t} \sum_{k=0}^{M} | b_k | \beta_k}{1 - 2^{-t} \sum_{k=1}^{N} | a_k | \alpha_k \max_\omega | D(e^{i\omega})^{-1} |} \langle x\rangle_K$$

$$+ \frac{\max_\omega | D(e^{i\omega})^{-1} | \left(\frac{1}{K+1} \sum_{n=0}^{N-1} | q_n |^2\right)^{\frac{1}{2}}}{1 - 2^{-t} \sum_{k=1}^{N} | a_k | \alpha_k \max_\omega | D(e^{i\omega})^{-1} |},$$

which completes the proof.

3.4 *Derivation of Inequalities (15) and (16)*

We have, from (1) and (13),

$$\sum_{k=0}^{N} a_k(w_{n-k} - \bar{w}_{n-k}) = \xi_n , \qquad n \geqq 0 \tag{29}$$

in which $a_0 \triangleq 1$,

$$\xi_n = \sum_{k=0}^{M} b_k(x_{n-k} - \bar{x}_{n-k}), \qquad n \geqq N$$

and

$$\xi_n = 0, \qquad n = 0, \qquad 2, \cdots , (N - 1).$$

Since $\bar{x}_n = \operatorname{sgn}(\bar{x}_n)h2^b$ for some integer b and some $h \in [\frac{1}{2}, 1]$ (assuming that $\bar{x}_n \neq 0$), the magnitude of the error in approximating $\bar{x}_n$ by the closest machine number $x_n = \operatorname{sgn}(\bar{x}_n)a2^b$ is at most $\frac{1}{2}2^{-t}2^b = \frac{1}{2}2^{-t}a^{-1} | x_n | \leqq 2^{-t} | x_n |$. Therefore, for $n \geqq N$

$$| \xi_n | \leqq 2^{-t} \sum_{k=0}^{M} | b_k |\cdot| x_{n-k} |,$$

and by Propositions 1, 2, and 3

$$\langle w - \bar{w}\rangle_K \leqq \max_\omega | D(e^{i\omega})^{-1} | 2^{-t} \cdot \sum_{k=0}^{M} | b_k | \left(\frac{1}{K+1} \sum_{n=N-M}^{K} | x_n |^2\right)^{\frac{1}{2}}, \qquad K \geqq N. \tag{30}$$

From (30) and (27)

$$\langle w - \bar{w}\rangle_K \leqq 2^{-t} \sum_{k=0}^{M} | b_k | \max_\omega | D(e^{i\omega})^{-1} |$$

$$\cdot \frac{\max\limits_{\omega} | D(e^{i\omega})/N(e^{i\omega}) | + \max\limits_{\omega} | N(e^{i\omega})^{-1} | 2^{-t} \sum\limits_{k=1}^{N} | a_k | \alpha_k}{1 - 2^{-t} \sum\limits_{k=0}^{M} | b_k | \beta_k \max\limits_{\omega} | N(e^{i\omega})^{-1} |} \langle y \rangle_K$$

$$+ \frac{2^{-t} \sum\limits_{k=0}^{M} | b_k | \max\limits_{\omega} | D(e^{i\omega})^{-1} | \max\limits_{\omega} | N(e^{i\omega})^{-1} | \left(\frac{1}{K+1} \sum\limits_{n=0}^{N-1} | q_n |^2 \right)^{\frac{1}{2}}}{1 - 2^{-t} \sum\limits_{k=0}^{M} | b_k | \beta_k \max\limits_{\omega} | N(e^{i\omega})^{-1} |}$$

for all $K \geqq N$, provided that $N(z) \neq 0$ for $| z | \geqq 1$, $b_0 \neq 0$, and

$$\min_{\omega} | N(e^{i\omega}) | > 2^{-t} \sum_{k=0}^{M} | b_k | \beta_k .$$

This completes the derivation.

3.5 *Proposition 1:*

If

$$\sum_{l=0}^{L} c_l r_{n-l} = \sum_{l=0}^{L'} d_l s_{n-l} + f_n , \qquad n \geqq 0$$

with: $r_n = s_n = 0$ for $n < 0$, $c_0 \neq 0$, and $\sum_{l=0}^{L} c_l z^{-l} \neq 0$ for $| z | \geqq 1$, then

$$r_n = \sum_{k=0}^{n} u_{n-k} s_k + \sum_{k=0}^{n} v_{n-k} f_k , \qquad n \geqq 0$$

in which

$$\sum_{n=0}^{\infty} | u_n | < \infty, \qquad \sum_{n=0}^{\infty} | v_n | < \infty,$$

$$\sum_{n=0}^{\infty} u_n e^{-in\omega} = \sum_{l=0}^{L'} d_l e^{-il\omega} \Big/ \sum_{l=0}^{L} c_l e^{-il\omega},$$

and

$$\sum_{n=0}^{\infty} v_n e^{-in\omega} = 1 \Big/ \sum_{l=0}^{L} c_l e^{-il\omega}$$

for $0 \leqq \omega \leqq 2\pi$.

*Proof:**

* The proof of this result, although rather trivial, is included because the writer knows of no reference where it is proved without the assumption that the sequences $\{s_n\}$ and $\{f_n\}$ are z-transformable.

Let $M > 0$, and let

$$\begin{aligned} \hat{s}_n &= s_n \quad \text{for} \quad n \leqq M \\ &= 0 \quad \text{for} \quad n > M \\ \hat{f}_n &= f_n \quad \text{for} \quad n \leqq M \\ &= 0 \quad \text{for} \quad n > M. \end{aligned}$$

Then $r_n = \hat{r}_n$ for $n \leqq M$, with

$$\sum_{l=0}^{L} c_l \hat{r}_{n-l} = \sum_{l=0}^{L'} d_l \hat{s}_{n-l} + \hat{f}_n , \qquad n \geqq 0$$

and with $\{\hat{r}_n\}$, $\{\hat{s}_n\}$, and $\{\hat{f}_n\}$ z-transformable. Therefore, we have

$$\hat{R}(z) = \left(\sum_{l=0}^{L'} d_l z^{-l}\right)\left(\sum_{l=0}^{L} c_l z^{-l}\right)^{-1} \hat{S}(z) + \left(\sum_{l=0}^{L} c_l z^{-l}\right)^{-1} \hat{F}(z)$$

in which

$$\begin{aligned} \hat{R}(z) &= \sum_{n=0}^{\infty} \hat{r}_n z^{-n} \\ \hat{S}(z) &= \sum_{n=0}^{M} s_n z^{-n} \\ \hat{F}(z) &= \sum_{n=0}^{M} f_n z^{-n}. \end{aligned}$$

Thus,

$$\hat{r}_n = \sum_{k=0}^{n} u_{n-k} \hat{s}_k + \sum_{k=0}^{n} v_{n-k} \hat{f}_k , \qquad n \geqq 0$$

and hence

$$r_n = \sum_{k=0}^{n} u_{n-k} s_k + \sum_{k=0}^{n} v_{n-k} f_k , \tag{31}$$

for $n = 0, 1, \cdots, M$. However, since M is arbitrary, (31) is satisfied for all $n \geqq 0$. This proves Proposition 1.

3.6 *Proposition 2:*

If

$$f_n = \sum_{l=0}^{n} c_{n-l} g_l , \qquad n \geqq 0$$

with $\sum_{l=0}^{\infty} | c_l | < \infty$, then

$$\langle f \rangle_K \leqq \max_{0 \leqq \omega \leqq 2\pi} \left| \sum_{l=0}^{\infty} c_l e^{-il\omega} \right| \langle g \rangle_K$$

for all $K \geqq 0$.

Proof:

$$\begin{aligned} \sum_{n=0}^{K} | f_n |^2 &= \frac{1}{2\pi} \int_0^{2\pi} \left| \sum_{n=0}^{K} e^{-in\omega} \sum_{l=0}^{n} c_{n-l} g_l \right|^2 d\omega \\ &= \frac{1}{2\pi} \int_0^{2\pi} \left| \sum_{n=0}^{K} e^{-in\omega} \sum_{l=0}^{n} c_{n-l} \hat{g}_l \right|^2 d\omega \end{aligned}$$

in which

$$\begin{aligned} \hat{g}_l &= g_l, \qquad l = 0, 1, \cdots, K \\ &= 0, \qquad l > K. \end{aligned}$$

Thus,

$$\begin{aligned} \sum_{n=0}^{K} | f_n |^2 &\leqq \frac{1}{2\pi} \int_0^{2\pi} \left| \sum_{n=0}^{\infty} e^{-in\omega} \sum_{l=0}^{n} c_{n-l} \hat{g}_l \right|^2 d\omega \\ &\leqq \frac{1}{2\pi} \int_0^{2\pi} \left| \sum_{l=0}^{\infty} c_l e^{-il\omega} \sum_{n=0}^{\infty} e^{-in\omega} \hat{g}_n \right|^2 d\omega \\ &\leqq \max_{\omega} \left| \sum_{l=0}^{\infty} c_l e^{-il\omega} \right|^2 \frac{1}{2\pi} \int_0^{2\pi} \left| \sum_{n=0}^{\infty} e^{-in\omega} \hat{g}_n \right|^2 d\omega \\ &\leqq \max_{\omega} \left| \sum_{l=0}^{\infty} c_l e^{-il\omega} \right|^2 \sum_{n=0}^{K} | g_n |^2, \end{aligned}$$

which proves Proposition 2.

3.7 *Proposition 3:*

If

$$| f_n | \leqq \sum_{l=0}^{L} | g_l | \cdot | h_{n-l} |, \qquad n \geqq N$$

with $L \leqq N$, then

$$\left(\sum_{n=N}^{K} | f_n |^2 \right)^{\frac{1}{2}} \leqq \left(\sum_{l=0}^{L} | g_l | \right) \left(\sum_{n=N-L}^{K} | h_n |^2 \right)^{\frac{1}{2}}$$

for all $K \geqq N$.

Proof:

$$\sum_{n=N}^{K} | f_n |^2 \leqq \sum_{n=N}^{K} \left| \sum_{l=0}^{L} | g_l | \cdot | h_{n-l} | \right|^2$$

$$\leqq \sum_{n=N}^{K} \left| \sum_{l=0}^{L} | g_l |^{\frac{1}{2}} \, | g_l |^{\frac{1}{2}} \, | \hat{h}_{n-l} | \right|^2$$

in which

$$\begin{aligned} \hat{h}_n &= h_n \qquad n = 0, 1, 2, \cdots, K \\ &= 0 \qquad n > K. \end{aligned}$$

Therefore, by the Schwarz inequality,

$$\sum_{n=N}^{K} | f_n |^2 \leqq \sum_{n=N}^{K} \sum_{l=0}^{L} | g_l | \sum_{l=0}^{L} | g_l | \cdot | \hat{h}_{n-l} |^2$$

$$\leqq \sum_{l=0}^{L} | g_l | \sum_{l=0}^{L} \left(| g_l | \sum_{n=N}^{K} | \hat{h}_{n-l} |^2 \right)$$

$$\leqq \sum_{l=0}^{L} | g_l | \sum_{l=0}^{L} \left(| g_l | \sum_{m=N-l}^{K-l} | \hat{h}_m |^2 \right)$$

$$\leqq \left(\sum_{l=0}^{L} | g_l | \right)^2 \sum_{m=N-L}^{K} | h_m |^2.$$

This completes the proof.

IV. ACKNOWLEDGMENT

The writer is indebted to his colleague J. F. Kaiser for emphasizing the need for analytical results relating to problems of the type discussed here.

REFERENCES

1. Blackman, R. B. and Tukey, J. W., *The Measurement of Power Spectra from the Point of View of Communication Engineering,* Dover Press, 1959.
2. Blackman, R. B., *Linear Data-Smoothing and Prediction in Theory and Practice,* Addison-Wesley, Reading, Massachusetts, 1965.
3. Kaiser, J. F., Digital Filters, Chapter 7 of *System Analysis by Digital Computer,* edited by F. F. Kuo and J. F. Kaiser, John Wiley & Sons, Inc., New York, 1966.
4. Bennett, W. R., Spectra of Quantized Signals, B.S.T.J., *27,* July, 1948, pp. 446–472.
5. Knowles, J. B. and Edwards, R., Effects of a Finite-Word-Length Computer in a Sampled-Data Feedback System, Proc. IEE (London), *112,* June, 1965, pp. 1197–1207.
6. Wilkinson, J. H., *Rounding Errors in Algebraic Processes,* Prentice-Hall, Englewood Cliffs, New Jersey, 1963.

17

Reprinted from *Proc. IEEE,* **57**(10), 1735–1747 (1969)

Error Analysis of Digital Filters Realized with Floating-Point Arithmetic

BEDE LIU, MEMBER, IEEE, AND TOYOHISA KANEKO, STUDENT MEMBER, IEEE

Abstract—This paper calculates the error at the output of a digital filter using floating-point arithmetic operations due to roundoff accumulation and input quantization. Expressions are derived for the mean square error for each of the three canonical forms of realization: direct, cascade, and parallel. Upper bounds that are useful for a special class of filters are given. The theoretical result is found to be in good agreement with the experimental result.

I. Introduction

SIGNAL processing by digital means has become increasingly attractive with the rapid technological advancement in digital devices, circuits, and systems. The availability of both large-scale general purpose computers and special purpose hardware has made real-time digital filtering both practical and economical.

A digital filter transforms a sequence of numbers, called the input, into another sequence, called the output, by means of a computational algorithm. Digital filters may be designed to achieve the characteristics of conventional analog filters for spectral shaping such as low pass, high pass, etc. [1]–[3]. In addition to the more obvious applications to the simulation of dynamic systems on the digital computer, digital filters have been employed in a number of diverse fields such as power spectra measurement [4], processing of missile data [5], speech processing [6], [7], seismic signal processing [8], touch-tone receiver [9], electrocardiogram analysis [10], automatic frequency control [11], and radar MTI [12].

The two most outstanding advantages of digital filters are their flexibility and their reliability, due to the fact that 1) their characteristics can be changed simply by reading in a new set of coefficients and that 2) they can be constructed with two-state switching elements. In addition, the use of digital filters avoids a number of problems arising in analog filter design such as element tolerance, large physical size of components for low frequency operation, and difficulty in realizing delay.

Despite the advantages offered by digital networks, there is an inherent accuracy problem associated with digital filters, since each number is represented by a finite number of bits and the arithmetic operations must be carried out with an accuracy limited by this finite word length.

The class of digital filters considered in this paper is that of linear constant coefficient filter specified by the difference equation

$$w_n = \sum_{i=0}^{M} b_i x_{n-i} - \sum_{i=1}^{L} a_i w_{n-i} \tag{1}$$

where $\{x_n\}$ is the input sequence and $\{w_n\}$ is the output sequence. In terms of Z transforms (1) can be written as

$$W(z) = H(z)X(z) \tag{2}$$

where $X(z)$ and $W(z)$ are the Z transforms of the input and output sequences respectively, and $H(z)$ is the so-called pulse transfer function defined as

$$H(z) = \frac{\sum_{i=0}^{M} b_i z^{-i}}{1 + \sum_{i=1}^{L} a_i z^{-i}}. \tag{3}$$

There are three canonical forms of realizing a digital filter. If the output sequence is calculated by using (1), the digital filter is said to be realized in the direct form. If the order of the filter L is higher than 2, $H(z)$ may be written as a sum of partial fractions as

$$H(z) = \sum_{i=1}^{K} H_i(z),$$

and the entire filter can be visualized as the parallel connection of simpler filters $H_i(z)$ of lower order. In this case, the filter is said to be realized in the parallel form. Similarly, if $H(z)$ is written as a product of factors,

$$H(z) = \prod_{i=1}^{K} \bar{H}_i(z),$$

the filter may be visualized as a cascade of lower order filters. The original filter is then said to be realized in the cascade form.

There are three common sources of errors associated with the filter of (1), namely:

1) input quantization—caused by the quantization of the input signal into a set of discrete levels,
2) coefficient truncation—caused by the fact that the coefficients a_k and b_k are each realized with a finite number of bits,
3) roundoff accumulation—caused by the accumulation of errors committed at each arithmetic operation because these operations are carried out with only finite bit accuracy.

Manuscript received November 4, 1968; revised June 6, 1969. This research is supported by the Air Force Office of Scientific Research under AFOSR Grant 1333-67 and by the National Science Foundation under Grant GK-1439.

The authors are with the Department of Electrical Engineering, Princeton University, Princeton, N. J. 08540.

Therefore, for the digital filter of (1) the actual computed output reference is in general different from $\{w_n\}$. We shall denote the actual output by $\{y_n\}$ and call $\{w_n\}$ the ideal output sequence.

For the filter of (1) realized with fixed-point arithmetic the above error problems have been studied extensively [13], [14]. Recently, Sandberg [15] studied the error problem for floating-point arithmetic and derived an absolute upper bound of the error accumulation due to roundoff.

In this paper we present a general approach to the error analysis problem of digital filters using floating-point arithmetic. The basic assumptions are stated in Section II. In Section III, explicit expressions for the mean square error due to roundoff accumulation are given for the three canonical filter forms. The derivations are presented in Section IV. The error depends on the ideal filter characteristics, the input signal spectrum, and the roundoff algorithm of the arithmetic operation. The effect of input quantization is analyzed in Section V. An upper bound on the error-to-signal ratio at the output of the filter is derived in Section VI. In Section VII two digital filters, one of second order and one of eighth order, are simulated on the computer and good agreement between theoretical and experimental results are obtained.

II. BASIC ASSUMPTIONS

Throughout this paper we shall be concerned with binary machines. Thus, a floating-point number is stored in the computer in the form of $(\text{sign}) \cdot \mu \cdot 2^\nu$, where μ and ν each have a fixed number of bits. The number ν, called the exponent, is an integer, and the number μ, called the mantissa or the fractional part, is between $\frac{1}{2}$ and 1. Usually zero is also permitted for μ. It is assumed that enough bits are allowed for the exponent so that no computed number will lie outside the permissable range. The notation $fl(\cdot)$ [16] is used to denote the machine number resulting from performing the arithmetic operation specified by the parenthesis in some designated order. For example, $fl(x+y)$ is the calculated sum of x and y; $fl(ax+by)$ is the calculated sum of two terms, one is the calculated product of a and x, the other is the calculated product of b and y.

When two floating-point machine numbers x and y each with a t-bit mantissa are multiplied, their exact product in floating point representation has in general a fractional part of $2t$ or $(2t-1)$ bits. To represent the product in the computer, only the most significant t bits of the mantissa can be retained. This can usually be achieved either by truncation or by rounding. In truncation, those bits beyond the most significant t bits are simply dropped. In rounding a 1 or a 0 is added to the tth bit in accordance with whether the $(t+1)$th bit is 1 or 0. In practice truncation is encountered more often than rounding. Similarly, truncation or rounding is needed in general for the addition of two floating-point numbers.

It is known [16] that for a machine with double precision accumulator,

$$fl(x+y) = (x+y)(1+\varepsilon) \qquad (4)$$

and

$$fl(x \cdot y) = (x \cdot y)(1+\delta) \qquad (5)$$

with $-2^{-t} \leq \varepsilon, \delta \leq 2^{-t}$ for rounding and $-2^{-t+1} \leq \varepsilon, \delta \leq 0$ for truncation, where t is the number of bits of the mantissa.

Since the errors ε and δ arise from roundoff, we will assume that they are random variables and are independent of x and y. Furthermore, we shall assume that the error committed at each arithmetic step is independent from those at other arithmetic steps. If rounding is used, the error is assumed to be uniformly distributed in $(-2^{-t}, 2^{-t})$ and if truncation is used, the error is assumed to be uniformly distributed in $(-2^{-t+1}, 0)$. These assumptions are discussed further in Appendix I. As will be seen, they lead to theoretical results which are in good agreement with the experimental result obtained by actual simulation.

III. SUMMARY OF RESULTS

In this section the principal results for roundoff accumulation are summarized. Their derivation is given in Section IV; the effect of input quantization is treated in Section V.

The quantities a_n, b_n, and x_n in (1) are machine numbers. The number of bits allowed to the mantissa is t. We assume that the input $\{x_n\}$ is zero mean and wide-sense stationary with autocorrelation function $R_{xx}(n)$ and power spectral density $\Phi_{xx}(z)$.

The error at the output at the nth sample is the difference between the actual output and the ideal output

$$e_n = y_n - w_n. \qquad (6)$$

It is shown in Section IV that the sequence $\{e_n\}$ is wide-sense stationary with zero mean for the three canonical forms of filter realization: direct, parallel, and cascade. For each form, the error power spectral density $\Phi_{ee}^r(z)$ is derived for rounding arithmetic and the error power spectral density $\Phi_{ee}^t(z)$ is derived for truncation arithmetic. The mean square value of e_n is related to $\Phi_{ee}(z)$ by

$$E\{e_n^2\} = \frac{1}{2\pi j} \oint \Phi_{ee}(z) \frac{dz}{z} \qquad (7)$$

where the contour integration is around $|z|=1$, taken counter clockwise.

A. Direct Form

1) *Rounding:* The power spectral density of the error sequence $\{e_n\}$ for this case is given by[1]

$$\Phi_{ee}^r(z) = \frac{\sigma_u^2}{|D(z)|^2} \qquad (8)$$

where $D(z)$ is the denominator of the pulse transfer function $H(z)$ and σ_u^2 is given by

---

[1] Equation (8) holds only for $|z|=1$, and $|D(z)|^2$ should be correctly written as $D(z)D(1/z)$. This oversimplification of notation also appears in a number of other equations in this paper. One may of course replace $|H_i(z)|^2$ by $H_i(z)H_i(1/z)$, $|N(z)|^2$ by $N(z)N(1/z)$ etc. in these equations. But some of these equations would become unduly long. Since in most cases one is interested in $|z|=1$ only, this oversimplified notation is adopted.

$$\sigma_u^2 = \left(\frac{2^{-2t}}{3}\right)\frac{1}{2\pi j}\oint\left[|B(z)|^2 + \left|A(z)\frac{N(z)}{D(z)}\right|^2 - 2C(z)\frac{N(z)}{D(z)}\right]\Phi_{xx}(z)\frac{dz}{z}. \tag{9}$$

In (9) t is the number of bits of the mantissa, $N(z)$ is the numerator of the pulse transfer function $H(z)$. The functions $|A(z)|^2$, $|B(z)|^2$ and $C(z)$ are defined as

$$\begin{aligned}|A(z)|^2 &= \sum_{k=1}^{L}\sum_{i=1}^{L} a_k a_i A_{k,i} z^{k-i}\\ |B(z)|^2 &= \sum_{k=0}^{M}\sum_{i=0}^{M} b_k b_i B_{k,i} z^{k-i}\\ C(z) &= \sum_{k=0}^{M}\sum_{i=1}^{L} b_k a_i z^{k-i} = N(1/z)[D(z) - 1].\end{aligned} \tag{10}$$

where[2]

$$A_{k,i} = \begin{cases} L + 2 - \max(k, i) & k \neq i \quad \text{or} \quad k = i = 1\\ L + 3 - k & k = i \neq 1\end{cases}$$

$$B_{k,i} = \begin{cases} M + 2 - \max(k, i) & k \neq i \quad \text{or} \quad k - i - 0\\ M + 3 - k & k = i \neq 0.\end{cases}$$

2) *Truncation:* The power spectral density of the error in this case is

$$\Phi_{ee}^t(z) = \Phi_{ee}^r(z) + \frac{2^{-2t}}{|D(z)|^2}\left|B'(z) - A'(z)\frac{N(z)}{D(z)}\right|^2\Phi_{xx}(z) \tag{11}$$

where

$$\begin{aligned}A'(z) &= \sum_{k=1}^{L}\alpha_k a_k z^{-k}\\ B'(z) &= \sum_{k=0}^{M}\beta_k b_k z^{-k}\\ \alpha_k &= \begin{cases} L + 1, & k = 1\\ L + 3 - k, & k \geq 2\end{cases}\\ \beta_k &= \begin{cases} M + 2, & k = 0\\ M + 3 - k, & k \geq 1\end{cases}\end{aligned} \tag{12}$$

and $\Phi_{ee}^r(z)$ is the spectrum of the error for the rounding case given in (8).

B. Parallel Form

To implement the digital filter in parallel form, $H(z)$ is written as

$$H(z) = \sum_{i=1}^{K} H_i(z) \tag{13}$$

where

$$H_i(z) = \frac{N_i(z)}{D_i(z)} = \frac{f_i + g_i z^{-1}}{1 + c_i z^{-1} + d_i z^{-2}}. \tag{14}$$

[2] We assume that none of the filter coefficients is zero or one, otherwise the value for $A_{k,i}$ or $B_{k,i}$ needs modification. See Section IV.

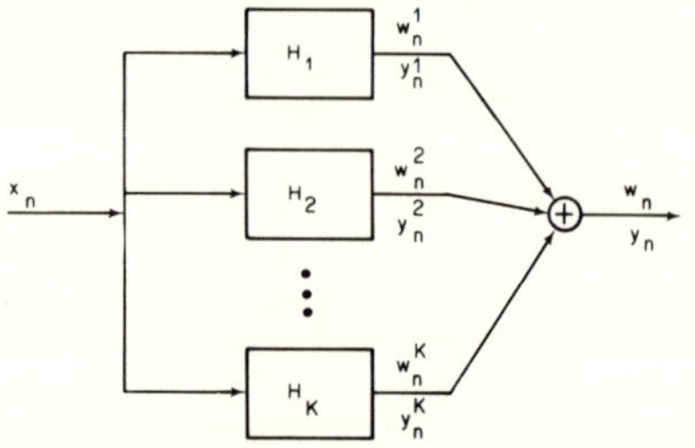

Fig. 1. Implementation of parallel programming.

The above expression includes the case of a real pole or a constant by setting $g_i = d_i = 0$ or $g_i = c_i = d_i = 0$, respectively. The implementation of a digital filter in parallel form is shown in Fig. 1, where K-intermediate outputs $\{y_n^i\}$, $i = 1, 2, \cdots, K$ are first calculated and then summed to form the total output $\{y_n\}$. The error is defined as in (6),

1) *Rounding:* The error spectrum is given by

$$\Phi_{ee}^r(z) = \frac{2^{-2t}}{3}\Phi_{xx}(z)\sum_{i=1}^{K}\sum_{j=1}^{K} F_{ij}\frac{N_i(z)N_j(1/z)}{D_i(z)D_j(1/z)} + \sum_{i=1}^{K}\left(1 + \frac{2^{-2t}}{3}F_{ii}\right)\frac{\sigma_i^2}{|D_i(z)|^2} \tag{15}$$

where

$$F_{ij} = \begin{cases} K - 1 & i = j = 1\\ K + 1 - \max(i, j) & \text{otherwise}\end{cases}$$

$$\begin{aligned}\sigma_i^2 &= \frac{2^{-2t}}{3}\frac{1}{2\pi j}\oint\left[|B_i(z)|^2 + \left|A_i(z)\frac{N_i(z)}{D_i(z)}\right|^2 - 2C_i(z)\frac{N_i(z)}{D_i(z)}\right]\Phi_{xx}(z)\frac{dz}{z}\\ |A_i(z)|^2 &= 3(c_i^2 + d_i^2) + 2c_i d_i(z + z^{-1})\\ |B_i(z)|^2 &= 3(f_i^2 + g^2) + 2f_i g_i(z + z^{-1})\\ C_i(z) &= g_i c_i + (f_i c_i + g_i d_i)z^{-1} + f_i d_i z^{-2}.\end{aligned} \tag{16}$$

2) *Truncation:* The error spectrum for this case is

$$\Phi_{ee}^t(z) = \Phi_{ee}^r(z) + 2^{-2t}\left|\sum_{i=1}^{K}\left\{P_i\frac{N_i(z)}{D_i(z)} + T_i(z)\right\}\right|^2\Phi_{xx}(z) \tag{17}$$

where

$$\begin{aligned}P_i &= \begin{cases} K - 1 & i = 1\\ K - i + 1 & i \geq 2\end{cases}\\ T_i(z) &= [B_i'(z)D_i(z) - A_i'(z)N_i(z)]/|D_i(z)|^2\\ A_i'(z) &= 3(c_i z^{-1} + d_i z^{-2})\\ B_i'(z) &= 3(f_i + g_i z^{-1})\end{aligned} \tag{18}$$

and $\Phi_{ee}^r(z)$ is given by (15).

C. Cascade Form

For the cascade realization form the pulse transfer function is written as a product of quadratic factors.

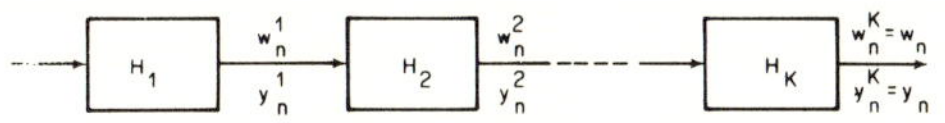

Fig. 2. Implementation of cascade programming.

$$H(z) = c \prod_{i=1}^{K} H_i(z) \tag{19}$$

where c is a constant and[3]

$$H_i(z) = \frac{N_i(z)}{D_i(z)} = \frac{1 + f_i z^{-1} + g_i z^{-2}}{1 + c_i z^{-1} + d_i z^{-2}}. \tag{20}$$

Again the expression for $H_i(z)$ includes the possibility of a real pole or a real zero by simply setting $d_i=0$ or $g_i=0$ respectively. The implementation of a digital filter in cascade form is shown in Fig. 2. From the input $\{x_n\}$, the intermediate output $\{y_n^1\}$ is first calculated, and this is the input to the second filter. Continuing in this manner, the final output $y_n^K = y_n$ is calculated.

1) *Rounding:* The spectrum of the error sequence $\{e_n\}$ is given by[4]

$$\begin{aligned}\Phi_{ee}^r(z) = \left|\frac{N_2 \cdots N_K}{D_2 \cdots D_K}\right|^2 \frac{\sigma_1^2}{|D_1|^2} + \left|\frac{N_3 \cdots N_K}{D_3 \cdots D_K}\right|^2 \frac{\sigma_2^2}{|D_2|^2} \\ + \cdots + \frac{|N_K|^2}{|D_K|} \frac{\sigma_{K-1}^2}{|D_{K-1}|^2} + \frac{\sigma_K^2}{|D_K|^2}\end{aligned} \tag{21}$$

where

$$\begin{aligned}\sigma_i^2 &= \left(\frac{2^{-2t}}{3}\right)\frac{1}{2\pi j}\oint \Phi_{xx}\left[|B_i|^2 + \left|A_i \frac{N_i}{D_i}\right|^2 \right. \\ &\qquad \left. - 2C_i \frac{N_i}{D_i}\right] \prod_{j=1}^{i-1} |H_i|^2 \frac{dz}{z} \\ |A_i|^2 &= 3(c_i^2 + d_i^2) + 2c_i d_i(z + z^{-1}) \\ |B_i|^2 &= (3 + 4f_i^2 + 3g_i^2) + (3f_i + 2f_i g_i)(z + z^{-1}) \\ C_i &= (1 + f_i z + g_i z^2)(c_i z^{-1} + d_i z^{-2}).\end{aligned} \tag{22}$$

2) *Truncation:* We have in this case,

$$\Phi_{ee}^t(z) = \Phi_{ee}^r(z) + 2^{-2t}|H(z)|^2 \left|\sum_{i=1}^{K} \frac{D_i(z)T_i(z)}{N_i(z)}\right|^2 \tag{23}$$

where

$$\begin{aligned}T_i(z) &= [B_i'(z)D_i(z) - A_i'(z)N_i(z)]/|D_i(z)|^2 \\ A_i'(z) &= 3(c_i z^{-1} + d_i z^{-2}) \\ B_i'(z) &= 3 + 4f_i z^{-1} + 3g_i z^{-2}\end{aligned} \tag{24}$$

and $\Phi_{ee}^r(z)$ is given by (21).

It should be pointed out that there is a design flexibility for the cascade realization form in the different ways of pairing of $N_i(z)$ and $D_i(z)$ and of ordering of $H_i(z)$. The mean square error would be different for each realization.

[3] The numbers f_i and g_i are, of course, different from those of (14).

[4] For simplicity the argument "z" will be omitted sometimes. That is, $N_k = N_k(z)$, $\Phi_{ee} = \Phi_{ee}(z)$, etc.

IV. Derivation of Results

In this section the results summarized in the previous section are derived.

To illustrate the approach of this paper, consider first a second-order filter specified by

$$w_n = b_0 x_n - (a_1 w_{n-1} + a_2 w_{n-2}). \tag{25}$$

The actual computed output is

$$y_n = fl[b_0 x_n - (a_1 y_{n-1} + a_2 y_{n-2})]. \tag{26}$$

The calculation is to be performed in the following manner. First, the products $a_1 y_{n-1}$, $a_2 y_{n-2}$, and $b_0 x_n$ are calculated separately. Then $a_1 y_{n-1}$ and $a_2 y_{n-2}$ are added. Finally, this sum is subtracted from $b_0 x_n$ to obtain y_n.

Following Sandberg [15], a flow diagram of Fig. 3 may be drawn by using (4) and (5). The quantities $\delta_{n,0}$, $\varepsilon_{n,1}$, $\varepsilon_{n,2}$, η_n, ξ_n are errors caused by roundoff at each arithmetic step, and are assumed to be identically distributed, independent random variables. Therefore, the actual output $\{y_n\}$ is seen to be given explicitly by

$$y_n = b_0 \theta_{n,0} x_n - \sum_{k=1}^{2} a_k \phi_{n,k} y_{n-k} \tag{27}$$

where

$$\begin{aligned}\theta_{n,0} &= (1 + \delta_{n,0})(1 + \xi_n) \\ \phi_{n,1} &= (1 + \varepsilon_{n,1})(1 + \eta_n)(1 + \xi_n) \\ \phi_{n,2} &= (1 + \varepsilon_{n,2})(1 + \eta_n)(1 + \xi_n).\end{aligned}$$

We need to calculate the $\{y_n\}$ sequence from (27) and compare it with the ideal output sequence $\{w_n\}$ specified by (25) to obtain the error $e_n = y_n - w_n$.

We shall consider first the case of the direct form; the result will be used for the two other canonical realization forms.

A. Direct Form

It is clear from the above discussion that for the digital filter of (1) realized in direct form, we have

$$y_n = fl\left(\sum_{k=0}^{M} b_k x_{n-k} - \sum_{k=1}^{L} a_k y_{n-k}\right). \tag{28}$$

The corresponding flow graph showing the effect of roundoff error is given by Fig. 4[5] which also indicates the order of the calculation. The actual output sequence is therefore given by

$$y_n = \sum_{k=0}^{M} b_k \theta_{n,k} x_{n-k} - \sum_{k=1}^{L} a_k \phi_{n,k} y_{n-k} \tag{29}$$

where

$$\theta_{n,0} = (1 + \xi_n)(1 + \delta_{n,0}) \prod_{i=1}^{M} (1 + \zeta_{n,i}) \tag{30}$$

$$\theta_{n,j} = (1 + \xi_n)(1 + \delta_{n,j}) \prod_{i=j}^{M} (1 + \zeta_{n,i}) \quad j = 1, 2, \cdots, M$$

[5] If $b_k=0$, then $\delta_{n,k} = \zeta_{n,k} = 0$. If $b_k=1$, then $\delta_{n,k}=0$. Similarly, $\varepsilon_{n,k} = \eta_{n,k} = 0$ if $a_k=0$, and $\varepsilon_{n,k}=0$ if $a_k=1$.

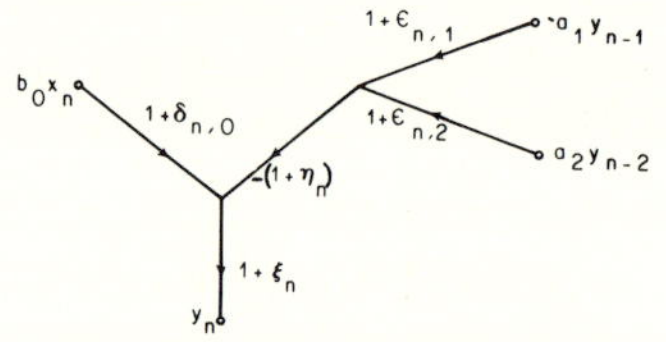

Fig. 3. Flow graph for second-order filter.

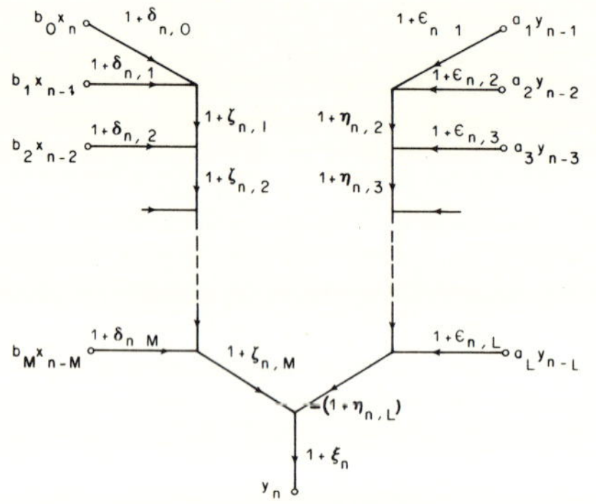

Fig. 4. Flow graph for Lth-order filter.

$$\phi_{n,1} = (1 + \xi_n)(1 + \varepsilon_{n,1}) \prod_{i=2}^{L} (1 + \eta_{n,i})$$

$$\phi_{n,j} = (1 + \xi_n)(1 + \varepsilon_{n,j}) \prod_{i=j}^{L} (1 + \eta_{n,i}) \quad j = 2, 3, \cdots, L.$$

By defining $a_0 = 1$ and $\phi_{n,0} = 1$ we may write (29) as

$$\sum_{k=0}^{L} a_k \phi_{n,k} y_{n-k} = \sum_{k=0}^{M} b_k \theta_{n,k} x_{n-k} \tag{31}$$

where $\xi_n, \varepsilon_{n,k}, \delta_{n,k}, \zeta_{n,k}, \eta_{n,k}$ are independent random variables. As discussed in Section II, we assume these random variables are uniformly distributed over the interval $(-2^{-t}, 2^{-t})$ or $(-2^{-t+1}, 0)$ depending on whether a rounding or a truncation arithmetic is used.

Equation (31) cannot be solved simply by the usual transform techniques since it has the time varying random coefficients $\phi_{n,k}$ and $\theta_{n,k}$.

To solve (31), we define $y_n', y_n'', y_n''', \cdots, y^{(p)}, \cdots$, by

$$\sum_{k=0}^{L} a_k \bar{\phi}_k y'_{n-k} = \sum_{k=0}^{M} b_k \bar{\theta}_k x_{n-k} \tag{32}$$

$$\sum_{k=0}^{L} a_k \bar{\phi}_k y''_{n-k} = \sum_{k=0}^{M} b_k(\theta_{n,k} - \bar{\theta}_k) x_{n-k} - \sum_{k=0}^{L} a_k(\phi_{n,k} - \bar{\phi}_k) y'_{n-k} \tag{33}$$

$$\sum_{k=0}^{L} a_k \bar{\phi}_k y^{(p)}_{n-k} = -\sum_{k=0}^{L} a_k(\phi_{n,k} - \bar{\phi}_k) y^{(p-1)}_{n-k}, \quad p = 3, 4, 5, \cdots \tag{34}$$

where $\bar{\phi}_k = E\{\phi_{n,k}\}$ and $\bar{\theta}_k = E\{\theta_{n,k}\}$ and E denotes the operation of taking the expectation. These two averages are independent of n (see Appendix III). Substitution of (32), (33), and (34) into (31), yields

$$y_n = y_n' + y_n'' + y_n''' + \cdots. \tag{35}$$

Since in (33) $(\theta_{n,k} - \bar{\theta}_k)$ and $(\phi_{n,k} - \bar{\phi}_k)$ are of the order of 2^{-t}, the order of magnitude of y_n'' is expected to be much smaller than that of y_n' and x_n.

In general, the magnitude of $y_n^{(p)}$ is expected to be much smaller than that of $y_n^{(p-1)}$. This is shown in Appendix II to be indeed the case if t is not too small. Therefore, only the $\{y_n'\}$ and the $\{y_n''\}$ sequences are significant. The error is given by

$$\begin{aligned} e_n &= y_n - w_n \\ &= y_n' + y_n'' - w_n. \end{aligned} \tag{36}$$

It is worth noting that in the case of rounding it can be shown that $\bar{\phi}_k$ and $\bar{\theta}_k$ are both equal to unity. By comparing (32) with (1), we see $y_n' = w_n$. Thus the $\{y_n''\}$ sequence is the same as the error sequence $\{e_n\}$.

Assume that the input $\{x_n\}$ is zero mean, and wide-sense stationary (WSS) with autocorrelation function $R_{xx}(n)$ and power spectral density $\Phi_{xx}(z)$. Then it is clear from (32) that y_n' is zero mean and WSS. Its power spectral density $\Phi_{y'y'}(z)$ is related to that of $\{x_n\}$ by

$$\Phi_{y'y'}(z) = \frac{N'(z)N'(1/z)}{D'(z)D'(1/z)} \Phi_{xx}(z) \tag{37}$$

and its autocorrelation function $R_{y'y'}(n)$ is given by

$$R_{y'y'}(n) = \frac{1}{2\pi j} \oint \Phi_{y'y'}(z) z^n \frac{dz}{z}$$

where

$$N'(z) = \sum_{k=0}^{M} b_k \bar{\theta}_k z^{-k},$$

and

$$D'(z) = \sum_{k=0}^{L} a_k \bar{\phi}_k z^{-k}. \tag{38}$$

Furthermore, y_n' is independent of $\theta_{m,j}$ and $\phi_{m,j}$ for all m and j. From (1), (33), and (36) we have

$$\sum_{k=0}^{L} a_k \bar{\phi}_k e_{n-k} = \sum_{k=0}^{L} a_k \bar{\phi}_k (y'_{n-k} - w_{n-k}) + u_n \tag{39}$$

with

$$u_n = \sum_{k=0}^{M} b_k(\theta_{n,k} - \bar{\theta}_k) x_{n-k} - \sum_{k=0}^{L} a_k(\phi_{n,k} - \bar{\phi}_k) y'_{n-k}. \tag{40}$$

To calculate the statistics of u_n, we need the statistics of $\theta_{n,k}$ and $\phi_{n,k}$. These can be evaluated in a straightforward manner from (30) and the result is summarized in Appendix III. It can be shown easily then that

$$E\{u_n\} = 0 \tag{41}$$

$$E\{u_n u_m\} = 0, \qquad n \neq m \tag{42}$$

$$E\{u_n^2\} = q^2 \sum_{k=0}^{M} \sum_{i=0}^{M} b_k b_i B_{k,i} R_{xx}(k-i) + q^2 \sum_{k=1}^{L} \sum_{i=1}^{L} a_k a_i A_{k,i} R_{y'y'}(k-i) - 2q^2 \sum_{k=0}^{M} \sum_{i=1}^{L} b_k a_i R_{xy'}(k-i) \quad (43)$$

where $q^2 = 2^{-2t}/3$ is the variance of a random variable uniformly distributed in $(-2^{-t}, 2^{-t})$ or in $(-2^{-t+1}, 0)$, and $R_{xy'}(k-i)$ is the cross-correlation function between $\{x_n\}$ and $\{y_n'\}$. $A_{k,i}$, $B_{k,i}$, and $C_{k,i}$ are defined by

$$\begin{aligned} A_{k,i} &= E\{(\phi_{n,k} - \bar{\phi}_k)(\phi_{n,i} - \bar{\phi}_i)\}/q^2 \\ B_{k,i} &= E\{(\theta_{n,k} - \bar{\theta}_k)(\theta_{n,i} - \bar{\theta}_i)\}/q^2 \\ C_{k,i} &= E\{(\theta_{n,k} - \bar{\theta}_k)(\phi_{n,i} - \bar{\phi}_i)\}/q^2. \end{aligned} \quad (44)$$

These quantities can be calculated from the result given in Appendix III. Since $(1+q^2)^k \approx 1 + kq^2$ if t is not too small, we have the excellent approximation

$$A_{k,i} \approx \begin{cases} L + 2 - \max(k, i) & k \neq i \text{ or } k = i = 1 \\ L + 3 - k & \text{otherwise} \end{cases}$$

$$B_{k,i} \approx \begin{cases} M + 2 - \max(k, i) & k \neq i \text{ or } k = i = 0 \\ M + 3 - k & \text{otherwise} \end{cases} \quad (45)$$

$$C_{k,i} \approx 1.$$

Equation (45) is valid if none of the filter coefficients is one or zero.[5] Thus we see $\{u_n\}$ is white and wide-sense stationary with variance given by (43) which may be rewritten as

$$\sigma_u^2 = E\{u_n^2\} = \left(\frac{2^{-2t}}{3}\right)\frac{1}{2\pi j}\oint \left[|B(z)|^2 \Phi_{xx}(z) + |A(z)|^2 \Phi_{y'y'}(z) - 2C(z)\Phi_{xy'}(z)\right]\frac{dz}{z} \quad (46)$$

where

$$|A(z)|^2 = \sum_{k=1}^{L} \sum_{i=1}^{L} a_k a_i A_{k,i} z^{k-i}$$

$$|B(z)|^2 = \sum_{k=0}^{M} \sum_{i=0}^{M} b_k b_i B_{k,i} z^{k-i} \quad (47)$$

$$C(z) = \sum_{k=0}^{M} \sum_{i=1}^{L} b_k a_i C_{k,i} z^{k-i}.$$

By using (46) and the relationship

$$\Phi_{xy'}(z) = \frac{N'(z)}{D'(z)} \Phi_{xx}(z) \quad (48)$$

we have

$$\sigma_u^2 = \left(\frac{2^{-2t}}{3}\right)\frac{1}{2\pi j}\oint \left[|B(z)|^2 + \left|A(z)\frac{N'(z)}{D'(z)}\right|^2 - 2C(z)\frac{N'(z)}{D'(z)}\right]\Phi_{xx}(z)\frac{dz}{z}. \quad (49)$$

Since $\bar{\phi}_k$ and $\bar{\theta}_k$ are either one or very close to one, the $N'(z)$ and $D'(z)$ inside the above integral can be replaced by $N(z)$ and $D(z)$ respectively.

From (39), we see that $\{e_n\}$ has zero mean and is wide-sense stationary. Its power spectral density $\Phi_{ee}(z)$ is given by

$$\Phi_{ee}(z) = \left|\frac{N(z)}{D(z)} - \frac{N'(z)}{D'(z)}\right|^2 \Phi_{xx}(z) + \frac{\sigma_u^2}{|D'(z)|^2}. \quad (50)$$

For the case of rounding $N'(z) = N(z)$, and $D'(z) = D(z)$. Equation (50) reduces to (8) given in Section III. For the case of truncation (50) reduces to (11) of Section III after

$$\left|\frac{N}{D} - \frac{N'}{D'}\right|^2$$

is expanded and second-order terms are neglected.

Equation (50) shows that the actual filter may be represented as the transfer function $N'(z)/D'(z)$ connected in parallel with another filter with transfer function $1/D'(z)$ and white noise $\{u_n\}$ as its input. This is illustrated in Fig. 5. It can be shown that the sequence $\{u_n\}$ defined by (40) is uncorrelated with $\{w_n\}$. We therefore have from Fig. 5,

$$\Phi_{ew}(z) = \Phi_{e''w}(z) = \Phi_{xx}(z)\frac{N(z)}{D(z)}\left[\frac{N'(z)}{D'(z)} - \frac{N(z)}{D(z)}\right]^* \quad (51)$$

and

$$\Phi_{yy}(z) = \left|\frac{N'(z)}{D'(z)}\right|^2 \Phi_{xx}(z) + \frac{\sigma_u^2}{|D'(z)|^2} \quad (52)$$

where the asterisk * denotes the complex conjugate. This result will be used later to derive the error for cascade and parallel forms of realization.

B. Parallel Form

For parallel form of realization (13), let $\{w_n^i\}$ and $\{y_n^i\}$, $i = 1, 2, \cdots, K$, be the ideal and the actual output sequences of the filter $H_i(z)$. For the input sequence $\{x_n\}$ we have

$$w_n^i = f_i x_n + g_i x_{n-1} - c_i w_{n-1}^i - d_i w_{n-2}^i \quad (53)$$

$$y_n^i = fl\left[f_i x_n + g_i x_{n-1} - c_i y_{n-1}^i - d_i y_{n-2}^i\right]. \quad (54)$$

Fig. 6 shows the corresponding flow graph. Since each unit $H_i(z)$ is implemented in the direct form, the spectral density function of the error at the ith intermediate output, $\Phi_{ee}^i(z)$ is obtained by using (46) and (50),

$$\Phi_{ee}^i(z) = \left|\frac{N_i}{D_i} - \frac{N_i'}{D_i'}\right|^2 \Phi_{xx} + \frac{\sigma_i^2}{|D_i'|^2} \quad (55)$$

where

$$\begin{aligned} \sigma_i^2 &= \frac{q^2}{2\pi j}\oint\left[|B_i|^2 + \left|A_i\frac{N_i}{D_i}\right|^2 - 2C_i\frac{N_i}{D_i}\right]\Phi_{xx}\frac{dz}{z} \\ |A_i|^2 &= 3(c_i^2 + d_i^2) + 2c_i d_i(z + z^{-1}) \\ |B_i|^2 &= 3(f_i^2 + g_i^2) + 2f_i g_i(z + z^{-1}) \\ C_i &= g_i c_i + (f_i c_i + g_i d_i)z^{-1} + f_i d_i z^{-2}. \end{aligned} \quad (56)$$

For rounding

$$N_i'(z) = N_i(z)$$
$$D_i'(z) = D_i(z)$$

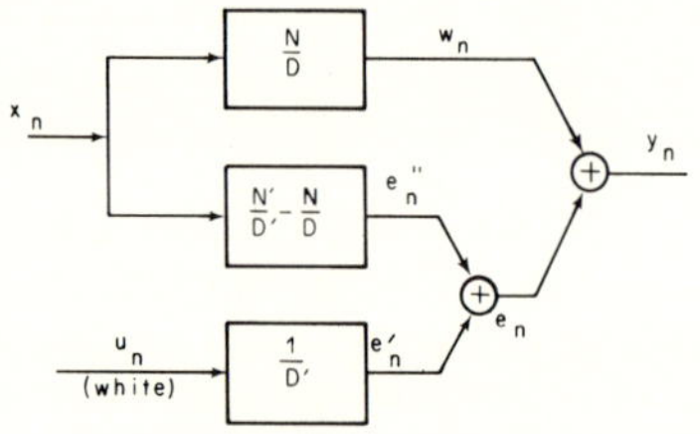

Fig. 5. Equivalent block diagram for direct programming showing effect of roundoff error accumulation.

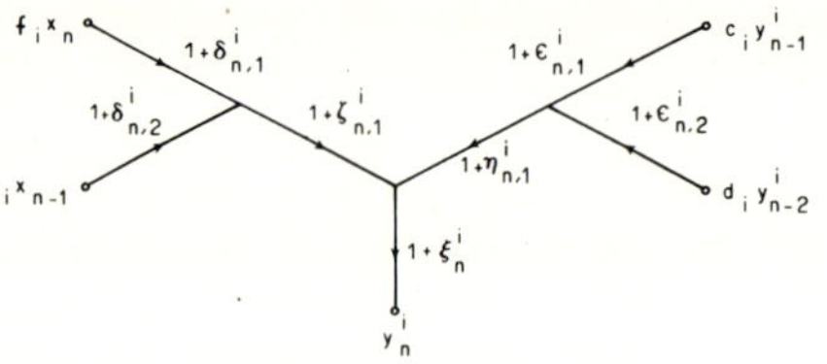

Fig. 6. Flow graph for ith parallel path

and for truncation

$$N_i'(z) = (f_i + g_i z^{-1})(1 - 2^{-t})^3$$
$$D_i'(z) = 1 + (c_i z^{-1} + d_i z^{-2})(1 - 2^{-t})^3.$$

N_i and D_i are given by (14).

The ideal and the actual outputs of the entire filter $H(z)$, $\{w_n\}$ and $\{y_n\}$, are related to $\{w_n^i\}$ and $\{y_n^i\}$ by

$$w_n = w_n^1 + w_n^2 + \cdots + w_n^K \tag{57}$$

$$y_n = fl\left[y_n^1 + y_n^2 + \cdots + y_n^K\right]. \tag{58}$$

If the summation of (58) is carried out from the left to the right, a corresponding flow graph can be drawn (Fig. 7). Thus

$$y_n = \sum_{i=1}^{K} \psi_{n,i} y_n^i \tag{59}$$

$$\psi_{n,i} = \begin{cases} \prod_{j=2}^{K} (1 + \zeta_{n,j}), & i = 1 \\ \prod_{j=i}^{K} (1 + \zeta_{n,j}), & i \geq 2. \end{cases} \tag{60}$$

Defining $e_n^i = y_n^i - w_n^i$ and $e_n = y_n - w_n$ to be the error of $H_i(z)$ and $H(z)$ respectively, we can rewrite (59) as

$$\begin{aligned} y_n &= \sum_{i=1}^{K} \psi_{n,i}(w_n^i + e_n^i) \\ &= \sum_{i=1}^{K} w_n^i + \sum_{i=1}^{K} (\psi_{n,i} - \bar{\psi}_i)(w_n^i + e_n^i) \\ &\quad + \sum_{i=1}^{K} (\bar{\psi}_i - 1)w_n^i + \sum_{i=1}^{K} \bar{\psi}_i e_n^i \end{aligned} \tag{61}$$

where $\bar{\psi}_i = E\{\psi_{n,i}\}$. The first term of (61) is exactly the ideal output w_n [see (57)]; the remaining terms represent the error e_n,

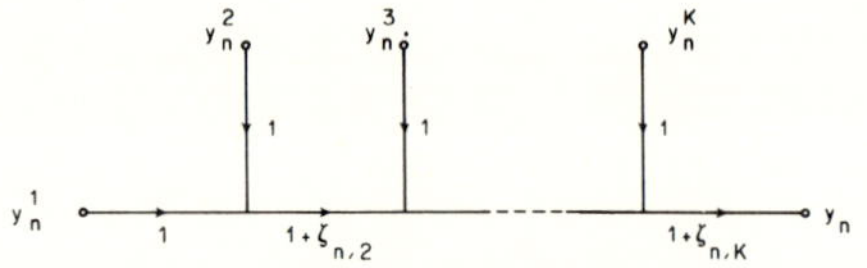

Fig. 7. Flow graph for parallel programming.

$$\begin{aligned} e_n = \sum_{i=1}^{K} (\psi_{n,i} - \bar{\psi}_i)(w_n^i + e_n^i) + \sum_{i=1}^{K} (\bar{\psi}_i - 1)w_n^i \\ + \sum_{i=1}^{K} \bar{\psi}_i e_n^i. \end{aligned} \tag{62}$$

The power spectral density of $\{e_n\}$ can be computed in a straightforward manner. The result is

$$\begin{aligned} \Phi_{ee}(z) &= q^2 \cdot \Phi_{xx} \sum_{i=1}^{K} \sum_{j=1}^{K} F_{ij} \frac{N_i' N_j'^*}{D_i' D_j'^*} \\ &\quad + \sum_{i=1}^{K} \left(1 + \frac{2^{-2t}}{3} F_{ii}\right) \frac{\sigma_i^2}{|D_i'|^2} \\ &\quad + \Phi_{xx} \left| \sum_{i=1}^{K} \left\{ (\bar{\psi}_i - 1)\frac{N_i}{D_i} + \bar{\psi}_i \left[\frac{N_i'}{D_i'} - \frac{N_i}{D_i}\right] \right\} \right|^2 \end{aligned} \tag{63}$$

where

$$\begin{aligned} F_{ij} &= E\{(\psi_{n,i} - \bar{\psi}_i)(\psi_{n,j} - \bar{\psi}_j)\}/q^2 \\ &= \begin{cases} [(1 + q^2)^{K-1} - 1]/q^2 & i = j = 1 \\ [(1 + q^2)^{K+1-\max(i,j)} - 1]/q^2 & \text{otherwise} \end{cases} \end{aligned} \tag{64}$$

$$F_{ij} \approx \begin{cases} K - 1 & i = j = 1 \\ K + 1 - \max(i,j) & \text{otherwise.} \end{cases}$$

For rounding $\bar{\psi}_i = 1$; thus the last term of (63) is zero and the first two terms agree with (15) of Section III.

For truncation

$$\bar{\psi}_i = \begin{cases} (1 - 2^{-t})^{K-1} & i = 1 \\ (1 - 2^{-t})^{K+1-i} & i \geq 2. \end{cases} \tag{65}$$

By neglecting second-order terms, it can be shown that (63) reduces to (17) of Section III after some algebra.

C. Cascade Form

Let $\{w_n^i\}$ and $\{y_n^i\}$ be the ideal and the actual output of the ith section respectively. Since the output of the ith transfer function $H_i(z)$ is the input of the $(i+1)$-th transfer function $H_{i+1}(z)$, the following equations hold.

$$w_n^{i+1} = w_n^i + f_i w_{n-1}^i + g_i w_{n-2}^i - c_i w_{n-1}^{i+1} - d_i w_{n-2}^{i+1} \tag{66}$$

$$y_n^{i+1} = fl[y_n^i + f_i y_{n-1}^i + g_i y_{n-2}^i - c_i y_{n-1}^{i+1} - d_i y_{n-2}^{i+1}]. \tag{67}$$

If we define $y_n^0 = x_n$, then (66) and (67) are valid for $i = 0, 1, 2, \cdots, K-1$. Fig. 8 shows the flow graph of (67). Note that there is no roundoff error for y_n^i since the constant term of $N_i(z)$ is unity. It is assumed that coefficients f_i, g_i, c_i, d_i are neither zero nor unity.

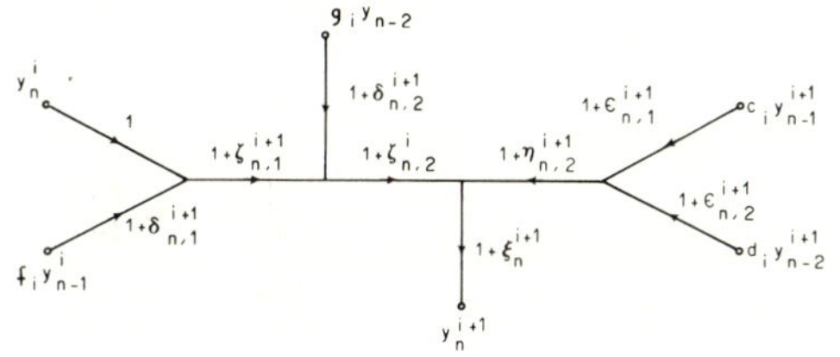

Fig. 8. Flow graph for ith section of cascade programming.

Denote the spectral density at the output of the ith section by $\Phi_{yy}^i(z)$. Then from (46) and (52), we have

$$\Phi_{yy}^i(z) = \left|\frac{N_i'(z)}{D_i'(z)}\right|^2 \Phi_{yy}^{i-1}(z) + \frac{\sigma_i^2}{|D_i'(z)|^2} \quad i = 1, 2, \cdots, K \quad (68)$$

$$\Phi_{yy}^0(z) = \Phi_{xx}(z).$$

For rounding

$$N_i'(z) = N_i(z)$$
$$D_i'(z) = D_i(z)$$

and for truncation

$$N_i'(z) = (1 + g_i z^{-2})(1 - 2^{-t})^3 + f_i z^{-1}(1 - 2^{-t})^4$$

and

$$D_i'(z) = 1 + (c_i z^{-1} + d_i z^{-2})(1 - 2^{-t})^3.$$

Here N_i and D_i are defined by (20),

$$\sigma_i^2 = \frac{q^2}{2\pi j}\oint G_i(z)\Phi_{yy}^i(z)\frac{dz}{z}$$

$$G_i(z) = |B_i|^2 + \left|A_i\frac{N_i}{D_i}\right|^2 - 2C_i\frac{N_i}{D_i} \quad (69)$$

$$|A_i|^2 \approx 3(c_i^2 + d_i^2) + 2c_i d_i(z + z^{-1})$$

$$|B_i|^2 \approx (3 + 4f_i^2 + 3g_i^2) + (3f_i + 2f_i g_i)(z + z^{-1})$$

$$C_i = (1 + f_i z + g_i z^2)(c_i z^{-1} + d_i z^{-2}).$$

By repeated substitution we can obtain σ_i^2 more explicitly.

$$\sigma_1^2 = \frac{q^2}{2\pi j}\oint \Phi_{xx} G_1 \frac{dz}{z}$$

$$\vdots$$

$$\sigma_i^2 = \frac{q^2}{2\pi j}\oint \Phi_{xx} G_i \prod_{j=1}^{i-1} |H_j'|^2 \frac{dz}{z} + \frac{q^2\sigma_{i-1}^2}{2\pi j}\oint \frac{G_i}{|D_{i-1}'|^2}\frac{dz}{z}$$
$$+ \frac{q^2\sigma_{i-2}^2}{2\pi j}\oint \frac{G_i|H_{i-1}'|^2}{|D_{i-2}'|^2}\frac{dz}{z} \quad (70)$$
$$+ \cdots + \frac{q^2\sigma_1^2}{2\pi j}\oint \frac{G_i \prod_{j=1}^{i-1}|H_j'|^2}{|D_1'|^2}\frac{dz}{z}$$

$$H_i' = \frac{N_i'}{D_i'}.$$

In practice, however, all terms except the first term in (70) are much smaller compared with the first term provided the word length t is not too short. Thus for all practical purposes, we have

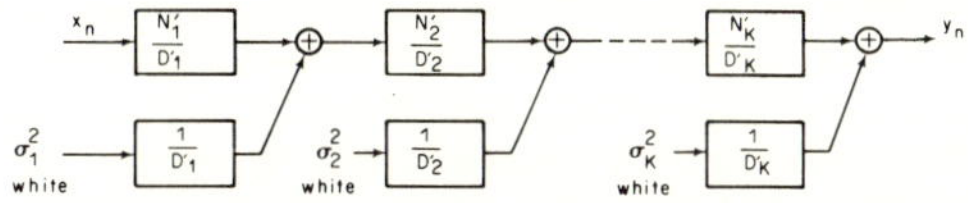

Fig. 9. Equivalent block diagram for cascade programming showing effect of roundoff error accumulation.

$$\sigma_i^2 = \frac{q^2}{2\pi j}\oint \Phi_{xx} G_i \prod_{j=1}^{i-1} |H_i'|^2 \frac{dz}{z}, \qquad i = 2, 3, \cdots, K. \quad (71)$$

At the output of the final stage the total spectrum is (see Fig. 9)

$$\Phi_{yy}^K = \Phi_{yy} = \left|\frac{N_1'N_2'\cdots N_K'}{D_1'D_2'\cdots D_K'}\right|^2 \Phi_{xx} + \left|\frac{N_2'\cdots N_K'}{D_2'\cdots D_K'}\right| \frac{\sigma_1^2}{|D_1'|^2}$$
$$+ \cdots + \left|\frac{N_K'}{D_K'}\right|^2 \frac{\sigma_{K-1}^2}{|D_{K-1}'|^2} + \frac{\sigma_K^2}{|D_K'|^2}. \quad (72)$$

The spectral density of the ideal output $\Phi_{ww}(z)$ is given by $\Phi_{ww}(z) = |H(z)|^2\Phi_{xx}(z)$ and the cross-spectral density between the actual and ideal output signals is given by

$$\Phi_{yw}(z) = \Phi_{xx}(z)H'(z)H^*(z) \quad (73)$$

where

$$H'(z) = \prod_{i=1}^{K} [N_i'(z)/D_i'(z)]$$

since the error is uncorrelated with the ideal output. Therefore, the spectral density of the error can be computed.

$$\Phi_{ee}(z) = \Phi_{yy}(z) + \Phi_{ww}(z) - \Phi_{yw}(z) - \Phi_{wy}(z)$$
$$= |H'(z) - H(z)|^2\Phi_{xx}(z)$$
$$+ \sum_{i=1}^{K-1} \frac{\sigma_i^2}{|D_i'(z)|^2} \prod_{j=i+1}^{K} |H_j'(z)|^2 + \frac{\sigma_K^2}{|D_K'(z)|^2}. \quad (74)$$

For rounding $N_i'(z) = N_i(z)$ and $D_i'(z) = D_i(z)$ and (74) reduces to (21) of Section III. For truncation, we may expand $|H(z) - H'(z)|^2$ in (74) and neglect the higher order terms. The result is that given by (23) of Section III.

V. Input Quantization

In Sections III and IV we have assumed that the input sequence $\{x_n\}$ are machine numbers. It will be shown in this section that the technique developed can also be used to treat the effect of input quantization.

The quantization of x_n may occur at two places. It may occur when the data is being taken; here the error is introduced by the quantizer at the analog-to-digital converter. It is well known that such an error can be represented by an additive noise component which is uncorrelated with the signal and has a zero mean and an approximately white spectrum [17], [18]. The analysis of this effect is therefore straightforward. The second place that input quantization may occur is in the computer where each number is represented in floating point form and the mantissa is truncated or rounded to t bits. The effect of such a quantization is considered here. Our discussion is restricted to the direct form of filter realization with rounding arithmetic. Extension to truncation arithmetic and to other two canonical forms of realization is straightforward.

Denote the rounded input by x'_n; we have

$$x'_n = x_n(1 + \gamma_n) \tag{75}$$

where γ_n is a random variable introduced to represent the roundoff error. Thus, γ_n is uniformly distributed in $(-2^{-t}, 2^{-t})$. Following the same approach of Section IV, (32) and (33) become in this case

$$\sum_{k=0}^{L} a_k y'_{n-k} = \sum_{k=0}^{M} b_k x_{n-k} \tag{76}$$

$$\sum_{k=0}^{L} a_k y''_{n-k} = \sum_{k=0}^{M} b_k(\theta_{n,k} - 1)x_{n-k} - \sum_{k=0}^{L} a_k(\phi_{n,k} - 1)y'_{n-k} + \sum_{k=0}^{M} b_k\gamma_{n-k}x_{n-k}. \tag{77}$$

From (76) $y'_n = w_n$. Thus $y''_n = e_n$. The first two terms on the right side of (77) are due to roundoff and the last term is due to the input quantization. It is easy to show that the final result is

$$\Phi_{ee}(z) = \Phi^r_{ee}(z) + \frac{2^{-2t}}{3} \cdot |H(z)|^2 \cdot \Phi_{xx}(z) \tag{78}$$

where $\Phi^r_{ee}(z)$ is the error spectrum due to roundoff alone and is given by (8).

VI. Output Error-To-Signal Ratio and Bounds

Quite often one is interested in the error-to-signal ratio at the output. We shall consider only the case of direct programming. Extension to other cases can be treated in a similar manner.

For rounding from (8), (9), and (36),

$$\frac{E\{e_n^2\}}{E\{w_n^2\}} = \frac{1}{2\pi j}\oint \frac{1}{|D(z)|^2}\frac{dz}{z} \cdot \frac{\dfrac{q^2}{2\pi j}\displaystyle\oint\left[|B(z)|^2 + \left|A(z)\frac{N(z)}{D(z)}\right|^2 - 2C(z)\frac{N(z)}{D(z)}\right]\Phi_{xx}(z)\frac{dz}{z}}{\dfrac{1}{2\pi j}\displaystyle\oint\left|\frac{N(z)}{D(z)}\right|^2\Phi_{xx}(z)\frac{dz}{z}}. \tag{79}$$

We have used the fact that the spectrum of the ideal output $\{w_n\}$ is

$$\Phi_{ww}(z) = \left|\frac{N(z)}{D(z)}\right|^2 \Phi_{xx}(z) \tag{80}$$

and

$$E\{w_n^2\} = \frac{1}{2\pi j}\oint \Phi_{ww}(z)\frac{dz}{z}. \tag{81}$$

This ratio depends on the input spectrum Φ_{xx}. An upper bound independent of Φ_{xx} may be derived from (79) as

$$\frac{E\{e_n^2\}}{E\{w_n^2\}} \le \frac{q^2}{2\pi j}\oint\frac{dz}{z|D(z)|^2} \cdot \max_{|z|=1}\left\{\left|\frac{D(z)}{N(z)}\right|^2 \cdot \left[|B(z)|^2 + \left|A(z)\frac{N(z)}{D(z)}\right|^2 - 2C(z)\frac{N(z)}{D(z)}\right]\right\}. \tag{82}$$

Similarly for truncation from (11) we have

$$\frac{E\{e_n^2\}}{E\{w_n^2\}} \le [E\{e_n^2\}/E\{w_n^2\}]^r + 2^{-2t}\max_{|z|=1}\left[\frac{1}{|N(z)|^2}\left|B'(z) - A'(z)\frac{N(z)}{D(z)}\right|^2\right] \tag{83}$$

where $[E\{e_n^2\}/E\{w_n^2\}]^r$ is the bound given by the right-hand side of (82).

VII. Examples

For the first example we take the second-order filter considered by Sandberg [15]:

$$w_n = x_n - (a_1 w_{n-1} + a_2 w_{n-2})$$

with $a_1 = -\sqrt{2}(1-0.001)$ and $a_2 = (1-0.001)^2$. The error-to-signal ratio at the output is given by (79) for rounding arithmetics. Suppose $\Phi_{xx}(z) = 1$ and $t = 27$. The result is $\sqrt{E[e_n^2]/E[w_n^2]} = 2.1 \times 10^{-7}$. For truncation arithmetic the error-to-signal ratio can be calculated from (11) and (81). The result is 1.1×10^{-5}. The upper bound derived in Section VI can be calculated as follows: $N(z) = 1$, $D(z) = 1 + a_1 z^{-1} + a_2 z^{-2}$, $|B(z)|^2 = 1$, $|A(z)|^2 = 3(a_1^2 + a_2^2) + 2a_1a_2(z + z^{-1})$, and $C(z) = a_1 z^{-1} + a_2 z^{-2}$. Thus,

$$\left|\frac{D}{N}\right|^2\left(|B|^2 + \left|\frac{AN}{D}\right|^2 - C\frac{N}{D} - C^*\frac{N^*}{D^*}\right) = 1 + 2(a_1^2 + a_2^2 + a_1a_2(z^{-1} + z).$$

Its maximum on $|z| = 1$ is calculated to be approximately $(7 + 2\sqrt{2})$. The value of the integral $(1/2\pi j)\oint(dz/z|D(z)|^2)$ is 0.5×10^3. With $t = 27$ the upper bound of $\sqrt{E[e_n^2]/E[w_n^2]}$ calculated from (82) to be 3.3×10^{-7}. The upper bound given by Sandberg is 5.8×10^{-5}. It should be pointed out that the bound given by Sandberg is a deterministic absolute bound whereas our results are statistical and the bound is on the ratio of the rms value of the error to that of the output.

If truncation arithmetic is used, we have the second term of (83) as

$$\max_{|z|=1}\left\{\left|\frac{D(z)}{N(z)}\right|^2 \frac{|1 - (a_1 z_1^{-1} + a_2 z^{-2})|^2}{|D(z)|^4}\right\}.$$

The upper bound of $\sqrt{E[e_n^2]/E[w_n^2]}$ for truncation is calculated to be 1.1×10^{-5}. It is possible to modify Sandberg's work for truncation arithmetic. The deterministic upper bound turns out to be twice as large as that for rounding arithmetics. For this example it is 1.2×10^{-4}.

To check the theoretical results derived in this paper, two digital filters are simulated on IBM 7094 computer, first using double precision, and then using single precision with both rounding and truncation arithmetics. The output from the single precision calculations are compared with

TABLE I

ρ	Theoretical	Experimental
0.5	2.54×10^{-8}	1.4×10^{-8}
0.6	3.18×10^{-8}	1.6×10^{-8}
0.7	4.09×10^{-8}	2.0×10^{-8}
0.8	5.85×10^{-8}	2.8×10^{-8}
0.9	1.13×10^{-7}	5.7×10^{-8}
0.92	1.40×10^{-7}	6.9×10^{-8}
0.95	2.24×10^{-7}	1.1×10^{-7}
0.97	3.73×10^{-7}	1.9×10^{-7}
0.98	5.59×10^{-7}	3.1×10^{-7}
0.99	1.12×10^{-6}	6.5×10^{-7}

that obtained from the double precision calculation so as to obtain the error. The first such example is a second-order filter with a lower Q than that of the previous example in order to prevent a long waiting time for the starting transient to settle down. The digital filter, specified by $w_n = x_n - (-\sqrt{2}\rho w_{n-1} + \rho^2 w_{n-2})$, has poles at $\rho(1 \pm j)/\sqrt{2}$. When rounding arithmetic is used, it has been reported recently [18] that there is good agreement between the theoretical and experimental results, at least for $\rho \leq 0.9$. At higher values of ρ we found that there is slight disagreement, and it is caused by the fact mentioned in Appendix I that the addition error in rounding does not have exactly zero mean. The result for truncation arithmetic is presented in Table I.

The second example for checking the theoretical result against the experimental result is an eighth-order low-pass elliptical filter function with 0.1 dB ripple in the passband and a minimum loss of 36.15 dB beyond 1.04974 times the cutoff frequency [20] the poles of this filter are at

$$\begin{aligned} &0.015\,536\,903\,2 \pm j\,1.010\,477\,744\,2 \\ &-0.076\,027\,165\,6 \pm j\,0.979\,887\,569\,3 \\ &-0.257\,843\,410\,8 \pm j\,0.844\,653\,718\,9 \\ &-0.605\,843\,851\,5 \pm j\,0.372\,426\,884\,1. \end{aligned}$$

The zeros are

$$\begin{aligned} &\pm j\,1.055\,576\,753\,3 \\ &\pm j\,1.121\,816\,272\,1 \\ &\pm j\,1.442\,940\,986\,8. \end{aligned}$$

This filter is converted to a digital filter by using a bilinear transformation with

$$\frac{\text{cutoff frequency}}{\text{sampling frequency}} = 0.2.$$

Table II gives the coefficients of direct, cascade, and parallel programmings. With $\Phi_{xx}(z) = 1$ as the input spectral density and $t = 27$ as the length of the mantissa, theoretical and experimental results for the noise-to-signal ratio,

$$\left[\frac{E\{e_n^2\}}{E\{w_n^2\}}\right]^{1/2}$$

are summarized in Table III.

From (11), (17), and (23) it is seen that the error for truncation arithmetic is larger than that for rounding arithmetic by an extra term. This extra term, caused by the fact that the mean of the error committed at each arithmetic step is not zero, can be significant or even dominant if the filter has high Q poles. The measured error for truncation arithmetic is about half as large as the theoretical value. This fact is to be anticipated since the assumption that the error distribution is uniform in $(-2^{-t+1}, 0)$ is a pessimistic one. It can be shown that the expression derived in this paper for truncation arithmetic is always larger than the actual error.

From Table III it is also seen that the roundoff error accumulation for the direct form is the largest of the three canonical forms of realization and the difference between the cascade and the parallel forms is not significant; a result known for fixed-point arithmetic realization of digital filters.

TABLE II

COEFFICIENTS OF FILTER USED IN EXAMPLE

Direct Programming	Cascade Programming	Parallel Programming
$a_0 = 1.000000000$	$f_1 = 2.000000000$	$f_1 = -2.122555509$
$a_1 = -2.436585057$	$f_2 = 0.09438004947$	$f_2 = -12.15445655$
$a_2 = 5.005281390$	$f_3 = -0.4034100191$	$f_3 = 4.070898703$
$a_3 = -5.998719282$	$f_4 = -0.5186245808$	$f_4 = -0.2014641232$
$a_4 = 5.683103860$		$f_5 = 11.40757748$
$a_5 = -3.706488054$	$g_i = 1.000000000$	
$a_6 = 1.809113610$	$i = 1, 2, 3, 4$	$g_1 = 17.02530132$
$a_7 = -0.5465520431$		$g_2 = -4.913407259$
$a_8 = 0.08766103068$	$c_1 = -0.6827461732$	$g_3 = -2.080689432$
	$c_2 = -0.6586646160$	$g_4 = 0.7889095393$
$b_0 = 0.04021459800$	$c_3 = -0.6049294717$	
$b_1 = 0.04714540096$	$c_4 = -0.5902447964$	$c_1 = -0.6827461732$
$b_2 = 0.09920489233$		$c_2 = -0.6586646160$
$b_3 = 0.1520584627$	$d_1 = 0.1800503899$	$c_3 = -0.6049294717$
$b_4 = 0.1195687466$	$d_2 = 0.5805234892$	$c_4 = -0.5902447964$
$b_5 = 0.1520584627$	$d_3 = 0.8636435133$	
$b_6 = 0.09920489233$	$d_4 = 0.9710871201$	$d_1 = 0.1800503899$
$b_7 = 0.04714540096$		$d_2 = 0.5805234892$
$b_8 = 0.04021459800$		$d_3 = 0.8636435133$
		$d_4 = 0.9710871201$
		$g_5 = c_5 = d_5 = 0$

TABLE III

Realization	Rounding		Truncation	
	Theoretical	Experimental	Theoretical	Experimental
Direct	4.30×10^{-6}	3.2×10^{-6}	9.09×10^{-6}	5.3×10^{-6}
Parallel	2.25×10^{-8}	2.6×10^{-8}	1.65×10^{-7}	8.7×10^{-8}
Cascade	3.43×10^{-8}	4.0×10^{-8}	1.83×10^{-7}	8.8×10^{-8}

VIII. CONCLUSION

Despite the outstanding advantages of digital filters, there is an inherent accuracy problem caused by the finite word length. For digital filters simulated on a large-scale general purpose machine the main problem is to calculate the error. For filters built with actual digital hardware the problem is to relate the error to the word length so that a choice of the word length can be made in order to insure a desired degree of accuracy.

This paper has presented a general approach to the statistical analysis of errors of linear constant coefficient digital filters using floating-point arithmetic. The effects of roundoff accumulation and of input data quantization are studied. The effect of coefficient truncation, which can be treated as a coefficient sensitivity problem [1], [22], is not considered in this paper. However, if the order of the filter is not too low, it is possible to analyze this effect statistically as was done in the fixed-point arithmetic case [14], and the approach presented in this paper would then be applicable.

Expressions are derived for the mean square error due to roundoff accumulation and input quantization. Results are given for both rounding and truncation arithmetics, and for each of the three canonical realization forms: direct, parallel, and cascade. The mean square error in all cases is proportional to 2^{-2t} where t is the length of the matissa. Good agreement between the theory and experiments are obtained.

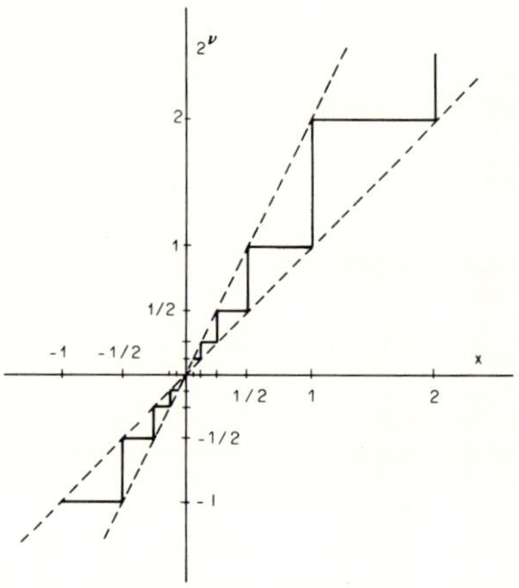

Fig. 10. 2^v as a function of x.

Appendix I

In Section II it is assumed that the product and sum of two floating-point numbers x and y are given by

$$fl(xy) = xy(1 + \delta)$$

$$fl(x + y) = (x + y)(1 + \varepsilon)$$

respectively, where ε and δ are random variables uniformly distributed over the interval $(-2^{-t}, 2^{-t})$ for rounding and over $(-2^{-t+1}, 0)$ for truncation. In this appendix we elaborate on the meaning of this assumption. We shall give an intuitive reason leading to the assumption, present some experimental data, and discuss the meaning of the results summarized in Section III as implied by our assumption.

Consider the number x represented in floating-point form, (sgn) $2^v \mu$. The exponent is given by

$$v = \lceil \log_2 x \rceil$$

where the brackets $\lceil\ \rceil$ denote the smallest integer exceeding the quantity inside the brackets. The mantissa is therefore

$$\mu = x/2^v.$$

For convenience x is taken to be positive. If only t bits is allowed to the mantissa, μ must be so truncated or rounded, thus comitting an error e given by

$$\begin{aligned} e &= x - 2^v \mu_t \\ &= 2^v(\mu - \mu_t) \end{aligned} \tag{84}$$

where μ_t is the truncated or the rounded version of μ. Denote by e' the difference $\mu - \mu_t$. It is clear that e' is bounded by

$$-2^{-t-1} \leq e' \leq 2^{-t-1}$$

for rounding and

$$-2^{-t} \leq e' \leq 0$$

for truncation.

Since e' is due to quantizing, it is reasonable to assume that it is uniformly distributed in its range. The quantity e would then be uniformly distributed over the range $(-2^v 2^{-t-1}, 2^v 2^{-t-1})$ or $(-2^v 2^{-t}, 0)$. The function 2^v is a piecewise constant function of x and is sketched in Fig. 10. It is seen that

$$x \leq 2^v \leq 2x$$

or

$$1 \leq (2^v/x) \leq 2. \tag{85}$$

The above characterization of e and e' is not easy to use in analysis because of the nonlinear dependence of 2^v on x. Equation (84) may be rewritten as

$$e = xe'' \tag{86}$$

with

$$e'' = (\mu - \mu_t)(2^v/x). \tag{87}$$

It is clear from (87) that

$$-2^{-t} \leq e'' \leq 2^{-t}$$

for rounding and

$$-2^{-t+1} \leq e'' \leq 0$$

for truncation.

The assumption used in this paper is that e'' is uniformly distributed in $(-2^{-t}, 2^{-t})$ for rounding and in $(-2^{-t+1}, 0)$ for truncation.

Consider now the multiplication of two floating-point numbers x and y each with a t bit mantissa. Their product will in general have a fractional part of $2t$ or $(2t-1)$ bits before rounding or truncation. This leads to our assumption of Section II

$$fl(xy) = xy(1 + \delta)$$

where δ is uniformly distributed in $(-2^{-t}, 2^{-t})$ or in $(-2^{-t+1}, 0)$. Because of (85), it is also reasonable to assume

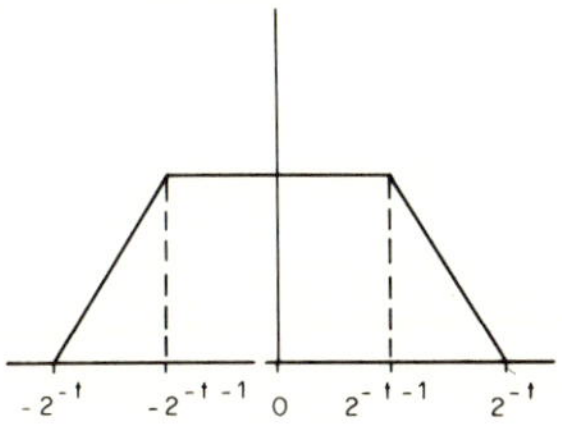

Fig. 11. Trapezoidal distribution of roundoff error.

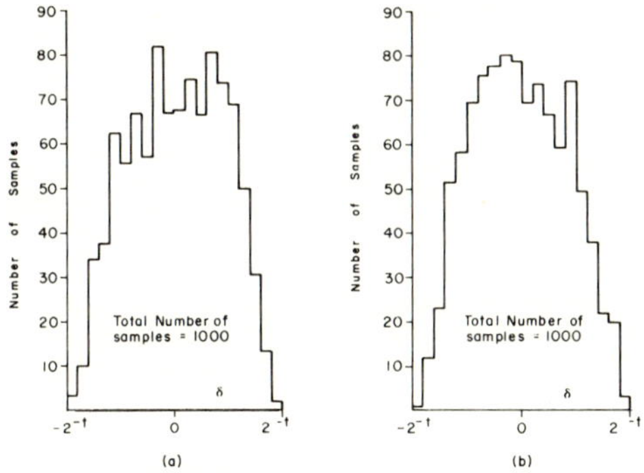

Fig. 12. Two typical histograms of multiplication error.

that the probability density of δ is not uniform but of a trapezoidal shape as shown in Fig. 11.

In the computer simulation reported in Section VII some statistics have been taken. Fig. 12 shows two typical histograms of the multiplicative error δ. It is seen that a trapezoidal density agrees with the data very well and that a uniform density agrees with the data reasonably well.

For addition the assumption made in this paper is

$$fl(x + y) = (x + y)(1 + \varepsilon)$$

where ε is uniformly distributed in $(-2^{-t}, 2^{-t})$ for rounding and in $(-2^{-t+1}, 0)$ for truncation. The actual situation, however, is more involved. If the magnitudes of x and y are about the same, the fractional part of the sum $x + y$ may be no more than t bits and the roundoff error would be zero. This situation is encountered frequently in calculating the sum

$$\sum_{k=1}^{L} a_k y_{n-k}$$

for filters with high Q poles. Also, in floating-point addition it can be shown that even with rounding arithmetic, the error in general does not have zero mean for a binary machine. Fig. 13 shows two histograms of addition error. It is seen that the agreement with our assumption is not too good.

It should be pointed out that in our analysis, only the

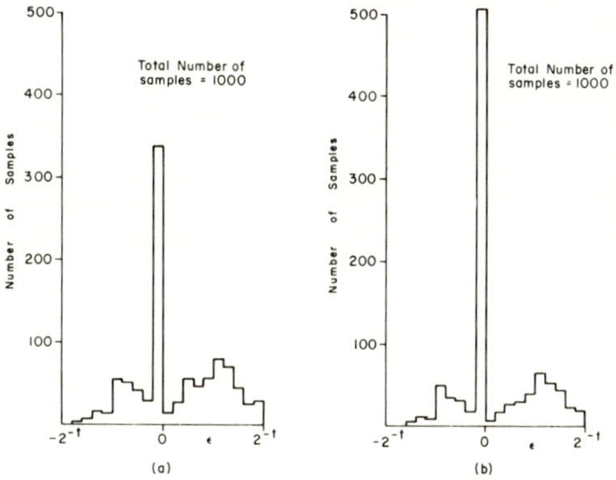

Fig. 13. Two typical histograms of addition error.

mean and variance of the errors are needed. Thus the exact density of these errors does not affect too much our result as long as the first and second moments remain the same.

Appendix II

In Section IV each term in the $\{y_n\}$ is written as

$$y_n = y_n' + y_n'' + y_n''' + \cdots. \tag{88}$$

It was then argued that $y_n', y_n'', y_n''', y_n'''' \cdots$ are in decreasing order of magnitude. Thus only the first two terms in (88) are of significance. We now show that this is indeed the case.

First, from (34), we note that $y_n^{(p)}$ and $y_n^{(p-1)}$ for $p \geq 3$ are related by the following inequality which is derived by one of Sandberg's results [15].

$$\frac{\langle y^{(p)}\rangle^{1/2}}{\langle y^{(p-1)}\rangle^{1/2}} \leq 2^{-t} \operatorname*{Max}_{|z|=1}\left\{\frac{1}{|D'(z)|}\right\} \sum_{k=0}^{L} |a_k \alpha_k| \tag{89}$$

$$\langle y^{(p)}\rangle = \lim_{K\to\infty} \frac{1}{K+1} \sum_{i=0}^{K} [y_i^{(p)}]^2$$

$$\langle y^{(p-1)}\rangle = \lim_{K\to\infty} \frac{1}{K+1} \sum_{i=0}^{K} [y_i^{(p-1)}]^2$$

$$\alpha_k = \begin{cases} 1.06(L+1) & k = 0, 1 \\ 1.06(L+3-k) & k = 2, 3, \cdots, L \end{cases}$$

$$D'(z) = \sum_{k=0}^{L} a_n \bar{\phi}_k z^{-k}.$$

Next, let

$$z_n = y_n''' + y_n'''' + \cdots.$$

From (34) we have

$$\sum_{k=0}^{L} a_k \bar{\phi}_k z_{n-k} = -\sum_{k=0}^{L} a_k(\phi_{n,k} - \bar{\phi}_k)(z_{n-k} + y_{n-k}''). \tag{90}$$

By using the inequality again, we obtain

$$\langle z\rangle^{1/2} \le 2^{-t}\mathcal{M}\langle z + y''\rangle^{1/2} \tag{91}$$

where

$$\langle z\rangle^{1/2} = \lim_{K\to\infty} \frac{1}{K+1}\sum_{k=0}^{K} (y_k''' + y_k'''' + \cdots)^2$$

and

$$\mathcal{M} = \operatorname*{Max}_{|z|=1}\left\{\frac{1}{|D'(z)|}\right\}\cdot\sum_{k=0}^{L}\cdot|a_k\alpha_k|.$$

It should also be pointed out that if t is not too small, then the question of deadband effect [21] will not arise. Since

$$\langle z + y''\rangle^{1/2} \le \sqrt{2}[\langle z\rangle^{1/2} + \langle y''\rangle^{1/2}]$$

we have

$$\frac{\langle z\rangle^{1/2}}{\langle y''\rangle^{1/2}} \le \frac{\mathcal{M}2^{-t}}{1 - \sqrt{2}\mathcal{M}2^{-t}} \tag{92}$$

It is seen that for moderately high t, the right sides of (89) and (92) are very small and can be neglected as was done in Section IV.

Appendix III

The first and second moments of $\theta_{n,k}$ and $\phi_{n,k}$ as given by (30) can be evaluated straightforwardly. The result is tabulated below with $q^2 = 2^{-2t}/3$.

1) Rounding:

$$E\{\theta_{n,j}\} = E\{\phi_{n,j}\} = 1 \qquad \text{for all } n \text{ and } j$$

$$E\{\theta_{n,j}^2\} = \begin{cases}(1+q^2)^{M+2} & j=0\\ (1+q^2)^{M+3-j} & j\ge 1\end{cases}$$

$$E\{\theta_{n,j}\theta_{n,k}\} = (1+q^2)^{M+2-j} \qquad M\ge j>k\ge 0$$

$$E\{\phi_{n,j}^2\} = \begin{cases}(1+q^2)^{L+1} & j=1\\ (1+q^2)^{L+3-j} & j\ge 2\end{cases}$$

$$E\{\phi_{n,j}\phi_{n,k}\} = (1+q^2)^{L+2-j} \qquad L\ge j>k\ge 1$$

$$E\{\theta_{n,j}\phi_{n,k}\} = \begin{cases}1+q^2 & k\ne 0\\ 1 & k=0\end{cases}$$

$$E\{\theta_{n,j}\theta_{m,k}\} = E\{\phi_{n,j}\phi_{m,k}\} = E\{\phi_{n,j}\theta_{m,k}\} = 1 \qquad n\ne m.$$

2) Truncation:

$$E\{\theta_{n,j}\} = \bar\theta_j = \begin{cases}(1-2^{-t})^{M+2} & j=0\\ (1-2^{-t})^{M+3-j} & j\ge 1\end{cases}$$

$$E\{\phi_{n,j}\} = \bar\phi_j = \begin{cases}(1-2^{-t})^{L+1} & j=1\\ (1-2^{-t})^{L+3-j} & j\ge 2\end{cases}$$

$$E\{\theta_{n,j}^2\} = \begin{cases}[(1-2^{-t})^2+q^2]^{M+2} & j=0\\ [(1-2^{-t})^2+q^2]^{M+3-j} & j\ge 1\end{cases}$$

$$E\{\theta_{n,j}\theta_{n,k}\} = \begin{cases}[(1-2^{-t})^2+q^2]^{M+2-j}(1-2^{-t})^{2+j-k} & M\ge j>k\ge 1\\ [(1-2^{-t})^2+q^2]^{M+2-j}(1-2^{-t})^{1+j} & M\ge j>k=0\end{cases}$$

$$E\{\phi_{n,j}^2\} = \begin{cases}[(1-2^{-t})^2+q^2]^{L+1} & j=1\\ [(1-2^{-t})^2+q^2]^{L+3-j} & j\ge 2\end{cases}$$

$$E\{\phi_{n,j}\phi_{n,k}\} = \begin{cases}[(1-2^{-t})^2+q^2]^{L+2-j}(1-2^{-t})^{2+j-k} & L\ge j>k\ge 2\\ [(1-2^{-t})^2+q^2]^{L+2-j}(1-2^{-t})^{j} & L\ge j>k=1\end{cases}$$

$$E\{\theta_{n,j}\phi_{n,k}\} = [(1-2^{-t})^2+q^2](1-2^{-t})^{L+M+4-j-k}.$$

References

[1] J. F. Kaiser, "Digital Filters," in *System Analysis by Digital Computer*, F. F. Kuo and J. F. Kaiser, Eds. New York: Wiley, 1966, pp. 218–285.

[2] R. M. Golden and J. F. Kaiser, "Design of wideband sampled-data filters," *Bell Sys. Tech. J.*, vol. 48, pp. 1533–1546, July 1964.

[3] C. M. Radar and B. Gold, "Digital filter design techniques in the frequency domain," *Proc. IEEE*, vol. 55, pp. 149–171, February 1967.

[4] R. B. Blackman and J. W. Tukey, *The Measurement of Power Spectra from the Point of View of Communication Engineering*. New York: Dover, 1959.

[5] J. F. A. Ormsby, "Design of numerical filters with applications to missile data processing," *J. ACM*, vol. 8, pp. 440–446, October 1961.

[6] R. M. Golden, "Digital computer simulation of a sampled-data voice excited vocoder," *J. Acoust. Soc. Am.*, vol. 35, pp. 1358–1366, 1963.

[7] C. M. Rader, "Study of vocoder filters by computer simulation," *J. Acoust. Soc. Am.*, vol. 36, p. 1023, 1964.

[8] E. J. Kelly and M. J. Levin, "Signal parameter estimation for seismometer arrays," MIT Lincoln Laboratory, Lexington, Mass., Tech. Rept. 339, January 8, 1964.

[9] L. B. Jackson, J. F. Kaiser, and H. S. McDonald, "An approach to implementation of digital filters," *IEEE Trans. Audio and Electroacoustics*, vol. AU-16, pp. 413–421, September 1968.

[10] C. S. Weaver, J. von der Groeben, P. E. Mantey, J. G. Toole, C. A. Cole, J. W. Fitzgerald, and R. W. Lawrence, "Digital filtering with applications to electrocardiogram processing," *IEEE Trans. Audio and Electroacoustics*, vol. AU-16, pp. 350–389, September 1968.

[11] M. J. Ferguson and P. E. Mantey, "Automatic frequency control via digital filtering," *IEEE Trans. Audio and Electroacoustics*, vol. AU-16, pp. 392–397, September 1968.

[12] A. I. Zverev, "Digital MTI radar filters," *IEEE Trans. Audio and Electroacoustics*, vol. AU-16, pp. 422–432, September 1968.

[13] J. B. Knowles and R. Edwards, "Effect of a finite-word-length computer in a sampled-data feedback system," *Proc. IEE* (London), vol. 112, pp. 1197–1207, June 1965.

[14] J. B. Knowles and E. M. Olcayto, "Coefficient Accuracy in Digital Filter Response," *IEEE Trans. Circuit Theory*, vol. CT-15, March 1968, pp. 31–41.

[15] I. W. Sandberg, "Floating-Point Round-off Accumulation in Digital Filter Realization," *Bell Syst. Tech. Jr.*, vol. 46, October 1967, pp. 1775–1791.

[16] J. H. Wilkinson, *Rounding Errors in Algebraic Processes*, Prentice-Hall, Englewood Cliffs, New Jersey 1963.

[17] B. Widrow, "Statistical analysis of amplitude-quantized sampled-data systems," *AIEE Trans. (Application and Industry)*, vol. 59, pp. 555–568, 1960 (January 1961 section).

[18] J. Katzenelson, "On errors introduced by combined sampling and quantization," *IRE Trans. Automatic Control*, vol. AC-7, pp. 58–68, April 1962.

[19] C. Weinstein and A. V. Oppenheim, "A comparison of roundoff noise in floating point and fixed point digital filter realizations," *Proc. IEEE* (Letters), vol. 57, pp. 1181–1183, June 1969.

[20] E. Christian and E. Eisenmann, *Filter Design Tables and Graphs*. New York: Wiley, 1966, p. 249.

[21] R. B. Blackman, *Linear Data—Smoothing and Prediction in Theory and in Practice*. Reading, Mass.: Addison-Wesley, 1965, pp. 75–81.

[22] P. E. Mantey, "Eigenvalue sensitivity and state-variable selection," *IEEE Trans. Automatic Control*, vol. AC-13, pp. 263–269, June 1968.

18

Reprinted from *IEEE Trans. Audio Electroacoustics,* **AU-20**(3), 206–212 (1972)

On the Design of Digital Filters with Coefficients of Limited Word Length

ERNST AVENHAUS

Abstract

The cost of a digital filter, if implemented as a special-purpose computer, depends heavily on the word length of the coefficients. Therefore, it should be reduced as much as possible. On the other hand, a small word length causes large coefficient deviations that impair the wanted performance of the digital filter. The necessary word length may be reduced by choosing a suitable structure for the filter. Two methods for doing this will be presented, taking selective filters as a model. A further reduction of the word length may be won by optimizing the rounded filter coefficients in the discrete parameter space. A description of a modified univariate search will be given.

I. Introduction

Let the transfer function of a digital filter with constant coefficients be described by

$$H(z) = \sum_{\mu=0}^{m} b_\mu z^\mu \Big/ \sum_{\nu=0}^{n} c_\nu z^\nu, \qquad c_n = 1 \tag{1}$$

with

$$z = e^{sT} = e^{(\sigma + j\omega)T}.$$

We consider the magnitude of this function on the unit circle in the z plane

$$|H(\Omega)| = |H(z)|\,|_{z=e^{j\Omega}}. \tag{2}$$

A wanted magnitude function $|H_w(\Omega)|$ and the tolerated deviation $\delta(\Omega)$ may be prescribed. As an example, Fig. 1 shows a tolerance scheme of a bandpass described by

$$|H_w(\Omega)| = \begin{cases} 1 & \text{in the passband (PB)} \\ 0 & \text{in the stopband (SB)} \end{cases} \tag{3}$$

and

$$\delta(\Omega) = \begin{cases} \delta_P & \text{in PB} \\ \delta_S & \text{in SB.} \end{cases} \tag{4}$$

A transfer function $|H(\Omega)|$ satisfies the tolerance conditions, if

$$\big|\,|H_w(\Omega)| - |H(\Omega)|\,\big| \le \delta(\Omega), \qquad \forall\Omega \tag{4a}$$

Manuscript received November 18, 1971.
The author is with the Institut für Nachrichtentechnik, Erlangen, Germany.

or

$$\epsilon(\Omega) = \frac{1}{\delta(\Omega)}\,\big|\,|H_w(\Omega)| - |H(\Omega)|\,\big| \le 1, \qquad \forall\Omega \tag{4b}$$

holds. In our example the normalized error function $\epsilon(\Omega)$ becomes

$$\epsilon(\Omega) = \begin{cases} \dfrac{1}{\delta_P}\,\big|1 - |H_N(\Omega)|\,\big|, & \text{in PB} \\ \dfrac{1}{\delta_S}\cdot |H_N(\Omega)|, & \text{in SB,} \end{cases} \tag{5}$$

where

$$H_N(z) = K \cdot H(z) \tag{6}$$

is the normalized transfer function. K can be arbitrarily chosen such that

$$\max \epsilon = \max_{\Omega} \epsilon(\Omega) \tag{7}$$

becomes a minimum. Since there are no constraints in the transition ranges between passband and stopbands, $\epsilon(\Omega)$ is not defined in these intervals. In the following max ϵ, defined as described above, will be considered as a figure of merit for judging the performance of an actual transfer function $H_N(z)$. At first it will be used to state the problem at hand in more detail.

According to (4b) max ϵ has to be smaller than, or at most equal to, one for satisfaction of the prescribed tolerance scheme. By minimizing max ϵ with an appropriate approximation procedure we get a set of coefficients of a filter with equal ripple behavior in the passband as well as in the stopband. In practical cases the parameters of this filter turn out to have about eight decimal digits accuracy, and the frequency response $|H_N(\Omega)|$ uses only a minimum fraction of the tolerance scheme, i.e., the performance will be described by a value max $\epsilon_0 < 1$, indicating that there is a certain margin which can be used for a reduction of the word length. In the case of the bandpass filter of Fig. 1, a filter of at least the eighth degree is necessary, leading to max $\epsilon_0 = 0.526054$. By reducing the word length of the coefficients step-by-step, a minimum increment $Q = 2^{-9}$ has been found to be necessary, in this particular example, for satisfaction of the tolerance scheme and we get max $\epsilon = 0.87$. A larger increment does not lead to an acceptable solution as has been indicated in Fig. 1.

In a more general approach we now investigate the word length sufficient to satisfy the specifications using the three well-known canonical forms shown in Figs. 2–4. In the direct form (Fig. 2), the coefficients b_μ, c_ν of (1) will be used. For the cascade form (Fig. 3), the transfer function has to be written in product form as

$$H(z) = K \cdot \prod_{\lambda=1}^{l} \frac{b_{2\lambda} z^2 + b_{1\lambda} z + b_{0\lambda}}{z^2 + c_{1\lambda} z + c_{0\lambda}} \tag{8}$$

with

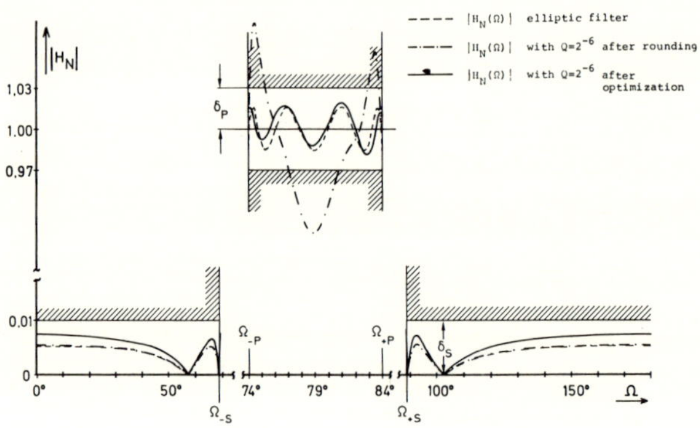

Fig. 1. The influence of rounding and optimization on the frequency response.

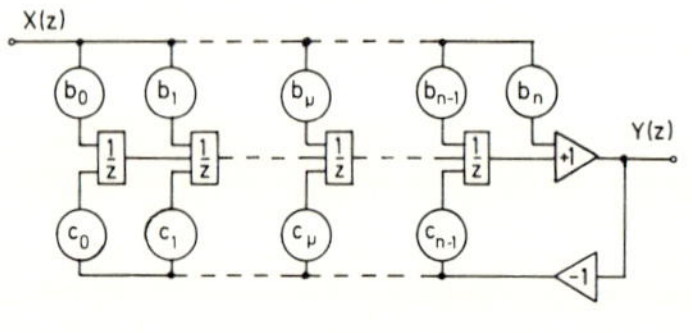

Fig. 2. Direct programming of a digital filter.

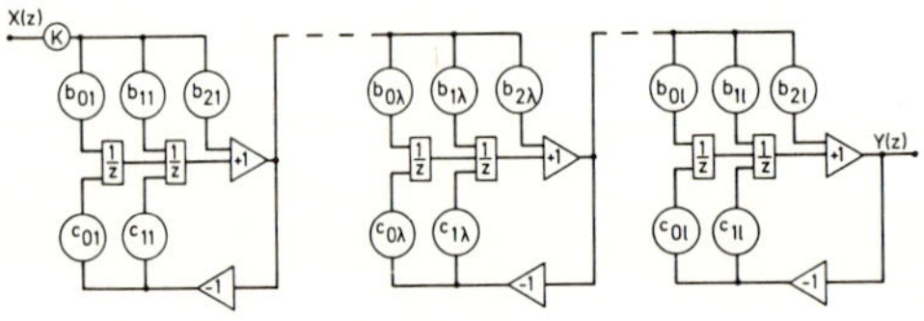

Fig. 3. Cascade programming of a digital filter.

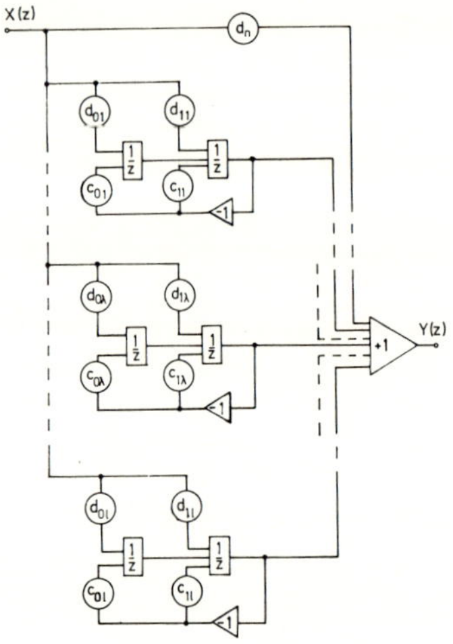

Fig. 4. Parallel programming of a digital filter.

$$l = \begin{cases} \dfrac{n}{2}, & \text{if } n \text{ is even} \\ \dfrac{n+1}{2}, & \text{if } n \text{ is odd.} \end{cases}$$

The coefficients of the parallel form (Fig. 4) will be won by the partial fraction expansion of (1) as

$$H(z) = d_0 + \sum_{\lambda=1}^{l} \frac{d_{1\lambda}z + d_{0\lambda}}{z^2 + c_{1\lambda}z + c_{0\lambda}}, \tag{9}$$

if we assume for simplicity all poles of $H(z)$ to be distinct.

The coefficients a_ν may be stored as fixed-point binary numbers as

$$a_\nu = \pm \sum_{i=i_M}^{i_L} \alpha_{\nu i} 2^i \tag{10}$$

where $\alpha_{\nu i}$ are 0 or 1, $\alpha_{\nu i_L}$ is the least significant bit, and $\alpha_{\nu i_M}$ is the most significant bit. Obviously, the range of coefficients is

$$-2 \cdot 2^{i_M} < a_\nu < 2 \cdot 2^{i_M}, \qquad \forall \nu \tag{11}$$

while the minimum increment, i.e., the quantization step size is defined as

$$Q = 2^{i_L}, \tag{12}$$

and the roundoff error is

$$|E| \leq \frac{Q}{2}. \tag{13}$$

Taking the one bit for the sign into account, the word length of the coefficients turns out to be

$$W = 2 + i_M - i_L. \tag{14}$$

II. Valuation of the Sufficient Word Length for Different Filter Structures

If the deviation of the magnitude function holds the inequality relation

$$\Delta|H(\Omega)| = \big||H(\Omega) + \Delta H(\Omega)| - |H(\Omega)|\big| \leq \delta(\Omega)(1 - \max \epsilon_0), \qquad \forall\Omega, \quad (15)$$

the tolerance conditions will be satisfied. According to (13) all coefficient errors are restricted by

$$|\Delta a_\nu| \leq \frac{Q}{2}, \qquad \forall \nu. \quad (16)$$

For a fixed frequency Ω, let

$$\max_{\Delta a_\nu} \{|H(\Omega) + \Delta H(\Omega, \Delta a_\nu)|\} \quad (17)$$

be the maximum magnitude of the transfer function if the coefficients are changed within the range limited by unequality (16) and let

$$\min_{\Delta a_\nu} \{|H(\Omega) + \Delta H(\Omega, \Delta a_\nu)|\} \quad (18)$$

be the minimum magnitude at Ω in dependance of Δa_ν. The deviation $\Delta|H(\Omega)|$ (15) is then bounded by

$$\Delta|H(\Omega)| \leq \max\{(\max_{\Delta a_\nu}|H + \Delta H| - |H|), (|H| - \min_{\Delta a_\nu}|H + \Delta H|)\}. \quad (19)$$

Now let

$$H(\Omega) = \frac{N(\Omega)}{D(\Omega)}$$

be the transfer function, $|\Delta_N|$ be an upper bound of the deviation of the numerator, and $|\Delta_D|$ be an upper bound of the deviation of the denominator. Then the unequalities

$$\max_{\Delta a_\nu} |H(\Omega) + \Delta H(\Omega, \Delta a_\nu)| \leq \frac{|N(\Omega)| + |\Delta_N|}{|D(\Omega)| - |\Delta_D|} \quad (20a)$$

and

$$\min_{\Delta a_\nu} |H(\Omega) + \Delta H(\Omega, \Delta a_\nu)| \geq \frac{|N(\Omega)| - |\Delta_N|}{|D(\Omega)| + |\Delta_D|} \quad (20b)$$

are valid. With these results (19) becomes

$$\Delta|H(\Omega)| \leq \frac{|N(\Omega)| + |\Delta_N|}{|D(\Omega)| - |\Delta_D|} - |H(\Omega)|. \quad (21)$$

This can be verified easily, if $|D(\Omega)| > |\Delta_D|$, i.e., if the system stays stable after changing the coefficients. Since $|\Delta_D|$ and $|\Delta_N|$ depend on the quantization, with (15) and (21) we get an implicit expression for the sufficient value Q_s for each particular frequency Ω

$$\frac{|N(\Omega)| + |\Delta_N(Q_s(\Omega))|}{|D(\Omega)| - |\Delta_D(Q_s(\Omega))|} - |H(\Omega)| = \delta(\Omega)(1 - \max \epsilon_0). \quad (22)$$

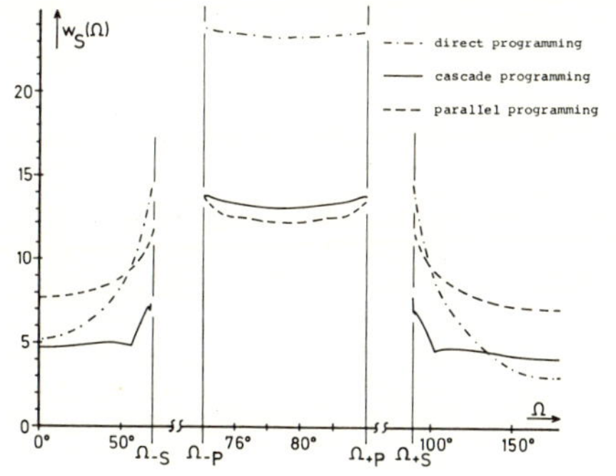

Fig. 5. Valuation of filter structures by sufficient word length functions.

The corresponding expressions for the cascade and parallel form are

$$\prod_{\lambda=1}^{l} \frac{|N_\lambda(\Omega)| + |\Delta_{N\lambda}(Q_s(\Omega))|}{|D_\lambda(\Omega)| - |\Delta_{D\lambda}(Q_s(\Omega))|} - \prod_{\lambda=1}^{l} \frac{|N_\lambda(\Omega)|}{|D_\lambda(\Omega)|} = \delta(\Omega)(1 - \max \epsilon_0) \quad (23)$$

and

$$\sum_{\lambda=1}^{l} \frac{|N_\lambda(\Omega)| + |\Delta_{N\lambda}(Q_s(\Omega))|}{|D_\lambda(\Omega)| - |\Delta_{D\lambda}(Q_s(\Omega))|} - \sum_{\lambda=1}^{l} \frac{|N_\lambda(\Omega)|}{|D_\lambda(\Omega)|} = \delta(\Omega)(1 - \max \epsilon_0). \quad (24)$$

Using $Q_s(\Omega)$, we define a word length function according to (14)

$$w_s(\Omega) = 2 + i_M - \mathrm{ld} Q_s(\Omega), \quad (25)$$

the maximum of which will be the sufficient word length for satisfying the tolerance scheme

$$W_s = \max_{\Omega} \{w_s(\Omega)\}. \quad (26)$$

For example, in the direct form

$$|\Delta_N| = \tfrac{1}{2} M_N \cdot Q_s(\Omega)$$

and

$$|\Delta_D| = \tfrac{1}{2} M_D \cdot Q_s(\Omega) \quad (27)$$

are upper bounds of the magnitude deviation in numerator and denominator, where M_N and M_D are the numbers of the coefficients in the numerator and denominator, respectively. With (22) we get

$$Q_s(\Omega) = 2 \frac{\left|\sum_{\nu=0}^{n} c_\nu e^{j\nu\Omega}\right| \delta(\Omega)(1 - \max \epsilon_0)}{M_N + M_D |H(\Omega)| + M_D \delta(\Omega)(1 - \max \epsilon_0)}. \quad (28)$$

Fig. 5 shows $w_s(\Omega)$ for the three canonical forms, using the bandpass of the eighth degree as an example. We see that for the direct form the sufficient word length is 24 bits, while 14 bits will suffice for the cascade or parallel form.

III. A Statistical Method for the Estimation of the Word Length

A better estimation of the word length may be won by handling the coefficient errors as random variables. This approach leads to an ensemble of filters with coefficients being statistically distributed in a properly chosen range. The variation range of the filter performance of this ensemble will be calculated [2], [3]. The influence of coefficient errors on the transmission property has been investigated in the frequency domain by Knowles and Olcayto [2]. For convenience the approach will be recalled briefly. After rounding, the coefficients a_ν° lie in the range $\pm\frac{1}{2}Q$ around the original values a_ν

$$a_\nu - \tfrac{1}{2}Q \le a_\nu^\circ \le a_\nu + \tfrac{1}{2}Q, \tag{29}$$

leading to a transfer function $H^\circ(\Omega)$. An ensemble of transfer functions $\tilde{H}(\Omega)$ with coefficients $\tilde{a}_\nu$ being equally distributed in the range

$$a_\nu - \frac{Q}{2} \le \tilde{a}_\nu = a_\nu + \Delta\tilde{a}_\nu < a_\nu + \frac{Q}{2} \tag{30}$$

obviously contains the transfer function $H^\circ(\Omega)$. So by investigating the ensemble $\tilde{H}(\Omega)$, we get a statement on $H^\circ(\Omega)$.

Knowles and Olcayto used the mean value

$$\sigma_\Omega^2 = \overline{\frac{1}{2\pi}\int_0^{2\pi} |\tilde{H}(\Omega) - H(\Omega)|^2 d\Omega} \tag{31}$$

as a measure, but this expression seems to be less appropriate for filters with high attenuation required in the stopband, as will be demonstrated later by an example. Here we use the mean value and the variance of the power transfer functions

$$\tilde{P}(\Omega) = \tilde{H}(\Omega)\cdot\tilde{H}^*(\Omega) \tag{32}$$

instead, where $\tilde{H}^*(\Omega)$ is the complex conjugate of $\tilde{H}(\Omega)$. $\tilde{P}(\Omega)$ can be expressed approximately by the power transfer function $P(\Omega)$ of the original filter and the sensitivity functions. It is

$$\begin{aligned}\tilde{P}(\Omega) &= P(\Omega) + \Delta\tilde{P}(\Omega)\\ &\approx P(\Omega) + \sum_{\nu=1}^{\eta_e} \frac{\partial P(\Omega)}{\partial a_\nu}\Delta\tilde{a}_\nu,\end{aligned} \tag{33}$$

where η_e is the total number of erroneous coefficients. An elliptic filter satisfies the unequality

$$\begin{aligned}&(|H_w(\Omega)| - \delta(\Omega)\cdot\max\epsilon_0)^2\\ &\le P(\Omega) \le (|H_w(\Omega)| + \delta(\Omega)\cdot\max\epsilon_0)^2, \qquad \forall\Omega.\end{aligned} \tag{34}$$

A member of the ensemble $\tilde{H}(\Omega)$ satisfies the tolerance constraints if the unequality

$$\begin{aligned}&(|H_w(\Omega)| - \delta(\Omega))^2 \le \tilde{P}(\Omega) = P(\Omega) + \Delta\tilde{P}(\Omega)\\ &\le (|H_w(\Omega)| + \delta(\Omega))^2, \qquad \forall\Omega\end{aligned} \tag{35}$$

holds. This condition is accomplished, if the unequality

$$\begin{aligned}|\Delta\tilde{P}(\Omega)| \le |&-2|H_w(\Omega)|\cdot\delta(\Omega)(1-\max\epsilon_0)\\ &+\delta^2(\Omega)(1-\max^2\epsilon_0)|, \qquad \forall\Omega\end{aligned} \tag{36}$$

is valid.

The mean value of the random variable $\tilde{P}(\Omega)$ is defined as

$$\mu_p(\Omega) = \int_{-\infty}^{+\infty} \tilde{P}(\Omega)\cdot p(\tilde{P}(\Omega))\, d\tilde{P}(\Omega) \tag{37}$$

with $p(\tilde{P}(\Omega))$ being the probability density function of $\tilde{P}(\Omega)$ at the fixed frequency Ω. In correspondence, the variance of the power transfer functions is

$$\sigma_p^2(\Omega) = \int_{-\infty}^{+\infty} (\tilde{P}(\Omega) - \mu_p(\Omega))^2\cdot p(\tilde{P}(\Omega))\, d\tilde{P}(\Omega). \tag{38}$$

Equations (37) and (38) may be evaluated by means of the probability density functions

$$p_\nu(\Delta\tilde{a}_\nu) = \begin{cases}\dfrac{1}{Q}, & \text{if } -\dfrac{Q}{2} \le \Delta\tilde{a}_\nu \le \dfrac{Q}{2}\\ 0, & \text{otherwise}\end{cases} \tag{39}$$

of the random variables $\Delta\tilde{a}_\nu$, if the functions

$$\begin{aligned}\tilde{P}_\nu(\Omega) = \tilde{P}(\Omega, \Delta a_1, \cdots, \Delta a_\nu, \cdots,\\ \Delta a_\kappa, \cdots, \Delta a_{\eta_e})\big|_{\Delta a_\kappa = \text{const},\ \kappa\ne\nu}\end{aligned} \tag{40}$$

are monotonous. This assumption holds, if $\tilde{P}(\Omega)$ is estimated by (33). $\mu_p(\Omega)$ and $\sigma_p(\Omega)$ can be shown to be [4]

$$\mu_p(\Omega) \approx P(\Omega)$$

and

$$\sigma_p^2(\Omega) \approx \frac{Q^2}{12}\sum_{\nu=1}^{\eta_e}\left(\frac{\partial P(\Omega)}{\partial a_\nu}\right)^2. \tag{41}$$

According to the finite value theorem [5], $p(\tilde{P}(\Omega))$ will approach a normal probability density function if η_e increases, since $\Delta\tilde{P}(\Omega)$ is approximately a sum of equally distributed random variables. By this we can state that 95 percent of the power transfer functions $\tilde{P}(\Omega)$ vary less than $2\cdot\sigma_p(\Omega, Q)$ at Ω.

$$|\Delta\tilde{P}(\Omega)| \le 2\sigma_p(\Omega, Q). \tag{42}$$

According to (36) and (42) we get a condition, which if satisfied, guarantees that 95 percent of the filters with the transfer functions $\tilde{H}(\Omega)$ meet the specifications at Ω

$$\begin{aligned}2\sigma_p(\Omega, Q) \le |&-2|H_w(\Omega)|\cdot\delta(\Omega)(1-\max\epsilon_0)\\ &+\delta^2(\Omega)(1-\max^2\epsilon_0)|.\end{aligned} \tag{43}$$

For estimating the percentage of filters satisfying the tolerance scheme for all Ω it would be necessary to evaluate the conditional probability of hurting the tolerance limits as a function of Ω. But the function $Q_p(\Omega)$, already won by (41) and (43), will allow us to judge a filter structure.

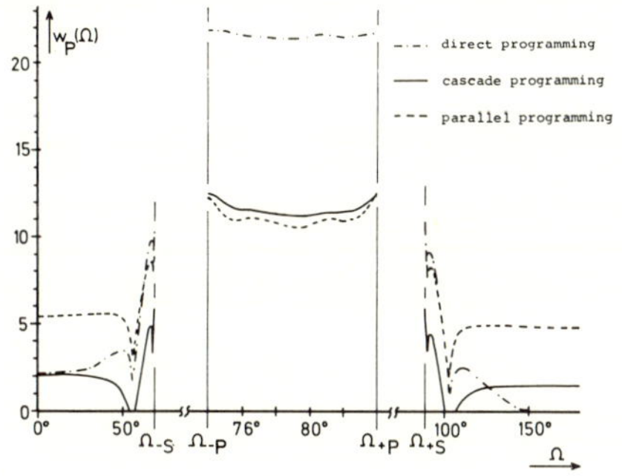

Fig. 6. Valuation of filter structures by statistical word length functions.

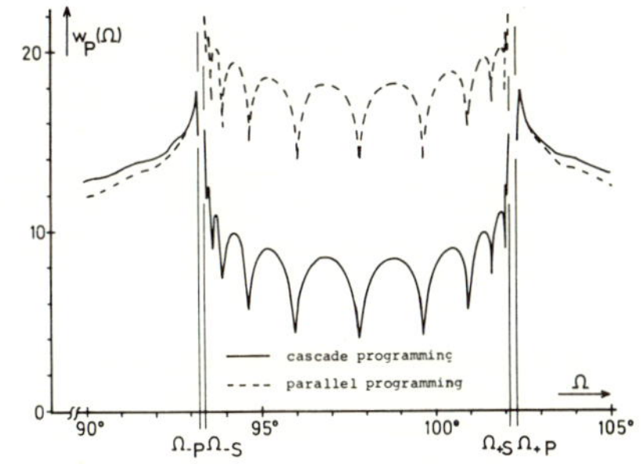

Fig. 7. Statistical word length function of a band rejection filter of the 22nd degree.

$$Q_p(\Omega) = \sqrt{3}\,\big|\, -2\,|H_w|\cdot\delta\cdot(1 - \max\epsilon_0) + \delta^2(1 - \max{}^2\epsilon_0)\,\big|\left(\sqrt{\sum_{\nu=1}^{n_e}\left(\frac{\partial P}{\partial a_\nu}\right)^2}\right)^{-1}. \tag{44}$$

With (14) we get the word length function

$$w_p(\Omega) = 2 + i_M - \operatorname{ld} Q_p(\Omega)$$

to estimate the word length

$$W_p = \max_{\Omega} w_p(\Omega). \tag{45}$$

Using the bandpass example, Fig. 6 shows the word length function $w_p(\Omega)$ of three canonical forms. According to this figure the word length estimations are reduced by about 1.5 bits in comparison to Fig. 5.

Fig. 7 shows the word length function $w_p(\Omega)$ of a band rejection filter of the 22nd degree first used by Kaiser [6] as an example. The tolerance scheme is given by

$$\Omega_{-P} = 93.20^\circ \quad \Omega_{-s} = 93.43^\circ \quad \Omega_{+s} = 102.10^\circ$$

$$\Omega_{+P} = 102.33^\circ \quad \delta_s = 2\cdot 10^{-4} \quad \delta_P = 0.037.$$

In the cascade structure the word length is determined by the passband behavior, while in the parallel structure the stopband requirements determine the word length. The difference is about four bits. In contrast to Knowles' and Olcayto's result, the cascade form should be favorable to parallel programming in this case.

IV. Direct Calculation of a Sufficient Word Length

For particular examples $|H(\Omega)|$, and according to (7) and (5), max ϵ can be calculated using a set of coefficients of certain word length W. In order to determine max ϵ, the extremal values of $|H(\Omega)|$ have to be calculated, and in addition the transfer function must be determined at the cutoff frequencies. Of course for each new set of coefficients the extremal values will be located at different frequencies. By checking whether max ϵ_w satisfies (4b) a sufficient word length can be calculated most easily. We introduce the function

$$d(W) = \operatorname{ld}\frac{\max\epsilon_W - \max\epsilon_0}{1 - \max\epsilon_0}. \tag{46}$$

Using $d(W)$, the condition (4b) will be transformed into

$$d(W) \le 0. \tag{47}$$

$d(W)$ has been calculated for the bandpass of the eighth degree, again using the three canonical structures (Fig. 8). It turns out that for the direct form a word length of 22 bits, for the two other forms 11 bits, is sufficient. Notice that $d(W)$ does not decrease monotonously. The results found by the two other methods are shown for comparison. The merits of the estimate are to be seen.

V. Optimization in Discrete Parameter Space

A. The Optimization Problem

If a magnitude function with rounded coefficients does not satisfy the tolerance scheme, we can try to vary the coefficients by an optimization procedure such that the figure of merit max ϵ becomes less than one. In doing so, only discrete coefficient values can be accepted.

$$a_\nu = l_\nu\cdot Q \tag{48}$$

where l_ν is an integer.

Generally speaking the normalized transfer function $H_N(\Omega)$ is determined by a set of

$$M_F = m + n \tag{49}$$

independent coefficients. Since in the case of selective filters the zeros are restricted to lie on the unit circle, the number of free parameters reduces to

$$M_F = \begin{cases} \dfrac{m}{2} + n, & m \text{ even} \\[2ex] \dfrac{m-1}{2} + n, & m \text{ odd.} \end{cases} \tag{50}$$

First, all optimization procedures based on evaluating the gradient can not be applied in this case. An approximate solution by integer linear programming is not possible

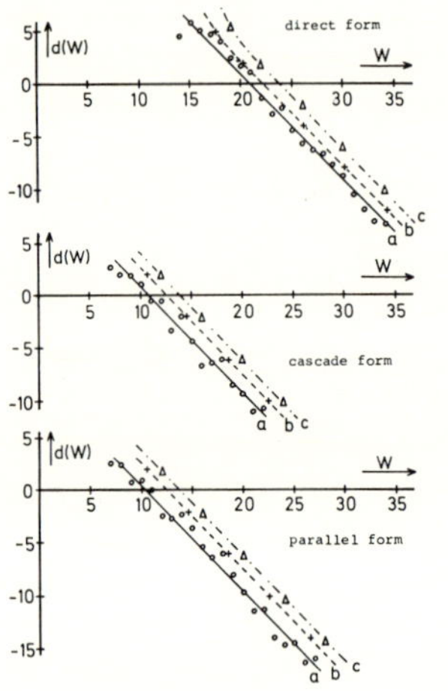

Fig. 8. Valuation of different structures. Line a: direct calculation $d(W)$; line b: statistical method; line c: upper bound method.

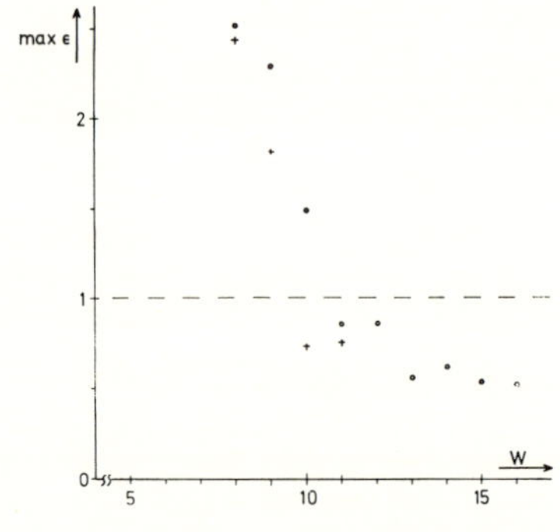

Fig. 9. The figure of merit as a function of the word length. 0: after rounding; +: after univariate search.

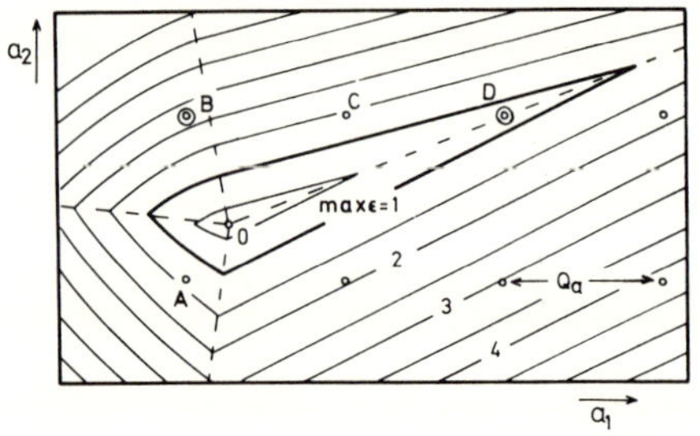

Fig. 10. The optimization problem in discrete parameter space.

either. The procedure of Gomory leads to a degenerate dual problem [7]. Three optimization procedures that have been investigated more thoroughly are described in the following section.

B. Global Search

The global minimum of max ϵ for coefficients of prescribed word length will be found in any case, if all points of the grid will be controlled in a sufficiently large range around the best continuous parameter set $\mathbf{a}$. But by choosing an M_F-dimensional cube with length Q of one side, we have to investigate 2^{M_F}-parameter sets. The bandpass of the eighth degree with $M_F=12$ free parameters, e.g., requires 4096 times the calculation of max ϵ, which needs 45 min computing time on the CD 3300. Nevertheless, the optimum has not been found, since the chosen cube was too small. Therefore, this procedure is not appropriate due to the prohibitively long computing time.

C. Univariate Search

We choose an initial parameter set $\mathbf{a}_i'$ in the discrete parameter space, which may be found by rounding the coefficients of the elliptic filter, and compute max ϵ_i (Fig. 1). The procedure is characterized by always varying only one parameter a_ν' in both directions by step-size Δa_ν, while the other parameters remain fixed. If we find a better parameter set by this way, it will become the new initial parameter set. The procedure is finished if all parameters have been varied without finding a better figure of merit. If we choose Δa_ν equal to the width of quantization Q, all parameter sets being investigated are points of the grid.

Fig. 9 shows max ϵ of the bandpass example after optimization at four different word lengths. In comparison with systems of rounded coefficients we observe an improvement. Nevertheless, we cannot find the global optimum by the aid of the univariate search.

D. A Modified Univariate Search

A difficulty one has to expect in the univariate search will be explained in a two-dimensional parameter space by Fig. 10. The closed lines connect spots with max ϵ being constant. Moreover, the points of the two-dimensional grid have been drawn in the neighborhood of the optimal continuous parameter set $\mathbf{a}$ (point 0). The central range contains all solutions of the continuous parameter space with max $\epsilon \leq 1$.

Rounding the coefficients of $\mathbf{a}$, point A will become the initial solution of the univariate search. The global optimum (point D) will not be found in discrete parameter space, since variation of only one coefficient by $\Delta a_\nu = Q$ will not lead from point A to a better figure of merit. Neither does the global search, if we choose a square with edge length Q.

On the other hand, if we reduce the word length by one bit, B may be the initial point, and both procedures will lead to the global optimum. This explains the nonmonotonous dependence on word length and max ϵ mentioned in Section IV.

An optimization procedure suitable in discrete parameter space, which is based on the univariate search, has to overstep points in the grid with unfavorable max ϵ, say from point A via B and C to point D.

The univariate search has been modified such that in addition to the initial solution, some ten parameter sets with the

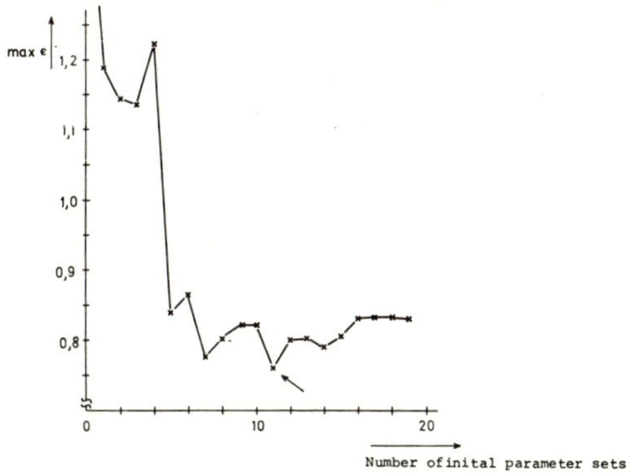

Fig. 11. Typical course of the figure of merit as a function of subsequent initial parameter sets.

next unfavorable max ε have been stored, not having been initial solutions so far. Now if no parameter variation of the initial solution will cause an improvement, the procedure will not stop; the parameter set with the next unfavorable figure of merit will become the new initial solution. By this way we can overstep points with higher levels. But since the stop criterion was lost, we have to fix a maximum number of initial points, say 20, which stops the procedure. Fig. 11 shows a typical course of max ε as a function of subsequent initial points.

It has been found that the procedure becomes much more efficient by combining it with the global search. We generate a matrix S the elements of which are

$$S_{K\nu} = \frac{\Delta\epsilon(\Omega_K)}{\Delta a_\nu}\bigg|_{\Delta a_\nu = Q}. \tag{51}$$

Here the Ω_K are the cutoff frequencies or the points, where $|H(\Omega)|$ has its extremal values. The $S_{K\nu}$ will be calculated during the univariate search anyway.

Using S we arrange the parameters a_ν according to their influence on max ε. Following this sequence we choose step-by-step four subsequent parameters and span a four-dimensional subspace for the global search with 2 Q being the length of one side of the cube. To save computing time the matrix S will be used to find the best max ε′ in the subspace by

$$\epsilon'(\Omega_\kappa) = \epsilon(\Omega_\kappa) + S\Delta a_\nu(\Omega_\kappa). \tag{52}$$

The result will be checked afterwards by computing the error function exactly.

Although it cannot be guaranteed that this method finds the global optimum, it turns out to be a great improvement if compared to the univariate search. It has been used to optimize the coefficients of ten different bandpass filters, having the same values $(\Omega_P - \Omega_{-P})$ and $(\Omega_S - \Omega_{-S})$ as the example in Fig. 1, but with different values Ω_P. Fig. 12 indicates the results for rounding to a step $Q = 2^{-7}$ and for optimization by the univariate search with and without modification. By the new procedure, a satisfying transfer

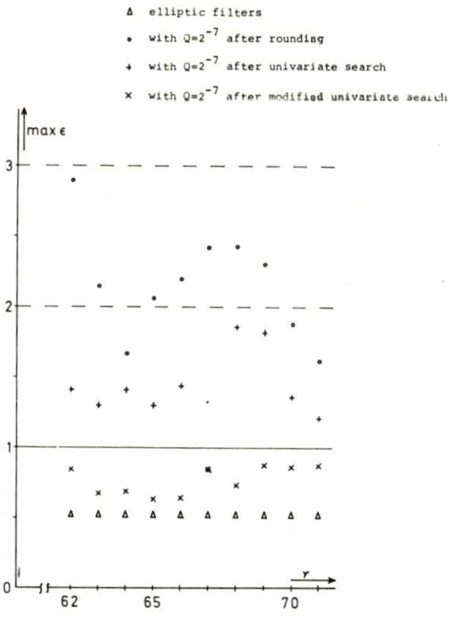

Fig. 12. Figures of merit of different bandpass filters.

function has been found in all examples, while the univariate search finds it only in one case.

Finally, Fig. 1 indicates how much can be won by this optimization technique. While $Q = 2^{-9}(W = 11)$ is needed, if the coefficients are rounded, only $Q = 2^{-6}(W = 8)$, i.e., three bits less are necessary, if the coefficients are optimized. The optimization of the bandpass filter takes about 10 min computing time.

Acknowledgment

The author wishes to thank Prof. W. Schüßler for stimulating this investigation, for numerous valuable suggestions, and for his help with the preparation of the manuscript. Calculations were made by means of the computer CD 3300 of the computation center of the Erlangen-Nürnberg University.

References

[1] W. Schüßler, "Zur allgemeinen Theorie der Verzweigungsnetzwerke," *Arch. Elek. Übertragung*, vol. 22, pp. 361–367, 1968.
[2] J. B. Knowles and E. M. Olcayto, "Coefficient accuracy and digital filter response," *IEEE Trans. Circuit Theory*, vol. CT-15, pp. 31–41, Mar. 1968.
[3] O. Mildenberger, "Die Genauigkeit der Lösung von Differenzengleichungen in Abhängigkeit von der Stellenzahl des verwendeten Rechners," *Arch. Elek. Übertragung*, vol. 25, pp. 90–100, Feb. 1971.
[4] E. Avenhaus, "Zum Entwurf digitaler Filter mit minimaler Speicherwortlänge für Koeffizienten und Zustandsgrößen," in *Ausgewählte Arbeiten über Nachrichtensysteme*, no. 13. Erlangen, Germany: Erlangen Univ., 1971.
[5] M. Fisz, *Wahrscheinlichkeitsrechnung und mathematische Statistik*. Berlin: VEB Deutscher Verlag der Wissenschaften, 1966, p. 181.
[6] R. M. Golden and J. F. Kaiser, "Design of wideband sampled-data filters," *Bell Syst. Tech. J.*, vol. 43, pp. 1533–1546, July 1964.
[7] S. J. Gass, *Linear Programming*. New York: McGraw-Hill, 1964 pp. 154–166.

Reprinted from *IEEE Trans. Audio Electroacoustics,* **AU-18**(4), 456–463 (1970)

Instability Thresholds in Digital Filters Due to Coefficient Rounding

ROBERT K. OTNES, Member, IEEE
University Software Systems
Los Angeles, Calif. 90024

LAWRENCE P. McNAMEE, Member, IEEE
Department of Computer Science
University of California
Los Angeles, Calif. 90024

Abstract

Asymptotic expressions are developed to determine the threshold frequency at which sine and tangent Butterworth digital low-pass recursive filters become unstable because of coefficient rounding. Both floating point and fixed point arithmetic is considered. The theoretical threshold values are compared with those obtained through actual operation of the filter on data.

Manuscript received March 28, 1970; revised June 11, 1970.

I. Introduction

Many of the basic design problems of digital filters have been solved using the techniques discussed in [1], [3], [5], [6], and [14].

If a computer were available for which the arithmetic was perfect, i.e., computer word lengths were infinite and there would be no overflow, underflow, or other errors when arithmetic operations are performed, then digital filters could be implemented whose action on data would exactly duplicate such desired design criterion.

Naturally, all real computers truncate both the filter coefficients and the data. The investigation of the resulting degradation of the filter performance currently falls into three areas of investigations.

1) Measurement of the deviation of the implemented filter from the ideal. For example, the mean-square error criterion is employed in [11] and [13].

2) Regarding the underflow (due to additions in the fixed point case or to addition and multiplication in the floating point case) as a noise source, and computing the spectrum of the output noise. This approach has been used in [2], [7]–[10], and more recently, [17].

3) Examination of both the computer arithmetic and word format on the one hand, and the filter and its method of implementation on the other, for the purpose of determining at what point these variables will cause a filter (designed to be stable) to become unstable. It is this last area that this paper explores.

Kaiser in [6] and [12] and others [16], discuss the stability threshold problem for the combined filter. Kaiser's solution is of the form

$$m_b > 1 + \left[-\log_2 \left(\frac{5\sqrt{N}}{2^{N+2}} \prod_{k=1}^{N} p_k T \right) \right] \tag{1}$$

where

m_b = the minimum number of binary digits required to represent the word for a stable filter of the low-pass type
N = the number of poles of the filter
T = the sampling interval in seconds
p_k = the kth pole of the transfer function in radians per second.

Equation (1) has the disadvantage of not giving a result directly in terms of bandwidth. The exact definition of bandwidth will be given in (25). Essentially, it is a parameter which tells the filter user which frequencies are transmitted by the filter; in the low-pass case, sinusoids with frequencies less than the bandwidth tend to be passed without change, while those with frequencies greater than the bandwidth are attenuated. Bandwidth is one of the most commonly used filter design characteristics. As will be shown, a threshold for instability on a floating point computer for the combined filter can be given in the form

$$m_b = B_N - N \log_2(2\pi BT) \tag{2}$$

where

B = minimum attainable bandwidth in Hz

β_N = the integer i such that

$$2^{i-1} \leq \binom{N}{[N/2]} < 2^i$$

where the brackets [] indicate the integer part of the enclosed quantity.

In the cascade or serial form of implementation where $N=2$, the filter has stages containing at most two poles, so that (2) applies in the form

$$B_{\text{cascade}} \approx \frac{2^{-(m_b-2)/2}}{2\pi T} \tag{3}$$

for the cascade filter. It is thus possible to go much lower in filter bandwidth in the cascade implementation than with the combined.

II. Terminology

In keeping with current notational usage, the input to the filter is a sequence $\{x_i\}$ and the output is the sequence $\{y_i\}$. The filtering action for the combined case is defined by the difference equation

$$y_i = \sum_{k=0}^{N} a_k x_{i-k} - \sum_{k=1}^{N} b_k y_{i-k}. \tag{4}$$

The filter is determined by the sets of coefficients $\{a_j\}$ and $\{b_j\}$. The cascade form of the same filter normally would have $[N/2]$ or $([N/2]+1)$ stages, most of them quadratic and having the same form as (4) with $N=2$ for each stage (and possibly one stage with $N=1$).

The Fourier transform of $\{x_i\}$, denoted by $X(f)$, is found from

$$X(f) = \sum_{i=-\infty}^{\infty} x_i \exp(-j2\pi f i T) \tag{5}$$

where f is in Hz. Employing a similar definition for $\{y_i\}$, the transfer function, the filter described by (4) may be found to be

$$H(f) = \frac{Y(f)}{X(f)} = \frac{\sum_{k=0}^{N} a_k \exp(-j2\pi f k T)}{1 + \sum_{k=1}^{N} b_k \exp(-j2\pi f k T)}. \tag{6}$$

As shown by Holtz and Leondes [4], $|H(f)|^2$ can always be reduced to a rational function in powers of $\sin^2(\pi f T)$.

III. Second-Order Filters

Without the loss of generality, the second-order filter given by

$$y_i = a_o x_i - b_1 y_{i-1} - b_2 y_{i-2} \tag{7}$$

may be considered; additional $\{x_i\}$ terms will not affect the stability of the filter. The investigation of instability will begin with the assumption that (7) is implemented on a perfect computer, that is, a computer with infinite word-length and with no underflow or overflow when computations are performed. The transfer function of the filter defined by (7) is

$$H(f) = \frac{a_o}{1 + b_1 \exp(-j2\pi f T) + b_2 \exp(-j4\pi f T)}. \tag{8}$$

The poles of the denominator of (8) are examined by finding solutions to

$$1 + b_1 \exp(-j2\pi f T) + b_2 \exp(-j4\pi f T) = 0. \tag{9}$$

The roots of (9) are

$$\{\exp(j2\pi f_1 T), \exp(j2\pi f_2 T)\} = \frac{-b_1 \pm \sqrt{b_1^2 - 4b_2}}{2} \tag{10}$$

or

$$\{f_1, f_2\} = \frac{1}{j2\pi T} \ln\left[\frac{-b_1 \pm \sqrt{b_1^2 - 4b_2}}{2}\right] \tag{11}$$

choosing the plus sign for f_1 and the minus sign for f_2. These roots are periodic and the variable f can take on complex values. Therefore, defining

$$f = \nu + j\sigma \tag{12}$$

where both ν and σ are real, two stable regions can be determined. If

$$b_1^2 < 4b_2 \tag{13}$$

then corresponding to complex roots,

$$\nu = \pm \frac{1}{2\pi T} \arctan \sqrt{4b_2/b_1^2 - 1}$$

and

$$\sigma = -\frac{1}{4\pi T} \ln b_2. \tag{14}$$

If

$$b_1^2 \geq 4b_2 \tag{15}$$

then corresponding to real roots,

$$\nu = 0 \quad \text{or} \quad \frac{1}{2T}$$

and

$$\sigma = -\frac{1}{2\pi T} \ln\left[\frac{-b_1 \pm \sqrt{b_1^2 - 4b_2}}{2}\right]. \tag{16}$$

The areas in the (b_1, b_2) plane corresponding to either the real or complex roots are shown in Fig. 1. Second-order filters are stable only within the triangular area shown in the figure. That is, σ is greater than zero within the triangle. In the frequency plane, the triangle discussed maps into the upper half of the (ν, σ) plane. This is shown in Fig. 2.

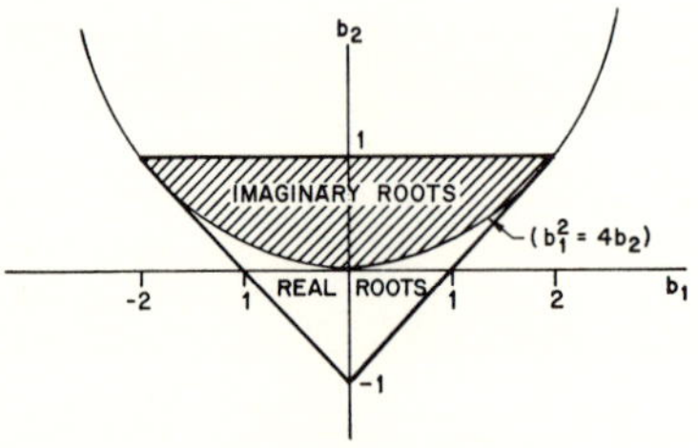

Fig. 1. Second-order coefficient space. All coefficient pairs within the triangle are stable.

Fig. 2. Points within the triangle of stability in the (b_1, b_2) plane map into the shaded area. The shaded portion represents the principal area, and repeats every $1/T$ units along the ν axis.

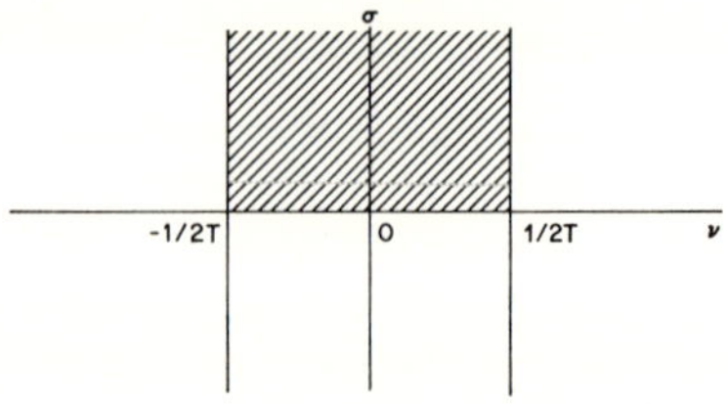

The threshold model accounts for the manner in which the filtering becomes unstable by simple rounding of the filter weights only. Assume that b_1 and b_2 are represented as *binary* numbers which have been rounded to m fractional digits:

$$\cdots XXXX \cdot \underbrace{XX \cdots XX}_{m \text{ binary digits}} 00000.$$

Rounding to m binary places is defined as adding $2^{-(m+1)}$ (a bit in position $m+1$) and then discarding all bits beyond position m. All of the other calculations for the filtering are the same. This model is useful whenever it is not necessary to discuss the roundoff error as noise.

When b_1 and b_2 are rounded, the position of the poles in the (ν, σ) plane shifts. In some cases this shift is enough to cause instability. Although it would appear fruitful to look for an end result in the (ν, σ) plane, as such a result will be given in terms of frequency, the effects of rounding can be clarified by first examining the (b_1, b_2) plane. Fig. 3 shows the $(b_1 b_2)$ plane for the case where m is equal to 2. That is, the two weights have been rounded to two binary positions. The previously infinite number of possible, stable two weight filters has been mapped into a discrete set of 81 weights, 49 of which correspond to stable filters, and 32 of which correspond to unstable filters.

The upper, left-hand corner of the triangle corresponds to low-pass filter weights. The shaded area corresponds to the two weight filters which were stable before rounding but unstable afterwards. This area, for the low-

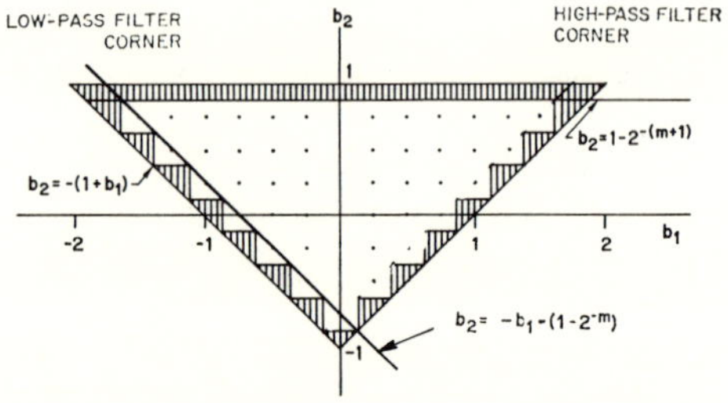

Fig. 3. Within the stability triangle, all perfect computer second-order filters are stable. On the other hand, for a computer model for which there is truncation of the weights with $m=2$, the weights must originally have been within the area bounded by the shaded portion to be stable, and can in fact only take on values corresponding to the points shown in the figure.

pass case, is bounded by two sets of equations. The first set is

$$b_2 = 1$$

and

$$b_2 = 1 - 2^{-(m+1)}. \tag{17}$$

These equations define the strip at the top of the triangle with a width of $2^{-(m+1)}$ units. The area along the left leg of the triangle is more difficult to describe. The sawtooth edge would be tedious to delineate mathematically. However, it does lie within the strip defined by the following set of equations.

$$b_2 = -b_1 - 1$$

and

$$b_2 = -b_1 - (1 - 2^{-m}). \tag{18}$$

Any point (pair of weights) falling within the strip defined by (18) has about an even chance of being unstable. These equations will be used to define the instability threshold.

Consider next the relationship of (17) and (18) to the frequency plane (ν, σ). For $b_1^2 < 4b_2$, (17) maps into:

$$\nu = \frac{1}{2\pi T} \arctan \sqrt{4(1 - 2^{-(m+1)})/b_1^2 - 1} \tag{19}$$

and

$$\sigma = -\frac{1}{4\pi T} \ln (1 - 2^{-(m+1)}). \tag{20}$$

For large m, (20) becomes

$$\sigma \approx -\frac{1}{4\pi T} [-2^{-(m+1)}] = \frac{2^{-(m+2)}}{2\pi T} \cdot \tag{21a}$$

On the other hand, (18), the sawtooth bounding equation, has two intersections which must be looked at: 1) the intersection of (18) and $b_2 = 1$, labeled P_1; and 2) the

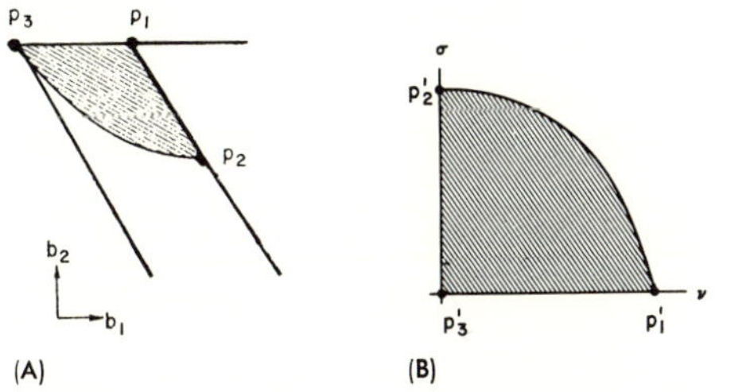

Fig. 4. The points p_1, p_2, and p_3 in the coefficient plane shown in (A) are mapped into the points p'_1, p'_2, and p'_3, respectively, in the frequency plane (B). The straight line segment p_1-p_2 maps (approximately) into a quarter of a circle in the frequency plane.

intersection of (18) and $b_1^2=4b_2$, labeled P_2. In the frequency plane (ν, σ) these become points P'_1 and P'_2: for P'_1

$$\begin{aligned}\sigma &= 0\\ \nu &= \frac{1}{2\pi T}\arctan\sqrt{4/(2-2^{-m})^2-1}\\ &= \frac{1}{2\pi T}\arctan\sqrt{(4\cdot 2^{-m}-2^{-2m})/(2-2^{-m})^2}\\ &\approx \frac{1}{2\pi T}\arctan 2^{-m/2}\\ &\approx \frac{2^{-m/2}}{2\pi T}\end{aligned} \tag{22}$$

for large enough m.

The second intersection yields for point P'_2

$$\begin{aligned}\nu &= 0\\ \sigma &= -\frac{1}{4\pi T}\ln(1-2\cdot 2^{-m/2}+2^{-m})\\ &\approx \frac{2^{-m/2}}{2\pi T}.\end{aligned} \tag{23}$$

These points are shown in Fig. 4.

A general result can be obtained; for large m, the equation

$$\nu^2+\sigma^2=\frac{2^{-m}}{(2\pi T)^2}, \qquad (\sigma>0) \tag{24a}$$

approximates the segment of (18) in the frequency plane between P'_1 and P'_2 (see Fig. 4). Define

$$B^2=\nu^2+\sigma^2. \tag{25}$$

Then

$$B=\frac{2^{-m/2}}{2(\pi T)}. \tag{26}$$

As

$$\sin\alpha \approx \tan\alpha \approx \alpha \tag{27}$$

for small α, B can be interpreted as the threshold frequency in Hz for which the filter becomes unstable. That is, if the poles are any closer to the (0, 0) point in the (ν, σ) plane than B Hz, the filter tends to be unstable. Note that this stability criterion in terms of B is larger than the criterion for σ given by (21a).

In summary, there are two thresholds for the low-pass filter:

$$B=\sqrt{\nu^2+\sigma^2}=\frac{2^{-m/2}}{(2\pi T)^2} \tag{24b}$$

and

$$\sigma=\frac{2^{-(m+2)}}{2\pi T}. \tag{21b}$$

In most practical situations, (24b) is the larger of the two bounds and will be encountered first.

As a filter made up of a cascade of stable filters is itself stable, (24b) can be regarded as the threshold of stability for the Butterworth low-pass filter if it is implemented in cascade form. This is discussed more fully in the following sections.

IV. Higher Order Low-Pass Filters

The results in the preceding section can be extended for higher order filters. The filter of chief interest is the Butterworth, low-pass, sine. Its transfer function has an absolute value squared of the form

$$|H(f)|^2=\frac{1}{1+\left(\dfrac{\sin(\pi f T)}{\sin(\pi B T)}\right)^{2N}}. \tag{28}$$

Here, B is the *cutoff frequency*, i.e., that frequency for which the power is reduced by one half. It has only one $\{a_j\}$ coefficient namely, a_o. The transfer function therefore has the form

$$H(f)=\frac{a_o}{1+\sum_{k=1}^{N} b_k\exp(-j2\pi f k T)}. \tag{29}$$

As it is required that

$$H(o)=1 \tag{30}$$

then

$$a_o=1+\sum_{k=1}^{N} b_k. \tag{31}$$

As noted by Kaiser [6], the right-hand side of (31) is very important. If it is *effectively* zero on a given digital computer, then that filter has a pole at 0 Hz, and thus becomes unstable. Therefore, it is necessary to find the circumstances for which

$$\left|1+\sum_{k=1}^{N} b_k\right| < 2^{-m}. \tag{32}$$

This will be accomplished by first finding an expression for (31), which will be referred to as a_o. As an intermediate step it is necessary to examine b_N, which may be shown to be

$$b_N = (-1)^N \exp\left(-\sum_{i=1}^{K} \lambda_i\right) \tag{33}$$

where the $|\lambda_i|$ terms are related to the damping ratios. In the Butterworth case under discussion, they are given by

$$\lambda_i = 2 \operatorname{arcsinh}\left[\sqrt{\frac{1}{2}\left\{-\cos^2\left(\frac{\lambda_B}{2}\right)+\sqrt{\cos^4\left(\frac{\lambda_B}{2}\right)+4\sin^2\left(\frac{\lambda_B}{2}\right)\sin^2(\theta_i)}\right\}}\right] \tag{34}$$

where

$$\theta_i = \frac{\pi}{2N}(N+1-2i), \qquad i = 1, \cdots, N. \tag{35}$$

Note that

$$\begin{aligned} |H(f)|^2 &= \frac{a_o}{(1+b_1z^{-1}+\cdots+b_Nz^{-N})}\cdot\frac{a_o}{(1+b_1z+\cdots+b_Nz^N)} \\ &= \frac{a_o^2}{[b_Nz^{-N}+\cdots+b_Nz^N]} \end{aligned} \tag{36}$$

where as usual

$$z = \exp(j2\pi fT). \tag{37}$$

Equation (28) may be arranged and expressed as

$$\begin{aligned} |H(f)|^2 &= \frac{\sin^{2N}(\pi BT)}{\sin^{2N}(\pi BT)+\sin^{2N}(\pi fT)} \\ &= \frac{[2\sin(\pi BT)]^{2N}}{[2\sin(\pi BT)]^{2N}+(-1)^N\sum_{k=0}^{2N}\binom{2N}{k}(-1)^k z^{-N+k}}. \end{aligned} \tag{38}$$

Equating the coefficients of equal powers of z^N in (36) and (38), it follows that

$$a_o^2 = (-1)^N b_N [2\sin(\pi BT)]^{2N} \tag{39}$$

so that

$$a_o = |b_N|^{1/2}[2\sin(\pi BT)]^N. \tag{40}$$

The stability threshold frequency B_m is defined by

$$[(-1)^N b_N]^{1/2}[2\sin(\pi B_m T)]^N = 2^{-m}. \tag{41}$$

For B_m less than this frequency, the condition defined by (32) holds and the filter is unstable.

For small BT,

$$[(-1)^N b_N]^{1/2N} \approx 1$$

and

$$\sin(\pi B_m T) \approx \pi B_m T \tag{42}$$

so that (41) can be rewritten as

$$B_m \approx \frac{2^{-m/N}}{2\pi T}. \tag{43}$$

When N is equal to 2, (43) is the same as (26).

This expression is plotted for various N in Fig. 5, with $T=0.005$. The curves show the drastic increase in the threshold as N increases. The result may also be used for the tangent Butterworth low-pass filter whose transfer function is of the form

$$H(f) = \frac{1}{1+\left(\dfrac{\tan(\pi fT)}{\tan(\pi BT)}\right)^{2N}}. \tag{44}$$

This may be seen by the following argument. For small BT, (27) holds and the poles of the two filters are approximately equal. When the poles are approximately equal, the filters will be unstable at about the same bandwidth. More formally, suppose that sine and tangent filters of the same order and same cutoff frequency are examined, and that r is one of the roots of unity to be used in the construction of the two filters, and that λ_s and λ_t are the sine and tangent poles, respectively, corresponding to r. That is

$$\sin\left(\frac{\lambda_s}{2}\right) = r\sin\left(\frac{\lambda_B}{2}\right) \tag{45}$$

and

$$\tan\left(\frac{\lambda_t}{2}\right) = r\tan\left(\frac{\lambda_B}{2}\right) \tag{46}$$

where

$$\lambda = 2\pi Tf. \tag{47}$$

Consider the ratio R defined b

$$R = \frac{\sin\left(\frac{\lambda_s}{2}\right)}{\sin\left(\frac{\lambda_t}{2}\right)} = \frac{r\sin\left(\frac{\lambda_B}{2}\right)}{\left\{\dfrac{r^2\tan^2\left(\frac{\lambda_B}{2}\right)}{1+r^2\tan^2\left(\frac{\lambda_B}{2}\right)}\right\}^{1/2}} = \sqrt{\cos^2\left(\frac{\lambda_B}{2}\right)+r^2\sin^2\left(\frac{\lambda_B}{2}\right)}. \tag{48}$$

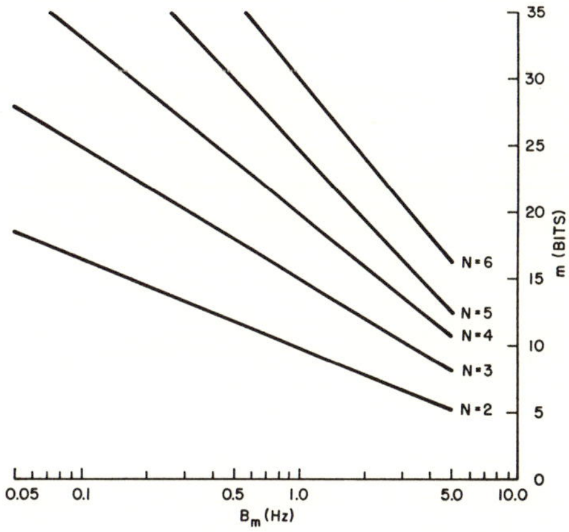

Fig. 5. Threshold frequencies for several values of N; $T=0.005$ is used so that B_m is effectively the percentage of the Nyquist folding frequency.

Suppose that r is defined by

$$r = \cos\theta + j\sin\theta. \tag{49}$$

Then

$$r^2 = \cos 2\theta + j\sin 2\theta \tag{50}$$

and the ratio squared can be written as

$$R^2 = \cos^2\left(\frac{\lambda_B}{2}\right) + [\cos 2\theta + j\sin 2\theta]\sin^2\left(\frac{\lambda_B}{2}\right). \tag{51}$$

Taking the absolute value squared of this,

$$|R^2|^4 = 1 - \cos^2\lambda_B\sin^2 2\theta \leq 1 \tag{52}$$

from which it follows that

$$\left|\sin\left(\frac{\lambda_s}{2}\right)\right| \leq \left|\sin\left(\frac{\lambda_t}{2}\right)\right|. \tag{53}$$

As the basic instability expression (41) is given in terms of sin ($\lambda/2$) rather than $\lambda/2$, it follows that the tangent form tends to be more stable than the sine version. In any event, they are asymptotically equal, so that it is reasonable to use (41) for both cases.

V. Theoretical and Actual Instability with Fixed and Floating Point Arithmetic

Equation (43) can be employed on actual machines with two additions to the model.

1) The number of binary digits m, is replaced with $(m_B-\beta_N)$, where m_B is the number of bits in the fixed point word or the number of bits in the mantissa of the floating point word on a given computer, and β_B is a corrective term; usually it is not possible to make use of all m_b bits; the corrective factor varies from machine to machine.

2) On an actual machine there will be underflow; in the case where fixed point arithmetic is used, it may occur when numbers are multiplied; in the floating point arithmetic case, it may occur both for addition and multiplication. In either case, this underflow may be regarded as a noise term. This sort of problem has been investigated by previously mentioned authors and is outside the scope of this paper.

The fixed point case is the easiest to describe. Suppose that the word on the fixed point machine has t bits as follows.

$$\overbrace{\underbrace{\alpha_0\alpha_1\alpha_2\cdots\alpha_{t-m-1}}_{t-m}\;\underbrace{\alpha_{t-m}\cdots\alpha_{t-1}}_{m}}^{t} \tag{54}$$

(with α_0 marked "sign bit", and the arrow between the two groups marking "location of the binary point")

There are various methods for taking into account the position of the binary point when filtering is performed. Here it will be assumed that the maximum value for m is chosen that can be employed without danger of *overflow*, and that m does not vary during the calculations.

The question then becomes, what is the minimum number of bits which must be set aside for the integer portion of the computer word. A partial answer arises from the following. Many times the data being processed comes from an analog-to-digital converter having q bits. It will be assumed that $q \leq m-1$, if q includes a sign bit. Until it is converted to engineering units, these digitized quantities can be regarded as being on the range $\pm(1-2^{-m})$. The filtered numbers can be larger than one; for example, a square wave alternating between plus and minus the maximum input value may rise to nearly $4/\pi$ of that value after being acted upon a low-pass filter which eliminates the higher harmonics.

Assume, however, that overshoot is not a problem. What remains is the question of how large m may be made as a function of the size of the weights. That is, the $(t-m-1)$ bits used to express that part of the number greater than unity must be at least large enough to express the largest coefficient. It can be readily demonstrated that the coefficients of the low-pass filter are bounded as follows.

$$|b_k| < \binom{N}{k} \tag{55}$$

with equality being achieved for zero bandwidth. Define β_N by

$$2^{(\beta_N-1)} < \binom{N}{\left[\frac{N}{2}\right]} \leq 2^{\beta_N}. \tag{56}$$

Then the number of bits needed to express the maximum coefficient in magnitude is less than or equal to β_N. Near the threshold, (45) nearly becomes an equality, so that the threshold itself becomes

TABLE I

Stability Threshold for some *t*-Bit Computers $T=0.005$

t (bits)	$B_{t,2}$ (Hz)	$B_{t,6}$ (Hz)
7	7.9	28
12	1.5	16
16	0.35	9.9
18	0.17	4.0

$$B_{t,N} \approx \frac{2^{-(t-1-\beta_N)/N}}{2\pi T}. \tag{57}$$

This approximately represents the lowest bound to which the threshold can be lowered on a t bit machine. If the cascade form is implemented, this becomes

$$B_{t,2} \approx \frac{2^{-(t-3)/2}}{2\pi T} \tag{58}$$

Some values are shown in Table I.

The floating point case can have unusual characteristics. The floating point word is made up of three parts: a sign, characteristic, and mantissa, usually in the following format.

$$\overbrace{\underset{\substack{\uparrow\\ \text{sign}\\ \text{bit}}}{\sigma} \quad \underbrace{k_1k_2\cdots k_P}_{\substack{\text{characteristic}\\ \text{bits}}} \quad \underbrace{\mu_1\mu_2\cdots\mu_Q}_{\substack{\text{mantissa}\\ \text{bits}}}}^{t\,=\,1+P+Q\ \text{bits}} \tag{59}$$

On the IBM 704, 709, 7090, etc., t is 36 and the data word defined by (59) would be

$$(-1)^\sigma \left(\sum_{q=1}^{27} \mu_q 2^{-q} \right) 2^{\left[\sum_{p=1}^{8} k_p 2^{8-p} - 128\right]}. \tag{60}$$

On the IBM 360 series there are 32 bits and there is a subtle difference. The data word has the form

$$(-1)^\sigma \left(\sum_{q=1}^{24} \mu_q 2^{-q} \right) 16^{\left[\sum_{p=1}^{7} k_p 2^{7-p} - 64\right]} \tag{61}$$

The fact that a power of 16 rather than a power of 2 is used as a scaling factor on the IBM 360 causes some difficulties.

The case for the IBM 704 format is a little easier to examine, so it will be discused first. There are 27 bits used to express the number. As at least two of these must be used for the portion of the number greater than zero, the largest possible number of bits for the fractional portion is 25. For higher order filters

$$m \geq 27 - \beta_N \tag{62}$$

and the threshold becomes

$$B_{\text{IBM } 704,N} \approx \frac{2^{-(27-\beta_N)/N}}{2\pi T}. \tag{63}$$

TABLE II

Threshold Values in Hz for Several Values of N for Two Computer Floating Point Word Formats

N	$\binom{N}{[N/2]}$	IBM 704 Format $m_{\max}$	$B_{m_{\max}}$	B_{actual}	IBM 360 Format $m_{\max}$	$B_{m_{\max}}$	B_{actual}
2	2	25	0.0054	—	20	0.032	—
3	3	24	0.13	0.10	20	0.32	0.15
4	6	23	0.6	0.4	20	1.0	1.0
5	10	23	1.3	1.3	20	1.0	2.0
6	20	22	2.4	2.4	16–20	3.2–5.0*	3.8

* For $N=6$ the threshold is difficult to calculate as $|b_3| \approx 16$, thus causing a rapid change in the effective number of bits.

TABLE III

Threshold Values in Hz for Fixed Point Arithmetic

N	$\binom{N}{[N/2]}$	Simulated $m=10$ B_{theory}	B_{actual}	Simulated $m=12$ B_{theory}	B_{actual}
2	2	1.0	0.9	0.5	0.4
3	3	3.12	3.2	2.0	1.8
4	6	5.66	5.7	3.97	3.8
5	10	7.95	8.2	6.10	6.1
6	20	10.0	10.4	8.95	9.1

The IBM 360 cannot be simplified this easily. On the IBM 704, μ_1 is always a one bit because floating point numbers are always normalized, i.e., shifted so that the most significant digit is at position μ_1 and the characteristic changed accordingly. In the IBM 360 floating point word μ_1, μ_2, and μ_3 may be zero as the normalizing is done in groups of four bits. This is consistent with using a power of 16 rather than 2. It thus turns out that p_N is a multiple of 4.

This difference shows up in Table II, which has values of $m_{\max}$ tabulated for the two computers along with theoretical and actual thresholds.

The actual thresholds were found by filtering unit step functions and observing the output of the filter. If the output differed significantly from unity, the filter was assumed to be unstable. As $m_{\max}$ is generally greater than the actual number of bits required to represent the portion of the largest coefficient which is greater than 1, the theoretical results are generally conservative.

Table III lists some fixed point results obtained from a fixed point arithmetic simulation program on an IBM 360 computer. Again, agreement is on the conservative side.

VI. Conclusions and Comments

A threshold of stability as a function of the number of bits in the fraction portion of the filter coefficient has been derived for the sine and tangent Butterworth low-pass filters. The resulting formula is more specific than that of Kaiser given in [6] in the sense that it establishes a stability threshold directly in terms of bandwidth.

These results can be extended to other filter types. Thresholds for the Chebyshev, Gaussian, and Bessel filters can possibly be found in terms of multiples of the result for the sine-Butterworth case.

The extension to the high-pass case is obvious. The bandpass and band reject thresholds appear to be somewhat more complicated.

References

[1] L. D. Enochson and R. K. Otnes, *Programming and Analysis of Digital Time Series Data*, SVM-3. Washington, D. C.: U. S. Printing Office, 1968.

[2] B. Gold and C. M. Rader, "Effects of quantization noise in digital filters," in *1966 Spring Joint Computer Conf.*, vol. 28. Washington, D. C.: Spartan, 1966, pp. 213–219.

[3] R. M. Golden, "Digital filter synthesis by sampled-data transformation," *IEEE Trans. Audio Electroacoust.*, vol. AU-16, pp. 321–329, September 1968.

[4] H. Holtz and C. T. Leondes, "The synthesis of recursive digital filters," *J. Ass. Comput. Mach.*, vol. 13, pp. 262–280, April 1966.

[5] J. F. Kaiser, "Design methods for sampled data filters," *Proc. 1st Allerton Conf. Circuit Syst. Theory* (Allerton, Ill.), pp. 221–236, November 1963.

[6] —, "Some practical considerations in the realization of linear digital filters," *Proc. 3rd Allerton Conf. Circuit Syst. Theory* (Monticello, Ill.), pp. 621–633, October 1965.

[7] J. B. Knowles and R. Edwards, "Complex cascade programming and associated computational errors," *Electron. Lett.*, vol. 1, pp. 160–161, August 1965.

[8] —, "Effect of a finite-word-length computer in a sampled-data feedback system," *Proc. Inst. Elec. Eng.*, vol. 112, pp. 1197–1207, June 1965.

[9] —, "Finite word-length effects in multirate direct digital control systems," *Proc. Inst. Elec. Eng.*, vol. 112, pp. 2376–2384, December 1965.

[10] —, "Simplified analysis of computational errors in a feedback system incorporating a digital computer," presented at Society of Instrument Testing Symp. Direct Digital Control (London), April 22, 1965.

[11] J. B. Knowles and E. M. Olcayto, "Coefficient accuracy and digital filter response," *IEEE Trans. Circuit Theory*, vol. CT-15, pp. 31–41, March 1968.

[12] F. F. Kuo and J. F. Kaiser, Eds., *Systems Analysis by Digital Computer*. New York: Wiley, 1966.

[13] B. Liu and T. Kaneko, "Error analysis of digital filters realized with floating-point arithmetic," *Proc. IEEE*, vol. 57, pp. 1735–1747, October 1969.

[14] R. K. Otnes, "An elementary design procedure for digital filters," *IEEE Trans. Audio Electroacoust.*, vol. AU-16, pp. 330–335, September 1968.

[15] C. M. Rader and B. Gold, "Digital filter design techniques in the frequency domain," *Proc. IEEE*, vol. 55, pp. 149–171, February 1967.

[16] C. S. Weaver, J. von der Groeben, P. E. Mantey, J. G. Toole, C. A. Cole, Jr., J. W. Fitzgerald, and R. W. Lawrence, "Digital filtering with applications to electrocardiogram processing," *IEEE Trans. Audio Electroacoust.*, vol. AU-16, pp. 350–391, September 1968.

[17] C. Weinstein and A. V. Oppenheim, "A comparison of roundoff noise in floating point and fixed point digital filter realizations," *Proc. IEEE* (Lett.), vol. 57, pp. 1181–1183, June 1969.

20

Reprinted from *IEEE Trans. Audio Electroacoustics,* **AU-21**(1), 27–30 (1973)

An Absolute Bound on Limit Cycles Due to Roundoff Errors in Digital Filters

JAMES L. LONG and TIMOTHY N. TRICK

***Abstract*—An absolute bound on limit cycle oscillations in fixed-point digital filter implementations due to roundoff errors is presented. Periodicity of the limit cycles is assumed in the derivation. Useful design results are explicitly given for the case of second-order filter sections. In addition it is shown that this bound is equal to the rms bound of Sandberg and Kaiser for real roots, and is never more than a factor of two greater than the rms bound for complex roots for second-order filters.**

I. Introduction

In the design of fixed-point digital filters one is particularly interested in determining the minimum bit-per-word requirement for the internal arithmetic. Bounds on the maximum value of the error due to roundoff induced limit cycles are of prime importance in making that determination. In this paper, a new bound on the maximum amplitude of limit cycles in fixed-point digital filters is presented.

II. Derivation of an Absolute Bound for Periodic Limit Cycles

Consider the Nth-order nonlinear difference equation (1) representing the direct form digital-filter realization

$$y(n) = -[\alpha_1 y(n-1)]_r - [\alpha_2 y(n-2)]_r - \cdots - [\alpha_N y(n-N)]_r + x(n) \quad (1)$$

where $[\cdot]_r$ denotes the standard quantization function. The coefficients $\alpha_1, \alpha_2, \cdots, \alpha_N$ and the numbers $y(-N), y(-N+1), \cdots, y(-1), y(0), \cdots$ are assumed to be fixed-point machine numbers. Roundoff after multiplication of these numbers accounts for the need of quantization functions in the otherwise linear equation (1). Each product is rounded before addition takes place, as is often the case in practical filters, to alleviate the need for double precision registers. Throughout this paper the independent forcing term $x(n)$ will be taken to be zero and the coefficients in (1) are such that the equation without quantizers is asymptotically stable.

Equation (1) may be replaced by a linear difference equation driven by roundoff induced error sequences $e_1(n), e_2(n), \cdots, e_N(n)$ as follows:

$$y(n) = -\alpha_1 y(n-1) + e_1(n) - \alpha_2 y(n-2) + e_2(n) - \cdots - \alpha_N y(n-N) + e_N(n). \quad (2)$$

Assuming a normalized quantization width of 1, the error terms in (2) are bounded in magnitude of $\frac{1}{2}$. A single error term may be defined as follows:

$$e(n) \triangleq e_1(n) + e_2(n) + \cdots + e_N(n) \quad (3)$$

where $|e(n)| \leq N/2$.

Now suppose that the limit cycle is periodic of period M; then the error input sequence has the property that

$$e(k) = e(k+M) \quad (4)$$

for all k. The steady-state output due to the roundoff induced error $e(n)$ is given by the discrete convolution

$$y(n) = \sum_{k=-\infty}^{n} h(n-k)e(k) \quad (5)$$

which can be decomposed in a manner directly analogous to the method of Trick and Anderson [3] for the case of a continuous convolution. First, (5) is expressed as a double summation

$$y(n) = \sum_{i=0}^{\infty}\left[\sum_{k=n-(i+1)M+1}^{k=n-iM} h(n-k)e(k)\right]. \quad (6)$$

Letting $p \triangleq n-iM-k$ one obtains

$$y(n) = \sum_{i=0}^{\infty}\left[\sum_{p=0}^{p=M-1} h(p+iM)e(n-iM-p)\right]. \quad (7)$$

Interchanging the summation in (7) and noting that $e(n-p-iM) = e(n-p)$ we have

$$y(n) = \sum_{p=0}^{M-1} e(n-p)\left[\sum_{i=0}^{\infty} h(p+iM)\right]. \quad (8)$$

The desired bound follows directly from (8) as

$$|y(n)|_{max} = \frac{N}{2}\sum_{p=0}^{M-1}\left|\sum_{i=0}^{\infty} h(p+iM)\right| \quad (9)$$

for any value of n.

A bound on the right-hand side of (9) for any M and the asymptotic behavior for large M are of particular interest. Considering (9) it is obvious that

$$\sum_{p=0}^{M-1}\left|\sum_{i=0}^{\infty} h(p+iM)\right| \leq \sum_{p=0}^{M-1}\sum_{i=0}^{\infty} |h(p+iM)|. \quad (10)$$

Letting $k \triangleq p+iM$ and interchanging order of summation in (10) gives

Manuscript received July 28, 1972; revised September 28, 1972. This work was supported by the Joint Services Electronics Program (U. S. Army, U. S. Navy, and U. S. Air Force) under Contract DAAB-07-67-C-0199.

The authors are with the Coordinated Science Laboratory, University of Illinois, Urbana, Ill. 61801.

$$\sum_{p=0}^{M-1}\sum_{i=0}^{\infty}|h(p+iM)| = \sum_{i=0}^{\infty}\sum_{k=iM}^{k=(i+1)M-1}|h(k)| = \sum_{k=0}^{\infty}|h(k)|. \quad (11)$$

Combining (9) and (11) one gets

$$|y(n)|_{\max} \le \frac{N}{2}\sum_{k=0}^{\infty}|h(k)| \quad (12)$$

which is independent of the value of n and M. Utilizing the asymptotic stability property assumed on $h(p)$ one can show that

$$\lim_{M\to\infty}\sum_{p=0}^{M-1}\left|\sum_{i=0}^{\infty}h(p+iM)\right| = \sum_{p=0}^{\infty}|h(p)|. \quad (13)$$

III. Absolute Bound for Second-Order Filters

Equation (9) can be reduced to an explicit function of the filter coefficients for the commonly used second-order filter. Consider the second-order transfer function

$$H(z) = \frac{z^2}{z^2+az+b}, \quad (14)$$

where letters a and b are used in place of α_1 and α_2 to agree with standard notation. The inversion formula for (14) is

$$h(p) = \frac{1}{2\pi j}\int_c \frac{z^2}{z^2+az+b}\cdot z^{p-1}dz \quad (15)$$

where c is taken to be the unit circle; therefore,

$$\sum_{i=0}^{\infty}h(p+iM) = \sum_{i=0}^{\infty}\left\{\frac{1}{2\pi j}\int_c \frac{z^{p+iM+1}}{z^2+az+b}dz\right\}. \quad (16)$$

Combining the results obtained by evaluating the right-hand side of (16) by the residue theorem with (9), one gets for the case of distinct roots of the equation z^2+az+b

$$|y(n)|_{\max} = \sum_{p=0}^{M-1}\left|\frac{1}{2\sqrt{\left(\frac{a}{2}\right)^2-b}}\left[\frac{\left(-\frac{a}{2}+\sqrt{\left(\frac{a}{2}\right)^2-b}\right)^{p+1}}{1-\left(-\frac{a}{2}+\sqrt{\left(\frac{a}{2}\right)^2-b}\right)^M} - \frac{\left(-\frac{a}{2}-\sqrt{\left(\frac{a}{2}\right)^2-b}\right)^{p+1}}{1-\left(-\frac{a}{2}-\sqrt{\left(\frac{a}{2}\right)^2-b}\right)^M}\right]\right| \quad (17)$$

and for the repeated root case

$$|y(n)|_{\max} = \sum_{p=0}^{M-1}\left|\left(-\frac{a}{2}\right)^p\left[\frac{p}{\left[1-\left(-\frac{a}{2}\right)^M\right]} + \frac{1+\left(-\frac{a}{2}\right)^M(M-1)}{\left[1-\left(-\frac{a}{2}\right)^M\right]^2}\right]\right|. \quad (18)$$

Note that (17) and (18) are computationally more attractive than the "matrix" method of Parker and Hess [2].

Two important special cases of the "periodic" bound (17) are when $M=1$ and $M=2$, corresponding to frequencies of dc and one-half the sampling frequency:

$$|y(n)|_{\max} = \frac{1}{1+a+b}, \qquad M=1 \quad (19)$$

$$|y(n)|_{\max} = \frac{1}{1+b-|a|}, \qquad M=2. \quad (20)$$

These bounds are the same as those derived by Jackson [5] for each of the above special cases, and they are also identical to the results obtained by Parker and Hess [2] with their matrix bound and by Kaiser and Sandberg [1] with their rms bound.

In general, the period of the limit cycle is not known, which makes the summation of the absolute value of the impulse response an attractive bound since it is the least upper bound on (17) independent of M. The evaluation of $\sum_{n=0}^{\infty}|h(n)|$ can be done exactly for the cases of real roots by straightforward summation of terms yielding the following:

$$\sum_{n=0}^{\infty}|h(n)| = \frac{1}{1-|a|+b} \quad (21)$$

for distinct real roots, and

$$\sum_{n=0}^{\infty}|h(n)| = \frac{1}{(1-\sqrt{b})^2} \quad (22)$$

for a double real root where $a^2=4b$.

Obtaining an exact expression for the case of complex roots is difficult, since the impulse response of a second-order filter with complex roots has the form

$$h(n) = \frac{r^n}{\sin\theta}\sin(n+1)\theta \quad (23)$$

where $r=\sqrt{b}$, $\theta=\cos^{-1}(-a/2\sqrt{b})$, and $4b>a^2$. In this case the least upper bound on the absolute value of the roundoff error is

$$\sum_{n=0}^{\infty}|h(n)| = \sum_{n=0}^{\infty}\left|\frac{r^n}{\sin\theta}\sin(n+1)\theta\right|. \quad (24)$$

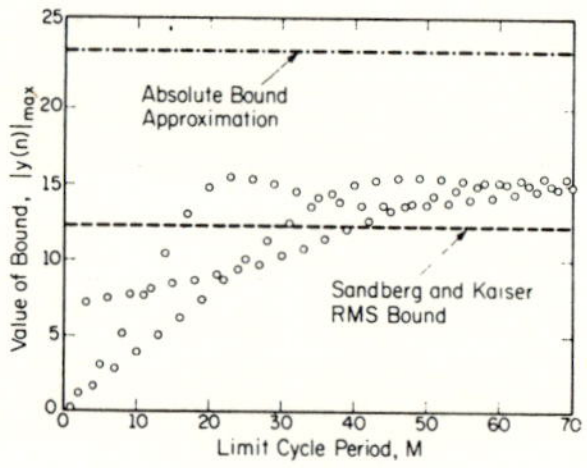

Fig. 1. Amplitude bound for $a=1.09$ and $b=0.9$.

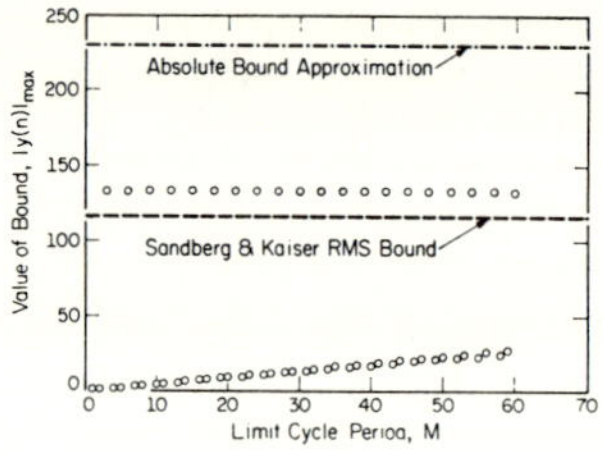

Fig. 2. Amplitude bound for $a=1.0$ and $b=0.99$.

The following two estimates of (24) can be useful for design purposes:

$$\sum_{n=0}^{\infty} | h(n) | \leq \sum_{n=0}^{\infty} \left| \frac{r^n}{\sin \theta} \right| = \frac{1+\sqrt{b}}{(1-b)\sqrt{1-\dfrac{a^2}{4b}}} \tag{25}$$

or

$$\sum_{n=0}^{\infty} | h(n) | \leq \sum_{n=0}^{\infty} (n+1) r^n = \frac{1}{(1-\sqrt{b})^2}. \tag{26}$$

Thus the absolute value of the roundoff error is bounded by

$$| y(n) |_{\max} \leq \begin{cases} \dfrac{1}{1-|a|+b}, & \text{for either } b \leq 0 \text{ or } b > 0 \text{ and } 2\sqrt{b} \leq |a| \quad (27) \\ \dfrac{1}{(1-\sqrt{b})^2}, & \text{for } b > 0 \text{ and } 2b\sqrt{\dfrac{2}{\sqrt{b}}} - 1 \leq |a| \leq 2\sqrt{b} \quad (28) \\ \dfrac{(1+\sqrt{b})}{(1-b)\sqrt{1-\dfrac{a^2}{4b}}}, & \text{for } b > 0 \text{ and } |a| \leq 2b\sqrt{\dfrac{2}{\sqrt{b}}} - 1. \quad (29) \end{cases}$$

IV. Comparison of Results

A computer program was used to evaluate the bound (17) for the case $M=7$ and $b=0.9$ while the parameter a was stepped from -1.7 to 1.7. The results of this computation were identical to those displayed graphically for the same parameter values by Hess [4] using his "matrix" method. Figs. 1 and 2 were also constructed using (17). Fixed values of a and b were used while M was varied. The rms bound of Kaiser and Sandberg [1] as well as the absolute bound derived in this paper are displayed on the graphs. One can see by the asymptotic behavior for large M that the absolute bound for complex roots is not the least upper bound due to the approximations made in (25) and (26). Values of the bounds in Figs. 1 and 2 can also be compared to the heuristic bound

$$| y(n) |_{\max} \leq \text{integer part of} \left[\frac{0.5}{1-b} \right] \tag{30}$$

which was derived by Jackson [5] using an "effective-value model." This bound yields numerical values of 5 for the example illustrated in Fig. 1 and 50 for the example illustrated in Fig. 2. However, Parker and Hess [2] have shown by example that (30) does not always bound the limit cycle, whereas (27)–(29) do.

Interesting comparisons can also be made between the absolute bounds (27)–(29) and the rms bounds obtained by Sandberg and Kaiser. Their results are as follows (using notation consistent with this paper):

$$\left(\frac{1}{M}\sum_{n=0}^{M-1} y^2(n)\right)^{1/2} \leq \rho(a, b) \tag{31}$$

for any $M>0$ and where

$$\rho(a, b) = \begin{cases} \dfrac{1}{1+b-|a|}, & \text{for either } b \leq 0 \text{ or } b > 0 \text{ and } |a| \geq \dfrac{4b}{1+b} \quad (32) \\ \dfrac{1}{(1-b)\left(1-\dfrac{a^2}{4b}\right)^{1/2}}, & \text{for } b > 0 \text{ and } |a| < \dfrac{4b}{1+b}. \quad (33) \end{cases}$$

One should make the important observation that (32) is identical to (27) over the region where $a^2-4b\geq 0$. Therefore, the rms bound (32) is also an absolute bound over the region $a^2-4b\geq 0$. Comparison of (29) and (33) over the region $b>0$ and $|a|<4b/(1+b)$ shows that they differ by a factor of $(1+\sqrt{b})$. Note that (29) cannot exceed (33) by more than a factor of two over the specified region and for small values of b they are approximately the same. The following important result can be shown by careful comparison of the bounds:

$$\frac{|y(n)|_{\max}}{\rho(a, b)} \leq 2 \tag{34}$$

for any a and b for which the linear filter is asymptotically stable. Realizing this fact, the designer using the rms bound can be certain that the maximum value of the limit cycle will not exceed this bound by a factor of more than two.

References

[1] I. W. Sandberg and J. F. Kaiser, "A bound on limit cycles in fixed-point implementations of digital filters," *IEEE Trans. Audio Electroacoust.*, vol. AU-20, pp. 110–112, June 1972.

[2] S. R. Parker and S. F. Hess, "Limit-cycle oscillations in digital filters," *IEEE Trans. Circuit Theory*, vol. CT-18, pp. 687–697, Nov. 1971.

[3] T. N. Trick and D. R. Anderson, "Stability of nonlinear networks," *IEEE Trans. Circuit Theory*, vol. CT-16, pp. 302–311, Aug. 1969.

[4] S. F. Hess, "A deterministic analysis of limit cycle oscillations in recursive digital filters due to quantization," Ph.D. dissertation, Naval Postgraduate School, Monterey, Calif., Dec. 1970.

[5] L. B. Jackson, "An analysis of limit cycles due to multiplication rounding in recursive digital (sub) filters," in *Proc. 7th Annual Allerton Conf. Circuit and System Theory*, pp. 69–78, Oct. 1969.

21

Reprinted from *IEEE Trans. Audio Electroacoustics,* **AU-20**(2), 110–112 (1972)

A Bound on Limit Cycles in Fixed-Point Implementations of Digital Filters

IRWIN W. SANDBERG, Member, IEEE
JAMES F. KAISER, Senior Member, IEEE
Bell Telephone Lab., Inc.
Murray Hill, N. J. 07974

Abstract

We present an upper bound on the rms value of self-sustained limit cycles in fixed-point implementations of digital filters. The bound can be easily evaluated for the important special case of sections of order two, and simulation results for this case show that the bound is a useful tool.

I. Introduction

The primary purpose of this paper is to present an upper bound on the rms value of self-sustained limit cycles due to roundoff in fixed-point implementations of digital filters. We consider in greatest detail the important practical case of second-order sections. However, our main result, the theorem proved in the Appendix, is applicable to sections of arbitrary order.

II. Second-Order Sections

If $\{y_n\}_{-2}^{\infty}$ is the sequence of machine numbers that corresponds to the zero-input response of a second-order digital filter, then

$$y_n = -ay_{n-1} - by_{n-2} + \delta_n, \qquad n \geq 0 \tag{1}$$

in which a and b are fixed real constants and $\{\delta_n\}_0^{\infty}$ takes into account the effects of roundoff. If, for example, $q(\cdot)$ denotes the standard quantization function defined by the condition that for all real x, $q(x)$ is equal to the integer part of $[x+0.5 \text{ sign } (x)]$, and if

$$y_n = q(-ay_{n-1}) + q(-by_{n-2}), \qquad n \geq 0 \tag{2}$$

(i.e., if each product is rounded and the rounded products are added) then $|\delta_n| \leq 1$ for all $n \geq 0$. Alternatively, if

$$y_n = q(-ay_{n-1} - by_{n-2}), \qquad n \geq 0, \tag{3}$$

then it is clear that $|\delta_n| \leq 0.5$ for all $n \geq 0$. It is assumed throughout that the infinite-precision counterpart of the digital filter is asymptotically stable (i.e., that $1+az^{-1}+bz^{-2} \neq 0$ for all $|z| \geq 1$).

Let us suppose that the numbers $y_{-2}, y_{-1}, y_0, y_1, \cdots$ of (1) approach[1] a limit cycle of period $(K+1)$ as $n \rightarrow \infty$, in which K is an arbitrary nonnegative integer. More explicitly, let $(y_n - p_n) \rightarrow 0$ as $n \rightarrow \infty$ and $(\delta_n - s_n) \rightarrow 0$ as $n \rightarrow \infty$ in which $\{p_n\}_{-\infty}^{\infty}$ and $\{s_n\}_{-\infty}^{\infty}$ satisfy $p_n = p_{n+K+1}$ and $s_n = s_{n+K+1}$ for all n. Let $\|P\|$ and $\|S\|$ denote, respectively,

$$\left(\frac{1}{K+1}\sum_{n=0}^{K} p_n^2\right)^{1/2} \quad \text{and} \quad \left(\frac{1}{K+1}\sum_{n=0}^{K} s_n^2\right)^{1/2}$$

the root-mean-square values of the response sequence $\{p_n\}_{-\infty}^{\infty}$ and the roundoff sequence $\{s_n\}_{-\infty}^{\infty}$.

According to a special case of the theorem proved in the Appendix, we have

$$\min_l \phi(l)\|S\| \leq \|P\| \leq \max_l \phi(l)\|S\| \tag{4}$$

in which l varies over $0, 1, \cdots, K$ and

$$\phi(l) = |H\{\exp[i2\pi l/(K+1)]\}|,$$

and $H(z) = (1+az^{-1}+bz^{-2})^{-1}$ for all $|z| = 1$. The left inequality of (4) is included for completeness. We shall not discuss it further here.

For the simple special cases in which $K=0$ and $K=1$ (i.e., for the dc and $\frac{1}{2}f_s$ situations), respectively,

$$\max_l \phi(l) = (1+b+a)^{-1} \tag{5}$$

and

$$\max_l \phi(l) = (1+b-|a|)^{-1}. \tag{6}$$

The expressions that appear on the right sides of (5) and (6) appear also in related upper bounds [1], [2] on the maximum absolute value of p_n for the cases $K=0$ and $K=1$. While bounds for such special situations are of interest, it is ordinarily not the case that the period $(K+1)$ of the limit cycle is known at the outset. Therefore, we proceed as follows.

Since it is certainly true that

$$\max_l \phi(l) \leq \max_{\theta \in [0,2\pi]} |H(e^{i\theta})|$$

for all K, and since it can be shown that

$$\max_{\theta \in [0,2\pi]} |H(e^{i\theta})| = \rho(a, b)$$

in which

$$\rho(a,b) = \begin{cases} \dfrac{1}{(1-b)\left(1-\dfrac{a^2}{4b}\right)^{1/2}}, & \text{for } b>0 \text{ and } |a| \leq \dfrac{4b}{1+b} \\[2ex] \dfrac{1}{1+b-|a|}, & \text{for either } b \leq 0 \text{ or } b>0 \text{ and } |a| \geq \dfrac{4b}{1+b} \end{cases}$$

we see that

$$\|P\| \leq \rho(a,b)\|S\|, \tag{7}$$

which is our main result concerning second-order sections. Referring to (1), if δ is the smallest positive number such that

Manuscript received August 27, 1971.

[1] Of course, since the digital filter is a finite-state machine, the limit cycle will in fact be reached in a finite number of sampling periods.

$|\delta_n| \leq \delta$ for all $n \geq 0$, then obviously $\|S\| \leq \delta$, and hence[2]

$$\|P\| \leq \delta\rho(a, b). \tag{8}$$

By considering certain simulation results reported by Jackson [1] and by Parker and Hess [3], we can obtain some understanding of the extent to which our bound (7) is not grossly pessimistic. For the case of complex-conjugate poles of $H(z)$, $\delta = 1$, and rounding after each multiplication as in (2), Jackson, using an "effective-value model," was led to the "bound"

$$\max_n |p_n| \leq \text{integer part of } [0.5(1-b)^{-1}] \tag{9}$$

which he reported was never exceeded in his simulations by more than unity (i.e., by more than one level). Recently Parker and Hess [3] have presented an example in which the right side of (9) is 11 and $\max_n |p_n|$ is 17. They have suggested, on the basis of their simulation studies, that (9) should be replaced by

$$\max_n |p_n| \leq 1.5(1-b)^{-1}. \tag{10}$$

The ratio of the right side of (8) for $b > 0$ and $|a| \leq 4b(1+b)^{-1}$ and with $\delta = 1$ to the right side of (10) is

$$(1.5)^{-1}\left(1 - \frac{a^2}{4b}\right)^{-1/2}. \tag{11}$$

This ratio of two different types of bounds does not differ from unity by more than a factor of, say, three or four for a large range of useful design values of $a^2/4b$. Thus, on the strength of this basically qualitative argument, we see that our bound is a useful bound for at least a large class of design parameters.

Appendix

The Theorem

Notation and Definitions

Let m be a positive integer, and let $c_1, c_2, \cdots, c_m$ be (real or complex) constants such that

$$1 + \sum_{k=1}^{m} c_k z^{-k} \neq 0, \qquad \text{for } |z| \geq 1. \tag{12}$$

Let $\{y_n\}_{-m}^{\infty}$ and $\{\delta_n\}_0^{\infty}$ be sequences of constants such that

$$y_n + \sum_{k=1}^{m} c_k y_{n-k} = \delta_n, \qquad n \geq 0 \tag{13}$$

and $(\delta_n - s_n) \to 0$ as $n \to \infty$ for some sequence $\{s_n\}_{-\infty}^{\infty}$ with the property that $s_n = s_{n+K+1}$ for $n = 0, \pm 1, \pm 2, \cdots$, in which K is a nonnegative constant.

Let $\{p_n\}_{-\infty}^{\infty}$ denote the sequence with the properties that $p_n = p_{n+K+1}$ for $n = 0, \pm 1, \pm 2, \cdots$ and $(y_n - p_n) \to 0$ as $n \to \infty$, and let $\|S\|$ and $\|P\|$ denote

$$\left(\frac{1}{K+1}\sum_{n=0}^{K} |s_n|^2\right)^{1/2} \quad \text{and} \quad \left(\frac{1}{K+1}\sum_{n=0}^{K} |p_n|^2\right)^{1/2},$$

respectively. Finally, let $\mathcal{R} = \{0, 1, 2, \cdots, K\}$, and

$$H(e^{i\theta}) = \sum_{n=0}^{\infty} h_n e^{-in\theta}$$

for $\theta \in [0, 2\pi]$ in which $\{h_n\}_0^{\infty}$ is the inverse z transform of

$$\left(1 + \sum_{k=1}^{m} c_k z^{-k}\right)^{-1}.$$

Theorem:

$$\min_{l \in \mathcal{R}} \phi(l)\|S\| \leq \|P\| \leq \max_{l \in \mathcal{R}} \phi(l)\|S\|$$

in which $\phi(l) = |H\{\exp[i2\pi l/(K+1)]\}|$.

Proof:[3]

From (12) and (13) it is clear that

$$y_n = r_n + \sum_{k=0}^{n} h_{n-k}\delta_k, \qquad n \geq 0 \tag{14}$$

in which $r_n \to 0$ as $n \to \infty$. Since $s_n = s_{n+K+1}$ for $n = 0, \pm 1, \pm 2, \cdots$, we have

$$s_k = \sum_{l=0}^{K} \hat{s}_l \exp\left(\frac{i2\pi lk}{K+1}\right)$$

with

$$\hat{s}_l = (K+1)^{-1} \sum_{n=0}^{K} s_n \exp\left(-\frac{i2\pi ln}{K+1}\right)$$

and

$$\sum_{n=0}^{K} |s_n|^2 = (K+1) \sum_{n=0}^{K} |\hat{s}_n|^2.$$

Moreover, we can show, using (12) and (14), that

$$p_n = \sum_{k=-\infty}^{n} h_{n-k} s_k$$

for all n.[4] Therefore

$$\begin{aligned} p_n &= \sum_{k=-\infty}^{n} h_{n-k} \sum_{l=0}^{K} \hat{s}_l \exp\left(\frac{i2\pi lk}{K+1}\right) \\ &= \sum_{l=0}^{K} \hat{s}_l \sum_{k=-\infty}^{n} h_{n-k} \exp\left(\frac{i2\pi lk}{K+1}\right) \\ &= \sum_{l=0}^{K} \hat{s}_l \exp\left(\frac{i2\pi ln}{K+1}\right) \sum_{m=0}^{\infty} h_m \exp\left(-\frac{i2\pi lm}{K+1}\right) \\ &= \sum_{l=0}^{K} H\left[\exp\left(\frac{i2\pi l}{K+1}\right)\right] \hat{s}_l \exp\left(\frac{i2\pi ln}{K+1}\right). \end{aligned}$$

Thus

[2] If the sequence $\{s_n\}_{-\infty}^{\infty}$ is approximately the closely spaced samples of a sinusoid (as is often observed in simulations), then $\|S\|$ is approximately equal to $(0.707)\delta$. This estimate of $\|S\|$ can often be used to improve the engineering utility of (7).

[3] The proof given here is similar to a portion of a proof given in [4, proof of Theorem 2]. Our theorem is similar to results proved earlier [5].

[4] We note in passing that

$$\max_n |p_n| \leq \max_n |s_n| \sum_{n=0}^{\infty} |h_n|,$$

which might be of use in other work.

$$\|P\|^2 = \sum_{l=0}^{K} \phi(l)^2 |\hat{s}_l|^2$$

$$= \max_{l \in \mathfrak{R}} \phi(l)^2 \sum_{l=0}^{K} |\hat{s}_l|^2$$

$$\leq \max_{l \in \mathfrak{R}} \phi(l)^2 \|S\|^2$$

and, similarly,

$$\|P\|^2 \geq \min_{l \in \mathfrak{R}} \phi(l)^2 \|S\|^2$$

which proves the theorem.

References

[1] L. B. Jackson, "An analysis of limit cycles due to multiplication rounding in recursive digital (sub) filters," in *Proc. 7th Annu. Allerton Conf. Circuit and Syst. Theory*, Oct. 1969, pp. 69–78.

[2] F. Bonzanigo, "Constant-input behavior of recursive digital filters," presented at the IEEE Arden House Workshop on Digital Filtering, Harriman, N. Y., Jan. 1970.

[3] S. R. Parker and S. F. Hess, "Limit-cycle oscillations in digital filters," *IEEE Trans. Circuit Theory*, vol. CT-18, pp. 687–697, Nov. 1971.

[4] I. W. Sandberg, "Some properties of a classic numerical integration formula," *Bell Syst. Tech. J.*, vol. 46, pp. 2061–2080, Nov. 1967.

[5] ——, "Two theorems on the accuracy of numerical solutions of systems of ordinary differential equations," *Bell Syst. Tech. J.*, vol. 46, pp. 1243–1266, July–Aug. 1967.

22

Reprinted from *Bell Syst. Tech. J.*, **48**(9), 2999–3020 (1969)

Overflow Oscillations in Digital Filters

By P. M. EBERT, JAMES E. MAZO,
AND MICHAEL G. TAYLOR

(Manuscript received May 9, 1969)

The cascade and parallel realizations of an arbitrary digital filter are both formed using second order sections as building blocks. This simple recursive filter is commonly implemented using 2's complement arithmetic for the addition operation. Overflow can then occur at the adder and the resulting nonlinearity causes self-oscillations in the filter. The character of the resulting oscillations for the second order section are here analyzed in some detail. A simple necessary and sufficient condition on the feedback tap gains to insure stability, even with the presence of the nonlinearity, is given although for many desired designs this will be too restrictive. A second question studied is the effect of modifying the "arithmetic" in order to quench the oscillations. In particular it is proven that if the 2's complement adder is modified so that it "saturates" when overflow occurs, then no self-oscillations will be present.

I. INTRODUCTION

A digital filter using idealized operations can easily be designed to be stable.[1] Nevertheless, in actual implementations, the output of such a stable filter can display large oscillations even when no input is present.* A known cause of this phenomenon is the fact that the digital filter realization of the required addition operation can cause overflow, thereby creating a severe nonlinearity.† Our purpose here is twofold. The first is to give a somewhat detailed analysis of the character of the oscillations when the filter is a simple second order recursive section with two feedback taps. This unit is the fundamental building block for the cascade and the parallel realization of digital filters, and as such is worthy of some scrutiny.[2] A simple conclusion which one can draw from

* To the best of our knowledge, these oscillations were first observed and diagnosed by L. B. Jackson of Bell Telephone Laboratories.

† In the present work rounding errors in multiplication or storage are neglected and therefore so are the little-understood oscillations attendant upon these nonlinearities.

the analysis is that the design of many useful filters requires using values of feedback coefficients such that the threat of oscillations is always present (with 2's complement arithmetic). Optimum solutions that cope with this state of affairs are still unknown. Some recent proposals include observing when overflow at the adder is to occur and then taking appropriate action. Our second purpose, then, is to discuss the effectiveness of some of these ideas, and to give a proof that modifying 2's complement arithmetic so that the adder "saturates" is an effective way to eliminate the oscillations. Questions of how this nonlinearity will affect the desired outputs from a particular ensemble of input signals are not yet answered however, and perhaps for some applications other solutions need be considered.

II. PROBLEM FORMULATION AND GENERAL DISCUSSION

As explained in the introduction, this paper deals primarily with the simple structure shown in Fig. 1. The outputs of the registers, which are storage elements with one unit of delay, are multiplied by coefficients a and b respectively, fed back, and "added" to the input in the accumulator. No round-off error is considered either in multiplication or storage, but overflow of the accumulator is not neglected. In other words, the accumulator will perform as a true adder if the sum of its inputs is in some range; otherwise a nonlinear behavior is observed.

Figure 2 shows the instantaneous input-output characteristic $f(v)$ of the device motivated by using 2's complement arithmetic. It is also important to note that there is no memory of the accumulator for past outputs; that is, the device is zeroed after the generation of each output.

If we let $x(t)$ be the input signal to the device, $y(t)$ the output, and

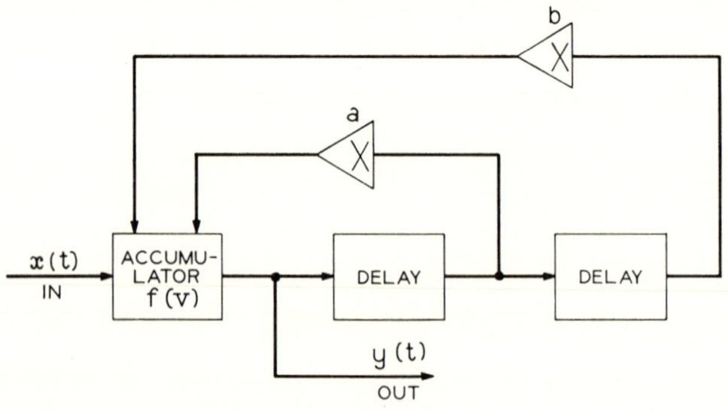

Fig. 1 — Basic configuration for the digital filter $\cdot y_{k+2} = f[ay_{k+1} + by_k + x_{k+2}]$.

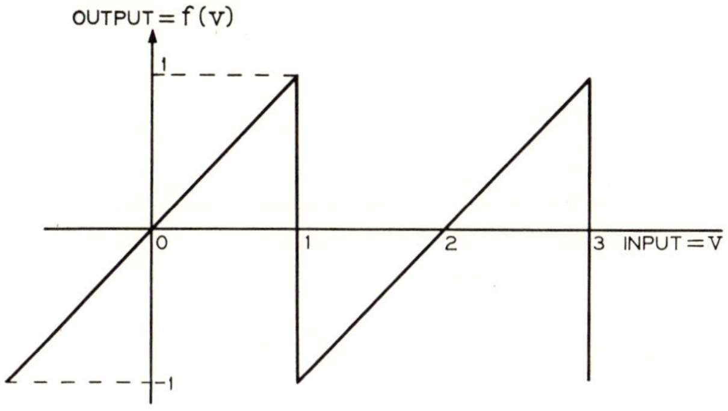

Fig. 2 — Instantaneous transfer function of the accumulator.

$f(\cdot)$ the nonlinear characteristic of the accumulator, we have the basic equation

$$y(t + 2) = f[ay(t + 1) + by(t) + x(t + 2)]. \tag{1}$$

We shall be concerned with the self-sustaining oscillations of the device that are observed even when no input is present [$x(t) = 0$], and when linear theory would predict the device to be stable.

By making this linear approximation $f(v) = v$, the linearized version of equation (1) becomes, with no driving term in the equation,

$$y(t + 2) - ay(t + 1) - by(t) = 0. \tag{2}$$

The roots of the characteristic equation for equation (2) are

$$\rho_{1,2} = \frac{a \pm (a^2 + 4b)^{\frac{1}{2}}}{2} \tag{3}$$

and the region of linear stability corresponds to the requirement that $| \rho_i | < 1$. This region is depicted as a subset of the a–b plane in Fig. 3. One has $| \rho_i | < 1$ if and only if one is within the large triangle shown in Fig. 3. For this situation any solution of (2) will damp out to zero after a sufficient period of time. Now note that (2) is not necessarily a valid reduction of (1) even when $x(t) = 0$. The output, by choice of f, has been assumed to be constrained to be less than unity, but this is not sufficient to guarantee that the argument of the function f is less than unity. For this to be the case we require

$$| ay(t + 1) + by(t) | < 1. \tag{4}$$

Since $| y(t) | < 1$, equation (4) will always be satisfied provided that

$$| a | + | b | < 1. \tag{5}$$

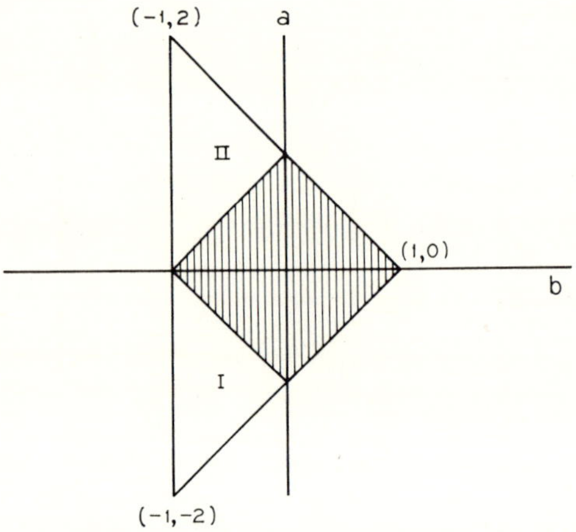

Fig. 3—Some interesting regions in the "space" of feedback tap weights. The hatching indicates stability even with the nonlinearity.

The subset of the a–b plane for which (5) is true is shown in Fig. 3 with vertical hatching, and is a subset of the region of linear stability. It is shown in this Section that if (5) is not satisfied there always exist self-sustained oscillations of the digital filter and hence (5) is both a necessary and sufficient condition for absence of self-sustained oscillations.* One way to avoid the oscillations in question is simply to impose the requirement (5). This trick has its limitations, however, for it clearly restricts design capabilities. The region of the s-plane which is shaded in Fig. 4 shows the allowable pole positions. Roughly speaking, one concludes that there are desirable filter characteristics that can be realized with this restriction and there are desirable characteristics that cannot.

It is not our purpose here to outline those applications for which (5) will not be restrictive; we proceed to sketch the situation when $|a| + |b| > 1$ and the threat of oscillation is present. Sections III and IV contain, we believe, a novel and interesting mathematical treatment of the general problem of classifying the self-oscillations of the nonlinear difference equation (1). However, for the user of digital filters a simple proof of the $|a| + |b| > 1$ being sufficient for threat of oscillations is of more immediate interest. After reading the simple proof of this fact given next in the present section, such a reader may wish to proceed directly to Section V.

Consider the possibility of undriven nonlinear operation giving a dc

* I. W. Sandberg has informed the authors that the necessity and sufficiency of (5) holding for absence of oscillations has also been obtained jointly by him and L. B. Jackson.

output, that is, $y_k \equiv y$ for all k. Equation (1), with $x(t) = 0$ becomes $y = f[(a + b)y]$. Assuming for definitness that $y > 0$, we can easily see from Fig. 2 that the above equation will be true if $(a + b)y = y - 2$, which implies $y = 2/(1 - a - b)$. One can show (see discussion following equation 17), that this y will have magnitude < 1 provided only that the tap values a and b lie in the region labeled I in Fig. 3. Thus a consistent dc oscillation is always possible for all (a, b) pairs in this region. Next consider the possibility of a period 2 oscillation. This amounts to finding a consistent solution to $y = f[(b - a)y]$. Proceeding as before we obtain

$$y = \frac{2}{1 + a - b}.$$

Thus y_k will be given by $(-1)^k y$, and will have magnitude less than unity if the (a, b) pair lies anywhere in region II of Fig. 3.

III. FURTHER ANALYSIS OF THE OSCILLATIONS

To analyze equation (1) in greater detail, it is very convenient to write it in the form similar to (2),

$$y(t + 2) - ay(t + 1) - by(t) = \sum_n a_n u(t + 2 - n), \qquad (6)$$

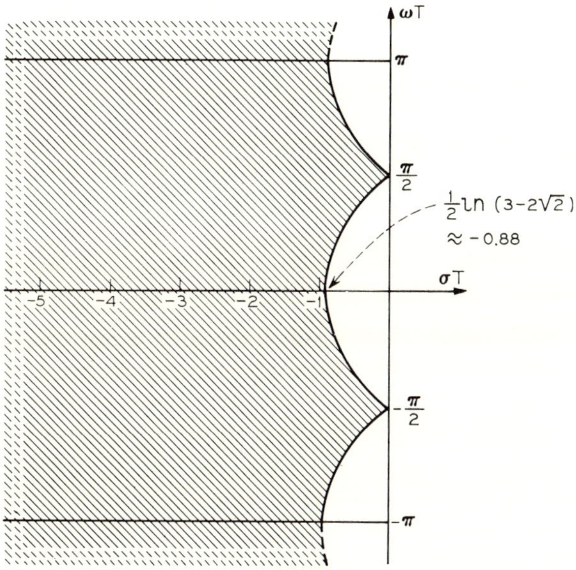

Fig. 4 — Pole locations in the s-plane (shaded region) realizable under the constraint that $|a| + |b| < 1$.

where $u(t)$ is a square pulse of unit height that one may conveniently think of as lasting from $t = 0$ until $t = 1$. This, of course, means that one interprets the solution of (6) to be a piecewise constant function like the actual output of the digital filter. For mathematical manipulations it is sometimes desirable to also interpret (6) as a difference equation, defined only for integer t. In this case one would write that $u(t - n) = \delta_{tn}$ where δ_{tn} is the familiar Kroneker symbol.

The point of the right side of (6) is simply to keep $| f(v) | < 1$ regardless of what value v has. From Fig. 2 we see that if $| v | < 1$, this added term is not needed and we take $a_n = 0$. If $1 < v < 3$ then we take $a_n = -2$, and if $-3 < v < -1$ we take $a_n = +2$. Since we have that $| y(t) | < 1$ and that linear stability (see Fig. 3) implies $| a | < 2$, $| b | < 1$, we need not consider further values of $| v |$. Thus in (6) $a_n = 0, \pm 2$ depending on whether or not $v(t) \equiv ay(t + 1) + by(t)$ crosses the lines $v = \pm 1$. It will be convenient to have a word for such crossings; we shall call them "clicks", borrowing a favorite word from FM theory. Then $a_n = 0, \pm 2$ depending on whether or not a click does not, or does, occur.

Note if one knew what the click sequence $\{a_n\}$ was, one could solve (6) simply by using the clicks to be the driving term for a linear equation. We are mainly interested in describing the self-sustained steady state oscillations of arbitrary period N. Hence initial conditions will play no essential role for us, for while they determine which oscillating mode appears as $t \to \infty$, they play no role in describing the modes. Our procedure will be as follows:

(*i*) Assume a click sequence of period N;

$$\begin{aligned} &a_0 , a_1 , a_2 , \cdots , a_{N-1} \\ &a_{lN+k} = a_k . \qquad l = 0, 1, \cdots \\ &\qquad\qquad\qquad 0 \leq k < N - 1. \end{aligned} \tag{7}$$

(*ii*) Using the assumed $\{a_n\}$, find the steady state solution of (6). However, only solutions that have $| y(t) | < 1$ for all t are allowed. (*iii*) Check that this steady state solution actually generates the assumed click sequence.

In carrying out the above program for some simple cases we observed that step *iii* never seemed to yield anything new. Indeed, surprising as it seems at first glance, step *iii* never has to be carried out. If one obtains a solution with $| y(t) | < 1$, this solution is consistent. That is, it automatically generates the assumed click sequence. The proof is simple.

One calculates the argument of the function f from (6):

$$ay(t+1) + by(t) = y(t+2) - \sum a_n u(t+2-n). \qquad (8)$$

We have a click at time $t + 2 = m$ if $| ay(m-2) + by(m-1) | > 1$. From (8),

$$| ay(m-2) + by(m-1) | = | y(m) - a_m |. \qquad (9)$$

Note then if in (9) $a_m = 0$, then $| ay(m-2) + by(m-1) | = | y(m) | < 1$; thus if there is no click at a particular time in the assumed click sequence the "solution" will not generate one. Next assume $a_m = +2$; then

$$ay(m-2) + by(m-1) = y(m) - 2 < -1, \qquad (10)$$

where we use $| y(t) | < 1$ again. Equation (10) says if a positive click is present in the assumed click sequence then the solution obtained from the linear equation (6), given by this click sequence, will reproduce the positive click. Obviously the same argument holds for a negative click, $a_m = -2$, and the proof of this point is complete.

The steady-state solution of our fundamental equation (6) for an arbitrary click sequence $\{a_m\}$ of period N is derived in the appendix. If we define

$$A_{N-1}\left(\frac{1}{z}\right) \equiv \sum_{n=0}^{N-1} a_n z^{-n} \qquad (11)$$

and

$$D(z) \equiv z^2 - az - b, \qquad (12)$$

and let r_i, $i = 1, \cdots, N$, be the N Nth roots of unity, then the (periodic) output values are given by

$$y_k = \frac{1}{N} \sum_{i=1}^{N} \frac{A_{N-1}\left(\frac{1}{r_i}\right)}{D(r_i)} r_i^k . \qquad (13)$$

The above expression gives the $\{y_k\}$ output sequence for any click sequence. We emphasize, however, that it is only a solution corresponding to a self-sustained oscillation of the digital filter if we have $| y_k | < 1$, all k. Whether or not this is true depends on the particular click sequence assumed.

Another form of the solution can be obtained by manipulation of (13). To write this down, define

$$b_n^{(k)} \equiv (\bar{a}_{k-1-n} + \bar{a}_{k-1-n+N})/2, \qquad (14)$$

where we understand $\bar{a}_j \equiv 0$ if j does not lie between 0 and $N - 1$, inclusive, and $\bar{a}_j \equiv a_j$ if it does. One of the $\bar{a}$'s in (14) will thus always be zero and $b_n^{(k)}$ has values of $\pm 1, 0$. The other form of the solution is then

$$y_k = \frac{2}{\rho_1 - \rho_2} \sum_{n=0}^{N-1} b_n^{(k)} \left[\frac{\rho_1^n}{1 - \rho_1^N} - \frac{\rho_2^n}{1 - \rho_2^N} \right] \qquad k = 0, 1, \cdots, N-1 \qquad (15)$$

where ρ_i are given in (3).

In (15) we have N vectors of dimension N, namely the $\{b_n^{(k)}\}$ $k = 0, 1, 2, \cdots, N - 1$. Note from (14), however, that they are all cyclic permutations of one another. Hence we may refer to the b vector, $\mathbf{b}$, of a solution, understanding that the $\mathbf{b}$ and all its cyclic permutations generate a solution in the sense of (15). Note that a cyclic permutation of the y_k has no real significance here; it simply changes the origin of time.

An interesting property of the solutions which we have written down follows from the fact that if we transform the point (a, b) in the ab-plane into another point by

$$\begin{aligned} a \rightarrow a' &= -a \\ b \rightarrow b' &= b \end{aligned} \qquad (16a)$$

then under this transformation

$$\begin{aligned} \rho_1 \rightarrow \rho_1' &= -\rho_2 \\ \rho_2 \rightarrow \rho_2' &= -\rho_1 . \end{aligned} \qquad (16b)$$

The property is this: Let N be an even integer and let $\mathbf{b} = (b_0, b_1, \cdots, b_{N-1})$ be a click vector generating a solution at point (a, b). Then the vector $\mathbf{b}' = (b_0, -b_1, b_2, -b_3, \cdots, b_{N-1})$ generates a solution at reflected point $(-a, b)$. The proof is simple. Note from (15),

$$\begin{aligned} y'^{(k)} &= \frac{2}{\rho_1' - \rho_2'} \sum_n b_n'^{(k)} \left[\frac{\rho_1'^n}{1 - \rho_1'^N} - \frac{\rho_2'^n}{1 - \rho_2'^N} \right] \\ &= \frac{2}{-\rho_2 + \rho_1} \sum_n (-1)^{k+n} b_n \left[\frac{(-\rho_2)^n}{1 - \rho_2^N} - \frac{(-\rho_1)^n}{1 - \rho_1^N} \right] = (-1)^k y^{(k)}. \end{aligned}$$

Hence if $| y^{(k)} | < 1$ then $| y'^{(k)} | < 1$. Note that the proof also supplies the value for $y'^{(k)}$ in terms of $y^{(k)}$. This theorem will be used later to generate new solutions from old ones.

Before leaving this general discussion in favor of exhibiting some solutions in the next section, we list a few more observations related to the click vector **b**. The click vector **b**, whose only allowed component values are ± 1, 0, completely characterizes the associated oscillation. Clearly there can then only be a finite number of oscillations of given period N. This number is upper bounded by 3^N, but will generally be much less. Also note that a cyclic permutation of the components of b cyclically permutates the output values y^k, and this latter is merely a shift in time. The permutated values are not physically distinct.

Also note that if we perform $\mathbf{b} \rightarrow -\mathbf{b}$ then $\mathbf{y} \rightarrow -\mathbf{y}$, and a solution of opposite sign is obtained. While this may often be distinguishable from the first solution, it is trivially related to it. Finally if one were to count the number **b** vectors of dimension N that yield new information, one would wish to exclude subperiods of N. Thus if $(+, 0, 0)$ is an generating **b** vector for period 3, $(+, 0, 0, +, 0, 0)$ generates a period 6 oscillation but this is not new information. We have not solved the problem of counting how many of the 3^N vectors are left after we impose the requirements of cyclic shifts, sign changes, and subperiods. At any rate, it is essential to test the ones that remain to check that they generate allowed solutions, $| y^k | < 1$.

IV. SOME EXPLICIT PERIODS AND REGIONS OF OSCILLATION

Now for a few explicit solutions. Consider the possibility of a dc "oscillation", namely, set $N = 1$. The only nontrivial click vector is $\mathbf{b} = (+)$. The solution is more immediate if we use (13). We have

$$y = \frac{2}{1 - a - b} \tag{17}$$

for the dc value of output. For what values of a and b within the triangle of Fig. 3 will we have $| y | < 1$? We require

$$| 1 - a - b | > 2 \tag{18}$$

which is equivalent to either

$$1 - a - b > 2 \tag{19a}$$

or

$$-1 + a + b > 2. \tag{19b}$$

Inequality (19a) (coupled with the linear stability requirement) defines the triangle labeled "I" in Fig. 3, while (19b) is outside the stability region and needs no further consideration. Thus any portion of the region $a < 0$ that we have not excluded from oscillations has now been shown to have them. They are of period 1; other period oscillations may (and do) occur in this region.

At this point it is amusing to use an earlier remark on the possibility of generating new solutions from an even period one by "reflection". Letting $N = 2$, the click vector $\mathbf{b} = (+, +)$ certainly generates a period 2 oscillation (albeit one with subperiods) in region I. Then the click vector $\mathbf{b} = (+, -)$ generates something really new: a period 2 oscillation in the region labeled II in Fig. 3. The amplitudes of the output are

$$y^{(k)} = (-1)^k \frac{2}{1 + a - b}, \qquad a > 0. \tag{20}$$

One more possibility of a click vector exists for period 2, and that is $\mathbf{b} = (+, 0)$. From (13) we write for possible output values

$$\begin{aligned} y_0 &= \frac{1}{1 - a - b} + \frac{1}{1 + a - b} \\ y_1 &= \frac{1}{1 - a - b} - \frac{1}{1 + a - b}. \end{aligned} \tag{21}$$

After a little uninteresting analysis one can conclude that we cannot have $| y_0 | < 1$, $| y_1 | < 1$ in (21) for any allowed values of a and b. Thus there are no other period 2 oscillations.

On to period 3. Now there are four click vectors which must be considered. These are $(+00)$, $(++0)$, $(+-0)$, $(++-)$. Even in this case an exhaustive check that the "solutions" generated are legitimate ones is trying. Therefore, we resort to a trick; we look for periods which may exist in the immediate neighborhood of the point $(a = 0, b = 1)$. This means $\rho_1 = i$, $\rho_2 = -i$. In this immediate neighborhood $\rho_2 = \rho_1^*$, and (15) reads

$$y = \frac{2}{\operatorname{Im} z} \operatorname{Im} \sum_{n=0}^{N-1} \frac{b_n z^n}{1 - z^N}, \tag{22}$$

where we have let $z = \rho_1$. Letting $N = 3$, $z = i$ gives

$$\begin{aligned} y_0 &= -b_0 + b_1 + b_2 \\ y_1 &= -b_1 + b_2 + b_0 \\ y_2 &= -b_2 + b_0 + b_1 . \end{aligned} \tag{23}$$

We now require $y_k = \pm 1$ as a test for the click vector $\mathbf{b}$. We see that only $(+00)$ qualifies as possibly yielding a solution in the neighborhood of $(a = 0, b = -1)$. A computer study shows that indeed the solution extends into the interior of the triangle and the region found is shown in Fig. 5. This immediately implies existence of the period 6 oscillation generated by $(+00-00)$ in the reflected region. Similarly, a period 5 oscillation region (with the concomitant period 10) generated by $(+0000)$ is shown in Fig. 6.

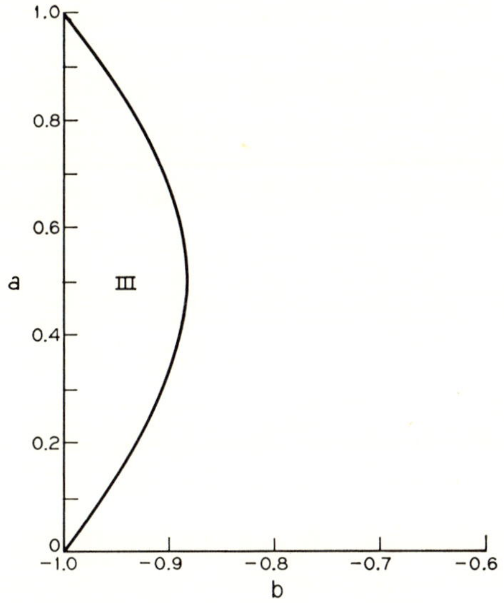

Fig. 5 — A region for period 3 oscillations.

It is very tempting to conjecture that the point ($a = 0, b = -1$) is a boundary point of any allowed region of oscillation. If this is true, a procedure like that used above may eliminate some otherwise very respectable **b** vectors from consideration. Note that for $N = 2$, $b = (+, 0)$ satisfies the required condition at $\rho_1 = i$, but we have shown this

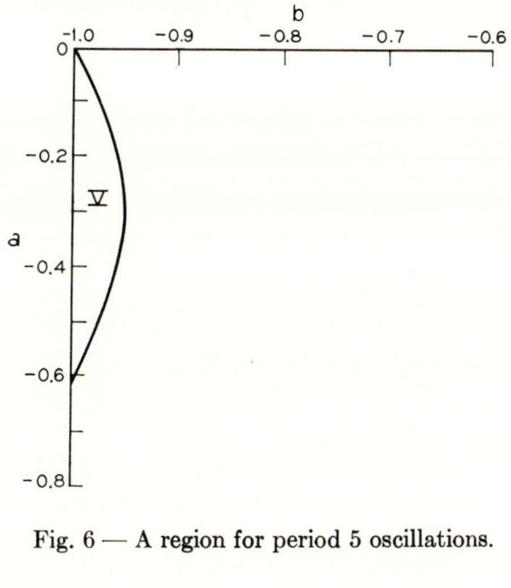

Fig. 6 — A region for period 5 oscillations.

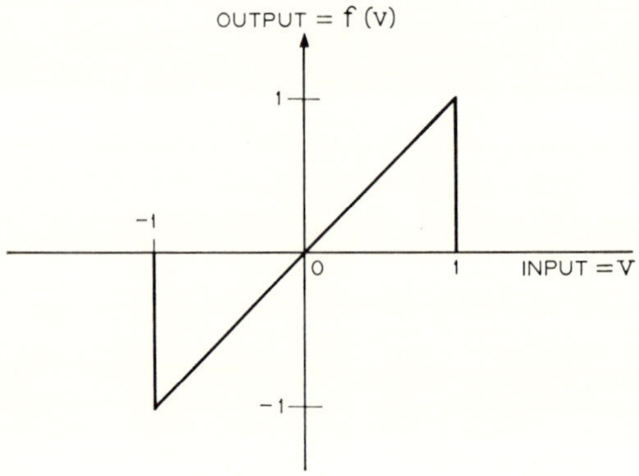

Fig. 7 — Zeroing arithmetic, shown above, also gives rise to oscillations.

is not extendable into the interior of the triangle. Hence existence at $z = i$ does not guarantee an allowed solution.

V. STABILITY WITH A MODIFIED ARITHMETIC

In an attempt to eliminate these oscillations, proposals have been made which rely on detecting overflow. One such suggestion dictates that when overflow occurs, the adder is directed to shift out zero. For refereace we call this zeroing arithmetic. The effective transfer function of the adder for zeroing arithmetic is given in Fig. 7. However, it can be shown by numerical example that such a procedure still leads to oscillations. Another possibility, "saturation arithmetic," is displayed in Fig. 8. Here a one (with the appropriate sign) is put out when overflow is detected. The remaining portion of this paper is devoted to proving that saturation arithmetic leads to stable operation whenever linear theory would predict it to be so.

To begin, we suppose for the moment that we ignore the fact that the digitally implemented adder is nonlinear. Then the second-order linear difference equation which governs the behavior of the undriven system has solutions y_k which may be described as follows:

Case 1: Complex roots for characteristic equation

$$y_k = \operatorname{Re} K_0 \exp(-\alpha k), \quad K_0 \text{ and } \alpha \text{ complex}, \quad \operatorname{Re} \alpha > 0.$$

$$k = 0, 1, 2, \cdots . \tag{24}$$

Case 2: Real but unequal roots

$$y_k = K_1 \exp(-\alpha k) + K_2 \exp(-\beta k). \quad K_i \text{ real}; \quad \alpha > 0, \quad \beta > 0. \tag{25}$$

Case 3: Real and equal roots

$$y_k = [K_1 + K_2 k] \exp(-\alpha k). \quad K_i \text{ real}; \quad \alpha > 0. \tag{26}$$

Using this information, coupled with knowledge of y_j and y_{j+1} for some j, it is easy to give a bound on the magnitudes of all future ($k \geqq j$) values of the output and to show this value goes to zero with increasing j. This is just another way to say that the solutions go to zero for the linear case. In the nonlinear case we cannot exclude the situation that some y_{k+1} will exceed unity and the nonlinearity will be operative. For saturation arithmetic the offending value must be set to unity if, for example, $y_{k+1} > +1$. We can, for conceptual purposes, regard this as a "squeezing" of the output from a value greater than unity down to the value one which is performed in a continuous fashion. The crux of the proof now comes in showing that the partial derivative of our bound (on future outputs) with respect to the most recent output y_{k+1} has, for saturation arithmetic, the same sign as y_{k+1}. Hence decreasing a value that is too large in magnitude will decrease the bound as well, and it will go to zero at least as fast as it does for the linear case.

To show how the above outline works, consider first the linear case with complex roots. From the form of the solution

$$y_k = \operatorname{Re} K_0 \exp(-\alpha k), \quad \operatorname{Re} \alpha > 0, \quad k = 0, 1, 2, \cdots,$$

it is clear that if we define

$$B_0 = |K_0|^2 \tag{27}$$

then $y_k^2 \leqq B_0$ for all $k \geqq 0$. We now express B_0 in terms of the values y_0, y_1 which are initially stored in the shift registers to yield

$$B_0 = y_0^2 + \frac{[y_1 - y_0 \operatorname{Re} \exp(-\alpha)]^2}{[\operatorname{Im} \exp(-\alpha)]^2}. \tag{28}$$

This suggests that one define the more general set of numbers

$$B_j = y_j^2 + \frac{[y_{j+1} - y_j \operatorname{Re} \exp(-\alpha)]^2}{[\operatorname{Im} \exp(-\alpha)]^2}. \tag{29}$$

Clearly, from the way that B_j is defined, we have that

$$y_k = \operatorname{Re} K_j \exp[-\alpha(k - j)], \qquad k \geqq j \tag{30}$$

where K_j is some appropriate complex number that satisfies

$$B_j = |K_j|^2. \tag{31}$$

From (30), the additional inequality that $y_k^2 \leqq B_j$ for all $k \geqq j$ follows,

Furthermore, one can see by comparing (30) and (24) that

$$| K_j |^2 = | K_0 |^2 \, | \exp(-\alpha j) |^2. \tag{32}$$

Hence, since the real part of α is positive, B_j goes monotonically to zero with increasing j.

To generalize the above arguments to a nonlinear situation of interest,* consider the following equation which follows from (29):

$$\frac{\partial B_j}{\partial y_{j+1}} = \frac{2}{[\text{Im} \exp(-\alpha)]^2} [y_{j+1} - y_j \text{ Re} \exp(-\alpha)]. \tag{33}$$

Now imagine B_{j-1} has been calculated from values stored in the registers. From *linear* theory we predict $y_{j+1}^{(L)}$ and $B_j^{(L)} \leqq B_{j-1} \exp(-2\alpha)$, by (32). Now if the $y_{j+1}^{(L)}$ generated by the linear equation were too large, say, then decreasing it to unity would, according to (33), *decrease* the bound B_j if we knew that

$$y_{j+1} - y_j \text{ Re} [\exp(-\alpha)] \geqq 0 \quad \text{for} \quad y_{j+1}^{(L)} \geqq y_{j+1} \geqq y_{j+1}^{(C)} \tag{34}$$

where $y_{j+1}^{(L)}$ is the linear prediction for y_{j+1} and $y_{j+1}^{(C)}$ is the correct value for the nonlinear circuit resulting from "squeezing" $y_{j+1}^{(L)}$ down. Since $| y_j | \leqq 1$ and Re $\exp(-\alpha) < 1$, (34) is always true for saturation arithmetic (see Fig. 8) because $y_{j+1}^{(C)} = +1$ (assuming $y_{j+1}^{(L)} > +1$) and (34) can never swing negative. Similar things happen, of course, if $y_{j+1} < -1$. Thus the bound decreases at least as fast as for the linear case (which is exponential) and stability is assured. For zeroing arithmetic $y_{j+1}^{(C)} = 0$, and thus the appropriate sign for (34) cannot be guaranteed which is in satisfying agreement with the known instability for this case.

For the next case of real but unequal roots, we now have reference to equation (25) and define our initial bound as

$$B_0 = 2(K_1^2 + K_2^2)$$

$$= 2\frac{[y_1 - \exp(-\alpha)y_0]^2 + [y_1 - \exp(-\beta)y_0]^2}{[\exp(-\alpha) - \exp(-\beta)]^2}. \tag{35}$$

The remaining details are too similar to those of the preceding case to warrant recording again; stability for saturation arithmetic holds here as well.

The last case to discuss occurs when we have real and equal roots.

* B_j calculated from (29) is a bound on future outputs for the nonlinear as well as the linear case. If $B_j \leq 1$ the two cases coincide, while of $B_j > 1$ the conclusion follows equally trivially since $|y_k| \leq 1$ for the nonlinear situation.

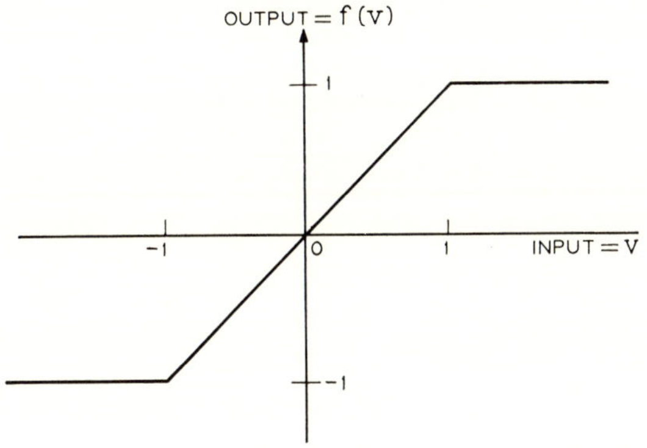

Fig. 8 — The above nonlinearity corresponds to saturation arithmetic and leads to stable behavior.

This situation, represented for the linear equation by equation (26), is more difficult to treat than the previous ones. The analog of (27) and (35) now is

$$B_0 = \max \begin{cases} 4K_1^2 \\ \dfrac{4K_2^2}{\alpha^2} \end{cases}. \tag{36}$$

That (36) yields a bound follows from the facts that (for $t \geqq 0$)

$$\begin{aligned} y_k^2 &\leqq \max_t \, [(K_1 + K_2 t) \exp(-\alpha t)]^2 \\ &\leqq 2 \max_t \, [K_1^2 + K_2^2 t^2] \exp(-2\alpha t) \\ &\leqq 4 \max \begin{cases} \max_t K_1^2 \exp(-2\alpha t) \\ \max_t K_2^2 t^2 \exp(-2\alpha t) \end{cases} \\ &= 4 \max \begin{cases} K_1^2 \\ \dfrac{K_2^2 \exp(-2)}{\alpha^2} \end{cases} \\ &\leqq 4 \max \begin{cases} K_1^2 \\ \dfrac{K_2^2}{\alpha^2} \end{cases}. \end{aligned}$$

Since

$$K_1^2 = y_0^2$$
$$\frac{K_2^2}{\alpha^2} = \frac{(y_1 \exp \alpha - y_0)^2}{\alpha^2}, \tag{37}$$

we define our general bound as

$$B_j = 4 \max \begin{cases} y_j^2 \\ \dfrac{(y_{j+1} \exp \alpha - y_j)^2}{\alpha^2}. \end{cases} \tag{38}$$

Using the solution $y_j = (K_1 + K_2 j) \exp(-\alpha j)$, we see that

$$\theta_j \equiv \frac{(y_{j+1} \exp \alpha - y_j)^2}{\alpha^2} \tag{39}$$

decreases by the multiplicative factor exp (-2α) for every unit increase of j. Further, suppose that $B_j = 4y_j^2$ for some j. That is, suppose

$$\frac{(y_{j+1} \exp \alpha - y_j)^2}{\alpha^2} < y_j^2 . \tag{40}$$

This implies

$$y_{j+1}^2 < y_j^2(1 + \alpha)^2 \exp(-2\alpha), \tag{41}$$

and so if next time $B_{j+1} = 4y_{j+1}^2$, then we have decreased by $(1 + \alpha)^2 \exp(-2\alpha) < 1$. On the other hand, if at the next step we have to choose $B_{j+1} = 4\theta_{j+1}$, we see

$$\frac{B_{j+1}}{B_j} = \frac{\theta_{j+1}}{y_j^2} \leqq \frac{\theta_{j+1}}{\theta_j} \leqq \exp(-2\alpha). \tag{42}$$

Likewise if we go from $4\theta_j$ to $4\theta_{j+1}$ we decrease by exp (-2α). Finally, a "transition" from $4\theta_j$ as a bound to $4y_{j+1}^2$ decreases the bound by a multiplicative factor of $(1 + \alpha)^2 \exp(-2\alpha)$. To see this we note that, by assumption,

$$B_j = \frac{4[y_{j+1} \exp \alpha - y_j]^2}{\alpha^2} \geqq 4y_j^2 . \tag{43}$$

Using the left-hand equality in (43) implies

$$| y_{j+1} | \exp \alpha \leqq \frac{\alpha(B_j)^{\frac{1}{2}}}{2} + | y_j |. \tag{44}$$

while $B_i \geqq 4y_i^2$ yields

$$| y_i | \leqq \frac{(B_i)^{\frac{1}{2}}}{2}. \tag{45}$$

Using (45) in (44) then allows us to deduce that

$$B_{i+1} = 4y_{i+1}^2 \leqq (1 + \alpha)^2 \exp(-2\alpha) B_i \tag{46}$$

as was claimed. To extend these arguments to the nonlinear case we again observe that

$$\frac{\partial B_i}{\partial y_{i+1}} \geqq 0 \tag{47}$$

for saturation arithmetic.

VI. GENERALIZATIONS TO OTHER STABLE NONLINEARITIES

Aside from the three nonlinearities already mentioned, there does not appear to be immediate engineering interest in seeing which other nonlinearities will or will not give rise to stable behavior of the filter. Having come this far, however, it is hard to resist asking if the method of proof we have used, or some slight extension of it, does suggest other nonlinearities for which stability will hold. The extension we consider is not to require

$$\frac{\partial B_i}{\partial y^{i+1}} \geqq 0$$

all during the "squeezing" operation, but merely that

$$B_i^L - B_i^C \geqq 0, \tag{48}$$

where B_i^L is the value of the bound using linear theory and B_i^C is the "correct" value. An inspection of the previous proofs shows that this is equivalent to

$$(y_{i+1}^L - ay_i)^2 - (y_{i+1}^C - ay_i)^2 > 0 \tag{49}$$

for all real a such that $| a | < 1$.

A little manipulation reduces (49) to

$$(y_{k+1}^L - y_{k+1}^C)(y_{k+1}^L + y_{k+1}^C - 2ay_k) \geqq 0. \tag{50}$$

Assuming $y_{k+1}^L > 0$, the first term in (50) to be nonnegative, and $| y_k | \leqq 1$, makes it apparent that

$$y_{k+1}^L + y_{k+1}^C \geqq 2 \tag{51}$$

is sufficient. The "stable nonlinearities" deduced from this kind of reasoning are outlined in Fig. 9. Thus any nonlinearity whose graph coincides with the identity function on the interval $[-1, 1]$ and whose remaining portions lie in the closed shaded region of Fig. 9 will be stable. The function in these regions need not be continuous and need not obey $f(-u) = -f(u)$.

An even higher degree of generality is achieved when we realize that nothing in our proofs required the nonlinearity $f(u)$ to be the same for successive values of the parameter k. This is tantamount to allowing the nonlinearity to be random in the following manner. Suppose a value of $y_{k+1}^L > 1$ has been predicted from linear theory (see Fig. 9). The perpendicular P to the v axis through y_{k+1}^L intersects the shaded region shown in Fig. 9 along a line segment. Choose randomly from this line segment the "value" of the nonlinearity to give y_{k+1}^C. The discussion in this Section shows that the solutions of the difference equation

$$y_{k+2} = f[ay_{k+1} + by_k] \tag{52}$$

which has the stochastic nonlinearity just described will be stable whenever the linear version has stable solutions.

APPENDIX

Derivation of the Steady-State Solution

We obtain the steady-state solution of our fundamental equation (6) using z-transforms. Recall that if one has a bounded sequence of number $\{a_n\}$, the z-transform is defined by

$$f(z) = \sum_{n=0}^{\infty} a_n z^{-n} \tag{53}$$

where (53) converges and is analytic outside the unit circle, $|z| > 1$. It is easy to show that if $\{a_n\}$ is periodic of period N, that is if $a_{N+n} = a_n$, then (53) becomes

$$f(z) = \frac{A_{N-1}\left(\frac{1}{z}\right)}{1 - z^{-N}} \tag{54}$$

where A_{N-1} is the polynomial of degree $(N - 1)$ in $1/z$ given by

$$A_{N-1}\left(\frac{1}{z}\right) = \sum_{n=0}^{N-1} a_n z^{-n}. \tag{55}$$

The N poles of $f(z)$ at the N roots of unity are apparent from (12), and there are no other poles.

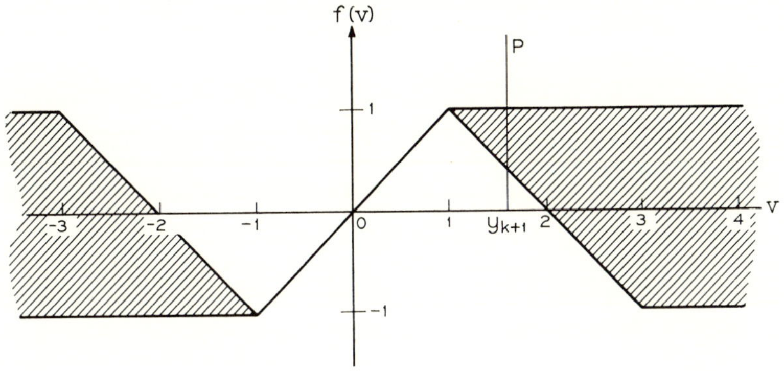

Fig. 9 — Any nonlinearity whose graph coincides with the identity function on the interval [−1, +1] and whose remaining portions lie in the (closed) shaded region will be stable. The possibility of generalizing this to a stochastic nonlinearity is also noted in the text.

Denoting by $Y(z)$ the z-transform of $y(t)$ *excluding* the additive terms involving initial conditions (since these will damp out because of linear stability) we have from (6) that

$$Y(z) = \frac{A_{N-1}\left(\frac{1}{z}\right)}{(z^2 - az - b)(1 - z^{-N})}. \tag{56}$$

The z-transform of the steady-state solution $\hat{Y}(z)$ must still be extracted from $Y(z)$. Since the unit circle $|z| = 1$ corresponds to the frequency axis if one were using Fourier transforms, we know, by analogy, the state steady-state portion of (56) will be the pole-terms. Let r_i, $i = 1, \cdots, N$ be the N Nth roots of unity and define

$$Q_i^{N-1}\left(\frac{1}{z}\right) \equiv \sum_{k=0}^{N-1} \left(\frac{1}{r_i}\right)^{N-1-k} \left(\frac{1}{z}\right)^k = \frac{1 - z^{-N}}{\frac{1}{r_i} - \frac{1}{z}}. \tag{57}$$

Note (57) implies

$$Q_i^{N-1}\left(\frac{1}{r_i}\right) = Nr_i . \tag{58}$$

Then from (56)–(58) we have

$$\hat{Y}(z) = \sum_{i=1}^{N} \frac{A_{N-1}\left(\frac{1}{r_i}\right)}{\left(\frac{1}{r_i} - \frac{1}{z}\right) \cdot Nr_i \cdot D(r_i)}, \tag{59}$$

where we have let

$$D(z) = z^2 - az - b. \tag{60}$$

Using (57) once more, the steady-state solution (59) may be written

$$\hat{Y}(z) = \frac{1}{1 - z^{-N}} \cdot \frac{1}{N} \sum_{i=1}^{N} \frac{A_{N-1}\left(\frac{1}{r_i}\right) Q_i^{N-1}\left(\frac{1}{z}\right)}{r_i D(r_i)}. \tag{61}$$

Referring back to the discussion at the beginning of this section, we see that (61) is the z-transform of a sequence $\{y_k\}$ of period N where

$$y_k = \text{coefficient of } z^{-k} \text{ in } \left\{ \frac{1}{N} \sum_{i=1}^{N} \frac{A_{N-1}\left(\frac{1}{r_i}\right) Q_i^{N-1}\left(\frac{1}{z}\right)}{r_i D(r_i)} \right\}$$

$$k = 0, 1, \cdots, N - 1. \tag{62}$$

Using (57) in (62) we obtain

$$y_k = \frac{1}{N} \sum_{i=1}^{N} \frac{A_{N-1}\left(\frac{1}{r_i}\right)}{D(r_i)} r_i^k, \tag{63}$$

where, in writing (63), we have used the fact that $r_i^N = 1$. Expression (63) thus gives the $\{y_k\}$ sequence for any click sequence. It is a solution corresponding to a self-sustained oscillation of the digital filter only if we have $|y_k| < 1$, all k.

Two sums appear in (63). The explicit one shown is the sum over the roots of unity; the hidden one is the polynomial $A_{N-1}(1/r_i)$. We will exhibit another form of solution (63) by explicitly doing the sum over the N roots. We begin by writing

$$A_{N-1}\left(\frac{1}{r_i}\right) = 2 \sum_{l=0}^{N-1} \frac{p_l}{r_i^l}, \qquad p_l = \pm 1, 0. \tag{64}$$

Thus p_l are the coefficients, except for the factor of 2, of the polynomial $A_{N-1}(z)$. We also write, by factoring $D(z)$ and expanding in partial fractions,

$$\frac{1}{D(z)} = \frac{1}{(z - \rho_1)(z - \rho_2)} = \frac{1}{\rho_1 - \rho_2} \left[\frac{1}{z - \rho_1} - \frac{1}{z - \rho_2} \right]. \tag{65}$$

Now note that if z is such a number than $z^N = 1$, we have (since $|\rho| < 1$ and $|z| = 1$)

$$\frac{1}{z-\rho} = \frac{1}{z}\sum_{n=0}^{\infty}\left(\frac{\rho}{z}\right)^n. \tag{66}$$

Let us look at the sum of the $n = 0, N, 2N$, etc., terms in the right side of (66), that is

$$1 + \frac{\rho^N}{z^N} + \frac{\rho^{2N}}{z^{2N}} + \frac{\rho^{3N}}{z^{3N}} + \cdots$$

$$= 1 + \rho^N + \rho^{2N} + \rho^{3N} + \cdots = \frac{1}{1-\rho^N}. \tag{67}$$

Treating the sum of terms

$$n = 1, N+1, 2N+1, \cdots$$

$$n = 2, N+2, 2N+2, \cdots$$

$$\vdots$$

$$n = N-1, N+(N-1), 2N+(N-1), \cdots$$

similarly, we have

$$\frac{1}{z-\rho} = \frac{1}{z}\cdot\frac{1}{1-\rho^N}\left[1 + \frac{\rho}{z} + \frac{\rho^2}{z^2} + \cdots + \frac{\rho^{N-1}}{z^{N-1}}\right]. \tag{68}$$

Finally letting $z = 1/r_i$ gives

$$\frac{1}{\frac{1}{r_i} - \rho} = \frac{r_i}{1-\rho^N}\sum_{n=0}^{N-1}[\rho r_i]^n. \tag{69}$$

Using (65) and (64) in (63) yields

$$y_k = \frac{1}{\rho_1 - \rho_2}\cdot\frac{2}{N}\sum_i r_i^k\left(\sum_{l=0}^{N-1}\frac{p_l}{r_i^l}\right)$$

$$\cdot\left[\frac{1}{r_i}\sum_{n=0}^{N-1}\frac{1}{r_i^n}\left(\frac{\rho_1^n}{1-\rho_1^n} - \frac{\rho_2^n}{1-\rho_2^n}\right)\right]. \tag{70}$$

Two sums in (70) are immediately done. First look at the sum over the roots of unity. This involves observing that

$$\sum_i r_i^{k-l-1-n} = \begin{cases} N & \text{if } k-l-1-n \equiv 0 \mod N, \\ 0 & \text{otherwise.}\end{cases} \tag{71}$$

The congruence indicated in (71) can only be satisfied here if $l = k -$

$1 - n$ or if $l = k - 1 - n + N$. Thus it is useful to define

$$2b_n^{(k)} \equiv \bar{a}_{k-1-n} + \bar{a}_{k-1-n+N}, \tag{72}$$

where we understand $\bar{a}_j \equiv 0$ if j does not lie between 0 and $N - 1$, inclusive, and $\bar{a}_j \equiv a_j$ if it does. One of the $\bar{a}$'s in (72) will thus always be zero and $b_n^{(k)}$ has values, like the p's, of $\pm 1, 0$. Using the discussion above surrounding equations (71) and (72) we perform next the sum over l and write another form of the solution:

$$y_k = \frac{2}{\rho_1 - \rho_2} \sum_{n=0}^{N-1} b_n^{(k)} \left[\frac{\rho_1^n}{1 - \rho_1^N} - \frac{\rho_2^n}{1 - \rho_2^N} \right] \qquad k = 0, 1, \cdots, N - 1. \tag{73}$$

REFERENCES

1. Rader, C. M., Gold, B., "Digital Filter Design Techniques in the Frequency Domain," Proc. IEEE, *55*, No. 2 (February 1967), pp. 149–171.
2. Jackson, L. B., Kaiser, J. F., and McDonald, H. S., "An Approach to the Implementation of Digital Filters," IEEE Trans. Audio and Electroacoustics, *AV-16*, No. 3 (September 1968), pp. 413–421.

23

Reprinted from *IEEE Trans. Audio Electroacoustics,* **AU-21**(2), 100–106 (1973)

Limit-Cycle Oscillations in Floating-Point Digital Filters

TOYOHISA KANEKO

Abstract—**In a digital filter realized with fixed-point arithmetic, there is a peculiar phenomenon known as limit-cycle oscillation, which is due to roundoff errors. For floating-point arithmetic, it has been conjectured that its amplitude is negligibly small, if it does exist. This paper shows that limit-cycle oscillations can exist in floating-point digital filters and that their amplitude can be large. Also, conditions for the existence of limit-cycle oscillations are derived.**

I. Introduction

Since all digital processors are implemented with a finite number of elements, there are inherent error problems due to the finite word length. The finite arithmetic involves rounding or truncating operations that are essentially nonlinear. Such a nonlinear operation will sometimes generate sustained oscillations known as limit-cycle oscillations if a feedback loop exists in an algorithm.

In digital filters, three sources of errors due to the finite word length have been identified: 1) input quantization error due to quantizing an input sequence, 2) coefficient truncation error due to truncating coefficients in the algorithm, and 3) roundoff error due to rounding intermediate results or final results. They have been studied extensively in the past. A recent paper by Liu [1] is an excellent review. Very recently, the phenomenon of limit-cycle oscillations has attracted much attention. A limit-cycle oscillation in a digital filter was first reported by Blackman [2], who called it a "deadband effect." It is a peculiar phenomenon caused by roundoff errors, and takes place only in a recursive filter. With a zero or constant value input, the output of a digital filter reaches and sustains a constant value different from the predicted value. Sometimes these are constant amplitude sinusoidal oscillations.

In the past, limit-cycle oscillations in digital filters have been studied only for fixed-point arithmetic. Jackson [3] derived bounds on the amplitude of zero-input limit-cycle oscillations by linearizing the nonlinear effect, and gave sufficient conditions on coefficients of first- and second-order digital filters to generate limit-cycle oscillations. A paper by Parker and Hess [4] derived three more exact bounds on the amplitude of limit-cycles using Lyapunov's direct method and a general matrix formulation.

There has been no work on limit-cycle oscillations in floating-point digital filters. It has been conjectured [1], [3] that their amplitude is negligibly small if they do exist. This conjecture seems to be founded on the fact that the floating-point quantizer has a very small

Manuscript received April 19, 1972.
The author is with the IBM T. J. Watson Research Center, Yorktown Heights, N. Y. 10598.

step size unlike a constant (not small) step size in the fixed-point quantizer.

This paper studies limit-cycle oscillations in floating-point digital filters. It will be shown that there can exist limit-cycle oscillations and that their amplitude can be large. Section II describes floating-point arithmetic and its properties involving roundoff errors. Limit-cycle oscillations in first- and second-order digital filters are analyzed in Sections III and IV, respectively. Section V describes some basic theorems on limit-cycle oscillations. Conditions for limit-cycle oscillations not to exist are described in Section VI. Section VII summarizes the results, and discusses similarities and differences between limit-cycle oscillations in fixed- and floating-point digital filters.

II. Floating-Point Arithmetic and Roundoff Errors

Before going into the analysis of limit-cycle oscillations in floating-point digital filters, we will review floating-point arithmetic and its roundoff error properties. A floating-point number is represented in the form

$$(\mathrm{sgn})\cdot(\alpha)\cdot\beta^m$$

where (sgn) is the sign, α is the mantissa or fractional part, β is the base, and m is the integer exponent. It is assumed throughout this paper that t digits are alloted to the mantissa, and that m is bounded as $-N<m\leq M$, and M and N are large enough so that there are neither overflows nor underflows.[1]

When two floating-point numbers with t-digit mantissas are multiplied, the mantissa of the exact product has a length of either $2t$ or $(2t-1)$ digits. The part beyond t significant digits is rounded, generating a roundoff error. There is, in general, a roundoff error in the addition of two floating-point numbers, or in shortening the length of the mantissa. It is shown by Wilkinson [5] that

$$fl(x\cdot y) = (x\cdot y)(1+\epsilon), \tag{1}$$

$$fl(x+y) = (x+y)(1+\delta), \tag{2}$$

$$fl(x) = x(1+\mu), \tag{3}$$

and

$$|\epsilon|,\ |\delta|,\ |\mu| \leq \tfrac{1}{2}\beta\cdot\beta^{-t} \tag{4}$$

where the notation $fl(\cdot)$ [5] denotes the machine number resulting from performing the arithmetic operation described in the parenthesis in some designated order, and $fl(x)$ is the result after shortening x. The statistical properties of the errors ϵ, δ, and μ were studied by Kaneko and Liu [6] recently. It is to be noted that a roundoff error due to the addition δ has a large probability at $\delta=0$, especially when the base β is large, say, $\beta=16$.

[1] Overflows and underflows occur in floating-point arithmetic when the exponents go beyond the upper bound, namely, M in the above example, and the lower bound, namely, $-N$, respectively.

III. Limit-Cycle Oscillations in First-Order Digital Filters

Limit-cycle oscillations in first- and second-order digital filters will be studied in this and subsequent sections, respectively. These filters are especially important because any higher order filter can be realized with a combination (cascade or parallel or both) of first- and second-order filters, and also because the arrangement is preferable in terms of roundoff error accumulation and coefficient truncation errors.

A first-order digital filter is specified by

$$w_n = x_n + aw_{n-1}, \qquad n > 0 \tag{5}$$

with $w_0=W$ where $\{x_n\}$ and $\{w_n\}$ are the input and output sequences, respectively, a is the coefficient of the filter, and W is an initial value. For the filter operation specified by (5) to be stable, we need

$$|a| < 1. \tag{6}$$

The so-called Q of the filter is defined by

$$Q = 1/(1-|a|). \tag{7}$$

When the filtering algorithm specified by (5) is carried out in a machine and the input is zero, i.e., $x_n=0$, we have

$$y_n = fl(ay_{n-1}) \tag{8}$$

where y_n is the actual output sequence in the machine, and the notation $fl(\ )$ is already defined in Section II. It is well known [7] that the filter specified by (8) has a limit-cycle oscillation of zero frequency if we have

$$u = fl(au) \tag{9}$$

for some $u(\neq 0)$. Equation (9) implies the existence of unity positive feedback in the operation. If it is negative, we have an alternating sign oscillation that will be discussed in Section V. Now we investigate solutions for u in (9) given a. When the machine has floating-point arithmetic with base β and length of the mantissa t, the filter[2] with the highest Q is given by

$$a = 1 - \beta^{-t}. \tag{10}$$

Although a is specified by the difference of two numbers 1 and β^{-t}, a is actually a number with the exponent zero and with the t-digit mantissa of the form

$$\underbrace{(\beta-1)(\beta-1)\cdots(\beta-1)}_{t\text{-digits}} \tag{11}$$

Using (1), (9) becomes

$$u = au(1+\epsilon)$$

or

[2] If one is allowed to use two numbers in specifying a value, a higher Q filter can be specified by, for example, the difference of 1 and β^{-m}:

$$a = 1 - \beta^{-m}$$

where m is the exponent and $m>t$. In this filter, the subsequent analysis favors the existence of a limit-cycle oscillation.

$$a = \frac{1}{1+\epsilon}, \qquad \text{if } u \neq 0$$

$$\epsilon = \frac{1-a}{a}. \tag{12}$$

If a is given by (10), then

$$\epsilon = \frac{\beta^{-t}}{1-\beta^{-t}} \approx \beta^{-t} + \beta^{-2t} + \beta^{-3t} + \cdots. \tag{13}$$

The right-hand side of (13) is larger than β^{-t}. However, $|\epsilon| \leq 2^{-t}$ for $\beta = 2$ from (4). Therefore, (13) does not hold with $\beta = 2$. That is, there are no limit cycles in a first-order digital filter for binary arithmetic.

Now we determine the amplitude of a limit-cycle oscillation for $\beta > 2$. Consider $u = (\beta^{-1}/2)\beta^m$, where m is an integer. Then for u to be a machine number implies that β must be an even integer:

$$fl(au) = fl[(1-\beta^{-t})(\beta^{-1}/2)\beta^m].$$

The mantissa of the exact product is

$$(\beta/2 - 1)\underbrace{(\beta-1)\cdots(\beta-1)}_{t\text{-digits}}\ \vdots\ (\beta/2)_{\,t+1}.$$

After rounding the $(t+1)$th digit, the above becomes $(\beta^{-1}/2)$ again. So

$$fl(au) = u.$$

Now take $u = (\alpha)\beta^m$ where α is the t-digit mantissa specified by

$$\alpha = .\alpha_1\alpha_2 \cdots \alpha_t$$

and

$$1 < \alpha_1 < \beta/2 \text{ or } \alpha_2\alpha_3 \cdots \alpha_t \neq 0, \qquad \text{if } \alpha_1 = 1.$$

The mantissa of the exact product au is

$$\begin{aligned}(1-\beta^{-t})\alpha &= .\alpha_1\alpha_2 \cdots \alpha_t - \beta^{-t}(\alpha_1\alpha_2 \cdots \alpha_t) \\ &= \underbrace{.\alpha_1\alpha_2 \cdots (\alpha_t - 1)}_{t\text{-digits}} \\ &\quad \cdot \underbrace{(\beta-1-\alpha_1)(\beta-1-\alpha_2) \cdots (\beta-\alpha_t)}_{t\text{-digits}}\end{aligned} \tag{14}$$

where $\alpha_t \geq 1$ is assumed. If $\alpha_t = 0$, borrow from the first nonzero α_i, $i = t-1, t-2, \cdots, 2$. Since the $(t+1)$th digit $(\beta - 1 - \alpha_1) \geq \beta/2$, the result after rounding is again

$$.\alpha_1\alpha_2 \cdots \alpha_t.$$

Therefore we have

$$fl(au) = u.$$

The behavior of the filter specified by (5) with $a = 1 - \beta^{-t}$ is as follows. If $(\beta^{-1}/2)\beta^m < W \leq \beta^m$, $\{y_n\}$ decreases until it is in $(\beta^{m-2}, (\beta^{-1}/2)\beta^m]$, and stays at the first value in the domain. If $\beta^{m-2} < W \leq (\beta^{-1}/2)\beta^m$, $\{y_n\}$ stays at the initial value W. This is illustrated in Fig. 1.

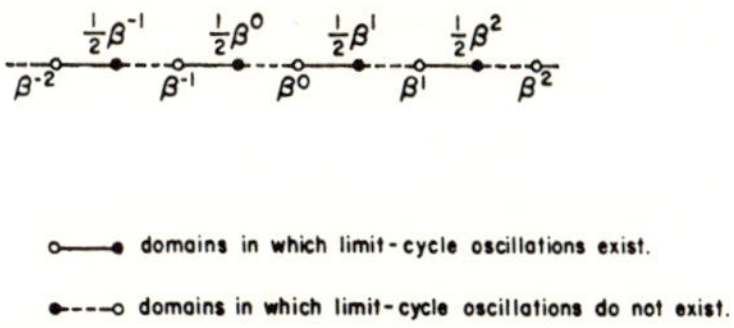

Fig. 1. Domains of limit-cycle oscillations in the first-order digital filter $w_n = (1-\beta^{-t})w_{n-1}$.

The number of domains where limit cycles exist is $2(M+N)$ if the exponent ranges as $-N < m \leq M$.

So far, only the filter specified by $a = 1 - \beta^{-t}$ has been studied. We now investigate a first-order filter specified by $a = 1 - k\beta^{-t}$ where k is a positive integer. With this filter, the $(t+1)$th digit of (14) becomes

$$\beta - 1 - k\alpha_1. \tag{15}$$

If $\beta - 1 - k\alpha_1 \geq \beta/2$, 1 is added to the tth digit after rounding, and we have $fl(au) = u$ again.

Therefore

$$\begin{aligned} k &\leq \beta/2 - 1, &\quad &\text{if } t \geq 2 \\ k &\leq \tfrac{1}{2}(\beta/2 - 1), &\quad &\text{if } t = 1 \end{aligned} \tag{16}$$

because α_1 can be 1 if $t \geq 2$, but $\alpha_1 \geq 2$ if $t = 1$. Then limit cycles exist in the domain

$$\left(\beta^{m-1}, \frac{\beta}{2k}\beta^{m-1}\right]. \tag{17}$$

The expression given in (17) implies that the range of the limit cycle domain decreases as k increases or as Q decreases.

Example: Take $\beta = 10$, $t = 2$. Then $k \leq 4$ from (16). Therefore $a = 0.96$ is the lowest Q filter with limit cycles. From (17), the domain of limit cycles for $a = 0.96$ is

$$(0.10 \times 10^m, 0.12 \times 10^m]$$

or

$$0.11 \times 10^m \quad \text{and} \quad 0.12 \times 10^m.$$

Also, a transient behavior is given in Table I using the above example and the initial value $W = 0.20$.

IV. Limit-Cycle Oscillations in Second-Order Digital Filters

A second-order digital filter is specified by

$$w_n = x_n - aw_{n-1} - bw_{n-2}, \qquad n \geq 2 \tag{18}$$

with

$$w_0 = W_0 \quad \text{and} \quad w_1 = W_1$$

where $\{x_n\}$ and $\{w_n\}$ are the input and output sequences, respectively, and a and b are the coefficients.

TABLE I
The Transient Behavior of a Filter $y_n = +ay_{n-1}$ where $a=0.96$, $\beta=10$, $t=2$, $W=y_0=0.20$

n	Exact ($a \times y_{n-1}$)	y_n
0		0.20
1	0.96×0.20=0.192	0.19
2	0.96×0.19=0.1824	0.18
3	0.96×0.18=0.1728	0.17
4	0.96×0.17=0.1632	0.16
5	0.96×0.16=0.1536	0.15
6	0.96×0.15=0.1440	0.14
7	0.96×0.14=0.1344	0.13
8	0.96×0.13=0.1248	0.12
9	0.96×0.12=0.1152	0.12
10	0.96×0.12=0.1152	0.12

The pulse transfer function $H(z)$ is given in the Z domain as

$$H(z) = 1/D(z)$$

and

$$D(z) = 1 + az^{-1} + bz^{-2}. \tag{19}$$

We will consider only a filter possessing two conjugate roots in $D(z)$. So

$$a^2 - 4b < 0. \tag{20}$$

For the filter to be stable, the two roots should be inside the unit circle:

$$0 < b < 1. \tag{21}$$

Also, we impose

$$a \neq 0 \tag{22}$$

because (18) with $a=0$ specifies only every other term and reduces to two decoupled first-order system already treated. The following two types of computational methods are commonly used for the filter specified by (18).

Type 1: The multiplications and addition in (18) are all carried out exactly with double precision accuracy, and the result is rounded into a t-digit floating-point number at the final step.

Type 2: The result at each arithmetic step in (18) is rounded into a t-digit floating-point number.

Using the notation $fl[\ ]$, the expression of (18) can be written as

$$y_n = fl[(-ay_{n-1}) + (-by_{n-2})] \text{ for type 1} \tag{23a}$$

$$y_n = fl[fl[(-a)y_{n-1}] + fl[(-b)y_{n-2}]] \text{ for type 2} \tag{23b}$$

where $\{y_n\}$ is the actual machine result.

Now let us first examine limit-cycle oscillations at zero frequency. The set $y_n = y_{n-1} = y_{n-2} = u$. For type 1, the same result can be obtained by multiplying $(-a-b)$ and u. Therefore, (23a) becomes

$$u = fl[\{-(a+b)\}u] \tag{24}$$

where is the same equation encountered in Section III. Therefore, $-(a+b)$ should satisfy (12).

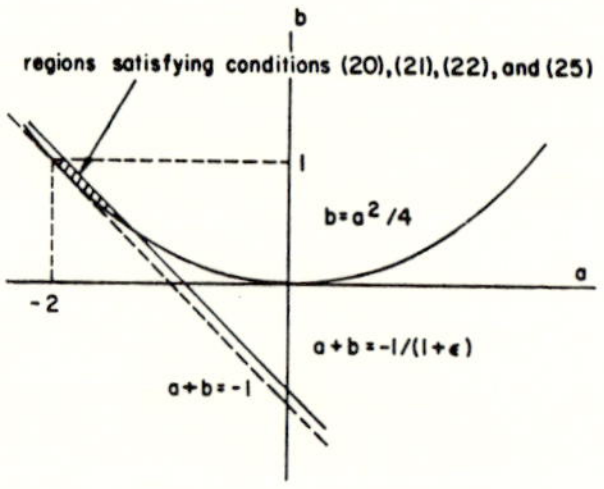

Fig. 2. Regions satisfying the conditions (20), (21), (22), and (25).

$$-a - b = \frac{1}{1+\epsilon}. \tag{25}$$

Conditions (20), (21), (22), and (25) are shown in Fig. 2.

As is seen in Fig. 2, the coefficients a and b satisfying the above conditions are situated close to -2 and 1, respectively. Hence, a and b may be written in the following form:

$$a = -(2 - k\beta^{-t+1})$$

and

$$b = 1 - l\beta^{-t} \tag{26}$$

where k and l are positive integers. Substituting a and b in (25), it can be shown that there is no ϵ satisfying (4). This is due to the fact that the relative accuracy of a is not as good as that of b because the first digit of a is assigned to be 1. To improve the accuracy of a, a second-order filter can be realized with[3]

$$w_n = x_n - 2(aw_{n-1}) - bw_{n-2} \tag{27}$$

which requires one more multiplication than (18). Then it can be easily shown that limit-cycle oscilations exist with $b = -a = 1 - k\beta^{-t}$ [k satisfies (16)].

Now let us examine zero frequency limit cycles for type 2. It is to be noted that a difference operation is carried out between the first and second terms on the right-hand side in (23b) because a is negative and b is positive. As was mentioned in Section II, roundoff error in the difference is often zero, especially for large β, say, 16. So roundoff error at this step is not considered. Let

$$fl(-au) = c_1 u$$

and

$$fl(bu) = c_2 u. \tag{28}$$

Then the condition that limit cycles exist is

$$c_1 - c_2 = 1. \tag{29}$$

An obvious solution is given by $c_2 = 1$ and $c_1 = 2$. But

[3] Also, the following form improves the accuracy of a.

$$w_n = x_n - (2-a)w_{n-1} - bw_{n-2}.$$

Existence of limit cycles can be shown for this form as well.

$c_1=2$ is not realizable in an a given in (26). In general, it is difficult to solve (28) for u, a, and b given c_1 and c_2. A somewhat heuristic method is illustrated by the following example.

Example: Take $\beta=10$ and $t=2$. First fix u and a or b. Say $a=-1.9$ and $u=0.33$. Then the b with which limit cycles exist should satisfy [from (23b)]

$$fl(b \times 0.33) = fl(1.9 \times 0.33) - 0.33 = 0.30.$$

Then $b=0.90$, 0.91, 0.92. $b=0.90$ should be excluded since it does not satisfy (20).

Example: Take $\beta=2$, $t=6$, $a=1.11011$, and $u=1.11000$. Then the b that generates a limit-cycle oscillation is given by

$$b = 0.110111$$

and it can be shown that $a^2-4b<0$.

As shown in this example, such limit cycles exist for a relatively low Q filter.

From (1) and (2), (23b) with $y_n=y_{n-1}=y_{n-2}=u$ can be written as

$$u = [-au(1+\xi) - bu(1+\eta)](1+\delta) \tag{30}$$

where

$$|\xi|, |\eta|, |\delta| \leq (\beta/2)\beta^{-t}.$$

Or

$$-a(1+\xi) - b(1+\eta) = \frac{1}{1+\delta}. \tag{31}$$

If a and b are given, it is possible to examine the existence of limit cycles using (30) and (31) [and (20) and (21)].

More interesting oscillation modes in a second-order digital filter occur for other than zero frequency. If the pulse transfer function is given in (19), a sufficient condition for the existence of a limit cycle is

$$b = 1.$$

In terms of the time domain, the coefficient of y_{n-2} is -1. For type 1, this condition is

$$b(1+\epsilon_n) = 1 \tag{32}$$

where ϵ_n is the relative rounding error of the result $(-ay_{n-1}-by_{n-2})$. Since b is fixed, ϵ_n should be constant as well. Therefore $(-ay_{n-1}-by_{n-2})$ can take values that result in the same rounding error. One class of values that satisfy the above condition is a set of three numbers $\{p, 0, -p\}$. Note that zero is always a trivial solution. Then the following limit cycle exists with $a=-b$:

$$p, p, 0, -p, -p, 0, p, p, \cdots.$$

Example: Take $\beta=10$, $t=1$, $W_0=W_1=0.2$, 0.3, 0.4, or 0.5. $w_n=0.9\,w_{n-1}-0.9\,w_{n-2}$.

Then

$$y_n = W_0, W_0, 0, -W_0, -W_0, 0, W_0, W_0, \cdots.$$

For type 2, condition (32) becomes

$$fl(by_{n-2}) = y_{n-2}, \tag{33}$$

which was studied extensively in Section III. That is, solutions exist for $b=1-k\beta^{-t}$ where k satisfies (16), and $\{y_n\}$ is in the domain specified by (17).

A significant difference for the solution of (33) between fixed- and floating-point arithmetics is as follows. The values of y_{n-2} that satisfy (33) are in the range $[-d, d]$ for fixed-point arithmetic, where d is the positive integer. However, for floating-point arithmetic, the solutions are in the multiple ranges specified by

$$(\beta^{m-1}, (\beta/2k)\beta^m] \tag{34}$$

or

$$[-(\beta/2k)\beta^m, -\beta^{m-1})$$

where $b=1-k\beta^{-t}$ and m is an integer. Also, zero is a trivial solution. Fig. 3(a) and (b) illustrate the above situations. For fixed-point arithmetic, the frequency of a limit-cycle oscillation is determined only by the coefficient a, and can go as low as zero frequency without any further restrictions. However, for floating-point arithmetic, $\{y_n\}$ cannot pass through domains other than those specified by (34). If this should be attempted, the amplitude is attenuated and the limit cycle diminishes. This peculiar situation implies that the frequency of a limit cycle has a lower bound. To determine a lower bound, first consider a situation shown in Fig. 4(a), where all points of a limit cycle belong to either $(\beta^{m-1}, (\beta/2k)\beta^{m-1}]$ or $[-(\beta/2k)\beta^{m-1}, -\beta^{m-1})$ for some m. As a matter of fact, m is determined by initial values. Consider a sinusoidal wave specified by

$$A \sin 2\pi f_0 s$$

where A is the amplitude, f_0 is the frequency, and s is the (continuous) time. For the limit cycle to remain in the domains specified by (34), a necessary condition is [see Fig. 4(a)]

$$A \sin 2\pi f_0(\tfrac{1}{2}) \geq \beta^{m-1}.$$

Here, the frequency of the sample sequence is normalized to 1. Therefore, we have

$$f_0 \geq \frac{1}{\pi} \sin^{-1}(\beta^{m-1}/A). \tag{35}$$

If $A=(\beta/2k)\beta^{m-1}$, then

$$f_0 \geq \frac{1}{\pi} \sin^{-1}(2k/\beta).$$

Another oscillation mode is shown in Fig. 4(b), where most points belong to the above domains, but the other points pass through some inside domains. Then the frequency can go lower than the bound given by (35). If

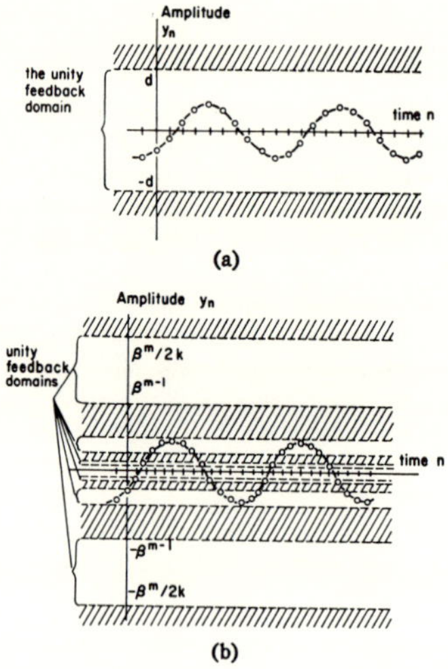

Fig. 3. (a) Limit-cycle oscillation in a second-order digital filter realized with fixed-point arithmetic. (b) Limit-cycle oscillation in a second-order digital filter realized with floating-point arithmetic.

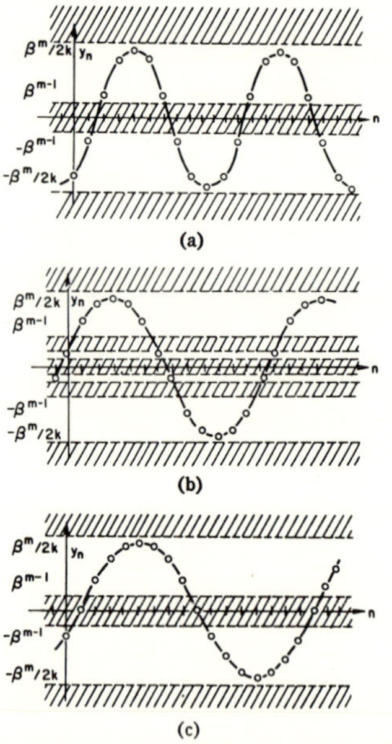

Fig. 4. (a) Limit-cycle oscillation existing only in two (positive and negative) domains. (b) Limit-cycle oscillation existing in more than two domains. (c) Limit-cycle oscillation that includes zero.

zero is included in the limit cycle, as is shown in Fig. 4(c), then the lower bound can be a half of (35):

$$f_0 \geq \frac{1}{2\pi} \sin^{-1} (\beta^{m-1}/A).$$

V. Some Basic Properties of Limit-Cycle Oscillations in Floating-Point Digital Filters

Consider an Nth order digital filter specified by

$$w_n = x_n - \sum_{i=1}^{N} a_i w_{n-i} \tag{36}$$

where $\{x_n\}$ and $\{w_n\}$ are the input and output sequences, respectively.

The pulse transfer function is given by

$$H(z) = 1/D(z) \tag{37}$$

and

$$D(z) = 1 + \sum_{i=1}^{N} a_i z^{-1}.$$

Theorem 1: When the digital filter specified by (36) utilizes floating-point arithmetic, a limit-cycle oscillation is described with a finite number of values $\{u_0, u_1, \cdots, u_{k-1}\}$ (therefore, the frequency is $1/k$). Then another limit-cycle oscillation exists that can be described by $\{u_0\beta^m, u_1\beta^m, \cdots, u_{k-1}\beta^m\}$ where m is an integer.

Proof: Since $\{u_0, \cdots, u_{k-1}\}$ are solutions, $\{y_n, y_{n+1}, \cdots, y_{n+k-1}\}$ can be computed by substituting those numbers where n is an arbitrary positive integer. Then multiply β^m in the equations. Because these multiplications do not change the relative amount of the roundoff errors, $\{u_0\beta^m, u_1\beta^m, \cdots, u_{k-1}\beta^m\}$ are also solutions.

Theorem 2: If u_0 is a limit-cycle solution of (36), then the filter specified by

$$w_n = x_n - \sum_{i=1}^{N} (-1)^i a_i w_{n-i}, \qquad x_n = 0 \tag{38}$$

generates the alternating sign oscillation

$$u_0, -u_0, u_0, -u_0, u_0, \cdots.$$

Proof: Again write the exact computational procedure including roundoff errors for (36) by substituting u_0. Since the equation $u_0 = -\sum a_i u_0$ holds, we have $(-1)^n u_0 = -\sum (-1)^i a_i (-1)^{n-i} u_0$, which implies the existence of the alternation sign oscillation $y_n = (-1)^n u_0$ in (38).

VI. Conditions for Limit-Cycle Oscillations not to Exist

From a designer's view, it is important to obtain conditions on coefficients that do not generate any limit-cycle oscillations. In general, limit-cycle oscillations do not exist in a low Q filter. More specifically, in a floating-point filter realized on a β-ary and t-digit machine, there are no limit-cycle oscillations for a first-order digital filter if (12) does not hold. So we have

$$\left|\frac{1-a}{a}\right| > \frac{1}{2}\beta^{1-t}. \quad (39)$$

Considering only a positive a, we have

$$a < 1/(1 + \tfrac{1}{2}\beta^{1-t}) \quad (40)$$

$$Q < (1 + \tfrac{1}{2}\beta^{1-t})2\beta^{t-1} \approx 2\beta^{t-1}. \quad (41)$$

Similarly, limit-cycle oscillations at zero frequency can be avoided in second-order digital filters of types 1 and 2 if (25) and (31), respectively, do not hold. Thus from (25)

$$-(a + b) < 1/(1 + \tfrac{1}{2}\beta^{1-t}). \quad (42)$$

Equation (31) can be rewritten as

$$-a\xi - b\eta + \frac{\delta}{1+\delta} = a + b + 1$$

where $|\xi|$, $|\eta|$, and $|\delta|$ are bounded by $\frac{1}{2}\beta^{1-t}$.

The above equality does not hold if

$$|a + b + 1| > (|a| + |b| + 1/(1 - \tfrac{1}{2}\beta^{1-t}))\tfrac{1}{2}\beta^{1-t}. \quad (43)$$

For limit-cycle oscillations at other than zero frequency, we obtain the following condition from (32):

$$|b| < 1/(1 + \tfrac{1}{2}\beta^{1-t}). \quad (44)$$

VII. Summary and Discussion

The existence of limit-cycle oscillations has been demonstrated for first- and second-order digital filters realized with floating-point arithmetic, whereas previously it had been conjectured that either no limit-cycle oscillations exist or, if they do, they are negligibly small. For first- and second-order digital filters realized with both fixed- and floating-point arithmetics, there exists a constant value limit-cycle oscillation(s). For the former arithmetic, there can be only one domain that includes zero. However, for the latter, there are many limit-cycle oscillations in separated domains. The zero value is an isolated point. In control theory terminology, limit cycles in fixed-point digital filters are stable, while those in floating-point filters are semistable.

For a second-order floating-point digital filter, there exist limit-cycle oscillations other than those of zero frequency or the alternative sign. For a limit cycle to be stable, its values should remain in domains where negative unity feedback can be obtained from y_{n-2}. This special condition imposes a lower bound for the frequency of the limit-cycle oscillation.

References

[1] B. Liu, "Effect of finite word length on the accuracy of digital filters—A review," *IEEE Trans. Circuit Theory*, vol. CT-18, pp. 670–677, Nov. 1971.

[2] R. B. Blackman, *Linear Data-Smoothing and Prediction in Theory and Practice*. Reading, Mass.: Addison-Wesley, 1965, pp. 75–79.

[3] L. B. Jackson, "An analysis of limit cycles due to multiplication rounding in recursive digital filters," in *Proc. 7th Annu. Allerton Conf. Circuit and Syst. Theory*, 1969, pp. 69–79.

[4] S. R. Parker and S. F. Hess, "Limit-cycle oscillations in digital filters," *IEEE Trans. Circuit Theory*, vol. CT-18, pp. 687–697, Nov. 1971.

[5] J. H. Wilkinson, *Rounding Errors in Algebraic Processors*. Englewood Cliffs, N. J.: Prentice-Hall, 1964.

[6] T. Kaneko and B. Liu, "On local round-off errors in floating-point arithmetic," *J. Ass. Comput. Mach.*, to be published.

[7] J. R. Ragazzini and G. F. Franklin, *Sampled-Data Control Systems*. New York: McGraw-Hill, 1958, pp. 93–105.

IV

The Fast Fourier Transform: Introduction

Editor's Comments on Paper 24

24 IEEE G-AE Subcommittee on Measurement Concepts: *What is the Fast Fourier Transform?*

Most engineers and applied scientists are undoubtedly familiar with some aspects of Fourier analysis. From the applications point of view, the basic theory is relatively simple. A problem arises, however, when one needs to calculate numerically the Fourier transform of a signal record that is not too short in its duration. By definition, the transform of a signal $x(t)$ is

$$X(f) = \int_{-\infty}^{\infty} x(t)e^{-2\pi ft}\, dt$$

One approach to the calculation would be to select an interval on the t axis outside of which $x(t)$ is negligible. Then, N points uniformly spaced in the interval are picked. With Δ_t denoting the separation between neighboring points, the Fourier integral can be approximated by a sum

$$X(f) = \sum_k x(k\Delta_t)e^{-2\pi fk\Delta_t}\, \Delta_t$$

There are N terms in the summation. Thus for each f, the calculation of $X(t)$ requires N complex multiplications and $(N - 1)$ complex additions. Suppose that $X(f)$ is to be calculated at N values of f, assumed to be uniformly spaced. The total calculation requires N^2 multiplications and $N(N - 1)$ additions.

The term "fast Fourier transform" or "FFT" refers to a family of algorithms that can perform such a calculation in $N \log N$ rather than N^2 arithmetic operations. The savings in computation effort by a factor of $N/\log N$ can be very important, because the amount of computing is often a determining factor in whether or not it is fruitful to pursue a certain course of investigation. With $N = 1024$, for example, a reduction by about two orders of magnitude is achieved by the use of FFT. This kind of improvement also makes it possible to construct dedicated hardware Fourier transform processors that operate in real time for some very important applications. The paper, written by a committee, serves as an excellent introduction to this subject.

24

Reprinted from *IEEE Trans. Audio Electroacoustics,* **AU-15**(2), 45–55 (1967)

What is the Fast Fourier Transform?

G-AE Subcommittee on Measurement Concepts

WILLIAM T. COCHRAN
JAMES W. COOLEY
DAVID L. FAVIN, MEMBER, IEEE
HOWARD D. HELMS, MEMBER, IEEE
REGINALD A. KAENEL, SENIOR MEMBER, IEEE
WILLIAM W. LANG, SENIOR MEMBER, IEEE
GEORGE C. MALING, JR., ASSOCIATE MEMBER, IEEE
DAVID E. NELSON, MEMBER, IEEE,
CHARLES M. RADER, MEMBER, IEEE
PETER D. WELCH

***Abstract*—The fast Fourier transform is a computational tool which facilitates signal analysis such as power spectrum analysis and filter simulation by means of digital computers. It is a method for efficiently computing the discrete Fourier transform of a series of data samples (referred to as a time series). In this paper, the discrete Fourier transform of a time series is defined, some of its properties are discussed, the associated fast method (fast Fourier transform) for computing this transform is derived, and some of the computational aspects of the method are presented. Examples are included to demonstrate the concepts involved.**

INTRODUCTION

An ALGORITHM for the computation of Fourier coefficients which requires much less computational effort than was required in the past was reported by Cooley and Tukey [1] in 1965. This method is now widely known as the "fast Fourier transform," and has produced major changes in computational techniques used in digital spectral analysis, filter simulation, and related fields. The technique has a long and interesting history that has been summarized by Cooley, Lewis, and Welch in this issue [2].

The fast Fourier transform (FFT) is a method for efficiently computing the discrete Fourier transform (DFT) of a time series (discrete data samples). The efficiency of this method is such that solutions to many problems can now be obtained substantially more economically than in the past. This is the reason for the very great current interest in this technique.

The discrete Fourier transform (DFT) is a transform in its own right such as the Fourier integral transform or the Fourier series transform. It is a powerful reversible mapping operation for time series. As the name implies, it has mathematical properties that are entirely analogous to those of the Fourier integral transform. In particular, it defines a spectrum of a time series; multiplication of the transform of two time series corresponds to convolving the time series.

If digital analysis techniques are to be used for analyzing a continuous waveform then it is necessary that the data be sampled (usually at equally spaced intervals of time) in order to produce a time series of discrete samples which can be fed into a digital computer. As is well known [6], such a time series completely represents the continuous waveform, provided this waveform is frequency band-limited and the samples are taken at a rate that is at least twice the highest frequency present in the waveform. When these samples are equally spaced they are known as Nyquist samples. It will be shown that the DFT of such a time series is closely related to the Fourier transform of the continuous waveform from which samples have been taken to form the time series. This makes the DFT particularly useful for power spectrum analysis and filter simulation on digital computers.

The fast Fourier transform (FFT), then, is a highly efficient procedure for computing the DFT of a time series. It takes advantage of the fact that the calculation of the coefficients of the DFT can be carried out iteratively, which results in a considerable savings of computation time. This manipulation is not intuitively obvious, perhaps explaining why this approach was overlooked for such a long time. Specifically, if the time series consists of $N=2^n$ samples, then about $2nN=2N\cdot\log_2 N$ arithmetic operations will be shown to be required to evaluate all N associated DFT coefficients. In comparison with the number of operations required for the calculation of the DFT coefficients with straightforward procedures (N^2), this number is so small when N is large as to completely change the computationally economical approach to various problems. For example, it has been reported that for $N=8192$ samples, the computations require about five seconds

Manuscript received March 10, 1967.

W. T. Cochran, D. L. Favin, and R. A. Kaenel are with Bell Telephone Laboratories, Inc., Murray Hill, N. J.

H. D. Helms is with Bell Telephone Laboratories, Inc., Whippany, N. J.

J. W. Cooley and P. D. Welch are with the IBM Research Center, Yorktown Heights, N. Y.

W. W. Lang and G. C. Maling are with the IBM Corporation, Poughkeepsie, N. Y.

C. M. Rader is with Lincoln Laboratory, Massachusetts Institute of Technology, Lexington, Mass. (Operated with support from the U. S. Air Force.)

D. E. Nelson is with the Electronics Division of the General Dynamics Corporation, Rochester, N. Y.

for the evaluation of all 8192 DFT coefficients on an IBM 7094 computer. Conventional procedures take on the order of half an hour.

The known applications where a substantial reduction in computation time has been achieved include: 1) computation of the power spectra and autocorrelation functions of sampled data [4]; 2) simulation of filters [5]; 3) pattern recognition by using a two-dimensional form of the DFT; 4) computation of bispectra, cross-covariance functions, cepstra and related functions; and 5) decomposing of convolved functions.

The Discrete Fourier Transform (DFT)

Definition of the DFT and its Inverse

Since the FFT is an efficient method for computing the DFT it is appropriate to begin by discussing the DFT and some of the properties that make it so useful a transformation. The DFT is defined by[1]

$$A_r = \sum_{k=0}^{N-1} X_k \exp(-2\pi jrk/N) \qquad r = 0, \cdots, N-1 \quad (1)$$

where A_r is the rth coefficient of the DFT and X_k denotes the kth sample of the time series which consists of N samples and $j=\sqrt{-1}$. The X_k's can be complex numbers and the A_r's are almost always complex. For notational convenience (1) is often written as

$$A_r = \sum_{k=0}^{N-1} (X_k) W^{rk} \qquad r = 0, \cdots, N-1 \quad (2)$$

where

$$W = \exp(-2\pi j/N). \quad (3)$$

Since the X_k's are often values of a function at discrete time points, the index r is sometimes called the "frequency" of the DFT. The DFT has also been called the "discrete Fourier transform" or the "discrete time, finite range Fourier transform."

There exists the usual inverse of the DFT and, because the form is very similar to that of the DFT, the FFT may be used to compute it.

The inverse of (2) is

$$X_l = (1/N) \sum_{r=0}^{N-1} A_r W^{-rl} \qquad l = 0, 1, \cdots, N-1. \quad (4)$$

This relationship is called the inverse discrete Fourier transform (IDFT). It is easy to show that this inversion is valid by inserting (2) into (4)

$$X_l = \sum_{r=0}^{N-1} \sum_{k=0}^{N-1} (X_k/N) W^{r(k-l)}. \quad (5)$$

Interchanging in (5) the order of summing over the indices r and k, and using the orthogonality relation

$$\sum_{r=0}^{N-1} \exp(2\pi j(n-m)r/N) = N, \text{ if } n \equiv m \bmod N$$
$$= 0, \text{ otherwise} \quad (6)$$

establishes that the right side of (5) is in fact equal to X_l.

It is useful to extend the range of definition of A_r to all integers (positive and negative). Within this definition it follows that

$$A_r = A_{N+r} = A_{2N+r} = \cdots \quad (7)$$

Similarly,

$$X_l = X_{N+l} = X_{2N+l} = \cdots. \quad (8)$$

Relationships between the DFT and the Fourier Transform of a Continuous Waveform

An important property that makes the DFT so eminently useful is the relationship between the DFT of a sequence of Nyquist samples and the Fourier transform of a continuous waveform, that is represented by the Nyquist samples. To recognize this relationship, consider a frequency band-limited waveform $g(t)$ whose Nyquist samples, X_k, vanish outside the time interval $0 \le t \le NT$

$$g(t) = \sum_{k=0}^{N-1} \frac{\sin(\pi(t-kT)/T)}{(\pi(t-kT)/T)} \cdot X_k \quad (9)$$

where T is the time spacing between the samples. A periodic repetition of $g(t)$ can be constructed that has identically the same Nyquist samples in the time interval $0 \le t \le NT$

$$g_p(t) = \sum_l \sum_{k=0}^{N-1} X_k \cdot \frac{\sin(\pi(t-kT-lNT)/T)}{(\pi(t-kT-lNT)/T)} \cdot \quad (10)$$

Let the Fourier transform of $g(t)$ be $G(f)$. As is well known [6], this transform is exactly specified at discrete frequencies by the complex Fourier series coefficients of $g_p(t)$. From this it follows:

$$\frac{G(n/NT)}{NT} = D_n$$
$$= (1/NT) \int_0^{NT} g_p(t) \cdot \exp(-2\pi jnt/NT) \cdot dt$$
$$= (1/NT) \sum_{k=0}^{N-1} X_k \cdot \exp(-2\pi jnkT/NT) \quad (11)$$

where $|n| \le N/2$ due to the spectral bandwidth limitation implicitly assumed by the sampling theorem underlying the validity of Nyquist samples.

Comparing (11) and (1) it is seen that they are exactly the same except for a factor of NT and (r, n) are both unbounded. That is,

$$N \cdot A_r = D_n \text{ for } r = n \text{ and } T = 1 \text{ second.} \quad (12)$$

The bounds specified for r and n require a correspondence which depends on (7)

[1] The definition of the DFT is not uniform in the literature. Some authors use A_r/N as the DFT coefficients, others use $A_r/\sqrt{N}$, still others use a positive exponent.

$$\frac{G(n/NT)}{NT} = D_n = N \cdot A_r$$

where

$$n = r \qquad \text{for } n = 0, 1, \cdots, q < N/2,$$

and

$$n = N - r \quad \text{for } n = -1, -2, \cdots, -q > -N/2 \quad (13)$$

and

$$\frac{G(n/NT)}{NT} = D_n = N \cdot A_r/2 \qquad \text{for } n = N/2. \quad (14)$$

Equations (13) and (14) give a direct relationship between the DFT coefficients and the Fourier transform at discrete frequencies for the waveform stipulated by (9). A one-to-one correspondence could have been obtained if the running variable r had been bounded by $\pm N/2$. This, however, would have required distinguishing between even and odd values of N, a distinction avoided by keeping r positive.

A waveform of the type considered by (9) is shown in Fig. 1(e). It is usually obtained as an approximation of a frequency band-limited source waveform [such as the one sketched in Fig. 1(a)] by truncating the Nyquist sample series of this waveform, and reconstructing the continuous waveform corresponding to the truncated Nyquist sample series [Fig. 1(b), (d), and (e)]. Notwithstanding the identity of the Nyquist samples of this reconstructed waveform and the frequency band-limited source waveform, these waveforms differ in the truncation interval [Fig. 1(c) and (e)]. The difference is usually referred to as aliasing distortion; the mechanics of this distortion is most apparent in the frequency domain [Fig. 1(c)–(e)]. It can be made negligibly small by choosing a sufficiently large product of the frequency bandwidth of the source waveform and the duration of the truncation interval [6] (e.g., N is greater than ten).

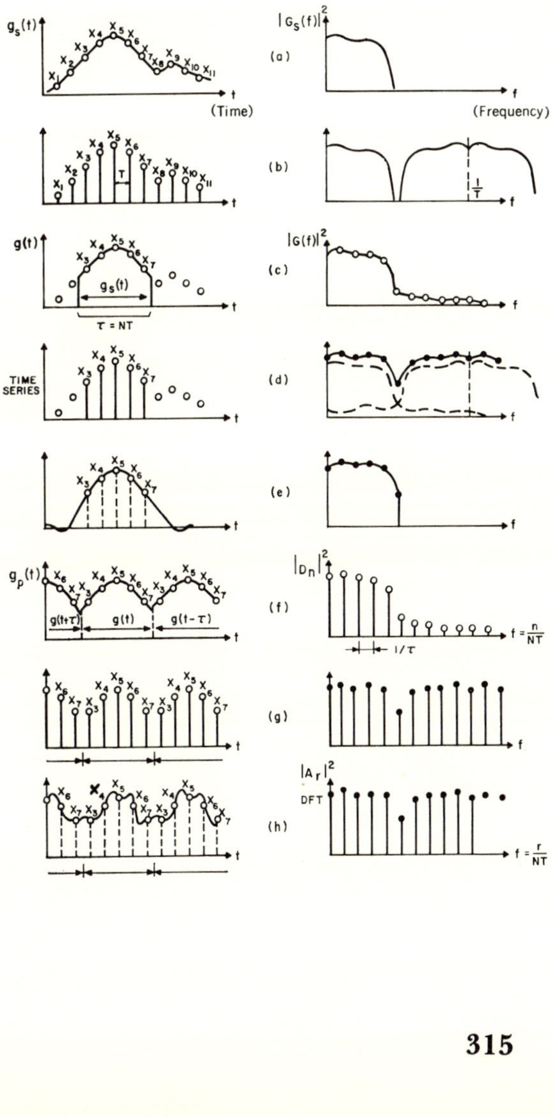

(a) Frequency-band-limited source waveform.
(b) Nyquist samples of the frequency band-limited source waveform.
(c) Truncated source waveform.
(d) Truncated series of Nyquist samples of the source waveform.
(e) Frequency-band-limited waveform whose Nyquist samples are identical to the truncated series of Nyquist samples of the source waveform.
(f) Periodic continuation of the truncated source waveform.
(g) Periodic continuation of the truncated series of Nyquist samples of the source waveform.
(h) DFT coefficients interpreted as Fourier series coefficients producing complex waveform.

Fig. 1. Related waveforms and their corresponding spectra as defined by the Fourier transforms (integral transforms for energy-limited waveforms; series transform for periodic waveforms).

These aliasing distortions are carried over directly to the discrete spectra of the periodically repeated waveforms [Fig. 1(f) and (g)], and appear correspondingly in the DFT of the truncated series of Nyquist samples [Fig. 1(h)]. It may be of interest to observe that the waveform corresponding to the DFT coefficients interpreted as Fourier series coefficients is complex [Fig. 1(h)].

Some Useful Properties of the DFT

Another property that makes the DFT eminently useful is the convolution relationship. That is, the IDFT of the product of two DFTs is the periodic mean convolution of the two time series of the DFTs. This relationship proves very useful when computing the filter output as a result of an input waveform; it becomes especially effective when computed by the FFT. A derivation of this property is given in Appendix A.

Other properties of the DFT are in agreement with the corresponding properties of the Fourier integral transform, perhaps with slight modifications. For example, the DFT of a time series circularly shifted by h is the DFT of the time series multiplied by W^{-rh}. Furthermore, the DFT of the sum of two functions is the sum of the DFT of the two functions. These properties are readily derived using the definition of the DFT. These and other properties have been compiled by Gentleman and Sande [7].

The Fast Fourier Transform

General Description of the FFT

As mentioned in the Introduction, the FFT is an algorithm that makes possible the computation of the DFT of a time series more rapidly than do other algorithms available. The possibility of computing the DFT by such a fast algorithm makes the DFT technique important. A comparison of the computational savings that may be achieved through use of the FFT is summarized in Table I for various computations that are frequently performed. It is important to add that the computational efforts listed represent comparable upper bounds; the actual efforts depend on the number N and the programming ingenuity applied [7].

It may be useful to point out that the FFT not only reduces the computation time; it also substantially reduces round-off errors associated with these computations. In fact, both computation time and round-off error essentially are reduced by a factor of $(\log_2 N)/N$ where N is the number of data samples in the time series. For example, if $N=1024=2^{10}$, then $N \cdot \log_2 N=10\,240$ [7], [9]. Conventional methods for computing (1) for $N=1024$ would require an effort proportional to $N^2=1\,048\,576$, more than 50 times that required with the FFT.

The FFT is a clever computational technique of sequentially combining progressively larger weighted sums of data samples so as to produce the DFT coefficients as defined by (2). The technique can be interpreted in terms of combining the DFTs of the individual data samples such that the occurrence times of these samples are taken into account sequentially and applied to the DFTs of progressively larger mutually exclusive subgroups of data samples, which are combined to ultimately produce the DFT of the complete series of data samples. The explanation of the FFT algorithm adopted in this paper is believed to be particularly descriptive for programming purposes.

TABLE I

COMPARISON OF THE NUMBER OF MULTIPLICATIONS REQUIRED USING "DIRECT" AND FFT METHODS

Operation	Formula	Approximate Number of Multiplications (upper comparable bounds)	
		Direct	FFT
Discrete Fourier Transform (DFT)	$\sum_{k=0}^{N-1} X_k e^{-2\pi jrk/N} \quad r=1, 2, \cdots, N-1$	N^2	$2N \log_2 N$
Filtering (Convolution)	$\sum_{k=0}^{N-1} X_k Y_{u-k} \quad u=0, 1, \cdots, N-1$	N^2	$3N \log_2 N$
Autocorrelation Functions	$\sum_{k=0}^{N-1-r} X_k X_{r+k} \quad r=0, 1, \cdots, N-1$	$\frac{N}{4}\left(\frac{N}{2}+3\right)$	$3N \log_2 N$
Two-Dimensional Fourier Transform (Pattern Analysis)	$\sum_{k=0}^{N-1}\sum_{l=0}^{N-1} X_{k,l} e^{-2\pi j(kq+r/N)} \quad r, q=0, 1, \cdots, N-1$	N^4	$4N^2 \log_2 N$
Two-Dimensional Filtering	$\sum_{k=0}^{N-1}\sum_{l=0}^{N-1} X_{k,l} Y_{q-k,r-l} \quad q, r=1, 2, \cdots, N-1$	N^4	$3N^2 \log_2 N$

Conventional Forms of the FFT

Decimation in Time: The DFT [as per (2)] and its inverse [see (4)] are of the same form so that a procedure, machine, or sub-routine capable of computing one can be used for computing the other by simply exchanging the roles of X_k and A_r, and making appropriate scale-factor and sign changes. The two basic forms of the FFT, each with its several modifications, are therefore equivalent. However, it is worth distinguishing between them and discussing them separately. Let us first consider the form used by Cooley and Tukey [1] which shall be called *decimation in time.* Reversing the roles of A_r and X_k gives the form called *decimation in frequency,* which will be considered afterwards.

Suppose a time series having N samples [such as X_k shown in Fig. 2(a)] is divided into two functions, Y_k and Z_k, each of which has only half as many points ($N/2$). The function Y_k is composed of the even-numbered points ($X_0, X_2, X_4 \cdots$), and Z_k is composed of the odd numbered points ($X_1, X_3, X_5 \cdots$). These functions are shown in Fig. 2(b) and (c), and we may write them formally as

$$\left.\begin{aligned} Y_k &= X_{2k} \\ Z_k &= X_{2k+1} \end{aligned}\right\} \quad k = 0, 1, 2, \cdots, \frac{N}{2} - 1. \tag{15}$$

Since Y_k and Z_k are sequences of $N/2$ points each, they have discrete Fourier transforms defined by

$$\left.\begin{aligned} B_r &= \sum_{k=0}^{(N/2)-1} Y_k \exp(-4\pi jrk/N) \\ C_r &= \sum_{k=0}^{(N/2)-1} Z_k \exp(-4\pi jrk/N) \end{aligned}\right\} \quad r = 0, 1, 2, \cdots, \frac{N}{2} - 1. \tag{16}$$

The discrete Fourier transform that we want is A_r, which we can write in terms of the odd- and even-numbered points

$$A_r = \sum_{k=0}^{(N/2)-1} \left\{ Y_k \exp(-4\pi jrk/N) + Z_k \exp\left(-\frac{2\pi jr}{N}[2k+1]\right)\right\} \quad r = 0, 1, 2, \cdots, N-1 \tag{17}$$

or

$$A_r = \sum_{k=0}^{(N/2)-1} Y_k \exp(-4\pi jrk/N) + \exp(-2\pi jr/N) \sum_{k=0}^{(N/2)-1} Z_k \exp(-4\pi jrk/N) \tag{18}$$

which, using (16), may be written in the following form:

$$A_r = B_r + \exp(-2\pi jr/N)C_r \quad 0 \le r < N/2. \tag{19}$$

For values of r greater than $N/2$, the discrete Fourier transforms B_r and C_r repeat periodically the values taken on when $r < N/2$. Therefore, substituting $r + N/2$ for r in (19), we obtain

$$\begin{aligned} A_{r+N/2} &= B_r + \exp\left(-2\pi j\left[r + \frac{N}{2}\right]/N\right)C_r \quad 0 \le r < N/2 \\ &= B_r - \exp(-2\pi jr/N)C_r \quad 0 \le r < N/2. \end{aligned} \tag{20}$$

By using (3), (19) and (20) may be written as

$$A_r = B_r + W^r C_r \quad 0 \le r < N/2 \tag{21}$$

$$A_{r+N/2} = B_r - W^r C_r \quad 0 \le r < N/2. \tag{22}$$

From (21) and (22), the first $N/2$ and last $N/2$ points of the discrete Fourier transform of X_k (a sequence having N samples) can be simply obtained from the DFT of Y_k and Z_k, both sequences of $N/2$ samples.

Assuming that we have a method which computes discrete Fourier transforms in a time proportional to the square of the number of samples, we can use this algorithm to compute the transforms of Y_k and Z_k, requiring a time proportional to $2(N/2)^2$, and use (21) and (22) to find A_r with additional N operations. This is illustrated in the signal flow graph of Fig. 3. The points on the left are the values of X_k (i.e., Y_k and Z_k), and the points on the right are the points of the discrete Fourier transform, A_r. For simplicity, Fig. 3 is drawn for the case where X_k is an eight-point function, and advantage is taken of the fact that $W^n = -W^{n-N/2}$, as per (3).

However, since Y_k and Z_k are to be transformed, and since we have shown that the computation of the DFT of N samples can be reduced to computing the DFTs of two sequences of $N/2$ samples each, the computation of B_k (or C_k) can be reduced to the computation of sequences of $N/4$ samples. These reductions can be carried out as long as each function has a number of samples that is divisible by 2. Thus, if $N = 2^n$ we can make n such reductions, applying (15), (21), and (22) first for N, then for $N/2, \cdots$, and finally for a two-point function. The discrete Fourier transform of a one-point function is, of course, the sample itself. The successive reduction of an eight-point discrete Fourier transform, begun in Fig. 3, is continued in Figs. 4 and 5. In Fig. 5 the operation has been completely reduced to complex multiplications and additions. From the signal flow graph there are 8 by 3 terminal nodes and 2 by 8 by 3 arrows, corresponding to 24 additions and 48 multiplications. Half of the multiplications can be omitted since the transmission indicated by the arrow is unity. Half of the remaining multiplications are also easily eliminated, as we shall see below. Thus, in general, $N \cdot \log_2 N$ complex additions and, at most, $\frac{1}{2}N \cdot \log_2 N$ complex multi-

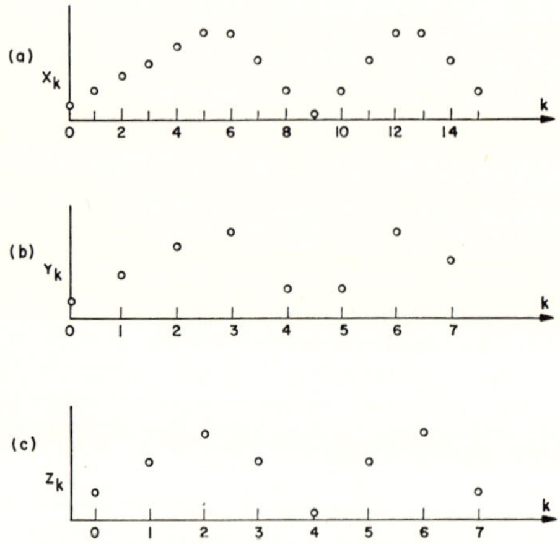

Fig. 2. Decomposition of a time series into two part-time series, each of which consists of half the samples.

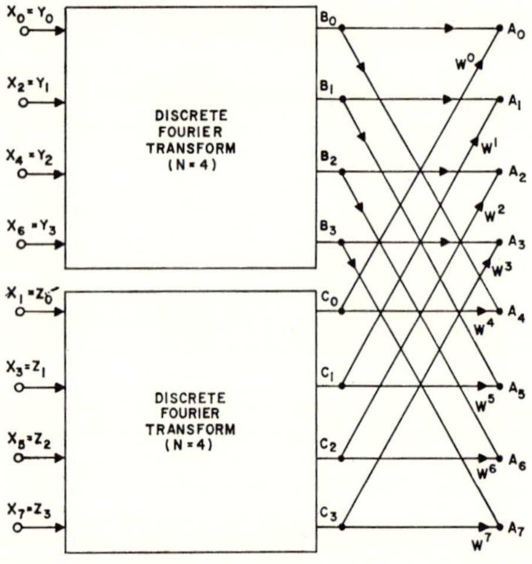

Fig. 3. Signal flow graph illustrating the reduction of endpoint DFT to two DFTs of $N/2$ points each, using decimation in time. The signal flow graph may be unfamiliar to some readers. Basically it is composed of dots (or nodes) and arrows (transmissions). Each node represents a variable, and the arrows terminating at that node originate at the nodes whose variables contribute to the value of the variable at that node. The contributions are additive, and the weight of each contribution, if other than unity, is indicated by the constant written close to the arrowhead of the transmission. Thus, in this example, the quantity A_7 at the bottom right node is equal to $B_3+W_7\times C_3$. Operations other than addition and constant multiplication must be clearly indicated by symbols other than · or ——→.

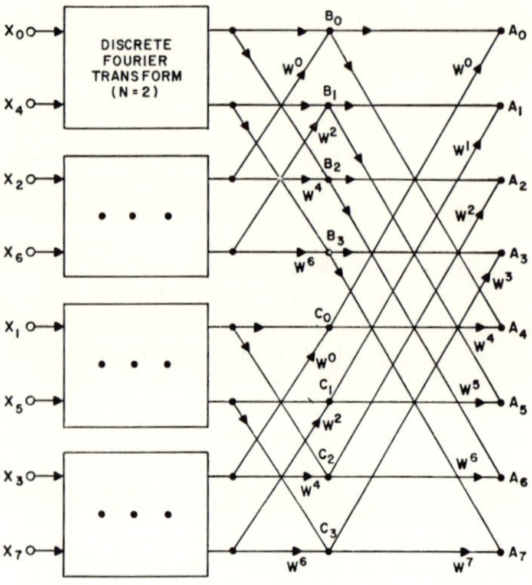

Fig. 4. Signal flow graph illustrating further reduction of the DFT computation suggested by Fig. 3.

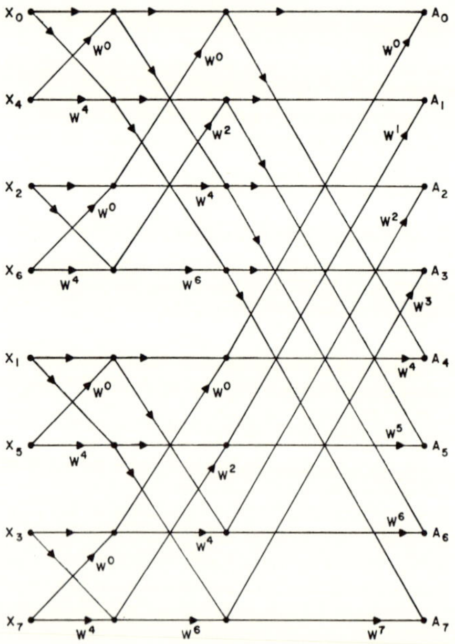

Fig. 5. Signal flow graph illustrating the computation of the DFT when the operations involved are completely reduced to multiplications and additions

plications are required for computation of the discrete Fourier transform of an N point sequence, where N is a power of 2.

When N is not a power of 2, but has a factor p, the development of equations analogous to (15) through (22) is possible by forming p different sequences, $Y_k^{(i)} = X_{pk+i}$, each having N/p samples. Each of these sequences has a DFT $B_r^{(i)}$, and the DFT of the sequence X_k can be computed from the p simpler DFTs with pN complex multiplications and additions. That is,

$$A_{r+m(N/p)} = \sum_{i=0}^{p-1} B_r^{(i)} W^{i[r+m(N/p)]}$$

$$m = 0, 1, 2, \cdots, p-1$$

$$r = 0, 1, 2, \cdots, \frac{N}{p} - 1. \quad (23)$$

The computation of the DFTs can be further simplified if N has additional prime factors.

Further information about the fast Fourier transform can be extracted from Fig. 5. For example, if the input sequence X_k is stored in computer memory in the order

$$X_0, X_4, X_2, X_6, X_1, X_5, X_3, X_7, \quad (24)$$

as in Fig. 5, the computation of the discrete Fourier transform may be done "in place," that is, by writing all intermediate results over the original data sequence, and writing the final answer over the intermediate results. Thus, no storage is needed beyond that required for the original N complex numbers. To see this, suppose that each node corresponds to two memory registers (the quantities to be stored are complex). The eight nodes farthest to the left in Fig. 5 then represent the registers containing the shuffled order input data. The first step in the computation is to compute the contents of the registers represented by the eight nodes just to the right of the input nodes. But each pair of input nodes affects only the corresponding pair of nodes immediately to the right, and if the computation deals with two nodes at a time, the newly computed quantities may be written into the registers from which the input values were taken, since the input values are no longer needed for further computation. The second step, computation of the quantities associated with the next vertical array of nodes to the right, also involves pairs of nodes although these pairs are now two locations apart instead of one. This fact does not change the property of "in place" computation, since each pair of nodes affects only the pair of nodes immediately to the right. After a new pair of results is computed, it may be stored in the registers which held the old results that are no longer needed. In the computation for the final array of nodes, corresponding to the values of the DFT, the computation involves pairs of nodes separated by four locations, but the "in place" property still holds.

For this version of the algorithm, the initial shuffling of the data sequence, X_k, was necessary for the "in place" computation. This shuffling is due to the repeated movement of odd-numbered members of a sequence to the end of the sequence during each stage of the reduction, as shown in Figs. 3, 4, and 5. This shuffling has been called *bit reversal*[2] because the samples are stored in bit-reversed order; i.e., $X_4 = X_{(100)_2}$ is stored in position $(011)_2 = 3$, etc. Note that the initial data shuffling can also be done "in place."

Variations of Decimation in Time: If one so desires, the signal flow graph shown in Fig. 5 can be manipulated to yield different forms of the *decimation in time* version of the algorithm. If one imagines that in Fig. 5 all the nodes on the same horizontal level as A_1 are interchanged with all the nodes on the same horizontal level as A_4, and all the nodes on the level of A_3 are interchanged with the nodes on the level of A_6, *with the arrows carried along with the nodes*, then one obtains a flow graph like that of Fig. 6.

For this rearrangement one need not shuffle the original data into the bit-reversed order, but the resulting spectrum needs to be *unshuffled*. An additional disadvantage might be that the powers of W needed in the computation are in bit-reversed order. Cooley's original description of the algorithm [1] corresponds to the flow graph of Fig. 6.

A somewhat more complicated rearrangement of Fig. 5 yields the signal flow graph of Fig. 7. For this case both the input data and the resulting spectrum are in "natural" order, and the coefficients in the computation are also used in a natural order. However, the computation may no longer be done "in place." Therefore, at least one other array of registers must be provided. This signal flow graph, and a procedure corresponding to it, are due to Stockham [8].

Decimation in Frequency: Let us now consider a second, quite distinct, form of the fast Fourier transform algorithm, *decimation in frequency*. This form was found independently by Sande [7], and Cooley and Stockham [8]. Let the time series X_k have a DFT A_r. The series and the DFT both contain N terms. As before, we divide X_k into two sequences having $N/2$ points each. However, the first sequence, Y_k, is now composed of the first $N/2$ points in X_k, and the second, Z_k, is composed of the last $N/2$ points in X_k. Formally, then

$$Y_k = X_k$$

$$k = 0, 1, 2, \cdots, \frac{N}{2} - 1. \quad (25)$$

$$Z_k = X_{k+N/2}$$

[2] This is a special case of digit reversal where the radix of the address is 2; more general digit reversals are available for transforms with other radices.

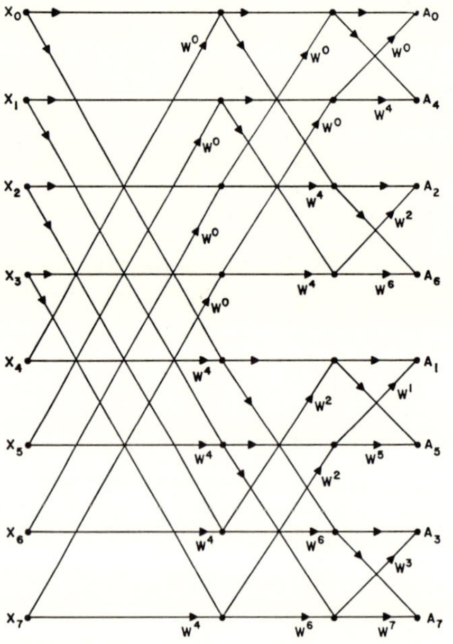

Fig. 6. Rearrangement of the flow graph of Fig. 5 illustrating the DFT computation from naturally ordered time samples.

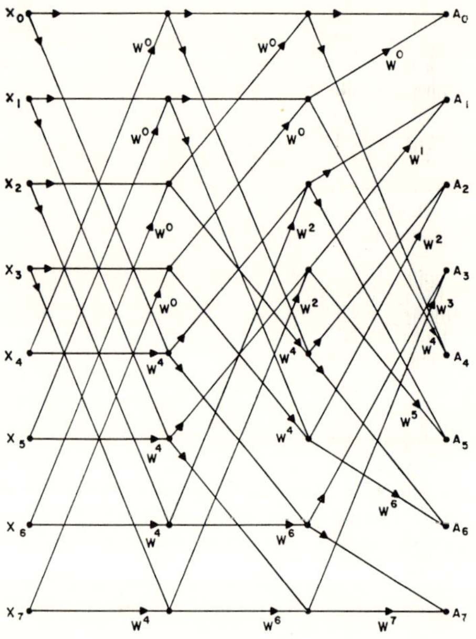

Fig. 7. Rearrangement of the flow graph of Fig. 5 illustrating the DFT computation without bit reversal.

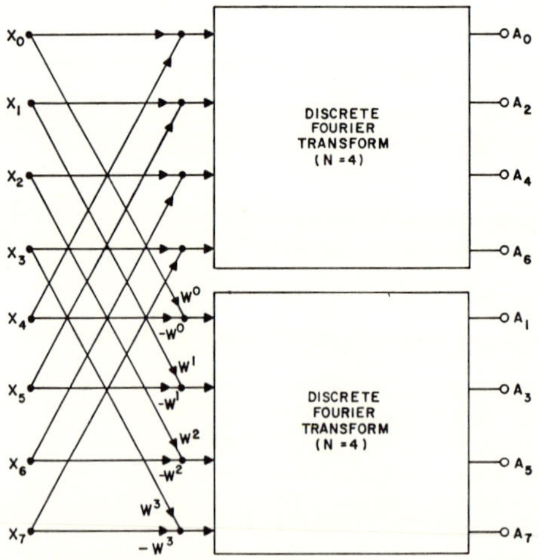

Fig. 8. Signal flow graph illustrating the reduction of endpoint DFT to two DFTs of $N/2$ points each, using decimation in frequency.

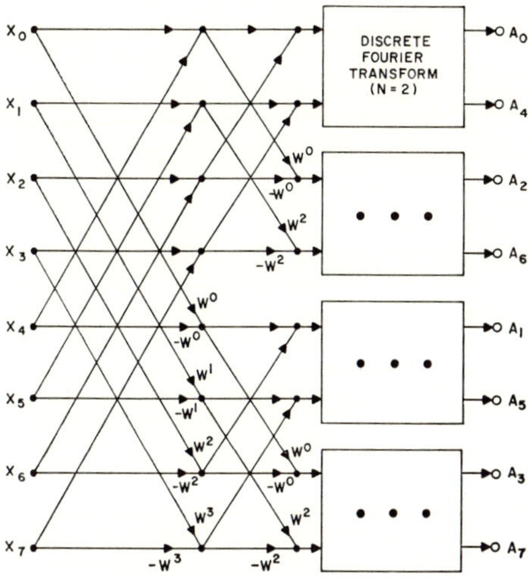

Fig. 9. Signal flow graph illustrating further reduction of the DFT computation suggested by Fig. 8.

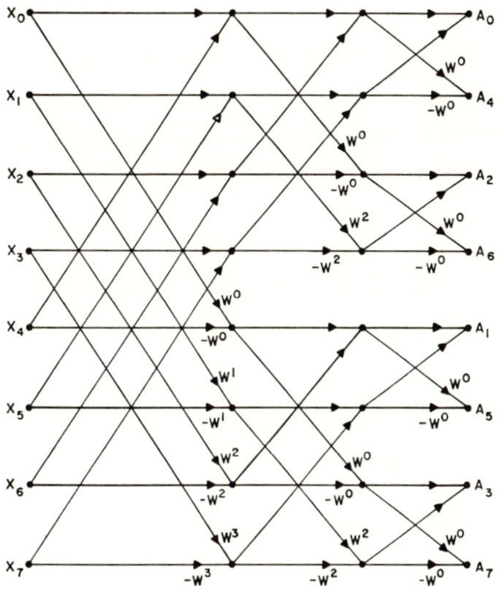

Fig. 10. Signal flow graph illustrating the computation of the DFT when the operations involved are completely reduced to multiplications and additions.

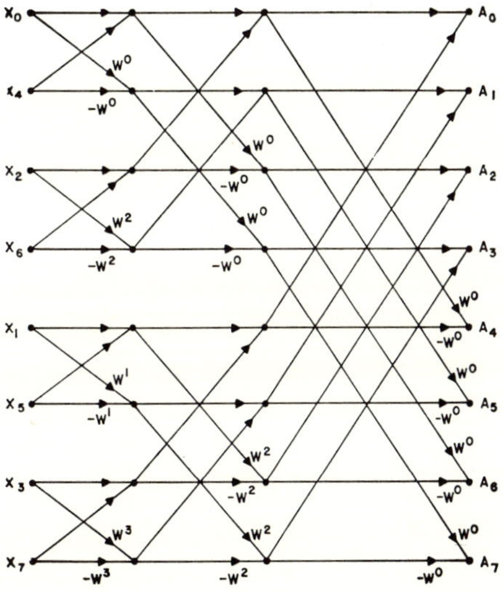

Fig. 11. Rearrangement of the flow graph of Fig. 10 illustrating the computation of the DFT to yield naturally ordered DFT coefficients.

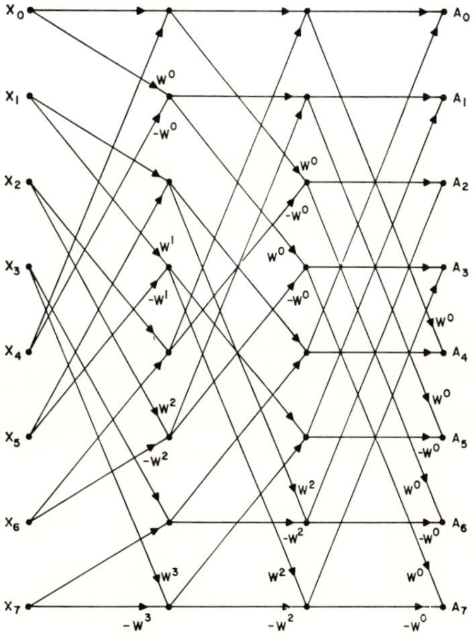

Fig. 12. Rearrangement of the flow graph of Fig. 10 illustrating the DFT computation without bit reversal.

The N point DFT of X_k may now be written in terms of Y_k and Z_k

$$A_r = \sum_{k=0}^{(N/2)-1} \left\{ Y_k \exp(-2\pi jrk/N) + Z_k \exp\left(-2\pi jr\left[k + \frac{N}{2}\right]/N\right)\right\} \quad (26)$$

$$A_r = \sum_{k=0}^{(N/2)-1} \{ Y_k + [\exp(-\pi jr)]Z_k\} \exp(-2\pi jrk/N). \quad (27)$$

Let us consider separately the even-numbered and odd-numbered points of the transform. Let the even-numbered points be R_r and the odd-numbered points be S_r, where

$$\begin{aligned} R_r &= A_{2r} \\ & \qquad\qquad 0 \le r < N/2. \\ S_r &= A_{2r+1} \end{aligned} \quad (28)$$

It is this step that may be called *decimation in frequency.* Note that for computing the even-numbered spectrum points, (27) becomes simply

$$R_r = A_{2r} = \sum_{k=0}^{(N/2)-1} \{ Y_k + Z_k\} e^{(-2\pi jrk)/(N/2)} \quad (29)$$

which we recognize as the $N/2$ point DFT of the function (Y_k+Z_k), the sum of the first $N/2$ and the last $N/2$ time samples. Similarly, for the odd-numbered spectrum points, (27) becomes

$$\begin{aligned} S_r = A_{2r+1} &= \sum_{k=0}^{(N/2)-1} \{ Y_k + Z_k \exp(-\pi j[2r+1])\} \cdot \exp(-2\pi j[2r+1]k/N) \\ &= \sum_{k=0}^{(N/2)-1} \{ Y_k - Z_k\} e^{(-2\pi jk)/N} e^{[(-2\pi jrk)/(N/2)]} \end{aligned} \quad (30)$$

which we recognize as the $N/2$ point DFT of the function $(Y_k-Z_k)\exp(-2\pi jk/N)$.

It can be concluded from (29) and (30) that the DFT of an N-sample sequence, X_k, may be determined as follows. For even-numbered transform points, it may be computed as an $N/2$ point DFT of a simple combination of the first $N/2$ and last $N/2$ samples of X_k. For odd-numbered transform points, it may be computed as another $N/2$ point DFT of a different simple combination of the first and last $N/2$ samples of X_k. This is illustrated in the signal flow graph of Fig. 8 for an eight-point function. W has been defined in (3).

As was the case with decimation in time, we can replace each of the DFTs indicated in Fig. 8 by two 2-point DFTs, and each of the 2-point DFTs by two 1-point transforms, these last being equivalency operations. These steps are indicated in Figs. 9 and 10.

Examination of Fig. 10 gives us much information about the method of decimation in frequency, and allows us to compare it with decimation in time. Both methods require $N/2 \cdot \log N$ complex additions, complex subtractions, and complex multiplications. Both computations can be done in place. If the coefficients in the computation are to be used in a "natural" rather than "bit-reversed" order, as in Figs. 5 and 10, then the decimation in frequency method works on time samples in unshuffled order and yields frequency samples in shuffled (bit-reversed) order. Recall that Fig. 5 yielded the opposite result.

We are also able to rearrange the nodes in Fig. 10 to obtain the signal flow graph, Fig. 11, which works on shuffled time samples and yields naturally ordered frequency samples, but the coefficients are needed by the computation in bit-reversed order. The geometry of this signal flow graph is identical to the geometry of Fig. 5, just as the geometry of Fig. 10 is identical to the geometry of Fig. 6. The differences lie in the transmissions.

A somewhat more complicated rearrangement of Fig. 10 (shown in Fig. 12) yields a signal flow graph that takes unshuffled samples of the time series and produces a set of Fourier coefficients that are ***not*** in bit-reversed order. The computation cannot, however, be done "in place," and at least one other array of registers must be provided. The method is similar to that shown in Fig. 7 for decimation in time. The forms of Figs. 5, 6, 7, 10, 11, and 12 constitute a set of what we might call canonic forms of the fast Fourier transform. We may choose among these forms to find an algorithm with the properties of "in place" computation, normally ordered input, normally ordered output, or normally ordered coefficients, but not all four at once. To achieve "in place" computation, we must deal with bit reversal, and to eliminate bit reversal we must give up "in place" computation. The two methods most effective when using homogeneous storage facilities are those providing in right order the sine and cosine coefficients needed in the computation. The other methods seem less desirable since they require wasteful tables. Still, all six methods have about equal usefulness, and the method used best will depend on the problem at hand. For example, the method shown in Fig. 10 may be used to transform from the time to the frequency domain, and the method shown in Fig. 4 may be used for the inverse transform. Any of the methods described above may be used for the inverse discrete Fourier transform if the coefficients are replaced by their complex conjugates, and if the result of the computation is multiplied by $1/N$.

The six forms mentioned are, in a sense, canonic, but one could also employ a combination of decimation in time and decimation in frequency at different stages in the reduction process, yielding a hybrid signal flow graph.

A Useful Computational Variation: It may be worth pointing out here how some programming simplicity is realized when the factors p and $q=N/p$ are relatively prime. As described by Cooley, Lewis, and Welch, [2], the "twiddle factor" W^{ir} of (23) can be eliminated by choosing subsequences of the X_k's that are different than those used before. The DFT computations are then conveniently performed in two stages.

1) Compute the q-point transforms

$$B_r^{(i)} = \sum_{k=0}^{q-1} Y_k^{(i)} \cdot W^{pkr} \qquad \begin{array}{l} i = 0, 1, \cdots, p-1 \\ r = 0, 1, \cdots, q-1 \end{array} \tag{31}$$

of each of the p sequences

$$Y_k^{(i)} = X_{pk+qi} \qquad \begin{array}{l} i = 0, 1, \cdots, p-1 \\ k = 0, 1, \cdots, q-1. \end{array} \tag{32}$$

2) Compute, then, the p-point transforms

$$A_s = \sum_{i=0}^{p-1} B_r^{(i)} \cdot W^{qim} \tag{33}$$

of the q sequences $B_r^{(i)}$, where

$$s = r \cdot p(p)_q^{-1} + m \cdot q(q)_p^{-1} \qquad (\text{mod } N, 0 \le s < N) \tag{34}$$

and the notation $(p)_q^{-1}$ is meant to represent the reciprocal of p, mod q, i.e., the solution of $p(p)_q^{-1}>1$ (mod q).

Conclusion

The integral transform method has been one of the foundations of analysis for many years because of the ease with which the transformed expressions may be manipulated, particularly in such diverse areas as acoustic wave propagation. speech transmission, linear network theory, transport phenomena, optics, and electromagnetic theory. Many problems which are particularly amenable to solution by integral transform methods have not been attacked by this method in the past because of the high cost of obtaining numerical results this way.

The fast Fourier transform has certainly modified the economics of solution by transform methods. Some new applications are presented in this special issue, and further interesting and profitable applications probably will be found during the next few years.

Appendix

As is well known, if the filter impulse response is frequency-band-limited to $1/2T$ Hz and is given by its Nyquist samples Y_h spaced T second apart, and furthermore, if the input waveform is also frequency-band-limited to $1/2T$ Hz and given by its Nyquist samples X_k spaced T second apart, then the filter output waveform is also frequency-band-limited to $1/2T$ Hz and completely specified by its Nyquist samples Z_s spaced T second apart

$$Z_s = \sum_{k=0}^{s} X_k \cdot Y_{s-k} = \sum_{l=0}^{s} X_{s-l} \cdot Y_l. \tag{35}$$

The convolution relationship facilitates computation of this equation.

To prove the convolution relationship, let the DFT of the X_k's be A_r and correspondingly the DFT of the Y_h's be B_r. The IDFT of the product of $A_r \cdot B_r$ then becomes [see (4)]

$$\begin{aligned} \left(\frac{1}{N^2}\right) \sum_{r=0}^{N-1} A_r B_r W^{-rs} &= \frac{1}{N^2} \sum_{r=0}^{N-1} \sum_{k=0}^{N-1} \sum_{l=0}^{N-1} X_k Y_l W^{r(k+l-s)} \\ &= \frac{1}{N} \sum_{k=0}^{N-1} \sum_{l=0}^{N-1} X_k Y_l \sum_{r=0}^{N-1} \frac{W^{r(k+l-s)}}{N} \\ &= \frac{1}{N} \sum_{k=0}^{s} X_k Y_{s-k} + \frac{1}{N} \sum_{k=s+1}^{N-1} X_k Y_{N+s-k} \\ &= \left(\frac{1}{N}\right) \cdot Z_s + \text{perturbation term.} \end{aligned} \tag{36}$$

If the first $N/2$ samples of each of the two time series (X_k) and (Y_h) are assumed to be identically zero, then the perturbation term of (36) is zero so that the IDFT of the product of the two DFTs multiplied by N is equal to the convolution product Z_s of (35). Since it is always possible to select the time series to be convolved such that half of the samples are zero, the convolution relationship for the DFT can be used to compute the convolution product [see (35)] of two time series.

It is useful to point out that if $A_r = B_r$, a periodic autocorrelation function emerges.

References

[1] J. W. Cooley and J. W. Tukey, "An algorithm for the machine calculation of complex Fourier series," *Math. of Comput.*, vol. 19, pp. 297–301, April 1965.

[2] J. W. Cooley, P. A. W. Lewis, and P. D. Welch, "Historical notes on the fast Fourier transform" this issue, p. 76–79.

[3] R. B. Blackman and J. W. Tukey, *The Measurement of Power Spectra.* New York: Dover, 1959.

[4] C. Bingham, M. D. Godfrey, and J. W. Tukey, "Modern techniques of power spectrum estimation," this issue, p. 56–66.

[5] T. G. Stockham, "High speed convolution and correlation," *1966 Spring Joint Computer Conf., AFIPS Proc.*, vol. 28. Washington, D. C.: Spartan, 1966, pp. 229–233.

[6] W. T. Cochran, J. J. Downing, D. L. Davin, H. D. Helms, R. A. Kaenel, W. W. Lang, and D. E. Nelson, "Burst measurements in the frequency domain," *Proc. IEEE*, vol. 54, pp. 830–841, June 1966.

[7] W. M. Gentleman and G. Sande, "Fast Fourier transforms—for fun and profit," *1966 Fall Joint Computer Conf. AFIPS Proc.*, vol. 29. Washington, D. C.: Spartan, 1966, pp. 563–578.

[8] Private communication.

[9] J. W. Cooley, "Applications of the fast Fourier transform method," *Proc. of the IBM Scientific Computing Symp.*, June 1966.

V

Algorithms, Hardware Implementation, and Some Applications of the Fast Fourier Transform

Editor's Comments on Papers 25 Through 32

25 Cooley and Tukey: *An Algorithm for the Machine Calculation of Complex Fourier Series*

26 Gentleman and Sande: *Fast Fourier Transforms—for Fun and Profit*

27 Singleton: *A Method for Computing the Fast Fourier Transform with Auxiliary Memory and Limited High-Speed Storage*

28 Gold and Bially: *Parallelism in Fast Fourier Transform Hardware*

29 Groginsky and Works: *A Pipeline Fast Fourier Transform*

30 Corinthios: *A Fast Fourier Transform for High-Speed Signal Processing*

31 Stockham: *High-Speed Convolution and Correlation*

32 Cooley, Lewis, and Welch: *Application of the Fast Fourier Transform to Computation of Fourier Integrals, Fourier Series, and Convolution Integrals*

The first three papers collected in this part deal with the various algorithms of FFT. There is an interesting history associated with FFT (Cooley et al., 1967, 1969). To many, however, the publication of the paper by Cooley and Tukey [25] marks the beginning of a surge of renewed interest in applying Fourier analysis to a variety of problems in many fields of engineering and science. The paper presents with classical elegance an algorithm for computing the Fourier transform from N discrete data points, when N is a highly composite number. Special attention is directed to the radix 2 case (i.e., $N = 2^m$, m an integer). Important features of the algorithm, such as in-place computation and bit reversal, are pointed out. The algorithm is later labeled the decimation-in-time algorithm.

In the next paper [26], Gentleman and Sande present a different algorithm. called the decimation-in-frequency algorithm by later workers. The two algorithms are compared in the paper, which also points out that a saving in computing can be achieved by using radix 4 with a possible radix 2 stage at the end. Other topics discussed in the paper include roundoff errors and a number of common applications of the FFT. Additional savings in computation may be possible by using still a higher radix, such as 8 or 16 (Bergland, 1968); programs for mixed radices has been reported (Singleton, 1969).

Since each complex number requires two storage locations, at least $2N$ data storage locations are needed to carry out the FFT algorithm for a sequence of length N. Some additional storage is also needed for the in-place computation. Thus, the size of high-speed (random-access) memory in a computer imposes a limit on the length of sequence for which the FFT algorithm can be applied directly. When the data exceed a certain length, auxiliary memory, such as tapes and disks, must be used. Gentleman and Sande [26] describe an approach to accomplish the transform of long sequences with virtually the same speed as if all computations are done with core storage. The number of files that must be set up on disks is $2^M/2^k$, where 2^M is the length of the sequence and 2^k is the length that can be fit into the core for the computer to take the FFT. In the paper by Singleton [27], an approach is presented to take the FFT of long sequences using only four serially organized memory files, such as magnetic tapes or serial disk files. The paper should be of particular interest to users of small computers for the processing of long records.

In applications that involve real-time processing or where speed is of importance, a special-purpose hardware FFT processor is usually preferred over the programming of a general-purpose computer. In such a hardware processor, one is no longer limited to a single arithmetic unit for carrying out the actual computation, as in the case of most general-purpose machines. It is therefore possible to explore the advantages offered by the parallel operation of arithmetic computation and/or memory, particularly when high-speed operations are considered. The paper by Gold and Bially [28] deals with this problem and proposes several structures that are highly efficient in computation and economical in hardware. The next two papers [29 and 30] deal more specifically with details of hardware processor. The pipeline structure proposed in Groginsky and Works' paper [29] has been applied in the processing of wideband radar signals. The reader is also referred to the paper by Peled and Liu [12] of Part III for a hardware implementation of the basic computation unit for FFT processors.

The FFT algorithms have been widely applied. We include in this part two papers that deal with specific applications. In Stockham's paper [31], a method is presented to calculate the convolution of two sequences with similar savings over the conventional approach by a factor proportional to $N/\log N$. The applications to the calculation of lag product, spectrum estimation, etc., are obvious. The reader may also recall the discussion in Gentleman and Sande's paper [26] on this topic. Stockham's method had been referred to as the overlap-add method. A similar method, discovered by Stockham and Helms independently, is known today as the select-and-save method (Helms, 1967). As the title indicates, the next paper [32], by Cooley, Lewis, and Welch, discusses in detail the computation of Fourier integrals, Fourier series, and other related problems.

References

Bergland, G. D. (1968). A Fast Fourier Transform Algorithm Using Base 8 Iterations, *Math. Comp.*, **22,** 275–279.

Cooley, J. W., P. A. Lewis, and P. D. Welch (1967). Historical Notes on the Fast Fourier Transform, *IEEE Trans. Audio Electroacoustics,* **AU-15,** 76–79.

———, R. L. Garwin, C. M. Rader, B. P. Bogert, and T. G. Stockham (1969). The 1968 Arden House Workshop on Fast Fourier Transform Processing, *IEEE Trans. Audio Electroacoustics,* **AU-17,** 66–75.

Helms, H. D. (1967). Fast Fourier Transform Method of Computing Difference Equations and Simulating Filters, *IEEE Trans. Audio Electroacoustics,* **AU-15,** 85–90.

Singleton, R. C. (1969). An Algorithm for Computing the Mixed Radix Fast Fourier Transform, *IEEE Trans. Audio Electroacoustics,* **AU-17,** 93–100.

25

Reprinted with permission of the publisher from *Math. Comp.*, **19**, 297–301 (1965)

An Algorithm for the Machine Calculation of Complex Fourier Series

By James W. Cooley and John W. Tukey

An efficient method for the calculation of the interactions of a 2^m factorial experiment was introduced by Yates and is widely known by his name. The generalization to 3^m was given by Box et al. [1]. Good [2] generalized these methods and gave elegant algorithms for which one class of applications is the calculation of Fourier series. In their full generality, Good's methods are applicable to certain problems in which one must multiply an N-vector by an $N \times N$ matrix which can be factored into m sparse matrices, where m is proportional to $\log N$. This results in a procedure requiring a number of operations proportional to $N \log N$ rather than N^2. These methods are applied here to the calculation of complex Fourier series. They are useful in situations where the number of data points is, or can be chosen to be, a highly composite number. The algorithm is here derived and presented in a rather different form. Attention is given to the choice of N. It is also shown how special advantage can be obtained in the use of a binary computer with $N = 2^m$ and how the entire calculation can be performed within the array of N data storage locations used for the given Fourier coefficients.

Consider the problem of calculating the complex Fourier series

$$X(j) = \sum_{k=0}^{N-1} A(k) \cdot W^{jk}, \qquad j = 0, 1, \cdots, N-1, \tag{1}$$

where the given Fourier coefficients $A(k)$ are complex and W is the principal Nth root of unity,

$$W = e^{2\pi i/N}. \tag{2}$$

A straightforward calculation using (1) would require N^2 operations where "operation" means, as it will throughout this note, a complex multiplication followed by a complex addition.

The algorithm described here iterates on the array of given complex Fourier amplitudes and yields the result in less than $2N \log_2 N$ operations without requiring more data storage than is required for the given array A. To derive the algorithm, suppose N is a composite, i.e., $N = r_1 \cdot r_2$. Then let the indices in (1) be expressed

$$\begin{aligned} j &= j_1 r_1 + j_0, \qquad j_0 = 0, 1, \cdots, r_1 - 1, \qquad j_1 = 0, 1, \cdots, r_2 - 1, \\ k &= k_1 r_2 + k_0, \qquad k_0 = 0, 1, \cdots, r_2 - 1, \qquad k_1 = 0, 1, \cdots, r_1 - 1. \end{aligned} \tag{3}$$

Then, one can write

$$X(j_1, j_0) = \sum_{k_0} \sum_{k_1} A(k_1, k_0) \cdot W^{jk_1 r_2} W^{jk_0}. \tag{4}$$

Received August 17, 1964. Research in part at Princeton University under the sponsorship of the Army Research Office (Durham). The authors wish to thank Richard Garwin for his essential role in communication and encouragement.

Since

$$W^{jk_1r_2} = W^{j_0k_1r_2}, \tag{5}$$

the inner sum, over k_1, depends only on j_0 and k_0 and can be defined as a new array,

$$A_1(j_0, k_0) = \sum_{k_1} A(k_1, k_0) \cdot W^{j_0k_1r_2}. \tag{6}$$

The result can then be written

$$X(j_1, j_0) = \sum_{k_0} A_1(j_0, k_0) \cdot W^{(j_1r_1+j_0)k_0}. \tag{7}$$

There are N elements in the array A_1, each requiring r_1 operations, giving a total of Nr_1 operations to obtain A_1. Similarly, it takes Nr_2 operations to calculate X from A_1. Therefore, this two-step algorithm, given by (6) and (7), requires a total of

$$T = N(r_1 + r_2) \tag{8}$$

operations.

It is easy to see how successive applications of the above procedure, starting with its application to (6), give an m-step algorithm requiring

$$T = N(r_1 + r_2 + \cdots + r_m) \tag{9}$$

operations, where

$$N = r_1 \cdot r_2 \cdots r_m. \tag{10}$$

If $r_j = s_j t_j$ with $s_j, t_j > 1$, then $s_j + t_j < r_j$ unless $s_j = t_j = 2$, when $s_j + t_j = r_j$. In general, then, using as many factors as possible provides a minimum to (9), but factors of 2 can be combined in pairs without loss. If we are able to choose N to be highly composite, we may make very real gains. If all r_j are equal to r, then, from (10) we have

$$m = \log_r N \tag{11}$$

and the total number of operations is

$$T(r) = rN \log_r N. \tag{12}$$

If $N = r^m s^n t^p \cdots$, then we find that

$$\begin{aligned} \frac{T}{N} &= m \cdot r + n \cdot s + p \cdot t + \cdots, \\ \log_2 N &= m \cdot \log_2 r + n \cdot \log_2 s + p \cdot \log_2 t + \cdots, \end{aligned} \tag{13}$$

so that

$$\frac{T}{N \log_2 N}$$

is a weighted mean of the quantities

$$\frac{r}{\log_2 r}, \quad \frac{s}{\log_2 s}, \quad \frac{t}{\log_2 t}, \cdots,$$

whose values run as follows

r	$\dfrac{r}{\log_2 r}$
2	2.00
3	1.88
4	2.00
5	2.15
6	2.31
7	2.49
8	2.67
9	2.82
10	3.01.

The use of $r_j = 3$ is formally most efficient, but the gain is only about 6% over the use of 2 or 4, which have other advantages. If necessary, the use of r_j up to 10 can increase the number of computations by no more than 50%. Accordingly, we can find "highly composite" values of N within a few percent of any given large number.

Whenever possible, the use of $N = r^m$ with $r = 2$ or 4 offers important advantages for computers with binary arithmetic, both in addressing and in multiplication economy.

The algorithm with $r = 2$ is derived by expressing the indices in the form

$$\begin{aligned} j &= j_{m-1}\cdot 2^{m-1} + \cdots + j_1\cdot 2 + j_0 , \\ k &= k_{m-1}\cdot 2^{m-1} + \cdots + k_1\cdot 2 + k_0 , \end{aligned} \tag{14}$$

where j_v and k_v are equal to 0 or 1 and are the contents of the respective bit positions in the binary representation of j and k. All arrays will now be written as functions of the bits of their indices. With this convention (1) is written

$$X(j_{m-1}, \cdots, j_0) = \sum_{k_0}\sum_{k_1}\cdots\sum_{k_{m-1}} A(k_{m-1}, \cdots, k_0)\cdot W^{jk_{m-1}\cdot 2^{m-1}+\cdots+jk_0}, \tag{15}$$

where the sums are over $k_v = 0, 1$. Since

$$W^{jk_{m-1}\cdot 2^{m-1}} = W^{j_0k_{m-1}\cdot 2^{m-1}}, \tag{16}$$

the innermost sum of (15), over k_{m-1}, depends only on $j_0, k_{m-2}, \cdots, k_0$ and can be written

$$A_1(j_0, k_{m-2}, \cdots, k_0) = \sum_{k_{m-1}} A(k_{m-1}, \cdots, k_0)\cdot W^{j_0k_{m-1}\cdot 2^{m-1}}. \tag{17}$$

Proceeding to the next innermost sum, over k_{m-2}, and so on, and using

$$W^{j\cdot k_{m-l}\cdot 2^{m-l}} = W^{(j_{l-1}\cdot 2^{l-1}+\cdots+j_0)k_{m-l}\cdot 2^{m-l}}, \tag{18}$$

one obtains successive arrays,

$$\begin{aligned} &A_l(j_0, \cdots, j_{l-1}, k_{m-l-1}, \cdots, k_0) \\ &\quad = \sum_{k_{m-l}} A_{l-1}(j_0, \cdots, j_{l-2}, k_{m-l}, \cdots, k_0)\cdot W^{(j_{l-1}\cdot 2^{l-1}+\cdots+j_0)\cdot k_{m-l}\cdot 2^{m-l}} \end{aligned} \tag{19}$$

for $l = 1, 2, \cdots, m$.

Writing out the sum this appears as

$$A_l(j_0, \cdots, j_{l-1}, k_{m-l-1}, \cdots, k_0)$$

$$(20) \quad \begin{aligned} &= A_{l-1}(j_0, \cdots, j_{l-2}, 0, k_{m-l-1}, \cdots, k_0) \\ &\quad + (-1)^{j_{l-1}} i^{j_{l-1} j_{l-2}} A_{l-1}(j_0, \cdots, j_{l-2}, 1, k_{m-l-1}, \cdots, k_0) \\ &\quad \cdot W^{(j_{l-3} \cdot 2^{l-3} + \cdots + j_0) \cdot 2^{m-l}}, \qquad j_{l-1} = 0, 1. \end{aligned}$$

According to the indexing convention, this is stored in a location whose index is

$$(21) \qquad j_0 \cdot 2^{m-1} + \cdots + j_{l-1} \cdot 2^{m-l} + k_{m-l-1} \cdot 2^{m-l-1} + \cdots + k_0 .$$

It can be seen in (20) that only the two storage locations with indices having 0 and 1 in the 2^{m-l} bit position are involved in the computation. Parallel computation is permitted since the operation described by (20) can be carried out with all values of $j_0, \cdots, j_{l-2}$, and $k_0, \cdots, k_{m-l-1}$ simultaneously. In some applications* it is convenient to use (20) to express A_l in terms of A_{l-2}, giving what is equivalent to an algorithm with $r = 4$.

The last array calculated gives the desired Fourier sums,

$$(22) \qquad X(j_{m-1}, \cdots, j_0) = A_m(j_0, \cdots, j_{m-1})$$

in such an order that the index of an X must have its binary bits put in reverse order to yield its index in the array A_m.

In some applications, where Fourier sums are to be evaluated twice, the above procedure could be programmed so that no bit-inversion is necessary. For example, consider the solution of the difference equation,

$$(23) \qquad aX(j + 1) + bX(j) + cX(j - 1) = F(j).$$

The present method could be first applied to calculate the Fourier amplitudes of $F(j)$ from the formula

$$(24) \qquad B(k) = \frac{1}{N} \sum_j F(j) W^{-jk}.$$

The Fourier amplitudes of the solution are, then,

$$(25) \qquad A(k) = \frac{B(k)}{aW^k + b + cW^{-k}}.$$

The $B(k)$ and $A(k)$ arrays are in bit-inverted order, but with an obvious modification of (20), $A(k)$ can be used to yield the solution with correct indexing.

A computer program for the IBM 7094 has been written which calculates three-dimensional Fourier sums by the above method. The computing time taken for computing three-dimensional $2^a \times 2^b \times 2^c$ arrays of data points was as follows:

---

* A multiple-processing circuit using this algorithm was designed by R. E. Miller and S. Winograd of the IBM Watson Research Center. In this case $r = 4$ was found to be most practical.

a	*b*	*c*	No. Pts.	Time (minutes)
4	4	3	2^{11}	.02
11	0	0	2^{11}	.02
4	4	4	2^{12}	.04
12	0	0	2^{12}	.07
5	4	4	2^{13}	.10
5	5	3	2^{13}	.12
13	0	0	2^{13}	.13

IBM Watson Research Center
Yorktown Heights, New York

Bell Telephone Laboratories,
Murray Hill, New Jersey

Princeton University
Princeton, New Jersey

1. G. E. P. Box, L. R. Connor, W. R. Cousins, O. L. Davies (Ed.), F. R. Hirnsworth & G. P. Silitto, *The Design and Analysis of Industrial Experiments*, Oliver & Boyd, Edinburgh, 1954.

2. I. J. Good, "The interaction algorithm and practical Fourier series," *J. Roy. Statist. Soc. Ser. B.*, v. 20, 1958, p. 361–372; Addendum, v. 22, 1960, p. 372–375. MR **21** #1674; MR **23** #A4231.

26

Reprinted from *AFIPS Proc. 1966 Fall Joint Computer Conf.*, Vol. 29, Spartan Books, 1966, pp. 563–578

FAST FOURIER TRANSFORMS—FOR FUN AND PROFIT

W. M. Gentleman

Bell Telephone Laboratories
Murray Hill, New Jersey

and

G. Sande *

Princeton University
Princeton, New Jersey

IMPLEMENTING FAST FOURIER TRANSFORMS

Definition and Elementary Properties of Fourier Transforms

The "Fast Fourier Transform" has now been widely known for about a year. During that time it has had a major effect on several areas of computing, the most striking example being techniques of numerical convolution, which have been completely revolutionized. What exactly is the "Fast Fourier Transform"?

In fact, the Fast Fourier Transform is nothing more than an algorithm whereby, for appropriate length sequences, the finite discrete Fourier transform of the sequence may be computed much more rapidly than by other available algorithms. The properties and uses of the finite discrete Fourier transform, which become practical when used with the fast algorithm, make the technique important. Let us therefore first consider some properties of such transforms.

The usual infinite Fourier integral transform is well known and widely used—the physicist when solving a partial differential equation, the communication engineer looking at noise and the statistician studying distributions may all resort to Fourier transforms, not only because the mathematics may be simplified, but because nature itself is often easier to understand in terms of frequency. What is less well known is that many of the properties of the usual Fourier transform also hold, perhaps with slight modification, for the Fourier transform defined on finite, equispaced, discrete sequences.

We see in Table I that the most significant change required to modify the usual theorems so that they apply to the finite discrete case is that indexing must be considered modulo N. A useful heuristic interpretation of this is to think of the sequence $X(t)$ as a function defined at equispaced points on a circle. This interpretation is to be contrasted with what is

*This work made use of computer facilities supported in part by National Science Foundation grant NSF–GP579. Research was partially supported by the Office of Naval Research under contract Nonr 1858(05) and by the National Research Council of Canada.

TABLE I

A Comparison of Usual and Finite Discrete Fourier Transforms

Usual	Finite Discrete
Definition	
$\hat{X}(\hat{t}) = \int_{-\infty}^{\infty} X(t) e^{2\pi i t \hat{t}} dt$	$\hat{X}(\hat{t}) = \sum_{t=0}^{N-1} X(t) e^{\frac{2\pi i t \hat{t}}{N}}$
Linearity	
The Fourier transform is a linear operator.	The Fourier transform is a linear operator.
Orthogonality	
$\int_{-\infty}^{\infty} e^{2\pi i t(\hat{t}-\hat{t}')} dt = \delta(\hat{t}-\hat{t}')$ where $\delta(\hat{t}-\hat{t}')$ is the Dirac delta function. That is, $\int_{b}^{a} f(\hat{t})\delta(\hat{t}) d\hat{t} = f(0)$ if 0 is in the interval (a,b), otherwise $\int_{a}^{b} f(\hat{t})\delta(\hat{t}) d\hat{t} = 0.$	$\sum_{t=0}^{N-1} e^{\frac{2\pi i t(\hat{t}-\hat{t}')}{N}} = N\delta_N(\hat{t}-\hat{t}')$ where δ_N is the Kronecker delta function with its argument being considered modulo N. That is, $\delta_N(kN) = 1$, for integer k, otherwise $\delta_N = 0$.
Inverse Transform	
If $\hat{X}(\hat{t}) = \int_{-\infty}^{\infty} X(t) e^{2\pi i t \hat{t}} dt$ then $X(t) = \int_{-\infty}^{\infty} \hat{X}(\hat{t}) e^{-2\pi i t \hat{t}} d\hat{t}$ which we observe can be considered as $X(t) =$ $\left\{ \int_{-\infty}^{\infty} \{\hat{X}(\hat{t})\}^* e^{2\pi i t \hat{t}} d\hat{t} \right\}^*$ or the complex conjugate of the Fourier transform of the complex conjugate of X.	If $\hat{X}(\hat{t}) = \sum_{t=0}^{N-1} X(t) e^{\frac{2\pi i t \hat{t}}{N}}$ then $X(t) = \frac{1}{N} \sum_{\hat{t}=0}^{N-1} \hat{X}(\hat{t}) e^{\frac{-2\pi i t \hat{t}}{N}}$ which we observe can be considered as $X(t) = \frac{1}{N} \times$ $\left\{ \sum_{\hat{t}=0}^{N-1} \{\hat{X}(\hat{t})\}^* e^{\frac{2\pi i t \hat{t}}{N}} \right\}^*$ or the complex conjugate of the Fourier transform of the complex conjugate of X, divided by N.
Convolution Theorem	
$\int_{-\infty}^{\infty} X(\tau) Y(t-\tau) d\tau$ $= \int_{-\infty}^{\infty} e^{-2\pi i t \hat{t}} \hat{X}(\hat{t}) \hat{Y}(\hat{t}) d\hat{t}$	$\sum_{\tau=0}^{N-1} X(\tau) Y(t-\tau) = \frac{1}{N} \times$ $\sum_{\hat{t}=0}^{N-1} e^{\frac{-2\pi i t \hat{t}}{N}} \hat{X}(\hat{t}) \hat{Y}(\hat{t})$

TABLE 1—(Continued)

Usual	Finite Discrete
that is, the inverse Fourier transform of the product of the Fourier transforms.	that is, the inverse Fourier transform of the product of the Fourier transforms. NOTE: The convolution here must be considered as cyclic, i.e., the indices of X and Y must be interpreted modulo N.
Operational Calculus	
An operational calculus can be defined, based on the property that $\int_{-\infty}^{\infty} \left(\frac{\partial}{\partial t} X(t) \right) e^{2\pi i t \hat{t}} dt = -2\pi i \hat{t} \hat{X}(\hat{t})$ i.e., the Fourier transform of the derivative of a function is the Fourier transform of the function, multiplied by $-2\pi i \hat{t}$.	An operational calculus can be defined, based on the property that $\sum_{t=0}^{N-1} (\Delta X(t)) e^{\frac{2\pi i t \hat{t}}{N}} = \left(e^{\frac{-2\pi i \hat{t}}{N}} - 1 \right) \hat{X}(\hat{t})$ i.e., the Fourier transform of the (forward) difference of a function is the Fourier transform of the function, multiplied by $\left(e^{\frac{-2\pi i \hat{t}}{N}} - 1 \right)$. NOTE: The difference here must be considered cyclically, so that $\Delta X(t) = X(t+1) - X(t)$ becomes $\Delta X(N-1) = X(N) - X(N-1) = X(0) - X(N-1)$ for the case of $t = N-1$.
Symmetries	
If X is real then $\hat{X}$ is hermitian symmetric, i.e., $\hat{X}(\hat{t}) = \{\hat{X}(-\hat{t})\}^*$; if X is hermitian symmetric then $\hat{X}$ is real. If Y is imaginary, then $\hat{Y}$ is hermitian antisymmetric, i.e., $\hat{Y}(\hat{t}) = -\{\hat{Y}(-\hat{t})\}^*$; if Y is hermitian antisymmetric then $\hat{Y}$ is imaginary.	If X is real then $\hat{X}$ is hermitian symmetric, i.e., $\hat{X}(\hat{t}) = \{\hat{X}(N-\hat{t})\}^*$; if X is hermitian symmetric, then $\hat{X}$ is real. If Y is imaginary, then $\hat{Y}$ is hermitian antisymmetric, i.e., $\hat{Y}(\hat{t}) = -\{\hat{Y}(N-\hat{t})\}^*$; if Y is hermitian antisymmetric then $\hat{Y}$ is imaginary. NOTE: The use of the terms hermitian symmetric and hermitian antisymmetric in the discrete case is consistent with that in the usual case if we interpret indices modulo N.
Shifting Theorems	
$\int_{-\infty}^{\infty} X(t+h) e^{2\pi i t \hat{t}} dt = e^{-2\pi i \hat{t} h} \hat{X}(\hat{t})$	$\sum_{t=0}^{N-1} X(t+h) e^{\frac{2\pi i t \hat{t}}{N}} = e^{\frac{-2\pi i \hat{t} h}{N}} \hat{X}(\hat{t})$

(These results are all readily proved from the definitions.)

often the natural interpretation—as sampled data from a continuous infinite function.

Basic Algebra of Fast Fourier Transforms

At this point we will define the notation $e(X) = e^{2\pi iX}$ so that we may write expressions such as those of Table I in a simpler form. The two most important properties of $e(X)$ are that

$$e(X+Y) = e(X)e(Y)$$

and

$$e(X) = 1 \quad \text{if } X \text{ is an integer.}$$

Using this notation, the finite discrete Fourier transform of the sequence $X(t)$ is

$$\hat{X}(\hat{t}) = \sum_{t=0}^{N-1} X(t)e\left(\frac{t\hat{t}}{N}\right)$$

Suppose N has factors A and B so that $N = AB$. Then writing $\hat{t} = \hat{a} + \hat{b}A$ and $t = b + aB$ where $a,\hat{a} = 0,1,\ldots,A-1$ and $b,\hat{b} = 0,1,\ldots,B-1$; we have

$$\hat{X}(\hat{a}+\hat{b}A) =$$

$$= \sum_{b=0}^{B-1}\sum_{a=0}^{A-1} X(b+aB)e\left(\frac{\{\hat{a}+\hat{b}A\}\{b+aB\}}{AB}\right)$$

$$= \sum_{b=0}^{B-1}\sum_{a=0}^{A-1} X(b+aB)e\left(\frac{\hat{a}b}{AB} + \frac{\hat{a}a}{A} + \frac{\hat{b}b}{B} + a\hat{b}\right)$$

$$= \sum_{b=0}^{B-1}\sum_{a=0}^{A-1} X(b+aB)e\left(\frac{\hat{a}b}{AB}\right)e\left(\frac{a\hat{a}}{A}\right)e\left(\frac{b\hat{b}}{B}\right)$$

as $a\hat{b}$ is integral implies $e(a\hat{b}) = 1$

$$= \sum_{b=0}^{B-1} e\left(\frac{b\hat{b}}{B}\right)\left\{e\left(\frac{\hat{a}b}{AB}\right)\sum_{a=0}^{A-1} X(b+aB)e\left(\frac{a\hat{a}}{A}\right)\right\}$$

If we define the B different sequences

$$W_b(a) = X(b+aH) \quad a = 0,1,\ldots,A-1$$

and the A different sequences

$$Z_{\hat{a}}(b) = e\left(\frac{\hat{a}b}{AB}\right)\sum_{a=0}^{A-1} X(b+aB)e\left(\frac{a\hat{a}}{A}\right)$$

$$b = 0,1,\ldots,B-1$$

we can write the above equation as

$$\hat{X}(\hat{a}+\hat{b}B) = \sum_{b=0}^{B-1} e\left(\frac{b\hat{b}}{B}\right) Z_{\hat{a}}(b)$$

where

$$Z_{\hat{a}}(b) = e\left(\frac{\hat{a}b}{AB}\right)\left\{\sum_{a=0}^{A-1} e\left(\frac{a\hat{a}}{A}\right) W_b(a)\right\}$$

We recognize

$$\sum_{b=0}^{B-1} e\left(\frac{b\hat{b}}{B}\right)Z_{\hat{a}}(b) \text{ and } \sum_{a=0}^{A-1} e\left(\frac{a\hat{a}}{A}\right) W_b(a)$$

as Fourier transforms themselves, applied to shorter sequences. Obtaining a subsequence by starting at the bth element and taking every Bth element thereafter, in the manner $W_b(a)$ is obtained from $X(b+aB)$, is called decimating by B. Observe that the sequence $Z_{\hat{a}}(b)$ are not quite the sequence of frequency $\hat{a}$ values from the transforms of these decimated sequences, but these values multiplied by a "twiddle factor" $e\left(\frac{\hat{a}b}{AB}\right)$.

The Fourier transform of the complete AB point sequence may thus be accomplished by doing the B different A point Fourier transforms, multiplying through by the appropriate twiddle factors, then doing the A different B point Fourier transforms. This is a recursive formula which defines the larger Fourier transform in terms of smaller ones. The total number of operations is now proportional to $AB(A+B)$ rather than $(AB)^2$ as it would be for a direct implementation of the definition, hence the name "Fast Fourier Transform".

Associated observations:

1. Attention is drawn to the special advantage of factors 2 or 4, in that since a Fourier transform of a 2 or 4 point sequence may be done using only additions and subtractions, one stage of complex arithmetic may be avoided.

2. Although the recursive algorithm above may be implemented directly, it is much better to simulate the recursion as described in the following sections, since one may avoid the unnecessary calculation of some complex exponentials. With our programs, for example, the simulating the recursion takes only 2/5 as long for a 2^{10} point transform.*

* All programs discussed in this paper were implemented using ALCOR Algol 60 on the IBM 7094 at Princeton University.

3. The multi-dimensional finite discrete Fourier transform

$$\hat{X}(\hat{t}_1,\ldots,\hat{t}_k) = \sum_{t_1=0}^{N_1-1} \cdots \sum_{t_k=0}^{N_k-1} X(t_1,\ldots,t_k)e\left(\frac{t_1\hat{t}_1}{N_1}\right) \ldots e\left(\frac{t_k\hat{t}_k}{N_k}\right)$$

can be computed efficiently by factorizing in each dimension separately, as above.

4. The algebra presented here is not quite that appearing in "An Algorithm for the Machine Calculations of Complex Fourier Series," by J. W. Cooley and J. W. Tukey.[2] The difference can be seen if we consider N as having three factors, $N = ABC$.

The Cooley-Tukey algebra then is

$$\hat{X}(\hat{c}+\hat{b}C+\hat{a}BC) = \sum_{a=0}^{A-1}\sum_{b=0}^{B-1}\sum_{c=0}^{C-1} X(a+bA+cAB)e\left(\frac{(\hat{c}+\hat{b}C+\hat{a}BC)\cdot(a+bA+cAB)}{ABC}\right)$$

$$= \sum_{a=0}^{A-1}\sum_{b=0}^{B-1}\sum_{c=0}^{C-1} X(a+bA+cAB)e\left(\frac{a\cdot(\hat{c}+\hat{b}C+\hat{a}BC)}{ABC}\right) \cdot e\left(\frac{b\cdot(\hat{c}+\hat{b}C+\hat{a}BC)}{BC}\right) e\left(\frac{c(\hat{c}+\hat{b}C+\hat{a}BC)}{C}\right)$$

$$= \sum_{a=0}^{A-1} e\left(\frac{a\cdot(\hat{c}+\hat{b}C+\hat{a}BC)}{ABC}\right)\sum_{b=0}^{B-1} e\left(\frac{b\cdot(\hat{c}+\hat{b}C)}{BC}\right) \sum_{c=0}^{C-1} e\left(\frac{c\hat{c}}{C}\right)X(a+bA+cAB)$$

If, rather than collecting on the unhatted variables as above, we choose to collect on the hatted variables, we obtain

$$\hat{X}(\hat{c}+\hat{b}C+\hat{a}BC) = \sum_{a}^{A-1}\sum_{b}^{B-1}\sum_{c}^{C-1} X(a+bA+cAB)e\left(\frac{\hat{a}\cdot(a+bA+cAB)}{A}\right) \cdot e\left(\frac{\hat{b}\cdot(a+bA+cAB)}{AB}\right) e\left(\frac{\hat{c}\cdot(a+bA+cAB)}{ABC}\right)$$

$$= \sum_{a=0}^{A-1} e\left(\frac{\hat{a}a}{A}\right)\sum_{b=0}^{B-1} e\left(\frac{\hat{b}\cdot(a+bA)}{AB}\right)\sum_{c=0}^{C-1} e\left(\frac{\hat{c}\cdot(a+bA+cAB)}{ABC}\right) X(a+bA+cAB)$$

In both cases we may factor the twiddle factor outside the summation. If we do this we obtain in the first case

$$\hat{X}(\hat{c}+\hat{b}C+\hat{a}BC) = \sum_{a=0}^{A-1} e\left(\frac{a\hat{a}}{A}\right) e\left(\frac{a\cdot[\hat{c}+\hat{b}C]}{ABC}\right) \sum_{b=0}^{B-1} e\left(\frac{b\hat{b}}{B}\right) e\left(\frac{b\hat{c}}{BC}\right)\sum_{c=0}^{C-1} e\left(\frac{c\hat{c}}{C}\right) X(a+bA+cAB)$$

(Cooley version)

In the second case we obtain

$$\hat{X}(\hat{c}+\hat{b}C+\hat{a}BC) = \sum_{a=0}^{A-1} e\left(\frac{a\hat{a}}{A}\right) e\left(\frac{a\hat{b}}{AB}\right)\sum_{b=0}^{B-1} e\left(\frac{b\hat{b}}{B}\right) e\left(\frac{\hat{c}(a+bA)}{ABC}\right)\sum_{c=0}^{C-1} e\left(\frac{c\hat{c}}{C}\right) X(a+bA+cAB)$$

(Sande version)

The difference between these two versions becomes important when we come to the details of implementing the fast Fourier transform. We go on to this next.

Fast Fourier Transforms Using Scratch Storage

Let us consider the Sande version applied to an X sequence stored in serial order. The summation over c represents a C point Fourier transform of points spaced AB apart. There is one such transform for each of the AB values of $a + bA$. The result of one such transform is a frequency sequence indexed by $\hat{c}$. If we look at the twiddle factor $e\left(\frac{\hat{c}(a+bA)}{ABC}\right)$ we see that it depends upon $\hat{c}$ and $a + bA$ in a very convenient manner. To introduce some descriptive terminology we may call $\hat{c}$ the frequency for this analysis and $a + bA$ the displacement for this analysis. At this stage there are no free indices corresponding to replications. When we store the intermediate results in the scratch storage area, placing each of the C point Fourier transforms in contiguous blocks indexed by the displacement leads to elements stored at $\hat{c} + C(a+bA)$. The intermediate summation over b represents a B point Fourier transform with points spaced AC apart; one such transform for each of the AC values of $\hat{c} + aC$. The result of any one of these transforms is a frequency sequence indexed by $\hat{b}$. The twiddle factor $e\left(\frac{a\hat{b}}{AB}\right)$ depends only upon a and $\hat{b}$, leaving $\hat{c}$ as a free index. Here we would call $\hat{b}$ the frequency for the analysis, a the displacement for the analysis and $\hat{c}$ the replication index. We would store the intermediate results at $\hat{c} + \hat{b}C + aBC$. In this case the contiguous blocks are spaced out by the replication factor C. The outer summation over a represents an A point Fourier transform with points spaced BC apart; one such transform for each of the BC values of $\hat{c} + \hat{b}C$. The result of any one of these transforms is a frequency sequence indexed by $\hat{a}$. There is no twiddle factor in this case. Here we would call $\hat{a}$ the frequency and $\hat{c} + \hat{b}C$ the replication index. There is no displacement at this stage. We would store the results at $\hat{c} + \hat{b}C + \hat{a}BC$. At this point we see that the results are stored in serial order of frequency.

When we compute one of the transforms we may wish to include the twiddle factor in the coefficients for the values of the replication index before computing new coefficients for a different displacement. If the Fourier transform is on two points there appears to be no advantage to either choice. For four points it is more economical to do the Fourier transform and then multiply by the twiddle factor. In the other cases it is more efficient if the twiddle factor is absorbed into the transform coefficients.

If we were to use the Cooley version on a sequence stored in serial order we would obtain an algorithm which differs only in minor details. The summation over c represents a C point Fourier transform of points spaced AB apart. The twiddle factor $e\left(\frac{b\hat{c}}{BC}\right)$ depends upon the frequency $\hat{c}$ and displacement b, leaving a free as a replication index. The elements are stored at $\hat{c} + aC + bAC$. The intermediate summation over b represents a B point Fourier transform of points spaced AC apart. The twiddle factor $e\left(\frac{a\cdot(\hat{c}+\hat{b}C)}{ABC}\right)$ depends upon the combined frequency $\hat{c} + \hat{b}C$ and the displacement a. There is no free index. The intermediate results are stored at $\hat{c} + \hat{b}C + aBC$. The outer summation represents an A point Fourier transform of results spaced BC apart. There is no twiddle factor in this case. The final results would be stored in serial order at $\hat{c} + \hat{b}C + \hat{a}BC$.

These two reductions are essentially identical from an algorithmic viewpoint when we use storage in this manner. We have only reversed the roles of the hatted and unhatted variables in the formation of the twiddle factors.

To obtain the general case of more than three factors we proceed to group our factors in three groups, for example $N = (P_1 \ldots P_{j-1})P_j(P_{j+1} \ldots P_n)$. If we identify

$$A = P_1 \ldots P_{j-1}$$
$$B = P_j$$
$$C = P_{j+1} \ldots P_n$$

and perform the above reductions we find that after two steps with this identification we arrive at exactly the same place as after only one step with the identification

$$A = P_1 \ldots P_{j-2}$$
$$B = P_{j-1}$$
$$C = P_j \ldots P_n$$

We can think of this as moving factors from the "A" part through the "B" part to the "C" part.

When we write a program to implement this, we set up a triply nested iteration. The outer loop selects the current factor being moved and sets up the limits and indexing parameters for the inner loops. The intermediate loop, in which we would compute the twiddle factor, corresponds to the displacement. The inner loop provides indexing over the replication variables.

Fast Fourier Transforms in Place

Let us reconsider the storage requirements for our algorithm. In particular, let us consider one of the small Fourier transforms on c for some value of $a + bA$. It is only during this transform that we will need this set of intermediate results. We have just vacated the C cells $a + bA + cAB$ (indexed by c) and are looking about for C cells in which to store our answers (indexed by $\hat{c}$). Why not use the cells that have just been vacated?

Having made this observation, let us reexamine the algorithm corresponding to the Sande factorization. The summation on c represents a C point Fourier transform of points spaced AB apart. The twiddle factor $e\left(\frac{\hat{c}\cdot(a+bA)}{ABC}\right)$ depends upon the frequency $\hat{c}$ and the displacement $a + bA$. There is no replication index. The intermediate results are stored at $a + bA + \hat{c}AB$. The summation on b represents a B point Fourier transform of points A apart. The twiddle factor $e\left(\frac{a\hat{b}}{AB}\right)$ depends upon the frequency $\hat{b}$ and the displacement a. The replication index is $\hat{c}$. We treat each of the blocks of length AB indexed by $\hat{c}$ in an exactly equivalent manner. The summation over a represents contiguous A point Fourier transforms. The frequency is $\hat{a}$ and the replication index is $\hat{b} + c\hat{B}$. There is no displacement. The final answers are then stored at $\hat{a} + \hat{b}A + \hat{c}AB$. The use of "displacement" and "replication" is now much more suggestive and natural than when we used these while describing the use of scratch storage.

The more general case of more than three factors is again obtained by grouping to obtain "A", "B", and "C" and then moving factors from the "A" part to the "C" part; the program being written as a triply nested iteration loop.

When we reexamine the Cooley factorization for serially stored data, we will see that it has an unpleasant feature. The summation over c represents a C point Fourier transformation of points spaced AB apart. The twiddle factor $e\left(\frac{b\hat{c}}{AB}\right)$ depends only upon b and $\hat{c}$, leaving a as a free index. The intermediate results are stored at $a + bA + \hat{c}AB$. The summation over b represents a B point Fourier transform of points spaced A apart. The twiddle factor $e\left(\frac{a \cdot (\hat{c}+\hat{b}c)}{ABC}\right)$ depends upon the displacement a and the combined frequency $\hat{c} + \hat{b}C$. For data which was originally stored serially, this is not convenient to compute. The intermediate results are stored at $a + \hat{b}A + \hat{c}AB$. The summation over a represents contiguous A point Fourier transforms. The final results are stored at $\hat{a} + \hat{b}A + \hat{c}AB$. More than three factors can be handled in the same manner as before.

The common unpleasant feature of both of these algorithms is that the Fourier coefficient for frequency $\hat{c} + \hat{b}C + \hat{a}BC$ is stored at $\hat{a} + \hat{b}A + \hat{c}AB$. Such a storage scheme may be described as storage with "digit reversed subscripts." Before we may use our Fourier coefficients we must unscramble them.* Unscrambling has been found to require only a small proportion of the total time for the algorithm. A very elegant solution corresponds to running two counters, one with normal digits (easily obtained by simply incrementing by one) and one with reversed digits (easily obtained by nesting loops with the innermost stepping by the largest increments) and recopying from one array to another where one counter is used for each array. If all of the factors are the same, the required interchanges become pairwise and it is possible to do the unscrambling in place. One may unscramble the real and imaginary parts separately, so scratch space could be generated by backing one or the other onto auxiliary store. A slower method of unscrambling is to follow the permutation cycles so that no scratch storage is needed.

*This problem has nothing to do with the representation of numbers in any particular machine, and unless the machine has a "reverse digit order" instruction (e.g., "reverse bit order" for a binary machine), no advantage can be taken of such representation.

TABLE II

TIME FOR A 1024 POINT TRANSFORM BY VARIOUS METHODS

Method	*Time*
radix 2	2.0 seconds
radix 4 (also radix 4 + 2 and mixed radices)	1.1 seconds
radix 2 (recursively implemented)	5.2 seconds
Goertzel's method	59.1 seconds

Two implementations, both for serially stored sequences, have been described. Alternately, two implementations, both for data stored with digit reversed subscripts, may be developed: in this instance the Cooley factorization has more convenient twiddle factors. For the last two, the final results are correctly stored—the "unscrambling" having been done first. Digit reversed subscripts may arise because we have

1. scrambled serially stored data,
2. not unscrambled after a previous Fourier transform,
3. generated data in scrambled order.

We have written various Fourier transform programs in the manner described above. These include one- and multi-dimensional versions of radix 2 (all factors equal to 2), radix 4 (all factors equal to 4), radix 4 + 2 (all factors equal to 4 except perhaps the last, which may be 2), and mixed radices (N is factored into as many 4's as possible, a 2 if necessary, then any 3's, 5's or other primes). Times required to transform a 1024 point sequence are given in Table II, including the time required by a recursive radix 2 transform and by the pre-fast Fourier transform "efficient" Goertzel's method[3](which still requires order N^2 operations) for comparison.

Our standard tool is the radix 4 + 2 Fourier transform which combines the speed of radix 4 with the flexibility of radix 2. The mixed radices Fourier transform is used when other factors, for example 10, are desired. The mixed radix is used less as it is more bulky (requiring scratch storage for unscrambling as well as being a larger program) and is less thoroughly optimized than the radix 4 + 2 Fourier transform.

Fast Fourier Transforms Using Hierarchical Store

As we become interested in doing larger and larger Fourier transforms, we reach a point where we may not be able to have all of the data in core simultaneously, and some must be kept in slower store such as drum, disk, or tape. At the other end of the scale, small amounts of very fast memory will be available with some of the "new generation" computers—memory much faster than the rest of core. Automatic systems for providing "virtual core" obscure the distinctions within this hierarchy of memories, but at great loss of efficiency for Fourier transforms. On the other hand, we can apply the basic recursion formula of the section *Basic Algebra of Fast Fourier Transforms* to employ hierarchical store in such a way as to operate on sequences large enough to require the slower store, yet to run little slower than if all the memory had the speed of the faster store.

In particular, when we factor N into the two factors A and B, we choose A as the largest Fourier transform that can be done in the faster store. We then compute the transforms (including unscrambling) of the B subsequences obtained by decimating the original sequence by B. We next multiply through by the appropriate twiddle factors, and do the B point transforms. Thus, other than decimating and recopying, which amounts to a matrix transpose, all the computations are done in the faster store, the loading and unloading of which can be overlapped. As an example, consider using a $32K$ machine with disk to Fourier transform a quarter million (2^{18}) point sequence (point here means complex number) initially on tape, returning the transform to tape. We will assume that we are willing to do transforms as large as 4096 points (2^{12}) in core.

The first step is to create 64 (2^6) files on disk, each of length 4096. The tape is read into core and the various decimated sequences read out onto disk—the zeroth point going into the zeroth file, the first into the first, etc. up to the sixty-third into the sixty-third, then the sixty-fourth point starting back in the zeroth file again, etc. Appropriately buffered, this can run almost at disk transfer speed.

The 64 files are now brought into core one at a time. When a file comes in, it is Fourier transformed using a standard in-core transform, then the points are multiplied by the twiddle factor $e(\hat{a}b/2^{18})$, where b is the number of the file (0 through 63) and $\hat{a}$ is the frequency index for the particular point. The file is then written out again. By choosing to use transforms of only 4096 points, we can fully buffer these operations, one file being Fourier transformed while the previous file is being written out and the succeeding file read in.

We now start on the second set of transforms. Here we make use of the random access feature of disk to read simultaneously from each of the 64 files, since there are 4096 subsequences of length 64, obtained by taking one element from each file. Each subsequence is brought in, transformed, and then returned to its previous position on the disk, that is, as one number in each of the files 0 through 63, according to the frequency index. Again we may completely buffer this operation so that the disk time comes free.

Finally we output our Fourier transform onto tape, sequentially writing out each of the files, 0 through 63.

Rough calculations for the IBM 7094 Mod I show that the initial decimating is limited by disk speed, and takes about 3 minutes, that the first 64 (4096 point) transforms will take a total of just under 6 minutes and are compute-bound so that disk time is free, that the last 4096 (64 point) transforms will take a total of about 3 minutes, again compute-bound so that disk time is free, and the final output is completely limited by disk speed, and takes about 3 minutes. Thus our quarter-million point Fourier transform has cost us only about 15 minutes, 9 minutes of which is computing, the same 9 minutes which would be required if the machine had enough core to do the entire problem internally, and 6 minutes of which is strictly tape to disk or disk to tape transmission that could be avoided if the sequence was initially available, and the final answers acceptable, in appropriate form or could at least be overlapped with other computing.

Roundoff Considerations

So far we have discussed the fast Fourier transform as though it could be done with complete accuracy. What happens when the algorithm is carried out in a real fixed word-length computer using floating point arithmetic? How does the error in doing this compare with the error from a direct application of the definition?

We may approach this question two ways—by strict bounds or by empirical values determined by experimentation. We shall derive bounds for the ratio of the Euclidean norm * of the error to the Euclidean norm of the data sequence, showing that whereas for direct application of the defining formula we can only bound this ratio by $1.06\sqrt{N}\,(2N)^{3/2}2^{-b}$. if we factor N into $n_1n_2\ldots n_k$ and use the fast Fourier transform we can bound the ratio by $1.06\sqrt{N}\left\{\sum_j (2n_j)^{3/2}\right\}2^{-b}$. In particular, if all the n_j are the same, so that $N = n^k$, then the ratio is less than $1.06\sqrt{N}\;k(2n)^{3/2}2^{-b}$, which is $k/n^{3/2(k-1)}$ times that for the direct calculation. (b here is the number of bits used in the mantissa of the floating-point representation). We shall then see from empirical results that the form of the bound is actually a fairly good description of the form of the observed errors, although the numerical constant is somewhat larger than is typical of actual trials.

Theorem: If we use the defining formula to Fourier transform a sequence $X(t)$ of length N in a machine using floating point arithmetic and a fixed word length with a b bit mantissa, then

$$||fl(\hat{X})-\hat{X}||_E < 1.06\sqrt{N}\,(2N)^{3/2}2^{-b}||\hat{X}||_E$$

Proof: We require the following lemma from Wilkinson [9] (p. 83):

Lemma: If A is a real p by q matrix and B is a real q by r matrix and if f (.) denotes the result of floating point computation then

$$||fl(AB)-AB||_E < 1.06q2^{-b}||A||_E||B||_E$$

q is the extent of the summation in the matrix multiplication.

The definition of the Fourier transform of $X(t)$ is also the definition of the multiplication of the vector $X(t)$ by the matrix $\{e(\hat{t}t/N)\}_{\hat{t}t}$.

Expressing this in real arithmetic, we multiply the real sequence $\{Re(X(t)),\ Im(X(t))\}$ by the real matrix $\begin{pmatrix} C & -S \\ S & C \end{pmatrix}$

where $C = \{cos\ 2\pi\hat{t}t/N\}_{\hat{t}t}$ and $S = \{sin\ 2\pi\hat{t}t/N\}_{\hat{t}t}$. The norm of this $2N\times2N$ real matrix is

$$\sum_{t=0}^{N-1}\sum_{\hat{t}=0}^{N-1}\{cos^2 2\pi\hat{t}t/N + sin^2 2\pi\hat{t}t/N + cos^2 2\pi\hat{t}t/N + sin^2 2\pi\hat{t}t/N\} = \sqrt{2N^2}$$

Since the extent of the summation is $2N$, we have by the lemma that $||fl(\hat{X})-\hat{X}||_E < 1.06\times2N\sqrt{2N^2}\,2^{-b}\,||X||_E$.

The proof is completed by observing that $||\hat{X}||_E = \sqrt{N}\,||X||_E$.

Theorem: If we use the fast Fourier transform, factoring N into $n_1n_2\ldots n_k$, to transform a sequence $X(t)$ in a machine using floating point arithmetic and a fixed word length with a b bit mantissa, then,

$$||fl(\hat{X})-\hat{X}||_E < 1.06\sqrt{N}\sum_{j=1}^{k}(2n_j)^{3/2}2^{-b}||\hat{X}||_E$$

Proof: Here we represent the transform as a sequence of matrix multiplies (one for each factor), and employ the above lemma. Specifically, the fast Fourier transform is equivalent to evaluating $\sqrt{N}\,M_kM_{k-1}\ldots M_1X$ where M_j is a matrix consisting of N/n_j disjoint n_j point Fourier transforms (including the twiddle factors for the next stage), rescaled to be orthogonal. Thus

*The Euclidean norm of a vector or sequence is the square root of the sum of the squares of the elements.

$$\begin{aligned} ||fl(\hat{X})-\hat{X}||_E &= \sqrt{N}\,||flM_kflM_{k-1}\ldots flM_1X - M_kM_{k-1}\ldots M_1X||_E \\ &= \sqrt{N}\,||flM_kflM_{k-1}\ldots flM_1X - M_kflM_{k-1}flM_{k-2}\ldots flM_1X \\ &\qquad + M_kflM_{k-1}\ldots flM_1X - M_kM_{k-1}flM_{k-2}\ldots flM_1X \\ &\qquad \ldots + M_kM_{k-1}\ldots flM_1X - M_kM_{k-1}\ldots M_1X||_E \\ &\le \sum_{j=1}^{k}\sqrt{N}\,||M_k\ldots M_{j+1}(flM_jflM_{j-1}\ldots flM_1X - M_jflM_{j-1}\ldots flM_1X)||_E \end{aligned}$$

but since the M_j are orthogonal

$$= \sqrt{N}\sum_{j=1}^{k}||flM_j(flM_{j-1}\ldots flM_1X) - M_j(flM_{j-1}\ldots flM_1X)||_E$$

When we compute the bound for $||fl(M_jY) - M_jY||_E$ we find that since M_j may be partitioned into N/n_j disjoint blocks, we can bound the error from each block separately, getting, since in real arithmetic each block is $2n_j$ square, and the norm of the block is $\sqrt{2n_j}$, a bound of $1.06(2n_j)^{3/2}2^{-b}||Y_S||_E$ where Y_S is the appropriate part of Y. By using Pythagoras' theorem this yields

$$||fl(M_jY) - M_jY||_E < 1.06(2n_j)^{3/2}2^{-b}||Y||_E$$

Finally we observe that except for error of order $N2^{-b}||X||_E$

$$||flM_{j-1}flM_{j-2}\ldots flM_1X||_E = ||M_j\ldots M_1X||_E$$

Immediately we have $||M_j\ldots M_1X||_E = ||X||_E$ because of the orthogonality, so

$$||fl(\hat{X}) - \hat{X}||_E < 1.06\sqrt{N}\sum_{j=1}^{k}(2n_j)^{3/2}2^{-b}||\hat{X}||_E$$

Corollary: If a sequence is transformed and then the inverse transform applied to the result, the norm of the difference between the initial and final sequences can be bounded by $2\times1.06(2N)^{3/2}2^{-b}||X||_E$ if we use the definition, or $2\times1.06\sum_j(2n_j)^{3/2}2^{-b}||X||_E$ if we use the fast Fourier transform.

With this corollary in mind, an experiment we can readily do is to take a sequence, transform, inverse transform, and then compute the norm of the difference between the original and resultant sequences, dividing this by the norm of the original sequence. Table III gives these ratios for random Gaussian sequences of length 2^k, $k = 1,\ldots,12$, when the transforms were computed using: (1) a radix 2 transform; (2) a radix 4 + 2 transform; (3) a radix 4 + 2 transform with rounding instead of truncation; (4) a Goertzel's method transform; and (5) a transform which directly implemented the definition. The results of three replications of the experiment are given. Figure 1 is a plot of the average ratio divided by $\log_2N$ against $\log_2N$ to show the essentially linear dependence (for the factored transforms) of the ratio on $\log_2N$. Attention is drawn to the common scaling factor of 10^{-8} and the consistency of the tabled ratios for the larger transforms.

TABLE III

OBSERVED RATIOS OF ERROR NORM TO SEQUENCE NORM FOR THREE RANDOM SEQUENCES AT EACH LENGTH.
(in units of 10^{-8})

Log_2N	Radix 2	Radix 4+2	Radix 4+2 with rounding	Goertzel's method	Defining formula
1	1.03	(1.03)	.46	2.60	4.27
	.53	(.53)	.13	3.91	1.54
	.57	(.57)	.00	2.73	4.14
2	3.43	1.09	.92	9.73	5.15
	2.23	.92	1.05	12.6	3.31
	1.89	1.96	1.07	15.8	4.14
3	4.04	4.99	2.65	32.8	9.75
	5.57	5.58	2.69	16.6	13.0
	4.51	5.01	2.39	34.4	7.83
4	9.14	7.11	2.12	98.9	17.7
	8.64	5.92	2.91	91.5	18.4
	7.76	6.62	2.71	121.	17.1
5	10.7	11.0	4.58	202.	33.8
	12.7	12.2	4.44	258.	36.7
	11.9	11.4	5.40	198.	36.2
6	13.2	11.2	3.38	548.	69.1
	14.4	12.0	3.70	787.	75.2
	14.0	10.2	3.59	806.	63.8
7	17.6	17.0	6.24	1290.	143.
	16.9	16.3	6.74	1990.	141.
	17.3	16.8	6.78	1900.	135.
8	20.0	16.3	4.82	7050.	286.
	20.1	16.6	5.09	3490.	288.
	20.7	16.2	4.66	5090.	269.
9	22.8	21.6	7.61	13700.	579.
	22.8	21.4	7.81	10700.	578.
	22.9	21.7	7.91	11500.	561.
10	25.2	21.0	5.82	32100.	1170.
	25.6	21.0	5.44	27600.	1160.
	25.8	21.0	5.44	29900.	1140.
11	28.1	26.2	8.70		
	28.1	26.1	8.56	—	—
	28.5	26.5	8.53		
12	30.7	25.5	6.30		
	30.7	25.5	6.21	—	—
	31.1	25.7	6.21		

USING FAST FOURIER TRANSFORMS

Classical Usage with Partial Differential Equations

One of the classical motivations for the study of Fourier transforms has been the application of finding solutions of partial differential equations. Hockney [4] gives a discussion of numerical solutions of partial differential equations. He considers the problem

$$\frac{\partial^2\Phi(x,y)}{\partial x^2} + \frac{\partial^2\Phi(x,y)}{\partial y^2} = \rho(x,y) \qquad 0 \le x \le l, \quad 0 \le y \le m$$

with boundary conditions $\Phi(x,y) = 0$ if $x = 0, l$ or $y = 0, m$. The solution is obtained by discretizing

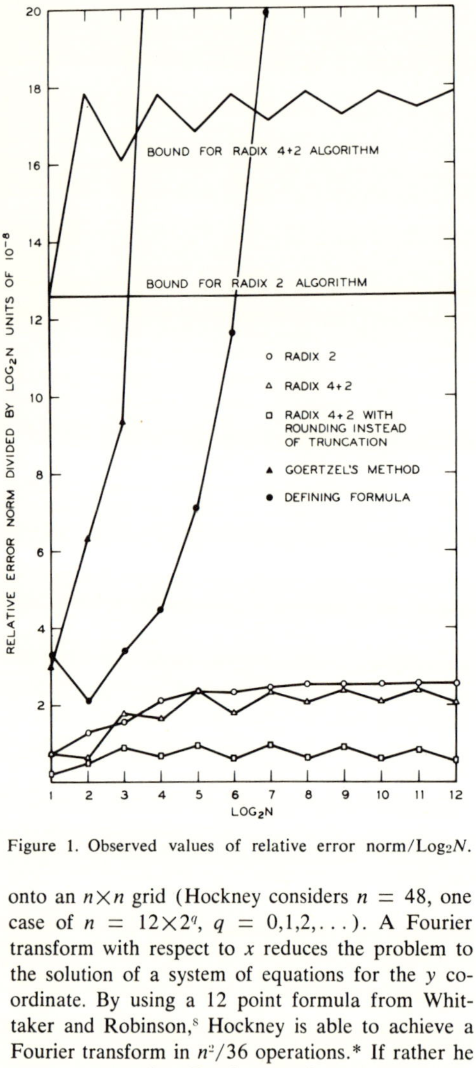

Figure 1. Observed values of relative error norm/$\text{Log}_2 N$.

onto an $n \times n$ grid (Hockney considers $n = 48$, one case of $n = 12 \times 2^q$, $q = 0,1,2,\ldots$). A Fourier transform with respect to x reduces the problem to the solution of a system of equations for the y coordinate. By using a 12 point formula from Whittaker and Robinson,[8] Hockney is able to achieve a Fourier transform in $n^2/36$ operations.* If rather he were to use a fast Fourier transform for $n = 2^p$, $p = 0,1,2,\ldots$ he could achieve a Fourier transform in approximately $\frac{3}{2} n \log_2 n$ operations. This estimate of operations leads to a net count of $6n^2 \log_2 n$ for Hockney's method. Under this accounting, Hockney's method is much superior to the other methods he considers. It must be cautioned that the factoring of 12×2^2 which Hockney uses for his Fourier transform will be equivalent to the fast Fourier transform. The apparent reduction only comes into play as the size, n, is increased past 48. As Hockney points out, the calculation of Fourier coefficients has long been recommended (and avoided because of cost) as a method of solving partial differential equations. Perhaps the fast Fourier transform will render the solution of partial differential equations both more economical and more straightforward.

*Operation in this section means a real operation of a multiplication and an addition.

Least Squares Approximation by Trigonometric Polynomials

Often one may want to approximate a given sequence by a band-limited sequence, that is, a sequence generated by a low order trigonometric polynomial. A possible example could be the determination of filter coefficients to approximate a desired transfer function.[5] The use of expansions in terms of orthogonal functions to obtain least square approximations is well known. The Fourier transform of a sequence gives its expansion in terms of the mutually orthogonal complex exponentials, hence to obtain the coefficients for the approximation, we simply retain the low order coefficients from the Fourier transform. We may compare the approximation to the original by looking at the inverse Fourier transform of the coefficients after replacing the unwanted coefficients by zeros.

If we had wanted a positive definite approximation, we could have approximated the square root of the original sequence. The square of the approximation to the root is then the positive approximation. Squaring a sequence like this has the effect of convolving the Fourier coefficients. Other variations in the approximating problem may be handled by similar specialized techniques.

A caution: Gibbs phenomenon is a persistent problem when using least squares, particularly in the case of trigonometric approximation. In some situations, the problem may be sufficiently bad to warrant the use of other approximation criteria.

Numerical Convolutions

To date, the most important uses of the fast Fourier transform have been in connection with the convolution theorem of Table I. Some uses of numerical convolutions are the following:

Auto- and Cross-Covariances: 1. Method and Timing Considerations. In the analysis of time series by digital spectrum analysis, there are many computational advantages to the "indirect" method of spectral estimation by computing the autocovariance and then Fourier transforming rather than the "direct" method by obtaining the periodogram and then smoothing.[1] For example, it is more practical to consider examining a spectrum through several different spectral windows by the indirect method. Similarly, when analyzing a pair of series with co- and quadrature-spectra, we often find the cross-covariance of the two series computationally convenient.

The first major use of the convolution theorem, by Sande,[6] was to compute the auto-covariance

$$R_{XX}(\tau) = \frac{1}{N} \sum \{X(t)\}^* X(t+\tau) \qquad \tau = 0,\pm 1,\pm 2,\ldots,\pm L$$

and cross-covariance

$$R_{XY}(\tau) = \frac{1}{N} \sum \{X(t)\}^* Y(t+\tau) \qquad \tau = 0,\pm 1,\pm 2,\ldots \pm L.$$

where the summations are overall values of t for which the products are defined. (These are not in the form of Table I but can be readily put into that form.) The convolution theorem may lead to improvements by a factor of 20 in computing time as compared to the summing of lagged products. The two major problems are that these are not defined cyclically and that N may not be convenient for use of the fast Fourier transform. Appending zeros to one or both ends of the series solves both problems.

To see this, extend X to length $N' \geq N + L$ by adding zeros so that

$$X'(t) = X(t) \qquad t = 0,1,\ldots,N-1$$
$$X'(t) = 0 \qquad t = N,\ldots,N'-1$$

Extend the Y series similarly. Fourier transform both new sequences to obtain

$$\hat{X}'(\hat{t}) = \sum_{t=0}^{N'-1} X'(t) e\left(\frac{t\hat{t}}{N'}\right)$$

$$\hat{Y}'(\hat{t}) = \sum_{s=0}^{N'-1} Y'(s) e\left(\frac{s\hat{t}}{N'}\right)$$

Inverse transform the sequence formed by the product of the Y' transform and the complex conjugate of the X' transform.

$$C(\tau) = \frac{1}{N'} \sum_{\hat{t}=0}^{N'-1} \{\hat{X}'(\hat{t})\}^* \hat{Y}'(\hat{t}) e\left(-\frac{\hat{t}\tau}{N'}\right)$$

$$= \frac{1}{N'} \sum_{\hat{t}=0}^{N'=1} \sum_{t=0}^{N'=1} \sum_{s=0}^{N'=1} \{X'(t)\}^* Y'(s) e\left(-\frac{\hat{t}\tau}{N'}\right) e\left(-\frac{t\hat{t}}{N'}\right) e\left(\frac{s\hat{t}}{N'}\right)$$

$$= \frac{1}{N'} \sum_{t=0}^{N'-1} \sum_{s=0}^{N'-1} \{X'(t)\}^* Y'(s) \sum_{\hat{t}=0}^{N'-1} e\left(\frac{\hat{t}(s-t-\tau)}{N'}\right)$$

which by the orthogonality condition of Table I gives

$$= \sum_{t=0}^{N'-1} \sum_{s=0}^{N'-1} \{X'(t)\}^* Y'(s) \delta_{N'}(s-t-\tau)$$

$$= \sum_{t=0}^{N'-1} \{X'(t)\}^* Y'(t+\tau)$$

Recall that the indices must be interpreted modulo N'. For $\tau < N' - N$, this expression is just N times the cross-covariance R_{XY} because then the extraneous products are precisely zero.

Note that the auto-covariance is precisely the cross-covariance of a series with itself, so that the outline just given also describes how to compute the auto-covariance. On the other hand, we may take advantage of the fact that in most time series problems the series are real to compute two auto-covariances at one time. We use one series as the real part and the other series as the imaginary part in forming the sequence which we Fourier transform, using the symmetry properties from Table I to separate them later.

Exactly how much faster the convolution theorem approach is than the lagged products depends not only on the length of series and proportion of the available lags desired, but also on details of the programs used. There are $\frac{(2N-L)\ (L+1)}{2}$ multiplications and additions, plus associated indexing operations, in summing lagged products. On the other hand, the number of operations required for the Fourier transforms (approximately proportional to $N' \log N'$) depends on our choice of N' which can vary according to the availability of Fourier transform routines, a routine which can handle mixed factors being much more flexible than a radix 4 + 2 routine which restricts N' to be a power of two. Figure 2 gives a comparison of direct summing and

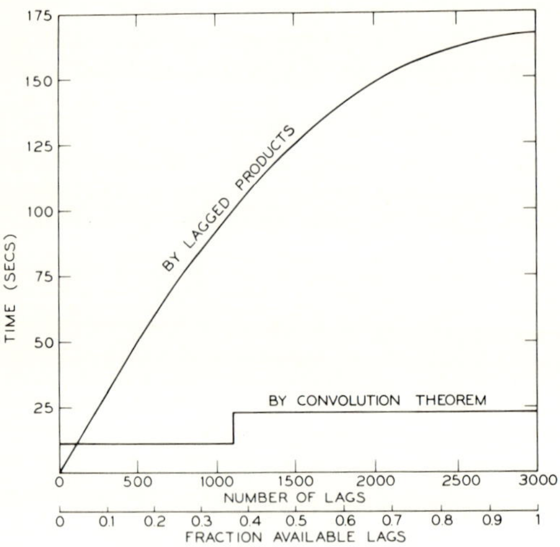

Figure 2. Computation times for auto-covariance of a 3000 point series, using convolution theorem or summing lagged products.

convolution theorem times using our radix 4 + 2 transform routine. Figure 3 shows (for this program) the regions in which either direct summing or the convolution theorem is more efficient as functions of N and L.

2. Extra-long Series. When we have to compute auto- or cross-covariances of extra-long series, we could just continue to use the method of the previous section, doing the oversize Fourier transforms by the method outlined in the section *Fast Fourier Transforms Using Hierarchical Store.* A simpler and occasionally more efficient method is also available. What we do instead is to partition the series X and Y into contiguous disjoint segments, and express the cross-covariance R_{XY} in terms of the cross-covariances between these various subseries.

A simple diagram will illustrate the method. In Fig. 4 we have plotted all the products of the elements of series Y against those of series X, the plotting position corresponding to the values of the respective indices. The cross-covariance $R_{XY}(\tau)$ then corresponds to $1/N$ times the sum over all elements on a line parallel to the main diagonal, but τ elements above it. Such a line is shown on the diagram.

As an example, we consider the problem of computing R_{XY} for $\tau = 0, \pm 1, \ldots \pm L$, when the largest cross-covariance we can compute by the method of the sub-section on Method and Timing Considerations is for series of length $N/5$. We are thus interested in those products in the hatched area of the diagram. The segments described above are indicated by the dotted lines; they are labeled $X_0, X_1, \ldots X_4$ and $Y_0, Y_1, \ldots Y_4$. We now see how $R_{XY}(\tau)$ may be calculated, by summing the appropriate cross-covariances of X_i with Y_j. For instance, for the particular value of τ illustrated,

$$R_{XY}(\tau) = \frac{1}{5}\Big\{ R_{X_0Y_0}(\tau) + R_{X_1Y_0}\left(\tau - \frac{N}{5}\right) + R_{X_1Y_1}(\tau) + R_{X_2Y_1}\left(\tau - \frac{N}{5}\right) + R_{X_2Y_2}(\tau) + R_{X_3Y_2}\left(\tau - \frac{N}{5}\right) + R_{X_3Y_3}(\tau) + R_{X_4Y_3}\left(\tau - \frac{N}{5}\right) + R_{X_4Y_4}(\tau)\Big\}$$

Filtering: Digital Filters. Often, when we have a set of data which is indexed by displacement or time, we wish to "smooth" our data to reduce the effect of unwanted "noise". If our original data were represented by $\{X(t)\}$ then forming

$$Y(t) = \frac{X(t)+X(t+1)+X(t+2)}{3}$$

would represent one method of smoothing our data. More generally, we could choose to use any arbitrary weights we pleased so that

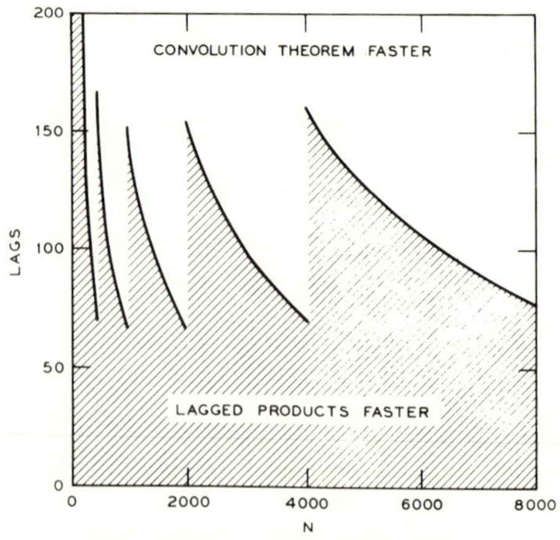

Figure 3. Region where using convolution theorem is faster than summing lagged products.

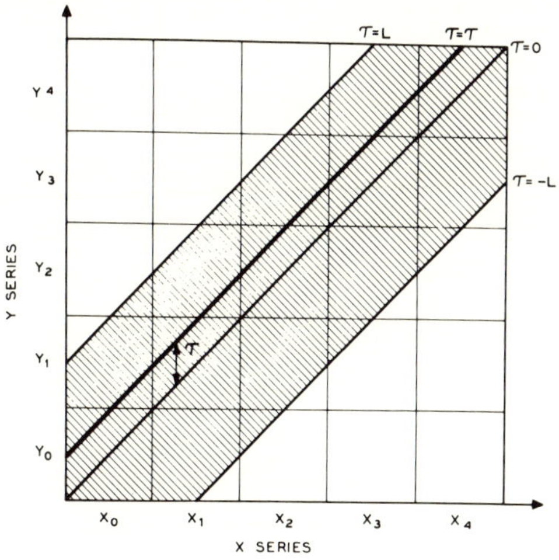

Figure 4. Diagram of cross products for computing cross-covariance.

$$Y(t) = c(k)X(t) + c(k-1)X(t+1) + \dots + c(o)X(t+k) = \sum_{j=0}^{k} c(k-j)X(t+j).$$

If, rather than smoothing, we had desired to estimate moving short-term integrals (possibly with a kernel) we would again be able to express the result as a summation over the data points weighted by suitable coefficients. The estimation of derivatives also leads to a similar expression.

Let us consider the application of such a moving average to a complex exponential of a fixed frequency. Thus our data would be $X(t) = e^{i\omega t} = e(ft)$ and the result of the moving average would be

$$\begin{aligned} Y(t) &= c(k)X(t) + \dots + c(o)X(t+k) \\ &= \sum_{j=0}^{k} c(k-j)X(t+j) \\ &= \sum_{j=0}^{k} c(k-j)e(f\times\{t+j\}) \\ &= e(tf) \sum_{j=0}^{k} c(k-j)e(fj). \end{aligned}$$

We can thus represent $Y(t)$ as the product of $e(ft)$ and $A(f) = \sum_{j=0}^{k} c(k-j)e(fj)$. $A(f)$ is called the (complex) gain of the moving average (or filter) determined by the coefficients $c(o),\dots,c(k)$. Moving averages are linear in the data, for example, the moving average of $U(t) + V(t)$ is the same as the sum of the separate moving averages of $U(t)$ and $V(t)$. Thus if we consider our data as being composed of many Fourier components, we see that the moving average affects each of the Fourier components separately.

All of these considerations can be compactly stated using the terminology we have at hand. Taking the moving average is a convolution with a set of weights. The complex gain of the moving average is the Fourier transform of the coefficients. Filtering corresponds to multiplying by the complex gain in the frequency domain.

This has given but a small introduction to the analysis of linear filters. It does not tell us how to design a filter to perform a desired task. For this problem, the reader is referred to general texts such as Hamming,[3] or Blackman and Tukey,[1] or to more specialized works such as Kaiser.[5]

2. Sectioning. After we have chosen the weights with which we would like to filter, we still must choose how to implement the filter. A typical situation might be to have 15,000 data points which we wish to filter with 50 weights. We could do this by computing the convolution directly. We might also extend our data to 16K by adding zeros and our filter to 16K weights by adding zeros and then doing two Fourier transforms to get our convolution. For this choice of numbers the decision is not dramatically in favor of one or the other method.

Applying a 16K Fourier transform to only 50 weights seems rather extravagant. Perhaps we could do better if we were to filter only sections of the original data at any one time as first suggested by Stockham.[7] The problem is then to find a sensible size for the various sections. Let us say that we have D data points and F filter weights. We want to find N (the section length) so as to minimize the total time. This scheme will take about $D/(N-F)$ Fourier transforms (transforming two real sequences at one time). The total time will then be $t \quad \dfrac{D}{N-F} \quad cN\ell n(N)$ where c may have a small dependency on N which we will ignore. By assuming that $F << D$ and that N is continuous we may find the minimum of t by differentiating with respect to N. Thus

$$\frac{\partial t}{\partial N} = Dc \frac{\partial}{\partial N}\left(\frac{N \ln(N)}{N-F}\right)$$

$$= Dc\left\{\frac{\ln(N)}{N-F} + \frac{N}{N-F}\frac{1}{N} - \frac{N\ln(N)}{(N-F)^2}\right\}$$

$$= \frac{Dc}{(N-F)^2}\{(N-F)\ln(N) + N - F - N\ln(N)\}$$

$$= \frac{Dc}{(N-F)^2}\{-F\ln(N) + N - F\}$$

$$\frac{\partial t}{\partial N} = 0 \quad \text{implies} \quad N - F(1 + \ln(N)) = 0$$

which yields

$$F = \frac{N}{1 + \ln(N)}.$$

For our case of 50 filter weights, this suggests using Fourier transforms of length about 300, for a reduction of the time required by a factor of two.

Interpolation. One of the standard operations in numerical computing is the interpolation in a table of function values to obtain approximate values on a grid at a finer spacing than the original.

1. Band-Limited Interpolation. The most obvious application of these ideas is to the case of band-limited interpolation, sometimes called trigonometric or cosine interpolation. This interpolation is done over the whole interval, by fitting a trigonometric polynomial (or complex exponential series) of sufficiently high order through all the given values, then evaluating it at the new points.[3] But the first step of this is exactly the finding of the Fourier transform of the original data. And the second step, as we shall see, can be accomplished by inverse transforming the sequence whose low order coefficients are the same as the coefficients of the transform of the original data, and whose high order coefficients are zero.

To justify this last statement, consider the case where the function is given at N equispaced grid points. We wish to interpolate in this sequence to obtain values at M times as many points. The original sequence of values, $X(t)$, has a Fourier transform $X(t)$ such that we have the representation

$$X(t) = \frac{1}{N}\sum_{\hat{t}=0}^{N-1} \hat{X}(\hat{t})e\left(-\frac{t\hat{t}}{N}\right)$$

Consider the augmented Fourier transform $\hat{Z}(\hat{s})$ $\hat{s} = 0,1,\ldots,N{\cdot}m-1$ such that

(a) $\hat{Z}(\hat{t}) = \hat{X}(\hat{t})\left(-\frac{N}{2}\right) < \hat{t} < \frac{N}{2}$

(b) $\hat{Z}(-N/2) = \hat{Z}(N/2) = 1/2\hat{X}(N/2)$ if N even

(c) $\hat{Z}(\hat{s}) = 0$ for all other values of $\hat{s}$.

When we recall that $\hat{X}(\hat{t})$ has period N and that $\hat{Z}(\hat{s})$ has period NM we identify

$$\hat{X}(-\hat{t}) = \hat{X}(N-\hat{t})$$

and

$$\hat{Z}(-\hat{s}) = \hat{Z}(NM-\hat{s})$$

The construction of $\hat{Z}$ corresponds to taking the circle on which the Fourier transform coefficients of $\hat{X}$ are defined, breaking it at $\frac{N}{2}$, and inserting $M(N-1)$ zeros so that the low frequency Fourier coefficients of the two sequences match, and the high frequency coefficients of $\hat{Z}(\hat{s})$ are zero. What is the inverse transform of $\hat{Z}(\hat{s})$?

$$Z(s) = \frac{1}{MN}\sum_{\hat{s}=0}^{MN-1} \hat{Z}(\hat{s})e\left(-\frac{(s/M)\hat{s}}{N}\right)$$

which, we note, is just the band-limited interpolation of $X(t)$, except for a factor $\frac{1}{M}$. For example, when $s = Mt$,

$$Z(Mt) = \frac{1}{MN}\sum_{\hat{s}=0}^{MN-1} \hat{Z}(\hat{s})e\left(-\frac{t\hat{s}}{N}\right)$$

$$= \frac{1}{M}X(t) \text{ by the definition of } \hat{Z}(\hat{s})$$

Applying this in the case where the original values are not at equispaced points merely makes the finding of the first transform, $\hat{X}(\hat{t})$, more complicated. Furthermore, the extension of this scheme to multiple dimensions is immediate.

We remark that a fundamental fact of band-limited interpolation is that it is periodic, so that the first points of a sequence influence the values interpolated between the last points. This property, and the long range effects of single values, may mean we should avoid band-limited interpolations for certain uses. It may, however, be extremely useful in other circumstances.

2. General Interpolation Rules as Convolutions. Knowing the limitations of band-limited interpolation, we often prefer to use a local formula such as Lagrangian polynomial interpolation. Using any interpolation rule such that the interpolated

values are linear combinations of the adjacent known values to subtabulate a table of equispaced data may very conveniently be done by convolutions. If we denote the value to be interpolated at $t + p$, $0 \leq p < 1$, by $Z_p(t)$, then the series $Z_p(t)$ is the convolution of the original series $X(t)$ with an appropriate series of weights:

$$Z_p(t) = \sum_{s=\alpha}^{\beta} W_p(s)X(t-s)$$

It is possible to arrange that the convolutions for all values of p be done simultaneously, but the computational effort may be increased over doing them separately. Since, typically, the series of weights will be short compared to the length of the series $X(t)$, it may be profitable to employ the sectioning described earlier. When applied to problems in more than one dimension, additional profit may be made if we use interpolation rules which are direct products or direct sums of lower dimensional rules, for example, a two dimensional interpolation rule such that the weights $W_{pq}(r,s)$ are either a product $W'_p(r)W''_q(s)$ or a sum $W'_p(r) + W''_q(s)$. The advantage of such rules is that the higher dimensional convolutions may then be done as a succession of lower dimensional convolutions or equivalently that the Fourier transforms of such sequences are the products or sums, respectively, of the Fourier transforms of the constituent parts.

Cauchy Products for Symbolic Polynomial Manipulation. The Cauchy product of two infinite series is a well-known result of college algebra:

$$\sum_{i=0}^{\infty} a_i x^i \cdot \sum_{j=0}^{\infty} b_j x^j = \sum_{k=0}^{\infty} c_k x^k \text{ where}$$

$$c_k = \sum_{n=0}^{k} a_k b_{n-k}$$

subject to some convergence criteria. We readily recognize that the $\{c_k\}$ are convolutions of the $\{a_i\}$ and $\{b_j\}$ and that we could use our techniques to evaluate these discrete convolutions.

We may arrive at this by considering an alternative viewpoint. Let us write $z = e(\theta)$ and recognize that $z^n = e(n\theta)$. With this notation the Fourier transform becomes

$$\hat{X}(\hat{t}) = \sum_{t=0}^{N-1} X(t)e(t\hat{t}/N) = \sum_{t=0}^{N-1} X(t)z^t \text{ where}$$

$$z = e(\hat{t}/N)$$

This is a truncated power series evaluated at points on the unit circle in the complex plane. Our convolution theorem is now a statement about multiplying polynomials and using a Cauchy product representation. The inverse Fourier transform gives us a way of evaluating coefficients of polynomials given as values equispaced on the unit circle.

Of the applications so far suggested, this most deserves the epithet "fun." It is doubtful if this technique will supplant conventional symbol manipulation techniques as it makes no use of special properties of the coefficient sequences. It does, however, admirably illustrate that maintaining an open mind may well bring forth new, surprising and occasionally beneficial applications of finite discrete Fourier analysis.

Considering Problems in Transform Space

It has long been a fact that some problems of physics are more tractable in transform space than they are in the regular coordinates. Our refound ability to switch from one to the other may mean that some previously intractable problems will now be solvable. In electromagnetic problems, the return signal can often be represented as a convolution with a kernel. Taking the Fourier transform can change the form of the non-linearity from convolution to multiplication and may make the problems more manageable. In pattern recognition problems, one may seek measures which are free from the effects of linear transformations. Taking the Fourier transform may greatly reduce the problem of finding such measures. It has been suggested that quantizing the Fourier transform of a signal rather than quantizing the signal itself may result in a higher quality transmission for communication.

There are certainly a wealth of problems in which Fourier analysis arises as a most natural tool to be employed. Unfortunately it has often been rejected because of its high computational cost. Perhaps the application of the fast Fourier transform could swing the economic balance to make this time-honored technique again competitive.

ACKNOWLEDGMENT

The authors would like to express their appreciation to those from whose work we have borrowed, and whose fruitful conversations have helped in the preparation of this paper. In particular we should like to thank C. Bingham, A. B. Langdon, C. L. Mallows, T. G. Stockham, and J. W. Tukey.

REFERENCES

1. R. B. Blackman and J. W. Tukey, *The Measurement of Power Spectra,* Dover, New York, 1958.

2. J. W. Cooley and J. W. Tukey, "An Algorithm for the Machine Calculation of Complex Fourier Series," *Mathematics of Computation,* vol. 19, no. 90, (1965) pp. 297–301.

3. R. W. Hamming, *Numerical Analysis for Scientists and Engineers,* McGraw-Hill, New York, 1962.

4. R. W. Hockney, "A Fast Direct Solution of Poisson's Equation Using Fourier Analysis," *Journal of the Association of Computing Machinery,* vol. 12, no. 1, (1965) pp. 95–113.

5. J. F. Kaiser, "Some Practical Considerations in the Realization of Linear Digital Filters," Proceedings of the Third Allerton Conference on Circuit and System Theory, Monticello, Illinois, October 1965.

6. G. Sande, "On an Alternative Method of Calculating Covariance Functions," unpublished, Princeton University, 1965.

7. T. G. Stockham, "High Speed Convolution and Correlation," *AFIPS, volume 28, 1966 Spring Joint Computer Conference,* Spartan Books, Washington, 1966.

8. Whittaker and Robinson, *Calculus of Observations,* Blackie & Son, London, 1944.

9. J. H. Wilkinson, *Rounding Errors in Algebraic Processes,* Prentice-Hall, Englewood Cliffs, New Jersey, 1963.

27

Reprinted from *IEEE Trans. Audio Electroacoustics,* **AU-15**(2), 91–98 (1967)

A Method for Computing the Fast Fourier Transform with Auxiliary Memory and Limited High-Speed Storage

RICHARD C. SINGLETON, SENIOR MEMBER, IEEE

Abstract—A method is given for computing the fast Fourier transform of arbitrarily large size using auxiliary memory files, such as magnetic tape or disk, for data storage. Four data files are used, two in and two out. A multivariate complex Fourier transform of $n=2^m$ data points is computed in m passes of the data, and the transformed result is permuted to normal order by $m-1$ additional passes. With buffered input-output, computing can be overlapped with reading and writing of data. Computing time is proportional to $n \log_2 n$. The method can be used with as few as three files, but file passing for permutation is reduced by using six or eight files. With eight files, the optimum number for a radix 2 transform, the transform is computed in m passes without need for additional permutation passes.

An ALGOL procedure for computing the complex Fourier transform with four, six, or eight files is listed, and timing and accuracy test results are given. This procedure allows an arbitrary number of variables, each dimension a power of 2.

Introduction

The fast Fourier transform is a method of computing the finite complex Fourier transform

$$\alpha_k = \frac{1}{n}\sum_{j=0}^{n-1} x_j \exp(i2\pi jk/n) \qquad \text{for } j = 0, 1, \cdots, n-1$$

or the inverse transform

$$x_j = \sum_{k=0}^{n-1} \alpha_k \exp(-i2\pi jk/n) \qquad \text{for } j = 0, 1, \cdots, n-1.$$

The key idea of the method, that of factoring n as n_1, $n_2, \cdots, n_m$, then decomposing the transform into n/n_j transforms of size n_j for $j=1, 2, \cdots, m$, was first proposed in 1942 by Danielson and Lanczos [1], [2]. Since the labor of computing a single transform of dimension n is of the order of n^2, the decomposition gives a considerable saving when n is not a prime, reducing computing to the order of $n(n_1+n_2+\cdots+n_m)$. Good [3] in 1958 formulated the fast Fourier transform for complex-valued and multivariate data. Cooley and Tukey [4] programmed Good's method with improvements, and their 1965 paper touched off the current surge of research on fast Fourier transform methods [5], [6] and applications [7], [8].

In this paper, we first give some general comments on programs for the fast Fourier transform, then show a method for computing radix 2 transforms of arbitrarily large size on any computer having four auxiliary files available for data storage. This method also has possible application in the design of special-purpose hardware for parallel computation of the fast transform. A B5500 ALGOL procedure for computing a multivariate Fourier transform using four, six, or eight auxiliary files is listed in the Appendix; results of testing this procedure for time and accuracy are also given.

Programs for the Fast Fourier Transform

The fast Fourier transform can be computed for any size n, but since the number of arithmetic operations is proportional to $n\sum n_j$ for $n=\Pi n_j$, we avoid large factors if possible. An ALGOL procedure for the general case of arbitrary factors is available [9] and other programs are known to exist [6].

The special case of $n=2^m$ has received considerable attention for reasons of efficiency and ease of programming. A version of Cooley's original program, written in a mixture of FORTRAN II and FAP and available through SHARE [10], computes complex Fourier transforms with up to three variables

$$\alpha_{jkl} = \frac{1}{n_1n_2n_3}\sum_{p=0}^{n_1-1}\sum_{q=0}^{n_2-1}\sum_{r=0}^{n_3-1} x_{pqr} \cdot\exp(i2\pi(pj/n_1 + qk/n_2 + rl/n_3)),$$

for the $n_1n_2n_3=2^m$ case. The author has written ALGOL procedures for the univariate case [11], [12]. A univariate transform can be used to compute multivariate transforms if desired; for example, a bivariate transform of data stored in a two-dimensional array is computed by first transforming the rows one by one, then the columns of the array. Other programs for $n=2^m$ have been written by Gentleman and Sande [6]. They favor factors of 4, with a final factor of 2 if m is odd, and claim a doubling of efficiency by this approach. The author's experience contradicts their claim; when computing with all the data in high-speed storage, a good

Manuscript received December 29, 1966; revised February 14, 1967. This work was supported by the Stanford Research Institute, out of Research and Development funds.

The author is with the Stanford Research Institute, Menlo Park, Calif.

radix 2 program is nearly as efficient as a radix 4 plus 2 program and is simpler. In computing the fast transform, a number of possibilities exist for reducing time by adding coding. However, the trade-off between efficiency and program size needs further study.

For a fast Fourier transform of size $n=2^m$, $n/4+1$ distinct sine function values, sin $(2\pi k/n)$ for $k=0, 1, \cdots, n/4$, are needed during the computation. In the Cooley program the sine values are tabled, while in the author's programs sine and cosine values are generated as needed. Tabling gives a moderate increase in efficiency but requires additional storage space. The difference between the two methods is small if trigonometric function values are generated recursively by a difference equation.

The fast Fourier transform can be done in place, with the results of each transform of size n_j replacing the values used to compute the transform. In this case a complex transform of size n requires $2n$ data storage locations, with each complex quantity taking two locations, plus 2 $\max_j n_j$ temporary storage locations. If n is a power of 2 and if the number of words of high-speed storage in the computer is also a power of 2, then n can be no larger than one quarter of the memory size for a complex Fourier transform computed without auxiliary memory. Thus, Cooley's program will compute a transform of size 8192 on an IBM 7094 with 32 768 words of core (with a computing time of about 0.13 minute).

The speed of the fast Fourier transform leads us to search for methods of computing transforms larger than can be accommodated within high-speed storage. On a system with virtual memory, we can organize the sequence of computations so as to reduce time lost through overlay of data storage [9], [11]; using this approach the author has computed complex transforms of size up to 65 536 on a Burroughs B5500 with 32 768 words of core. Another approach suggested by Cooley [8] and in greater detail by Gentleman and Sande [6], is that of breaking the problem up into transforms that will fit within memory, storing each result in a separate auxiliary memory file, then combining results at the end. In the next section we propose still another approach, a method of computing the fast Fourier transform using four auxiliary files and requiring negligible high-speed storage for data.

An Algorithm for Use of Auxiliary Memory

Here, we assume the availability of four serial-organized auxiliary memory files, such as magnetic tapes or serial disk files. The reader will note, however, that three tape units can be used at the cost of changing two reels per pass or of added tape copying.

In this method the $n=2^m$ complex data points are first written on two of the files, with the first $n/2$ complex values in sequence on the first file and the remaining $n/2$ on the second. The fast Fourier transform is computed while copying the data back and forth between the two pairs of files. As with the fast transform done in place within high-speed memory, this method also leaves the transformed result in reverse binary order, requiring a permutation to restore it to normal order. Again, this permutation can either follow the transform or precede it, using in the later case a fast transform procedure operating on data stored initially in reverse binary order. We consider both approaches.

If the data are originally in normal order, we read a pair of complex data elements, one from each input file, compute a two-by-two transform and write the two resulting complex values on the first output file. After $n/2$ entries have been written we switch to writing on the second output file. The two-by-two transform computed on the kth pass, where $k=1, 2, \cdots, m$, is

$$y_{2j} = x_j + x_{j+n/2}$$
$$y_{2j+1} = x_j - x_{j+n/2} \exp\left(i\pi(j \div 2^{k-1})/2^{m-k}\right)$$

for $j=0, 1, \cdots (n/2)-1$, where $\div$ represents integer division without remainder. The successive passes are identical, except for interchanging input and output files and using a different sequence of trigonometric function values. The sequence of steps for the $n=8$ case is shown in Figs. 1 and 2, where

$$c^j = \exp(i2\pi j/n).$$

We see that the result is in reverse binary order.

This method of computing the fast Fourier transform offers interesting possibilities for parallel computation. As with other methods, the two-by-two transforms at each step are independent and can be computed in parallel. However, this method has the additional property of having the same schedule of data access and storage at each of the m steps of an $n=2^m$ transform.[1] If we copy back and forth between two data arrays, we need only a single fixed set of data transmission paths in each direction for computing the transform, plus an additional set of data transmission paths in one direction to permute the final result to normal order. We must, of course, also provide a way of changing the set of trigonometric function values used as multipliers of the data at each step in the transform.

If the data are originally in reverse binary—for example, as the result of a previous transform—we reverse the steps outlined above. In each pass we take pairs of complex entries from the first input file until it is exhausted, then from the second input file, compute a two-by-two transform with each pair, and write one result on each of the two output files. The two-by-two transform computed on the kth pass, where $k=1, 2, \cdots, m$, is

[1] M. Pease of Stanford Research Institute independently noted the parallel computation possibilities of this arrangement of the fast Fourier transform, and is exploring the idea further.

α_0	α_4	$(\alpha_0+\alpha_4)$	$(\alpha_2+\alpha_6)$
α_1	α_5	$(\alpha_0-\alpha_4)$	$c^2(\alpha_2-\alpha_6)$
α_2	α_6	$(\alpha_1+\alpha_5)$	$(\alpha_3+\alpha_7)$
α_3	α_7	$c^1(\alpha_1-\alpha_5)$	$c^3(\alpha_3-\alpha_7)$
Input to Step One		*Input to Step Two*	

$[(\alpha_0+\alpha_4)+(\alpha_2+\alpha_6)]$	$[(\alpha_1+\alpha_5)+(\alpha_3+\alpha_7)]$
$[(\alpha_0+\alpha_4)-(\alpha_2+\alpha_6)]$	$c^2[(\alpha_1+\alpha_5)-(\alpha_3+\alpha_7)]$
$[(\alpha_0-\alpha_4)+c^2(\alpha_2-\alpha_6)]$	$[c^1(\alpha_1-\alpha_5)+c^3(\alpha_3-\alpha_7)]$
$[(\alpha_0-\alpha_4)-c^2(\alpha_2-\alpha_6)]$	$c^2[c^1(\alpha_1-\alpha_5)-c^3(\alpha_3-\alpha_7)]$

Input to Step Three

Fig. 1. Auxiliary memory input files for $N=8$, data in normal order.

x_0	x_1	x_0+x_4	x_0-x_4
x_4	x_5	x_2+x_6	x_2-x_6
x_2	x_3	x_1+x_5	x_1-x_5
x_6	x_7	x_3+x_7	x_3-x_7
Input to Step One		*Input to Step Two*	

$(x_0+x_4)+(x_2+x_6)$	$(x_0+x_4)-(x_2+x_6)$
$(x_1+x_5)+(x_3+x_7)$	$(x_1+x_5)-(x_3+x_7)$
$(x_0-x_4)+c^2(x_2-x_6)$	$(x_0-x_4)-c^2(x_2-x_6)$
$(x_1-x_5)+c^2(x_3-x_7)$	$(x_1-x_5)-c^2(x_3-x_7)$

Input to Step Three

Fig. 3. Auxiliary memory input files for $N=8$, data in reverse binary order.

$$[(\alpha_0+\alpha_4)+(\alpha_2+\alpha_6)]+[(\alpha_1+\alpha_5)+(\alpha_3+\alpha_7)]=x_0$$
$$[(\alpha_0+\alpha_4)+(\alpha_2+\alpha_6)]-[(\alpha_1+\alpha_5)+(\alpha_3+\alpha_7)]=x_4$$
$$[(\alpha_0+\alpha_4)-(\alpha_2+\alpha_6)]+c^2[(\alpha_1+\alpha_5)-(\alpha_3+\alpha_7)]=x_2$$
$$[(\alpha_0+\alpha_4)-(\alpha_2+\alpha_6)]-c^2[(\alpha_1+\alpha_5)-(\alpha_3+\alpha_7)]=x_6$$

First Output File at Step Three

$$[(\alpha_0-\alpha_4)+c^2(\alpha_2-\alpha_6)]+[c^1(\alpha_1-\alpha_5)+c^3(\alpha_3-\alpha_7)]=x_1$$
$$[(\alpha_0-\alpha_4)+c^2(\alpha_2-\alpha_6)]-[c^1(\alpha_1-\alpha_5)+c^3(\alpha_3-\alpha_7)]=x_5$$
$$[(\alpha_0-\alpha_4)-c^2(\alpha_2-\alpha_6)]+c^2[c^1(\alpha_1-\alpha_5)-c^3(\alpha_3-\alpha_7)]=x_3$$
$$[(\alpha_0-\alpha_4)-c^2(\alpha_2-\alpha_6)]-c^2[c^1(\alpha_1-\alpha_5)-c^3(\alpha_3-\alpha_7)]=x_7$$

Second Output File at Step Three

Fig. 2. Transform output files for $N=8$, data in normal order.

$$[(x_0+x_4)+(x_2+x_6)]+[(x_1+x_5)+(x_3+x_7)]=\alpha_0$$
$$[(x_0-x_4)+c^2(x_2-x_6)]+c^1[(x_1-x_5)+c^2(x_3-x_7)]=\alpha_1$$
$$[(x_0+x_4)-(x_2+x_6)]+c^2[(x_1+x_5)-(x_3+x_7)]=\alpha_2$$
$$[(x_0-x_4)-c^2(x_2-x_6)]+c^3[(x_1-x_5)-c^2(x_3-x_7)]=\alpha_3$$

First Output File at Step Three

$$[(x_0+x_4)+(x_2+x_6)]-[(x_1+x_5)+(x_3+x_7)]=\alpha_4$$
$$[(x_0-x_4)+c^2(x_2-x_6)]-c^1[(x_1-x_5)+c^2(x_3-x_7)]=\alpha_5$$
$$[(x_0+x_4)-(x_2+x_6)]-c^2[(x_1+x_5)-(x_3+x_7)]=\alpha_6$$
$$[(x_0-x_4)-c^2(x_2-x_6)]-c^3[(x_1-x_5)-c^2(x_3-x_7)]=\alpha_7$$

Second Output File at Step Three

Fig. 4. Transform output files for $N=8$, data in reverse binary order.

$$y_j = x_{2j} + x_{2j+1}\exp\left(i\pi(j \div 2^{m-k})/2^{k-1}\right)$$
$$y_{j+n/2} = x_{2j} - x_{2j+1}\exp\left(i\pi(j \div 2^{m-k})/2^{k-1}\right)$$

for $j=0, 1, \cdots, (n/2)1$. The successive passes are identical except for using a different sequence of trigonometric function values. The sequence of steps for the $n=8$ case is shown in Figs. 3 and 4. We see that the final result is in normal order.

To reorder the complex data entries in reverse binary order, we require $m-1$ passes of the file. On the first pass, we read complex entries alternately from the first and second input files and write them on the first output file until it is filled with $n/2$ items, and then write them on the second output file. On the second pass, we copy two entries from the first input file, then two from the second, and so forth, until the first output file is filled, then continue with the second output file. On the kth pass, we copy 2^{k-1} entries from the first input file alternately with 2^{k-1} entries from the second. On completion of the $(m-1)$th pass, the complex data entries are in reverse binary order. The permutation matrix of this reordering is symmetric; thus, applying the procedure a second time will restore the initial ordering. We can also permute between normal and reverse binary order by following the above steps in inverse order.

If, instead of having the assumed initial ordering, the data are written with the real components in the first file and the imaginary components in the second file, we add one step at the beginning of the reordering for a total of m steps. On this step we alternately read single real and imaginary values from the two files and write the complex pairs on the first output file until it is filled, then on the second output file.

This method has been tested on a Burroughs B5500 computer, using both serial disk and magnetic tapes for auxiliary memory. Computing times were comparable with those for other fast Fourier transform programs on the same computer; however, input-output channel times were about twice the computing time. In the next section, we consider ways of reducing input-output time to achieve a better time balance.

Modification to Reduce Input-Output

In the previous section we separated the computation into a permutation phase and a transform phase. In doing this we have essentially followed the plan of computing the transform in place. Although the data are also permuted during each of the m transform steps, this permutation has a cycle length of m, giving the same order at the end as at the beginning.

The first modification in the algorithm is to start with the data in normal order, and alternate computing and permutation steps in such a way that the final result is in normal order. This change by itself gives no speed-up in the transform, but if we then read and write the data in records of 2^r items, where $1 \leq r < m-1$, we can conveniently combine computing and permuting during the final $r+2$ steps, thus eliminating $r+1$ permutation passes. The total number of passes of the file is then $2m-r-2$, except that if $r=m-1$ the transform is completed in m passes. The transform is done in two phases, the first with $m-r$ computing passes and $m-r-2$ permutation passes, and the second with r computing passes.

In this method we start with the first $n/2$ complex entries in normal order in one input file and the remaining $n/2$ entries in a second input file. During the first phase a record is read from each of the two input files and stored in an array. Complex pairs of entries are then selected in sequence from the two arrays, a two-by-two transform computed, and the results stored in place of the original entries. No permutation is done. When computing with one pair of records is completed, we write one record on each output file and then refill the input arrays as before. At the end of the computing pass the files are rewound and a permutation pass is done. In the permutation pass, groups of records are read from the first input file, and alternately written on the first and second output files. When the first input file is exhausted the second is similarly recopied. On the first permutation pass, groups of 2^{m-r-2} records are copied, i.e., one half of the first input file is written on the first output file and the other half on the second, and similarly with the second input file. On the second permutation pass, the group size is one quarter of the size of each input file. On the final permutation pass, groups of two records each are alternately written on the two output files until the first input file is exhausted, then from the second input file. Logically, the next permutation would alternate single records; but we avoid a separate pass for this permutation by altering the input procedure on the next succeeding computing pass, reading pairs of records from the first input file into the two data arrays, computing the transform, then writing the records as before on the two output files. After the first input file is exhausted, we go on with the second.

We then enter the second phase of the computation for the final r steps of the transform. In each of the r passes of the file the input file is read one record at a time, from the beginning of the first input file to the end of the second. Thus, we can if necessary reduce the number of files to three in this phase by copying the second output file onto the end of the first output file at the end of each pass. In this phase three data arrays of size equal to the record size are used in the computer, one to hold an input record and two more to hold records being built up for the two output files. The permutation in the kth step, where $m-r+1 \leq k \leq m$, is done by selecting pairs of entries 2^{m-k} apart from the input record to compute each two-by-two transform, then storing the pair of results in sequence, one each in the two arrays associated with the output files. Using the locations the results are stored in as the location index, the two-by-two transforms computed in both the first and second phase are

$$y_j = x_j + x_{j+n/2} \exp\left(i\pi(j \div 2^{m-k})/2^{k-1}\right)$$
$$y_{j+n/2} = x_j - x_{j+n/2} \exp\left(i\pi(j \div 2^{m-k})/2^{k-1}\right)$$

for $k=1, 2, \cdots, m$ and $j=0, 1, \cdots, (n/2)-1$.

A B5500 ALGOL procedure for computing the fast transform in this way is listed in the Appendix, along with a driver program for testing the procedure. The reader familiar with ALGOL-60 should be able to trace in detail the steps of the algorithm, keeping in mind the following differences in symbols:

B5500 ALGOL	ALGOL-60
DIV	÷
*	↑
AND	∧
OR	∨
NOT	¬

The procedure READ (F, K, A [*]) reads a record of K words from file F into array A. The procedure WRITE is similar. The program as listed is set up for magnetic tape files, but only the four file declarations need be changed to use disk files.

Computing and input–output channel times were measured for transforms of $2^9, 2^{10}, \cdots, 2^{16}$ complex data points, using a buffer and array size of 512 words, and disk files for auxiliary memory. These times, shown in Fig. 5, increase somewhat faster than linearly with the number of data points. For larger problems, extrapolation beyond $n=2^{16}$ at a rate of $n \log_2 n$ should give good predictions of times. When magnetic tape files are used instead of disk, times increase by a small amount. At $n=2^{12}$ the ratio of computing time is 1.05 and the ratio of input–output channel times is 1.10, but these ratios decline gradually with increasing problem size. At $n=2^{16}$ the ratios for computing and input–output channel time are 1.02 and 1.05.

Another modification to reduce input–output channel time is to add two output files, using two files in and four files out, and copy two of the output files onto the ends of the other two at the end of each transform step during the first phase. This step replaces the permutation step and takes only half as long as the permutation.

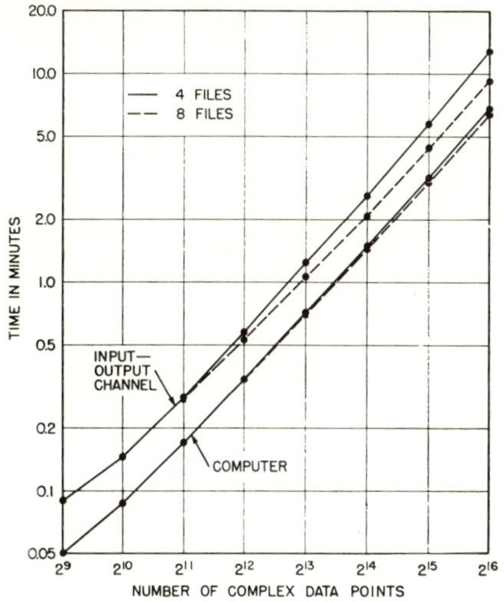

Fig. 5. Computer and input–output channel times for the fast Fourier transform.

As a further possibility, eight files can be used, four in and four out, and the permutation steps during the first phase avoided entirely. In this case the transform is computed in m passes of the files. The transform procedure in the Appendix runs with four, six, or eight files during the first phase, and closes the additional files, and runs with four during the second phase. Times for eight files are also shown in Fig. 5; times for six files lie about halfway between those for four and eight. For larger problems, the $n=16$ times listed in Table I can be extrapolated at an $n \log_2 n$ rate. Single variable transforms were computed during these timing tests. Multivariable transforms run slightly faster because of the fewer number of trigonometric function values and the smaller amount of indexing.

TABLE I

FAST FOURIER TRANSFORM TIMES FOR $N=2^{16}$

	Four Files	Six Files (time in minutes)	Eight Files
Computing	6.75	6.54	6.37
I–O Channel	12.71	10.60	9.25

Whatever the number of files used in the computation, the data are stored originally in two files, zero and two or one and three, and the output appears in one of these two pairs of files. In each pass the input file numbers are of one parity, odd or even, and the output files are of the other parity; the roles switch between passes. In the program listed in the Appendix, the switch file declaration is used as shown for four files, and is changed to

```
SWITCH FILE FT = FFT0, FFT1, FFT2, FFT3,
                 FFT4, FFT4, FFT5, FFT5
```

for six files, and to

```
SWITCH FILE FT = FFT0, FFT1, FFT2, FFT3,
                 FFT4, FFT5, FFT6, FFT7
```

for eight files. The value of the variable NF in the procedure call is also changed to agree with the number of files to be used.

The number of variables in the transform is given by the value of the parameter NV, and the log base 2 of the jth variable is given by the array entry DIM_j for $j=1, 2, \cdots, NV$. If, for example, NV = 3, and the three dimensions are $n_1=2^{m_1}$, $n_2=2^{m_2}$, and $n_3=2^{m_3}$, where $n=2^{m_1+m_2+m_3}$ is the total number of complex data points, we set $DIM_1=m_1$, $DIM_2=m_2$, and $DIM_3=m_3$, and store the data in locations

$$k_1 + k_2 n_1 + k_3 n_1 n_2 \qquad \text{for } 0 \le k_j \le n_j - 1,$$

viewing the input files as a single sequence from the beginning of the first to the end of the second. The transformed result is indexed in the same way. Since the number of variables is not limited, this procedure can be used to calculate interactions in a 2^m factorial experiment, as described by Good [3]; in this case $NV=m$ and $DIM_j=1$ for $j=1, 2, \cdots, m$.

To compute the single-variate transform of $n=2^m$ complex data values x_j for $j=0, 1, \cdots, n-1$, we set $NV=1$ and $DIM_1=m$. If negexp is false, then the transform

$$\alpha_k = \sum_{j=0}^{n-1} x_j \exp(i2\pi jk/n) \quad \text{for } k = 0, 1, \cdots, n-1$$

is computed; and if negexp is true,

$$\alpha_k = \sum_{j=0}^{n-1} x_j \exp(-i2\pi jk/n) \quad \text{for } k = 0, 1, \cdots, n-1$$

is computed. One followed by the other, in either order, gives n times the original data, except for roundoff errors.

With the radix 2 transform, no advantage is gained by using more than eight files. If, on the other hand, we consider using a radix 4 transform with eight files, four in and four out, a transform of $n=2^{2m}$ data points still takes $2m+1$ passes, including an initial pass to divide the data into four files and a final pass to consolidate the file, since a permutation step is needed between each transform step. To gain full advantage from a radix 4 transform we need 32 files, 16 in and 16 out; with this number of files an $n=4^m$ transform is done in $m+2$ passes, including an initial pass to divide the file and a final pass to consolidate. This plan looks practical for the B5500 computer, but has not been tried. In general, to compute a radix p transform without reordering between transform steps we need $2p^2$ files, p^2 in and p^2 out; however, only p of the input files need be open at any one time.

Accuracy

The arithmetic operations used in the fast Fourier transform procedure given here are unaltered by changes in the number of files. In testing for accuracy, we generated as data random normal deviates with zero mean and unit standard deviation. The transform results were first compared with those from another well-tested fast transform program to make sure the correct transform was being computed. In subsequent tests, the procedure was used to transform, then inverse transform, the data set, giving in theory n times the original data as a result. The squared differences of the original and $1/n$ times the result values were summed and divided by the sum of squares of the original values. The square root of this ratio, i.e., the ratio of rms error to rms data, was used as a measure of accuracy. Typical results are listed in Table II, with three trials for each transform size. In the floating point number representation used in the B5500 computer, an error in the final bit of a number x gives a difference of 14.6×10^{-12} for $1<|x|<8$, as is true for about one third of our data, and a difference of 1.8×10^{-12} for $\frac{1}{8}<|x|<1$, as is true of most of the remainder of our data, except for about 10 percent with $|x|<\frac{1}{8}$. By comparison, the observed errors appear satisfactorily small. Single variable transforms were used for these tests; however, limited tests with three variables show even lower errors, as should be expected from the fact that fewer iterative extrapolations of trigonometric function values are needed.

TABLE II
Accuracy Tests of Single Variable Transform

Number of Data Points	First Trial	Second Trial	Third Trial
	(rms error/rms data in units of 10^{-12}		
2^9	19.9	19.9	18.9
2^{10}	21.8	21.5	22.0
2^{11}	26.2	25.2	25.1
2^{12}	40.4	40.5	40.8
2^{13}	65.5	66.5	65.8
2^{14}	106.2	105.3	106.3
2^{15}	109.9	109.5	109.7
2^{16}	153.8	153.8	154.2

In the procedure listed in the Appendix, cosine and sine function values are computed during the first phase using the second difference relations

$$C_{k+1} = R \times \cos(k\theta) + C_k$$
$$\cos((k+1)\theta) = \cos(k\theta) + C_{k+1}, \quad \text{and}$$
$$S_{k+1} = R \times \sin(k\theta) + S_k$$
$$\sin((k+1)\theta) = \sin(k\theta) + S_{k+1},$$

where the constant multiplier is

$$R = -4 \times \sin^2(\theta/2)$$

and the initial values are

$$C_0 = 2 \times \sin^2(\theta/2),$$
$$S_0 = \sin(\theta),$$
$$\cos(0) = 1, \quad \text{and}$$
$$\sin(0) = 0.$$

This method gives very good accuracy. However, transform errors can be reduced by about one half (one binary place) by use of the following slightly slower method:

$$TEMP = \cos(k\theta)$$
$$\cos((k+1)\theta) = [CS \times \cos(k\theta) - SD \times \sin(k\theta)] + \cos(k\theta)$$
$$\sin((k+1)\theta) = [CS \times \sin(k\theta) + SD \times TEMP] + \sin(k\theta),$$

where the constant multipliers are

$$CS = -2 \times \sin^2(\theta/2) \quad \text{and}$$
$$SD = \sin(\theta).$$

Trigonometric values are computed in this way during the second phase of the transform procedure listed in the Appendix. This method gives transform results that are nearly as accurate as when the computer's library trigonometric function procedure is used to compute all values.

Conclusions

We have shown that the fast Fourier transform of $n=2^m$ complex data points can be computed using auxiliary memory for data storage with only a relatively small amount of high-speed memory required for temporary storage in addition to program. Using four auxiliary files, at most $2m-1$ passes of the data are required to compute the transform and restore the result to normal order. The logic of the transform calculation in this method may also find application in building special hardware for parallel computation of the fast Fourier transform.

Although the fast transform can be computed with as few as three auxiliary files, added efficiency results from increasing the number of files from four to six or eight. With eight auxiliary memory files the transform is computed in m passes of the data, with the final result in normal order. For the radix 2 transform there is no advantage gained by increasing the number of files beyond eight.

Appendix

A B5500 ALGOL Procedure for the Fast Fourier Transform with Auxiliary Memory

```
BEGIN COMMENT TEST AUXILIARY MEMORY TRANSFORM;
DEFINE KBF= 512#;
FILE FFT0 (2,512); FILE FFT1 (2,512);
FILE FFT2 (2,512); FILE FFT3 (2,512);
SWITCH FILE FT := FFT0,FFT1,FFT2,FFT3;
INTEGER NA,NB,NC,ND;
SAVE ARRAY AFA,AFB,AFC[0:KBF-1];

PROCEDURE FASTFOURIERAUX(NV,DIM,NEGEXP,NF);
    VALUE NV,NEGEXP,NF; INTEGER NV,NF;
    BOOLEAN NEGEXP; ARRAY DIM[0];
COMMENT NV IS THE NUMBER OF VARIABLES.
    DIM[J] IS THE LOG BASE 2 OF THE DIMENSION OF
        VARIABLE J.  THE INPUT FILE LOCATION OF THE
        COMPLEX DATA ENTRY WITH INDEX (K1,K2,...KNV) IS
        K1+K2×2*DIM[1]+...+KNV×2*(DIM[1]+...+DIM[NV-1]),
        CONSIDERING THE FILE AS A SINGLE SEQUENCE OF
        ENTRIES.
    NEGEXP=TRUE GIVES A NEGATIVE SIGN IN THE
        EXPONENTIAL AND FALSE GIVES A POSITIVE SIGN.
    NF IS THE NUMBER OF FILES, 4,6, OR 8, TO BE USED;
BEGIN INTEGER J,JJ,JK,K,KK,N,SPAN,KC,KD,KS,KR,NR,NS;
    REAL AB,BB,RE,IM,CN,SN,RAD,CD,SD,R,RS;
    BOOLEAN FIRST;
    LABEL L,L2,L3,L4,L5,L6,L7;
    PROCEDURE REWINDFILES;
    BEGIN FOR NB := 0,1,2,3 DO REWIND(FT[NB]);
        NA := NC; NC := 1-NC; ND := NC+2;
        NB := NA+2;
    END REWINDFILES;
    N := DIM[1]; J := JJ := K := KC := KR := 0;
    FOR JK := 2 STEP 1 UNTIL NV DO N := DIM[JK]+N;
    SPAN := N := 2*N; RAD := 6.28318530718/(4×N);
    IF NEGEXP THEN RAD := -RAD; JK := DIM[NV];
    NC := 1-NA; ND := NC+4; NB := NA+2;
    NR := JJ := N DIV (KBF+KBF); NS := N DIV KBF;
    IF JJ<2 OR NF<6 THEN
    BEGIN JJ := N; NF := 4; ND := NC+2 END ELSE
    IF JJ=2 OR NF<8 THEN NF := 6 ELSE NF := 8;
    IF NR=0 THEN FIRST := TRUE ELSE FIRST := FALSE;
    KD := KBF DIV 2; KS := N; GO TO L2;
    COMMENT THE FOLLOWING SECTION REORDERS RECORDS
        BETWEEN STEPS OF THE FIRST PHASE OF THE
        TRANSFORM WHEN 4 FILES ARE USED;
L:  READ(FT[NA],KBF,AFA[*]);
    WRITE(FT[NC],KBF,AFA[*]); KR := KR+1;
    K := K+KBF; IF K<KS THEN GO TO L; K := 0;
    IF NC<2 THEN NC := NC+2 ELSE NC := NC-2;
    IF KR<NS THEN GO TO L; KR := 0;
    IF NA<2 THEN
    BEGIN NA := NA+2; GO TO L END;
    REWINDFILES;
    COMMENT FIRST PHASE OF TRANSFORM STARTS HERE;
L2: CN := 1; SN := 0; SD := RS; RS := SIN(RAD×SPAN);
    R := -(RS+RS)*2; CD := -0.5×R;
    SPAN := SPAN DIV 2; KK := 0;
L3: READ(FT[NA],KBF,AFA[*]);
    READ(FT[NB],KBF,AFB[*]);
    FOR J := 0, J+1 WHILE J<KBF DO
    BEGIN AB := AFB[J]; BB := AFB[J+1];
        RE := CN×AB-SN×BB; IM := SN×AB+CN×BB;
        AFB[J] := AFA[J]-RE; AFA[J] := AFA[J]+RE;
        J := J+1;
        AFB[J] := AFA[J]-IM; AFA[J] := AFA[J]+IM;
    END;
    KC := KC+KD; IF KC=SPAN THEN
    BEGIN KC := 0;
        COMMENT COMPUTE THE NEXT COS AND SIN VALUES;
        CD := R×CN+CD; CN := CN+CD;
        SD := R×SN+SD; SN := SN+SD;
    END;
    WRITE(FT[NC],KBF,AFA[*]);
    WRITE(FT[ND],KBF,AFB[*]);
    KK := KK+1; IF KK=JJ THEN
    BEGIN KK := 0; IF NC<2 THEN NC := NC+2 ELSE
        NC := NC-2; ND := NC+4;
    END;
    KR := KR+1; IF KR<NR THEN GO TO L3; KR := 0;
    FIRST := NOT FIRST; IF FIRST THEN
    BEGIN IF KS=N THEN GO TO L3 ELSE
        IF KS=KBF THEN NB := NA := NA+2 ELSE
        IF NF=8 THEN
        BEGIN NA := NA+4; NB := NB+4 END;
        GO TO L3;
    END;
    IF NF=8 AND KS≠N THEN
    BEGIN IF JJ=2 OR JJ=N THEN
        BEGIN CLOSE(FT[NA]); CLOSE(FT[NB]) END ELSE
        BEGIN REWIND(FT[NA]); REWIND(FT[NB]) END;
    END;
    IF JJ<N THEN
    BEGIN REWIND(FT[NC+4]); REWIND(FT[NC+6]);
        IF JJ>2 THEN JJ := JJ DIV 2 ELSE
        BEGIN NF := 8; JJ := N END;
    END;
    COMMENT THE FOLLOWING SECTION REORDERS RECORDS
        BETWEEN STEPS OF THE FIRST PHASE OF THE
        TRANSFORM WHEN 6 FILES ARE USED;
    IF NF=6 THEN FOR NB := NC, NC+2 DO
    BEGIN ND := NB+4;
L4:     READ(FT[ND],KBF,AFA[*]);
        WRITE(FT[NB],KBF,AFA[*]); KR := KR+1;
        IF KR<NR THEN GO TO L4; KR := 0;
        REWIND(FT[ND]);
    END;
    REWINDFILES; IF JJ<N THEN ND := NC+4;
    JK := JK-1; IF JK=0 THEN
    BEGIN NV := NV-1; JK := DIM[NV]; SPAN := N END;
    KS := KS DIV 2; IF KS>KBF THEN
    BEGIN IF NF≥6 THEN GO TO L2 ELSE GO TO L END
    ELSE IF KS=KBF THEN
    BEGIN NB := NA; NF := 4; GO TO L2 END; J := 0;
    COMMENT SECOND PHASE OF TRANSFORM STARTS HERE;
L5: CN := 1; SN := 0; SD := RS; RS := SIN(RAD×SPAN);
    CD := -2×RS*2;
    SPAN := SPAN DIV 2; KS := KD; KD := KD DIV 2;
L6: READ(FT[NA],KBF,AFC[*]); KK := 0; K := KS;
L7: AB := AFC[KK+KS]; BB := AFC[KK+KS+1];
    RE := CN×AB-SN×BB; IM := SN×AB+CN×BB;
    AB := AFC[KK]; BB := AFC[KK+1]; KK := KK+2;
    AFA[J] := AB+RE; AFB[J] := AB-RE; J := J+1;
    AFA[J] := BB+IM; AFB[J] := BB-IM; J := J+1;
    IF KK<K THEN GO TO L7;
    KC := KC+KD; IF KC=SPAN THEN
    BEGIN KC := 0; AB := CN;
        COMMENT COMPUTE THE NEXT COS AND SIN VALUES;
        CN := (CD×CN-SD×SN)+CN;
        SN := (CD×SN+SD×AB)+SN;
    END;
    KK := KK+KS; IF KK<KBF THEN
    BEGIN K := KK+KS; GO TO L7 END;
    IF J=KBF THEN
    BEGIN WRITE(FT[NC],KBF,AFA[*]);
        WRITE(FT[ND],KBF,AFB[*]); J := 0;
    END;
    KR := KR+1; IF KR<NS THEN GO TO L6; KR := 0;
    IF NA<2 THEN
    BEGIN NA := NA+2; GO TO L6 END;
    REWINDFILES;
    JK := JK-1; IF JK>0 THEN GO TO L5; SPAN := N;
    NV := NV-1; JK := DIM[NV]; IF NV>0 THEN GO TO L5;
END FASTFOURIERAUX;

COMMENT DRIVER PROGRAM STARTS HERE;
INTEGER J,JJ,K,M,N,NV,LIM,RDM,RDN;
INTEGER ARRAY D[0:6];
REAL SS1,SS2,SS3,SS4,AA,BB;
FILE LP 4(1,15);
PROCEDURE NORMAL(RDM,RX,RY);
    INTEGER RDM; REAL RX,RY;
COMMENT COMPUTES A PAIR RX,RY OF RANDOM NORMAL
    DEVIATES WITH MEAN = 0, S.D. = 1.
    CHOOSE INITIAL VALUE OF RDM ODD AND <2*27;
BEGIN REAL R; LABEL LR;
LR: RDM := 3589×RDM; RDM := RDM-(RDM DIV 134217728)×134217728;
    RX := (RDM-67108864)/67108864;
    RDM := 3589×RDM; RDM := RDM-(RDM DIV 134217728)×134217728;
    RY := (RDM-67108864)/67108864;
    R := RX*2+RY*2; IF R≥1.0 THEN GO TO LR;
    R := SQRT(-2×LN(R)/R); RX := RX×R; RY := RY×R;
END NORMAL;
RDN := 41365;
FOR M := 9 STEP 1 UNTIL 12 DO BEGIN
WRITE(LP[DBL],<"M =",I2>,M);
N := 2*M; LIM := N DIV KBF; JJ := KBF-1;

NV := 1; D[1] := M;
NA := 0; NB := 2; NC := 1; ND := 3;
RDM := RDN; SS3 := SS4 := 0;
WRITE(LP[DBL],<"RDM =",I10>,RDM);
COMMENT WRITE DATA FILES;
FOR J := 1 STEP 1 UNTIL LIM DO
BEGIN FOR K := 0 STEP 1 UNTIL JJ DO
    BEGIN NORMAL(RDM,AA,BB); AFA[K] := AA;
        AFB[K] := BB; SS3 := AA*2+SS3; SS4 := BB*2+SS4
    END;
    WRITE(FT[NA],KBF,AFA[*]);
    WRITE(FT[NB],KBF,AFB[*]);
END;
REWIND(FT[NA]); REWIND(FT[NB]);
FASTFOURIERAUX(NV,D,FALSE,4);
FASTFOURIERAUX(NV,D,TRUE,4);
RDM := RDN; SS1 := SS2 := 0;
FOR J := 1 STEP 1 UNTIL LIM DO
BEGIN READ(FT[NA],KBF,AFA[*]);
    READ(FT[NB],KBF,AFB[*]);
    FOR K := 0 STEP 1 UNTIL JJ DO
    BEGIN NORMAL(RDM,AA,BB);
        SS1 := (AFA[K]/N-AA)*2+SS1;
        SS2 := (AFB[K]/N-BB)*2+SS2;
    END;
END;
REWIND(FT[NA]); REWIND(FT[NB]);
WRITE(LP[DBL],<3E16.6>,SQRT(SS1/SS3),SQRT(SS2/SS4),
    SQRT((SS1+SS2)/(SS3+SS4)));
RDN := RDM; WRITE(LP[PAGE]);
END;
END.
```

References

[1] G. C. Danielson and C. Lanczos, "Some improvements in practical Fourier analysis and their application to X-ray scattering from liquids," *J. Franklin Institute*, vol. 233, pp. 365–380 and pp. 435–452, April 1942.

[2] P. Rudnick, "Note on the calculation of Fourier series," *Math. of Comput.*, vol. 20, pp. 429–430, July 1966.

[3] I. J. Good, "The interaction algorithm and practical Fourier series," *J. Royal Statist. Soc.*, ser. B, vol. 20, pp. 361–372, 1958; Addendum, vol. 22, pp. 372–375, 1960.

[4] J. W. Cooley and J. W. Tukey, "An algorithm for the machine calculation of complex Fourier series," *Math. of Comput.*, vol. 19, pp. 297–301, April 1965.

[5] T. G. Stockham, "High-speed convolution and correlation," *1966 Spring Joint Computer Conf., AFIPS Proc.*, vol. 28. Washington, D. C.: Spartan, 1966, pp. 229–233.

[6] W. M. Gentleman and G. Sande, "Fast Fourier transforms—for fun and profit," *1966 Fall Joint Computer Conf., AFIPS Proc.*, vol. 29. Washington, D. C.: Spartan, 1966, pp. 563–578.

[7] R. C. Singleton and T. C. Poulter, "Spectral analysis of the call of the male killer whale," this issue, p. 104–113.

[8] L. E. Alsop and A. A. Nowroozi, "Fast Fourier analysis," *J. of Geophys. Res.*, vol. 71, p. 5482–5483, November 15, 1966.

[9] R. C. Singleton, "An ALGOL procedure for the fast Fourier transform with arbitrary factors," SRI Project 181531-132, Stanford Research Institute, Menlo Park, Calif., December 1966 (Defense Doc. Ctr. AD-643 997).

[10] J. W. Cooley, "Harmonic Analysis Complex Fourier Series," SHARE Program Library SDA 3425, February 7, 1966.

[11] R. C. Singleton, "ALGOL procedures for the fast Fourier transform," SRI Project 181531-132, Stanford Research Institute, Menlo Park, Calif., November 1966 (Defense Doc. Ctr. AD-643 996).

[12] R. C. Singleton, "An ALGOL convolution procedure based on the fast Fourier transform," SRI Project 181531-132, Stanford Research Institute, Menlo Park, Calif., January 1967 (Defense Doc. Ctr. AD-646 628).

Reprinted from *IEEE Trans. Audio Electroacoustics,* **AU-21**(1), 5–16 (1973)

Parallelism in Fast Fourier Transform Hardware

BEN GOLD and THEODORE BIALLY

***Abstract*—The fast Fourier transform algorithm is derived by means of successive fracturing of one-dimensional data strings into two-dimensional arrays. Using this formulation, a diagrammatic representation of mixed radix and highest radix FFT algorithms is derived. Using this representation, two broad classes of FFT hardware are explored, from the point of view of speed, parallelism, radix number, and type of memory.**

I. Introduction

Over the past few years, a large body of knowledge has been generated on the subject of fast Fourier transform algorithms [6]–[9], [12], [14]. While most of this work has been concerned with the radix-2 algorithm, substantial information has also been made available on higher radix and mixed radix algorithms [2], [16], [19]. All of these FFT algorithms have been used in both programming and hardware applications [1], [3]–[5], [10], [11], [17], [18].

FFT programs naturally have to be conditioned by the architecture of the particular computer. On the other hand, the designer of an FFT hardware device has the freedom to match the hardware design and the FFT algorithm. Such freedom is of particular advantage, for example, in designing high-speed systems, such as matched filters for wide bandwidth radar systems. The material in this paper is directed especially at such problems, where the designer wants to configure his digital structure so as to squeeze all the speed possible from his system.

The following example illustrates the difference between software and hardware methodology in FFT design. Consider first a general-purpose computer with a register length such that two registers are required to hold a single complex datum. Furthermore, the computer has the capability of performing a single arithmetic function at a time, such as an addition or multiplication. A radix-2 algorithm would require eight memory cycles, four multiply cycles, and six add cycles for each of $N/2\ \log_2 N$ elementary computations. Since all of these are performed sequentially, a very restrictive upper limit is established on the speed capability of such a computer performing the FFT.

Manuscript received May 4, 1972. This work was sponsored by the Department of the Air Force.
The authors are with M.I.T. Lincoln Laboratory, Lexington, Mass. 02173.

Now, consider a piece of special-purpose FFT hardware. It is possible to design such hardware so that memory and arithmetic cycles overlap. Furthermore, both memory and arithmetic can be paralleled in such a way as to cause each elementary computation to be performed in one cycle. Thus, it can be seen that hardware design can be specifically tailored to the FFT algorithm to give substantial speed increases. Obviously, a general-purpose computer can also be designed where special emphasis is given to FFT performance, and such computers have been built [20], but in these structures it is difficult to attain the optimum speed while maintaining the desired generality. In this paper, we concentrate on the representation of FFT structures with both memory and arithmetic parallelism plus control, which leads to optimum speed.

II. Representation of FFT Structures

The FFT computes the discrete Fourier transform (DFT) of N input variables $x(0), x(1), \cdots, x(N-1)$ where, by definition,

$$X(k) = \text{DFT}(x(n)) = \sum_{n=0}^{N-1} x(n)W^{nk},$$

$$W = \exp\left(-j\frac{2\pi}{N}\right), \qquad n = 0, 1, \cdots, N-1. \quad (1)$$

The FFT gains speed by factoring the composite number N into the product of integers:

$$N = r_1 r_2 \cdots r_m. \quad (2)$$

Equation (2) corresponds to fracturing the one-dimensional signal $x(n)$ into an m-dimensional signal. For example, if $N=64$ and $M=3$, we might have $r_1=r_2=r_3=4$, and the 64-point transform is translated into a three-dimensional $4\times4\times4$ transform. If N is a prime number so that factorization of N is not possible, the original signal can usually be augmented with zeros, and a new composite value of N is attained.

A convenient way of performing the FFT algorithm is as follows. Begin by representing N as the product of two numbers, say, $N=N_1N_2$. If N_1 and N_2 are both composite, they can further be broken down: $N_1=N_{11}N_{12}$, $N_2=N_{21}N_{22}$. In this way, each iteration consists of going through the mechanics of a two-dimensional transform to achieve a one-dimensional transform. The procedure for performing this latter operation is the following.

1) Arrange the one-dimensional data into a two-dimensional array, as shown for example, on the left of Fig. 1.

2) Do DFT's on each row individually.

3) Multiply each term of the resultant matrix by W_1^{ij} where i is the row index and j is the column index and $W_1=\exp(-j(2\pi/N))$.

4) Transform the resultant matrix column by column to obtain the result. Thus, if we demonstrate that a

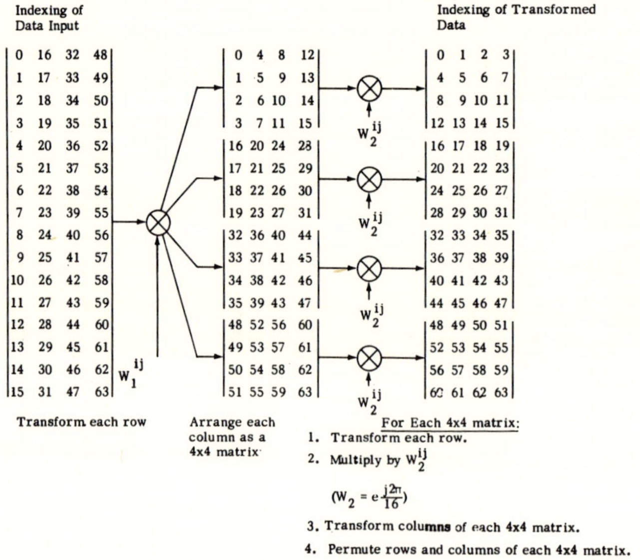

Fig. 1. Illustration of a 64-point DFT by successive fracturing of one-dimensional data into two dimensions.

given one-dimensional FFT can be reformulated in two dimensions, we will have effectively shown how to do the FFT in general. Beginning with (1), if N is composite, it can be written as $N=LM$. As an example, let $L=3$, $M=5$ so that, given $x(0)$, $x(1) \cdots x(14)$, we need to find $X(0)$, $X(1) \cdots X(14)$. Our plan is to first arrange the $x(n)$'s and $X(k)$'s into rectangular arrays as follows:

$$\begin{array}{cccccc} m \rightarrow & & & m = 0, 1 \cdots M-1 & & \\ l & x(0) & x(1) & x(2) & x(3) & x(4) \\ \downarrow & x(5) & x(6) & x(7) & x(8) & x(9) \\ l = 0, 1 \cdots & x(10) & x(11) & x(12) & x(13) & x(14) \\ L-1 & & & & & \end{array}$$

$$\begin{array}{cccccc} r \rightarrow & & & r = 0, 1 \cdots M-1 & & \\ s & X(0) & X(3) & X(6) & X(9) & X(12) \\ \downarrow & X(1) & X(4) & X(7) & X(10) & X(13) \\ s = 0, 1 \cdots & X(2) & X(5) & X(8) & X(11) & X(14). \\ L-1 & & & & & \end{array}$$

Notice that the orderings of the sequence and its transform are orthogonal. Given this, the variables n and k can be represented as

$$n = Ml + m, \quad k = Lr + s. \tag{3}$$

Substituting (3) into (1), we obtain

$$X(Lr + s) = \sum_{m=0}^{M-1} \sum_{l=0}^{L-1} x(Ml + m) W^{(Ml+m)(Lr+s)}. \tag{4}$$

Expanding $W^{(Ml+m)(Lr+s)}$, observing that $W^{MLrl} = W^{Nlr} = 1$, and properly associating the indices with the summation terms, we can rearrange (4):

$$X(Lr + s) = \sum_{m=0}^{M-1} W^{Lmr} W^{ms} \sum_{l=0}^{L-1} x(Ml + m) W^{Msl}. \tag{5}$$

Now, let us define

$$q(s, m) = \sum_{l=0}^{L-1} x(Ml + m)(W^M)^{sl}. \tag{6}$$

For any fixed m, $q(s, m)$ can be interpreted as a column transform on the $x(n)$ array. For example, if $m=1$, $M=5$, $L=3$, then

$$q(s, 1) = x(1)(W^5)^0 + x(6)(W^5)^1 + x(11)(W^5)^2, \tag{7}$$

which we observe to be the DFT of column 1 of the $x(n)$ array; W^5 is the correct kernel for this three-point transform because $W = \exp(j2\pi/15)$, so that $W^5 = \exp(j2\pi/3)$. Substituting (6) into (5) yields

$$X(Lr + s) = \sum_{m=0}^{M-1} q(m, s) W^{ms} (W^L)^{mr}. \tag{8}$$

Equation (8) is seen to be a row transform of the array of products $q(m, s)W^{ms}$, with W^L representing the kernel of this transform. Using the indexing arrangement shown next to the arrays, we can thus compute (4) by successive applications of (1) and (8), leading precisely to the procedure we initially outlined. The W^{ms} of (8) can be identified as the twiddle factors of the algorithm.

Accordingly, the complete FFT algorithm will consist of successively breaking down one-dimensional sig-

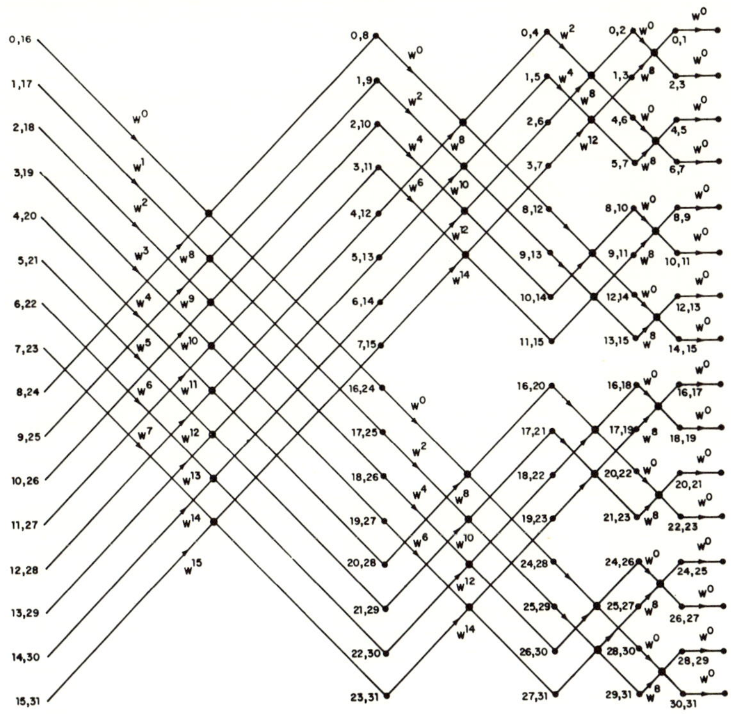

Fig. 2. Flow diagram for radix-2 32-point FFT with two complex words per register.

nals into their two-dimensional counterparts. An illustration of this technique is given in Fig. 1. We begin with $N=64$ and first represent it as a 4×16 array, as shown in Fig. 1. After performing steps 1), 2), and 3), we are ready for step 4). However, rather than explicitly performing four 16-point column transforms, we choose to represent each column by a 4×4 array. Now steps 1)–4) are performed on each individual 4×4 array. Notice that the second set of twiddle factors are based on 16-point transforms, so that $W_2=\exp(j2\pi)/16$. This particular illustration can be interpreted as a fracturing of $N=4\times4\times4$, since there are three levels of computation, each level corresponding to the computation of a set of 16 4-point DFT's plus twiddle factors. Such a computation is called a radix-4 computation. More generally, if N can be expressed as $N=r^m$, and if the algorithm is carried out via a succession of r-point transforms, the resultant FFT is called a radix-r FFT.

In a radix-r FFT, we denote an "elementary computation" (EC) by an r-point DFT followed by multiplication of each of the r results by the appropriate twiddle factor W^{ij}. It is easy to show that the number of EC's required is

$$C_r = \frac{N}{r}\log_r N. \qquad (9)$$

Equation (9) shows that C_r decreases as r increases; in fact, when $r=N$, $C_r=1$. But, of course, the complexity of an EC increases with increasing r. For example, for $r=2$, EC consists of a single complex multiplication and a pair of complex additions, while for $r=4$, EC consists of three complex multiplications and some extra additions. Thus, the "best" value of r is not easily arrived at, and depends greatly on the particular problem.

III. Nomenclature for FFT's Using Parallelism

Most FFT algorithms exist as programs on general-purpose computers where the memory word length usually contains one data sample. In discussing high-speed FFT hardware, it is important to include the possibility of parallelism both in the EC's and in the memory cycles. We therefore introduce a flow diagram nomenclature slightly different from the Rader diagrams of the radix-2 FFT usually found in the literature [12]. In this nomenclature, the memory parallelism is contained in the data assignments as seen in Fig. 2. The arrows and associated W's represent the parameters of the EC's, and the nodes (heavy dots) show how the outputs must be distributed in order to be read for the next computation level. Notice that there are two pairs of data points entering and leaving each node in the 32-point transform of Fig. 2. This derives from the fact

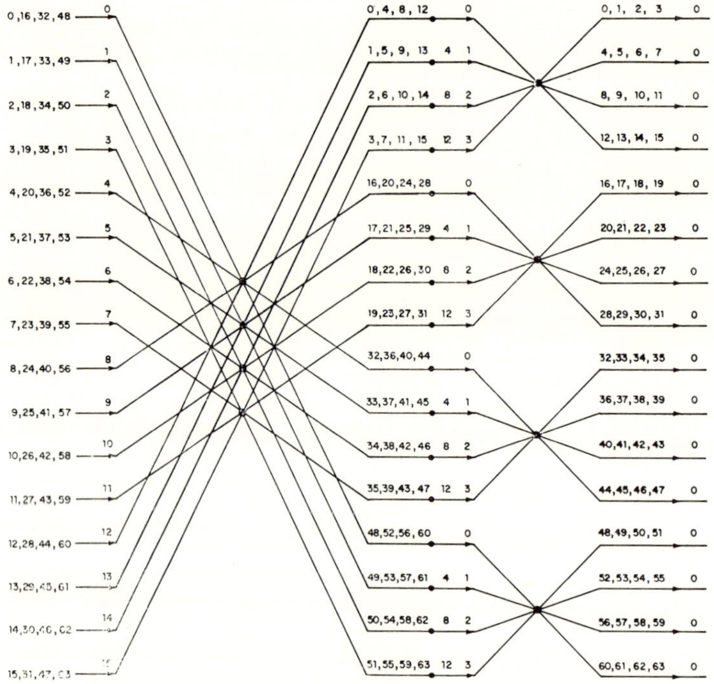

Fig. 3. 64-point FFT, radix-4 with quadrapole word memory.

that each line in the diagram represents a complete EC of the transform, and it implies that the computational model must effectively reorder (permute the rows and columns) of a 2×2 matrix after each EC. Here we have the in-place idea extended to four data points. This is the natural extension of the in-place algorithm in which two data points are computed, and the two output results are entered back into the original memory location.

The nomenclature of Fig. 2 can be extended to the radix-4 case of Fig. 1. The result is seen in Fig. 3. Now we have a quadruplet of inputs to an EC (represented by the arrow). The number near the arrow represents the four twiddle factors of each EC as follows. The number k means that the twiddle factors are W^0, W^k, W^{2k}, W^{3k}. Each node accepts a 4×4 matrix of EC outputs and distributes them via a row–column permutation.

IV. EC's for $r=2$ and $r=4$

For the "decimation in time" algorithm [12], the EC is

$$A' = A + CW^i$$
$$C' = A - CW^i \qquad (10)$$

where A and C are the complex data points, $W=\exp(j2\pi/N)$, and i as an index which is a function of how many steps have been completed in the algorithm.

For the "decimation in frequency" algorithm,

$$A' = A + C$$
$$C' = (A - C)W^i. \qquad (11)$$

Thus, both EC's always involve two input data points A and C, a coefficient W^i, and two output data points A' and C'.

A radix-4 EC corresponds to a four-point DFT followed by multiplication of each output by a complex "twiddle" factor. Thus there are four inputs, A, B, C, and D, and four outputs, A', B', C', and D', given by

$$A' = (A + B + C + D)W^0$$
$$B' = (A + jB + C - jD)W^k$$
$$C' = (A - B + C - D)W^{2k}$$
$$D' = (A - jB - C + jD)W^{3k} \qquad (12)$$

In (12), $j=\sqrt{-1}$ and k is an algorithm-position-dependent index; note, however, that the first "twiddle" is always $W^0=1$.

V. Parallel Memory—Pipeline Arithmetic Configuration of Radix-r FFT

In Section III we showed that a radix-r FFT can be implemented in a parallel way by performing a total of r

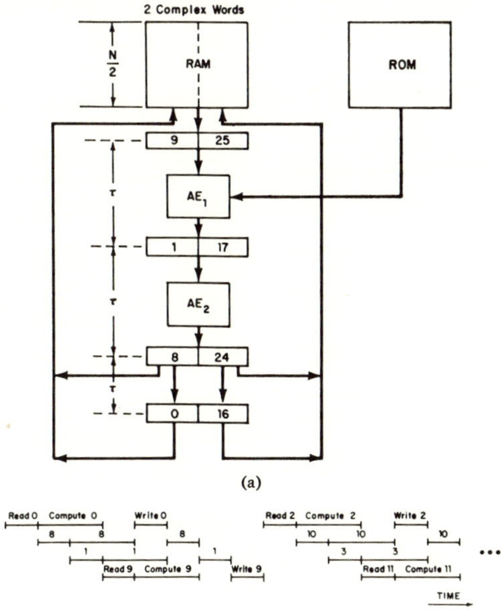

Fig. 4. Structure and timing diagram for radix-2 FFT with parallel memory. (a) Structure. (b) Timing.

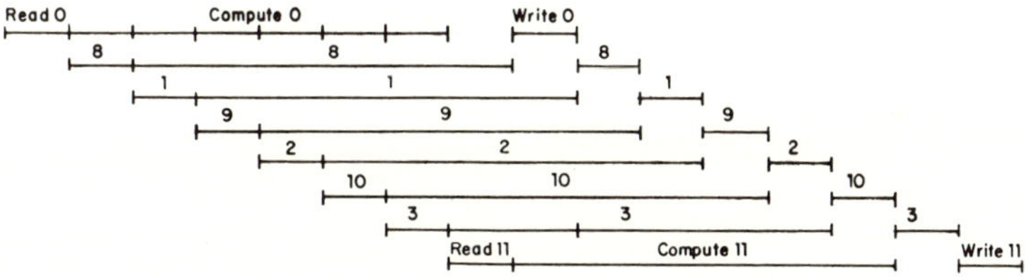

Fig. 5. Alternate timing diagram for structure of Fig. 4.

EC's, and then permuting the rows and columns of the result before doing the next set of r EC's. In this section, a specific computing structure that can perform such FFT's will be examined; this structure has the following features.

1) Memory cycles and computation cycles are completely overlapped; we can think of such a structure as "matched" in the sense that the system is designed to be both memory limited and arithmetically limited simultaneously.

2) The data are contained in a random access memory with a register length equal to r complex data words.

A radix-2 example of such a structure and its associated (partial) timing diagram is shown in Fig. 4. The RAM consists of $N/2$ registers, each of the length of two complex data words. The numbers in the timing diagram correspond to the 32-point transform depicted in Fig. 2, which also shows how the addressing is to be sequenced. In order to realize a matched system, the second read cycle must, without pause, follow the first read cycle, etc Thus, subsequent read cycles must be overlapped with the EC. Furthermore, since writing must follow the completion of at least two EC's, we can see from Fig. 4's timing diagram that four successive read cycles are required before writing can occur. This means that the AE *must* contain pipelining, that is, a second EC begins before the first is finished. Fig. 4 shows the pipelining needed when it is assumed that an EC takes two memory cycles; the state of the machine shown corresponds in time to the arrow in the timing diagram.

Should the designer be unable to construct an AE that performs its function in two memory cycles, a longer pipeline, of six memory cycles, will also result in a matched system; this timing is depicted in Fig. 5. The resulting structure will be a lengthened version of Fig. 4, with a total of six pipelining buffers entered between six partial arithmetic elements; if the computation can pro-

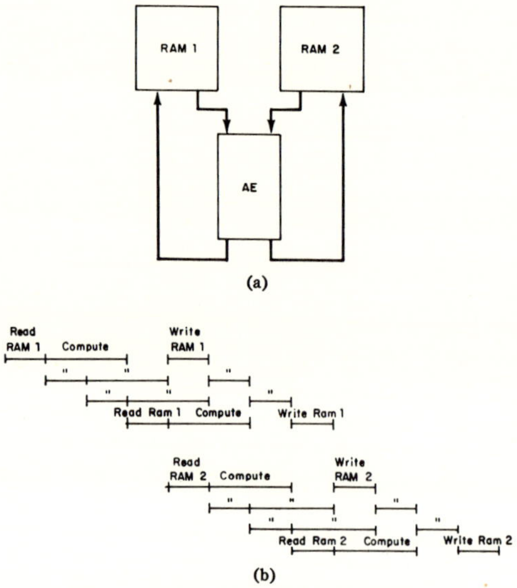

Fig. 6. Two RAM's doing two FFT's on one AE. (a) Structure. (b) Timing.

ceed more rapidly, some of these partial AE's will be "nulls," simply wires between adjacent buffers.

The timing diagram of Fig. 4 shows that whereas the memory is being continuously exercised, the AE is only working half the time; during memory read cycles, the AE pipeline buffer registers are being filled, and during the write cycles, these registers are being emptied. McHugh [2] has suggested that another memory can be serviced in an interleaved way by the one AE; the situation and timing is depicted in Fig. 6. While the results of the first four computations are being written into RAM 1, RAM 2 is keeping the pipeline filled by reading into the AE.

These ideas can be extended to higher radices; Fig. 7 shows a radix-4 and its associated timing. Here the computation is fractured into four units (1 unit = 1 memory cycle time) and eight reads are followed by eight writes. The pipeline culminates in a 4×4 permutation matrix; this matrix is represented in Fig. 7 by four registers, each containing the result of a radix-4 EC. Since these registers keep getting new information clocked in every unit, permutation clearly involves a form of buffer storage that we have not shown in the figure. In the same way as in the radix-2 case, two RAM's can be serviced by a single AE.

The structure just described satisfies (9) in the sense that the complete FFT takes exactly C_r memory cycle times. Table I shows the value of C_r for various N's and r's. We see that large time savings are possible by recourse to higher radix FFT's, but beyond $r = 4$, the price one pays in extra EC and permutation hardware appears to be prohibitive. As an example of the bandwidth attainable by these structures, assume that a memory cycle time is 50 ns. A radix-4 computation of a 256-point transform yields $C_r = 1024$; since both a read and write cycle are needed for each EC, we get 1024×100 ns $= 102.4$ μs, which corresponds to a 2.5-MHz data rate. It is interesting to note that the speed requirements on the hardware are specified almost exclusively by the input bandwidth and not the FFT size to be processed, whereas the memory size requirements are only a function of the FFT size. These comments are pertinent to the design of digital pulse compression systems with differing time and bandwidth requirements.

A variation on the above configurations can be made by increasing the parallelism beyond the radix size. Thus, in Fig. 8 we show a 32-point radix-2 transform wherein the memory register length is 4 rather than 2 (as in Fig. 2). The coefficients may be found by comparing with Fig. 2. In this case, a matched system is attained by employing two EC's in parallel and performing the permutations indicated in the flow diagram. For any degree of parallelism, radix-2 implies that no greater than two rows of results must be permuted. In this system, C_r is C_2, but since two EC's are done in parallel, the FFT time is halved; since $C_4 = 1/4\ C_2$, a radix-4 computation using the same memory format would be twice as fast.

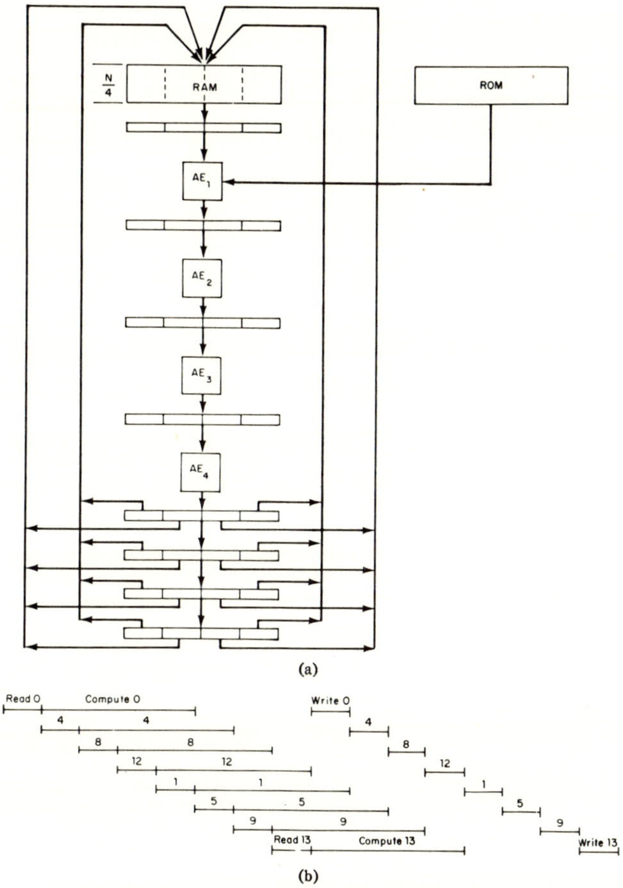

Fig. 7. Structure and timing for radix-4 FFT. (a) Structure. (b) Timing.

TABLE I

Tabulation of C_r for Various N and r

N \ r	2	4	8	16
64	192	48	16	—
128	448	—	—	—
256	1024	1024	—	32
512	2304	—	192	—
1024	5120	1280	—	—
2048	11 264	—	—	—
4096	25 576	6144	2048	768

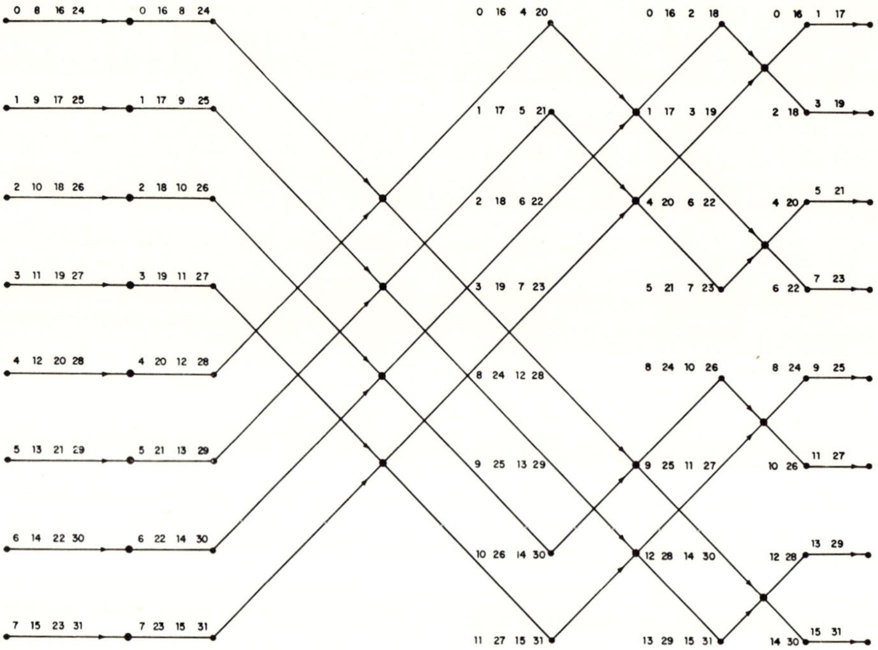

Fig. 8. Radix-2 32-point FFT with parallelism of 4.

VI. Pipeline FFT Structure of Radix-r

An FFT hardware architecture that is of particular interest is the pipeline processor [13]. The term "pipeline FFT" processor refers to the particular structure described in this section. It has also been referred to as a "cascade" processor [15]. It is unfortunate that we have also used the term "pipeline" in Section V to denote a particular AE structure. Although there is good reason to use the term "pipeline" in both cases, it is not intended that the structures discussed are thereby similar. In this device, log N identical arithmetic elements are operated in parallel in a way that requires surprisingly simple and straightforward control functions. Although a number of variations of the basic design exist, all of these implement the FFT via a radix-2 algorithm. Some reference has been made to higher radix pipeline structures in the literature [15], but the specific architecture of these machines has not been presented in any detail. In this section, we extend the FFT pipeline concept to the implementations of arbitrary radix, and derive some basic parameters of these processors. It will be shown that when several waveforms are to be processed simultaneously, one can achieve significant savings in both arithmetic hardware and total memory size through the use of higher radix pipeline structures.

The basic computational element of a radix-r pipeline processor is a device that computes an r-point DFT and effects an arbitrary complex weighting on $r-1$ of the resultant r output data (Fig. 9). The input sequence is applied to the computational element in a parallel format via r input ports, while the outputs are produced simultaneously, after a finite delay, at r output ports. Note that this computational element, a "CE," effects the basic EC of the radix-r transform. An alternate element could be constructed in which weighting (twiddle factors) is applied to the inputs instead of to the outputs. With the exception of the sequence in which coefficients will be required, the resultant machines will be completely equivalent.

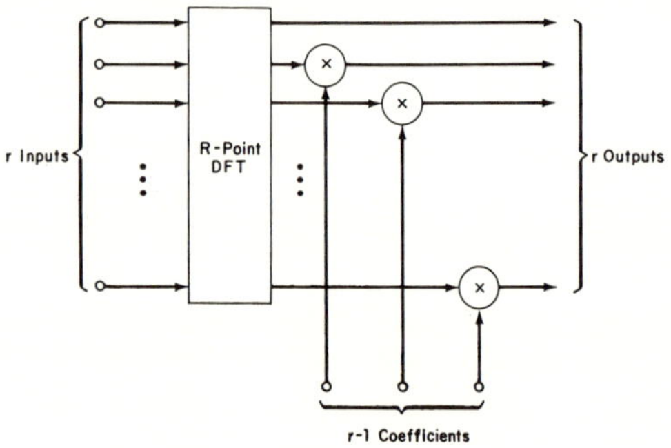

Fig. 9. Computational element for radix-5 pipeline processor.

We assume that the size N of the transform to be performed is an integral power of r, i.e.,

$$N = r^m, \qquad m = 1, 2, \cdots.$$

With this restriction, the transform is computed in m stages in which, at each stage, the computational element of Fig. 9 is exercised N/r times. In the pipeline processor, m separate computational elements are used, and each one is dedicated to a particular level of the algorithm.

Referring to the flow diagram of either Figs. 2 or 3, it is observed that in the first level of a radix-r computation, each EC is performed on points that in the original data set are separated by N/r locations. In general, the kth level consists of EC's that are performed on data that are separated by N/r^k positions. In the pipeline processor, these separations of input data points are obtained through the use of shift register memories, as indicated for the first level of the transform in Fig. 10. Assuming that the signal to be transformed is applied to the commutator switch at the left of the diagram, it is easy to verify that the correct data sequences will appear at the inputs to the CE if the commutator is advanced by one position after each block of N/r points is received. We assume for simplicity that the CE performs an EC instantaneously, so that the outputs occur simultaneously with the inputs. In practice, of course, some integral number of clock periods will elapse between inputs and outputs, but the throughput rate will remain the same.

Although the r inputs of the CE are in parallel, the element operates at the same rate at which incoming data are clocked. The timing is such that, as it is shown in Fig. 10, the processor remains idle during the time that the first $r-1/r\ N$ data occur, and then it computes the first level of the transform during the receipt of the last N/r points. Clearly, a more efficient mode of operation would use the CE continuously. This efficiency is possible if one were to simultaneously process r waveforms, where each is delayed from the next by N/r samples. In this case, an r-pole commutator, as indicated in Fig. 11, would replace the switch of Fig. 9. Fig. 12 illustrates the timing of this stage. Note that while each of the r inputs is devoted to a separate and independent data block, the r output terminals simultaneously deliver data from the same block. In the illustration, the ijth output segment is the jth portion of the output due to the ith input sequence.

Although the above discussion refers specifically to the first stage of the transform, it applies equally to all successive stages. Referring again to Figs. 2 or 3, it is easily observed that after having completed the first level, the remainder of the transform can be thought of as being composed of r completely separate N/r point FFT's, at least in terms of the flow paths in the diagram. Twiddle factors, of course, are chosen based on the N-point case. Further, it turns out that each N/r-point transform that originates in the second level is to be effected on a block of data that at the first stage output occurs at a single output terminal. That is to say, each of the blocks ij in Fig. 12 is to be treated as an independent block of data and transformed accordingly.

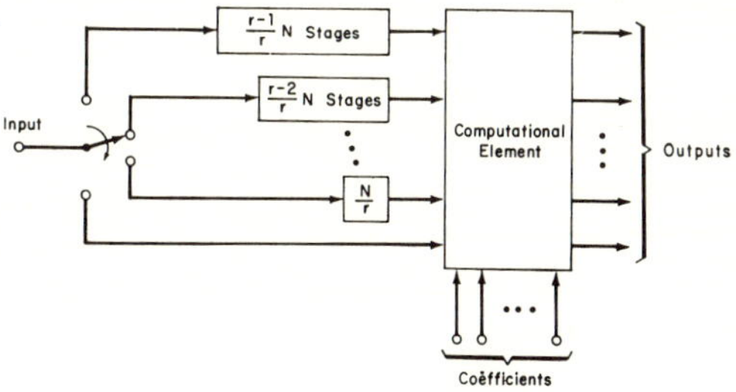

Fig. 10. First stage of radix-5 pipeline.

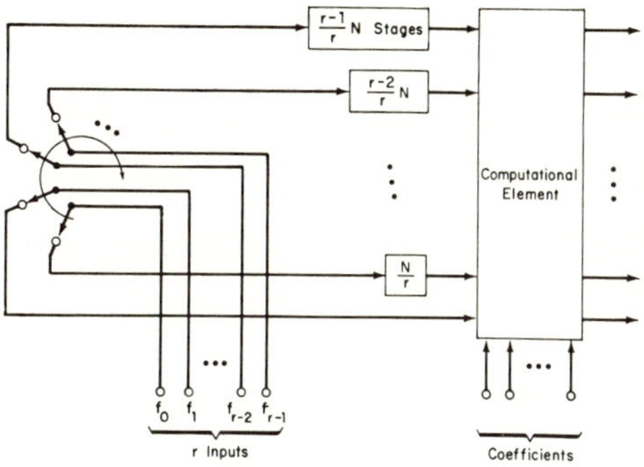

Fig. 11. Multiple input switching scheme.

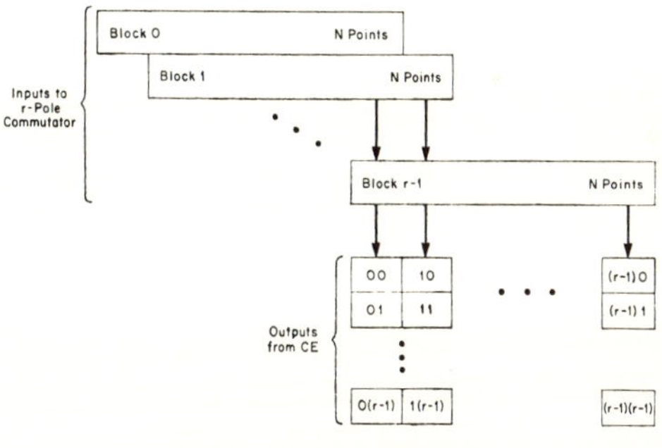

Fig. 12. Timing of first pipeline.

In effect, one can think of the second stage of the algorithm as being the first stage of a new set of FFT's of length N/r. Note that r of these N/r-point transforms are to be effected simultaneously. Our experience with the first level indicates that r transforms can be handled simultaneously if the inputs are skewed in time as in Fig. 12. Extending this concept to the second level of the algorithm, we follow the outputs of the first CE

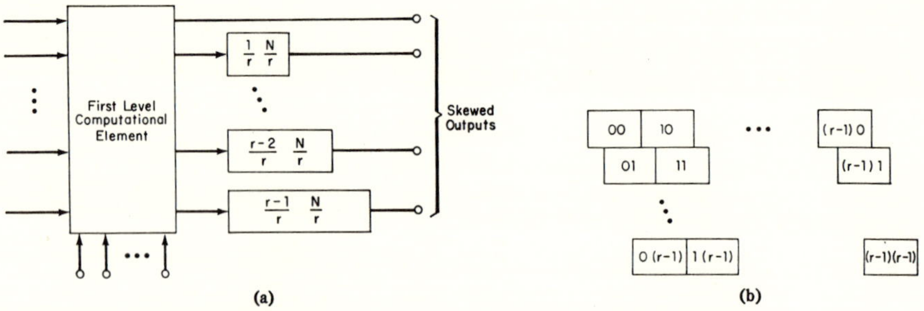

Fig. 13. Output memories. (a) Memory arrangement. (b) Time relationships.

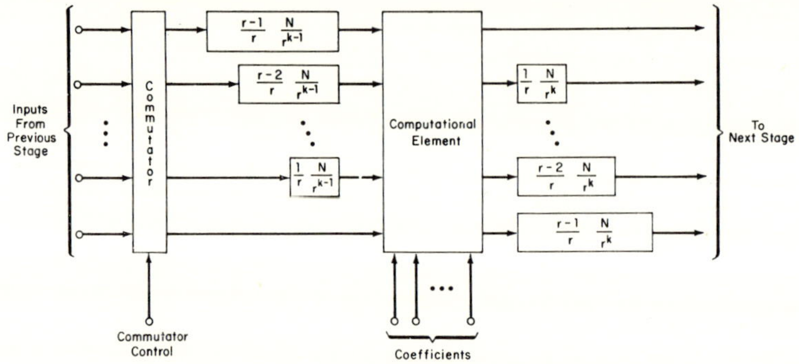

Fig. 14. kth coefficients stage of radix-r pipeline.

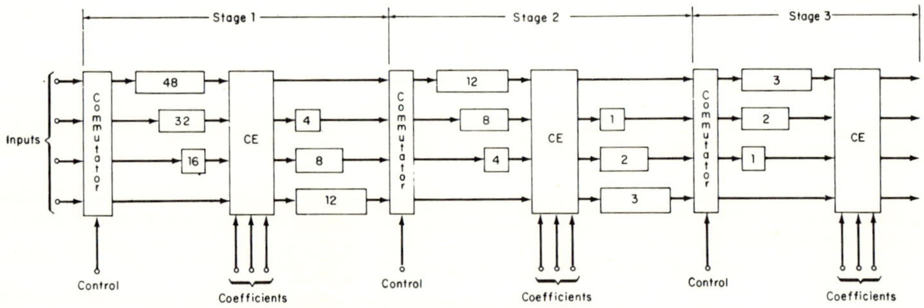

Fig. 15. Radix-4 pipeline $N=64$.

by a set of shift register delay lines that misalign the output groups (Fig. 13). Note that these shift registers are $1/r$ as long as those that are used at the input of the computational element. Merging Figs. 12 and 13, we obtain a structure that accepts an r block of data length N, skewed by N/r points, effects one level of an FFT on them, and delivers a succession of r blocks of data of length N/r, skewed by N/r^2. It is a simple matter to extend this concept to the kth stage of a radix-r FFT, in which one accepts r blocks of data of length N/r^{k-1}, skewed by N/r^k, and delivers a succession of r blocks of data of length N/r^k, skewed by N/r^{k+1}. Thus, the kth stage of a radix-r pipeline FFT processor has the structure shown in Fig. 14. A complete radix-r pipeline for a 64-point transform ($m=3$) is shown in Fig. 15. This device handles four simultaneous transforms.

It might be pointed out that the commutator switches are not operated at the same rate. Each successive switch is operated r times faster than the one before it, with the last switch changing its state at the system sampling rate. The last stage delivers r output points at once, with all r referring to the same input data block.

The N outputs for any one input sequence are delivered in N/r successive groups of r. Note that no provision for serializing these values has been included in Fig. 15. This could be accomplished with additional memory if necessary.

Several observations and generalizations can be made about pipeline FFT processors of the type described. First, it has already been noted that a radix-r machine will handle r transforms simultaneously. It is of interest to compare the amount of arithmetic hardware and total memory that the various implementations require per computed transform.

A. Arithmetic Hardware

A computational element for a pipeline processor consists of a number of adders and multipliers. Assuming that CE's are designed in as efficient a way as possible, the following is true.

Radix-2:	2	Adders
	1	Multiplier
Radix-4:	8	Adders
	3	Multipliers
Radix-8:	24	Adders
	9	Multipliers

In the above, all elements operate on complex quantities, i.e., an adder consists of two separate (real) adders, and a multiplier contains four real multipliers and two real adders. In the radix-8 case, two of the multipliers operate with fixed coefficients. This might result in some hardware economy. Since a radix-r pipeline requires $\log_r N$ CE's, we obtain the results given in Table II.

Of course, the fact that more transforms are performed in a higher radix pipeline means that the total amount of hardware that one uses increases with increasing radix. The economies of higher radix implementations become real only when the transforms have to be performed at sufficiently high rates that hardware parallelism would be required in implementations of lower radix. From Table II it is obvious that radix-4 pipeline structures offer a net savings of 25 percent in multiplier hardware. Radix-8 structures seem to offer little, if any, advantage over the radix-4 case.

B. Memory Requirements

From Fig. 13 it may be verified that the kth stage of a radix-r pipeline uses

$$\sum_{m=1}^{r-1} \frac{m}{r}\left(\frac{N}{r^{k-1}} + \frac{N}{r^k}\right) \tag{13}$$

stages of shift-register memory. Further, there are $\log_r N$ stages so that the total memory requirement M is given by

$$M = \sum_{k=1}^{\log_r N-1} \sum_{m=1}^{r-1} \frac{n}{r}\left(\frac{N}{r^{k-1}} + \frac{N}{r^k}\right)$$

or

TABLE II

N	r	Arithmetic requirements per transform
256	2	8 Adders, 4 Multipliers
1024	2	10 Adders, 5 Multipliers
4096	2	12 Adders, 6 Multipliers
256	4	8 Adders, 3 Multipliers
1024	4	10 Adders, $3\frac{3}{4}$ Multipliers
4096	4	12 Adders, $4\frac{1}{2}$ Multipliers
4096	8	12 Adders, $4\frac{1}{4}$ Multipliers

$$M = \frac{N(r+1)}{r}\left(\sum_{k=1}^{\log_r N-1} \frac{1}{r^k}\right)\left(\sum_{m=1}^{r-1} m\right). \tag{14}$$

Now

$$\sum_{m=1}^{r-1} m = \frac{r(r-1)}{2}, \tag{15}$$

and if the pipeline is reasonably long, then

$$\sum_{k=1}^{\log_r N-1} \frac{1}{r^k} \simeq \frac{r}{r-1} \tag{16}$$

so that

$$M = \frac{N(r+1)}{2}. \tag{17}$$

On a per-transform basis, the pipeline requires $N(r+1)/2r$ complex registers of memory. This is a monotonically decreasing function of r, equal to $3/4\ N$ for $r=2$ and approaching $N/2$ for large r. As in the arithmetic case, if several very-high-speed transforms are to be done simultaneously, there will be a net savings of memory if higher radix implementations are employed.

Appendix

Representation of a Mixed Radix FFT

Although the FFT representations shown in Figs. 2 and 3 were done for radix-2 and radix-4, similar representations can be drawn for mixed radices. An example is given in Fig. 16 of a 60-point FFT expressed as a $5\times3\times4$ matrix. The first stage of the process shows a 5×12 two-dimensional array. Each five-number row undergoes a five-point DFT, and the five results are multiplied by the corresponding row of twiddle factors. For example, row 3 is multiplied by W^0, W^3, W^6, W^9, W^{12} which is represented in the figure by an arrow with W^3. Since twiddles in the rth column have an exponent that is always r times the exponent in column 1, one exponent serves to identify all twiddles for that row. The inputs and outputs from the heavy node points express the fact that the data points from the input registers are redistributed among the output registers via row–column permutations. Thus, the 15 numbers in registers 0, 4, and 8 of the first stage become the 15 numbers of registers 0, 4, 8, 12, 16 of the second stage. To perform the 12-point column DFT's, each column is expressed as a 4×3 array. We then move on to the next

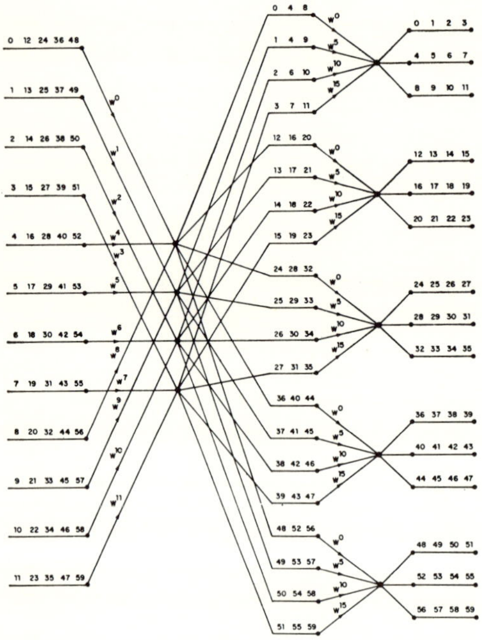

Fig. 16

stage wherein three-point DFT's are performed on the rows, then twiddled, and finally, four-point DFT's are performed on the columns (which are expressed as rows in the last stage). This structure results in successive row DFT's interspersed with twiddling.

Because of the mixed radix, the size of the rows and columns keep varying from stage to stage. This makes it awkward to perform the algorithm in-place, as we were able to do when a fixed radix was used.

References

[1] H. Andrews, "A high-speed algorithm for the computer generation of Fourier transforms," *IEEE Trans. Comput.* (Short Notes), vol. C-17, pp. 373–375, Apr. 1968.

[2] G. D. Bergland, "A fast Fourier transform algorithm using base 8 iterations," *Math. Comput.*, vol. 22, pp. 275–279, Apr. 1968.

[3] ——, "Fast Fourier transform hardware implementations—A survey," *IEEE Trans. Audio Electroacoust.*, vol. AU-17, pp. 109–119, June 1969.

[4] ——, "Fast Fourier transform hardware implementations—An overview," *IEEE Trans. Audio Electroacoust.*, vol. AU-17, pp. 104–108, June 1969.

[5] G. D. Bergland and D. E. Wilson, "An FFT algorithm for a global, highly parallel processor," *IEEE Trans. Audio Electroacoust.*, vol. AU-17, pp. 125–127, June 1969.

[6] W. T. Cochran *et al.*, "What is the fast Fourier transform?," *IEEE Trans. Audio Electroacoust.*, vol. AU-15, pp. 45–55, June 1967.

[7] J. W. Cooley and J. W. Tukey, "An algorithm for the machine computation of complex Fourier series," *Math. Comput.*, vol. 19, pp. 297–301, Apr. 1965.

[8] J. W. Cooley, P. Lewis, and P. Welch, "Application of the fast Fourier transform to computation of Fourier integrals, Fourier series, and convolution integrals," *IEEE Trans. Audio Electroacoust.*, vol. AU-15, pp. 79–84, June 1967.

[9] ——, "Historical notes on the fast Fourier transform," *IEEE Trans. Audio Electroacoust.*, vol. AU-15, pp. 76–79, June 1967.

[10] M J. Corinthios "The design of a class of fast Fourier transform computers," *IEEE Trans. Comput.*, vol. C-20, pp. 617–623, June 1971.

[11] ——, "A fast Fourier transform for high-speed signal processing," *IEEE Trans. Comput.*, vol. C-20, pp. 843–846, Aug. 1971.

[12] B. Gold and C. M. Rader, *Digital Processing of Signals*. New York: McGraw-Hill, 1969.

[13] H. L. Groginsky and G. A. Works, "A pipeline fast Fourier transform," *IEEE Trans. Comput.*, vol. C-19, pp. 1015–1019, Nov. 1970.

[14] R. Klahn, R. R. Shively, E. Gomez, and M. J. Gilmartin, "The time-saver: FFT hardware," *Electronics*, pp. 92–97, June 24, 1968.

[15] G. C. O'Leary, "Nonrecursive digital filtering using cascade fast Fourier transformers," *IEEE Trans. Audio Electroacoust.*, vol. AU-18, pp. 177–183, June 1970.

[16] M. C. Pease, "An adaptation of the fast Fourier transform for parallel processing," *J. Ass. Comput. Mach.*, vol. 15, pp. 252–264, Apr. 1968.

[17] R. R. Shively, "A digital processor to perform a fast Fourier transform," in *Proc. 1st IEEE Comput. Conf.*, 1967, pp. 21–24.

[18] ——, "A digital processor to generate spectra in real time," *IEEE Trans. Comput.*, vol. C-17, pp. 485–481, May 1968.

[19] R. C. Singleton, "An algorithm for computing the mixed radix fast Fourier transform," *IEEE Trans. Audio Electroacoust.*, vol. AU-17, pp. 93–103, June 1969.

[20] B. Gold, I. L. Lebow, P. G. McHugh, and C. M. Rader, "The FDP, A fast programmable signal processor," *IEEE Trans. Comput.*, vol. C-20, pp. 33–38, Jan. 1971.

29

Reprinted from *IEEE Trans. Computers,* **C-19**(11), 1015–1019 (1970)

A Pipeline Fast Fourier Transform

HERBERT L. GROGINSKY, SENIOR MEMBER, IEEE, AND GEORGE A. WORKS

Abstract—This paper describes a novel structure for a hardwired fast Fourier transform (FFT) signal processor that promises to permit digital spectrum analysis to achieve throughput rates consistent with extremely wide-band radars. The technique is based on the use of serial storage for data and intermediate results and multiple arithmetic units each of which carries out a sparse Fourier transform. Details of the system are described for data sample sizes that are binary multiples, but the technique is applicable to any composite number.

Index Terms—Cascade Fourier transform, digital signal processor, Doppler radar, fast Fourier transform, radar–sonar signal processor, radix-two fast Fourier transform, real-time signal processor.

Introduction

THIS paper describes a novel structure for a hardwired FFT signal processor that promises to permit digital spectrum analysis to achieve throughput rates consistent with extremely wide-band radars.

The processor consists of a number of modular units connected in cascade through switches that direct the flow of information from memory to arithmetic units. The switching required to carry out the process is simple and is controlled by a binary counter. The processor is similar to the binary analyzer described by Bergland and Hale [1], but employs only N complex words of storage to compute the FFT of N complex data samples.[1] Bergland [2] has listed many alternative organizations of FFT processors. Recently O'Leary [10] has also proposed a similar structure.

We show that the Cooley–Tukey algorithm does a natural interleaving of data gathered by the time multiplexing of a number of independent channels, typical of radars and sonars. In this concept, the successive stages or iterations of the fundamental algorithm are each carried out in the separate cascaded modules. Using shift registers as digital delay lines permits new data to be entered into the processor while the processing of earlier data blocks is carried out. In effect the overall delay required is equal to the time required to gather the analysis sample block N in each of the separate channels. As the Nth complex data sample is loaded into the digital delay line, the first analysis frequency appears at the output. The output appears in precisely the same channel sequence as the data when they were loaded into the delay line. The output frequencies, however, appear in the scrambled sequence associated with the algorithm.

The control device, namely the binary counter, yields a digital number identifying both the channel number and the frequency currently being outputted. In addition, it specifies the instants at which the separate modules are to be

Manuscript received November 7, 1969; revised April 27, 1970. This work was supported by Raytheon research and development funding. A patent has been filed on the basic structure of this signal processor. This paper was presented at EASCON'69 (Electronics and Aerospace Systems Convention), Washington, D. C., October 27–29, 1969.

The authors are with the Raytheon Company, Sudbury, Mass.

[1] Although the processor described here is cascade in structure, we prefer the pipeline designation used by computer designers [11] because this structure permits direct application of pipeline arithmetic techniques.

switched and a digital number identifying the sine/cosine values needed by each of the stages. This structure, although hardwired, does permit the flexible interchange of channels processed for data sample length per channel. Thus a system capable of processing N complex samples in a single channel is also capable of processing N/L samples in each of L channels provided L is a factor of N. The modular design of the device permits the duplicate arithmetic units to weight the input prior to the FFT operation and to present the output in magnitude. Furthermore, it allows computation of Fourier transforms at the rate at which new data can be inserted into the digital delay line. Fundamentally, the signal processing rate is independent of the data sample length.

A pipeline FFT configured to process radar data from a pulse Doppler tracking radar has been designed and tested. The system uses MOSFET shift registers as the digital delay lines, TTL in the arithmetic units and MOS LSI READ ONLY memory to store sine/cosine tables and filter shaping weight functions.

The system processes eight range channels taking 512 complex samples per channel. It is designed to obtain subclutter visibility of 60 dB and achieves this using 12-bit fixed-point internal operations in the arithmetic unit. The throughput rate achieved by the system is 128K samples per second.

Theory of Operation

In this section, the method of operation of the pipeline FFT is explained in terms of the fundamental mathematical operations that must be carried out. An extensive literature now exists [3] explaining the basic principles of the FFT. The discussion here emphasizes certain features of the analysis, permitting a hardwired realization of the algorithm to achieve the goals set forth in the introduction.

The discrete Fourier transform (DFT) is defined by

$$X_m = \sum_{n=0}^{N-1} x_n W_N^{mn} \tag{1}$$

where

$$W_N^z = e^{-j(2\pi z/N)},$$

x_n is a complex data sample, and X_m is the complex image of the data at frequency m/N.

Theory shows that when N is a composite number

$$N = \prod_{k=1}^{M} r_k \tag{2}$$

where the r_k are a set of integers (possibly with repeats), and (1) may be calculated iteratively in M stages as follows.

$$a_{\mu,\nu}^m = \sum_{l=0}^{r_m-1} W_{R_m}^{-\mu l} a_{p,q}^{m-1} \tag{3}$$

with

$$p = \mu \bmod R_{m-1}$$
$$q = \nu + lC_m$$

and the ranges

$$1 \leq m \leq M, \quad 0 \leq \mu < R_m, \quad 0 \leq \nu < C_m$$

where

$$R_m = \prod_{k=1}^{m} r_k$$
$$C_m = N/R_m$$
$$a_{\mu,0}^0 \doteq x_n$$

and

$$\mu \bmod r \triangleq \mu - \text{greatest integer in } \mu/r.$$

In fact, when this is done, the number of calculations drops from N^2 complex operations (MULTIPLY and ADD) to $N(r_1+r_2+\cdots+r_M)$ such operations. This iterative process both reduces the amount of hardware required to realize the operation, and provides the basic pipeline structure permitting new calculations to proceed before the results of earlier calculations are completed. This form of the algorithm is known as the Cooley–Tukey version.

When $r_k=2$ for all k, the algorithm is conveniently summarized by the flow diagram shown in Fig. 1. In this figure, the input data enter the left-hand column and each successive column corresponds to a later stage of the iterative process. The coefficients indicated at the input then correspond to the index of the input data x_n and give its order in the data stream. The figures shown at each later stage indicate the coefficients of the rotation vector W_{2^m} that must be applied to the lower branch entering each node.

A number of important features of the algorithm may be seen by examining Fig. 1. First, we observe that each stage needs only the data generated from the preceding stage. Second, if each stage is processed in order of arrival, the first stage examines data points displaced by half the data length ($N/2$), the second by one quarter of the data length ($N/4$), etc. Third, if the data were available in a continuous stream, the first stage could be processing one block of data while the second stage processed the next earlier block and so on through all M stages. Fourth, the rotation vector required, W_{2^m}, has the same periodicity as the data displacement interval. Finally, we note that the output appears in the usual scrambled order of frequency associated with this version of the algorithm.

The significance of these remarks is that each of the stages may be realized with a basic component whose general form is shown in Fig. 2. Any m module alternately transfers blocks of 2^m data samples into the delay line and into the arithmetic unit. When the data block just fills the delay line, the arithmetic unit obtains a rotation vector (from a READ ONLY memory) and begins its operation. The next block of 2^m input data samples are sent to the arithmetic unit that now produces two complex outputs in response to the two complex inputs it receives. One of the outputs is immediately transferred to the next stage while the other output is sent to the delay line. Thus in the interim period when the delay line is filled with fresh input data, the contents of the

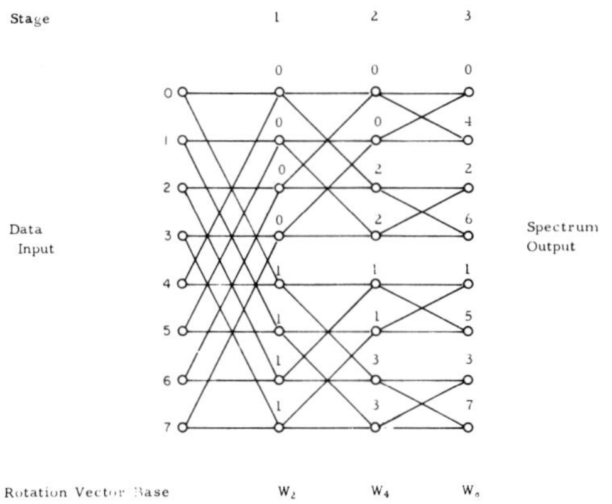

Fig. 1. Flow diagram of a Cooley–Tukey FFT.

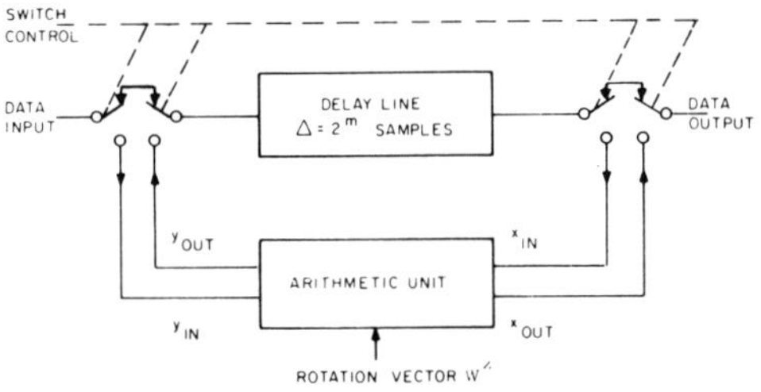

Fig. 2. Pipeline FFT m module.

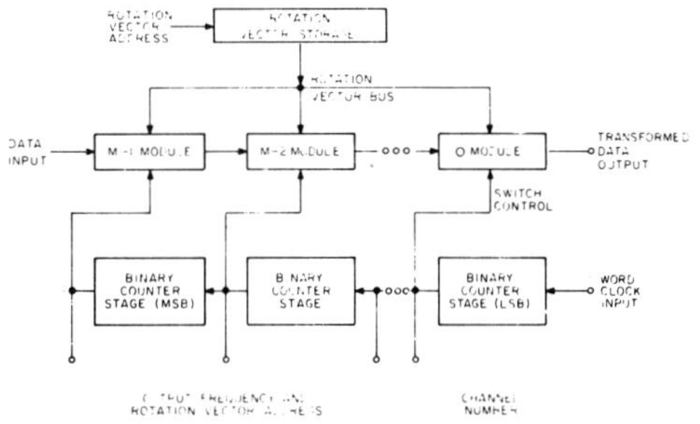

Fig. 3. Pipeline FFT processor.

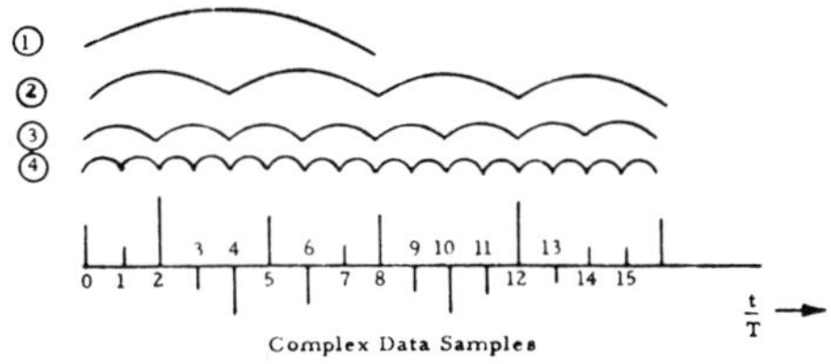

Fig. 4. FFT interleaving mechanism.

line containing the results of processing the earlier blocks are transferred to the next stage. The arithmetic unit, of course, computes the complex two-point transforms, shown below.

$$x_{\text{out}} = x_{\text{in}} + y_{\text{in}} W^{z}$$
$$y_{\text{out}} = x_{\text{in}} - y_{\text{in}} W^{z}. \tag{4}$$

This module design may be assembled into a system for computing DFTs in blocks of $N = 2^M$ samples as shown in Fig. 3. The rotation vector storage shown in the figure is a table of roots of unity, or sines and cosines, which is shared by all m modules. Fig. 1 shows that $N/2$ different rotation vectors are read to process one block of N samples. Indeed if they are produced in the order required for the last stage, the rotation vectors required for the earlier stages may be obtained by strobing this list at the proper instant in advance of its need. Thus the sines and cosines required for each stage may be obtained by providing a register for each arithmetic unit, all driven from a common bus. Note that exactly M arithmetic units and exactly $N-1$ complex data points of storage are needed in this system and that the first transform output is obtained immediately after the last data sample in the block of N is received.

The theory cited above leads to still another important observation, namely that the output at any intermediate stage of the process has a simple interpretation. Theory shows that

$$a^m_{\mu,\nu} = \sum_{k=0}^{R_m - 1} x_{\nu + kC_m} W^{\mu k}_{R_m}, \tag{5}$$

which is precisely the discrete Fourier transform of all groups of data points separated by an interval C_m. This may be better understood in reference to Fig. 4, which shows the natural FFT interleaving mechanism that results from the Cooley–Tukey algorithm. In terms of a sequential process, the output of the first stages results in $N/2$ independent two-point transforms; the output of the second stage yields $N/4$ independent four-point transforms, etc. Thus, if two independent streams of complex data were entered into the input interleaved with one another, the module 1 stage of the cascade processor would produce two independent DFTs of each data stream. The spectral component of each channel of data is outputted before the spectral frequency is changed. In particular for pulsed radar or sonar application, where the data for many range samples are received before a new sample may be taken, this system permits the data to be processed in order of arrival with no modification of the control circuits and without requiring the data to be reassembled first into consecutive (noninterleaved) data streams.

Equation (5) shows that the index μ, which is the rotation vector coefficient in (3), can also be regarded as the current frequency being outputted, namely μ/R_m. When these normalized frequencies (chosen to make R_m a unit period) are modified to account for the expanded sampling interval C_m, the current frequency being outputted may be regarded on absolute scales as the frequency $\mu/R_m C_m = \mu/N$. Thus, in Fig. 1, the coefficient given at every stage in the process indicates not only the coefficient of the rotation vector applied to the lower branch but the current frequency (in absolute terms) as well.

The control mechanism for this system is perhaps its most elegant feature. Fig. 1 shows clearly that a binary counter driven in synchronism with the data stream generates a signal designating the processing interval (i.e., the switch position) for each stage. However, if any output is taken at any intermediate stage, say at the $k+1$st stage, then the lower k bits of the counter, in normal order, give the channel number of the data currently being outputted, while the upper $M-k$ bits taken in bit-reversed order, give the frequency that is currently being outputted and, in addition, describes the address (θ) of the sine and cosines needed in the current computation of the $k+1$st stage. Thus, the control mechanism contains all the information needed to carry out the spectral analysis as well as to descramble the output data. In many applications, it is unnecessary to descramble the output data provided one can identify the frequency of any component.

Indeed the structure shown in Fig. 3 may be readily modified to calculate transforms when the data samples are given in scrambled order. This structure also permits the trade-off of channels processed for data length per channel by taking outputs at an intermediate stage. The modified structure produces the output sequence in natural order in both the channels and in time.

Computational Errors and Quantization

If the input signal to an FFT machine is obtained from an analog-to-digital converter, it must be sufficiently finely quantized so that quantization noise is uncorrelated to avoid distortion of the Fourier transform. Widrow [4] has shown that for signals in the presence of Gaussian noise, choosing the quantization grain or value of the least significant bit of the quantizer equal to three times the noise standard derivation provides a good approximation to this condition. The value of the most significant bit of the quantizer must be greater than or equal to one half the peak value of the input signal (without noise) to avoid peak limiting. The minimum number of bits Q required to represent a signal of peak value $\pm V_p$ with additive Gaussian noise of standard deviation σ_n is therefore given by

$$Q = \log_2(V_p/3\sigma_n) + 1. \tag{6}$$

One or more bits may be added to further reduce quantization noise.

The number of bits per data word required in an FFT processor is usually greater than the number of bits Q in the input word to reduce computational noise. The two-point transform (4), which is the building block of larger Fourier transforms, may be accomplished by four real multiplications and six real additions. Following each two-point transform, words may be truncated or rounded to maintain constant word size or allowed to grow.

Welch [5] has shown that rounding after each two-point transform leads to a relative rms output error ε, which is bounded above by

$$\varepsilon = (0.3)2^{-B+(M+3)/2}/\text{rms (input)} \tag{7}$$

for a transform of 2^M samples using B-bit arithmetic. The error may be reduced by allowing words to lengthen from stage to stage. This can be particularly attractive in a cascade machine because most of the storage in such a machine may be associated with the first few stages of the transform.

Alternate Configurations

The computing module circuitry may be configured in a number of different ways to meet different requirements. The arithmetic unit that computes two-point transforms may employ either four real multiplications and six real additions or three real multiplications and nine real additions. The multiplications may be performed either before data are stored in the shift register or after data emerge from the shift register. Goldstone [6] has shown that multipliers may be time shared between adjacent modules by performing half of the required multiplications as data enter the shift register and the other half as data leave. Some of the total shift register delay may be incorporated in a pipeline multiplier for high-speed operation.

In order to make use of the speed possible in a pipeline processor, one word delays must be inserted between the computing modules. These delays permit each module to begin computation at the start of a word time rather than to wait for the preceding modules to compute the input it requires. These intermodule delays do not appreciably complicate the control circuitry of the processor, since they may be compensated by delaying the control and rotation vector inputs to the module by a delay equal to the total data delay.

A cascade processor requires no multipliers in the first two modules. The rotation vectors (W_2^0 and W_4^1) used in these modules may be implemented by, at most, a switching circuit to interchange the real and imaginary part of the data word and invert the sign of the resulting real part when rotation by W_4^1 is required. The multipliers may be implemented if desired and used for data windowing.

Data windowing, or multiplying input data samples by a data window function, is a technique used to change the frequency response of the equivalent N filters whose output is computed by the FFT. If an FFT of N samples taken at equal intervals T is computed, a data window that is exactly zero everywhere except over sampling interval of duration NT must be used. If no data windowing is intentionally performed, input data samples are weighted by a data window function that is unity over the sampling interval and zero elsewhere.

The frequency response of the equivalent FFT filters is given by the discrete Fourier transform of the data window function. The frequency response of an FFT filter when no data windowing is performed is therefore the DFT of a unit-amplitude pulse of duration NT, which (normalized) is

$$H(f) = \frac{\sin \pi NTf}{N \sin \pi Tf} \tag{8}$$

The relatively slow decrease in amplitude with increasing frequency of this function makes it undesirable for Fourier analyzer applications, in which data windowing is commonly used. Frequently used data window functions include the Hamming [7], Hanning [7], Dolph Tchebyscheff [8], and Taylor [9] functions. The properties of these functions have been extensively described.

Implementation

The pipeline structure was used to implement a real-time FFT machine to provide spectral analysis for a tracking pulse-Doppler radar. This machine simultaneously processes eight channels of Doppler data at a sample rate of 16 000 complex samples per second per channel with a transform length of 512 complex samples and a quantization of 24 bits per complex sample. The FFT throughput rate is slightly over three million bits per second. A Taylor data window is employed to permit 60-dB subclutter visibility of high-speed targets.

The machine is composed of nine processing modules, a synchronizer, A/D converter, and display. The processing modules employ a total of 2600 TTL integrated circuits for switching and arithmetic functions and a total of 500 200-bit MOSFET shift registers for delay. Words are stored in bit-serial form in the shift registers, but arithmetic operations are performed in bit-parallel. Interstage delay is incorporated in the arithmetic units. Data windowing is performed by the "spare" multiplier in the first processing module. An eight complex sample input buffer following the A/D converter allows the FFT machine to operate at a constant word rate equal to eight times the 8–16 kHz radar PRF, independent of the ranges at which samples are collected. Fig. 5 shows a photograph of the complete breadboard FFT cascade processor.

Fig. 5. Pipeline processor machine.

Conclusions

The pipeline FFT processor has proven to be an effective tool to meet the real-time spectral analysis requirements of many radar and sonar systems. Its structure permits sufficient paralleling of operations, such that the processing time is limited solely by the time it takes to collect the data. It is efficient in storage requirement since it requires only as much storage as the number of samples to be processed. It is simple to control because basically all of the control information can be generated by a binary counter. Furthermore the use of this counter guarantees proper synchronization of the control function with the data stream passing through the device.

Perhaps its greatest disadvantage is the scrambled order in which the output appears. In the radar/sonar applications, for which the device was designed, this fault was transparent. It does, however, make certain operations in the frequency domain, such as smoothing over frequency, somewhat more difficult. Descrambling is possible in a pipeline processor at the cost of additional memory less than the number of words in the processed data block.

The remarkable flexibility with which the structure may be reconfigured to trade channels for data sample length and to carry out the inverse transform functions is one of its most satisfying properties. Properly utilized, the device is able to carry out auto- and cross-correlation, block convolutions, and cross-spectral density calculations.

References

[1] G. D. Bergland and H. W. Hale, "Digital real-time spectral analysis," *IEEE Trans. Electronic Computers*, vol. EC-16, pp. 180–185, April 1967.

[2] G. D. Bergland, "Fast Fourier transform hardware implementations—An overview," *IEEE Trans. Audio Electroacoust.*, vol. AU-17, pp. 104–108, June 1969.

[3] ——, "A guided tour of the fast Fourier transform," *IEEE Spectrum*, vol. 6, pp. 41–52, July 1969.

[4] B. Widrow, "Statistical analysis of amplitude-quantized sampled-data systems," *AIEE Trans.*, vol. 79, pt. 2, pp. 555–567, January 1961.

[5] P. D. Welch, "A fixed-point fast Fourier transform error analysis," *IEEE Trans. Audio Electroacoust.*, vol. AU-17, pp. 151–157, June 1969.

[6] B. J. Goldstone, "Serial FFT—More efficient utilization of the multiplier," Raytheon Co. internal memo BFX-R-29, October 1968.

[7] R. B. Blackman and J. W. Tukey, *The Measurement of Power Spectra.* New York: Dover, 1958.

[8] C. L. Dolph, "A current distribution for broadside arrays which optimizes the relationship between beam width and side-lobe level," *Proc. IRE*, vol. 34, pp. 335–348, June 1946.

[9] T. T. Taylor, "Design of line-source antennas for narrow beamwidth and low sidelobes," *IRE Trans. Antennas Propag.*, vol. AP-3, pp. 16–28, January 1955.

[10] G. O'Leary, "A high-speed cascade fast Fourier transformer," presented at IEEE Arden House Workshop on Digital Filtering, January 1970.

[11] W. R. Graham, "The parallel, pipeline and conventional computer," *Datamation*, vol. 16, pp. 68–71, April 1970.

30

Reprinted from *IEEE Trans. Computers,* **C-20**(8), 843–846 (1971)

A Fast Fourier Transform for High-Speed Signal Processing

MICHAEL J. CORINTHIOS, STUDENT MEMBER, IEEE

Abstract—An arbitrary-radix fast Fourier transform algorithm and a design of its implementing signal processing machine are introduced. The algorithm yields an implementation with a level of parallelism proportional to the radix r of factorization of the discrete Fourier transform, allows 100 percent utilization of the arithmetic unit, and yields properly ordered Fourier coefficients without the need for pre- or postordering of data.

It is shown that through a slight deviation from ideally hardwired machine architecture, the feedback time between input and output memories in a sequential processor as the one described in [6] can be completely eliminated. Elimination of the feedback time yields a faster machine and the speedup factor is related to the ratio of the speed of performing arithmetic to that of moving data in memory. If arithmetic can be performed at the data shifting speed, this speedup factor reaches the value (r+1).

Throughout the iterations of the algorithm, the operands are at least, or are always N/r^2 points apart, N being the number of data points. Partitioning of the data memory into several submemories in the form of queues, which can be simultaneously accessed for parallel operation, is shown to be possible. Effective use can thus be made of long shift registers or integrated circuit random access memories.

For 1024-word records it is shown that real-time power spectrum computation of signals sampled at a rate of 16 MHz is presently feasible. A radix-4 256-word machine such as the ones described here has been designed by the author and its construction is nearing completion.

Index Terms—Convolution, correlation, digital filtering, digital signal processing, fast Fourier transform, machine organization, special-purpose computer, spectral analysis, time-series analysis.

I. Introduction

IN previous suggestions for designing fast Fourier transform (FFT) signal processors [1], [2], machine organization ranged from a serial sequential type to a parallel cascade one with the latter requiring a large number of arithmetic units, which for the number of points usually encountered may prove to be prohibitively expensive to implement.

In [3] Pease has shown a design of a hardwired parallel processor. He then considered a compromise that yields a slower but more feasible machine design where emphasis is placed on the utilization of memories with relatively slow access time and where a number of arithmetic units are assumed to be operating in parallel to compute a radix-2 fast Fourier transform [4]. In [5] the author begins from the other end of the scale and arrives at a serial hardwired radix-2 machine employing a single complex multiplier. In [6] the same concept is generalized to yield an implementation with a level of parallelism proportional to the radix r of factorization of the DFT. The resulting machine is simple in organization and because it is hardwired, requires very little control.

We notice, however, that in this implementation each iteration calls for a feedback phase in which data are serially moved from the output to the input memory in an order determined by a permutation operator. Therefore, when r is increased the computation time is proportionally reduced while that of the feedback phase remains the same.

In an implementation in which the rate of moving data in memory is an order of magnitude, say, higher than the rate of performing arithmetic, such a machine organization yields a reasonable compromise between the cost of the machine and the processing speed. However, if the speed of performing arithmetic, particularly complex multiplication, approaches that of moving data in memory, the overhead time spent in the permutation phases becomes unreasonable. We thus search for an algorithm which would reduce or eliminate the feedback time; probably at the cost of a deviation from ideal hardwired conditions.

In this paper this compromise is arrived at. The feedback phase is completely eliminated and 100 percent utilization of the arithmetic unit obtained. The price paid, in the form of a need for some addressing or added gating, will be shown to be well justified.

A hardware-oriented FFT algorithm is first introduced, followed by the organization of the signal processor.

II. Ordered-Input Ordered-Output FFT Algorithms

We state here without proof one of four sets of algorithms that has been derived in [6]. Each set has been obtained by factoring the discrete Fourier transform to an arbitrary radix. We deal here with the ordered-input ordered-output (OIOO) algorithm, which yields an asymmetric-type machine. Application of the technique to the other algorithms in [6] is straightforward.

Employing the same notation as in [6] we can write the OIOO algorithm as follows:

$$T_N = \prod_{m=1}^{n} (p_m^{(r)} \mu_m^{(r)} S^{(r)}) \tag{1}$$

where T_N is the finite Fourier transformation matrix that when operating on a time series represented by a vector f yields the Fourier coefficients represented by the vector F, i.e., for a record length of $N = r^n$ samples we have

$$F = (1/N) T_N f. \tag{2}$$

Manuscript received September 1, 1970; revised March 5, 1971. This work was supported in part under National Research Council of Canada Grants A3148 and A3951.

The author is with the Department of Electrical Engineering, University of Toronto, Toronto, Ont., Canada.

The pertinent definitions of matrices as derived in [6] are stated in the following, where the symbol $\times$ stands for the Kronecker product of matrices [3] and $p_m^{(r)}$ is a permutation given by

$$p_i^{(r)} = I_{r^{n-i}} \times p_{r^i}^{(r)} \tag{3}$$

where the notation I_K denotes the indentity matrix of dimension K, and $P_K^{(r)}$ the ideal shuffle-base r permutation matrix operating on a vector of dimension K. $\mu_m^{(r)}$ is the weighting or twiddle operator and is given by

$$\mu_i^{(r)} = I_{r^{n-i}} \times D_{r^i}^{(r)} \tag{4}$$

where in general

$$D_{N/k}^{(r)} = \text{quasidiag}\,(I_{N/rk}, K_k, K_{2k}, K_{3k}, \cdots, K_{(r-1)k}) \tag{5}$$

and

$$K_m = \text{diag}\left(0, m, 2m, 3m, \cdots, \left(\frac{N}{rk} - 1\right)m\right); \tag{6}$$

the exponents rather than the actual values of the matrix coefficients are written for brevity of notation, as in [3], [5], [6]; $S^{(r)}$ is the preweighting r-point transform operator given by

$$S^{(r)} = (I_{N/r} \times T_r) \tag{7}$$

and

$$T_r = \begin{bmatrix} 0 & 0 & 0 & 0 \cdots & 0 \\ 0 & N/r & 2N/r & 3N/r \cdots & (r-1)N/r \\ 0 & 2N/r & 4N/r & 6N/r \cdots & 2(r-1)N/r \\ \cdot & \cdot & \cdot & \cdot \cdots & \cdot \\ 0 & (r-1)N/r & \cdot & \cdot \cdots & (r-1)^2 N/r \end{bmatrix}. \tag{8}$$

III. Modified Ordered-Input Ordered-Output Algorithm

The modification is simply performed as follows. We have

$$T_N = S^{(r)} p_2^{(r)} \mu_2^{(r)} S^{(r)} \cdots p_{n-1}^{(r)} \mu_{n-1}^{(r)} S^{(r)} p_n^{(r)} \mu_n^{(r)} S^{(r)} \tag{9}$$

which can be rewritten as:

$$T_N = S_1^{(r)} \mu_2^{(r)} S_2^{(r)} \mu_3^{(r)} \cdots S_{n-2}^{(r)} \mu_{n-1}^{(r)} S_{n-1}^{(r)} \mu_n^{(r)} S_n^{(r)} \tag{10}$$

that is,

$$T_N = \prod_{m=1}^{n} (\mu_m^{(r)} S_m^{(r)}) \tag{11}$$

where in general

$$S_{m-1}^{(r)} = S^{(r)} p_m^{(r)}; \qquad m = 2, 3, \cdots, n \tag{12}$$

$$S_n^{(r)} = S^{(r)} \tag{13}$$

and

$$\mu_1 = I_N. \tag{14}$$

We now show that the preweighting operator $S_m^{(r)}$ calls always for combining data that are at least N/r^2 words apart. Omitting the superscript (r), since it is known that operations are performed to base r, we have for $m \neq 1$

$$S_{m-1} = Sp_m = (I_{N/r} \times T_r)p_m \tag{15}$$

$$= p_m p_m^{-1}(I_{N/r} \times T_r)p_m \tag{16}$$

and we can easily show that

$$p_m^{-1}(I_{N/r} \times T_r)p_m = (I_{N/r^2} \times T_r \times I_r) \tag{17}$$

and therefore

$$S_{m-1} = p_m(I_{N/r^2} \times T_r \times I_r). \tag{18}$$

Thus we can see that the matrix I_{N/r^2} in the second factor causes the operator S_{m-1} to operate on data that are always N/r^2 words apart. In the first iteration, however, the operator S_n operates on data which are N/r words apart.

IV. Machine Organization

Fig. 1 shows a flow diagram of the modified OIOO algorithm for $N=8$ and $r=2$. We observe that in no iteration do we need to simultaneously access data that are closer than $N/r^2=2$ words apart.

Fig. 2 shows an example of a radix-4 machine organization. Generalization to other values of r should be straightforward. (For higher values of r, however, we may have to perform the operation of the matrix T_r iteratively; as we have done with T_N.)

As shown in Fig. 2 the machine includes two memories MEM1 and MEM2, each storing N words, an arithmetic unit (AU), a READ-ONLY memory or equivalent [6], and some switches. The AU includes preweighters and weighters and is identical to that described in [6]. Assuming arithmetic is to be performed at the data-shifting speed, this machine is $r+1=5$ times as fast as the hardwired one described in [6]. This gain in speed is achieved by eliminating the permutation operation in which N words were fed back from the output to the input memory after each iteration. The price paid is mainly the added gating.

Both MEM1 and MEM2 are divided into r submemories (SM), each of length N/r^2; and each SM is again divided into r queues. Each queue may consist of a long shift register for each bit of the data words. Switch $S2$ gates r words at each clock pulse to the AU. In all but the first iteration these words constitute the data at the tops of the r queues of a selected SM. In the first iteration the queues of each SM are connected to form a long queue, and the r words at the tops of the these formed queues are fed to AU through $S2$. (To simplify the diagram this connection is not shown explicitly in Fig. 2.) When the input data to the AU are selected from MEM1, $S3$ gates the output r words of the AU to MEM2 and vice versa. Switch $S1$ is obviously needed for gating new data into the machine. $S3$ is not really needed but is included in Fig. 2 to simplify the description of operation.

At each clock the data in the queues are shifted one bit to the right. For example, when MEM1 is empty and MEM2 full, $S2$ starts to access data from MEM2, which acts then as a data source and $S3$ feeds the output of AU to MEM1, which acts as a sink. When either MEM1 or MEM2 is in the sink mode, the r queues in each of its SMs are connected

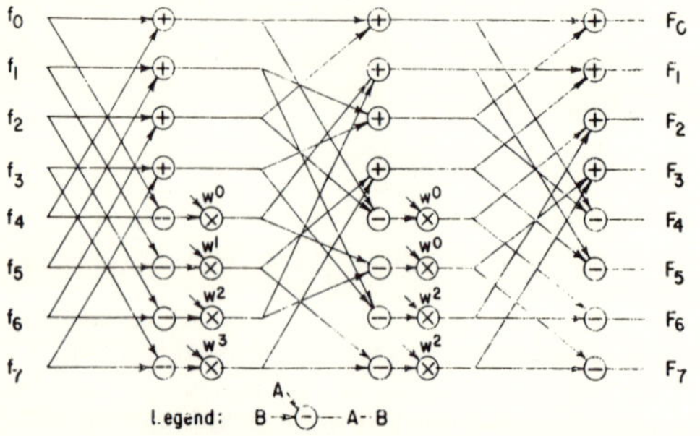

Fig. 1. Graphical representation of the modified ordered-input ordered-output algorithm for the case $N=8$ and $r=2$.

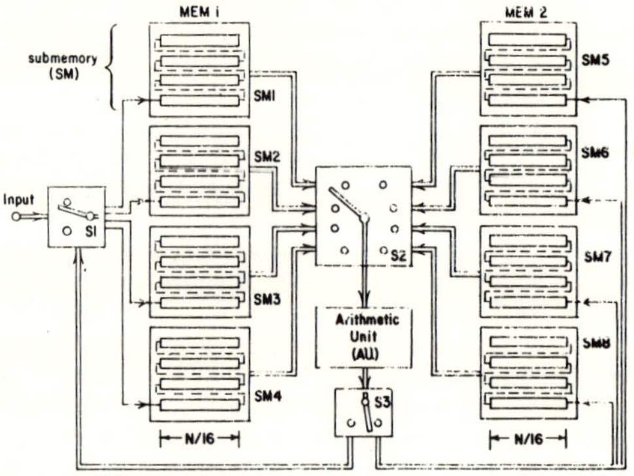

Fig. 2. Machine organization of the FFT signal processor for $r=4$. A single line represents one word; a double line four words.

to form one long queue, to the rear of which data are fed from $S3$.

The nth iteration calls for preweighting only; no multiplication is required. Thus during the nth iteration the data at the output of the preweighter are gated out of the processor, and the Fourier coefficients are in proper order. If the modification were performed on the postpermutation algorithm of [6], then all iterations after the first are identical. However, the output coefficients are not then in a properly ascending order.

We note that the data source during the nth iteration can be either MEM1 or MEM2 depending on whether n is odd or even, respectively. During this same iteration new data are unloaded from the input buffer (IB), which is needed for real-time operation, and stored in MEM1. If n is odd the IB may have to be partitioned into r queues, so that the new record would be all in MEM1 at the instant the last r points of the previous record leave MEM1. Alternatively, the IB may be partitioned into r-long queues and unloaded into MEM2 during the nth iteration; the choice being hardware dependent. If n is even the IB consists of r-long queues and is unloaded into MEM1 during the nth iteration.

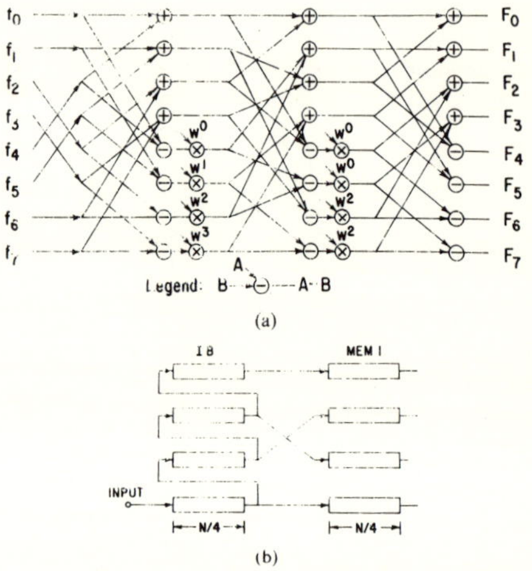

Fig. 3. Factoring of the first iteration to obtain more uniform iterations. (a) An example for the case $N=8$, $r=2$. (b) Schematic diagram of implementation in a radix-2 machine.

We also note that since the last iteration includes no multiplication ($\mu_1 = I_N$) the power spectrum can be evaluated during the nth iteration by making use of the otherwise idle multipliers. This is performed by gating the Fourier coefficients $a+jb$ at the outputs of the preweighters into the complex multipliers, and multiplying them by their conjugates to obtain the components (a^2+b^2) of the power spectrum. Power spectra can thus be computed in the same time as that required to perform the Fourier transformation. The power spectrum is also obtained in proper order which is advantageous in most applications.

V. Factoring of the First Iteration

We have noticed that the first iteration differs from subsequent ones in its call for operation on data which are N/r rather than N/r^2 points apart. To avoid this nonuniformity with its effect on the size of $S2$, we perform a permutation operation while unloading IB into MEM1. The r-point transform of the first iteration can then be made identical to one of the other $n-1$ iterations. One choice is to make the first iteration identical to the last, i.e.,

$$S_n = S_1 \tag{19}$$

and using (15) we have

$$S = S_1 p_2^{-1} \tag{20}$$

and

$$\begin{aligned} p_2^{-1} &= (I_{r^{n-2}} \times P_{r^2})^{-1} \\ &= I_{r^{n-2}} \times P_{r^2}^{-1} \\ &= I_{N/r^2} \times P_{r^2} = p_2. \end{aligned} \tag{21}$$

Fig. 3(a) shows a flow diagram of the modified algorithm

where the operator S_n is factored into a permutation operator p_2 followed by operator S_1. A schematic diagram of the interconnection between IB and MEM1 for a radix-2 machine is shown in Fig. 3(b). The permutation described by p_2 is thus performed while IB is unloaded into MEM1.

VI. Maximum Sampling Frequency

Let us denote by t_p the time of performing preweighting, i.e., of performing one r-point transform and by t_m the time of performing preweighting followed by weighting of r points. The time of Fourier transformation can then be computed assuming that the IB is loaded in MEM1 (or MEM2) during the nth iteration. Since the nth iteration involves preweighting only we have the time of one Fourier transformation equaling

$$(n-1)(N/r)t_m + (N/r)t_p. \tag{22}$$

To find the maximum sampling frequency for real-time Fourier transformation of a real-valued time series, we assume that arithmetic can be performed at the data shifting speed; say 15 MHz. (This may require pipelining in the multiplier and/or propagating the carries between its planes, rather than in the same plane as described in [6].) Moreover since the input time series is real-valued, it is possible to process two successive records simultaneously. Separating the two records at the output can be shown to be possible for this high-speed real-time operation. With these assumptions we have the time of Fourier transforming N real-valued data in real time equaling $(1/2)n(N/r)t_{sh}$ where t_{sh} is the minimum data shifting period.

For example, for $N = 1024$, $r = 4$, and $t_{sh} = 100$ ns, the time for Fourier transformation or power spectrum computation equals 64 μs, and the maximum sampling frequency for real-time operation equals 1024/64 equals 16 MHz.

VII. A Cascade Processor

At the present time, except for very small data arrays, a cascade processor may prove to be prohibitively expensive. However, within a decade this may not be the case. The conversion of the machine described into the cascade type is conceptually straightforward. Thus, instead of using an input buffer memory and oscillating data successively between two memories, we utilize a number of memory arrays equal to the number of stages (iterations) in the algorithm. The machine is thus a pipelined one and the effective time of processing is the time of processing one stage. The number of arithmetic units are thus increased by an amount equal to the number of stages. Moreover, simultaneous weighting coefficients have to be furnished for each stage.

(The maximum sampling speed if a cascade processor were involved is given by

$$\frac{N}{(1/2)(N/r)t_{sh}} = \frac{2r}{t_{sh}}$$

which is about 80 MHz for the example previously considered.)

VIII. Conclusion

We have seen that by partitioning the data memory into r^2 instead of r queues, it is possible to eliminate the need for the feedback permutation operations. Thus if arithmetic can be performed at the data shifting speed, then the processing time of an iteration is that corresponding to N/r shifting clock pulses, compared to $(N+N/r)$ in the hardwired machine described in [6]. The speed of processing is thus increased by a factor of $(r+1)$.

To achieve this high processing speed we made a compromise regarding the amount of hardwiring in the machine. The partitioning of the memories calls now for a larger number of shorter shift registers. However, this is not really significant since integrated circuit registers are still relatively limited in length. Most of the cost is in the added gating and some in the control. However, the cost increase may be justified by noting that the arithmetic unit, the ROM, and most other components of the system have not been changed; and therefore, the fractional increase in cost is still low compared to the speedup factor of $r+1$.

This paper illustrates again the effectiveness of using long shift registers for dynamic storage of data. The speed of moving data, elimination of addressing, and simplicity of machine organization constitute, in this context, some of their advantages over core-type memories. The algorithms are also well suited for an implementation using integrated-circuit random access memories (RAM). For example, a 256-word machine could be built of 16- by 4-bit RAMs, which have an access time of 50 ns. The suggestion for uniformity of iterations given in Section V allows full utilization of the wired-OR capability of these components.

Finally, we notice that in addition to the reduction in the number of multiplications offered by factorization to higher radices, the reduction of roundoff or truncation errors associated with them constitutes an additional advantage that can be significant in determining the word length and hence the cost of the implemented machine.

Acknowledgment

The author wishes to thank Prof. J. L. Yen for many helpful suggestions and discussions, and for his continuous encouragement and support and Prof. K. C. Smith for his instructive criticism and many valuable ideas.

References

[1] G. D. Bergland, "Fast Fourier transform hardware implementations—An overview," *IEEE Trans. Audio Electroacoust.*, vol. AU-17, June 1969, pp. 104–108.

[2] J. E. Whelchel, Jr., and D. F. Guinn, "FFT organizations for high-speed digital filtering," *IEEE Trans. Audio Electroacoust.*, vol. AU-18, June 1970, pp. 159–168.

[3] M. C. Pease, "An adaption of the fast Fourier transform for parallel processing," *J. Ass. Comput. Mach.*, vol. 15, Apr. 1968, pp. 252–264.

[4] ——, "Organization of large scale Fourier processors," *J. Ass. Comput. Mach.*, vol. 16, July 1969, pp. 474–482.

[5] M. J. Corinthios, *A Time Series Analyzer*, vol. 19, MRI Symposia Ser. New York: Polytechnic Press, 1969, pp. 47–61.

[6] ——, "The design of a class of fast Fourier transform computers," *IEEE Trans. Comput.*, vol. C-20, June 1971, pp. 617–623.

Reprinted from *AFIPS Proc. 1966 Spring Joint Computer Conf.*, Vol. 28, Spartan Books, 1966, pp. 229–233

HIGH-SPEED CONVOLUTION AND CORRELATION*

Thomas G. Stockham, Jr.
Massachusetts Institute of Technology, Project MAC
Cambridge, Massachusetts

INTRODUCTION

Cooley and Tukey[1] have disclosed a procedure for synthesizing and analyzing Fourier series for discrete periodic complex functions.† For functions of period N, where N is a power of 2, computation times are proportional to $N \log_2 N$ as expressed in Eq. (0).

$$T_{ct} = k_{ct} N \log_2 N \tag{0}$$

where k_{ct} is the constant of proportionality. For one realization for the IBM 7094, k_{ct} has been measured at 60 μsec. Normally the times required are proportional to N^2. For $N = 1000$ speed-up factors in the order of 50 have been realized! Eq. (1b) synthesizes the Fourier series in question. The complex Fourier coefficients are given by the analysis equation, Eq. (1a).

$$F(k) = \sum_{j=0}^{N-1} f(j) w^{-jk} \tag{1a}$$

$$f(j) = \frac{1}{N} \sum_{k=0}^{N-1} F(k) w^{jk} \tag{1b}$$

where $w = e^{2\pi i/N}$, the principal Nth root of unity. The functions f and F are said to form a discrete periodic complex transform pair. Both functions are of period N since

$$F(k) = F(k + cN) \tag{2a}$$

and

$$f(j) = f(j + cN) \tag{2b}$$

TRANSFORM PRODUCTS

Consider two functions g and h and their transforms G and H. Let G and H be multiplied to form the function C according to Eq. (3),

$$C(k) = G(k) \times H(k) \tag{3}$$

and consider the inverse transform $c(j)$. $c(j)$ is given by Eq. (4)

$$c(j) = \frac{1}{N} \sum_{J=0}^{N-1} g(J) h(j - J)$$

$$= \frac{1}{N} \sum_{J=0}^{N-1} h(J) g(j - J) \tag{4}$$

as a sum of lagged products where the lags are performed circularly. Those values that are shifted from one end of the summation interval are circulated into the other.

The time required to compute $c(j)$ from either form of Eq. (4) is proportional to N^2. If one computes the transforms of g and h, performs the multiplication of Eq. (3), and then computes the inverse

*Work reported herein was supported (in part) by Project MAC, an M.I.T. research program sponsored by the Advanced Research Projects Agency, Department of Defense, under Office of Naval Research Contract Number Nonr-4102(01).

†To be able to use this procedure the period must be a highly composite number.

transform of C, one requires a time given by Eq. (5)

$$T_{\rm circ} = 3\,k_{ct}\,N\log_2 N + k_m N$$
$$= k_{\rm circ}\,N(\log_2 N + \mu) \qquad (5)$$

where $k_{\rm circ} = 3k_{ct}$, $\mu = k_m/k_{\rm circ}$, and $k_m N$ is the time required to compute Eq. (3). Of course this assumes N is a power of 2. Similar savings would be possible provided N is a highly composite number.

APERIODIC CONVOLUTION

The circular lagged product discussed above can be alternately regarded as a convolution of periodic functions of equal period. Through suitable modification a periodic convolution can be used to compute an aperiodic convolution when each aperiodic function has zero value everywhere outside some single finite aperture.

Let the functions be called $d(j)$ and $s(j)$. Let the larger finite aperture contain M discrete points and let the smaller contain N discrete points. The result of convolving these functions can be obtained from the result of circularly convolving suitable augmented functions. Let these augmented functions be periodic of period L, where L is the smallest power of 2 greater than or equal to $M + N$. Let them be called $da(j)$ and $sa(j)$ respectively, and be formed as indicated by Eq. (6).

$$\begin{aligned} fa(j) &= f(j + j_0) && 0 \le j \le M - 1 \\ &= 0 && M \le j \le L - 1 \qquad (6) \\ &= fa(j + nL) && \text{otherwise} \end{aligned}$$

where j_0 symbolizes the first point in the aperture of the function in question. The intervals of zero values permit the two functions to be totally non-overlapped for at least one lagged product even though the lag is a circular one. Thus, while the result is itself a periodic function, each period is an exact replica of the desired aperiodic result.

The time required to compute this result is given in Eq. (7).

$$T_{\rm aper} = k_{\rm circ}\,L(\log_2 L + \mu) \qquad (7)$$

where $M + N \le L < 2(M + N)$. For this case, while L must be adjusted to a power of 2 so that the high-speed Fourier transform can be applied, no restrictions are placed upon the values of either M or N.

SECTIONING

Let us assume that M is the aperture of $d(j)$ and N is that of $s(j)$. In situations where M is considerably larger than N, the procedure may be further streamlined by sectioning $d(j)$ into pieces each of which contains P discrete points where $P + N = L$, a power of 2. We require K sections where

$$K = \text{least integer} \ge M/P \qquad (8)$$

Let the ith section of $d(j)$ be called $d_i(j)$. Each section is convolved aperiodically with $s(j)$ according to the discussion of the previous section, through the periodic convolution of the augmented sections, $da_i(j)$ and $sa(j)$.

Each result section, $r_i(j)$, has length $L = P + N$ and must be additively overlapped with its neighbors to form the composite result $r(j)$ which will be of length

$$KP + N \ge M + N \qquad (9a)$$

If $r_i(j)$ is regarded as an aperiodic function with zero value for arguments outside the range $0 \le j \le L - 1$, these overlapped additions may be expressed as

$$r(j) = \sum_{i=0}^{K-1} r_i(j - iP) \quad j = 0,1,\ldots KP + N - 1 \qquad (9b)$$

Each overlap margin has width N and there are $K - 1$ of them.

The time required for this aperiodic sectioned convolution is given in Eq. (10).

$$\begin{aligned} T_{\rm sect} &= k_{ct}(P + N)\log_2(P + N) \\ &\quad + 2Kk_{ct}(P + N)\log_2(P + N) \\ &\quad + Kk_{\rm aux}(P + N) \\ &= k_{ct}(2K + 1)(P + N)\log_2(P + N) \\ &\quad + Kk_{\rm aux}(P + N) \\ &\approx k_{ct}(2K + 1)(P + N)[\log_2(P + N) + \mu'] \end{aligned} \qquad (10)$$

where $\mu' = k_{\rm aux}/2k_{ct}$. $Kk_{\rm aux}(P + N)$ is the time required to complete auxiliary processes. These processes involve the multiplications of Eq. (3), the formation of the augmented sections $da_i(j)$, and the formation of $r(j)$ from the result sections $r_i(j)$. For the author's realization in which core memory was used for the secondary storage of input and output data, μ' was measured to be 1.5, which gives $k_{\rm aux} = 3k_{ct} \approx 300\,\mu$sec. If slower forms of auxiliary storage were employed, this figure would be enlarged slightly.

For a specific pair of values M and N, P should be chosen to minimize $T_{\rm sect}$. Since $P + N$ must be a power of 2, it is a simple matter to evaluate Eq. (10) for a few values of P that are compatible with this constraint and select the optimum choice. The size of available memory will place an additional constraint on how large $P + N$ may be allowed to become. Memory allocation considerations degrade the benefits of these methods when N becomes too large. In extreme cases one is forced to split the kernel, $s(j)$, into packets, each of which is considered separately. The results corresponding to all packets are then added together after each has been shifted by a suitable number of packet widths. For the author's realization N must be limited to occupy about 1/8 of the memory not used for the program or for the secondary storage of input/output data. For larger N, packets would be required.

COMBINATION OF SECTIONS IN PAIRS

If both functions to be convolved are real instead of complex, further time savings over Eq. (10) can be made by combining adjacent even and odd subscripted sections $da_i(j)$ into complex composites. Let even subscripted $da_i(j)$ be used as real parts and odd subscripted $da_{i+1}(j)$ be used as imaginary parts. Such a complex composite can then be transformed through the application of Eqs. (1a), (3), and (1b) to produce a complex composite result section. The desired even and odd subscripted result sections $r_i(j)$ and $r_{i+1}(j)$ are respectively the real and imaginary parts of that complex result section.

This device reduces the time required to perform the convolution by approximately a factor of 2. More precisely it modifies K by changing Eq. (8) to

$$K = \text{least integer} \geq M/2P \tag{11}$$

For very large numbers of sections, K, Eq. (10) can be simplified to a form involving M explicitly instead of implicitly through K. That form is given in Eq. (12)

$$T_{\text{fast}} \approx k_{ct} M((P + N)/P)[\log_2(P + N) + \mu'] \tag{12}$$

Since it makes no sense to choose $P < N$, for simple estimates of an approximate computation time we can write

$$T_{\text{fast}} \approx 2k_{ct}M[\log_2 N + \mu' + 1] \tag{13}$$

EMPIRICAL TIMES

The process for combined-sectioned-aperiodic convolution of real functions described above was implemented in the MAD language on the IBM 7094 Computer. Comparisons were made with a MAD language realization of a standard sum of lagged products for N = 16, 24, 32, 48, 64, 96, 128, 192, and 256. In each case M was selected to cause Eq. (11) to be fulfilled with the equal sign. This step favors the fast method by avoiding edge effects. However, P was not selected according to the optimization method described above (under "Sectioning Convolution"), but rather by selecting L as large as possible under the constraint.

$$\ln L \geq P/N \tag{14}$$

This choice can favor the standard method.

Table 1 compares for various N the actual computation times required in seconds as well as times in milliseconds per unit lag. Values of M, K, and L are also given.

Relative speed factors are shown in Table 2.

ACCURACY

The accuracy of the computational procedure described above is expected to be as good or better

Table 1. Comparative Convolution Times for Various N

N	16	24	32	48	64	96	128	192	256
M	192	208	384	416	768	832	1536	1664	3584
K	2	1	2	1	2	1	2	1	1
L	64	128	128	256	256	512	512	1024	2048
	Time in seconds								
T_{standard}	0.2	0.31	0.8	1.25	3.0	5.0	12	20	48
T_{fast}	0.3	0.4	0.6	0.8	1.3	1.8	3.0	3.8	8.0
	Time in milliseconds per unit lag								
$T_{\text{standard}/M}$	1.0	1.4	2.0	3.0	3.9	6.0	7.8	12.0	13.3
$T_{\text{fast}/M}$	1.5	1.9	1.5	1.9	1.6	2.1	1.9	2.2	2.2

Table 2. Speed Factors for Various N

N	16	24	32	48	64	96	128	192	256	512	1024	2048	4096
Speed factor	$\frac{2}{3}$	$\frac{3}{4}$	$\frac{4}{3}$	1.5	2.3	2.8	4.0	5.2	6	13*	24*	44*	80*

*Estimated values.

than that obtainable by summing products. Specific investigations of the accuracy of the program used to accumulate the data of Tables 1 and 2 are in process at the time of this writing. The above expectations are fostered by accuracy measurements made for floating-point data on the Cooley-Tukey procedure and a standard Fourier procedure. Since the standard Fourier procedure computes summed products, its accuracy characteristics are similar to those of a standard convolution which also computes summed products. Cases involving functions of period 64 and 256 were measured and it was discovered that two Cooley-Tukey transforms in cascade produced respectively as much, and half as much, error as a single standard Fourier transform. This data implies that the procedures disclosed here may yield more accurate results than standard methods with increasing relative accuracy for larger N.

APPLICATIONS

Today the major applications for the computation of lagged products are digital signal processing and spectral analysis.

Digital signal processing, or digital filtering as it is sometimes called, is often accomplished through the use of suitable difference equation techniques. For difference equations characterized by only a few parameters, computations may be performed in times short compared to those required for a standard lagged product or the method described here. However, in some cases, the desired filter characteristics are too complex to permit realization by a sufficiently simple difference equation. The most notable cases are those requiring high frequency selectivity coupled with short-duration impulse response and those in which the impulse response is found through physical measurements. In these situations it is desirable to employ the techniques described here either alone or cascaded with difference equation filters.

The standard methods for performing spectral analysis[2] involves the computation of lagged products of the form

$$F(j) = \sum_{J=0}^{N-j-1} x(J)y(J+j) \qquad (15)$$

which, in turn, after weighting by so-called spectral windows are Fourier transformed into power spectrum estimates. Speed advantages can be gained when Eq. (15) is evaluated in a manner similar to that outlined above (under "Aperiodic Convolution") except that in this case L is only required to exceed $N + \Omega$ where Ω is the number of lags to be considered. This relaxed requirement on L is possible because it is not necessary to avoid the effect of performing the lags circularly for all L lags but rather for only Ω of them. An additional constraint is that Ω be larger than a multiple of $\log_2 L$. The usual practice is to evaluate Eq. (15) for a number of lags equal to a substantial fraction of N. Since the typical situation involves values of N in the hundreds and thousands, the associated savings may be appreciable for this application.

Digital spatial filtering is becoming an increasingly important subject.[3,4] The principles discussed here are easily extended to the computation of lagged products across two or more dimensions. Time savings depend on the total number of data points contained within the entire data space in question, and they depend on this number in a manner similar to that characterizing the one-dimension case.

ACKNOWLEDGMENTS

The author is indebted to Charles M. Rader of the MIT Lincoln Laboratory for his ideas concerning the Cooley-Tukey algorithm and to Alan V. Oppenheim of the Electrical Engineering Department, MIT, for suggesting that high-speed convolutions might be realized through the utilization of that algorithm. During the preparation of this work the author became aware of the related independent efforts of Howard D. Helms, Bell Telephone Laboratories, and Gordon Sande, Jr., Princeton University.

REFERENCES

1. J. W. Cooley and J. W. Tukey, "An Algorithm for the Machine Calculation of Complex Fourier Series," *Mathematics of Computation*, vol. 19, no. 90, pp. 297–301, (Apr. 1965).

2. R. B. Blackman and J. W. Tukey, *The Measurement of Power Spectra*, Dover Publications, New York, 1959; also *Bell System Technical Journal*, Jan. and Mar. 1958.

3. T. S. Huang and O. J. Tretiak, "Research in Picture Processing," *Optical and Electro-Optical Information Processing*, J. Tippett et al, eds., MIT Press, Cambridge, Mass., 1965, Chap. 3.

4. T. S. Huang, "PCM Picture Transmission," *IEEE Spectrum*, vol. 2, no. 12, pp. 57–63 (Dec. 1965).

32

Reprinted from *IEEE Trans. Audio Electroacoustics,* **AU-15**(2), 79–84 (1967)

Application of the Fast Fourier Transform to Computation of Fourier Integrals, Fourier Series, and Convolution Integrals

JAMES W. COOLEY, PETER A. W. LEWIS, AND PETER D. WELCH

***Abstract*—The fast Fourier transform is a computational procedure for calculating the finite Fourier transform of a time series. In this paper, the properties of the finite Fourier transform are related to commonly used integral transforms including the Fourier transform and convolution integrals. The relationship between the finite Fourier transform and Fourier series is also discussed.**

Introduction

MANY PROBLEMS of current interest require the use of a variety of numerical methods for their solution. A technique that has found wide applicability is the integral transform method. However, numerical problems are generally solved with the aid of a digital computer which is not designed to handle the continuous waveforms that occur when integral transform methods are used. For this reason, it is necessary to convert continuous time series or other functions to a series of discrete data samples, and to perform numerical operations such as the finite Fourier transform on these samples. The fast Fourier transform (FFT) algorithm has reduced the time required to compute finite Fourier transforms by the fraction $\log_2 N/N$ where N is the number of discrete data samples. For large values of N, this reduction is important. It is therefore necessary to relate the properties of continuous integral transforms to the properties of the finite Fourier transform in order to take advantage of digital computers and the fast Fourier transform algorithm. In this paper, the correspondence between the finite Fourier transform and the Fourier integral is described, and a method that can be used to compute Fourier integrals is discussed in detail. Other problems covered are the computation of convolution integrals and integral equations of the convolution type. Finally, the relationship between the finite Fourier transform and the Fourier series is discussed in order to apply the FFT to problems that involve harmonic analysis and synthesis.

Manuscript received January 26, 1967; revised March 6, 1967.

The authors are with the IBM Research Center, Yorktown Heights, N. Y.

Relationship between the Finite Fourier Transform and the Fourier Transform: Use of the Algorithm to Calculate Fourier Integrals

Suppose we have a function $x(t)$ which has a Fourier transform

$$a(f) = \int_{-\infty}^{\infty} x(t)e^{-2\pi i f t}dt. \tag{1}$$

Then, the inverse relationship is

$$x(t) = \int_{-\infty}^{\infty} a(f)e^{2\pi i f t}df. \tag{2}$$

If $x(t)$ is sampled at intervals of length Δt, then (2) expressed at the sampling points $j\cdot\Delta t, j=0, \pm 1, \pm 2, \cdots,$ can be written

$$x(j\cdot\Delta t) = \int_{-\infty}^{\infty} a(f)e^{2\pi i j f/F}df$$
$$= \sum_{k=-\infty}^{\infty} \int_{kF}^{(k+1)F} a(f)e^{2\pi i j f/F}df \tag{3}$$

where $F=1/\Delta t$. F is twice the Nyquist frequency, $1/(2\Delta t)$. Now the exponential function $e^{2\pi i j f/F}$ is a periodic function of f with period F so that by changing the variable of integration one can put (3) in the form

$$x(j\cdot\Delta t) = \int_0^F a_p(f)e^{2\pi i j f/F}df \tag{4}$$

where

$$a_p(f) = \sum_{k=-\infty}^{\infty} a(f+kF). \tag{5}$$

Here, and in what follows, the subscript p on a function will denote the periodic function formed by the superposition of the nonperiodic function shifted by all multiples of a fundamental period. The function $a_p(f)$ is said to be an "aliased" version of $a(f)$, with the aliasing occurring relative to the Nyquist frequency $F/2$.

Since $a_p(f)$ is a periodic function of f it has a Fourier series expansion. Further, we see from (4) that the coefficients of this expansion are given by $1/F$ times the sequence $x(j\cdot\Delta t)$. Hence, (4) has the reciprocal equation

$$a_p(f) = \frac{1}{F}\sum_{j=-\infty}^{\infty} x(j\cdot\Delta t)e^{-2\pi i f j/F}. \tag{6}$$

In this relationship between $a_p(f)$ and $x(j\cdot\Delta t)$ the usual roles of time and frequency are interchanged; i.e., a periodic, continuous function of frequency corresponds to a sequence of time coefficients. Now, (6) is a discrete Fourier transform but it is not finite. However, if we consider the values of $a_p(f)$ at N equally spaced points between 0 and F, i.e., sample $a_p(f)$ at intervals $\Delta f = F/N = 1/(N\Delta t)$, we obtain

$$a_p(n\Delta f) = \frac{1}{F}\sum_{j=-\infty}^{\infty} x(j\Delta t)e^{-2\pi i j n/N}$$
$$= \frac{1}{F}\sum_{j=0}^{N-1}\left\{\sum_{l=-\infty}^{\infty} x(j\Delta t + lN\Delta t)\right\} e^{-2\pi i j n/N}. \tag{7}$$

The latter result follows from the fact that $e^{-2\pi i j n/N}$ is a periodic sequence of j with period N. Hence, we finally have

$$a_p(n\Delta f) = \frac{1}{F}\sum_{j=0}^{N-1} x_p(j\Delta t)e^{-2\pi i j n/N} \tag{8}$$

where

$$x_p(t) = \sum_{l=-\infty}^{\infty} x(t+lT) \tag{9}$$

is periodic of period $T=N\Delta t=1/\Delta F$. It is only a matter of a multiplicative constant to put (8) in the form of the finite Fourier transform and prove the following important theorem.

Theorem 1

If $x(t)$, $-\infty < t < \infty$, and $a(f)$, $-\infty < f < \infty$, are a Fourier integral transform pair,

$$x(t) \leftrightarrow a(f),$$

then $Tx_p(j\Delta t)$, $j=0, 1, \cdots, N-1$, and $a_p(n\Delta f)$, $n=0, 1, 2, \cdots, N-1$, are a finite Fourier transform pair,

$$Tx_p(j\Delta t) = T\sum_{l=-\infty}^{\infty} x(j\Delta t + lT) \leftrightarrow \sum_{k=-\infty}^{\infty} a(n\Delta f + kF)$$
$$= a_p(n\Delta f)$$

where $\Delta f = 1/(N\Delta t) = 1/T$.

It is to be remembered that the finite Fourier transform pair

$$X(j) \leftrightarrow A(n)$$

is defined by

$$X(j) = \sum_{n=0}^{N-1} A(n)W_N^{jn}, \qquad W_N = e^{2\pi i/N}.$$

Thus, if two functions are Fourier transforms of one another, then the sequences obtained from them by aliasing and sampling in this fashion are finite Fourier transforms of one another.

An interesting application of this theorem is to the problem of determining the vector $X(j)$ which corresponds to the normalized Gaussian function $x(t)=e^{-\pi t^2}$, the function having the unique property that it is its own Fourier transform, i.e.,

$$x(t) = e^{-\pi t^2} \leftrightarrow a(f) = e^{-\pi f^2}. \tag{11}$$

We must have $\Delta t = \Delta f$, which implies that $T=F=\sqrt{N}$. Theorem 1 now gives us Corollary 1.

Corollary 1:

$$X(j) = Tx_p(j\Delta t) = \sqrt{N}\sum_{k=-\infty}^{\infty} e^{-\pi(j+kN)^2/N} \tag{12}$$

and

$$A(n) = a_p(n\Delta f) = \frac{1}{\sqrt{N}}X(n), \quad j, n = 0, 1, \cdots, N-1,$$

form a finite Fourier transform pair.

The factor $\sqrt{N}$ arises because, for computational efficiency, we did not define the discrete Fourier transform so as to preserve the Euclidean norm in the transformation. Otherwise, if we had defined $X(j)$ and $A(n)$ with a factor $1/\sqrt{N}$ before the summation sign, we could have obtained a transform pair of identical functions which, for $-T/2<j\Delta t<T/2$, approximate the normalized Gaussian function (11) for large N.

A further word is in order at this point concerning the relationship between $a(f)$ and $a_p(f)$. Suppose Δt is chosen so that the aliasing involved in constructing $a_p(f)$ is negligible, i.e., $a(f)\approx 0$ for $|f|>\frac{1}{2}F$. As we said before, $a_p(f)$ is periodic of period F and if we view it from $-\frac{1}{2}F$ to $\frac{1}{2}F$ then from (5)

$$a(f) \approx a_p(f) \qquad \text{for } |f| \leq \tfrac{1}{2}F.$$

However, if we view it from 0 to F (as we do in the finite Fourier transform approach) then the negative f half of $a(f)$ is produced to the right of the positive f half of $a(f)$:

$$a(f) \approx a_p(f) \qquad 0 \leq f \leq \tfrac{1}{2}F$$
$$a(f-F) \approx a_p(f) \qquad \tfrac{1}{2}F \leq f \leq F.$$

In other words, $a_p(f)$ does not approximate $a(f)$ for $0\leq f\leq F$ but rather for $|f|\leq\frac{1}{2}F$. The relationship between $a_p(f)$ and $a(f)$ is illustrated in Fig. 1.

We will now discuss the application of Theorem 1 to the problem of using the fast Fourier transform algorithm for the evaluation of Fourier integrals. We will discuss the problem of evaluating $a(f)$ given $x(t)$, but the procedure for going in the other direction is a parallel one. Consider again the continuous function $x(t)$. In our formulation [see (7) and (8)] we have replaced the problem of calculating the frequency function $a(f)$ by the problem of calculating, from the sequence $x(j\Delta t)$, the periodic function, $a_p(f)$, constructed by adding the set of functions $a(f+kF)$ for $k=0$, ± 1, ± 2, $\cdots$. Hence, the error in approximating $a(f)$ by $a_p(f)$, in the range $-\frac{1}{2}F<f<\frac{1}{2}F$, is the sum of the $a(f+kF)$'s for $k\neq 0$. Thus, by choosing Δt small enough, $F=1/\Delta t$ can be made sufficiently large so that this error is negligible in the frequency range of interest. This proper choice of Δt is the first choice to make in applying the algorithm. If one is interested in the entire frequency range then Δt should be chosen so that the contributions of $a(f)$ for all $|f|>\frac{1}{2}F$ are negligible.

It is worthy of note here that (8) is precisely the formula one would have obtained in approximating the integral in (1) by the trapezoidal rule. The advantage of the present discussion is that the "integration error" is seen to be the difference between $a(f)$ and $a_p(f)$. Thus, if one has an estimate of the frequency function, one can predict the magnitude of the error introduced by the numerical integration.

With Δt chosen, one now has the freedom to choose the interval $\Delta f=1/(N\Delta t)$ at which one wants to sample

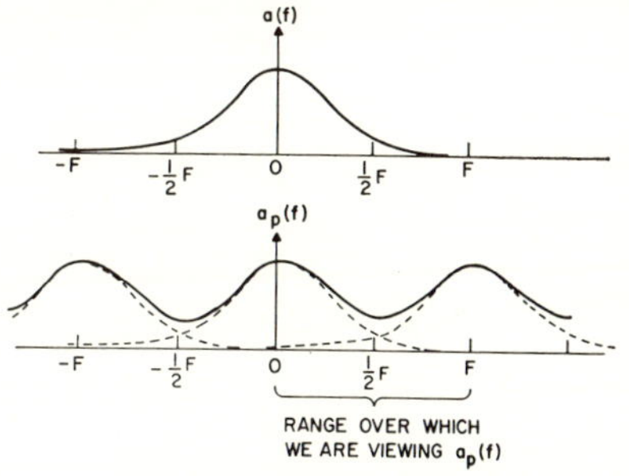

Fig. 1. Relationship between the frequency function $a(f)$ and the periodic aliased function $a_p(f)$.

$a_p(f)$. One does this by selecting N, and, for the sake of the fast Fourier transform algorithm, N should be highly composite (preferably a power of 2). In addition, N must be large enough so that the resulting Δf gives sufficiently fine spacing of the frequency estimates. With the choice of N we then form the aliased sequence $x_p(j\Delta t)=\sum_{l=-\infty}^{\infty} x(j\Delta t+lT)$ and apply the algorithm. In the application of the algorithm the sequence $X(j) = Tx_p(j\Delta t)$ would be formed. If we let $A(n)$ be the finite Fourier transform of $X(j)$ then $A(n)=a_p(n\Delta f)$.

With the formation of $x_p(j\Delta t)$, we have done in the time domain what was originally done in the frequency domain: we have formed a periodic sequence $x_p(j\Delta t)$ by displacing $x(j\Delta t)$ by 0, $\pm N$, $\pm 2N$, $\cdots$, points and adding the resulting sequences. The difference between $x_p(j\Delta t)$ and $a_p(n\Delta f)$ is that while Δt is chosen so as to make the frequency period $F=1/\Delta t$ just large enough to encompass the region where $a(f)$ is significantly different from zero, N need not be chosen so that $N\Delta t$ just encompasses the nonzero extent of $x(j\Delta t)$. Rather, N might be chosen to be less than or greater than the nonzero extent of $x(j\Delta t)$, depending on the frequency resolution desired. A choice of N for which $N\Delta t$ is less than the nonzero range of $x(j\Delta t)$ is illustrated in Fig. 2.

The method outlined above gives independent control over the upper frequency ($\frac{1}{2}F$) and frequency resolution ($\Delta f=1/T$) parameters. If you do not follow this method and do not do the aliasing of $x(j\Delta t)$ leading to $x_p(j\Delta t)$ you will, in some cases, obtain greater frequency resolution than you want and, hence, do more computing than necessary. Also note that the positioning of $x(t)$ in the interval $[0, N\Delta t=T]$ is automatically taken care of in the aliasing operation. Equation (9) places $x(t)$ in the interval $[0, T]$ without disturbing the phases of its components. Finally, stressing again a point made earlier, the finite Fourier transform will produce a sampling of the periodic function $a_p(f)$ from 0 to F, in which the negative f half is reproduced to the right of the positive f half. This might be confusing since experimenters are accustomed to viewing $a_p(f)$ from $-\frac{1}{2}F$ to $\frac{1}{2}F$.

In the preceding discussion we have assumed that

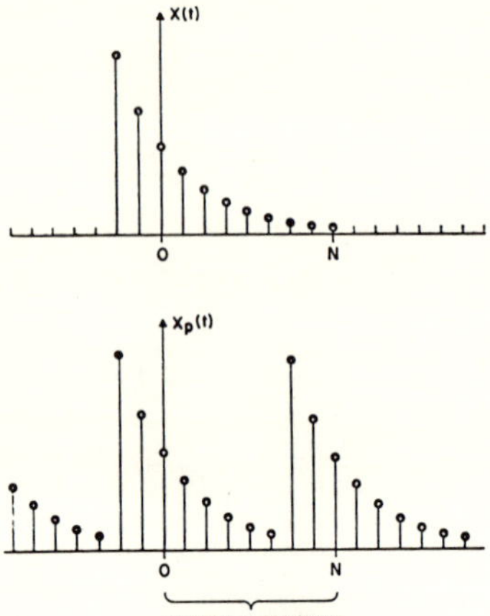

Fig. 2. Relationship between sampled values of the time function $x(t)$ and the periodic aliased function $x_p(t)$.

$x(t)$ is such that the infinite summation involved in obtaining $x_p(j\Delta t)$ could be performed [e.g., that $x(t)$ was nonzero only over a finite interval]. If the summation involved in $x_p(j\Delta t)$ must be truncated then the estimated frequency function will be convolved with a $\sin \pi f/\pi f$ shaped spectral window. This spectral window is not a desirable one and one might choose to apply the smoothing techniques used in spectral analysis for similarly truncated covariance functions.

Mention was also made above of the equivalence between trapezoidal rule integration and (7) for the discrete Fourier transform. Actually, trapezoidal rule integration yields better accuracy than one would anticipate for the type of functions considered here. To demonstrate this, consider the Euler–MacLaurin sum formula for integrating an arbitrary function $f(t)$:

$$\begin{aligned}\int_{t_0}^{t_N} f(t)dt &= \Delta t(\tfrac{1}{2}f_0 + f_1 + \cdots + f_{N-1} + \tfrac{1}{2}f_N) \\ &\quad - \frac{(\Delta t)^2}{12}(f_N' - f_0') + \frac{(\Delta t)^4}{720}(f_N''' - f_0''') \\ &\quad - \frac{(\Delta t)^6}{30240}(f_N^v - f_0^v) + \cdots, \end{aligned} \tag{13}$$

where primes denote derivatives. Here it can be seen that if the integrand and odd derivatives at the end points t_0, t_N are equal and/or negligible, then the discrete Fourier transform (7), terminated at indices corresponding to t_0 and t_N, becomes identical with the Euler–MacLaurin sum formula (13) applied to the integrand

$$f(t) = x(t)e^{-2\pi i t f}$$

with an integration interval Δt.

Relationship between Fourier Series and the Finite Fourier Transform: Application of the Algorithm to Harmonic Analysis and Synthesis

Suppose, we have a function $x(t)$ which is periodic of period T. Then $x(t)$ has a Fourier series expansion

$$x(t) = \sum_{n=-\infty}^{\infty} c(n)e^{2\pi i(nt/T)} \tag{14}$$

where the $c(n)$ are given by

$$c(n) = \frac{1}{T}\int_0^T x(t)e^{-2\pi i(nt/N)}dt. \tag{15}$$

Now, if we sample $x(t)$ at N equally spaced points between 0 and T, we generate the sequence $x(j\Delta t)$ where $\Delta t = T/N$. This sequence is periodic of period N; substituting in (14), we obtain

$$\begin{aligned} x(j\Delta t) = x(jT/N) &= \sum_{n=-\infty}^{\infty} c(n)e^{2\pi i(nj/N)} \\ &= \sum_{n=0}^{N-1}\left\{\sum_{l=-\infty}^{\infty} c(n+Nl)\right\} e^{2\pi i(nj/N)} \\ &= \sum_{n=1}^{N-1} c_p(n)e^{2\pi i(nj/N)}. \end{aligned} \tag{16}$$

Thus, we see that $x(j\Delta t)$ is the finite Fourier transform of

$$c_p(n) = \sum_{l=-\infty}^{\infty} c(n+Nl).$$

This is summarized by Theorem 2.

Theorem 2

If the periodic function $x(t)$ with period T has the Fourier series expansion $c(n)$,

$$x(t) \leftrightarrow c(n)$$

then the periodic sequence $x(j\Delta t)$ of period N, where $\Delta t = T/N$, has the finite Fourier transform $c_p(n)$:

$$x(j\Delta t) \leftrightarrow c_p(n) = \sum_{l=-\infty}^{\infty} c(n+lN).$$

From this we see that in using the algorithm for harmonic analysis we should pick an N such that the error due to aliasing in the approximation of $c(n)$ by $\sum_{l=-\infty}^{\infty} c(n+lN)$ is acceptable. Then let $\Delta t = T/N$ form $x(j\Delta t)$, and take its finite Fourier transform. Again, as with the Fourier transform, if we let

$$c_p(n) = \sum_{l=-\infty}^{\infty} c(n+lN)$$

$$\text{then } c_p(n) \approx c(n) \text{ for } n = 0, \cdots, N/2,$$

and $c_p(N-n) \approx c(-n)$ for $n = -1, \cdots, -N/2$.

The Calculation of Convolution Integrals

Consider the problem of computing the convolution integral

$$z(t) = \int_{-\infty}^{\infty} x(t-\tau)y(\tau)d\tau. \quad (17)$$

The convolution theorem of Fourier theory states that if

$$x(t) \leftrightarrow a(f)$$
$$y(t) \leftrightarrow b(f)$$
$$z(t) \leftrightarrow c(f)$$

then

$$c(f) = a(f)b(f). \quad (18)$$

The integral (17) may be thought of as describing a process which has an input function, or signal $x(t)$, which is averaged over points $t-\tau$ around each point t, with weights given by $y(\tau)$, the "impulse response function." The transform of $y(t)$, $b(f)$, is known as the transfer function and the output of the process is the convolution $z(t)$. For stable systems, the impulse response function $y(t)$ must decrease to zero or to negligible values for large $|t|$. Therefore, let us assume that $y(t)=0$ for $|t|>T_y$. In this case, (17) can be written

$$z(t) = \int_{-T_y}^{T_y} x(t-\tau)y(\tau)d\tau. \quad (19)$$

In a practical situation, one obtains signals such as $x(t)$ in segments of a finite length, say $0\leq t \leq T$. For such a segment, the output can be determined by (17) or (19) for $T_y\leq t\leq T-T_y$.

One can, of course, approximate (19) by numerical integration. The simplest and most convenient formula would be the trapezoidal rule which, applied to (19), gives

$$z(j\Delta t) \approx \Delta t \sum_{k=-K}^{k=K} x[(j-k)\Delta t]y(k\Delta t) \quad (20)$$

where $\Delta t = T/N$, $K=T_y/\Delta t$, and $j=0, 1, 2, \cdots, N-t$. From the convolution property of finite Fourier transforms we know that if $A(n)$ and $B(n)$ are the N-point Fourier transforms of $x(j\Delta t)$ and $y(j\Delta t)$, with indexes interpreted modulo N, then, the N-point Fourier transform of $z(j\Delta t)$ will be given, except for a wrap-around error, by

$$\Delta t N A(n)B(n). \quad (21)$$

This wrap-around error affects values of j in (20) where the index $j-k$ of $x[(j-k)\Delta t]$ goes out of the range $(0, N-1)$. The convolution, so computed, will have been computed as though $x(j\Delta t)$ repeated itself outside the $(0, N-1)$ interval; i.e., for $j<0$, the identity $x(j\Delta t) = x[(N+j)\Delta t]$ is assumed. Thus, $z(j\Delta t)$ is incorrect for $0\leq j<K$ and $N-K\leq j<N$. Valid results are obtained, however, for $K\leq j<N-K$. Therefore, if the next segment of signal is taken to start at $j=N-2K$, and the procedure is repeated, one will have valid outputs for a continuous set of $2(N-2K)$ points. This can be continued on such overlapping segments.

It is desirable to estimate the integration error in the use of the trapezoidal rule as was done earlier for Fourier integrals. To do so, we can assume that $x_p(t)$ and $y_p(t)$ are just periodic repetitions, with period T, of $x(t)$ and $y(t)$, respectively, in $[0, T]$ and that no "time-aliasing" is present. Next, we note that the values of $z_p(t)$, the "aliased" function $z(t)$ at sampling points $t_j=j\Delta t$, $j=0, 1, 2, \cdots, N-1$, are given by Theorem 1 as the Fourier synthesis of the "aliased" finite Fourier transform of $z(t)$,

$$Tz_p(j\Delta t) \leftrightarrow c_p(n\Delta f) \text{ (finite Fourier transform).} \quad (22)$$

The procedure considered here to obtain $z(t)$ would be to compute the finite Fourier transforms, $a_p(n\Delta f)$ and $b_p(n\Delta f)$,

$$Tx_p(j\Delta t) \leftrightarrow a_p(n\Delta f)$$
$$Ty_p(j\Delta t) \leftrightarrow b_p(n\Delta f), \quad (23)$$

and multiply them to get an approximation

$$\hat{c}_p(n\Delta f) = a_p(n\Delta f)b_p(n\Delta f) \quad (24)$$

to $c_P(f)$ at the points $f=n\Delta f$, $n=0, 1, \cdots, N-1$.

We define the aliasing error

$$e_a(f) = a_p(f) - a(f),$$

and similarly for $b(f)$ and $c(f)$. In most practical applications, one can assume that

$$e_a(f) \approx a(f+F) + a(f-F). \quad (25)$$

The error in the transform of the sequence $zp(j\Delta t)$ is

$$\hat{c}_p(f) - c_p(f) = [a(f)+e_a(f)][b(f)+e_b(f)] - c_p(f)$$
$$\approx a(f)e_b(f) + b(f)e_a(f), \quad (26)$$

where we neglect the product of the aliasing errors. This error is small if both $e_a(f)$ and $e_b(f)$ are small. Applying Parsevals theorem this yields a small mean-square error in the convolution.

The second source of error, in going from $c(f)$ to $z(t)$, was just shown to affect only the end regions of $z(t)$. It is eliminated by proper overlap.

The procedure just discussed may be summarized as follows:

1) Determine the sampling rate $\Delta t = 1/F$ such that the frequency domain aliasing, given by (26), is small.

2) Segment $x(t)$ into intervals of length T which yield efficient application of the algorithm.

3) Let $X(j)=x(j\Delta t)$ be the input to the FFT subroutine. The output will be

$$A(n) = \frac{1}{T} a_p(n\Delta f).$$

4) Let $Y(j) = y(j\Delta t)$ be the input to the FFT subroutine. The output will be

$$B(n) = \frac{1}{T} b_p(n\Delta f).$$

Perhaps some confusion can be avoided by pointing out again that indexes are to be interpreted modulo N so that $y(j\Delta t)$, for negative j, will appear in the input vector in the position corresponding to $Y(N+j)$, i.e., in the top half of the vector.

5) Compute

$$C(n) = TA(n)B(n) = \frac{1}{T} \hat{c}_p(n\Delta f).$$

6) With $C(n)$ as input to the FFT subroutine, do Fourier synthesis to get

$$Z(j) = z_p(j\Delta t), \qquad j = 0, 1, \cdots, N-1.$$

7) Let the next segment of signal start at $T - 2T_y$ and repeat the whole procedure. Note that the transform of the weight function $y(t)$ need not be recomputed.

This discussion applies to a system where the impulse response function $y(t)$ is nonzero for both positive and negative t. This is appropriate in cases where the input is a disturbance of a physical system or a spatial distribution of illumination, as in image processing. In the latter case, t is a space variable. When t is a time variable and the convolution describes a causal relationship between signal and response, $y(t)$ must be zero for $t<0$; i.e., cause must always precede effect. The only change to be made in the previous discussion is that $y(t) = 0$ for $t<0$ so that the lower limit of the integral in (17) is zero. Another effect of this is that there is no wraparound error for $N-K \le j < N$, and the output values of $Z(j)$ for these indices are correct. This means that the overlap of segments of signal need only be by an amount T_y rather than $2T_y$, or K instead of $2K$ sampling points.

Finally, one may indeed have a signal which has started at some point in time; i.e., $x(t) = 0$ for $t<0$. In this case, the upper limit of integration in (17) is t

$$Z(t) = \int_0^t x(t-\tau)y(\tau)d\tau. \tag{27}$$

This is the situation usually treated by Laplace transform methods. Here, it is a special case of the above procedure. Whether treating the first block of a segmented signal or taking the whole signal as a single segment, one would merely set

$$x(t) = 0 \qquad \text{for } -T_y \le t < 0$$

or, equivalently,

$$x(j\Delta t) = 0 \qquad \text{for } N-K \le j < N.$$

The Solution of Integral Equations of the Convolution Type

When dealing with processes described by the convolution integral (17) one often has one or both of two other associated problems: 1) to determine the impulse response $y(t)$ of a system when given an input $x(t)$ and an output $z(t)$; and 2) to determine the input $x(t)$ when given the output $z(t)$ and the impulse response.

These two problems are, in principle, the same, since $x(t)$ and $y(t)$ are interchangeable in the expression of the convolution. However, in practice, there is a difference in the time-domain aliasing which comes from the size of the regions over which $x(t)$ and $y(t)$ are significantly different from zero. Let us, for now, however, suppose that we are solving for $y(t)$ when given $x(t)$ and $z(t)$.

Again, there is a computational advantage is using Fourier transform methods. One computes the Fourier transform $a(f)$ and $c(f)$ and divides to get the transfer function

$$b(f) = c(f)/a(f). \tag{28}$$

If the impulse response function $y(t)$ is wanted, one need only perform Fourier synthesis to compute it from $b(f)$. However, it is often $b(f)$ that is really wanted for the subsequent signal processing.

Now, if $z(t)$ and $x(t)$ are given then we assume $z_p(t) = z(t)$ and $x_p(t) = x(t)$ in the interval $(0, T)$ and, by discrete Fourier transformations, we obtain $c_p(n\Delta f)$ and $a_p(n\Delta f)$, $n = 0, 1, \cdots, N-1$. The error in the frequency domain, between the desired transfer function $b_p(f)$ and the computed transfer function $c_p(f)/a_p(f)$, is

$$b_p(f) - \frac{c_p(f)}{a_p(f)} \approx \frac{b(f)e_a(f) + a(f)e_b(f)}{a_p(f)} \cdot \tag{29}$$

This result shows that the "integration error" amounts to an error in the frequency domain which is a linear function of the "overlap" of the frequency spectrum $a(f)$ of the input signal with the "aliasing" error $e_b(f)$ of the transfer function, the "overlap" of the transfer function $b(f)$ and the aliasing error $e_a(f)$ of the input signal, and the aliasing in $c_p(f)$. The error is also inversely proportional to the "aliased" input frequency function $a_p(f)$. If $a(f)$ is very small or zero for some f, then there is little or no input at the frequency f and the output signal can convey little or no information about the response of the system to such a frequency.

The result of this analysis is to evaluate the error incurred in converting the problem to a discrete one. It is seen that the numerator reflects the integration error—the aliasing error resulting from the size of Δt. The second cause of error is the smallness of the input frequency functions. The first of these yields criteria for selecting the sampling rate; the second yields criteria for evaluating or perhaps selecting an input function $x(t)$ for the purpose of accurately determining the transfer function of a system.

VI

Roundoff Error in Fast Fourier Transform

Editor's Comments on Papers 33 and 34

33 Welch: *A Fixed-Point Fast Fourier Transform Error Analysis*

34 Kaneko and Liu: *Accumulation of Round-Off Error in Fast Fourier Transforms*

The roundoff error in using the FFT algorithms depends on the algorithm, the type of arithmetic, the word length, and the radix. In the first paper [33], Welch presents an analysis for the fixed-point case. The array of numbers calculated at the successive intermediate stages increases on the average. Since the dynamic range of a fixed-point system is limited, a rescaling may be necessary at some or all of the stages in order to avoid overflow. There is, however, no way of predicting whether a rescaling is required beforehand. One has to resort to upper and lower bound on the mean-squared error. The paper also presents simulation results that correlate well with the analytical results. In the paper by Gentleman and Sande [26], Part V, an upper bound on the mean-squared error for floating-point FFT is presented. Weinstein calculates the mean-squared error for the special case of white-noise input (Weinstein, 1969). His simulation results show, however, that error expression agrees reasonable well when the input is a sinusoidal sequence. In the paper by Kaneko and Liu [34], a detailed analysis of the roundoff error in FTT using floating-point arithmetic is presented. Both rounding and chopping arithmetics are considered. Some special cases are included as examples that illustrate the general results. The reader should also refer to the paper by Oppenheim and Weinstein [13] in Part III for a discussion of the roundoff error in applying FFT to digital filtering.

Reference

Weinstein, C. J. (1969). Round-off Noise in Floating Point Fast Fourier Transform Computation, *IEEE Trans. Audio Electroacoustics,* **AU-17,** 209–215.

A Fixed-Point Fast Fourier Transform Error Analysis

PETER D. WELCH, Member, IEEE
IBM Watson Research Center
Yorktown Heights, N. Y.

Abstract

This paper contains an analysis of the fixed-point accuracy of the power of two, fast Fourier transform algorithm. This analysis leads to approximate upper and lower bounds on the root-mean-square error. Also included are the results of some accuracy experiments on a simulated fixed-point machine and their comparison with the error upper bound.

I. Introduction

In many situations there is interest in implementing the fast Fourier transform using fixed-point arithmetic. In this case the effect of the word size on the accuracy of the calculation is of obvious importance both with regard to the design of special-purpose machines and with regard to the accuracy attainable from existing machines. This paper contains an analysis of the fixed-point accuracy of the power of two, fast Fourier transform (FFT) algorithm. This analysis leads to approximate upper and lower bounds on the root-mean-square error. Also included are the results of some accuracy experiments on a simulated fixed-point machine and their comparison with the error upper bound.

II. The Finite Fourier Transform

If $X(j)$, $j=0, 1, \cdots, N-1$, is a sequence of complex numbers, then the finite Fourier transform of $X(j)$ is the sequence

$$A(n) = (1/N) \sum_{j=0}^{N-1} X(j) \exp{-2\pi i j n/N} \qquad (1)$$

$$n = 0, 1, \cdots, N-1.$$

The inverse transform is

$$X(j) = \sum_{n=0}^{N-1} A(n) \exp 2\pi i j n/N. \qquad (2)$$

In both of the above equations, $i=(-1)^{1/2}$. We will be considering a fixed-point calculation of these transforms using the fast Fourier transform algorithm [1], [2]. In connection with (1), we will consider the calculation of $NA(n)$ from $X(j)$. N^{-1} would then be included as an overall scale factor at the end. Now considering the calculation of $NA(n)$ from $X(j)$ or $X(j)$ from $A(n)$, Parseval's theorem states:

$$\sum_{j=0}^{N-1} |X(j)|^2 = N \sum_{n=0}^{N-1} |A(n)|^2$$

or

$$\sum_{n=0}^{N-1} |NA(n)|^2 = N \sum_{j=0}^{N-1} |X(j)|^2 \qquad (3)$$

and we see that the mean-square value of the result is N times the mean-square value of the initial sequence. This fact will be used below.

III. The Inner Loop of the Fast Fourier Transform Algorithm: Step-by-Step Scaling

The inner loop of the power of two FFT algorithm operates on two complex numbers from the sequence. It takes these two numbers and produces two new complex numbers which replace the original ones in the sequence.

Manuscript received February 26, 1969; revised April 9, 1969.
This work was supported in part by the Advanced Research Projects Agency, Dept. of Defense, Contract AF 19-67-C-0198.

Copyright © 1969 by the Institute of Electrical and Electronics Engineers, Inc.
Reprinted from *IEEE Trans. Audio Electroacoustics*, **AU-17**(2), 151–157 (1969)

Let $X_m(i)$ and $X_m(j)$ be the original complex numbers. Then, the new pair $X_{m+1}(i)$, $X_{m+1}(j)$ are given by

$$\begin{aligned} X_{m+1}(i) &= X_m(i) + X_m(j)W \\ X_{m+1}(j) &= X_m(i) - X_m(j)W \end{aligned} \tag{4}$$

where W is a complex root of unity. If we write these equations out in terms of their real and imaginary parts, we get

$$\begin{aligned} \mathrm{Re}\,\{X_{m+1}(i)\} &= \mathrm{Re}\,\{X_m(i)\} + \mathrm{Re}\,\{X_m(i)\}\ \mathrm{Re}\,\{W\} \\ &\quad - \mathrm{Im}\,\{X_m(i)\}\ \mathrm{Im}\,\{W\} \\ &\ \ \vdots \\ \mathrm{Im}\,\{X_{m+1}(j)\} &= \mathrm{Im}\,\{X_m(i)\} - \mathrm{Re}\,\{X_m(i)\}\ \mathrm{Im}\,\{W\} \\ &\quad - \mathrm{Im}\,\{X_m(j)\}\ \mathrm{Re}\,\{W\}. \end{aligned} \tag{5}$$

At each stage the algorithm goes through the entire sequence of N numbers in this fashion, two at a time. If $N=2^M$, then the number of such stages in the computation is M.

As we move from stage to stage through the calculation, the magnitudes of the numbers in the sequence generally increase which means that it can be kept properly scaled by right shifts. Consider first the root-mean square of the complex numbers. From (4) we have

$$\left[\frac{|X_{m+1}(i)|^2 + |X_{m+1}(j)|^2}{2}\right]^{1/2} = \sqrt{2}\left[\frac{|X_m(i)|^2 + |X_m(j)|^2}{2}\right]^{1/2}. \tag{6}$$

Hence, in the root-mean-square sense, the numbers (both real and complex) are increasing by $\sqrt{2}$ at each stage. Consider next the maximum modulus of the complex numbers. From (4) one can easily show that

$$\begin{aligned} &\max\{|X_m(i)|,\ |X_m(j)|\} \\ &\qquad \le \max\{|X_{m+1}(i)|,\ |X_{m+1}(j)|\} \\ &\qquad \le 2\max\{|X_m(i)|,\ |X_m(j)|\}. \end{aligned} \tag{7}$$

Hence the maximum modulus of the array of complex numbers is nondecreasing.

In what follows, we will assume that the numbers are scaled so that the binary point lies at the extreme left. With this assumption the relationships among the numbers is as shown in Fig. 1. The outside square gives the region of possible values, $\mathrm{Re}\{X_m(i)\}<1$ and $\mathrm{Im}\{X_m(i)\}<1$. The circle inscribed in this square gives the region $|X_m(i)|<1$. The inside square gives the region $\mathrm{Re}\{X_m(i)\}<1/2$, $\mathrm{Im}\{X_m(i)\}<1/2$. Finally, the circle inscribed in this latter square gives the region $|X_m(i)|<1/2$. Now if $X_m(i)$ and $X_m(j)$ are inside the smaller circle, then (7) tells us that $X_{m+1}(i)$ and $X_{m+1}(j)$ will be inside the larger circle and hence not result in an overflow. Consequently, if we control the sequence at the mth stage so that $|X_m(i)|<1/2$, we are certain we will have no overflow at the $m+1$st stage. However, if $X_m(i)$ and $X_m(j)$ are inside the smaller square, then it is possible for $X_{m+1}(i)$ or $X_{m+1}(j)$ to be outside the larger square and hence result in an overflow. Consequently, we cannot control the sequence to prevent overflow by keeping the absolute values of the real and imaginary parts less than one-half. Furthermore, the maximum absolute value of the real and imaginary parts can increase by more than a factor of two and hence a simple right shift is not a sufficient correction.

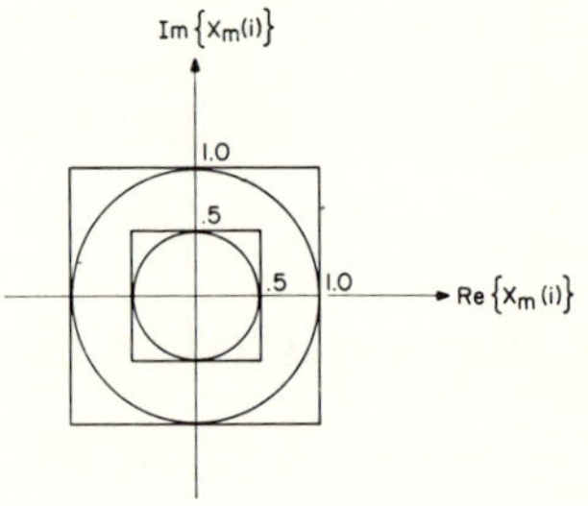

Fig. 1. Regions important to the rescaling of the sequence.

The above results and observations suggest a number of alternative ways of keeping the array properly scaled. The three that seem most reasonable are the following.

1) *Shifting Right One Bit At Every Iteration:* If the initial sequence, $X_0(i)$, is scaled so that $|X_0(i)|<1/2$ for all i and if there is a right shift of one bit after every iteration (excluding the last) then there will be no overflows.

2) *Controlling the Sequence so that* $|X_m(i)|<1/2$*:* Again assume the initial sequence is scaled so that $|X_0(i)|<1/2$ for all i. Then at each iteration we check $|X_m(i)|$ and if it is greater than one half for any i we shift right one bit before each calculation throughout the next iteration.

3) *Testing for an Overflow:* In this case the initial sequence is scaled so that $\mathrm{Re}\{X_0(i)\}<1$ and $\mathrm{Im}\{X_0(i)\}<1$. Whenever an overflow occurs in an iteration the entire sequence (part of which will be new results, part of which will be entries yet to be processed) is shifted right by one bit and the iteration is continued at the point at which the overflow occurred. In this case there could be two overflows during an iteration.

The first alternative is the simplest, but the least accurate. Since it is not generally necessary to rescale the sequence at each iteration, there is an unnecessary loss in accuracy. The second alternative is also not as accurate as possible because one less than the total number of bits available is being used for the representation of the sequence. This alternative also requires the computation of the modulus of every member of the sequence at each iteration. The third alternative is the most accurate. It has the disadvantage that one must process through the sequence an additional time whenever there is an overflow. The indexing for this processing is, however, straightforward. It would not be the complex indexing required for the algorithm. In comparing the speed of the second and third alternatives one would be comparing the speed of two overflow tests, two loads, two stores, and a transfer with that of the calculation or approximation of the modulus and a test of its magnitude. This comparison would depend greatly upon the particular machine and the particular approximation to the magnitude function.

A modification of the second alternative was adopted by Shively [3]. In this modification, if $|X_m(i)| > 1/2$, the right shift was made *after* each calculation in the next iteration. Provision was made for possible overflow. We will give an error analysis of the third alternative below. A microcoding performance study of this third alternative for the IBM 360/40 can be found in [4]. Although this error analysis applies to the third alternative it can be easily modified to apply to the second. In addition, the upper bound given applies directly to the first alternative. The analysis can also be modified for the power of four algorithm.

IV. A Fixed-Point Error Analysis

A. Introduction

We will assume, in this analysis, that the inputs [i.e., the real and imaginary parts of $X(j)$ or $A(n)$] are represented by B bits plus a sign. We assume the binary point lies to the left of the leftmost bit. We showed earlier that the magnitudes of the members of the sequence would generally increase as we moved from stage to stage in calculation. Hence, the method of operation is to test for overflow within the inner loop. If there is no overflow, the calculation proceeds as usual. If there is an overflow, then the two inputs producing the overflow are shifted right until there is no overflow. The amount of the shift is recorded (it will be either one or two bits) and the entire sequence is shifted right this same amount. In this scheme, we shift not only those elements we have already calculated but also those yet to be done. The total number of shifts is accumulated and the power of two, raised to the negative of this total number of shifts, constitutes an overall scale factor to be applied to the final sequence.

There are two operations which produce errors which are propagated through the calculation:

1) When two B bit numbers are multiplied together a $2B$ bit product results. If this product is rounded to B bits, an error whose variance is

$$\Delta_1^2 = 2^{-2B}/12 \tag{8}$$

is created. This error has a standard deviation of

$$\Delta_1 = 2^{-B}/\sqrt{12} \approx 0.3(2^{-B}). \tag{9}$$

2) When two B bit numbers are added together and there is an overflow, then the sum must be shifted right and a bit lost. If this bit is a zero, there is no error. If it is a one, there is an error of $\pm 2^{-B}$ depending upon whether the number is positive or negative. The variance of this error (it is unbiased assuming there are an equal number of positive and negative numbers) is

$$\Delta_2^2 = 2^{-2B}/2. \tag{10}$$

It has a standard deviation

$$\Delta_2 = 2^{-B-1/2} \approx 0.7(2^{-B}). \tag{11}$$

In addition, we will consider the effects of the propagation of errors present in the initial sequence. The variance of these errors we designate by δ^2. In the simplest case these errors would be the quantization errors resulting from the A/D conversion of an analog signal.

B. Upper Bound Analysis

In this section, we give an upper bound analysis of the ratio of the rms error to the rms of the completed transform. This upper bound is obtained by assuming that during each step of the calculation there is an overflow and a need to rescale. We let $X_k(j)$ be a typical real element at the kth stage (i.e., the real or imaginary part of a complex element) and let

$$\begin{aligned} V(X_k) &= \overline{\text{variance}\ \{X_k(j)\}} \\ &= \frac{1}{N}\sum_{j=0}^{N-1} \text{variance}\ \{X_k(j)\}. \end{aligned} \tag{12}$$

(This notation, a bar over the symbol indicating an average over the sequence, will be carried throughout the paper.) We will, in what follows, replace Δ_2^2 by $6\Delta_1^2$. We will also let $\Delta^2 = \Delta_1^2$.

Since the first stage gives an overflow, the original data must be rescaled or truncated by one bit. Hence,

$$V(X_0) = 6\Delta^2 + \delta^2. \tag{13}$$

In going from the original data to the results of the first stage, $W=1$ and, hence, there is no multiplication and we either add or subtract. Further, we assume that the next stage will result in an overflow and hence we will have to rescale. This gives

$$\begin{aligned} V(X_1) &= 2V(X_0) + 4\cdot 6\Delta^2 \\ V(X_1) &= 2(6\Delta^2) + 2\delta^2 + 4\cdot 6\Delta^2. \end{aligned} \tag{14}$$

In these expressions and this entire discussion, we are assuming all errors to be independent and, hence, that the variance of the sum is the sum of the variances. Going from the first stage to the second stage, we have $W=(-1)^{1/2}$ and again there are only additions and subtractions. Thus, with the rescaling,

$$\begin{aligned} V(X_2) &= 2V(X_1) + 4^2\cdot 6\Delta^2 \\ &= 2^2(6\Delta^2) + 2^2\delta^2 + 2(4\cdot 6\Delta^2) + 4^2\cdot 6\Delta^2. \end{aligned} \tag{15}$$

In going from the second stage to the third stage, we have multiplications and we have them in all subsequent stages. In generating the third stage, half the inner loops have multiplications. Consider the first equation of (5). All the other equations are identical in terms of error propagation. Remember that $X_3(i)$ is complex:

$$\operatorname{Re}\{X_3(i)\} = \operatorname{Re}\{X_2(i)\} + \operatorname{Re}\{X_2(j)\}\operatorname{Re}\{W\} - \operatorname{Im}\{X_2(j)\}\operatorname{Im}\{W\}. \tag{16}$$

Equation (16) yields, with rounding to B bits after the addition and with rescaling,

$$\begin{aligned} V'(X_3) &= V(X_2) + [\overline{\operatorname{Re}^2\{X_2(j)\}} + \overline{\operatorname{Im}^2\{X_2(j)\}}]V(W) \\ &\quad + [\operatorname{Re}^2(W) + \operatorname{Im}^2(W)]V(X_2) \\ &\quad + (4^3\Delta^2) + 4^3\cdot 6\Delta^2 \\ &= V(X_2) + \overline{|X_2(j)|^2}\Delta^2 + V(X_2) + (4^3\Delta^2) + 4^3\cdot 6\Delta^2. \end{aligned} \tag{17}$$

In (17), the first term is the variance of the first term of (16). The second and third terms of (17) are the variance of the full $2B$ bit products given by the second and third terms of (16). The fourth term of (17) is the result of rounding after the addition. The fifth term is the rescaling term. Finally, we saw in (6) that the average modulus squared of the complex numbers is increasing by a factor of 2 every stage. Hence, if we let K equal the average modulus squared of the initial array, i.e.,

$$K = \overline{|X_0(j)|^2} = \frac{1}{N}\sum_{j=0}^{N-1} |X_0(j)|^2,$$

then we have

$$V'(X_3) = 2V(X_2) + 2^2K\Delta^2 + 4^3\Delta^2 + 4^3\cdot 6\Delta^2. \tag{18}$$

Equation (18) would be correct for $V(X_3)$ if all the inner loops involved multiplications. However, at this stage only half of them do and, hence,

$$\begin{aligned} V(X_3) &= 2V(X_2) + 2K\Delta^2 + 4^3\Delta^2/2 + 4^3\cdot 6\Delta^2 \\ &= 2^3(6\Delta^2) + 2^3\delta^2 + 2^2(4\cdot 6\Delta^2) + 2(4^2\cdot 6\Delta^2) \\ &\quad + 4^3\cdot 6\Delta^2 + 2K\Delta^2 + 4^3\Delta^2/2. \end{aligned} \tag{19}$$

In the next stage, three quarters of the inner loops require multiplications and these multiplications get progressively more numerous as the stages increase. Hence, from here on, we will assume all stages have multiplications in all the inner loops. Thus, applying the above techniques, we get

$$\begin{aligned} V(X_4) &= 2^4(6\Delta^2) + 2^4\delta^2 + 2^3(6\cdot 4\Delta^2) + 2^2(6\cdot 4^2\Delta^2) \\ &\quad + 2(6\cdot 4^3\Delta^2) + 6\cdot 4^4\Delta^2 \\ &\quad + 2^2K\Delta^2 + 2^3K\Delta^2 + 4^3\Delta^2 + 4^4\Delta^2 \end{aligned} \tag{20}$$

and, generally, if M is the last stage

$$\begin{aligned} V(X_M) &= 2^M(6\Delta^2) + 2^M\delta^2 + 2^{M-1}(6\cdot 4\Delta^2) + \cdots \\ &\quad + 2(6\cdot 4^{M-1}\Delta^2) + 2^{M-2}K\Delta^2 + (M-3)2^{M-1}K\Delta^2 \\ &\quad + 2^{M-4}(4^3\Delta^2) + 2^{M-4}(4^4\Delta^2) + \cdots + (4^M\Delta^2) \\ &= (1.5)2^{M+2}\Delta^2(1 + 2 + \cdots + 2^{M-1}) + 2^M\delta^2 \\ &\quad + (M-2.5)2^{M-1}K\Delta^2 \\ &\quad + 2^{M+2}\Delta^2 + 2^{M+4}(1 + 2 \cdots + 2^{M-4})\Delta^2 \end{aligned} \tag{21}$$

or

$$\begin{aligned} V(X_M) &\approx (1.5)2^{2M+2}\Delta^2 + 2^M\delta^2 + (M-2.5)2^{M-1}K\Delta^2 \\ &\quad + 2^{M+2}\Delta^2 + 2^{2M+1}\Delta^2 \\ &\approx 2^{2M+3}\Delta^2 + 2^M\delta^2 + (M-2.5)2^{M-1}K\Delta^2 + 2^{M+2}\Delta^2. \end{aligned} \tag{22}$$

K is the average of the square of the absolute values of the initial complex array. Hence, applying Parseval's theorem (3), the average of the square of the absolute values of the final array will be 2^MK. What is most meaningful in this case, however, is the mean square of the real numbers, which is $2^MK/2$. Hence we have

$$\frac{V(X_M)}{2^MK/2} \approx \frac{2^{M+3}\Delta^2}{K/2} + \frac{2\delta^2}{K/2} + \frac{(M-2.5)\Delta^2/2}{1/2} + \frac{2^2\Delta^2}{K/2}, \tag{23}$$

and, finally, for large M,

$$\frac{\text{rms (error)}}{\text{rms (result)}} \approx \frac{2^{(M+3)/2}\Delta}{\sqrt{K/2}} \approx \frac{2^{(M+3)/2}2^{-B}(0.3)}{\text{rms (initial array)}}. \tag{24}$$

Equation (24) gives an approximate upper bound for the ratio of the rms of the error to the rms of the answer. Notice that this bound increases as the $\sqrt{N}$ or $\frac{1}{2}$ bit per stage.

C. Lower Bound Analysis

We will now obtain an approximate lower bound for the ratio of the rms of the error to the rms of the answer. We obtain this lower bound by assuming that there are no overflows in the calculation and, hence, no shifts of the array. In this case,

$$\begin{aligned} V(X_0) &= \delta^2 \\ V(X_1) &= 2\delta^2 \\ V(X_2) &= 2^2\delta^2. \end{aligned} \tag{25}$$

In the third stage, half of the inner loops involve a multiplication and, hence,

$$V(X_3) = (1/2)(2^2K\Delta^2) + 1/2(\Delta^2) + 2^3\delta^2. \tag{26}$$

This can be seen by considering the first term of (17). The first term of (26) comes from the second term of the first of equations (17). The second term of (26) is caused by the rounding to B bits. Now, as before,

$$\begin{aligned} V(X_4) &= 2V(X_3) + 2^3K\Delta^2 + \Delta^2 \\ &= 2^2K\Delta^2 + 2^3K\Delta^2 + \Delta^2 + \Delta^2 + 2^4\delta^2. \end{aligned} \tag{27}$$

Finally,

$$\begin{aligned} &V(X_M) \\ &= 2^{M-2}K\Delta^2 + (M-3)2^{M-1}K\Delta^2 + 2^{M-3}\Delta^2 + 2^{M-5}\Delta^2 \\ &\quad + 2^{M-6}\Delta^2 + \cdots + \Delta^2 + 2^M\delta^2 \\ &= (M-2.5)2^{M-1}K\Delta^2 + 2^{M-3}\Delta^2 \\ &\quad + (1 + \cdots + 2^{M-5})\Delta^2 + 2^M\delta^2 \\ &\approx (M-2.5)2^{M-1}K\Delta^2 + 2^{M-3}\Delta^2 + 2^{M-4}\Delta^2 + 2^M\delta^2. \end{aligned} \tag{28}$$

As in Section IV-B, the mean square of the final sequence of real numbers is $2^M \cdot K/2$. Hence, we have

$$\frac{V(X_M)}{2^MK/2} \approx (M-2.5)\Delta^2 + \frac{\Delta^2/8}{K/2} + \frac{\Delta^2/6}{K/2} + \frac{\delta^2}{K/2} \cdot \tag{29}$$

Now one has to be careful in interpreting (29) to obtain an approximate lower bound. In actuality, the only way to have a situation in which there are no shifts is to have a small K and, in fact, one which approaches zero as N (or M) becomes large. However, if we assume that the word size expands to the left as necessary rather than overflowing, then this analysis does provide a lower bound to the error. With this interpretation, as M becomes large, we have

$$\frac{\text{rms (error)}}{\text{rms (result)}} \approx (M-2.5)^{1/2}(.3)2^{-B}. \tag{30}$$

The lower bound increases as $M^{1/2} = \frac{1}{2}\log_2 N$. This is the rate of increase which has been observed for the floating-point calculation [5], [6].

D. Some Experimental Results

An IBM 7094 program was written to perform a fixed-point calculation using the fast Fourier transform algorithm, as described above. The program was capable of simulating a fixed-point machine of any word size up to 35 bits plus a sign. Experiments were run with fixed-point numbers of 17 bits plus a sign. This corresponds to $B=17$ in the analysis of Section IV-B and C.

We will now describe some experimental results. In these experiments we did not consider the propagation of the error present in the original sequence. Thus we considered the case where $\delta^2=0$. The experiments were performed as follows. Floating-point input was fixed to 17 bits plus a sign. This fixed input was then transformed with the fixed-point program. The fixed-point output was then floated. Next, the fixed-point 17-bit input was floated and a floating-point transform taken. Since this floating-point transform uses a floating-point word with a 27-bit mantissa, it was considered the correct answer. Finally, the rms of the difference between the fixed-point and floating-point answers was taken. We also obtained the maximum absolute error and average error.

Fig. 2 contains the result of transforming random numbers which lie between zero and one (placed in both the real and imaginary parts). In this and subsequent tests, three runs were made for every power of two from 8 to 2048. Since these random numbers have a dc component of one-half, the fixed-point program must rescale at least $N-1$ times. Hence, one would expect the error to lie close to the theoretical upper bound as given by (24). This theoretical upper bound is also plotted in Fig. 2 and the results are seen to lie slightly above it. The rms of the original array, $\sqrt{K/2}$, is approximately 0.58.

Fig. 3 contains the results of transforming three sine waves plus random numbers between zero and one-half in the real part and all zeros in the imaginary part. Specifically,

$$\begin{aligned} \text{Re}\{X(j)\} &= 1/2[Y(j) + (1/2)\sin(2\pi 8j/N) \\ &\quad + (1/4)\sin(2\pi 4j/N) \\ &\quad + (1/4)\sin(2\pi 8j/N)] \\ \text{Im}\{X(j)\} &= 0 \end{aligned}$$

where the $Y(j)$ are random numbers between zero and one. Again, there is a dc component of magnitude one-fourth and the array must be rescaled at least $N-2$ times. Thus, one would expect these results to be lower relative to the theoretical upper bound than the case depicted in Fig. 2. From Fig. 3 one can see that this is in fact the case. The rms of the original array $\sqrt{K/2}$ is, in this case, approximately 0.35. This is the reason the upper bound curve is higher than that of Fig. 2.

Fig. 4 contains the results of transforming random numbers from minus one to one (in both real and imaginary parts). In this case, the dc component is zero and there is no other strong component. The number of shifts should be approximately $(\log_2 N)/2$ or one-half shift per stage. Hence, one would expect the error curve to lie well below the theoretical upper bound, as is the case. In this case, $\sqrt{K/2}=0.58$.

Fig. 5 contains the results of an experiment identical to that used for Fig. 3, except that the random numbers are between $\pm\frac{1}{2}$. The results are as expected. In this case, $\sqrt{k/2}$-0.35.

Finally, Fig. 6 contains the results of transforming a sine wave in the real part and zero in the imaginary part. The sine wave was $\sin(2\pi j/8)$. Although in this case the array must be rescaled in at least $N-2$ times, the error is well below the upper bound. Here, $\sqrt{K/2}=0.5$.

In all these calculations the bias, as reflected by the average error, was negligible compared with the rms error. Furthermore, the maximum error was of the same order of magnitude as the rms error and hence the error was not due to the effect of a few, highly inaccurate terms.

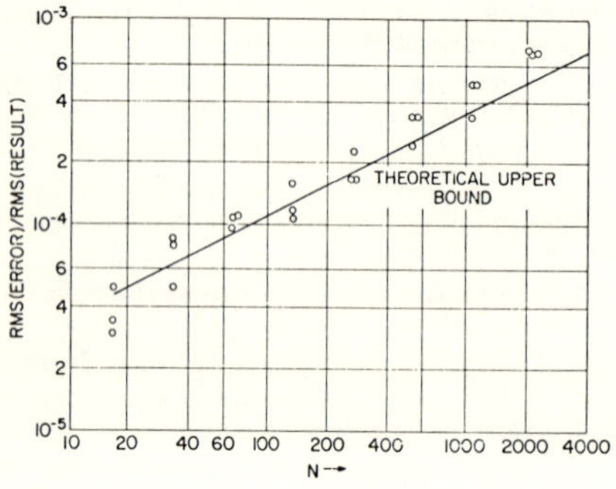

Fig. 2. Experimental error results: random numbers between 0 and 1; $B=17$.

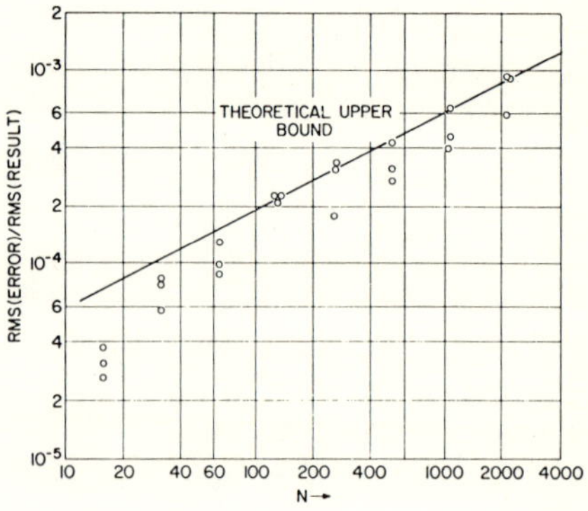

Fig. 3. Experimental error results: random numbers plus 3 sine waves; $0<\text{random numbers}<\frac{1}{2}$; $B=17$.

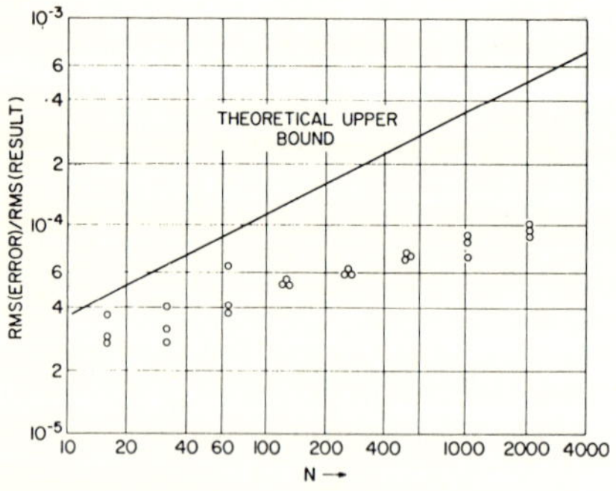

Fig. 4. Experimental error results: random numbers between -1 and 1; $B=17$.

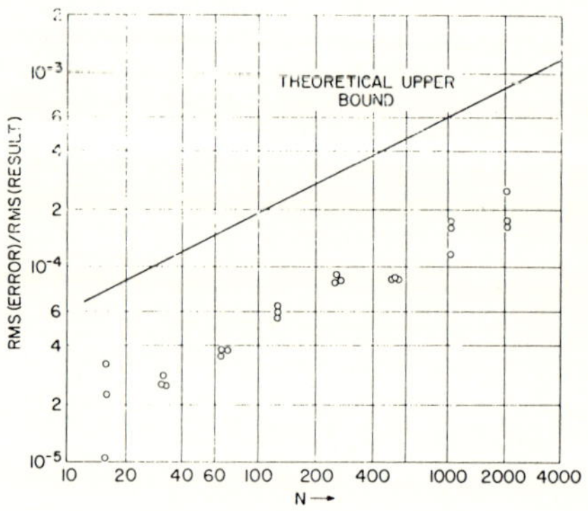

Fig. 5. Experimental error results: random numbers plus 3 sine waves; $-\frac{1}{2}<\text{random numbers}<\frac{1}{2}$; $B=17$.

Fig. 6. Experimental error results: single sine wave; $B=17$.

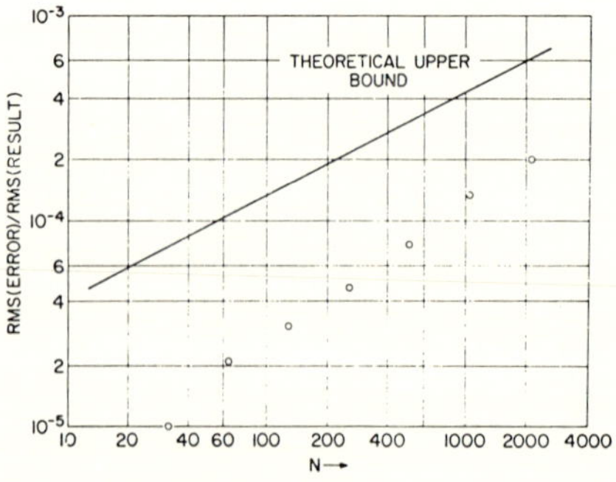

E. Conclusions and Additional Comments

The upper bound obtained in Section IV-B is of the form

$$\frac{\text{rms (error)}}{\text{rms (result)}} \leq \frac{2^{(M+3)/2}2^{-B}C}{\text{rms (initial sequence)}} \tag{31}$$

where $C=0.3$. On the basis of the experimental results we would recommend a bound with $C=0.4$.

We also carried through the analysis for a sign magnitude machine with truncation rather than rounding. In this case, the analytical upper bound was of the form given by (31) but with $C=0.4$. However, the experimental results were again higher and we would recommend a bound with $C=0.6$. The case of a twos-complement machine with truncation was not analyzed as analysis became exceedingly complex. However, experimental results indicated a bound of the form given by (31) with $C=0.9$.

It should be pointed out that if we are taking the transform to estimate spectra then we will be either averaging over frequency in a single periodogram or over time in a sequence of periodograms and this averaging will decrease the error discussed here as well as the usual statistical error. Finally, if we are taking a transform and then its inverse, Oppenheim and Weinstein have shown [7] that the errors in the two transforms are not independent.

Acknowledgment

The author would like to thank R. Ascher for assistance in programming the fixed-point calculations. He would also like to thank the referee for a number of corrections and valuable suggestions.

References

[1] J. W. Cooley and J. W. Tukey, "An algorithm for machine calculation of complex Fourier series," *Math. Comp.*, vol. 19, pp. 297–301, April 1965.

[2] J. W. Cooley, "Finite complex Fourier transform," SHARE Program Library: PK FORT, October 6, 1966.

[3] R. R. Shively, "A digital processor to generate spectra in real time," *1st Ann. IEEE Computer Conf., Digest of Papers*, pp. 21–24, 1967.

[4] "Experimental signal processing system," IBM Corp., 3rd Quart. Tech. Rept., under contract with the Directorate of Planning and Technology, Electronic Systems Div., AFSC, USAF, Hanscom Field, Bedford, Mass., Contract F19628-67-C-0198.

[5] J. W. Cooley, P. A. W. Lewis, and P. D. Welch, "The fast Fourier transform algorithm and its applications," IBM Corp., Res. Rept. RC 1743, February 9, 1967.

[6] W. M. Gentleman and G. Sande, "Fast Fourier transforms for fun and profit," *1966 Fall Joint Computer Conf., AFIPS Proc.*, vol. 29. Washington, D.C.: Spartan, 1966, pp. 563–578.

[7] A. V. Oppenheim and C. Weinstein, "A bound on the output of a circular convolution with application to digital filtering," this issue, pp. 120–124.

34

Reprinted from *J. Assoc. Computing Machinery,* **17**(4), 637–654 (1970)

Accumulation of Round-Off Error in Fast Fourier Transforms

TOYOHISA KANEKO* AND BEDE LIU

Princeton University,† Princeton, New Jersey

ABSTRACT. The fast Fourier transform (FFT) is an algorithm to compute the discrete Fourier coefficients with a substantial time saving over conventional methods. The finite word length used in the computer causes an error in computing the Fourier coefficients. This paper derives explicit expressions for the mean square error in the FFT when floating-point arithmetics are used. Upper and lower bounds for the total relative mean square error are given. The theoretical results are in good agreement with the actual error observed by taking the FFT of data sequences.

KEY WORDS AND PHRASES: FFT, finite word length effect, round-off accumulation, mean square error

CR CATEGORIES: 5.11

1. *Introduction*

The fast Fourier transform (FFT) [2, 3, 6] is an algorithm to compute the discrete Fourier coefficients efficiently. Because of the substantial time saving over the conventional methods, the fast Fourier transform has found important applications in a number of diverse fields such as the measurement of power spectra [1], the processing of speech signals [4], the simulation of and design of digital filters [7], and coherent optical signal processing [10].

There is an inherent accuracy problem in calculating the Fourier coefficients due to the finite word length used in digital computers. This problem has received little attention until recently. Gentleman and Sande [6] obtained an absolute upper bound on the sum of square errors due to round-off accumulation. Weinstein [12] derived the relative mean square error for the case of white noise data. The error problem of FFT using a fixed-point arithmetic was investigated by Welch [14].

This paper is concerned with the mean square error in the computation of the Fourier coefficients using the FFT method with floating-point arithmetic. We first derive an explicit expression for the error of each Fourier coefficient. This error depends on the number of bits used in the arithmetic, the accuracy of the data, as well as the data sequence itself. Two types of data, deterministic and random sequences, are considered, and the results are given for both rounding and chopping arithmetics. Simple upper and lower bounds for the error that are valid for all data sequences are given. The general expression for the error is then calculated for three special data sequences: a sinusoidal sequence, white noise, and random sequences with

* Present address: IBM, Thomas J. Watson Research Center, Yorktown Heights, New York.
† Department of Electrical Engineering. This work is supported by the Air Force Office of Scientific Research, Office of Aerospace Research, USAF, under Grant No. AFOSR-1333-67 and by the National Science Foundation under Grant GK-1439.

rational power spectral density. Good agreement between numerical experiments and the theory is obtained.

The organization of the paper is as follows. In Section 2 the problem is formulated and basic assumptions are stated. The main results are summarized in Sections 3 and 4. The derivation is given in Section 5. In Section 6 some results of the actual observed error are compared with the theoretical results. In Section 7 the effects of two other causes of the error of FFT are discussed.

2. *Formulation of the Problem*

The discrete Fourier transform of a sequence $\{x(n)\}_{n=0}^{N-1}$ is defined by [3] as

$$A(p) = \sum_{n=0}^{N-1} x(n) W^{np}, \qquad p = 0, 1, 2, \cdots, N-1, \tag{1}$$

where $W = e^{-j2\pi/N}$ and $j = \sqrt{-1}$. For simplicity, our discussion is restricted to the case where N is a power of 2; the results can be extended to radices other than 2. By using the FFT method, the Fourier coefficients $\{A(p)\}_{p=0}^{N-1}$ can be calculated in $m = \log_2 N$ iterative steps. At each step, an array of N complex numbers is generated by using only the numbers in the previous array. To explain the FFT algorithm, let each integer p, $p = 0, 1, 2, \cdots, N-1$, be expanded into a binary form as

$$p = 2^{m-1}p_0 + 2^{m-2}p_1 + \cdots + 2p_{m-2} + p_{m-1}, \qquad p_i = 0 \quad \text{or} \quad 1, \tag{2}$$

and let $\{A_k(p)\}_{p=0}^{N-1}$ denote the N complex numbers calculated at the kth step. Then the FFT algorithm[1] may be expressed as

$$A_{k+1}(p) = \begin{cases} A_k(p) + A_k(p + 2^{m-1-k}) & \text{if } p_k = 0, \qquad \text{(3a)} \\ [A_k(p - 2^{m-1-k}) - A_k(p)]w_k(p) & \text{if } p_k = 1, \qquad \text{(3b)} \end{cases}$$

where $w_k(p)$ is a power of W given by

$$w_k(p) = W^z,$$
$$\text{where } z = [2^k(2^{m-1-k}p_k + 2^{m-2-k}p_{k+1} + \cdots + 2p_{m-2} + p_{m-1}) - 2^{m-1}p_k].$$

It should be pointed out that the argument $p + 2^{m-1-k}$ in eq. (3a) and the argument $p - 2^{m-1-k}$ in eq. (3b) can be obtained from the binary expansion, eq. (2), by simply replacing p_k by its 1-complement. This fact will be used later in Section 5. Equation (3) is carried out for $k = 0, 1, 2, \cdots, m-1$, with $A_0(p) = x(p)$. It can be shown [6] that at the last step $\{A_m(p)\}_{p=0}^{N-1}$ are the discrete Fourier coefficients in a rearranged order. Specifically, $A_m(p) = A(p^*)$ with

$$p^* = 2_m^{-1}p_{m-1} + 2^{m-2}p_{m-2} + \cdots + 2p_1 + p_0 . \tag{4}$$

Figure 1 shows the flowgraph[2] of the actual computation for the case $N = 2^4$.

Due to the finite word length used in digital computers, all numbers can only be realized with finite bit accuracy and the results of all arithmetic operations must be

[1] This particular algorithm is known as decimation-in-frequency [6]. Another commonly used algorithm is known as decimation-in-time [3], which is not discussed in this paper.

[2] For readers not familiar with the notion of flowgraph, it consists of nodes and directed branches. Each node represents a variable which is the weighted sum of the variables at the originating nodes of all the branches that terminate on that node. The weights, if other than unity, are shown for each branch.

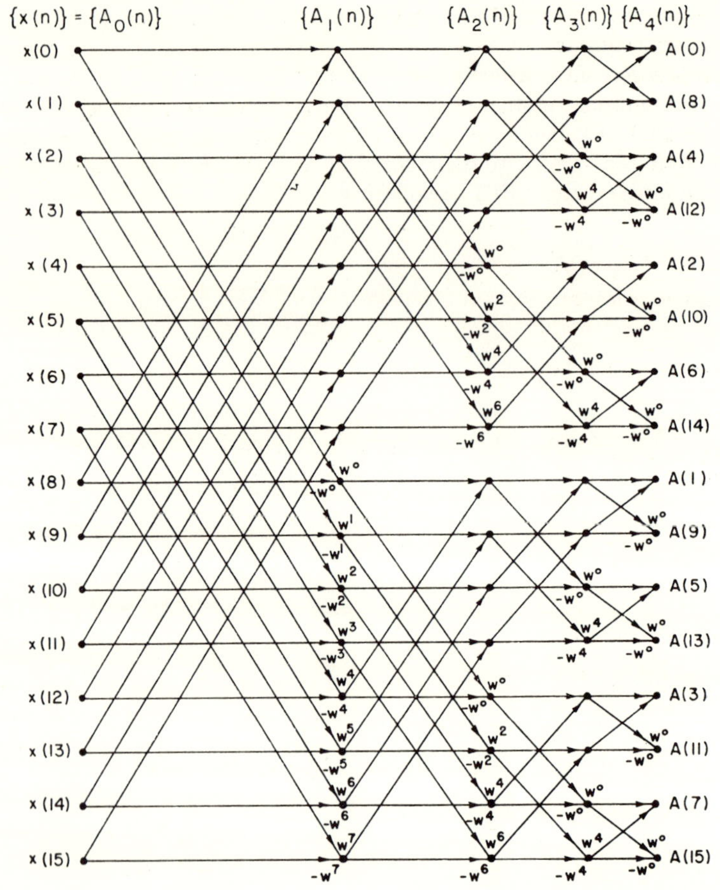

FIG. 1. Flowgraph of FFT (decimation in frequency), $N = 2^4$, $W^0 = 1$, $W^4 = j$

rounded. Therefore the actual array computed by using eq. (3) is in general different from $\{A_k(p)\}_{p=0}^{N-1}$. Let the actual computed array be denoted by $\{A_k'(p)\}_{p=0}^{N-1}$ and let $e_k(p)$ be the error of the pth element at step k,

$$e_k(p) = A_k'(p) - A_k(p). \tag{5}$$

Then

$$e(p) = e_m(p^*) \tag{6}$$

is the error committed in the computation of the Fourier coefficients $\{A(p)\}_{p=0}^{N-1}$ by using the FFT method.

Throughout this paper, we are concerned with binary machines using floating-point arithmetic with a double precision accumulator. Thus each machine number is expressed in the form of (sign) $\cdot 2^a \cdot b$, where the exponent a is an integer and the mantissa or the fractional part b is between $\frac{1}{2}$ and 1. We shall assume that t bits have been allotted to the mantissa and that enough bits have been assigned to the exponent so that no computed numbers will lie outside the permissible range.

The notation fl (·) [15] will be used to denote the machine number resulting from the arithmetic operation specified by the expression inside the parentheses in some designated order. Two standard round-off procedures for arithmetic operations [5, 15], rounding and chopping, will be considered. In rounding, a 1 or a 0 is added to the tth bit in accordance with whether the $(t+1)$-th bit of the result of an arithmetic operation is 1 or 0. In chopping, those bits beyond the most significant t bits are simply dropped.

It is known [5, 15] that, if the mantissa has t bits, then

$$\mathrm{fl}(x+y) = (x+y)(1+\alpha), \tag{7}$$

$$\mathrm{fl}(x \cdot y) = xy(1+\beta), \tag{8}$$

where the errors α and β satisfy $-2^{-t} \leq \alpha, \beta \leq 2^{-t}$ for rounding and $-2^{-t+1} \leq \alpha$, $\beta \leq 0$ for chopping.

Since the errors α and β arise from round-off, we may assume that they are random variables and independent of x and y, provided that the length of the mantissa t is not too short [8, 11]. Moreover, they are assumed to be uniformly distributed in $(-2^{-t}, 2^{-t})$ for rounding and in $(-2^{-t+1}, 0)$ for chopping. These assumptions have been shown to be in good agreement with experimental data [9, 13].[3]

3. *Summary of Results*

The main results of this paper are summarized in this section; the derivation is given in Section 5. As in Section 2, let $e(p)$ be the error in computing the Fourier coefficient $A(p)$ from $N = 2^m$ data samples $\{x(n)\}_{n=0}^{N-1}$ by using the FFT decimation-in-frequency algorithm, and let p be expanded in a binary form

$$\begin{aligned} p &= 2^{m-1}p_0 + 2^{m-2}p_1 + \cdots + 2p_{m-2} + p_{m-1}, \\ p_i &= 1 \quad \text{or} \quad 0, \quad i = 0, 1, \cdots, m-1. \end{aligned} \tag{2}$$

(a) *Rounding arithmetic.* When rounding arithmetic is used, it is shown in Section 5 that the mean of $e(p)$ is zero,

$$\mathcal{E}\{e(p)\}^R = 0 \tag{9}$$

and the mean square value of $e(p)$ is given by

$$\mathcal{E}\{|e(p)|^2\}^R = \frac{2^{-2t}}{3}\left[|A(I_{m-1,p})|^2 + \sum_{k=0}^{m-2} c_k 2^{-m+k+1} \sum_{i_{k+1}=0}^{1} \cdots \sum_{i_{m-1}=0}^{1} |A(I_{k,p})|^2\right], \tag{10}$$

where the superscript R denotes "rounding," t is the number of bits in the mantissa,

[3] For nonbinary machines such as the IBM 360 series where base 16 is used, the assumption that α and β are uniformly distributed is not appropriate. However, as will be seen, the results derived in this paper depend only on the mean and variances of the random errors α and β due to round-off at each arithmetic step. Thus the results can be extended straightforwardly to nonbinary machines provided only the mean and variance of the individual round-off errors can be either calculated or measured.

$\mathcal{E}$ denotes the expectation, and

$$c_k = \begin{cases} 1 + 2p_{m-1-k}, & 0 \le k \le m-3, \\ 1, & k = m-2,\ m-1. \end{cases} \tag{11}$$

The index $I_{k,p}$ is an integer between 0 and $N - 1$, and depends on the binary numbers $i_{k+1}, i_{k+2}, \cdots, i_{m-1}$ through the relationship

$$\begin{aligned} I_{k,p} = {} & 2^{m-1} i_{m-1} + 2^{m-2} i_{m-2} + \cdots + 2^{k+1} i_{k+1} \\ & + 2^k p_{m-1-k} + \cdots + 2p_{m-2} + p_{m-1}. \end{aligned} \tag{12}$$

(b) *Chopping arithmetic.* When chopping arithmetic is used, the mean of $e(p)$ is given by

$$\mathcal{E}\{e(p)\}^C = -2^{-t} \sum_{k=0}^{m-1} c_k A(p) \tag{13}$$

and the mean square value of $e(p)$ is given by

$$\mathcal{E}\{| e(p) |^2\}^C = 2^{-2t} \left(\sum_{k=0}^{m-1} c_k \right)^2 | A(p) |^2 + \mathcal{E}\{| e(p) |^2\}^R, \tag{14}$$

where c_k is defined in eq. (11) and $\mathcal{E}\{| e(p) |^2\}^R$ is the mean square error for rounding given by eq. (10). Here the superscript C denotes "chopping."

(c) *Random* $\{x(n)\}$. If the sequence $\{x(n)\}_{n=0}^{N-1}$ consists of N consecutive samples taken from a discrete parameter random process with autocorrelation function $R_{xx}(k)$, the mean square value of the error $e(p)$ may be expressed as

$$\begin{aligned} \mathcal{E}\{| e(p) |^2\}^R = {} & \frac{2^{-2t}}{3} \sum_{k=0}^{m-1} c_k 2^{m-k-1} \sum_{n=-(2^{k+1}-1)}^{(2^{k+1}-1)} (2^{k+1} - | n |) \\ & \times R_{xx}(n \cdot 2^{m-1-k}) W^{n2^{m-1-k} J_{k,p}} \end{aligned} \tag{15}$$

where $W = e^{-j2\pi/N}$ and

$$J_{k,p} = 2^k p_{m-1-k} + 2^{k-1} p_{m-2-k} + \cdots + 2p_{m-2} + p_{m-1} \tag{16}$$

and

$$\begin{aligned} \mathcal{E}\{| e(p) |^2\}^C = {} & 2^{-2t} \left(\sum_{k=0}^{m-1} c_k \right)^2 \times \sum_{n=-(N-1)}^{(N-1)} (N - | n |) R_{xx}(n) W^{pn} \\ & + \mathcal{E}\{| e(p) |^2\}^R, \end{aligned} \tag{17}$$

where $\mathcal{E}\{| e(p) |^2\}^R$ is given by eq. (15).

(d) *Total relative mean square error.* Quite often, one is interested in an overall measure of the error rather than the error in each Fourier coefficient. It can be shown that the total relative mean square error is bounded by

$$m \cdot \frac{2^{-2t}}{3} \le \frac{\sum_{p=0}^{N-1} \mathcal{E}\{| e(p) |^2\}^R}{\sum_{p=0}^{N-1} | A(p) |^2} < 3m \cdot \frac{2^{-2t}}{3} \tag{18}$$

for rounding arithmetic and

$$m^2 2^{-2t} + m\cdot\frac{2^{-2t}}{3} \leq \frac{\sum_{p=0}^{N-1} \mathcal{E}\{| e(p) |^2\}^C}{\sum_{p=0}^{N-1} | A(p) |^2} < 9m^2\cdot 2^{-2t} + 3m\cdot\frac{2^{-2t}}{3} \tag{19}$$

for chopping arithmetic. For a random data sequence, $| A(p) |^2$ in the denominator should be replaced by $\mathcal{E} | A(p) |^2$.

4. *Special Cases*

It is interesting to calculate the mean square error given in Section 3 for some special data sequences.

(a) *Sinusoidal data.* Consider the data

$$x(n) = e^{j(\omega nT+\varphi)}$$

where ω is the angular frequency, T is the sampling period, and φ is the phase. Let $\theta = \omega T$ and assume $\theta \neq 2\pi p 2^{-m}$, $p = 0, \cdot, \cdots, 2^m - 1$. Then the Fourier coefficients $\{A(p)\}$ can be found straightforwardly by using eq. (1),

$$A(p) = e^{j\varphi}(1 - \exp j\theta\cdot 2^m)/[1 - \exp j(\theta - 2\pi p\cdot 2^{-m})].$$

It can be shown, after considerable algebra, that eq. (10) becomes

$$\mathcal{E}\{| e(p) |^2\}^R = \frac{2^{-2t}}{3}\sum_{k=0}^{m-1} c_k\, 2^{m-k-1} \frac{| 1 - e^{j(\theta-2\pi p 2^{-m})} |^2}{| 1 - e^{j(\theta-2\pi 2^{-m}J_{k,p})2^{m-k-1}} |^2} \tag{20}$$

and $J_{k,p}$ is given by eq. (16). The error for chopping arithmetic can be calculated easily from eq. (14) by using the expression for $A(p)$ just derived.

(b) *White noise data.* The simplest case of random data sequence is white noise and its autocorrelation function is

$$R_{xx}(n) = \begin{cases} 1, & n = 0, \\ 0, & n \neq 0. \end{cases}$$

Substituting this autocorrelation function into eq. (15) we have

$$\mathcal{E}\{| e(p) |^2\}^R = \frac{2^{-2t}}{3} 2^m \sum_{k=0}^{m-1} c_k. \tag{21}$$

Also by taking the summation of eq. (21) over all p and on using the relationship

$$\sum_{p=0}^{N-1}\sum_{k=0}^{m-1} c_k = m2^m + 2\cdot 2^{m-1}(m - 2),$$

we have

$$\sum_{p=0}^{N-1} \mathcal{E}\{| e(p) |^2\}^R = \frac{2^{-2t}}{3} 2^{2m}\cdot 2(m - 1).$$

Since the sum of squared Fourier coefficient is

$$\sum_{p=0}^{N-1} \mathcal{E} | A(p) |^2 = 2^{2m},$$

the error-to-signal ratio is given by

$$\sum_{p=0}^{N-1} \mathcal{E}\{|e(p)|^2\}^R \Big/ \sum_{p=0}^{N-1} \mathcal{E}|A(p)|^2 = 2(m-1)\cdot\frac{2^{-2t}}{3} \tag{22}$$

which is similar to a result given by Weinstein [12].

For chopping arithmetic, it can be shown that eq. (17) reduces in this case to

$$\mathcal{E}\{|e(p)|^2\}^C = 2^{-2t}2^m\left(\sum_{k=0}^{m-1} c_k\right)^2 + \frac{2^{-2t}}{3}2^m\sum_{k=0}^{m-1} c_k \tag{23}$$

and the total error-to-signal ratio is

$$\sum_{p=0}^{N-1} \mathcal{E}\{|e(p)|^2\}^C \Big/ \sum_{p=0}^{N-1} \mathcal{E}|A(p)|^2 = 2^{-2t}(4m^2 - 7m + 2) + \frac{2^{-2t}}{3}2(m-1). \tag{24}$$

(c) *Random data sequence with rational power spectral density.* Consider the wide sense stationary random sequence $\{x(n)\}$ with autocorrelation function given by

$$R_{xx}(n) = \sum_{i=1}^{L} \rho_i e^{-\nu_i|n|}\cos 2\pi\xi_i n,$$

where ρ_i, $\nu_i \geq 0$ for $i = 1, 2, \cdots, L$. This class of random sequence has rational power spectral density and is encountered frequently in practice. For example, the case with $L = 1$, and $\xi_1 = 0$ gives the familiar Markov sequence. Equation (15) in this case becomes

$$\mathcal{E}\{|e(p)|^2\}^R = \frac{2^{-2t}}{3}2^{m-1}\sum_{i=1}^{L}\rho_i\sum_{k=0}^{m-1} c_k \operatorname{Re}\left\{\frac{1+\Gamma_{1i}}{1-\Gamma_{1i}} + \frac{1+\Gamma_{2i}}{1-\Gamma_{2i}} - 2^{-k}\left[\frac{\Gamma_{1i}(1-\Gamma_{1i}^{2^{k+1}})}{(1-\Gamma_{1i})^2} + \frac{\Gamma_{2i}(1-\Gamma_{2i}^{2^{k+1}})}{(1-\Gamma_{2i})^2}\right]\right\}, \tag{25}$$

where

$$\Gamma_{1i} = \exp\left[-2^{m-k-1}\nu_i - 2\pi j\, 2^{-k-1}(J_{k,p} + 2^m\xi_i)\right],$$

$$\Gamma_{2i} = \exp\left[-2^{m-k-1}\nu_i - 2\pi j\, 2^{-k-1}(J_{k,p} - 2^m\xi_i)\right]$$

and $J_{k,p}$ is given by eq. (16). The result for chopping arithmetic is

$$\mathcal{E}\{|e(p)|^2\}^C = 2^{-2t}\left(\sum_{k=1}^{m-1} c_k\right)^2\sum_{i=1}^{L}\rho_i \operatorname{Re}\left\{2^{m-1}\left[\frac{1+\Delta_{1i}}{1-\Delta_{1i}} + \frac{1+\Delta_{2i}}{1-\Delta_{2i}}\right] - \frac{\Delta_{1i}(1-\Delta_{1i}^{2^m})}{(1-\Delta_{1i})^2} - \frac{\Delta_{2i}(1-\Delta_{2i}^{2^m})}{(1-\Delta_{2i})^2}\right\} + \mathcal{E}\{|e(p)|^2\}^R \tag{26}$$

where

$$\Delta_{1i} = \exp\left[-\nu_i - 2\pi j(2^{-m}p + \xi_i)\right], \qquad \Delta_{2i} = \exp\left[-\nu_i - 2\pi j(2^{-m}p - \xi_i)\right]$$

and $\mathcal{E}\{|e(p)|^2\}^R$ is given by eq. (25).

5. *Derivation of Results*

In eq. (3) the $\{A_k(p)\}$ are complex numbers, so their real and imaginary parts are calculated separately. Let

$$\begin{aligned} B_k(p) &= \mathrm{Re}[A_k(p)], \qquad C_k(p) = \mathrm{Im}[A_k(p)], \\ U_k(p) &= \mathrm{Re}[w_k(p)], \qquad V_k(p) = \mathrm{Im}[w_k(p)], \end{aligned} \tag{27}$$

where the notations Re[·] and Im[·] denote, respectively, the real and imaginary parts of the quantity inside the bracket [·]. Equation (3) can be rewritten as

$$\begin{aligned} B_{k+1}(p) &= B_k(p) + B_k(q), \\ C_{k+1}(p) &= C_k(p) + C_k(q) \qquad \text{if } p_k = 0, \\ B_{k+1}(p) &= [B_k(r) - B_k(p)]U_k(p) - [C_k(r) - C_k(p)]V_k(p), \\ C_{k+1}(p) &= [C_k(r) - C_k(p)]U_k(p) + [B_k(r) - B_k(p)]V_k(p) \qquad \text{if } p_k = 1, \end{aligned} \tag{28}$$

where $q = p + 2^{m-1-k}$ and $r = p - 2^{m-1-k}$. On using the prime to denote the actual calculated results, the real and imaginary parts of $A'_{k+1}(p)$ are given by

$$\begin{aligned} B'_{k+1}(p) &= \mathrm{fl}\{B_k'(p) + B_k'(q)\}, \\ C'_{k+1}(p) &= \mathrm{fl}\{C_k'(p) + C_k'(q)\} \qquad \text{if } p_k = 0, \\ B'_{k+1}(p) &= \mathrm{fl}\{[B_k'(r) - B_k'(p)]U_k(p) - [C_k'(r) - C_k'(p)]V_k(p)\}, \\ C'_{k+1}(p) &= \mathrm{fl}\{[C_k'(r) - C_k'(p)]U_k(p) + [B_k'(r) - B_k'(p)]V_k(p)\} \\ &\qquad \text{if } p_k = 1. \end{aligned} \tag{29}$$

By introducing an error for each of the arithmetic steps in accordance with eqs. (7) and (8), we may write eq. (29) (see Figure 2) as

$$\begin{aligned} B'_{k+1}(p) &= [B_k'(p) + B_k'(q)](1 + \gamma'_{k,p}), \\ C'_{k+1}(p) &= [C_k'(p) + C_k'(q)](1 + \gamma''_{k,p}) \qquad \text{if } p_k = 0, \\ B'_{k+1}(p) &= [B_k'(r) - B_k'(p)]U_k(p)(1 + \delta'_{k,p})(1 + \zeta'_{k,p})(1 + \lambda'_{k,p}) \\ &\quad - [C_k'(r) - C_k'(p)]V_k(p)(1 + \delta''_{k,p})(1 + \zeta''_{k,p})(1 + \lambda'_{k,p}) \\ C'_{k+1}(p) &= [C_k'(r) - C_k'(p)]U_k(p)(1 + \epsilon'_{k,p})(1 + \eta'_{k,p})(1 + \lambda''_{k,p}) \\ &\quad + [B_k'(r) - B_k'(p)]V_k'(p)(1 + \epsilon''_{k,p})(1 + \eta''_{k,p})(1 + \lambda''_{k,p}) \\ &\qquad \text{if } p_k = 1, \end{aligned} \tag{30}$$

where the quantities $\gamma'_{k,p}$, $\gamma''_{k,p}$, $\delta'_{k,p}$, $\delta''_{k,p}$, $\epsilon'_{k,p}$, $\epsilon''_{k,p}$, $\zeta'_{k,p}$, $\zeta''_{k,p}$, $\eta'_{k,p}$, $\eta''_{k,p}$, $\lambda'_{k,p}$, $\lambda''_{k,p}$ are independent random variables uniformly distributed in $(-2^{-t}, 2^{-t})$ or in $(-2^{-t+1}, 0)$ depending whether rounding or chopping arithmetic is used.

It should be noted, however, that when $w_k(p)$ is 1 or j, the multiplication by $w_k(p)$ introduces no error, and therefore the multiplication errors $\eta'_{k,p}$, $\eta''_{k,p}$, $\zeta'_{k,p}$, and $\zeta''_{k,p}$ are all equal to zero. Also, the addition error $\lambda'_{k,p}$ and $\lambda''_{k,p}$ are zero in this case, since either $U_k(p) = 0$ or $V_k(p) = 0$. This happens for $k = m - 1$ and $k = m - 2$ where all $w_k(p)$ are either 1 or j. Thus $\eta'_{k,p} = \eta''_{k,p} = \zeta'_{k,p} = \zeta''_{k,p} = \lambda'_{k,p} =$

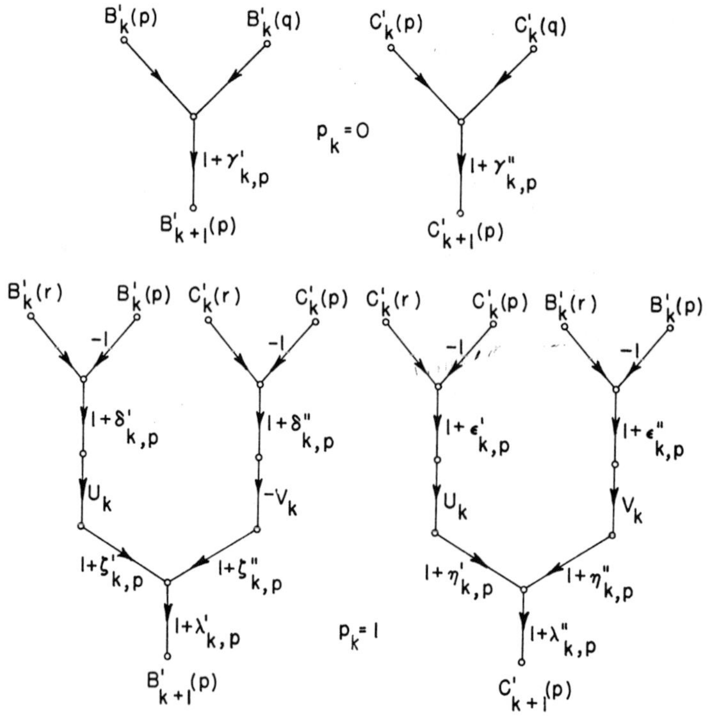

FIG. 2. Flowgraph of errors in computing $A_{k+1}(p)$

$\lambda''_{k,p} = 0$ for $k = m - 1$ and $k = m - 2$. For each $k \le m - 3$, all but two of the 2^{m-1-k} different values of $w_k(p)$ are neither 1 nor j. However, the analysis to be presented shall still assign an error for the multiplication by those two $w_k(p)$. Although it is straightforward to set the appropriate error to zero, the final result becomes rather a complicated function of the intermediate values $A_k(p)$ and is not too useful. Therefore, multiplication errors will be assumed for all $w_k(p)$ except for $k = m - 1$ and $m - 2$ where the multiplication errors will be taken to be zero. This represents a slightly pessimistic approach, and our final result will be larger than the actual error by a small amount.

The error $e_k(p)$ defined in eq. (5) is complex and can be written as

$$e_k(p) = B_k'(p) - B_k(p) + j[C_k'(p) - C_k(p)], \qquad k = 1, 2, \cdots, m, \quad p = 0, 1, \cdots, N-1 \tag{31}$$

with

$$e_0(p) = 0, \qquad p = 0, 1, \cdots, N-1. \tag{32}$$

From eqs. (28), (30), and (31), we have

$$e_{k+1}(p) = \begin{cases} e_k(p) + e_k(q) + f_k(p) & \text{if } p_k = 0, \quad (33a) \\ [e_k(r) - e_k(p)]w_k(p) + f_k(p) & \text{if } p_k = 1, \quad (33b) \end{cases}$$

where $f_k(p)$ is given by

$$
\begin{aligned}
f_k(p) &= \gamma'_{k,p}[B_k'(p) + B_k'(q)] + j\gamma''_{k,p}[C_k'(p) + C_k'(q)] \qquad \text{if } p_k = 0,\\
&= [(1 + \delta'_{k,p})(1 + \zeta'_{k,p})(1 + \lambda'_{k,p}) - 1][B_k'(r) - B_k'(p)]U_k(p)\\
&\quad - [(1 + \delta''_{k,p})(1 + \zeta''_{k,p})(1 + \lambda'_{k,p}) - 1][C_k'(r) - C_k'(p)]V_k(p)\\
&\quad + j[(1 + \epsilon'_{k,p})(1 + \eta'_{k,p})(1 + \lambda''_{k,p}) - 1][C_k'(r) - C_k'(p)]U_k(p)\\
&\quad + j[(1 + \epsilon''_{k,p})(1 + \eta''_{k,p})(1 + \lambda''_{k,p}) - 1][B_k'(r) - B_k'(p)]V_k(p)\\
&\qquad \text{if } p_k = 1.
\end{aligned} \tag{34}
$$

Thus we see the accumulation of round-off error is determined by the recursive eqs. (33) and (34) with initial conditions given by eq. (32). To calculate the mean square value of the error $e(p) = e_m(p^*)$, we proceed separately for the case of rounding arithmetic and chopping arithmetic.

(a) *Rounding arithmetic.* We first calculate some statistics of $f_k(p)$. On using the fact that the random variables $\gamma'_{k,p}$, $\gamma''_{k,p}$, $\delta'_{k,p}$, $\delta''_{k,p}$, $\epsilon'_{k,p}$, $\epsilon''_{k,p}$, $\zeta'_{k,p}$, $\zeta''_{k,p}$, $\eta'_{k,p}$, $\eta''_{k,p}$, $\lambda'_{k,p}$, $\lambda''_{k,p}$ are independent and uniformly distributed in $(-2^{-t}, 2^{-t})$, it can be readily shown that

$$\mathcal{E}\{f_k(p)\} = 0, \tag{35}$$

$$\mathcal{E}\{f_k(p)\bar{f}_i(s)\} = 0, \qquad k \neq i \quad \text{or} \quad p \neq s, \tag{36}$$

$$\mathcal{E}\{|f_k(p)|^2\} = \begin{cases} b_k(2^{-2t}/3)\,|A_k'(p) + A_k'(q)|^2 & \text{if } p_k = 0, \quad (37a)\\ [(1 + 2^{-2t}/3)b_k - 1]\,|A_k'(p) - A_k'(r)|^2 & \text{if } p_k = 1, \quad (37b)\end{cases}$$

where $\bar{f}_i(s)$ is the complex conjugate of $f_i(s)$, $2^{-2t}/3$ is the variance of a random variable uniformly distributed in $(-2^{-t}, 2^{-t})$, and

$$b_k = \begin{cases} 1 + 2p_k, & k = 0, 1, \cdots, m-3,\\ 1, & k = m-2, \quad m-1. \end{cases} \tag{38}$$

If t is not too small, then $(1 + 2^{-2t}/3)^{b_k} \approx 1 + b_k 2^{-2t}/3$ and $A_k'(p) \approx A_k(p)$. Equation (37) becomes, by using eq. (3),

$$\mathcal{E}\{|f_k(p)|^2\} = (2^{-2t}/3)b_k\,|A_{k+1}(p)|^2. \tag{39}$$

From eqs. (32), (33), and (35), it is seen that

$$\mathcal{E}\{e_k(p)\} = 0. \tag{40}$$

Next we notice that the small random errors associated with the three terms $e_k(p)$, $e_k(q)$, and $f_k(p)$ in eq. (33a) and with the three terms $e_k(p)$, $e_k(r)$, and $f_k(p)$ in eq. (33b) are all different, and hence independent. From eqs. (33), (35), and (36), we have

$$
\begin{aligned}
\mathcal{E}\{|e_{k+1}(p)|^2\} &= \mathcal{E}\{|e_k(p)|^2\} + \mathcal{E}\{|e_k(q)|^2\} + \mathcal{E}\{|f_k(p)|^2\}, \qquad p_k = 0,\\
&= \mathcal{E}\{|e_k(r)|^2\} + \mathcal{E}\{|e_k(p)|^2\} + \mathcal{E}\{|f_k(p)|^2\}, \qquad p_k = 1.
\end{aligned} \tag{41}
$$

Equation (41) becomes, with $k = m - 1$,

$$\mathcal{E}\{|e_m(p)|^2\} = \mathcal{E}\{|f_{m-1}(p)|^2\} + \mathcal{E}\{|e_{m-1}(p)|^2\} + \mathcal{E}\{|e_{m-1}(q)|^2\} \tag{42}$$

or

$$\mathcal{E}\{|e_m(p)|^2\} = \mathcal{E}\{|f_{m-1}(p)|^2\} + \mathcal{E}\{|e_{m-1}(p)|^2\} + \mathcal{E}\{|e_{m-1}(r)|^2\}.$$

The two arguments in the two $e_{m-1}(\ )$ terms, p and q or p and r, can be obtained from eq. (2) by putting p_{m-1} equal to 1 and equal to 0. On using eq. (41) with $k = m - 2$, each of the $\mathcal{E}\{|e_{m-1}(\ )|^2\}$ terms in eq. (42) can be expressed as a sum of one $\mathcal{E}\{|f_{m-2}(\ )|^2\}$ term and two $\mathcal{E}\{|e_{m-2}(\cdot)|^2\}$ terms. Thus $\mathcal{E}\{|e_m(p)|^2\}$ of eq. (42) can be expressed as the sum of one term of $\mathcal{E}\{|f_{m-1}(p)|^2\}$, two terms of $\mathcal{E}\{|f_{m-2}(\ )|^2\}$, and four terms of $\mathcal{E}\{|e_{m-2}(\ )|^2\}$. The argument in the two $\mathcal{E}\{|f_{m-2}(\ )|^2\}$ can be obtained from eq. (2) by putting p_{m-1} equal to 1 and 0. The arguments in the four terms of $\mathcal{E}\{|e_{m-2}(\ )|^2\}$ can be obtained from eq. (2) by putting p_{m-1} equal to 1 and 0 and p_{m-2} equal to 1 and 0. Continuing in this manner, we finally arrive at

$$\begin{aligned}\mathcal{E}\{|e_m(p)|^2\} &= \mathcal{E}\{|f_{m-1}(s_{m-1,p})|^2\} + \sum_{i_{m-1}=0}^{1} \mathcal{E}\{|f_{m-2}(s_{m-2,p})|^2\} \\ &\quad + \cdots + \sum_{i_2=0}^{1} \cdots \sum_{i_{m-1}=0}^{1} \mathcal{E}\{|f_1(s_{1,p})|^2\} + \sum_{i_1=0}^{1} \cdots \sum_{i_{m-1}=0}^{1} \mathcal{E}\{|f_0(s_{0,p})|^2\}\end{aligned} \tag{43}$$

where

$$\begin{aligned}s_{k,p} &= 2^{m-1}p_0 + 2^{m-2}p_1 + \cdots + 2^{m-1-k}p_k \\ &\quad + 2^{m-2-k}i_{k+1} + \cdots + 2i_{m-2} + i_{m-1}, \quad k = 0, 1, \cdots, m-1.\end{aligned} \tag{44}$$

In each term of eq. (43), the summation of $\mathcal{E}\{|f_k(s_{k,p})|^2\}$ can be expressed in terms of the summation of $|A_{k+1}(s_{k,p})|^2$ in view of eq. (39). Furthermore, since the 2^{m-1-k} numbers $\{A_m(s_{k,p})\}$ with $i_{m-1}, i_{m-2}, \cdots, i_{k+1} = 0, 1$ may be regarded as the discrete Fourier transform of the "data" $\{A_k(s_{k,p})\}$, $i_{m-1}, i_{m-2}, \cdots, i_{k+1} = 1, 0$, Parseval's theorem applies:

$$\sum_{i_{k+1}=0}^{1} \cdots \sum_{i_{m-1}=0}^{1} |A_k(s_{k,p})|^2 = 2^{-(m-k-1)} \sum_{i_{k+1}=0}^{1} \cdots \sum_{i_{m-1}=0}^{1} |A_m(s_{k,p})|^2. \tag{45}$$

Thus the mean square value of the error $e_m(p)$ is given by

$$\begin{aligned}&\mathcal{E}\{|e_m(p)|^2\}^R \\ &= \frac{2^{-2t}}{3}\left[|A_m(s_{m-1,p})|^2 + \sum_{k=0}^{m-2} b_k 2^{-m+k+1} \sum_{i_{k+1}=0}^{1} \cdots \sum_{i_{m-1}=0}^{1} |A_m(s_{k,p})|^2\right].\end{aligned} \tag{46}$$

By using eq. (6) and rearranging the indices of eq. (46), we obtain eq. (10) given in Section 3.

(b) *Chopping arithmetic.* When chopping is used for round-off, the random variables $\gamma'_{k,p}$, $\gamma''_{k,p}$, $\delta'_{k,p}$, $\delta''_{k,p}$, $\epsilon'_{k,p}$, $\epsilon''_{k,p}$, $\zeta'_{k,p}$, $\zeta''_{k,p}$, $\eta'_{k,p}$, $\eta''_{k,p}$, $\lambda'_{k,p}$, and $\lambda''_{k,p}$ all have nonzero mean $\mu = -2^{-t}$, although their variances are still equal to $2^{-2t}/3$. The mean of $f_k(p)$ of eq. (34) is $\mu b_k[A_k'(p) + A_k'(q)]$ if $p_k = 0$, and is $\mu b_k[A_k'(r) - A_k'(p)]w_k(p)$ if $p_k = 1$. By rewriting eq. (34) as the sum of its mean and a term with a zero mean, we have

$$[f_k(p)]^C = \begin{cases}\mu b_k[A_k'(p) + A_k'(q)] + [f_k(p)]^R & \text{if} \quad p_k = 0, \\ \mu b_k[A_k'(r) - A_k'(p)]w_k(p) + [f_k(p)]^R & \text{if} \quad p_k = 1,\end{cases} \tag{47}$$

where the superscripts C and R refer to chopping and rounding, respectively, and

b_k is given by eq. (38). From eqs. (30), (34), and (47), we have

$$A'_{k+1}(p) = \begin{cases} (1 + \mu b_k)[A_k'(p) + A_k'(q)] + [f_k(p)]^R & \text{if } p_k = 0, \\ (1 + \mu b_k)[A_k'(r) - A_k'(p)]w_k(p) + [f_k(p)]^R & \text{if } p_k = 1. \end{cases} \tag{48}$$

To solve for $A_k'(p)$ from eq. (48), we define $A_k''(p)$, $A_k'''(p)$, $\cdots$, $A_k^{(v)}(p)$, $\cdots$ by

$$A''_{k+1}(p) = \begin{cases} (1 + \mu b_k)[A_k''(p) + A_k''(q)] & \text{if } p_k = 0, \\ (1 + \mu b_k)[A_k''(r) - A_k''(p)]w_k(p) & \text{if } p_k = 1, \end{cases} \tag{49}$$

$$A'''_{k+1}(p) = \begin{cases} A_k'''(p) + A_k'''(q) + [f_k(p)]^R & \text{if } p_k = 0, \\ [A_k'''(r) - A_k'''(p)]w_k(p) + [f_k(p)]^R & \text{if } p_k = 1, \end{cases} \tag{50}$$

$$A^{(v)}_{k+1}(p) = \begin{cases} A_k^{(v)}(p) + A_k^{(v)}(q) + \mu b_k[A_k^{(v-1)}(p) + A_k^{(v-1)}(q)] \\ \qquad \text{if } p_k = 0, \\ \{A_k^{(v)}(r) - A_k^{(v)}(p) + \mu b_k[A_k^{(v-1)}(r) - A_k^{(v-1)}(p)]\}w_k(p) \\ \qquad \text{if } p_k = 1, \end{cases} \tag{51}$$
$$v = 3, 4, \cdots,$$

with the initial condition $A_0''(p) = x(p)$, $A_0^{(v)}(p) = 0$, $v \geq 3$. By summing these equations and comparing with eq. (48), it is seen that

$$A_k'(p) = A''(p) + A_k'''(p) + A_k''''(p) + \cdots. \tag{52}$$

It can be readily shown from eq. (49), by using eq. (3) and the initial condition $A_0(p) = x(p)$, that

$$A_m''(p) = A_m(p) \prod_{k=0}^{m-1} (1 + \mu b_k). \tag{53}$$

Equation (50) is identical to eq. (33) with $f_k(p) = [f_k(p)]^R$. Therefore, we have

$$A_m'''(p) = [e_m(p)]^R, \tag{54}$$

where the superscript R refers to "rounding." Because of the factor μ in eq. (51), it is clear that, for each $v = 3, 4, \cdots$, the order of magnitude of $A_k^{(v)}(p)$ is smaller than that of $A_k^{(v-1)}(p)$ by a factor of approximately μ. Thus, for all practical purposes, the terms beyond the first two on the right side of eq. (52) can be ignored, and we have

$$A_k'(p) = A_k''(p) + A_k'''(p). \tag{55}$$

From eqs. (53), (54), and (55), we have for chopping arithmetic

$$A_m'(p) = A_m(p) \prod_{k=0}^{m-1} (1 + \mu b_k) + [e_m(p)]^R. \tag{56}$$

Therefore,

$$\begin{aligned} [e_m(p)]^C &= A_m'(p) - A_m(p), \\ &= [e_m(p)]^R + A_m(p)\left[\prod_{k=0}^{m-1} (1 + \mu b_k) - 1\right]. \end{aligned} \tag{57}$$

If the mantissa t is not too short, the approximation

$$\prod_{k=0}^{m-1} (1 + \mu b_k) \approx 1 + \mu \sum_{k=0}^{m-1} b_k \tag{58}$$

is excellent. So

$$[e_m(p)]^C = [e_m(p)]^R + A_m(p)\mu \sum_{k=0}^{m-1} b_k \,. \tag{59}$$

It has been shown earlier that $[e_m(p)]^R$ has zero mean and its variance is given by eq. (46). Equations (13) and (14) follow readily from eq. (59).

Frequently, the data $\{x(n)\}_{n=0}^{N-1}$ represents a segment taken from a discrete parameter stochastic process, and several segments may be used to calculate the Fourier coefficients for such purposes as spectral estimation. It is possible in this case to relate the mean square error derived earlier to the autocorrelation function of the stochastic process.

Consider the case of rounding arithmetic first. The $|A(I_{k,p})|^2$ in eq. (10) are now random variables and should be replaced by $\mathcal{E}\{|A(I_{k,p})|^2\}$ where the expectation $\mathcal{E}$ is to be taken with respect to the random process $x(n)$. From eq. (1), we have

$$\begin{aligned}\mathcal{E}\{|A(p)|^2\} &= \mathcal{E}\sum_{n=0}^{N-1}\sum_{k=0}^{N-1} x(n)\bar{x}(k)W^{(n-k)p} \\ &= \sum_{n=0}^{N-1}\sum_{k=0}^{N-1} R_{xx}(n-k)W^{(n-k)p} \\ &= \sum_{k=-(N-1)}^{(N-1)} (N - |k|)R_{xx}(k)W^{pk}.\end{aligned} \tag{60}$$

On replacing $|A(p)|^2$ in eq. (46) by this $\mathcal{E}\{|A(p)|^2\}$, one obtains eq. (15). The derivation of eq. (17) is similar and will not be repeated here.

To derive the bounds given in inequalities (18) and (19), we sum eqs. (10) and (14) over all p, making use of the observations that

$$\sum_{p=0}^{N-1}\sum_{i_{k+1}=0}^{1}\cdots\sum_{i_{m-1}=0}^{1} |A(I_{k,p})|^2 = 2^{m-1-k}\sum_{p=0}^{N-1}|A(p)|^2$$

and that $1 \le c_i < 3$, $i = 0, \cdots, 2, \cdots, m-1$.

6. *Experimental Results*

A time series $\{x(n)\}$ with autocorrelation function $R_{xx}(n) = e^{-\nu|n|}$ is generated by the recursive relation

$$x(n+1) = z(n) + e^{-\nu}x(n), \tag{61}$$

where $z(n)$ is a sequence of independent random numbers. A number of segments consisting of N sample values of $\{x(n)\}$ are taken, and for each segment two FFT are calculated. In the first one, double precision arithmetics are used and the Fourier coefficients obtained are considered as ideal. In the second, the FFT is calculated with 27-bit rounding and with 27-bit chopping arithmetics. By comparing the results of the calculations we obtain the error $e(p)$. Figure 3 shows the

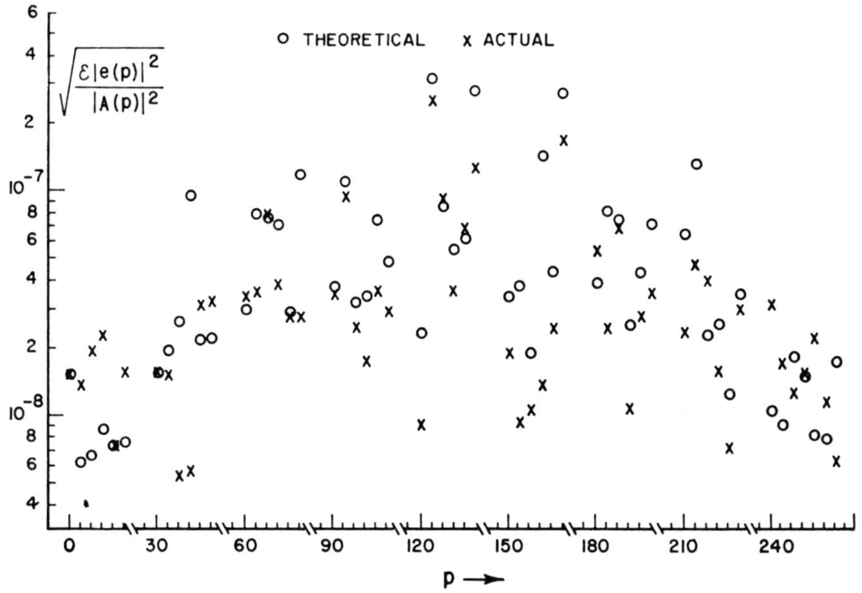

FIG. 3. Comparison of actually observed errors with theoretical calculations using eq. (10) (rounding arithmetic, $t = 27$, $m = 8$, deterministic)

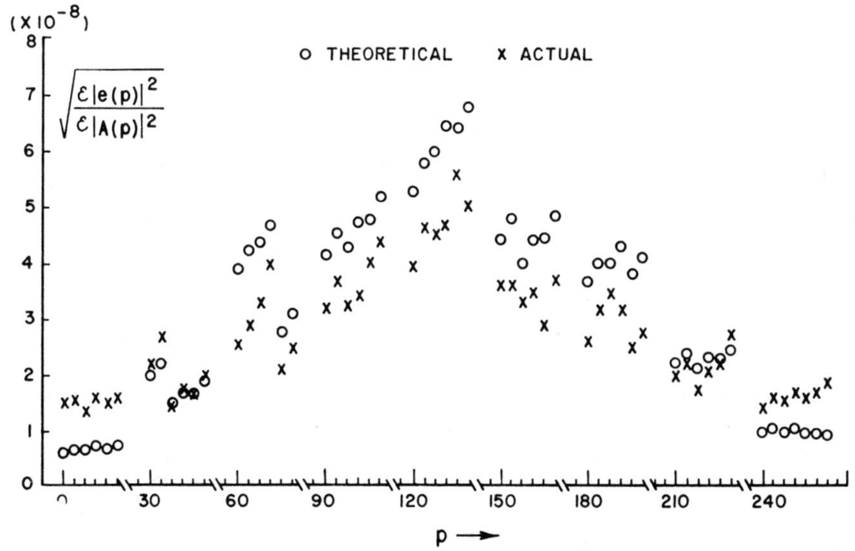

FIG. 4. Comparison of actually observed errors with theoretical calculations using eq. (15) or (26) (rounding arithmetic, $t = 27$, $m = 8$, random data)

comparison of the theoretical result, eq. (10), with the actual measured result for a single segment with $N = 2^8$ and $\nu = 0.2$.

Next, 50 segments are taken to obtain the sample average $\mathcal{E}\{|A(p)|^2\}$, and the results of eqs. (15) and (17) are compared with the actual mean square error.

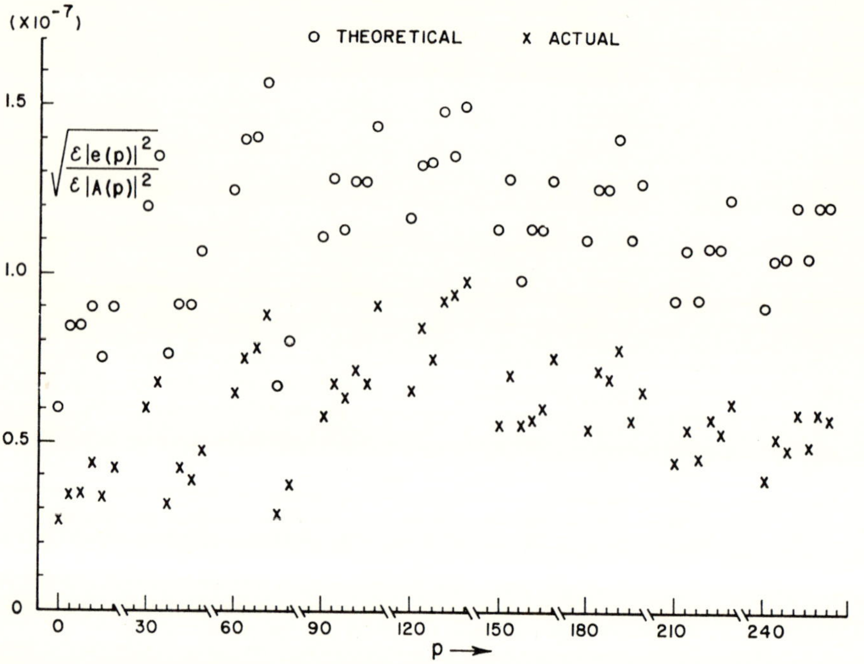

FIG. 5. Comparison of actually observed errors with theoretical calculations using eq. (17) (chopping arithmetic, $t = 27$, $m = 8$, random data)

Figure 4 shows the comparison for rounding arithmetic, and Figure 5 shows the comparison for chopping arithmetic. In both cases, good agreement is seen.

It is noticed that for chopping arithmetic the theoretical values are almost twice as large as the experimental result. This is due to the fact that the assumption that the small random errors have mean -2^{-t} for chopping is known from experience to be slightly pessimistic [9, 13]. The actual measured mean is about -0.6×2^{-t}.

Calculations are made also for different values of the mantissa length t and different values of the length of the segment $N = 2^m$. Good agreement is obtained for t larger than 12 and m up to 10; the results are not presented here.

The total relative mean square error is also measured to test the validity of inequalities (18) and (19). The results are shown in Figure 6. It is seen that the bounds for rounding arithmetic are good, but for chopping arithmetic, the lower bound given by (19) is above the measured error. This is again due to the fact that the mean of the small random errors is not exactly -2^{-t} as used in the theoretical calculation. The upper and lower bounds using -0.6×2^{-t} as the mean are also shown in Figure 6 as corrected bounds and the actual observed error is seen to lie within the bounds.

7. *Effect of Quantization of $\{x(n)\}$ and Inaccuracy of $w_k(p)$*

The discussion presented in previous sections concerns the round-off accumulation effect only. There are two other common causes of error due to the finite word

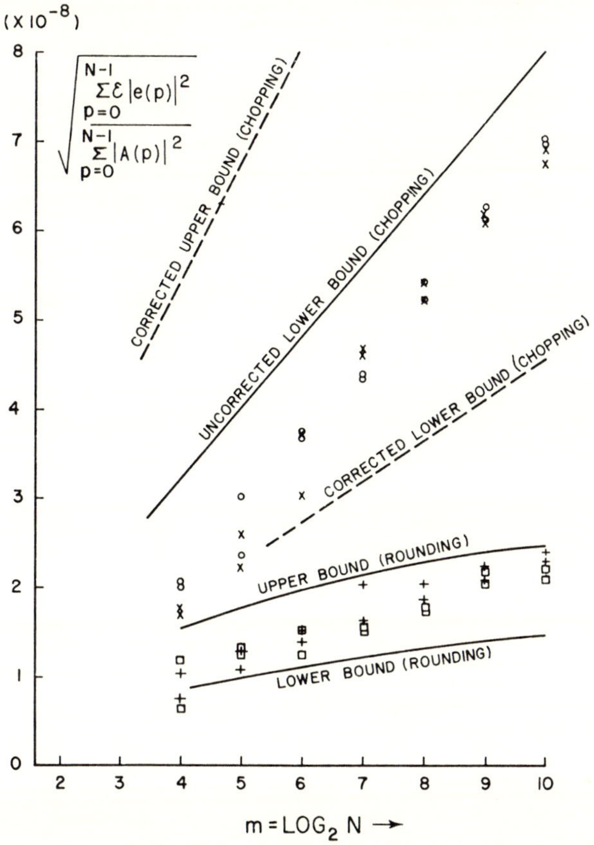

FIG. 6. Comparison of actually observed relative rms error with the bounds given by inequalities (18) and (19) ($t = 27$, deterministic, "○" and "□" are white noise data, "×" and "+" are random data generated by eq. (60))

length in computing the Fourier coefficients. They are the quantization of the data $\{x(n)\}$ and the inaccuracy of the numbers $w_k(p)$.

The effect of the quantization of $\{x(n)\}$ can be treated readily. Let $x'(n)$ be obtained from $x(n)$ either by rounding or by chopping the mantissa to t' bits.[4] Then from the discussion of Section 2

$$\mathrm{Re}[x'(n)] = (1 + \theta_n)\,\mathrm{Re}[x(n)], \qquad \mathrm{Im}[x'(n)] = (1 + \xi_n)\,\mathrm{Im}[x(n)], \tag{62}$$

where θ_n and ξ_n are random variables distributed uniformly in $(-2^{-t'}, 2^{-t'})$ or in $(-2^{-t'+1}, 0)$ depending whether rounding or chopping is used. The effect of eq. (62) is to modify the initial condition, eq. (32), to

$$e_0(n) = \theta_n\,\mathrm{Re}[x(n)] + j\xi_n\,\mathrm{Im}[x(n)]. \tag{63}$$

Thus

$$\mathcal{E}\{|\,e_0(n)\,|^2\} = \begin{cases} (2^{-2t'}/3)\;|\,x(n)\,|^2 & \text{for rounding,} \\ \frac{4}{3}\,2^{-2t'}\;|\,x(n)\,|^2 & \text{for chopping.} \end{cases} \tag{64}$$

[4] Usually $t' < t$. But if the data $x(n)$ are generated in the machine, we may have $t' = t$.

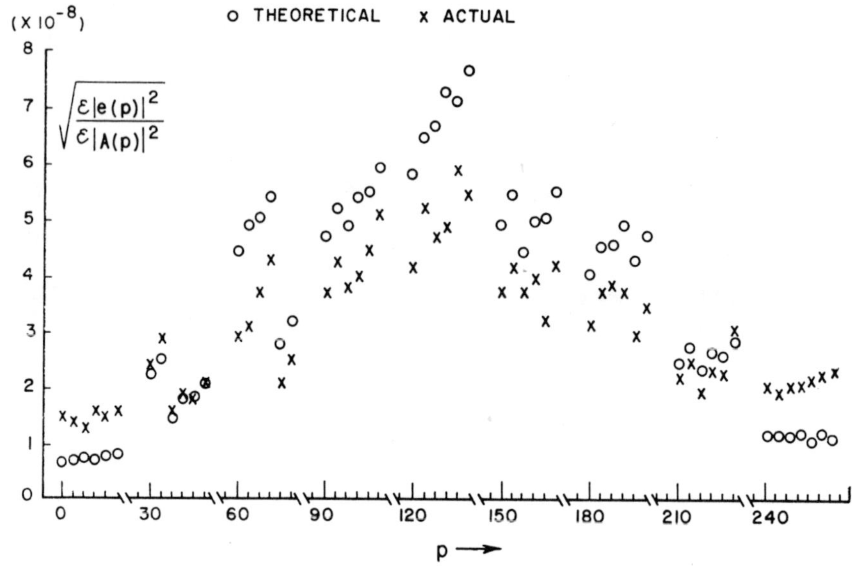

FIG. 7. Comparison of observed errors with theoretical calculations using eq. (67), including effect of inaccuracy in $w_k(p)$ (rounding arithmetic, $t = 27$, $m = 8$, random data)

With this modification, the final result of eqs. (46) and (59) can be shown to remain the same except for an addition term of

$$\frac{2^{-2t'}}{3} 2^{-m} \sum_{k=0}^{N-1} |A(k)|^2 \quad \text{for rounding}$$

or (65)

$$\frac{4}{3} 2^{-2t'} 2^{-m} \sum_{k=0}^{N-1} |A(k)|^2 \quad \text{for chopping}$$

which is independent of p.

Another cause for error that has been neglected in the treatment of the previous sections is the fact that the coefficients $w_k(p)$ can only be represented in the machine with an accuracy of t bits. The experimental result reported in Section 6 was obtained with t-bit accuracy for the coefficients $w_k(p)$ in both the double precision calculation and the t-bit calculation. If double precision $w_k(p)$ is used, then the difference between the results obtained from the double precision FFT calculation and those obtained from the t-bit FFT calculation would be the error due to the combined effect of the inaccuracy of $w_k(p)$ and the round-off accumulation. This is shown in Figure 7 by the "×" for the case of rounding arithmetic.

By comparing these errors with those due to round-off accumulation alone shown in Figure 4, we see that the amount by which the error is increased due to the coefficient inaccuracy is not too significant, about 15 percent.

It is possible to analyze the effect of the inaccuracy of $w_k(p)$ as follows. Since the coefficients $w_k(p)$ are realized in the machine with only t-bit mantissa, we have, as before,

$$\begin{aligned} U_k'(p) &= (1 + \varphi_{k,p}) U_k(p), \\ V_k'(p) &= (1 + \psi_{k,p}) V_k(p), \end{aligned} \tag{66}$$

where $U_k'(p)$ and $V_k'(p)$ denote the t-bit machine numbers for $U_k(p)$ and $V_k(p)$, respectively, and the errors $\varphi_{k,p}$ and $\psi_{k,p}$ are random variables distributed uniformly in either $(-2^{-t}, 2^{-t})$ or $(-2^{-t+1}, 0)$ depending whether rounding or chopping is used. One may now proceed with the analysis of Section 5 by adding the factors $(1 + \varphi_{k,p})$ and $(1 + \psi_{k,p})$ in the appropriate places in eq. (34). The difficulty now lies in the fact that for different values of k and p, $w_k(p)$ may be the same. This means that the errors $\varphi_{k,p}$ and $\psi_{k,p}$ are not necessarily independent for different values of k and p. However, if this point is disregarded by treating $\varphi_{k,p}$ and $\psi_{k,p}$ still as independent random variables for different k and p, one arrives at, for rounding arithmetic,

$$\mathcal{E}\,|\, f_k(p)\,|^2 = (2^{-2t}/3)b_k'\,|\, A_{k+1}(p)\,|^2, \tag{67}$$

where

$$b_k' = \begin{cases} 1 + 3p_k, & k = 0, 1, \cdots, m-3, \\ 1, & k = m-2, m-1, \end{cases} \tag{68}$$

instead of eq. (37). The final result is identical with eq. (10) with c_k replaced by c_k', which is obtained from b_k' by replacing p_k by p_{m-1-k}. For chopping arithmetic, the final result is the same as eq. (14) with c_k replaced by c_k'. The details are omitted.

The theoretical $\mathcal{E}\{|\, e(p)\,|^2\}^R$ is calculated for the $x(n)$ sequence of Section 6 and the results are shown in Figure 7 as "○". It is seen that the agreement between the theoretical and experimental results is quite good.

REFERENCES

1. Bingham, C., Godfrey, M. D., and Tukey, J. W. Modern techniques of power spectrum estimation. *IEEE Trans. AU-15* (June 1967), 56–66.
2. Cochran, W. T., et al. What is the fast Fourier transform? *Proc. IEEE 55*, 10 (Oct. 1967), 1644–1673.
3. Cooley, J. W., and Tukey, J. W. An algorithm for the machine calculation of complex Fourier series. *Math. Comput. 19* (April 1965), 297–301.
4. Flanagan, J. L. Spectrum analysis in speech coding. *IEEE Trans. AU-15* (June 1967), 66–69.
5. Forsythe, G., and Moler, C. B. *Computer Solution of Linear Algebraic Systems.* Prentice-Hall, Englewood Cliffs, N. J., 1967.
6. Gentleman, W. W., and Sande, G. Fast Fourier transform for fun and profit. Proc. AFIPS 1966 Fall Joint Comput. Conf., Vol. 29, Spartan Books, New York, pp. 563–578.
7. Helms, H. D. Fast Fourier transform method of computing difference equations and simulating filters. *IEEE Trans. AU-15* (June 1967), 85–90.
8. Katzenelson, J. On errors introduced by combined sampling and quantization. *IRE Trans. AC-7* (April 1962), 58–68.
9. Liu, B., and Kaneko, T. Error analysis of digital filters realized with floating-point arithmetic. *Proc. IEEE 57* (Oct. 1969), 1735–1747.
10. Lohmann, A. W., and Paris, D. P. Binary Fraunhofer holograms generated by computer. *Appl. Optics 6* (1967) 1739–1748.
11. Watts, D. J., and Katzenelson, J. Discussion of: On errors introduced by combined sampling and quantization, Vol. AC-7. *IEEE Trans. AC-8* (April 1963), 187–188.
12. Weinstein, C. J. Roundoff noise in floating point fast Fourier transform computation. *IEEE Trans. AU-17* (Sept. 1969), 209–215.
13. Weinstein, C., and Oppenheim, A. V. A comparison of rounding noise in floating point and fixed point digital filter realization. *Proc. IEEE 57*, 6 (June 1969), 1181–1183.
14. Welch, P. D. A fixed-point fast Fourier transform error analysis. *IEEE Trans. AU-17* (June 1969), 151–157.
15. Wilkinson, J. H. *Rounding Errors in Algebraic Processes.* Prentice-Hall, Englewood Cliffs, N. J., 1963.

RECEIVED AUGUST, 1969; REVISED SEPTEMBER, 1969

Author Citation Index

Subject Index

The Editor

BEDE LIU is Professor of Electrical Engineering at Princeton University. He holds a B.S.E.E. degree from the National Taiwan University, and his M.E.E. and D.E.E. degrees from the Polytechnic Institute of Brooklyn. Prior to joining the Princeton faculty in 1962, he was associated with Bell Laboratories, Allen B. DuMont Laboratory, and Western Electric Company. He has also served as technical consultant to various companies. During 1970–1971, he was a Visiting Professor at the National Taiwan University and at the University of California at Berkeley.

A fellow of the Institute of Electrical and Electronics Engineers (IEEE), Professor Liu's current research interests include digital signal processing, computer holograph, and various aspects of communication systems.